Human-Computer Interaction in Intelligent Environments

This book offers readers a holistic understanding of intelligent environments, encompassing their definition, design, interaction paradigms, the role of artificial intelligence (AI), and the associated broader philosophical and procedural aspects.

This book:

- Elaborates on AI research and the creation of intelligent environments.
- Zooms in on designing interactions with the IoT, intelligent agents, and robots.
- Discusses overarching topics for the design of intelligent environments, including user interface adaptation, design for all, sustainability, cybersecurity, privacy, and trust.
- Provides insights into the intricacies of various intelligent environment contexts, such as in automotive, urban interfaces, smart cities, and beyond.

This book has been written for individuals interested in Human-Computer Interaction research and applications.

Human-Computer Interaction in Intelligent Environments

Edited by
Constantine Stephanidis and Gavriel Salvendy

CRC Press
Taylor & Francis Group
Boca Raton London New York

CRC Press is an imprint of the
Taylor & Francis Group, an **informa** business

Designed cover image: shutterstock

First edition published 2025
by CRC Press
6000 Broken Sound Parkway NW, Suite 300, Boca Raton, FL 33487-2742

and by CRC Press
4 Park Square, Milton Park, Abingdon, Oxon, OX14 4RN

CRC Press is an imprint of Taylor & Francis Group, LLC

© 2025 selection and editorial matter, Constantine Stephanidis and Gavriel Salvendy individual chapters, the contributors

Reasonable efforts have been made to publish reliable data and information, but the author and publisher cannot assume responsibility for the validity of all materials or the consequences of their use. The authors and publishers have attempted to trace the copyright holders of all material reproduced in this publication and apologize to copyright holders if permission to publish in this form has not been obtained. If any copyright material has not been acknowledged please write and let us know so we may rectify in any future reprint.

Except as permitted under U.S. Copyright Law, no part of this book may be reprinted, reproduced, transmitted, or utilized in any form by any electronic, mechanical, or other means, now known or hereafter invented, including photocopying, microfilming, and recording, or in any information storage or retrieval system, without written permission from the publishers.

For permission to photocopy or use material electronically from this work, access www.copyright.com or contact the Copyright Clearance Center, Inc. (CCC), 222 Rosewood Drive, Danvers, MA 01923, 978-750-8400. For works that are not available on CCC please contact mpkbookspermissions@tandf.co.uk

Trademark notice: Product or corporate names may be trademarks or registered trademarks and are used only for identification and explanation without intent to infringe.

ISBN: 978-1-032-37004-0 (hbk)
ISBN: 978-1-032-79138-8 (pbk)
ISBN: 978-1-003-49068-5 (ebk)

DOI: 10.1201/9781003490685

Typeset in Times
by codeMantra

Contents

Preface vii
Editors viii
List of Contributors ix

Chapter 1 Design for Intelligent Environments 1

Constantine Stephanidis, Asterios Leonidis, Maria Korozi, Vasilis Kouroumalis, Ilia Adami, and Stavroula Ntoa

Chapter 2 User Interface Adaptation and Design for All 44

Constantine Stephanidis

Chapter 3 Human Behavior in Cybersecurity Privacy and Trust 77

Abbas Moallem

Chapter 4 Human-Centered AI 108

Brent Winslow and Ozlem Ozmen Garibay

Chapter 5 AI in HCI Design and User Experience 141

Wei Xu

Chapter 6 Interacting with the Internet of Things 171

Fulvio Corno, Luigi De Russis, Alberto Monge Roffarello, and Juan Pablo Sáenz Moreno

Chapter 7 Conversational Agents 201

Ana Paula Chaves, Charlotte van Hooijdonk, Christine Liebrecht, Guilherme Corredato Guerino, Heloisa Candello, Minha Lee, Matthias Kraus, and Marco Aurelio Gerosa

Chapter 8 Artificial Intelligence for Customer Services 241

Christian Matt and Helena Weith

Chapter 9 Automotive User Interfaces: Enhancing Future Mobility 263

Myounghoon Jeon, Chihab Nadri, Manhua Wang, and Abhraneil Dam

Chapter 10 Human–Robot Interaction 305

Connor Esterwood, Qiaoning Zhang, X. Jessie Yang, and Lionel P. Robert

Chapter 11 Human-Agent Teaming 333

    Anthony L. Baker, Shan G. Lakhmani, and Jessie Y. C. Chen

Chapter 12 Ambient Assisted Living Solutions 364

    Martina Ziefle, Julia Offermann, and Wiktoria Wilkowska

Chapter 13 Urban Interfaces 384

    Luke Hespanhol, Joel Fredericks, Martin Tomitsch, and Andrew Vande Moere

Chapter 14 Sustainability and Citizen Science 408

    Yao Sun and Ann Majchrzak

Index 439

Preface

Human-Computer Interaction (HCI) is a multidisciplinary field exhibiting increasing significance in the modern world, shaping our interactions with technology and transforming our daily lives. HCI plays a critical role in bridging the gap between human capabilities and technological breakthroughs. As technology becomes progressively intertwined with our daily lives, it plays a significant role in enhancing human interaction with advanced information technologies.

We are pleased to present six books in the series titled 'Human-Computer Interaction: Foundations and Advances', which provide extensive coverage of the HCI field. Collectively, this book series addresses the theoretical and application aspects of HCI and thus provides important resources for HCI researchers, practitioners, and students. Each book addresses theoretical and practical aspects of HCI as well as state-of-the-art technological advancements, aiming to illuminate how innovations reshape this field, redefine user interactions, and elevate the overall user experience. The six book titles in the series are:

- FOUNDATIONS AND FUNDAMENTALS IN HUMAN-COMPUTER INTERACTION
- DESIGNING FOR USABILITY, INCLUSION AND SUSTAINABILITY IN HUMAN-COMPUTER INTERACTION
- USER EXPERIENCE METHODS AND TOOLS IN HUMAN-COMPUTER INTERACTION
- INTERACTION TECHNIQUES AND TECHNOLOGIES IN HUMAN-COMPUTER INTERACTION
- HUMAN-COMPUTER INTERACTION IN INTELLIGENT ENVIRONMENTS
- HUMAN-COMPUTER INTERACTION IN VARIOUS APPLICATION DOMAINS

Within this series, readers will discover a wealth of information encompassing the foundational elements, the established and emerging domains, and the cutting-edge technological advancements, as well as invaluable insights into key concepts, challenges, and the latest and most important research and practices. Each book provides a thorough examination and a standalone source of information.

To further enhance this exploration, the books also showcase the practical applications of HCI approaches and methodologies, illustrating their utility. Insightful figures, tables, and references supplement the above, while an in-depth discussion and future directions in each chapter aim to engage readers in spearheading their research and application endeavors in the field of HCI.

This book, titled *Human-Computer Interaction in Intelligent Environments*, explores the dynamic relationship between humans and intelligent environments, with a specific emphasis on the role of artificial intelligence (AI) and the Internet of Things (IoT). This book offers readers a holistic understanding of intelligent environments, encompassing their definition, design, interaction paradigms, the role of AI, and the associated broader philosophical and procedural aspects. It takes a comprehensive HCI perspective in the context of AI research and the creation of intelligent environments, elaborating on the design process, human-centered AI, AI in HCI design and user experience, ambient assisted living, and AI for customer services. It zooms in on designing interactions with the IoT, conversing and teaming with intelligent agents and interacting with robots, and zooms out to discuss the overarching principles that design should aim to adhere to, such as user interface adaptation, design for all, sustainability, cybersecurity, privacy, and trust. This book also provides insights into the intricacies of various intelligent environment contexts, ranging from automotive to room-based systems, to entire cities, to urban interfaces, and beyond.

This book contains 14 chapters, written by 44 authors. This book includes 80 figures, 36 tables, and 2,378 references for documenting and providing supporting data to the presented and discussed information.

Constantine Stephanidis, Gavriel Salvendy
November 2023

Editors

Constantine Stephanidis is Professor at the Department of Computer Science of the University of Crete, past Director of the Institute of Computer Science of FORTH and Founding Head of its Human-Computer Interaction (HCI) Laboratory and its Ambient Intelligence Program. He has been the Principal Investigator for over 180 European Commission, National and Industry funded projects. He is the Founding Editor of the International Journal *Universal Access in the Information Society*, co-Editor of the *International Journal of Human-Computer Interaction* and General Chair of the HCI International Conference. He is the President of the Council for Research and Innovation of the Region of Crete and the President of the Hellenic National Accessibility Authority.

Gavriel Salvendy is University Distinguished Professor at the College of Engineering and Computer Science at the University of Central Florida and Founding President of the Academy of Science, Engineering, and Medicine of Florida. He is Professor Emeritus of Industrial Engineering at Purdue University and Chair Professor Emeritus and Founding Head of the Department of Industrial Engineering at Tsinghua University, Beijing, P.R. China. He is a member of the National Academy of Engineering and the recipient of the John Fritz Medal which is frequently referred to as the Nobel Prize in Engineering. He is the Founding Editor of the *International Journal of Human-Computer Interaction* and the Founder of the International Conference on Human-Computer Interaction - now in its 40th year.

Contributors

Ilia Adami
Foundation for Research and Technology – Hellas (FORTH)
Institute of Computer Science Heraklion, Crete, Greece

Anthony L. Baker
DEVCOM Army Research Laboratory
Aberdeen Proving Ground, Maryland

Heloisa Candello
IBM Research
São Paulo, Brazil

Ana Paula Chaves
College of Engineering, Informatics, and Applied Sciences, School of Informatics, Computing, and Cyber Systems
Northern Arizona University
Flagstaff, Arizona

Jessie Y. C. Chen
DEVCOM Army Research Laboratory
Aberdeen Proving Ground, Maryland

Fulvio Corno
Department of Computer and Control Engineering (DAUIN)
Politecnico di Torino
Torino, Italy

Abhraneil Dam
Grado Department of Industrial and Systems Engineering
Virginia Tech
Blacksburg, Virginia

Luigi De Russis
Department of Computer and Control Engineering (DAUIN)
Politecnico di Torino
Torino, Italy

Connor Esterwood
School of Information
University of Michigan
Ann Arbor, Michigan

Joel Fredericks
Design Lab – School of Architecture, Design and Planning
The University of Sydney
Sydney, Australia

Ozlem Ozmen Garibay
University of Central Florida
Orlando, Florida

Marco Aurelio Gerosa
College of Engineering, Informatics, and Applied Sciences, School of Informatics, Computing, and Cyber Systems
Northern Arizona University
Flagstaff, Arizona

Guilherme Corredato Guerino
Departamento de Informática
State University of Maringá
Maringá, Paraná, Brazil

Luke Hespanhol
Design Lab – School of Architecture, Design and Planning
The University of Sydney
Sydney, Australia

Myounghoon Jeon
Grado Department of Industrial and Systems Engineering
Virginia Tech
Blacksburg, Virginia

Maria Korozi
Foundation for Research and Technology – Hellas (FORTH)
Institute of Computer Science Heraklion, Crete, Greece

Vassilis Kouroumalis
Foundation for Research and Technology –
 Hellas (FORTH)
Institute of Computer Science Heraklion,
 Crete, Greece

Matthias Kraus
Institute of Computer Science
Augsburg University
Augsburg, Germany

Shan G. Lakhmani
DEVCOM Army Research Laboratory
Aberdeen Proving Ground, Maryland

Minha Lee
Industrial Design Department
Eindhoven University of Technology
Eindhoven, The Netherlands

Asterios Leonidis
Foundation for Research and
 Technology – Hellas (FORTH)
Institute of Computer Science
Heraklion, Crete, Greece

Christine Liebrecht
Department of Communication and Cognition,
 Tilburg School of Humanities and Digital
 Sciences
Tilburg University
Tilburg, the Netherlands

Ann Majchrzak
University of Southern California
Marshall School of Business
Los Angeles, California

Christian Matt
University of Bern
Institute of Information Systems
Bern, Switzerland

Abbas Moallem
Department of Industrial & Systems
 Engineering
San Jose State University
San Jose, California

Juan Pablo Saénz Moreno
Department of Computer and Control
 Engineering (DAUIN)
Politecnico di Torino
Torino, Italy

Chihab Nadri
Grado Department of Industrial and Systems
 Engineering
Virginia Tech
Blacksburg, Virginia

Stavroula Ntoa
Foundation for Research and Technology –
 Hellas (FORTH)
Institute of Computer Science Heraklion,
 Crete, Greece

Julia Offermann
Chair of Communication Science,
 Human-Computer Interaction Center
RWTH Aachen University
Aachen, Germany

Lionel P. Robert
School of Information and Robotics
 Department
University of Michigan
Ann Arbor, Michigan

Alberto Monge Roffarello
Department of Computer and Control
 Engineering (DAUIN)
Politecnico di Torino
Torino, Italy

Constantine Stephanidis
University of Crete, Computer Science
 Department
and
Foundation for Research and Technology -
 Hellas (FORTH)
Institute of Computer Science
Heraklion, Crete, Greece

Yao Sun
New Jersey Institute of Technology
University Heights, Newark

Contributors

Martin Tomitsch
TD School
University of Technology Sydney
Sydney, Australia

Charlotte van Hooijdonk
Humanities, Department of Languages,
   Literature and Communication
Utrecht University
Utrecht, the Netherlands

Andrew Vande Moere
Research[x]Design, Department of Architecture
KU Leuven
Leuven, Belgium

Manhua Wang
Grado Department of Industrial and Systems
   Engineering
Virginia Tech
Blacksburg, Virginia

Helena Weith
University of Bern
Institute of Information Systems
Bern, Switzerland

Wiktoria Wilkowska
Chair of Communication Science,
   Human-Computer Interaction Center
RWTH Aachen University
Aachen, Germany

Brent Winslow
Google, LLC
San Francisco, California

Wei Xu
Zhejiang University
Zhejiang, P.R. China

X. Jessie Yang
School of Information and
Department of Industrial and Operations
   Engineering
University of Michigan
Ann Arbor, Michigan

Qiaoning Zhang
School of Information
University of Michigan
Ann Arbor, Michigan

Martina Ziefle
Chair of Communication Science,
   Human-Computer Interaction Center
RWTH Aachen University
Aachen, Germany

1 Design for Intelligent Environments

Constantine Stephanidis, Asterios Leonidis, Maria Korozi, Vasilis Kouroumalis, Ilia Adami, and Stavroula Ntoa

1.1 INTRODUCTION TO DESIGNING INTELLIGENT ENVIRONMENTS AND DESIGN PROCESS

This chapter aims to introduce the reader to the concept of Intelligent Environments (IEs) (Brumitt et al., 2000) and the challenges of designing them. Along with the introduction of concepts, disambiguation of terms, and presentation of the current state of research, this chapter focuses on the description of the *AmI Design Process*, a design process and a framework with supporting tools, developed in the context of the Ambient Intelligence Programme[1] of ICS-FORTH[2]. The framework can be used as a reference or a guide to researchers and practitioners on how to tackle the complexity of designing IEs and managing the design process based on the *user-centered design* principles (Abras et al., 2004) and the *Design Thinking* process (Brown, 2008; Park & McKilligan, 2018) offering a holistic approach to the challenge that is to solving problems. This chapter also provides a use case of the design of an IE that was carried out with the described process and tools.

1.1.1 IEs and Their Objectives

An IE encompasses both physical and digital spaces enhanced with cutting-edge technologies and includes sensors, actuators, data processing, networking capabilities, artificial intelligence (AI), and a range of input and output devices (Augusto et al., 2013). Their integration enables the environment to perceive, adjust to, and react to changes in its surroundings and to user interactions.

The goals and objectives of an IE across various domains revolve around creating a responsive, user-centric, and adaptable environment that enhances the quality of life and promotes well-being, improves efficiency, and provides personalized experiences for individuals within it while adhering to ethical and sustainable principles. To achieve these aims, IEs gather and analyze data from diverse sources, including sensors and user inputs, toward making well-informed decisions and adapting to changing contexts. Table 1.1 summarizes the predominant objectives of IEs as depicted in the relevant literature.

1.1.2 IEs and Relevant Concepts

IEs, the Internet of Things (IoT), cyber-physical systems (CPS), ubiquitous environments, and ambient intelligence (AmI) are interconnected concepts within the broader domain of smart and responsive environments. In summary, IEs are a specific application of them as they leverage these concepts (and associated principles and technologies) toward their instantiation. Below, a clarification of these concepts is provided, along with an explanation of their contributions to the development of IEs:

IoT: Atzori et al. (2010) refer to a network of interconnected physical objects or devices (things) that are equipped with sensors, actuators, and communication capabilities. IoT plays a fundamental role in IEs by establishing the infrastructure for collecting data from diverse sensors and devices. It further enables real-time decision-making and control through the seamless exchange of data and information.

TABLE 1.1
Objectives of Intelligent Environments

Objectives	Description
Enhanced user experience (Emiliani & Stephanidis, 2005; Stephanidis et al., 2021)	• Create an environment that prioritizes user satisfaction and comfort. • Tailor services and interactions to meet individual user needs and preferences. • Employ Human-Centered Design to ensure ease of use, unobtrusiveness, privacy, and personalization • Design for human-technology symbiosis
Context awareness (Perera et al., 2013; Zhang & Tao, 2020)	• Develop an environment capable of sensing and understanding the context in which it operates. • Utilize sensors, data, and AI to adapt to changing conditions and user contexts.
Automation and efficiency (Vijayan et al., 2020)	• Implement automation to streamline tasks and processes within the environment. • Improve efficiency, reduce manual interventions, and optimize resource usage.
Safety and security (Dunne et al., 2021)	• Ensure the safety and security of users and their data within the environment. • Implement robust security measures and privacy protection mechanisms.
Personalization (Bianchi et al., 2019; Kontogianni & Alepis, 2020; Libanori et al., 2022; Peng et al., 2019; Quijano-Sánchez et al., 2020)	• Provide personalized services, recommendations, and interactions based on user profiles and historical data. • Adapt content and features to suit individual preferences.
Energy efficiency and sustainability (H. Kim et al., 2021; Robinson et al., 2015; Scorpio et al., 2020)	• Optimize energy consumption within the environment to reduce environmental impact and lower energy costs. • Promote sustainable practices and algorithms to optimize resource usage, reduce waste, and minimize environmental impact.
Adaptation to user needs (J. C. Augusto & Muñoz, 2019)	• Continuously adapt and evolve the environment based on user feedback and changing requirements. • Support dynamic customization of services and configurations.
Accessibility and inclusivity (Burzagli & Emiliani, 2017; Ntoa et al., 2022; Stephanidis, 2023)	• Ensure that the environment is accessible to users with diverse needs, including those with disabilities. • Design with inclusivity in mind to cater to a wide range of users. • Employ multimodal input and multimedia output, inherent in the IE, to cater to the needs of diverse target users • Design for explainability, transparency, and fairness
Data-driven decision-making (Bousdekis et al., 2021)	• Utilize data analytics and AI to make informed decisions and provide actionable insights. • Support data-driven optimization and resource allocation.
Seamless interactions (Belk et al., 2019; K. Kim et al., 2023; Rubio-Drosdov et al., 2017; Sun et al., 2019)	• Facilitate seamless interactions among users, devices, and systems within the environment. • Ensure that interactions are intuitive and user-friendly.

CPS: CPS are systems characterized by the seamless integration of computational and physical components, facilitating their interaction with the physical world and enabling real-time responses to changes (Baheti & Gill, 2011). In the context of IEs, they are primarily employed for monitoring, controlling, and managing physical processes and devices, as well as overseeing critical infrastructures, transportation systems, and industrial/production procedures.

Ubiquitous environments: Lyytinen & Yoo (2002) emphasized the concept of pervasive computing, wherein computing resources and capabilities seamlessly integrate into daily life. IEs extend this idea by embracing the notion of computing's omnipresence, offering support to users wherever they go. This fosters a seamless and uninterrupted user experience (UX) across diverse settings and contexts, ensuring continuity as users transition between them.

AmI: Cook et al. (2009) refer to the ability of an environment to be aware of and responsive to the presence and needs of individuals within it. This entails the integration of AI, context awareness, and adaptive systems to create environments capable of anticipating user needs and delivering proactive assistance. AmI technologies are frequently used to enhance the UX and automate tasks within IEs.

1.1.3 Application Domains and Acceptance of IEs

IEs have versatile applications spanning a wide spectrum of domains and industries, including smart homes, offices, healthcare facilities, and urban environments (Table 1.2). In these contexts, advanced technologies are harnessed to develop responsive, user-centric, and adaptable solutions

TABLE 1.2
Application Domains of Intelligent Environments

Smart homes (Stojkoska & Trivodaliev, 2017; Marikyan et al., 2019)	• Solutions mostly refer to home automation systems that control lighting, heating, security, and entertainment based on user preferences and sensor data. • Smart homes also incorporate voice assistants and AI-powered devices for convenience and energy efficiency.
Smart cities (Yin et al., 2015; Kirimtat et al., 2020)	• IoT sensors and data analytics are used to optimize transportation, energy consumption, waste management, and public safety. • They aim to improve urban living by enhancing infrastructure and services. • Energy-efficient buildings and grid management can contribute to energy consumption optimization, carbon footprint reductions, and sustainability enhancement.
Healthcare (Selvaraj & Sundaravaradhan, 2020)	• Focus on patient monitoring, remote health management, and personalized healthcare solutions. • Wearable devices, medical sensors, and AI are usually used to support healthcare professionals and improve the overall results or consequences of medical treatment or care for patients. • In hospitals and healthcare facilities IEs are used to improve patient care, asset tracking, and facility management. • RFID tags, real-time location systems, and patient monitoring devices are commonly used.
Smart offices (Papagiannidis & Marikyan, 2020)	• Enhanced workplace productivity, comfort, and energy efficiency. • Solutions mostly include smart lighting, climate control, occupancy sensors, and conference room management systems.
Industrial automation (Deshpande et al., 2016; Peres et al., 2020)	• In industrial settings, IEs with the use of CPS and automation aim to optimize manufacturing processes, reduce downtime, and enhance safety. • Predictive maintenance systems and robotic automation are usually employed.
Retail and marketing (Dlamini & Johnston, 2016; Weber & Schütte, 2019)	• AI and data analytics are used to provide personalized shopping experiences, optimize inventory management, and improve customer engagement. • Beacon technology and customer tracking are also common in retail environments.
Education (Korozi et al., 2019)	• IEs in education focus on creating smart classrooms and campuses with interactive learning technologies, adaptive content delivery, and IoT-based campus management.
Transportation (Zantalis et al., 2019)	• In the transportation sector, IEs include smart transportation systems that optimize traffic flow, improve public transportation, and enhance road safety. • They use real-time data and sensors to manage traffic and monitor vehicle conditions.
Agriculture (Stratakis et al., 2022)	• IEs are applied in precision agriculture, where sensors and data analytics optimize crop management, irrigation, and livestock monitoring. • They help increase agricultural productivity while conserving resources.
Entertainment and Media (Leonidis et al., 2021a)	• In the context of entertainment, IEs provide personalized content recommendations, immersive experiences, and interactive media using AI and augmented reality.
Hospitality (Leonidis et al., 2013; Nadkarni et al., 2020)	• In the hospitality industry, IEs offer smart hotel rooms, keyless entry, room automation, and guest services for an enhanced guest experience.

that offer benefits such as automation, energy efficiency, safety enhancements, and personalized services. Below, a non-exhaustive list of domains where IEs are commonly encountered and making a significant impact is presented, while as technology continues to advance, their applications are likely to expand into new ones.

While IEs have demonstrated significant potential (Weber et al., 2005) and have found applications across various domains (Dunne et al., 2021), their rapid growth in recent years (Kotha & Gupta, 2018) and optimistic future projections (Kawamoto et al., 2014) have not translated into widespread acceptance and popularity, as initially anticipated.

Alongside the evident factors, including the absence of standardization (Delsing & Lindgren, 2005), the need for technical expertise (Leist & Ferring, 2012), resistance to change (Ben Allouch et al., 2009), privacy and security concerns (Kirchbuchner et al., 2015), several critical ones have played a pivotal role in driving this deviation. These factors encompass: (1) the substantial cost of the initial investment (Song et al., 2020), (2) limited awareness of the value, benefits, and capabilities of IEs (Augusto et al., 2013; Emiliani & Stephanidis, 2005; Prandi et al., 2020), (3) challenges related to reliability and compatibility with existing infrastructure and legacy systems (Perumal et al., 2010), (4) the complexity of designing, deploying, and maintaining environments integrating diverse technologies (Aarts, 2004), and (5) the necessity for successful implementation to involve collaboration (Remagnino & Foresti, 2005) among experts from various disciplines, including Engineering, Computer Science, Social sciences, and Design.

The former challenges are expected to become easier to overcome over time as prices decrease due to increased availability, people become more accustomed to and accepting of the new reality, and regulations providing stronger privacy safeguards. Addressing the latter challenges can be achieved through the implementation of a formalized design process tailored to IEs that would: (1) emphasize the inclusion of essential functionalities to minimize unnecessary expenses, (2) provide effective solutions to real-world problems, (3) facilitate early testing with legacy systems to enhance interoperability, and (4) address the issues of complexity and collaboration through well-defined procedures.

The *AmI Design Process* described in this chapter has emerged as a result of ongoing efforts to design and develop highly interactive IEs, carried out within the broader context of the long-term interdisciplinary AmI RTD Programme of ICS-FORTH dedicated to the development and application of cutting-edge user-centric AmI technologies and smart environments. Moreover, it has been driven by the observation of lack of a comprehensive framework for constructing such solutions. The proposed design framework draws upon our extensive experience gained from developing in-vitro intelligent spaces and publicly available interactive systems across many sites for a period of over a decade.[3] Our work has spanned diverse application domains,[4] including but not limited to 'Education' (Korozi et al., 2019; Prinianakis et al., 2021), 'Hospitality' (Leonidis et al., 2013), 'Sports' (Leonidis et al., 2021a), 'Entertainment' (Leonidis et al., 2019), 'Well-being' (Leonidis et al., 2021b), 'Agriculture' (Stratakis et al., 2022), etc.

This chapter is organized as follows: In Section 1.2, we delve into the challenges, problems, and barriers associated with design for IEs explore similar design approaches, and elaborate on the motivation and approach. Section 1.3 provides a comprehensive overview of the design framework, while Section 1.4 outlines the developed supporting tools. Moving on to Section 1.5, we showcase the application of the *AmI Design Process* to a specific use case, and finally, in Section 1.6, we draw conclusions and recommendations.

1.2 DESIGN FOR IEs: CHALLENGES, PROBLEMS, AND BARRIERS

Designing any interactive space (e.g., a living room) within a complex environment, such as an 'Intelligent Home,' is far from a straightforward endeavor presenting a multitude of challenges (Augusto et al., 2018; Salem et al., 2017). Notably, the specific requirements for even a single intelligent application often necessitate fundamental alterations to the environment's design, resulting in a significant escalation of the context's complexity. The undertaking is inherently multidisciplinary, requiring expertise across diverse domains, including electrical and mechanical engineering, computer science, control theory, human factors engineering, architecture, etc. Such multidisciplinarity adds layers of complexity to the design process, often demanding collaborative efforts from multiple teams from different disciplines, and while at first glance, the design process for IEs shares numerous similarities with traditional digital systems, undeniably it introduces additional challenges and intricacies (e.g., safety, reliability, security, and privacy).

Moreover, IEs often comprise a wide array of devices and systems that must harmoniously collaborate. Hence, it is crucial to tackle interoperability challenges as an integral part of the design process. Along the same lines, these environments entail the integration of physical and digital components, introducing new facets to the design process that necessitate attention. This integration adds new dimensions to the design process that must be considered, such as the physical environment in which the system operates, the sensors and actuators used to interact with that environment, and the human factors associated with the system's use. Additionally, careful consideration should be taken with regard to scalability, requiring systems that can flexibly adapt to changes in size or scope over time.

Finally, considerations for human factors such as learnability, usability, and accessibility must be seamlessly integrated throughout the design process. Therefore, involving end users in the design process of such environments is of paramount importance. However, this can be more challenging in comparison to traditional digital systems, as often end users may possess limited experience or familiarity with the technology, thereby potentially requiring additional training or support to effectively contribute to the process.

1.2.1 Design Aspects Specific to IEs

In the field of Human-Computer Interaction (HCI), it has long been recognized that a formalized design process plays a crucial role in mitigating risks, enhancing solution quality, improving user satisfaction, and boosting the chances of successful adoption. This approach is even more pronounced within the domain of IEs. Given their inherent complexities (some of them were outlined above), and the need to secure their acceptance and successful implementation, a design methodology distinct from the traditional approach becomes imperative. Such a tailored method is essential for addressing the unique challenges and considerations associated with crafting and designing complex, context-aware environments. The key aspects that should be covered by such a method aiming to offer a suitable foundation for creating IEs are summarized in Table 1.3:

1.2.2 Designing Intelligent Artifacts vs. Designing Intelligent Services

The term "Designing intelligent artifacts" pertains to the process of developing physical or digital objects, devices, or systems that integrate intelligence and demonstrate intelligent behaviors, capabilities, or functionalities. These intelligent artifacts are intentionally designed to engage with their environment, dynamically adapt to changing conditions to enhance their usefulness and functionality, autonomously execute tasks or collaborate with human guidance, and operate either independently or in seamless harmony with other artifacts, services, or even entire environments (Gomes et al., 2014). On the other hand, "Designing intelligent services" encompasses the creation of service-oriented solutions that leverage technology to offer users enhanced and intelligent experiences (Tapia et al., 2010). This approach focuses on delivering intelligent capabilities, insights, and interactions within the context of services, rather than focusing on physical or digital artifacts; this distinction sets it apart from the process of designing intelligent artifacts.

However, it is important to note that within IEs, both artifacts and services coexist and their collaboration is what transforms ordinary daily environments into smart ones capable of anticipating and proactively supporting users with their activities. Their design requires interdisciplinary collaboration among designers, engineers, AI specialists, ethicists, UX experts, and more, so as to produce solutions that are not only technologically advanced but also user-friendly, all while adhering to ethical and practical considerations (e.g., efficiency) while increasing convenience, or providing novel experiences, thereby meeting user needs and preferences. Table 1.4 summarizes

TABLE 1.3
Design Aspects of Intelligent Environments

Environment or artifact definition	• Provide a systematic and creative process to conceive, plan, and shape physical or digital environments. • Support the definition of environments or individual artifacts integrated with smart technologies aimed at enhancing user experiences, well-being, and efficiency. • Enable a conceptualization process spanning from modest extensions to existing environments to the meticulous creation of entirely new, yet feasible, IEs. • Ensure versatility to allow for the proposal of solutions of varying sizes, ranging from a smart kettle to a smart city. • Allow for the exploration of alternative solutions to address limitations posed by various factors (e.g., financial resources).
Technology integration	• Accommodate the seamless integration of diverse technologies, including sensors, IoT devices, AI algorithms, and connectivity solutions, contributing to the development of intelligent and responsive environments. • Ensure compatibility with existing infrastructure, legacy systems, and interoperability with diverse devices and platforms.
User-centered design	• Emphasize the user-centric approach and the fundamental importance of understanding user needs, behaviors, and preferences at the heart of the process. • Acknowledge the pivotal role of user-related activities (e.g., user research, personas, user testing, etc.). • Prioritize user needs, preferences, and experiences.
Interdisciplinary nature	• Emphasize that the design of IEs is an interdisciplinary endeavor that involves stakeholders from multiple fields, including architecture, industrial design, human-computer interaction, engineering, artist, healthcare experts, artificial intelligence, etc. • Procure active collaboration among experts from these domains.
Iterative process	• Promote the iterative nature of the process, where prototypes and concepts are continuously refined based on user feedback, data analysis, and changing requirements, to enhance functionality, user satisfaction, and efficiency.
Context awareness	• Account for the fact that IEs should be designed to be context-aware, enabling them to adapt to evolving circumstances and user contexts.
Real-world applications	• Facilitate the process for either enhancing existing infrastructures (e.g., homes, offices, healthcare facilities, urban environments) through the implementation of Ambient Intelligence Technologies or for application in the development of such environments from the ground up. • Recognize and address the existence of inevitable technical constraints linked to the limitations of available technologies and infrastructure (e.g., capabilities of sensors and devices, network bandwidth, and computational resources).
Ethical considerations	• Acknowledge the ethical dimensions of designing IEs, including privacy, security, fairness, accessibility, transparency, and legal constraints. • Ethical design principles should guide decision-making throughout the process.
Challenges and considerations	• Allow for adjustments to meet the demands of interoperability, scalability, sustainability, and the ongoing need for maintenance and updates, as the challenges associated with designing IEs can emerge at any point. • Acknowledge that budgetary constraints exert a significant impact on the design and implementation of IEs, as the availability of financial resources can influence the selection of technologies and the overall project scope. • Incorporate cost-effective methods to evaluate the effectiveness and efficiency of design decisions before committing to their potentially costly implementation.
Future directions	• Do not confine the process to existing technical solutions; instead, allow for the integration of new components or emerging trends and future directions, such as augmented and mixed reality, AI-driven automation, smart cities, sustainable living, and more. • Provide the essential hooks to facilitate the seamless introduction of intelligent components into established environments and their subsequent testing, optimization, and improvement.

TABLE 1.4
Design Aspects of Intelligent Artifacts vs. Intelligent Services

Aspect	1. Nature of the Solution	Impact on Design Process
Artifact	Focuses on creating physical or digital objects, devices, or systems incorporating intelligence. *Examples*: smart devices, robots, autonomous vehicles.	In both cases, the ability to conceptualize and prototype remains equally important. In artifact design, the process should empower designers to evaluate the artifact within its actual operational conditions. In service design, the process should be flexible enough to incorporate activities related to the core functionality of the service, like the ability to evaluate how the new service integrates into the existing space or assess its added value in different contexts.
Service	Centers on services that offer intelligent functionalities, data-driven insights, and personalized experiences. Examples: virtual assistants, recommendation systems, and data analytics services.	

Aspect	2. User Interaction	Impact on Design Process
Artifact	Often involves physical interfaces, sensors, and direct manipulation of the artifact (e.g., voice commands to a smart speaker).	Encourage active user engagement and facilitate real-time testing and assessment to ensure the solutions delivered, are both user-friendly (e.g., appropriateness of interaction paradigm, intuitiveness of commands, touch interfaces, gestures, or other modalities) and meaningful within their intended context of use.
Service	Typically occurs through digital interfaces, software applications, or online platforms (e.g., interacting with a chatbot on a website or receiving personalized recommendations).	

Aspect	3. Function-specific vs. Generic Task	Impact on Design Process
Artifact	Primarily designed for specific functions or tasks (e.g., a robotic vacuum cleaner designed for cleaning).	The process should encompass activities aimed at gathering user feedback regarding the functions or tasks the artifact should perform in the case of an artifact. For services, the process should guide designers to adopt a broader perspective to understand how the service aligns with the overall context.
Service	Designed to fulfill broader user needs and offer a range of features, capabilities, or information (e.g., a virtual assistant that provides weather updates, schedules appointments, and answers questions).	

(Continued)

TABLE 1.4 (*Continued*)
Design Aspects of Intelligent Artifacts vs. Intelligent Services

Aspect	4. Data Utilization	Impact on Design Process
Artifact	Data is collected and processed by the artifact itself or its associated devices to enhance its functionality (e.g., a self-driving car using sensor data to navigate).	For artifact design, it should be ensured that the algorithms and mechanisms underpinning its functionality align with the initial conceptualization (respond to stimuli or user inputs, learn and adapt from user interaction). For service design, the process from the designer's perspective should include: a. Defining the necessary functionality of the service to enhance the user experience within these environments. b. Evaluating realistic scenarios to understand how the service interacts effectively. c. Experiencing firsthand how the service seamlessly integrates with other components and services within the environment.
Service	Rely on data analysis and AI algorithms to deliver personalized recommendations, predictions, or insights (e.g., a dinner recommending service based on user preferences and level of daily activity).	

Aspect	5. Ethical Considerations	Impact on Design Process
Artifact	Concerns related to the artifact's behavior, safety, and impact on the physical environment.	From the perspective of an artifact, the process should empower designers to verify whether the object functions as expected without posing any risk to its users within its intended context of use. Likewise, from a service perspective, the process should offer guidance to ensure that the service being developed not only adheres to its relevant ethical considerations independently but also when integrated into a technologically advanced environment, does not present any security threats.
Service	Encompasses issues such as user privacy, data security, algorithmic bias, and transparency in service operations.	

their differences and elucidates how the design process should be tailored to encompass both their shared and distinctive aspects.

While both service and artifact design involve the incorporation of intelligence, they diverge in their nature, user interaction, and the range of functionalities and services they provide. Nonetheless, an effective design process should be adaptable to encompass both approaches.

1.2.3 Similar Design Approaches

Much work has been done for the software engineering (i.e., technical) part of the development, and the literature has quite a few examples of attempts to tackle specific parts of the process, such as requirements elicitation or definition (Alawairdhi & Aleisa, 2011; Evans et al., 2014). From a theoretical perspective as well, there are many different approaches and frameworks that can be used to design IEs, like (1) user-centered design (UCD) (DIS, 2009) in an attempt to understand and meet the needs and preferences of users, (2) the concept of affordances (Norman, 2004), and (3) solutions stemming from the fields of AI and machine learning (Ramos et al., 2008) such as reinforcement learning, natural language processing, etc. that can be helpful in designing IEs. Overall though, designing IEs requires a multidisciplinary approach (Augusto et al., 2013) that draws on a wide range of theoretical perspectives and practical considerations, in order to create systems that are effective, efficient, and user-friendly.

Over the years, many notable attempts have been made to apply or even modify existing design frameworks to design IEs. Examples of such attempts can be classified under three main topics: processes to design artifacts or applications for IEs, processes toward designing smart spaces, and processes designing even larger environments like smart cities. We will present some of these examples next.

1.2.3.1 Smart Artifacts

Starting with individual intelligent artifacts or services, researchers have made significant progress in the requirements definition task by introducing sub-loops, improving the requirements collection process, and resulting in the development of highly usable smart artifacts (Kopetz et al., 2019; Psarommatis & May, 2023). To achieve this, various methods have been employed to increase user involvement and collaboration across cross-disciplinary stakeholders (Wang & Moulden, 2021), including focus groups, interviews, and surveys with prospective end users (Alkhafaji et al., 2020), individual conversations with the elderly (Imbesi & Mincolelli, 2020), or observations and workshops with experts (Crovari et al., 2019).

1.2.3.2 Intelligent Spaces

Next, with respect to the development of intelligent spaces, user involvement remains the central focus. Researchers aim to make users stakeholders in IoT products (Chin et al., 2019) and engage them in the early stages of development (Gil et al., 2022). This approach allows for more integrated perspectives from relevant disciplines, especially when smart spaces may not meet user expectations (Mitchell Finnigan & Clear, 2020). To foster agency and participation, users are encouraged to develop mental models of building experiences, and designers explore innovative environments using "design fictions" (creative ways to elaborate on speculative, but realistic visions of the future) (Reisinger et al., 2019). Additionally, researchers have proposed solutions for conducting realistic and immersive experiments (Gil et al., 2022) with "cheaper" prototypes (Gullà et al., 2019), anticipating issues and failures that may arise in real-world scenarios. For example, ARENA aims to enhance manufacturing productivity in smart factories (Piardi et al., 2019), and a mixed reality (MR)-based reference framework introduced by Dasgupta et al. (2019) helps represent, visualize, and model data, digital components, and interaction scenarios in smart building environments (SBE). Lastly, some advanced approaches have been suggested to tackle the design flow of such systems. These approaches may involve a refinement process starting from high-level definitions and progressing to the full formal synthesis of the underlying architecture (Zeng et al., 2020). Alternatively, they may introduce sub-methods to emphasize and identify highly ethical considerations or gather detailed requirements tailored to specific contexts (Augusto et al., 2020).

1.2.3.3 Larger IEs (e.g., Smart Cities)

In larger-scale projects, the focus is on supporting designers in making informed decisions through experimentation. For instance, the work of Sánchez Sepúlveda et al. (2019) assists experts in construction and design fields with urban design activities using virtual reality (VR), while the system in Scorpio et al. (2020) highlights the current use of VR in outdoor lighting design. Moreover, Sanaeipoor & Emami (2020) suggested adopting innovative augmented reality (AR) technologies to advance smart cities, enabling experimentation and transformation of how designers perceive and interact with the physical world. Finally, close collaboration between different disciplines, knowledge sharing, and various forms of co-design are emphasized throughout the research (Catalano et al., 2021; Matz & Götz, 2020; Sanaeipoor & Emami, 2020).

1.2.4 MOTIVATION AND APPROACH

Our team has been implementing IEs for almost two decades and has been engaged in the design, development, deployment in-vivo, and maintenance of hundreds of innovative interactive artifacts. We have also conceptualized and created various intelligent spaces, including: (1) a smart home consisting of intelligent spaces such as the living room (Leonidis et al., 2019), kitchen (Gavaletakis et al., 2022), bathroom and bedroom (Leonidis et al., 2021b), and dining area (Margetis et al., 2013), (2) an intelligent greenhouse (Stratakis et al., 2022), (3) a smart classroom (Korozi et al., 2019; Prinianakis et al., 2021), (4) a smart exhibition space (Grammenos et al., 2010; Zabulis et al., 2013), (5) an innovative multi-purpose room known as the 'Whiteroom' (Leonidis et al., 2020; Partarakis et al., 2017), and (6) a smart hotel room (Leonidis et al., 2013).

Additionally, we have developed numerous small- and medium-scale applications to complement these intelligent spaces.

Over the years, we have come to recognize that even the development of smaller applications or artifacts, let alone planning the creation of an entirely new space, demands a substantial amount of work and coordination. Designing an entire environment is an inherently ambitious undertaking, far more complex than 'simple' application design. Furthermore, even the design of seemingly simple artifacts requires the collaboration of experts from multiple disciplines. Various existing processes and approaches discussed in Section 1.2.3 were not considered fully comprehensive as they did not cover all the core stages of design (i.e., problem definition, relevant research, prototyping, implementation, and testing), nor were they agnostic of the type of IE, nor did they accommodate the need for the process to be able to connect the design of the large environments (general) to that of a specific solution or artifact that would then need to be integrated into the larger environment.

Given that IEs inherently aim to support their users and that the user should be the focus of attention when designing such applications, one possible and convenient solution is to employ *Design Thinking* (Brown, 2008; Park & McKilligan, 2018). It is a user-centered approach to problem-solving and innovation that emphasizes empathy, creativity, and iterative development, which has several benefits when applied to the development of user-oriented applications and services like placing the user at the center of the design process, encouraging designers to empathize with users and gain a deep understanding of their perspectives, fosters a creative mindset, encourages cross-functional collaboration and accommodates changes and adaptations as new insights and feedback emerge among others.

Nevertheless, while design thinking provides a strong foundation for user-oriented design, its application to IEs may require some adaptations and additional considerations since, among others: (1) IEs are often more complex than standalone applications, (2) collaboration among experts from diverse fields is often required, (3) integration of various technologies and interoperability are significant challenges, (4) data privacy and security are not clearly covered, (5) IEs should plan for scalability and adaptability to changing user needs and technological advancements, (6) IEs should consider the long-term impact of on users' behaviors, habits, and well-being, as they gradually become integral parts of users' lives. As a result, we have developed a process (Section 1.3) to tackle the aforementioned challenges and cover the design aspects described in Section 1.2.1.

1.3 DESIGN FRAMEWORK FOR IEs

This section presents the *AmI Design Process* (Figure 1.1), which aims to address the following high-level objectives:

Incorporate human-centered design and design thinking: by establishing a structured process that aligns with their design principles and practices (i.e., involves end users in the design process to ensure that the final result is effective, efficient, and aligned with users' needs, includes comprehensive assessments of performance, usability, effectiveness, and impact, among other factors).

Foster flexibility: by providing a generic process that is agnostic of the type and size of the environment and accommodating both the top-down development of entire environments and the integration of solutions/artifacts into existing ones.

Facilitate the identification of essential disciplines and their responsibilities: by determining the right blend of disciplines for inclusion in the process, defining the precise roles and duties for each discipline, and deciding when and how a discipline's contributions should be incorporated into the overall process.

Promote innovation and research: by encouraging the involvement of researchers and their innovative ideas, mitigating bias, promoting the introduction of fresh concepts, and providing valuable research and design experience while exposing them to cutting-edge technologies.

Design for Intelligent Environments

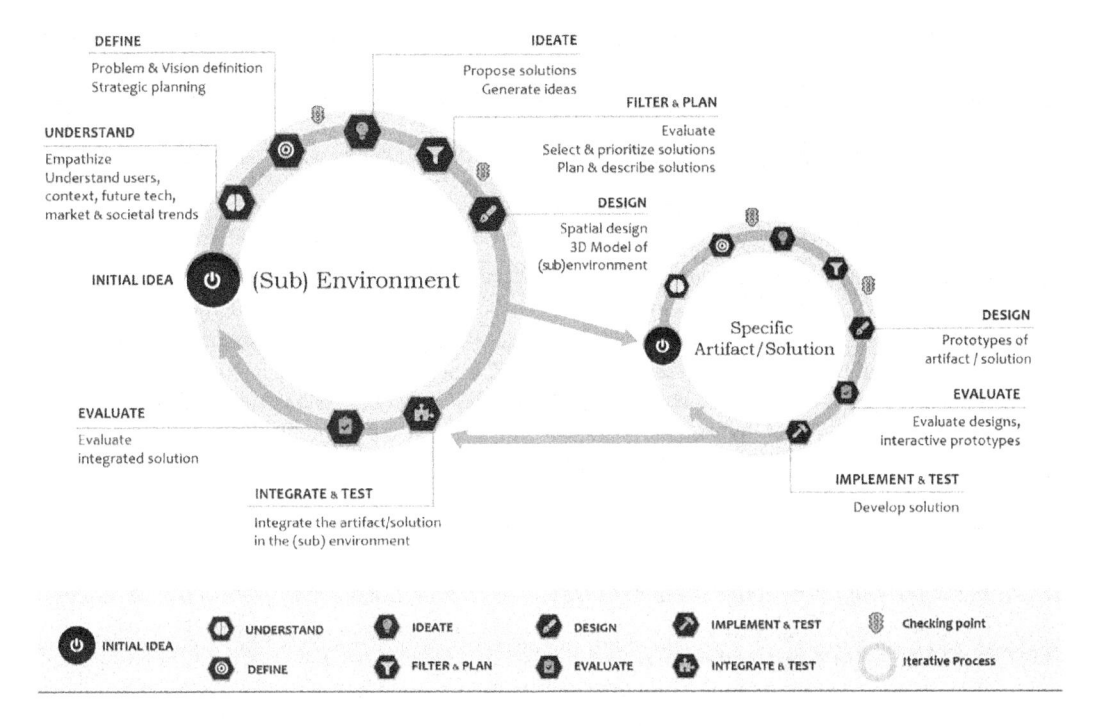

FIGURE 1.1 The AmI Design Process

Facilitate project selection, planning, and management: by employing filtering and decision-making mechanisms for the optimal selection of problems and solutions to focus upon.

Balance innovation and feasibility: by evaluating the technical and economic viability of proposed solutions, taking into account market potential, cost-effectiveness, and prioritizing the development of products that effectively address real-world challenges while improving their potential of acceptance.

Provide a collection of IE-specific tools: that address the unique challenges of complex environments.

The framework follows an iterative process for designing a large-scale environment (e.g., a smart home) or a sub-environment (e.g., a smart living room inside a smart home), or a specific artifact or solution to be integrated into an IE (e.g., a smart sofa for a smart living room).

As depicted in Figure 1.1, the process starts at the large, outer, circle as we are exploring the needs and requirements of an entire environment or sub-environment. The starting point of the process is the initial idea or goal. It can be the eureka moment or a goal formulating for years before becoming an actual idea. It can be something small, like "why not design a smart kettle for the smart kitchen?" or something entirely grand and ambitious. "We will build a smart home". Or "we will design a small smart city from scratch". In any case, an idea or goal must be had. After completing the first four stages of understanding, defining, filtering, and ideating for that environment, we then move to the small circle as the focus shifts to the designing of specific artifacts and solutions for that environment. In the figure, this is depicted with the connecting line that goes from the design stage of the outer circle to the understanding stage of the inner circle. Once inside the small circle, the team follows the same stages as before only this time the stages are applied to the specific solution or artifact. After the implementation and evaluation of the specific artifact or solution, we move back to the outer circle in order to integrate and evaluate it in the context of the entire environment. We can also move to an inner circle with the same steps, addressing a sub-target, as many times as needed, depending on the project scope and objectives. This is depicted in the figure with the connective line from the integrate stage of the small circle to the integrate stage of the outer circle.

The two circles represent the concept that environments and sub-environments are larger conceptual and physical entities that contain other systems and applications that are intrinsically linked and therefore cannot and should not be designed in isolation from each other. Therefore, as a team we want to think and plan from the general to the specific, from the outer larger environment to the specific solution or artifact. The purpose is to have a structured process that addresses the large and small challenges within this context of development.

Furthermore, as mentioned earlier, one of the biggest challenges of designing IEs is the high complexity of the endeavor and the large number of resources, i.e., budget, time, and human resources, needed. This reality makes the need for a good project management plan imperative. To ensure that the project team remains within the scope of the project and that they focus their efforts on the right solutions, we have employed two checking points in the process (depicted as traffic lights in Figure 1.1). The first checking point is after the Define stage and the second is after the Filter and Plan stage. These are the moments in time when critical decisions must be made, i.e., the project stakeholders must decide to go ahead and commit to the project (or the selected solution) and adjust it as appropriate or abandon it altogether. The decision is made based on the information collected and provided during these stages. Factors that will determine if a project moves forward, in our experience, have been budgetary concerns, matters of technical feasibility, idea deemed uninteresting, too risky (large investment for unknown results), lack of resources, and in the recent past, COVID restrictions (the quarantine effectively delayed a lot of activities, including meeting and designing within the physical space of interest, or evaluating ideas in situ).

As shown in Section 1.2, any system developed needs to be well-designed and evaluated before the actual development begins. This saves time and costs, despite appearing to be more time-consuming. This is especially true for IEs, where the costs and resources needed, as well as the complexity of the task, are many times more, not only in the sense of involving the development of multiple systems but also the design and building of an entire physical space.

The framework is intentionally designed to be flexible. This means that while we provide explicit guidance on who should be involved and what should be produced at each stage, we leave the selection of the techniques to be used to complete each respective stage (the how-to) to the discretion of the design and development team. Nonetheless, we do provide a list of recommended methods and techniques drawn from our past experience, which we consider essential for the effective implementation of the process. In other words, the *AmI Design Process* model is flexible in the 'HOW' section of the various phases and explicit in the 'WHAT,' 'WHY,' and 'RESULT' sections.

It is important to note here that as the process is based on the UCD principles and the *Design Thinking* process, it is fully iterative, meaning that the results of any of the stages, either in the large or in the small circle, may lead the team to a previous stage.

Finally, while forward-thinking, we must emphasize that we regard the topics of 'privacy,' 'security,' 'accessibility' and 'ethics' as highly critical. They are of paramount importance, and we highlight multiple instances where they must be taken into consideration. However, given the context and constraints of this chapter, we will not delve into addressing these issues in much detail.

The stages of the process are explained and analyzed below.

1.3.1 Stage 1: Understand

WHAT: In the first stage of the *AmI Design Process*, the focus is divided into three main goals: (1) to understand the users of the environment we are designing for, (2) to understand the context of use of this environment, and (3) to understand the state-of-the-art regarding the technologies that are used in the environment, as well as future technological and market trends.

The collected knowledge will then be analyzed and translated into the initial draft of the user-requirements document which is the output of this stage.

1.3.1.1 Understand and Empathize with Users

Just like in other HCI design approaches such as design thinking, user-centered design (UCD), and UX design, the first step of the *AmI Design Process* is to identify the target audience of the environment we are designing for, and gain a deep understanding of their respective needs, preferences, and opinions about it. By target audience, we mean the people (also referred to as users) who have a direct stake or interest in the environment and any other peripheral stakeholders. In addition, in this stage, we also want to explore the users' insights, experiences, and opinions on smart environments and technologies in general and what requirements they must fulfill in order to be regarded as beneficial and supportive of their needs. In the example of designing a broad smart environment such as a smart home for a typical family, the identified target audience could be divided into different groups: (1) parents, (2) children of various ages, (3) elderly relatives that may live in the house, (4) other stakeholders such as family members and family friends that occasionally visit and interact with the family. Representative users from all these groups should be included in the efforts to understand and empathize with their needs and experiences as these may differ from group to group.

1.3.1.2 Understand the Context of Use

The second goal of this stage is to understand the context of use of the environment, i.e., understand the when and how this environment is utilized. This means to identify the type of activities, events, and routines performed in that environment by the target audience. The description of these activities can be general or very specific, but in any case, it is good to note who performs each activity, who participates in it, and how often it takes place. The goal is to record a typical day in that environment. In the example of the smart home environment, the identified activities may be cooking breakfast, working from home, studying homework, cooking lunch or dinner, watching TV, etc. The activities may be organized in meaningful themes, i.e., morning routine, evening routine, dinner, breakfast, etc. For each one of those, we would want to gather more data regarding who performs or participates in each activity, what is needed to complete the activity, what are the common issues they face during the activity, what they like or dislike about it, etc.

1.3.1.3 Understand Technological and Market Trends

The third goal of this stage is to understand technology, market, and social trends regarding the environment we want to design. ICTs and the IoT technologies are constantly evolving creating new opportunities and possibilities for the development of new smart products, applications, and services for various settings, such as assisted living, households, e-health, education, etc. Smart technologies are already being used in everyday living devices, such as mobile phones, cars, TVs, refrigerators, lighting systems, vacuums, etc. Advancements in IoT technologies and communication protocols are a constant force for expanding the possibilities of the integration of smart technologies in everyday devices. For example, recently a new connectivity protocol, called Matter,[5] was launched by Amazon, Apple, Google, Samsung, and many other manufacturers promising improved device compatibility and security in the hopes to increase the rate of adoption of such solutions. The design team of a smart environment should be aware and have a good understanding of these trends. Data regarding the state-of-the-art, implementation challenges, tested use case scenarios, types of available commercially-ready products and applications, and social trends that may affect their adoption rates is crucial for making strategic decisions regarding the smart applications and products we want to focus on.

HOW: There are many HCI research tools and techniques that can be used to achieve the aforementioned goals. Below, we describe just a few of those that we think are of great value, however, the design team can pick and choose other techniques and tools that best fit their needs and their resources.

1.3.1.4 Scientific Research

Conducting scientific research on published studies about IEs, technologies, and use cases will help the design team collect data regarding all three goals of the first stage. We can learn a lot from reading the latest publications on technological innovations, applications, and use cases of smart technologies. The research should also be expanded to include studies that followed a user-centric approach focusing on exploring the user needs, preferences, and motivations regarding such technologies and environments. Furthermore, the research can be expanded to include domains other than the one we are designing for. For example, if we are designing a smart home environment, we may want to expand the research on other environments such as assisted-living environments, health-related environments, or educational environments. The results of good scientific research may become a good source of inspiration and ideation for solutions.

1.3.1.5 Market Research

Market research is another type of desktop research that can help the design team collect valuable data regarding smart technologies. The goal of this type of research is to look for already available commercial plug-and-play artifacts, software, and hardware that can be used in smart environments with little or no configuration required. This will allow the team to understand what is already possible and available and identify opportunities and needs for new services and applications. Off-the-shelf smart technologies, artifacts, and applications can be reused and repurposed to fit the needs of the environment they are designing. This can save us valuable resources down the line. Market research will help us acquire information mainly regarding the third goal, but can also be useful for the other two goals as well.

1.3.1.6 Questionnaires

One of the best ways to find out what users think or believe about a subject is to conduct a survey in the form of a questionnaire. In our context, the questionnaire's goal is to collect data regarding users' experiences, perceptions, and attitudes toward smart technologies in general and toward technologies specific to the home environment. Parameters such as perceived usefulness, positive and negative views regarding smart technologies and environments, and user expectations are worth investigating. A well-planned and well-constructed questionnaire can help the team understand the user better and provide valuable information that will feed into the user requirements document. There are many resources online on how to build a good questionnaire, e.g., (Lee, 2006; Müller et al., 2014; Stantcheva, 2023). It is important to note that from the scientific research, described above, we can draw ideas and suggestions for questions to include in the questionnaire.

1.3.1.7 Interviews

Just like questionnaires, interviews are another form of survey that is also very commonly used to collect data on users. The main difference is that interviews provide qualitative data, whereas questionnaires can provide quantitative data. Interviews are also a great tool to further deepen the understanding of the user's attitude and perception toward smart technologies. Depending on the experience of the user with smart technologies, it may be necessary to explain the basic concepts regarding such technologies at the beginning of the interview to ensure that the user understands the context of the questions. There are many online resources on how to conduct fruitful interviews, such as the Interaction-Design Foundation,[6] the Usability gov,[7] and the Design Kit[8] websites. More information is also available from the works of Taherdoost (2022).

1.3.1.8 Field Study-Observations

A field study in the form of observations is a research method that involves studying users in the context of their own environment. It allows the team to observe how users behave, act, and interact with each other, and use technology in the context of an environment. It is a great way to collect data that will help the team deepen their understanding of users' needs, habits, routines, issues, goals, and activities. It is a relatively easy method to apply[9] and it can be combined with in-depth interviews for more comprehensive results.

1.3.1.9 Personas

Personas is a very useful and commonly used technique in HCI that helps the design team identify and empathize with users' needs, goals, behaviors, worries, aspirations, preferences, etc. Personas are fictitious characters that are created from data collected during user research with actual users (see techniques above)

Design for Intelligent Environments

and represent the different user types that might use the environment or the artifact that we are designing. It is a great way to enable the design team to step out of their own bias and assumptions and think through the user's perspective. More information is available from the works of (Nielsen, 2013) as well as from other of the numerous resources available online and in the bibliography. Personas can also be combined with empathy mapping, another useful technique for gaining a deeper understanding of the users by specifying what the user says, thinks, does, and feels. More information on empathy mapping is available from the Nielsen Norman Group[10] website and UX Booth[11] among others.

1.3.1.10 Affinity Diagrams

Affinity diagrams, also known as the KJ method originated by Jiro Kawakita (Kawakita, 1975), can also help the team organize and consolidate big sets of collected data related to problems, user needs, opinions, attitudes, etc. into meaningful relationships. This process is great for making sense of insights gathered during research and can be used in the ideation stage as well, for organizing the generated ideas.[12]

> *WHO:* In the research activities that take place in the first stage of the *AmI Design Process*, it is important for the following project stakeholders to participate:
> - *Project managers*: the persons who will oversee the entire design and development process from start to finish. They must be involved in all stages to coordinate the activities of each stage, manage timelines and resources, and oversee the entire production lifecycle. Therefore, it is imperative that they have a good understanding of the requirements of the environment and its users from the very beginning.
> - *Interaction/UX designers*: the persons who will design the interactive applications that will be included in the smart environment. They need to be involved from the very beginning of the process as their expertise and experience can help not only in the user requirements elicitation phase but also in the analysis phase of the collected data.
> - *Software engineers*: the persons who will develop the applications and systems for the smart environment. The developers of the team should be involved in this stage as well, since they too need to be able to empathize with the user's needs regarding smart technologies.
> - *End users (users)*: representative users from the identified target audience of the environment. They have the central focus in the design process and it is their voice that needs to be heard. In this stage, they will be involved as participants to user requirements elicitation activities, i.e., questionnaires, surveys, observations, focus groups, etc. However, it is important to cultivate a long-term collaborative relationship with them, by asking them if they would like to be notified to participate in future project activities.
> - *Domain experts*: This refers to people who have specific expertise and experience regarding the environment we are designing. For example, if we are designing a healthcare facility, we will need to involve medical staff i.e., doctors, nurses, etc., administrative support staff, and other stakeholders from the management of the facility.
> - *Architect, interior designer*: persons who will be consulted in the integration of the systems and applications in the space of the smart environment. Designing for an IE involves designing a solution, system, or artifact that will probably need to be integrated either in an existing physical space or in a new one. To this end, the expertise of an architect or interior designer will be very valuable.
>
> *RESULT:* At the end of this stage, after organizing and analyzing the collected data, the design team should be able to draft the initial list of user requirements and the use case scenarios document for the environment. This is a living document meaning that it will be continuously adjusted and updated as needed and as the team's understanding of the environment and user needs deepens. The user requirement document can also be accompanied by a state-of-the-art report that will provide an overview of what technologies are currently available for smart environments. This document is also to be kept updated throughout the process as we need to keep an eye on new technology advancements.

1.3.2 Stage 2: Define

WHAT: In this stage, we have two main goals. The first goal is to identify and define specific problems and user needs of the environment we are designing, as these emerge from the analysis of the collected data from the previous stage. The second goal is to decide as a team which of the identified problems we would like to narrow our focus on. To make this decision, the team should analyze various parameters of each problem separately and strategically pick the ones that they want to focus on. Parameters can be the maturity of the technologies involved, technical and resource-related constraints, the impact value of the solution, the research interest and opportunity for innovation, etc. The criteria for this decision should be defined by the team in advance and based on information collected from the previous stage. For example, the team may decide to focus on a problem that creates multiple opportunities for innovation, or on a problem whose solution would have a very high impact on multiple user groups. Whatever the criteria, it is important for the entire team by the end of this stage to have formulated a strategy for the project with a clear vision and objectives for what they want to accomplish and why. This is an important exercise that will facilitate the overall management of the project, i.e., timelines and required resources (budget and human resources) and will ensure that the team stays committed, motivated, and focused on achieving the set goals and objectives.

HOW: There are many tools and techniques in the HCI bibliography and online resources that can help a team with these two goals. Below we provide a list of some of the techniques that we have used in the past and that we have found them to be easy to apply and effective in this context.

1.3.2.1 Frame the Design Challenge

This is a simple technique[13] that allows the team to articulate each problem that has been identified, taking into consideration the impact of its solution, the constraints, and contextual factors that may affect the outcome. It also allows the team to start thinking of potential solutions. In our experience, this is a great tool because it provides a structured template to describe the problems in a uniform way. At this point, there is no need to be very specific in the description of the solutions as this will be done later on at the ideation stage. The important thing is to identify and define problems based on the collected user requirements.

1.3.2.2 How Might We

This exercise[14] can be used in combination with the above technique and after we have identified the problems that have emerged from the previous stage. It is an excellent way for the design team to turn the identified problems into opportunities for design. Basically, the team can take each one of the problems that have been described using the previous technique and brainstorm ideas for solutions. Again, there is no need for the ideas to be very specific and detailed, as this will be done at a later stage. At this point, we are more interested in exploring design opportunities.

1.3.2.3 Top Five

This is another simple technique[15] that can help the members of the team identify the five top themes and ideas that jump out from the analysis of the collected data as the most important or more interesting. The technique can help the team strategize, uncover themes, isolate key ideas, and reveal opportunities for design. It can be used after the other two techniques described above for the team to sort out the top five problems described and rank them based on various criteria, i.e., highest research interest, highest impact, lowest technical constraints, etc. This technique will be useful if the team has identified a large list of problems from the user requirements to choose from.

1.3.2.4 SWOT

This is a powerful technique borrowed from the business management discipline, which has been used as a tool for strategic decision-making and has been developed in various contexts of use and disciplines

(Benzaghta et al., 2021). It is a simple technique that helps a team to identify strengths, weaknesses, opportunities, and threats of new ideas for products and services in relation to business competition or project planning. In our context, this technique can facilitate the decision-making process of which of the identified problems the team would like to focus on. One can find examples of how the SWOT analysis can be implemented in HCI projects in the works of (Engelbrecht et al., 2019; Syed et al., 2021).

> *WHO:* Just like in the previous stage, the following groups of stakeholders should participate in the problem definition phase:
> - *Project managers*: As managers of the project, they are responsible for managing the project's goals, needs, resources, and scope. They also have to make sure that the project resources available (time, budget, and human resources) are used wisely so that the project does not derail from its goals. They are the key stakeholders for any decision-making and strategic planning activity of the team and they have the ultimate authority on final decisions.
> - *Interaction/UX designers*: designers have experience and expertise in identifying patterns and themes in the collected data and analyzing it form user requirements which is the main goal of this stage.
> - *Software engineers/domain experts*: their expertise on technologies and software development is valuable in the initial framing of the project's goals, identification of user-requirements, and possibilities for solutions.
>
> *RESULT:* At the end of this stage, we highly recommend gathering all the information acquired during this stage in a single report that can be reviewed and revisited at any point down the line and by any team member. This report should include the vision and the objectives that the project wishes to accomplish, the initial list of the identified problems and their resulted analysis (e.g., impact, constraints, opportunities for design, etc.—see first described technique above), and the list of the problems that the team decided to pursue solving. The latter should be accompanied by a justification for their selection based on information such as the SWOT analysis or any other decision-making process used that led to their selection. Having a clear purpose in mind now the team can move to the next stage which is to ideate for possible solutions.

1.3.3 STAGE 3: IDEATE

> *WHAT:* By applying the various methods, outlined in the previous stage, to clearly define the problem to be solved, the team should now have a solid idea of what they want to create, outlined in the proposal produced in the previous stage, which should contain among other things a clear problem definition. Thus, we reach the most creative part of the process, where the ultimate goal is to foster innovation and apply it to the defined problem statement. In a sense, this is the heart of the design thinking process that the framework is based upon, customized to suit the needs of designing smart environments and artifacts. The key points at this stage of the process are to involve a diverse mix of stakeholders, apply ideation methods and prepare them to be presented to group experts for the next stage (Filter and Plan).
>
> *HOW:* There is a plethora of methods currently used in the industry and market, as well as an evolving and ever-expanding literature on how to generate ideas and foster innovation, which is the major focus of the design thinking process/mindset.

Conducting brainstorming sessions is the most essential part of this stage, where all stakeholders meet with the sole purpose of generating ideas and solutions to the clearly defined problem. Any brainstorm session should begin by presenting the problem statement and the goal of the session, in effect communicating to the people involved the outcomes of the previous stages and the purpose of the session. Therefore, whoever is hosting the session should prepare an introduction for that purpose that is brief and succinct and would bring up to speed members of the team that have not participated in the previous stages. It should be noted that it might be a good idea to not share the state-of-the-art report or any commentary of the technology available, as this might shift the participants' ideation towards very specific solutions and prevent them from thinking outside the box. Another thing we applied in our sessions was to explicitly state to the participants that any privacy, security, and even ethics issues should be ignored at this stage, not because they have been resolved or that they do not matter (quite the contrary) but because dwelling on them would inevitably foster hesitation and hinder the ideation process. These significant issues should be dealt with at the next stage (Filter and Plan).

During the brainstorming sessions, it is possible to apply different ideation methods, whichever suits any particular team. Different brainstorming sessions can be conducted with a different focus each time, depending on the strategic planning of the team. A list of methods that are suitable can be found in Section 1.3.9, in the summary Figure 1.2. Our team has successfully employed storyboarding and journey maps, which are both particularly suited for generating ideas for environments and sub-environments. Other methods include mind mapping, SCAMPER, crazy 8s, card sorting, "top five" and more.

The sessions can be conducted at any place but we found conducting them at the space of interest (e.g., the intelligent home) to be helpful to inspire participants to imagine their ideas applied inside the space they are currently in.

Work is being done to enhance an in-house tool (Wizard of AmI – see Section 1.4.4) to provide participants the option to meet in a virtual environment that can be manipulated on the spot (adding or removing elements in the environment – effectively rapid prototyping in 3D) and include members of the team that cannot be physically present.

Rapid prototyping, especially at a napkin-level of fidelity, can be extremely effective in communicating ideas that can then be built upon and further explored and enhanced by other participants. These can be produced on the spot, with pen and paper available for each participant or by participants taking turns to demonstrate an idea on the whiteboard. How these are done and discussed is up to the host of the session but in our experience, it is convenient to allow some time for users to sketch mockups or storyboards for any particular problem statement and then take turns to present them, personally or by the host. This will vary depending on the team dynamics and personalities of the people involved; the optimal procedure for each team will eventually present itself over time.

The same procedure (allowing time and then present) can also be applied with post-it notes to record single ideas and then discuss them by bundling them on the board. The categorization of ideas also helps a lot to identify key services and contexts of use for smart environments that can then be analyzed to see if they should be translated to actual services provided by the environment. For example, when brainstorming ideas for the smart kitchen environment, immediately the services for food storage and food preparation came to the surface and were shown to be tightly interlinked.

WHO: The essential roles of people to participate in this stage of the process is:
- *Interaction/UX designers*: During this stage, their role is to facilitate and host the brainstorming sessions, lead and guide the rapid prototyping activities, and any other ideation method employed by the project/research team.
- *Domain experts*: Domain experts should be included in this phase, particularly when it involves environments and sub-environments with domain particularities (such as healthcare or education). It should be noted however that at this stage they are not to provide negative feedback or point out flaws in the ideation process but rather facilitate the interaction/UX designers with feedback to organize the ideation sessions. One or more meetings/interviews before and after major sessions would be a good idea in order to pinpoint the focus of each session and facilitate the recording of the ideas and their categorization.
- *Users*: The process is a user-centered approach (Abras et al., 2004), based on the design thinking process (Brown, 2008), which places the user at the heart of the design, focusing on their needs. Every method listed in the ideation stage should involve users. When tackling general environments (such as a home or a public space), the diversity

of the users involved should be maximum, while at more specific environments or proposed artifact solutions, the selection can be narrower based on the market research, personas, and the domain experts' feedback.
- *Software engineers*: While software engineers are not typical users, it would be a good idea to involve them in these processes, not only to gain their feedback and ideas but more importantly to give them a chance to empathize with potential users and get a better glimpse of their perspective. It is up to the project manager and UX designers to prepare them and decide who to include specifically.

RESULT: At the end of this stage, all the ideas generated should be recorded and categorized. One specific format will be described in detail in Section 1.4.8, tentatively called the *Book of Ideas*. Essentially, each idea is cataloged with a unique ID number, including a brief description, any prototypes that were produced (more can be added later) and any existing or potential services of the environment that are required for the idea to work if identified. Related ideas/services could also be included (e.g., if an idea depends on the information or outcome of another idea). The Book of Ideas can be a single entity for an entire environment/project divided into sections (sub-environments) that contain subsections of other sub-environments or smart artifacts or services, which themselves can be viewed/analyzed as a smaller booklet, so to speak, of ideas.

This is the primary outcome that will be used for the next stage. A "Book of Ideas" application is currently in the works, which will allow easy creation, editing, and sharing of ideas, including the ability to interconnect with the *Wizard of AmI* (i.e., exporting Wizard of AmI representations of ideas directly into the Book and vice versa). However, the Book of Ideas can be a simple word or excel file or even a physical book with drawings.

1.3.4 STAGE 4: FILTER AND PLAN

At this stage, the ideas generated from the ideation stage are investigated and analyzed with the help of various experts. The goal of this stage is to filter out ideas that are unfeasible for any reason (cost-prohibitive, technologically impossible, resource-intensive, etc.) or just plain uninteresting or bad. This will depend upon many factors of course, most important of which being the actual focus of each respective project. In any case, each idea in the Book of Ideas after this process, should be enriched with more comments from the people giving feedback and marked as feasible or infeasible (in the short-term, long-term or in the realm of science fiction!) and how good or valuable and worth pursuing.

It is basically a revisit of the feasibility study and the SWOT analysis, only this time with much more concrete and numerous ideas to analyze, including a variety of domain experts to weigh on. After this stage is over, the project team will have the ideas prioritized and a much clearer view on how to move forward and with what.

One more thing to note is that at this stage the issues of privacy and security complications of the ideas are assessed both from a legal and ethical standpoint. Any problems, risks or challenges should be noted down and be considered at every consequent stage of the process.

HOW: A small team (or even a single person depending on the project size) should conduct informal interviews with domain experts and any person who understands the problem and has a unique perspective to bring. Depending on the size of the task (i.e., the number of ideas to be filtered and the number of people to be interviewed), work can be divided between members of the team that will be conducting these interviews. It is essential that the team meets up in the end to coordinate and update the Book of Ideas accordingly.

The interview process itself is fairly informal; the expert is presented with a single idea at the time and asked to evaluate, comment on it and provide feedback regarding feasibility, how interesting the idea is and what potential it has (from a research point of view or the market) and perhaps add upon it, either with comments or further rapid prototyping to be added to the Book entry. Potential pitfalls, risks and challenges are identified as well as a very rough estimate of the potential cost, if possible. Eventually, through multiple interviews it should be obvious that some ideas have more appeal across different people and those are prioritized and selected for further design and development. It should be expected that not all experts will agree, depending on their domain, and it is up to the team and project manager to discuss the outcome of this process before moving forward to the next stage.

In our experience, some of these interviews have been very useful to discuss the ideas and expand on the larger repercussions of the project and are generally excellent for fostering inspiration and food for thought. In a few instances, some ideas have been broken down into more detailed ideas or even produced entirely new ideas, all of which should be cataloged in the manner described in the previous stage and identified with a unique ID. In some cases, especially in relation to larger projects (like environments), it probably will be necessary to conduct more than one interview, possibly focused on certain ideas that have been significantly commented upon (i.e., stand out and added upon).

A good practice is to have a meeting at the end of the stage with the project manager presiding, to make sure everyone is on the same page and discuss which ideas will be greenlit or generally make any plans for the immediate future and how to move forward.

The latter is the most important outcome of this stage, at the end of which, the team should have a clear idea if the project examined, whether an environment or an artifact, should move forward to the next stage, i.e., design (hence the traffic light in the diagram). This is important because from this point forward a decision has been made if the idea is worth pursuing and resources are committed to it.

WHO:
- *Project manager*: The final decisions rest with the project manager after soliciting with the team members. Their role is to review the revised Book of Ideas and decide if a project should be greenlighted or remain in the Book until further notice taking into consideration the availability of resources (budget, timelines, and human resources).
- *Domain experts*: Their role is to evaluate, comment and/or expand on the ideas presented to them, primarily from the perspective of their expertise.
- *Software engineers*: Their role is similar to those of the domain experts, which means that their input should be focused on their estimation of the feasibility of an idea resource-wise (cost, time, people).
- *Interaction/UX designers*: Their role at this point is to envision the ideas discussed in the larger picture and try to envision their part in the larger picture of the UX. Is the idea taking away or adding to the UX? Does it burden the user with actions that were not necessary before this idea? And so on.

RESULT: At the end of this stage, the *Book of Ideas* has been updated and the ideas worth pursuing have been identified. In the case of specific artifacts and solutions, the whole process will provide enough feedback to refine the requirements document.

1.3.5 STAGE 5: DESIGN

WHAT: At this point, the team has developed a solid concept of the environment (large circle) or artifact (small circle) to be designed and implemented. In the case of an environment, the Design stage involves HCI designers working with architects or other disciplines related to building spaces, either working on an existing space or working together to design and build an entirely new environment. The result of this stage should be any representation of the final environment design (blueprints, 3D spatial models, etc.).

In the case of an artifact or system, using the process described, starting with a basic entry in the Book of Ideas at the ideation stage, after going through filtering and planning and getting greenlit by management, that entry should have been enriched with the basic and high-level requirements (now a separate document, if convenient) that have been identified and verified by the stakeholders. Lo-fi mockups might have been added as well. In any case, at this stage designs can now be produced to flesh out the actual solution. In the case of specific solutions or artifacts, this is the entry point to the small circle of the process.

This stage is tightly linked with the next stage ("Evaluate") and in UCD fashion stresses the need for iteration between stages until the desired goal has been reached. For a period of time, depending on size, scope, and ambition of each project, these two stages effectively run concurrently, as one larger stage, until the Evaluate stage concludes that the Design stage is over. There are many design activities and methods that can be employed in this stage. It is beyond the scope of this chapter to enlist and discuss them all so we have selected the most fundamental ones to include in the description, but every design team may use (in addition to prototyping at least) those that suit them.

At the end of this stage, visual prototypes of varying degrees of fidelity should have been produced and used to further refine the requirements through iterations of evaluations until a detailed plan of the environment or hi-fi, pixel-perfect prototypes of the artifact can be handed to the development team.

HOW: Lo-fi and Hi-fi prototypes

This is the most fundamental design practice and it is irreplaceable. Starting with low-fidelity prototypes, some of which may have been produced in previous stages, then moving on to higher fidelity prototypes as the design-evaluate iterative cycle progresses. In the case of environments, it is very beneficial to produce physical architectural models or design virtual 3D models of spaces. The latter may be used in a VR setting that the team may visit for discussion, ideation, or evaluation purposes.

These prototypes should be produced and discussed with the help of actual users, not just designers and engineers.

During the Design stage of the larger circle of the process, i.e., when producing design prototypes for a smart environment, with multiple interaction points and screens, it is more complicated than designing for a single device or even an ecosystem of devices. For supporting this activity, the Wizard of AmI tool (see Section 1.4.4) has been developed, in order to assign mockups to different screens and coordinate their appearance according to different states of the environment. This is extremely helpful for the evaluation stage that follows. Future work for this tool includes combining the VR models mentioned with the ability to rapidly add or remove screens or interactive points, as well as assign interface mockups for the various scenario states. That, in turn, will allow the design team to roleplay scenarios in the virtual environment, which can be edited and adapted at will.

Participatory/co-design workshop: Co-design or participatory design activities involve users early on in the design process, allowing them to provide input for solutions that will be used by them. This is a vital human-centered design activity that is more necessary when tackling smart environments, in the sense that people should have a direct say about where they are essentially going to live. The co-design workshop should be designed and coordinated by the UX experts. Efforts should be made to plan, recruit, and inform/orient participants, especially those with no particular experience or expertise in these design activities, which should be the case for most users of an environment. If available and possible it is recommended to conduct these co-design workshops in the actual environment the team is designing for. This can be very helpful to visualize design solutions in the actual space.

Further information, as well as resources and tips on how to conduct these workshops, can be found in Book 'USER EXPERIENCE METHODS AND TOOLS IN HUMAN-COMPUTER INTERACTION', Chapter 10 of this Book Series.

Product Design and Industrial Design: These are design disciplines that are associated with physical products or solutions. When designing for an environment, which can be anything from a room to a larger public space, it is wise to follow practices from these two domains, depending on the relevance of the task at hand. For example, when our team was developing the smart bedroom, we had to tackle designs for a physical, intelligent wardrobe. Similarly, when designing the intelligent living room, design decisions regarding the smart sofa and table had to be taken, with the help of engineers.

Storyboarding is an old, well-tested, and invaluable method of visualizing and communicating design ideas regarding the UX and the flow of interaction. They can be created and used during co-design workshops or as presentation tools for the team to discuss design ideas.

WHO: Interaction/UX/UI experts and designers

Interaction/UX and user interface (UI) designers should take the lead in the design activities, producing prototypes and coordinating co-design activities. These roles might be distinct people (different people might design interactions and others design interfaces and others still plan and host workshops) or more than one can be assumed by an expert with the appropriate experience in those roles. It is important that the design team includes expertise in all these areas.

Architects and interior designers: Designing environments or buildings will inevitably require the expertise of architects and interior designers. It is highly recommended to include such expertise early on (as seen in previous stages) so their input can be combined with those of the UX designers and the users involved. In the case of specific artifacts or solutions (small circle) it is not necessary to include them, unless the specific artifact is part of the living space. For example, the kitchen table tops, the intelligent living room, and so on.

Users: Depending on the project's characteristics and the specific target audience, the recruitment process should be tailored accordingly.

Software engineers: Software engineers are usually excluded from the design stage as it is a different role and mindset that is usually needed. In the case of complex environment design, it can be difficult to proceed without feedback from the software engineers, particularly when tackling using interactive prototypes in any given environment. Operating the *Wizard of AmI* tool and handling on-the-fly scripting for impromptu rapid prototyping are two tasks that needed the support of engineers.

Industrial engineers: Like software engineers, industrial engineers can also be helpful during the design phase, during co-design workshops, handling issues with physical objects in the smart environment (sofas, tables, etc.) and contributing ideas from their perspective in the design activities. It is not as important to include them when designing for a specific artifact or service, depending on the context of use.

RESULT: At the end of this stage mockups (lo-fi, hi-fi) that can be evaluated are produced. Virtual or physical 3D models of the envisioned environments. Scenarios depicted in storyboards. As seen in Figure 1.2 heavy iteration is expected between the design and evaluation stages. The models and mockups are discussed and improved, informing and fine-tuning the requirements in the process until the user requirements document is finalized (early iterations) and mockups evolve to high-fidelity or interactive prototypes (final iterations) that can be further evaluated and finally passed on to the development stage.

Design for Intelligent Environments 23

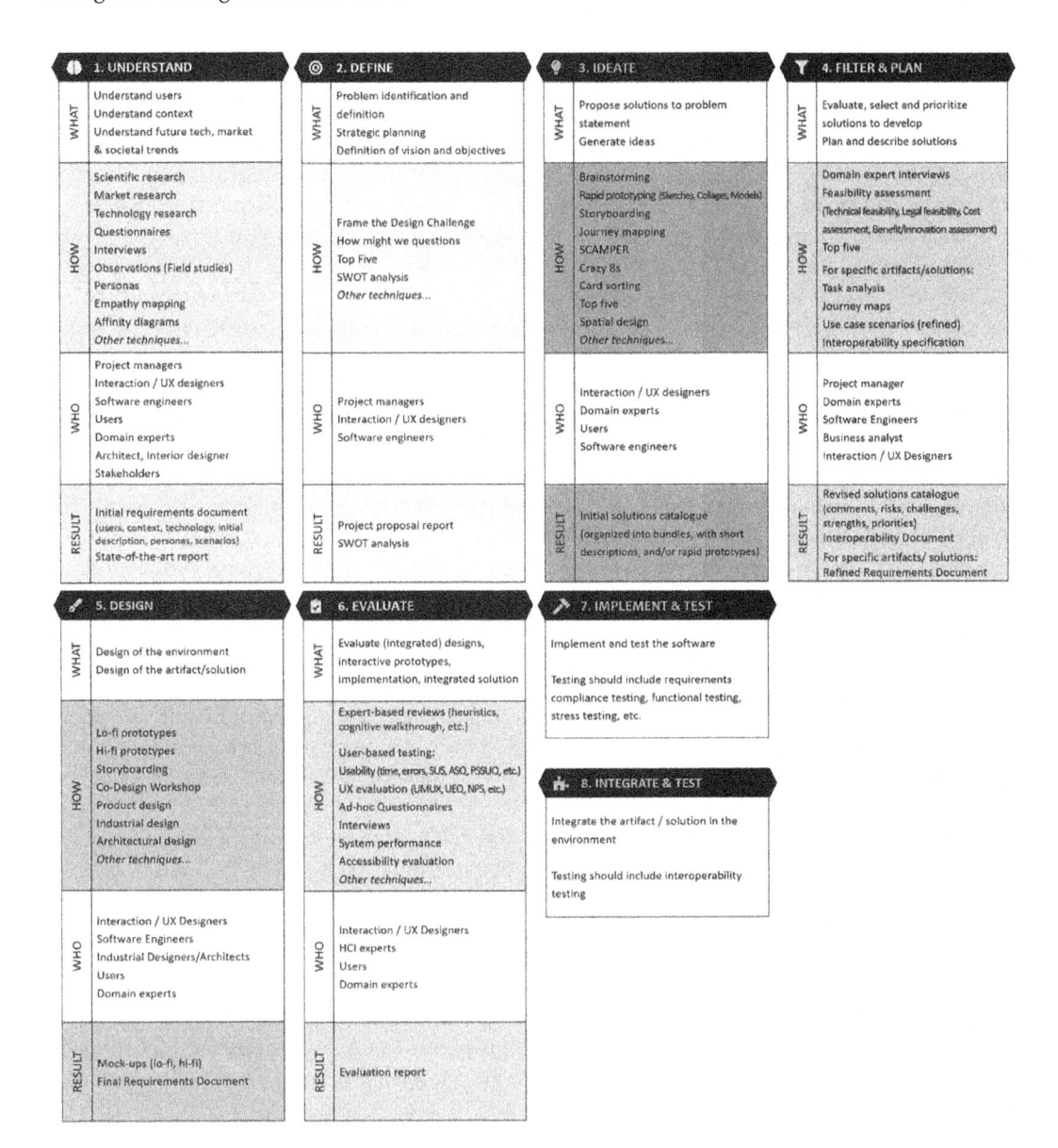

	1. UNDERSTAND	2. DEFINE	3. IDEATE	4. FILTER & PLAN
WHAT	Understand users Understand context Understand future tech, market & societal trends	Problem identification and definition Strategic planning Definition of vision and objectives	Propose solutions to problem statement Generate ideas	Evaluate, select and prioritize solutions to develop Plan and describe solutions
HOW	Scientific research Market research Technology research Questionnaires Interviews Observations (Field studies) Personas Empathy mapping Affinity diagrams Other techniques...	Frame the Design Challenge How might we questions Top Five SWOT analysis Other techniques...	Brainstorming Rapid prototyping (Sketches, Collages, Models) Storyboarding Journey mapping SCAMPER Crazy 8s Card sorting Top five Spatial design Other techniques...	Domain expert Interviews Feasibility assessment (Technical feasibility, Legal feasibility, Cost assessment, Benefit/Innovation assessment) Top five For specific artifacts/solutions: Task analysis Journey maps Use case scenarios (refined) Interoperability specification
WHO	Project managers Interaction / UX designers Software engineers Users Domain experts Architect, Interior designer Stakeholders	Project managers Interaction / UX designers Software engineers	Interaction / UX designers Domain experts Users Software engineers	Project manager Domain experts Software Engineers Business analyst Interaction / UX Designers
RESULT	Initial requirements document (users, context, technology, initial description, personas, scenarios) State-of-the-art report	Project proposal report SWOT analysis	Initial solutions catalogue (organized into bundles, with short descriptions, and/or rapid prototypes)	Revised solutions catalogue (comments, risks, challenges, strengths, priorities) Interoperability Document For specific artifacts/ solutions: Refined Requirements Document

	5. DESIGN	6. EVALUATE	7. IMPLEMENT & TEST
WHAT	Design of the environment Design of the artifact/solution	Evaluate (integrated) designs, interactive prototypes, implementation, integrated solution	Implement and test the software Testing should include requirements compliance testing, functional testing, stress testing, etc.
HOW	Lo-fi prototypes Hi-fi prototypes Storyboarding Co-Design Workshop Product design Industrial design Architectural design Other techniques...	Expert-based reviews (heuristics, cognitive walkthrough, etc.) User-based testing: Usability (time, errors, SUS, ASQ, PSSUQ, etc.) UX evaluation (UMUX, UEQ, NPS, etc.) Ad-hoc Questionnaires Interviews System performance Accessibility evaluation Other techniques...	**8. INTEGRATE & TEST** Integrate the artifact / solution in the environment Testing should include interoperability testing
WHO	Interaction / UX Designers Software Engineers Industrial Designers/Architects Users Domain experts	Interaction / UX Designers HCI experts Users Domain experts	
RESULT	Mock-ups (lo-fi, hi-fi) Final Requirements Document	Evaluation report	

FIGURE 1.2 Overview of each stage and recommended methods for its execution.

1.3.6 Stage 6: Evaluate Design

WHAT: This is the stage where we are now ready to evaluate the results of the *AmI Design Process* up to this point, i.e., the prototypes of the solutions we have created based on the outputs of the previous stages. The main goal of testing is to see whether our prototypes meet the defined user and technical requirements and fulfill their purpose. Just like all iterative design processes, the evaluation stage of the *AmI Design Process* is not to be performed only once, but as many times as needed and at any point in the design process. It is important to note that the results of this stage may lead the team back to previous stages, i.e., to redefine the problem, to further investigate user needs, or to ideate and redesign the solution. Testing does not have to be very elaborate and formal with hundreds of users in order to be effective. Often, iterative "quick and dirty" testing techniques can yield high returns with much less overhead. However, for a more holistic assessment of whether the designed solution fulfills the user needs, it is best to test for both usability and UX. Apart from the abovementioned types of evaluation, when the prototype has reached a mature enough level, we can also conduct other specialized evaluations, such as system performance and functionality evaluation, accessibility evaluation, security performance, etc. Furthermore, once the prototype has been fully implemented, it then needs to be integrated and tested within the context of the IE that it is designed for (see stages 7 and 8 further down).

HOW: There are many evaluation methods for assessing usability and UX in the HCI field to choose from. In general, we can divide the inspection methods into two main categories, expert-based inspection methods and user-based inspection methods. In the user-based inspections category, there are methods that measure strictly usability parameters, i.e., effectiveness, efficiency, learnability, and satisfaction, and there are methods that measure UX parameters, i.e., hedonic and pragmatic attributes such as attractiveness, novelty, innovation, ease of use, etc. Also, there are techniques that produce qualitative data and techniques that produce quantitative data. It is up to the design team to decide which evaluation method to use for each case scenario or prototype they want to evaluate. Sometimes it is necessary to use multiple methods. There are no strict rules on which method to use as long as some form of evaluation occurs. We discuss some of the most popular evaluation methods below.

1.3.6.1 Expert-Based Evaluations

Expert-based evaluation is a commonly used method in the HCI field especially in early design iterations because it is effective, quick to produce results, and does not require many resources (Nielsen, 1994b). In addition, it can, in principle, involve a wider range of tasks than in user-based evaluation, thus allowing the team to examine thoroughly all the features and functionality of the prototype. Expert-based evaluation is best conducted at the early stages to hush out any major design shortcomings and usability issues before introducing them to user-based evaluation with actual users. In expert-based evaluations, the inspection is ideally conducted by HCI or UX usability experts who look at the application from the perspective of the end-user. They perform common tasks while noting any areas in the design that may cause problems for the user. The evaluators base their judgments on prior experiences and knowledge of common human factors and ergonomics guidelines, principles, and standards. However, in technology-specific domains, the evaluation can also be performed by a domain expert with specialized knowledge of the technology employed.

There are a few inspection techniques available for expert-based evaluation, but the two most common ones are cognitive walkthroughs and heuristics analysis.

1.3.6.2 Cognitive Walkthroughs and Heuristics

In cognitive walkthroughs, the evaluator examines the prototype, through the eyes of the user, performing typical tasks and identifying areas in the design or the functionality that could potentially cause confusion or user errors (Rubin, 1995). Specifically, in cognitive walkthroughs, the evaluator has the following questions in the mind while examining the application or system (Cockton et al., 2009):

1. Does the user know what to do? Is it the correct action?
2. Will the user notice that the correct action is available? Is the action visible? Will users recognize it?
3. Will the user associate the correct action with the effect to be achieved? The action may be visible, but will the user understand it?
4. If the correct action is performed, will the user see that progress is being made toward the solution of the task? Is there system feedback to inform the user of progress? Will they see it? Will they understand it?

The results of the walkthrough are aggregated in a final single report and disseminated to the design team for rectification.

In a similar way, heuristics analysis also involves the inspection of the prototype by usability or HCI experts. Only this time, the experts inspect the prototype against a set of 10 usability guidelines and principles introduced by usability expert Jacob Nielsen (Nielsen, 1994a). Each expert produces a report with their findings and a severity score of 1–4 with 1 being assigned to a minor cosmetic error and 4 being assigned to a critical error. The results can then be aggregated in a single report and disseminated to the design team.

1.3.6.3 User-Based Evaluation

In user-based evaluations, the design prototype is evaluated by a selected set of representative end users of the application. There are many user-based evaluation techniques and choosing one depends mainly on what we want to measure or examine. For example, if we want to measure usability, we can use metrics such as time on task, number of errors, user success rate for task completion, and standardized post-study questionnaires. Below we provide a list of the post-study questionnaires for the assessment of usability and UX that we have used in most of our usability evaluations because they are easy to administer, have high internal reliability and sensitivity, and are free of charge. However, there are more options in the bibliography available to choose from.

Examples of post-study questionnaires:

- *System usability questionnaire (SUS)*: (Brooke, 1996) a widely used simple, ten-item Likert scale that gives a global view of subjective assessments of usability, such as perceived ease of use, effectiveness, efficiency, and satisfaction.
- *After-task questionnaire (ASQ)*: (Lewis, 1995) developed by IBM, a simple standardized three-question scale questionnaire that can be used right after the completion of a specific user task to assess how difficult the user perceived it to be. It uses a seven-item Likert scale, anchored at the endpoints with the terms "Strongly agree" for 1 and "Strongly disagree" for 7.
- *Post-study system usability questionnaire (PSSUQ)*: a post-study standardized questionnaire also developed by IBM (Lewis, 1992) that measures system usefulness, quality of information, and quality of interface.
- *Usability metric for user experience (UMUX)*: is a four-item Likert scale used for measuring the perceived usability of an application. It is designed to provide results similar to those obtained with the ten-item system usability scale and is organized around the ISO 9241-11 definition of usability (Finstad, 2010). Its lite version, UMUX-Lite, includes only two questions. This instrument is also free of charge.
- *UEQ (long and short versions):* The long version and the short version of this questionnaire (Hinderks et al., 2019) measure the UX of interactive products and applications. The included scales cover both hedonic i.e., attractiveness, novelty, stimulation, etc., and pragmatic i.e., efficiency, perspicuity, and dependability aspects regarding UX. The questionnaire is free of charge and very easy to use. It is an excellent instrument, highly reliable and it is provided in various languages free of charge, along with its own scoring template which can be downloaded from their website.[16]
- *Net promoter score (NPS)*: (Reichheld, 2003) is a loyalty metric typically used in customer surveys that aims to quantify customers' overall satisfaction with a product or service and the likelihood of someone recommending the product, or service to a friend or colleague. The metric comprises a single-item question, "How likely are you to recommend this application/installation to a relative or a close friend" and it is measured on a scale from 0 to 10. In recent years this single-item question has been used in the HCI field as a complementary metric to other usability metrics (Lewis, 2018).

- *Post study interview*: At the end of the study, the evaluator can have a de-briefing interview with the participant to further discuss their insights on the tested application. The interview can be semi-structured including specific questions that can then be expanded as it seems fit.

The reader can look into the work of (Hodrien & Fernando, 2021) for a comparative analysis of all types of post-study usability and UX questionnaires.

WHO: In this stage, the following three groups of project stakeholders should participate:
- *End users*: These are the selected sample of representative users that will participate in the evaluations. The selection criteria of the participants and the number of participants should be carefully planned by the entire team prior to the evaluation.
- *HCI experts*: These are people who have experience and expertise in conducting expert-based evaluations and/or in planning and carrying out user-based evaluations.
- *Interaction/UX Designers*: The designers can participate in the evaluation stage mainly as observers and assistants during the evaluation process, and of course as recipients of the results of the evaluation. Sometimes when the project team is small, the evaluations may be carried out by the design team itself. However, in this case, extra caution should be given so as to not introduce any bias in the study and to not let the participants know of the designers' capacity. In some early informal design iterations, designers may also carry out small tests with users selected informally (i.e., colleagues, friends, family, etc.) to hush out any major design flaws.
- *Software engineers*: The software engineers can also participate in this stage as observers and/or assistants during the evaluation process. When the evaluation setting permits remote observations, it is a great opportunity for the developers to experience first-hand the users' reactions to the application or system.

RESULT: The output of this stage is an aggregate report with the results of the evaluation of the prototypes. The report should include any statistical analysis, the scores of the metrics used, and a description of the findings. This report should be disseminated to the entire project team, i.e., project manager and design team for further analysis. Based on the analysis of results, the team must then discuss how to continue from thereon, i.e., what changes need to be implemented in the prototypes, whether more testing is needed, whether there is a need to go back to any of the previous stages either to elicit more user requirements, or to redesign the concept. If it is found that the solution prototype doesn't need any adjustments, the team can move to the next stage which is the final implementation and integration stage that will be discussed in the following section.

1.3.7 STAGE 7: (IMPLEMENT - INTEGRATE) & TEST

At this stage, the refined prototypes are implemented. Depending on the perspective, whether it's a small-scale artifact/solution or a large-scale environment, the objectives of this stage vary. In the case of a small-scale artifact/solution, this stage involves a range of physical activities, including software development, sensor installation, casing construction, microelectronics soldering, and comprehensive testing, encompassing requirements compliance testing, functional testing, and stress testing, among others. The result is a fully operational component, whether physical, virtual, or a combination of both, ready for deployment within an environment.

Conversely, in the context of a large-scale environment, this stage pertains to the activities required for deploying the new artifact/solution within the environment. This can encompass tasks as straightforward as installing a new application on a computer or as extensive as reconfiguring an entire room from the ground up while verifying the expected functionality and interoperability of the new component. The outcome is a fully operational system seamlessly integrated into the environment.

The team collaborating on this stage includes software engineers, architects, mechanical engineers, electrical engineers, carpenters, blacksmiths, and other domain experts essential to the prototype's creation.

1.3.8 Stage 8: Evaluate Integrated Solution

At this stage, we have returned to the larger circle, following the successful deployment of an artifact/solution within the environment. Conducting tests in real-world deployment settings is crucial to validate the UX under authentic usage conditions. The execution of this stage closely resembles that of Stage 6, with the key distinction being that evaluation now occurs on a fully functioning system which is installed in its actual setup and interacts with the other intelligent services and artifacts already present in the environment.

1.3.9 Summary

The process described in the previous sections can be summarized as an expanded and modified version of a *Design Thinking* process, tailored to meet the needs of the overall research vision and the particular needs of the team at every stage. It features some well-known activities, placed in the appropriate time, along with a number of suggestions and supporting tools and calls for specific roles and outcomes from each stage.

It is an ongoing effort that will keep getting refined, as more experience is gained and newer methods or tools become available. Figure 1.2 provides an overview of each stage, along with some recommended methods.

1.4 SUPPORTING TOOLS

Numerous tools are available to support the design process of IEs. **Design tools** aid designers in visualizing and refining their concepts before advancing to prototyping; **Simulation tools** generate digital simulations of the IE, enabling the testing of different scenarios and configurations before creating a physical prototype; **Prototyping tools** allow designers to assess the functionality of the environment and identify any technical issues before implementing the final design; **Data visualization tools** are employed to present data collected from the IE, helping designers identify patterns and insights that inform the environment's design; and, **Collaboration tools** facilitate communication among team members, ensuring that everyone remains aligned and the project progresses according to schedule.

The *AmI Design Process* is complemented by a suite of tools, which, when combined with other available commercial options, constitute a comprehensive set that designers can leverage when applying this design framework.

1.4.1 AmI Solertis: Streamlining IE Behavior Definition

Creating services for IEs entails the utilization of diverse technologies and protocols by the various technological components in order to define and expose their functionality. The choice of a particular protocol is determined by a combination of factors, including technical capabilities, hardware considerations (such as network interfaces, processing power, and battery-based operation), and software aspects (like the operating system and runtime environment). Additionally, this decision may be influenced by prospective standards and guidelines, which advocate certain approaches (Bandyopadhyay & Sen, 2011; Sheng et al., 2013).

AmI-Solertis (Leonidis et al., 2017) offers a streamlined solution for seamlessly integrating external services (i.e., AmI artifacts), regardless of their type, whether they are back-end, front-end, or hybrid services adhering to the Software-as-a-Service paradigm (Turner et al., 2003)). Additionally, it facilitates the creation of scripts, namely *AmI Scripts*, which specify the behavior of technological facilities (i.e., business logic) in crafting comprehensive, intelligent, and personalized environment experiences by combining various components. The *AmI-Solertis* system is built using a micro-service architecture style that enables it to be used as a backbone (Newman, 2015) across a wide range of ubiquitous systems (Chen et al., 2004) and IEs, each with distinct objectives. These applications may range from composing new compound services using existing ones to defining the behavior of a smart hotel room (Leonidis et al., 2013), controlling a smart home, or building an intelligent management system for a smart city (Batty et al., 2012).

Leveraging the advantages of asynchronous and event-based communication (Gamma et al., 2005; Michelson, 2006, p.; Van Kesteren & Jackson, 2007), *AmI-Solertis* introduces a unified hybrid communication protocol. This protocol combines the widely adopted Representational State Transfer (REST) (Fielding & Taylor, 2000) and the OpenAPI Specification (OAS)[17] with asynchronous and event-based communication capabilities, facilitating the integration of diverse services in a standardized yet agnostic manner (as depicted in Figure 1.3). Thus, AmI artifacts and AmI scripts provide a REST interface for handling incoming calls and convey their intentions to the IE by emitting relevant events through the AmI-Solertis Event Federator.

AmI-Solertis simplifies the intricacies of configuring and executing remote calls through automatically generated proxies, thereby alleviating the challenges of distributed programming. This is achieved by: (1) masking remote operations into local methods and (2) allowing consumers to express their interest in events originating from a remote component without the need to specify underlying topology and infrastructure details.

Beyond reducing code complexity, these proxies empower *AmI-Solertis* to dynamically adapt and modify the invocation process to meet evolving requirements. This includes actions like rerouting a call to a replicated host for load balancing, immediate call termination in case of an unavailable remote endpoint, call interception for logging relevant QoS-related metrics, and substituting a target endpoint with another offering semantically similar functionality.

In addition to managing AmI components (such as AmI artifacts or scripts), *AmI-Solertis* provides an online integrated development environment (IDE) known as *AmI-Solertis Studio*. This tool is designed to assist developers in creating, exploring, deploying, and optimizing AmI scripts (i.e., programs) responsible for controlling the behavior of an IE. These scripts accomplish this by combining and orchestrating various AmI artifacts, or other AmI scripts within the ecosystem.

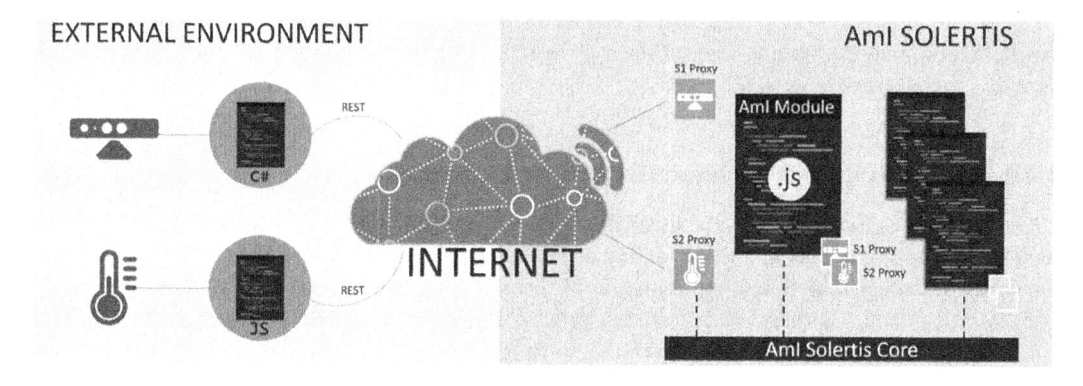

FIGURE 1.3 The AmI-Solertis hybrid communication protocol. (Leonidis et al., 2017).

1.4.2 Frameworks for Dictating the Behavior of IEs

LECTOR (Korozi et al., 2017b) is a framework designed to make use of existing ambient capabilities within IEs (i.e., in stages 7 and 8 of the proposed model), to identify when users require assistance or support and intervene to enhance their quality of life. It follows the trigger-action model (Ur et al., 2014), a prevalent approach for programming IEs using straightforward "if-then" rules. *LECTOR* introduces a three-step process for linking behaviors with interventions. The initial step involves defining a behavior, followed by describing the conditions that trigger the behavior, and finally connecting it to an intervention.

While this decomposition increases the number of steps required to connect a trigger to an intervention, it offers scalability and improved rule management. Each of the three essential elements (behavior, trigger, and intervention) is defined independently and only linked based on their outcomes. Therefore, any element can be modified without affecting the others, as long as their outcomes remain consistent. This approach not only minimizes unintended consequences but also fosters collaboration, as new rules can be easily created by different users, given that their "connection points" will always be their outcomes. This concept is inspired by how an Application Programming Interface (API) simplifies programming and allows computer programs to evolve independently by abstracting the underlying implementation and exposing only the necessary objects.

The core concepts of this rule-based approach are explained as follows:

- A **Rule** is a model that connects a behavior with an intervention through a trigger.
- **Behavior** represents the actions of a user or device (e.g., a user shouting or sleeping).
- A **Trigger** is a high-level behavior model capable of initiating an intervention.
- **Interventions** are system-guided actions aimed at assisting or supporting users in their daily activities.
- **Intervention Hosts** are environmental elements that can either display an application with carefully curated content or control the physical environment.

Both developers and individuals without technical backgrounds can effortlessly craft these rules through the user-friendly *LECTORstudio* (Korozi et al., 2017a) authoring tool. *LECTORstudio* features an intuitive interface that allows developers to seamlessly integrate the essential elements for programming IEs. Additionally, it empowers residents to fashion their own scenarios and tailor LECTOR's decision-making process to align with their specific needs. As an alternative to *LECTORstudio*, which employs a graphical user interface (GUI) for creating "if-then" rules to program IEs, an exploration into conversational interfaces (CIs) led to the development of *ParlAmI* (Stefanidi et al., 2019), a versatile multimodal conversational interface. *ParlAmI* introduces a hybrid approach that combines Natural Language Understanding (NLU) with semantic reasoning and service-oriented engineering to create a multimodal CI that helps users define the behavior of IEs. This user-friendly approach is particularly advantageous for those with limited or no programming experience, as it involves natural language interactions with a context-aware intelligent virtual agent, effectively functioning as a chatbot.

1.4.3 Validating IEs Behavior

AmITest (Louloudakis et al., 2016) is a specialized testing framework designed for IEs (i.e., in stages 6, 7, and 8 of the proposed model). These environments pose unique challenges due to their architectural and computational complexity (as they incorporate a lot of sensors, actuators, and interactive devices), necessitating not only user-friendly programming but also rigorous testing and validation of their behavior.

AmITest is tailored to support users responsible for defining the behavior of IEs. These users span a wide spectrum, from motivated users seeking to enhance their environments to seasoned IT professionals meticulously configuring the environment's behavior and features. To accommodate this diversity while recognizing that many users are not programming experts, *AmITest* streamlines the testing process by providing a visual programming tool for end users to define their tests. The *AmITest* framework is an integral part of the AmI Solertis framework, utilizing the dynamic scripting language of *AmI Solertis* to test the artifacts within the SLE being evaluated.

AmITest comprises two key components:

1. *AmIScript testing agents*: This framework focuses on testing and validating the behavior of IEs defined in an imperative domain-specific language called AmIScript.
2. *Tests management and deployment suite*: This suite empowers end users to visually program various tests related to AmI environment behavior, using a building block-based GUI. These components collaborate to provide users with flexible programming options, enabling them to configure aspects of the system's behavior and validate whether it aligns with their expectations.

Challenges in testing arise from the inherent complexity of IE. These systems feature numerous distributed, interrelated components that often operate asynchronously within themselves and with each other. Treating individual artifacts in isolation during testing is inadequate due to the high level of interdependency among them. *AmITest* addresses this challenge by focusing on isolating and verifying the values of artifacts. For example, in a classroom scenario, the system should respond appropriately if a student becomes distracted during a lecture. Validating such behavior requires monitoring and asserting the values of various artifacts, like the teacher's workstation transitioning from "classroom overview" to "inattention detected" and finally to "mini-game launched," or ensuring the student's intelligent desk (Prinianakis et al., 2021) restricts interaction to the teacher-initiated mini-game. To support these tests, *AmITest* employs a sophisticated monitoring mechanism, ensuring that value checks (expectations) occur only after the necessary handling actions (promises) have been completed.

This framework's significance lies in its ability to streamline the testing of complex IEs while accommodating the diverse range of users responsible for defining and fine-tuning their behavior.

1.4.4 Wizard of AmI

The *Wizard of AmI* (Arampatzis & Stephanidis, 2021) is an online platform available on both desktop and mobile devices, purpose-built to facilitate the design of applications for IEs by empowering designers to create interactive, prototype-based scenarios. In recognition of the evolving landscape of IEs, which are continually enriched with interactive displays, smart devices, and diverse sensors capable of recognizing multiple users and their activities, the *Wizard of AmI* distinguishes itself from existing interactive prototyping tools. While conventional tools primarily aid in designing applications tailored for a single device, the *Wizard of AmI* takes a forward-looking approach. It caters to applications destined to span various devices within the environment, whose behavior is influenced by unpredictable factors, including implicit or explicit user actions and the ever-changing state of the environment.

Within this context, the *Wizard of AmI* empowers designers by offering a multifaceted approach to application development for IEs. Initially, designers utilize the platform to craft user scenarios and map a series of mockups onto the IE environment, effectively choosing what content to present and where. Subsequently, through an innovative graph-inspired interface, they can define the flow and interactivity of these scenarios, pinpointing precisely when and how they should unfold. Finally, designers have the capability to execute these scenarios either in a virtual setting or on-site, enabling them to comprehensively evaluate the overall UX and conduct user-based assessments. During on-site scenario enactment, following the well-established 'Wizard of Oz' technique, the mobile companion application provides designers with real-time control over the execution flow, responding to unexpected factors in the dynamic IE environment.

In practice, the *Wizard of AmI* proves to be an invaluable asset for design teams navigating the complex realm of IEs. This tool serves as a comprehensive solution, facilitating not only the creation of interactive prototypes within intelligent spaces but also fostering interaction with these prototypes in both real-world and simulated environments. Moreover, it empowers designers to critically assess selected interaction modalities and the behavior of various artifacts during specific tasks, enabling refinement and optimization. Most remarkably, it sparks innovation by encouraging exploration of ideas for parts or artifacts within the environment that has yet to materialize. In essence, the *Wizard of AmI* stands as a versatile ally in the journey to craft intuitive and effective applications for the ever-evolving landscape of IEs.

1.4.5 ARGUS DESIGNER

The *ARgus Designer* (H. Stefanidi et al., 2022) is an innovative system designed to aid evaluators in conducting user studies (i.e., in stages 6 and 8 of the proposed model), with a dual focus:

a. Evaluating the usability and UX of AR and MR applications.
b. Assessing applications targeting IEs that may still be in developmental stages, lacking elements such as smart devices or even furniture.

Over recent years, AR/MR applications have surged in popularity among researchers, developers, enterprises, and consumers alike. With a continuous influx of new MR devices and accessories entering the market, such as HoloLens, Magic Leap, Varjo, and forthcoming products from Meta and Apple, the AR/MR field is poised to revolutionize the world of computing. As this field continues to mature, there is an increasingly urgent need for comprehensive evaluation tools tailored to these applications. While numerous approaches have emerged to tackle the challenges of evaluating AR/MR applications, with some even leveraging these technologies as part of the evaluation process, only a handful of attempts have been made to develop tools that comprehensively support the evaluation process from start to finish.

To address this need, the *ARgus Designer* tool enriches the information provided to experts in real-time during testing, all while they are wearing an MR device, notably the HoloLens 2, with the aim of streamlining the evaluation process. This system primarily targets the domain of prototyping for IEs, where its initial instantiation enhances users' surroundings by overlaying virtual interactive artifacts, 3D objects, and digital UIs onto the real-world environment. Users gain the ability to craft their own virtual experiment scenes, seamlessly integrating digital objects into the MR world. Furthermore, they have the flexibility to select specific metrics and features to include in the MR environment during the evaluation process. This versatile evaluation tool supports multiple users, fostering collaborative efforts among experts as they immerse themselves fully in the MR space.

During runtime, evaluators not only observe various metrics and features visualized in AR but also gain access to a virtual workstation, known as the *ARgus Workstation*. This workstation empowers evaluators to control the flow of the experiment and collect a range of runtime measures and data, including performance indicators and critical events of interest. An additional notable feature of the system is the incorporation of virtual questionnaires, which not only enhances expert supervision but, more importantly, minimize disruptions in the immersive experience.

Finally, *ARgus Designer* offers valuable support for post-study analysis, allowing experts to review raw data and replay evaluation experiments within the MR environment. Across all these stages, this tool endeavors to provide an efficient, all-encompassing approach to conducting user-based evaluations with the aid of AR/MR technologies.

1.4.6 UINIFY

UInify (Barka et al., 2017) is a framework designed to assist designers of IEs in delivering cohesive UXs to end users (i.e., in stages 5–8 of the proposed model). Specifically, it provides: (1) a suite of tools that enable the visual combination of several individual UIs towards introducing new UI compositions, and (2) a universal style guide ensuring that the resulting UIs maintain a consistent look and feel across all devices.

UInify leverages concepts from UI Mashup web technologies (Daniel et al., 2007) and UI composition (Daniel & Matera, 2014) to create unified applications that bring together various UI components into a single presentation layer for all connected devices and services. Additionally, *UInify*'s capability to control multiple connected devices simultaneously enables seamless multimodal UXs.

By using *UInify*, an IE can integrate distinct software components developed independently by different experts, all under a common roof. This approach offers several advantages

Maximum reusability: Developers can compose desired UIs using existing functionality from separate web applications, eliminating the need to build everything from scratch.

Reduced development costs: Repurposing existing software significantly reduces development time and costs compared to re-implementing everything.

Enhanced configurability: *UInify* allows the creation of unlimited compositions that can be activated in a context-sensitive manner. This flexibility ensures that the same components can address different needs.

End-user customization: *UInify* empowers end users to create their own custom compositions based on the available components to match their unique preferences and needs.

1.4.7 UXAmI Observer

In the context of evaluating UX in IEs (i.e., in stages 6 and 8 of the proposed model), researchers and practitioners can be greatly assisted by tools for automating the evaluation process. Such a specialized tool created to help evaluators in user-centered assessments within IEs is *UXAmI Observer* (Ntoa et al., 2019), which helps in planning and running user-based experiments and supports evaluators through visualizations of the data acquired. In more detail, considering the large number of attributes that should be studied to assess UX in IEs (Ntoa et al., 2021), *UXAmI Observer* aims to automate – to the extent possible – the evaluation process. In this respect, it aggregates experimental data and provides an analysis of the results of experiments, incorporates information regarding the context of use and fosters the objectivity of recordings by acquiring data directly from the infrastructure of the environment.

Evaluators can easily set up an experiment by defining (1) the evaluation targets which may be one or more artifacts or applications in the IE; (2) the study participants by providing demographic information, such as age, gender, technology expertise; and (3) the evaluation tasks, which is optional and pertains specifically to task-based experiments. Then, for each experiment, they can monitor in real time how it evolves or proceed with post-experiment processing. As a result, an experiment encompasses one or more evaluation targets, tasks (if any), participants, an identifying name and a description of its objectives, as well as experiment sessions, that is, usages of the system by one or more participants concurrently. Furthermore, an experiment features aggregated statistics, providing an overview of the success rate achieved, interaction modalities employed by the users, as well as interaction accuracy achieved (see Figure 1.4).

Information regarding each session features manual, semi-automated, and automated attributes. Manually annotated metrics for each session include user input errors, interaction errors, and help requests. Additional attributes provided by IE include eye-tracking data and other psychophysiological measurements (if available), adaptations or decisions introduced by the system (e.g., turn on the room lights), as well as adaptations or decisions rejected by the user (e.g., lights turned back off), input errors, and implicit interactions. The data collected for a session are presented to the experimenter organized into four main categories: (1) session timeline, a horizontal timeline with all the recorded or manually inserted by the experimenter points of interest; (2) interaction timeline, a vertical timeline with ordered according to the time of their occurrence indication of task start, task end, user input commands, implicit interactions, system responses, and introduced adaptations; (3) a diagram with all the system responses ordered in a path; and (4) interaction statistics, featuring

Design for Intelligent Environments

FIGURE 1.4 Experiment screen in UXAmI Observer.

information per session participant(s) regarding the aforementioned metrics. Additional insights are offered by the tool as aggregated statistics, also associated with the floorplan of the IE. Finally, an option for comparing experiments is also integrated in *UXAmI Observer*, allowing the evaluator to verify if UX was improved across various evaluation iterations.

Overall, the tool can facilitate and automate to the extent possible the evaluation process for one or more systems and artifacts in an IE, providing rich insights to evaluators regarding the tasks executed by users, the system responses and decisions, their acceptance by users, as well as other user-related information collected by sensors of the IE. Furthermore, the tool facilitates annotation and editing of automatically collected metrics by experimenters themselves, allowing them to be in control of the entire process and the collected results.

1.4.8 The "Book of Ideas"

As mentioned in Sections 1.3.3–1.3.5, when many ideas are being analyzed and refined, our team is using and continuously improving a tool to catalog proposed ideas and solutions for the environment and for specific artifacts or services. This repository, in a display of literalism, is informally called the *Book of Ideas*.

The *Book of Ideas* at its core is simply a catalog of ideas that are generated at the first stages of the framework. It can be a simple document or spreadsheet containing a list of ideas, as was its first incarnation in our team, or it can be a repository tool that can work with other tools, as is the version we are developing now. The *Book of Ideas* can be a single entity for an entire environment/project divided into sections (sub-environments) that contain subsections of other sub-environments or smart artifacts or services, which themselves can be viewed/analyzed as a smaller booklet, so to speak, of ideas. Ideally, a hypertext web of ideas allows for easy navigation and reference between ideas. Each idea should have specific information; specifically, a unique ID, a title and description, any services from the IE that are needed for the idea to work (existing or not) and if possible, one or more mockups.

As the stages of the process progress, so can the ideas be included in the book. Mockups are added as the team discusses and brainstorms on a specific idea. In the stage "Filter and Plan", the ideas from the "Book" are presented to different domain experts or any relevant stakeholders to be commented upon or criticized. After that process, more comments should be available for each idea discussed, including thoughts on how good and worth pursuing an idea is. Some sort of grading can be applied here, with the goal to have a ranked list of ideas, where the good ones are marked as such. Additionally, issues of compatibility or gaps in the services available to IEs, issues of environmental or architectural requirements, and other technical concerns should be addressed and commented upon in the appropriate place.

What each team will do with the repository of ideas is up to them. It is up to the team to decide who has access to the Book of Ideas and who has editing privileges, such as adding comments and mockups.

A fruitful practice is to use this repository during meetings of the design team, as a starting point for discussions about the research vision and the overall scope of what the team actually wants to achieve, discuss requirements of the environments (in terms of physical artifacts or software services) or simply work more on designs.

These design team meetings foster team building, focusing on vision and sharing a common goal. They also allow for creativity and foster innovation. Finally, out of these meetings, a tentative roadmap of what the team wants to accomplish can be formed. Development teams or people with roles other than design can and should use the Book of Ideas, for similar reasons as mentioned above but also to plan ahead and coordinate on technical issues and concerns regarding development.

Currently, further work is needed to expand the initial version of the Book of Ideas into a tool that will work closely with the *Wizard of AmI* tool and will include some presentation features for the easier and more effective presentation of the ideas included. These ideas ideally should be able to be exported in the *Wizard of AmI* tool, including the virtual or AR feature being developed now.

1.5 USE CASE: THE INTELLIGENT LIVING ROOM

This section demonstrates how we applied the presented process to the design of an intelligent living room. The intelligent living room (Leonidis et al., 2019) is a simulation space, part of the "Intelligent Home" located in the AmI Facility[18] of ICS-FORTH. Aligned with the core objectives of IEs, the **initial idea** was to establish a smart ecosystem that not only improves overall quality of life but also offers continuous support to residents in their daily activities, concurrently enhancing leisure activities and entertainment options. Designing an entire room within a complex environment such as an Intelligent Home was not a straightforward endeavor, but following the *AmI Design Process* allowed for a systematic and methodical approach, resulting in a well-integrated IE.

During a series of meetings involving the development team, which included analysts, designers, interior designers, and programmers, as well as engaging with several potential end users, the design team successfully crafted scenarios and personas for the "**Understand**" and "**Define**" stages of the *AmI Design Process*. This collaborative effort culminated in the development of an initial requirements document outlining the high-level functionality of the envisioned intelligent living room.

As a subsequent step, we conducted several brainstorming sessions for the "**Ideate**" stage. These sessions commenced with a clear directive: (1) to encourage the free expression of ideas, regardless of their cost, complexity, or feasibility and (2) to concentrate exclusively on interaction aspects while hypothetically assuming that all ethical, privacy, and security concerns had been resolved, even though recent literature suggests that this may not be the case (Bettini & Riboni, 2015; Campbell et al., 2003). These sessions generated a multitude of ideas, all of which were incorporated into the **initial solutions catalog** (e.g., sleep monitoring, step-by-step cooking assistance, outfit recommendation, and children safety monitoring).

Next, during the "**Filter & Plan**" stage, the initial solutions were subjected to a rigorous evaluation process, involving interviews with domain experts, including computer vision specialists, industrial engineers, automation and robotics experts, architects, and interior designers. This systematic assessment led to the identification and elimination of ideas that were currently unfeasible, notably those involving advanced robotics and intricate object recognition. Additionally, experienced interaction designers and developers scrutinized the ideas, offering valuable insights, comments, and preferences regarding which concepts held the greatest potential in terms of innovation, research interest, and acceptance by end users (prioritizing ideas that exhibited attractiveness, novelty, and engagement). Ultimately, this process yielded a curated set of ideas for the intelligent living room, each accompanied by a detailed description, potential challenges, relevant comments from domain experts, if available, and associated services. During this stage, it became evident that the intelligent living room required the integration of numerous intelligent artifacts. These encompassed a sofa able to detect user presence and relay information about the user's weight and seating posture, an intelligent coffee table able to serve as an expansive display area for presenting secondary information, and a solution, namely *SurroundWall*, that transforms the wall around the TV into a secondary large display. For each of these artifacts, the design team embarked on individualized design processes, iterating through the stages of the *AmI Design Process* from the very outset.

The "**Design**" stage of the intelligent living room started by creating a realistic 3D representation of the space (Figure 1.5a). In accordance with the practice followed in similar use cases (Roalter et al., 2011), such a prototype would be of outmost importance as it can assist designers in selecting the best solution by encouraging reflection in design, permitting exploration of design ideas and imagining the ramifications of design decisions. Given that the available space was empty (with no furniture), there was the opportunity to conceptualize different alternatives to better fulfill the requirements selected from the previous stages. In more details, a collaborative effort involving UX experts and specialists from various pertinent fields, such as computer vision, industrial design, automation and robotics, architecture, and interior design, was undertaken during multiple focus group sessions. During these sessions, multiple design alternatives were created for the intelligent living room and were subjected to evaluation during the "**Evaluate**" stage through a variety of methods. These methods included computer simulations, role-playing scenarios, and interviews with potential end users or experts. The final model specified the type and placement of furniture and detailed the arrangement of technological equipment.

Prior to progressing to the "**Implement & Test**" phase for the entire room, it was imperative that the "**Design**" stage for each conceptualized intelligent artifact (such as sofa, coffee table, and SurroundWall) was completed, and that all specifications were readily available for the construction of the required furniture and installation of the necessary equipment.

Upon reviewing the final designs of the living room and its integrated artifacts, UI and UX experts embarked on distinct design processes for the most promising applications, as identified during the "Ideation" stage. At this juncture, it is pertinent to delve into the "**Evaluate**" stage of these applications, which took place following the "**Implementation**" of the intelligent living room (Figure 1.5b), wherein all artifacts and equipment had been positioned in their final locations. During that stage, the challenge of quickly evaluating low- or high-fidelity design mockups soon emerged. Since we were no longer designing for a single screen (or just for screens, since we include different interaction modalities such as speech interaction and air gestures), it was very difficult to assess the UX of different parts of the room as we were unsure how second (or even third in some cases) screens should behave. For that purpose, the *The Wizard of AmI* (Section 1.4.4) was used in order to help the design team toward creating interactive prototypes for the intelligent space. Following an iterative design process, before proceeding with the implementation of the application, the developed prototypes were "**Evaluated**" (with the help of *The Wizard of AmI*) by UX experts and end users in order to gain valuable feedback and identify usability and UX-related issues early in the design process. The revised prototypes were subsequently "**Implemented**" by the development team using the *AmI Solertis* (Section 1.4.1), *LECTOR* (Section 1.4.2) and *UInify* (Section 1.4.6) frameworks.

Revisiting the Design Process for the entire living room, with regard to the "**Integration & Test**" stage, once all artifacts were positioned and the envisioned functionality, including applications and

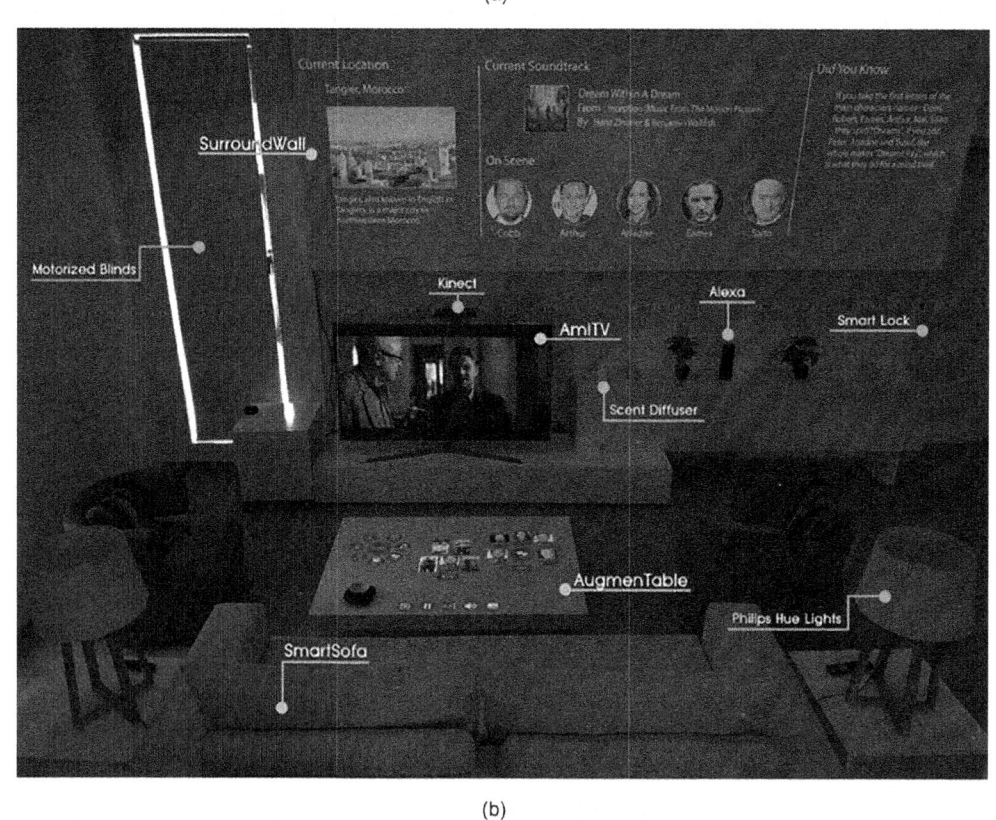

FIGURE 1.5 (a) The 3D representation of the living room as it resulted from the "Design" Stage (Leonidis et al., 2019) i; (b) The current "Intelligent Living Room" setup (Leonidis et al., 2017).

services, was prepared, the operation of each component and the overall environment's behavior were verified through *AmITest* (see Section 1.4.3).

Finally, a series of user-based evaluation experiments were conducted with the help *UXAmI Observer* (Section 1.4.7) and *ARgus Designer* (Section 1.4.5) in order to draw insights by observing the users interacting with the living room environment.

1.6 CONCLUSIONS

The *AmI Design Process* presented in this chapter is both a design process and a framework to design and develop IEs and intelligent artifacts. It is a process with explicit stages, defined outcomes and it is also a living framework with a goal, mindset, and supporting tools, to create hospitable and enjoyable IEs. It was born out of necessity, part of the main vision behind ICS-FORTH's Intelligent Home as the practical means to define and handle the research goals, activities and management of the design and development of any project. It is based on the UCD principles and the design thinking process, modified and expanded to suit the needs of (our) IEs' design.

One of the greatest challenges in IEs is the acceptance of the new technologies, particularly due to mistrust on the side of users on issues like privacy and security. The *AmI Design Process* does not explicitly address these issues because they ultimately depend on many factors and differ between projects. We explicitly choose not to consider them during the ideation phase as to not hinder creativity and place stumbling blocks on the participants' mind flow. However, during the Filter & Plan stages and onwards, special care should be given to address all such concerns, including legal and ethical. It is planned that in the near future an updated version of the *AmI Design Process* will more explicitly and directly address these currently missing features.

Another aspect that is not explicitly addressed in the various descriptions of the stages is the complexity of running the process. As can be seen, there are a lot of people involved during each phase, with a diverse set of roles and many team activities that need to be coordinated. In essence, the success of the process depends on factors such as management effectiveness, coordination, team dynamics, and effective management of time and resources.

While a plethora of design tools are readily available, it has become evident that the process of creating IEs is significantly more intricate and challenging than initially anticipated. The design process and the framework detailed in this chapter are expected to undergo continuous refinement, particularly with the maturation of emerging technologies like virtual rapid prototyping, which is anticipated to have a significant impact on the design and evaluation phases of the *AmI Design Process*. Additionally, we expect that even more tools and methodologies will be gradually introduced to aid design activities at various stages of this complex journal, thus enhancing the landscape of available third-party tools, many of which are AI-based (e.g., those provided by Google[19]). Some of these tools show great promise in terms of reducing the time and effort required for research, filtering, and planning; therefore, any organization venturing into the design of intelligent spaces will undoubtedly be influenced by the ongoing evolution and maturation of AI technologies. Finally, the incorporation of ethical and privacy considerations will need to evolve into inherent functionality rather than being treated as optional add-ons. As the design landscape for IEs continues to advance, addressing these crucial aspects will be integral to creating more responsible and effective solutions for the future.

NOTES

1. https://ami.ics.forth.gr/en/home/
2. https://www.ics.forth.gr
3. AmI Programme Installations - https://ami.ics.forth.gr/en/installation-category/
4. AmI Programme Domains - https://ami.ics.forth.gr/en/domain/
5. https://www.consumerreports.org/home-garden/smart-home/matter-smart-home-standard-faq-a9475777045/
6. https://www.interaction-design.org/literature/article/how-to-conduct-user-interviews
7. https://www.usability.gov/how-to-and-tools/methods/individual-interviews.html
8. https://www.designkit.org/methods/interview.html
9. https://www.nngroup.com/articles/field-studies/
10. https://www.nngroup.com/articles/empathy-mapping/
11. https://uxbooth.com/articles/empathy-mapping-a-guide-to-getting-inside-a-users-head/
12. https://www.interaction-design.org/literature/article/affinity-diagrams-learn-how-to-cluster-and-bundle-ideas-and-facts

13 https://www.designkit.org/methods/frame-your-design-challenge.html
14 https://www.designkit.org/methods/how-might-we.html
15 https://www.designkit.org/methods/top-five.html
16 https://www.ueq-online.org/
17 https://swagger.io/specification/
18 https://ami.ics.forth.gr/en/domain/home/
19 https://blog.google/technology/ai/notebooklm-google-ai/, https://arxiv.org/abs/2205.11487, https://cloud.google.com/duet-ai

REFERENCES

Aarts, E. (2004). Ambient intelligence: A multimedia perspective. *IEEE Multimedia*, *11*(1), 12–19.

Abras, C., Maloney-Krichmar, D., & Preece, J. (2004). User-centered design. In W. Bainbridge (Ed), *Encyclopedia of Human-Computer Interaction* (Vol. 37, No. 4, pp. 445–456). Thousand Oaks: Sage Publications.

Alawairdhi, M., & Aleisa, E. (2011). A scenario-based approach for requirements elicitation for software systems complying with the utilization of ubiquitous computing technologies. In *Computer Software and Applications Conference Workshops (COMPSACW), 2011 IEEE 35th Annual* (pp. 341–344).

Alkhafaji, A., Fallahkhair, S., & Haig, E. (2020). A theoretical framework for designing smart and ubiquitous learning environments for outdoor cultural heritage. *Journal of Cultural Heritage*, *46*, 244–258.

Arampatzis, D., & Stephanidis, C. (2021). *Wizard of AmI: A System for Building and Enacting Interactive Prototypes in Intelligent Environments*. Heraklion: University of Crete.

Atzori, L., Iera, A., & Morabito, G. (2010). The internet of things: A survey. *Computer Networks*, *54*(15), 2787–2805.

Augusto, J. C., & Muñoz, A. (2019). User preferences in intelligent environments. *Applied Artificial Intelligence*, *33*(12), 1069–1091.

Augusto, J. C., Callaghan, V., Cook, D., Kameas, A., & Satoh, I. (2013). Intelligent environments: A manifesto. *Human-Centric Computing and Information Sciences*, *3*, 1–18.

Augusto, J., Giménez-Manuel, J., Quinde, M., Oguego, C., Ali, M., & James-Reynolds, C. (2020). A smart environments architecture (search). *Applied Artificial Intelligence*, *34*(2), 155–186.

Augusto, J., Kramer, D., Alegre, U., Covaci, A., & Santokhee, A. (2018). The user-centred intelligent environments development process as a guide to co-create smart technology for people with special needs. *Universal Access in the Information Society*, *17*(1), 115–130.

Baheti, R., & Gill, H. (2011). Cyber-physical systems. *The Impact of Control Technology*, *12*(1), 161–166.

Bandyopadhyay, D., & Sen, J. (2011). Internet of things: Applications and challenges in technology and standardization. *Wireless Personal Communications*, *58*(1), 49–69.

Barka, A., Leonidis, A., Antona, M., & Stephanidis, C. (2017). A Unified Interactive System for Controlling a Smart Home. In *Proceedings of the ACM Europe Celebration of Women in Computing (*womENcourage 2017*)* (pp. 6–8).

Batty, M., Axhausen, K. W., Giannotti, F., Pozdnoukhov, A., Bazzani, A., Wachowicz, M., Ouzounis, G., & Portugali, Y. (2012). Smart cities of the future. *The European Physical Journal Special Topics*, *214*(1), 481–518.

Belk, M., Fidas, C., & Pitsillides, A. (2019). FlexPass: Symbiosis of seamless user authentication schemes in IoT. In *Extended Abstracts of the 2019 CHI Conference on Human Factors in Computing Systems* (pp. 1–6).

Ben Allouch, S., van Dijk, J. A., & Peters, O. (2009). The acceptance of domestic ambient intelligence appliances by prospective users. In *Pervasive Computing: 7th International Conference, Pervasive 2009, Nara, Japan, May 11-14, 2009. Proceedings 7* (pp. 77–94).

Benzaghta, M. A., Elwalda, A., Mousa, M. M., Erkan, I., & Rahman, M. (2021). SWOT analysis applications: An integrative literature review. *Journal of Global Business Insights*, *6*(1), 55–73.

Bettini, C., & Riboni, D. (2015). Privacy protection in pervasive systems: State of the art and technical challenges. *Pervasive and Mobile Computing*, *17*, 159–174.

Bianchi, V., Bassoli, M., Lombardo, G., Fornacciari, P., Mordonini, M., & De Munari, I. (2019). IoT wearable sensor and deep learning: An integrated approach for personalized human activity recognition in a smart home environment. *IEEE Internet of Things Journal*, *6*(5), 8553–8562.

Bousdekis, A., Lepenioti, K., Apostolou, D., & Mentzas, G. (2021). A review of data-driven decision-making methods for industry 4.0 maintenance applications. *Electronics*, *10*(7), 828.

Brooke, J. (1996). SUS-A quick and dirty usability scale. In P. W. Jordan, B. Thomas, I. L. McClelland, & B. Weerdmeester (Eds), *Usability Evaluation in Industry* (vol. 189, no. 194, pp. 4–7). London: CRC Press.

Brown, T. (2008). Design thinking. *Harvard Business Review*, *86*(6), 84.

Brumitt, B., Meyers, B., Krumm, J., Kern, A., & Shafer, S. (2000). Easyliving: Technologies for intelligent environments. In *International Symposium on Handheld and Ubiquitous Computing* (pp. 12–29).

Burzagli, L., & Emiliani, P. L. (2017). Universal design in ambient intelligent environments. In *Universal Access in Human-Computer Interaction. Design and Development Approaches and Methods: 11th International Conference, UAHCI 2017, Held as Part of HCI International 2017, Vancouver, BC, Canada, July 9-14, 2017, Proceedings, Part I* 11 (pp. 31–42).

Campbell, R., Al-Muhtadi, J., Naldurg, P., Sampemane, G., & Mickunas, M. D. (2003). Towards security and privacy for pervasive computing. In *Software Security-Theories and Systems* (pp. 1–15). Berlin, Heidelberg: Springer.

Catalano, C., Meslec, M., Boileau, J., Guarino, R., Aurich, I., Baumann, N., Chartier, F., Dalix, P., Deramond, S., Laube, P., & others. (2021). Smart sustainable cities of the new millennium: Towards design for nature. *Circular Economy and Sustainability*, *1*(3), 1053–1086.

Chen, H., Perich, F., Finin, T., & Joshi, A. (2004). Soupa: Standard ontology for ubiquitous and pervasive applications. In *The First Annual International Conference on Mobile and Ubiquitous Systems: Networking and Services, 2004. MOBIQUITOUS 2004* (pp. 258–267).

Chin, J., Callaghan, V., & Allouch, S. B. (2019). The Internet-of-Things: Reflections on the past, present and future from a user-centered and smart environment perspective. *Journal of Ambient Intelligence and Smart Environments*, *11*(1), 45–69.

Cockton, G., Woolrych, A., & Lavery, D. (2009). Inspection-based evaluations. In A. Sears & J. A. Jacko (Eds), *Human-Computer Interaction* (pp. 289–308). Boca Raton, FL: CRC Press.

Cook, D. J., Augusto, J. C., & Jakkula, V. R. (2009). Ambient intelligence: Technologies, applications, and opportunities. *Pervasive and Mobile Computing*, *5*(4), 4.

Crovari, P., Gianotti, M., Riccardi, F., & Garzotto, F. (2019). Designing a smart toy: Guidelines from the experience with smart dolphin "SAM". In *Proceedings of the 13th Biannual Conference of the Italian SIGCHI Chapter: Designing the Next Interaction* (pp. 1–10).

Daniel, F., & Matera, M. (2014). *Mashups: Concepts, Models and Architectures*. Berlin, Heidelberg: Springer.

Daniel, F., Yu, J., Benatallah, B., Casati, F., Matera, M., & Saint-Paul, R. (2007). Understanding ui integration: A survey of problems, technologies, and opportunities. *IEEE Internet Computing*, *11*(3), 59–66.

Dasgupta, A., Handosa, M., Manuel, M., & Gračanin, D. (2019). A user-centric design framework for smart built environments: A mixed reality perspective. In *Distributed, Ambient and Pervasive Interactions: 7th International Conference, DAPI 2019, Held as Part of the 21st HCI International Conference, HCII 2019, Orlando, FL, USA, July 26-31, 2019, Proceedings 21* (pp. 124–143).

Delsing, J., & Lindgren, P. (2005). Sensor communication technology towards ambient intelligence. *Measurement Science and Technology*, *16*(4), R37.

Deshpande, A., Pitale, P., & Sanap, S. (2016). Industrial automation using Internet of Things (IOT). *International Journal of Advanced Research in Computer Engineering & Technology (IJARCET)*, *5*(2), 266–269.

DIS, I. (2009). 9241-210: 2010. Ergonomics of human system interaction-Part 210: Human-centred design for interactive systems. In *International Standardization Organization (ISO)*. Switzerland.

Dlamini, N. N., & Johnston, K. (2016). The use, benefits and challenges of using the Internet of Things (IoT) in retail businesses: A literature review. In *2016 International Conference on Advances in Computing and Communication Engineering (ICACCE)* (pp. 430–436).

Dunne, R., Morris, T., & Harper, S. (2021). A survey of ambient intelligence. *ACM Computing Surveys (CSUR)*, *54*(4), 1–27.

Emiliani, P. L., & Stephanidis, C. (2005). Universal access to ambient intelligence environments: Opportunities and challenges for people with disabilities. *IBM Systems Journal*, *44*(3), 605–619.

Engelbrecht, H., Lindeman, R. W., & Hoermann, S. (2019). A SWOT analysis of the field of virtual reality for firefighter training. *Frontiers in Robotics and AI*, *6*, 101.

Evans, C., Brodie, L., & Augusto, J. C. (2014). Requirements engineering for intelligent environments. In *2014 International Conference on Intelligent Environments* (pp. 154–161).

Fielding, R. T., & Taylor, R. N. (2000). *Architectural Styles and the Design of Network-Based Software Architectures* (Vol. 7). Irvine: University of California.

Finstad, K. (2010). The usability metric for user experience. *Interacting with Computers*, *22*(5), 323–327.

Gavaletakis, M., Leonidis, A., Stivaktakis, N. M., Korozi, M., Roulios, M., & Stephanidis, C. (2022). An accessible smart kitchen cupboard. In *Proceedings of the 24th International ACM SIGACCESS Conference on Computers and Accessibility* (pp. 1–4).

Gil, M., Albert, M., Fons, J., & Pelechano, V. (2022). Modeling and "smart" prototyping human-in-the-loop interactions for AmI environments. *Personal and Ubiquitous Computing, 26*(6), 1413–1444.

Gomes, M., Carneiro, D., Pimenta, A., Nunes, M., Novais, P., & Neves, J. (2014). Improving modularity, interoperability and extensibility in ambient intelligence. In *Ambient Intelligence-Software and Applications: 5th International Symposium on Ambient Intelligence* (pp. 63–70).

Grammenos, D., Michel, D., Zabulis, X., & Argyros, A. A. (2010). PaperView: Augmenting physical surfaces with location-aware digital information. In *Proceedings of the Fifth International Conference on Tangible, Embedded, and Embodied Interaction* (pp. 57–60).

Gullà, F., Menghi, R., Papetti, A., Carulli, M., Bordegoni, M., Gaggioli, A., & Germani, M. (2019). Prototyping adaptive systems in smart environments using virtual reality. *International Journal on Interactive Design and Manufacturing (IJIDeM), 13*, 597–616.

Hinderks, A., Schrepp, M., Mayo, F. J. D., Escalona, M. J., & Thomaschewski, J. (2019). Developing a UX KPI based on the user experience questionnaire. *Computer Standards & Interfaces, 65*, 38–44.

Hodrien, A., & Fernando, T. (2021). A review of post-study and post-task subjective questionnaires to guide assessment of system usability. *Journal of Usability Studies, 16*(3), 203–232.

Imbesi, S., & Mincolelli, G. (2020). Design of smart devices for older people: A user centered approach for the collection of users' needs. In *Intelligent Human Systems Integration 2020: Proceedings of the 3rd International Conference on Intelligent Human Systems Integration (IHSI 2020): Integrating People and Intelligent Systems,* February 19-21, 2020, *Modena, Italy* (pp. 860–864).

Kawakita, J. (1975). *The KJ Method-A Scientific Approach to Problem Solving*. Tokyo: Kawakita Research Institute.

Kawamoto, Y., Nishiyama, H., Kato, N., Yoshimura, N., & Yamamoto, S. (2014). Internet of things (IoT): Present state and future prospects. *IEICE Transactions on Information and Systems, 97*(10), 2568–2575.

Kim, H., Choi, H., Kang, H., An, J., Yeom, S., & Hong, T. (2021). A systematic review of the smart energy conservation system: From smart homes to sustainable smart cities. *Renewable and Sustainable Energy Reviews, 140*, 110755.

Kim, K., Norouzi, N., Jo, D., Bruder, G., & Welch, G. F. (2023). The augmented reality internet of things: Opportunities of embodied interactions in transreality. In *Springer Handbook of Augmented Reality* (pp. 797–829). Berlin, Heidelberg: Springer.

Kirchbuchner, F., Grosse-Puppendahl, T., Hastall, M. R., Distler, M., & Kuijper, A. (2015). Ambient intelligence from senior citizens' perspectives: Understanding privacy concerns, technology acceptance, and expectations. In *Ambient Intelligence: 12th European Conference, AmI 2015, Athens, Greece,* November 11-13, 2015, *Proceedings 12* (pp. 48–59).

Kirimtat, A., Krejcar, O., Kertesz, A., & Tasgetiren, M. F. (2020). Future trends and current state of smart city concepts: A survey. *IEEE Access, 8*, 86448–86467.

Kontogianni, A., & Alepis, E. (2020). Smart tourism: State of the art and literature review for the last six years. *Array, 6*, 100020.

Kopetz, J. P., Wessel, D., & Jochems, N. (2019). User-centered development of smart glasses support for skills training in nursing education. *I-Com, 18*(3), 287–299.

Korozi, M., Antona, M., Ntagianta, A., Leonidis, A., & Stephanidis, C. (2017a). LECTORstudio: *Creating inattention alarms and interventions to reengage the students in the educational process*. In *Proceedings of the 10th Annual International Conference of Education, Research and Innovation*.

Korozi, M., Leonidis, A., Antona, M., & Stephanidis, C. (2017b). Lector: Towards reengaging students in the educational process inside smart classrooms. In *9th International Conference on Intelligent Human Computer Interaction*.

Korozi, M., Stefanidi, E., Samaritaki, G., Prinianakis, A., Katzourakis, A., Leonidis, A., & Antona, M. (2019). Shaping the intelligent classroom of the future. In *International Conference on Human-Computer Interaction* (pp. 200–212).

Kotha, H. D., & Gupta, V. M. (2018). IoT application: A survey. *International Journal of Engineering and Technology (IJET), 7*(2.7), 891–896.

Gamma, E., Helm, R., Johnson, R., & Vlissides, J. (1995). *Design Patterns: Elements of Reusable Object-Oriented Software*. Pearson Deutschland GmbH.

Lee, S. H. (2006). Constructing effective questionnaires. In *Handbook of Human Performance Technology* (pp. 760–779). Hoboken, NJ: Pfeiffer Wiley.

Leist, A., & Ferring, D. (2012). Technology and aging: Inhibiting and facilitating factors in ICT use. In *Constructing Ambient Intelligence: AmI 2011 Workshops, Amsterdam, The Netherlands, November 16-18, 2011. Revised Selected Papers 2* (pp. 166–169).

Leonidis, A., Arampatzis, D., Louloudakis, N., & Stephanidis, C. (2017). The AmI-Solertis system: Creating user experiences in smart environments. In *Proceedings of the 13th IEEE International Conference on Wireless and Mobile Computing, Networking and Communications*.

Leonidis, A., Korozi, M., Kouroumalis, V., Adamakis, E., Milathianakis, D., & Stephanidis, C. (2021a). Going beyond second screens: Applications for the multi-display intelligent living room. In *ACM International Conference on Interactive Media Experiences* (pp. 187–193).

Leonidis, A., Korozi, M., Kouroumalis, V., Poutouris, E., Stefanidi, E., Arampatzis, D., Sykianaki, E., Anyfantis, N., Kalligiannakis, E., & Nicodemou, V. C. (2019). Ambient intelligence in the living room. *Sensors, 19*(22), 5011.

Leonidis, A., Korozi, M., Margetis, G., Grammenos, D., & Stephanidis, C. (2013). An intelligent hotel room. In *International Joint Conference on Ambient Intelligence* (pp. 241–246).

Leonidis, A., Korozi, M., Nikitakis, G., Ntagianta, A., Dimopoulos, A., Zidianakis, E., Stefanidi, E., & Antona, M. (2020). CognitOS board: A wall-sized board to support presentations in intelligent environments. *Technologies, 8*(4), 66.

Leonidis, A., Korozi, M., Sykianaki, E., Tsolakou, E., Kouroumalis, V., Ioannidi, D., Stavridakis, A., Antona, M., & Stephanidis, C. (2021b). Improving stress management and sleep hygiene in intelligent homes. *Sensors, 21*(7), 2398.

Lewis, J. R. (1992). Psychometric evaluation of the post-study system usability questionnaire: The PSSUQ. *Proceedings of the Human Factors Society Annual Meeting, 36*(16), 1259–1260.

Lewis, J. R. (1995). IBM computer usability satisfaction questionnaires: Psychometric evaluation and instructions for use. *International Journal of Human-Computer Interaction, 7*(1), 57–78.

Lewis, J. R. (2018). The system usability scale: Past, present, and future. *International Journal of Human-Computer Interaction, 34*(7), 577–590.

Libanori, A., Chen, G., Zhao, X., Zhou, Y., & Chen, J. (2022). Smart textiles for personalized healthcare. *Nature Electronics, 5*(3), 142–156.

Louloudakis, N., Leonidis, A., & Stephanidis, C. (2016). AmITest: A testing framework for ambient intelligence learning applications. *eLmL, 2016*, 87.

Lyytinen, K., & Yoo, Y. (2002). Ubiquitous computing. *Communications of the ACM, 45*(12), 63–96.

Margetis, G., Grammenos, D., Zabulis, X., & Stephanidis, C. (2013). iEat: An interactive table for restaurant customers' experience enhancement. In *HCI International 2013-Posters' Extended Abstracts: International Conference, HCI International 2013, Las Vegas, NV, USA, July 21-26, 2013, Proceedings, Part II 15* (pp. 666–670).

Marikyan, D., Papagiannidis, S., & Alamanos, E. (2019). A systematic review of the smart home literature: A user perspective. *Technological Forecasting and Social Change, 138*, 139–154.

Matz, A., & Götz, C. (2020). Designing human-centered interactions for smart environments based on heterogeneous, interrelated systems: A user research method for the "age of services" (URSERVe). In *Design, User Experience, and Usability. Design for Contemporary Interactive Environments: 9th International Conference, DUXU 2020, Held as Part of the 22nd HCI International Conference, HCII 2020, Copenhagen, Denmark, July 19-24, 2020, Proceedings, Part II 22* (pp. 99–116).

Michelson, B. M. (2006). Event-driven architecture overview. *Patricia Seybold Group, 2*(12), 10–1571.

Mitchell Finnigan, S., & Clear, A. K. (2020). "No powers, man!": A Student Perspective on Designing University Smart Building Interactions. In *Proceedings of the 2020 CHI Conference on Human Factors in Computing Systems* (pp. 1–14).

Müller, H., Sedley, A., & Ferrall-Nunge, E. (2014). Survey research in HCI. In J. Olson & W. Kellogg (Eds), *Ways of Knowing in HCI* (pp. 229–266). New York: Springer.

Nadkarni, S., Kriechbaumer, F., Rothenberger, M., & Christodoulidou, N. (2020). The path to the Hotel of Things: Internet of Things and Big Data converging in hospitality. *Journal of Hospitality and Tourism Technology, 11*(1), 93–107.

Newman, S. (2015). *Building Microservices: Designing Fine-Grained Systems*. California: O'Reilly Media, Inc.

Nielsen, J. (1994a). Heuristic evaluation. In J. Nielsen & R. L. Mack (Eds), *Usability Inspection Methods*. New York: John Wiley & Sons.

Nielsen, J. (1994b). *Usability Engineering*. San Francisco, CA: Morgan Kaufmann.

Nielsen, L. (2013). *Personas-User Focused Design* (Vol. 15). London: Springer.

Norman, D. (2004). *Affordances and Design* (p. 14). Unpublished Article. https://jnd.org/affordances-and-design/.

Ntoa, S., Margetis, G., Antona, M., & Stephanidis, C. (2019). UXAmI observer: An automated user experience evaluation tool for ambient intelligence environments. In *Intelligent Systems and Applications: Proceedings of the 2018 Intelligent Systems Conference (IntelliSys) Volume 1*, (pp. 1350–1370).

Ntoa, S., Margetis, G., Antona, M., & Stephanidis, C. (2021). User experience evaluation in intelligent environments: A comprehensive framework. *Technologies, 9*(2), 41.

Ntoa, S., Margetis, G., Antona, M., & Stephanidis, C. (2022). Digital accessibility in intelligent environments. In *Human-Automation Interaction: Manufacturing, Services and User Experience* (pp. 453–475). Cham: Springer International Publishing.

Papagiannidis, S., & Marikyan, D. (2020). Smart offices: A productivity and well-being perspective. *International Journal of Information Management*, *51*, 102027.

Park, H., & McKilligan, S. (2018). A systematic literature review for human-computer interaction and design thinking process integration. In *Design, User Experience, and Usability: Theory and Practice: 7th International Conference, DUXU 2018, Held as Part of HCI International 2018, Las Vegas, NV, USA, July 15-20, 2018, Proceedings, Part I 7* (pp. 725–740).

Partarakis, N., Grammenos, D., Margetis, G., Zidianakis, E., Drossis, G., Leonidis, A., Metaxakis, G., Antona, M., & Stephanidis, C. (2017). Digital cultural heritage experience in ambient intelligence. In M. Ioannides, N. Magnenat-Thalmann, & G. Papagiannakis (Eds), *Mixed Reality and Gamification for Cultural Heritage* (pp. 473–505). Cham: Springer.

Peng, H., Ma, S., & Spector, J. M. (2019). Personalized adaptive learning: An emerging pedagogical approach enabled by a smart learning environment. *Smart Learning Environments*, *6*(1), 1–14.

Perera, C., Zaslavsky, A., Christen, P., & Georgakopoulos, D. (2013). Context aware computing for the internet of things: A survey. *IEEE Communications Surveys & Tutorials*, *16*(1), 414–454.

Peres, R. S., Jia, X., Lee, J., Sun, K., Colombo, A. W., & Barata, J. (2020). Industrial artificial intelligence in industry 4.0-systematic review, challenges and outlook. *IEEE Access*, *8*, 220121–220139.

Perumal, T., Leong, C. Y., Samsudin, K., Mansor, S., & others. (2010). Middleware for heterogeneous subsystems interoperability in intelligent buildings. *Automation in Construction*, *19*(2), 160–168.

Piardi, L., Kalempa, V. C., Limeira, M., de Oliveira, A. S., & Leitão, P. (2019). Arena-Augmented reality to enhanced experimentation in smart warehouses. *Sensors*, *19*(19), 4308.

Prandi, C., Monti, L., Ceccarini, C., & Salomoni, P. (2020). Smart campus: Fostering the community awareness through an intelligent environment. *Mobile Networks and Applications*, *25*, 945–952.

Prinianakis, A., Stefanidi, H., Leonidis, A., Korozi, M., Katzourakis, A., Stamatakis, E., & Antona, M. (2021). An intelligent modular student desk. In *INTED2021 Proceedings* (pp. 9508–9518).

Psarommatis, F., & May, G. (2023). A literature review and design methodology for digital twins in the era of zero defect manufacturing. *International Journal of Production Research*, *61*(16), 5723–5743.

Quijano-Sánchez, L., Cantador, I., Cortés-Cediel, M. E., & Gil, O. (2020). Recommender systems for smart cities. *Information Systems*, *92*, 101545.

Ramos, C., Augusto, J. C., & Shapiro, D. (2008). Ambient intelligence—The next step for artificial intelligence. *IEEE Intelligent Systems*, *23*(2), 15–18.

Reichheld, F. F. (2003). The one number you need to grow. *Harvard Business Review*, *81*(12), 46–55.

Reisinger, M. R., Prost, S., Schrammel, J., & Fröhlich, P. (2019). User requirements for the design of smart homes: Dimensions and goals. In *Ambient Intelligence: 15th European Conference, AmI 2019, Rome, Italy, November 13-15, 2019, Proceedings 15* (pp. 41–57).

Remagnino, P., & Foresti, G. L. (2005). Ambient intelligence: A new multidisciplinary paradigm. *IEEE Transactions on Systems, Man, and Cybernetics - Part A: Systems and Humans*, *35*(1), 1–6. https://doi.org/10.1109/TSMCA.2004.838456

Roalter, L., Moller, A., Diewald, S., & Kranz, M. (2011). Developing intelligent environments: A development tool chain for creation, testing and simulation of smart and intelligent environments. In *2011 Seventh International Conference on Intelligent Environments* (pp. 214–221).

Robinson, D. C., Sanders, D. A., & Mazharsolook, E. (2015). Ambient intelligence for optimal manufacturing and energy efficiency. *Assembly Automation*, *35*(3), 234–248.

Rubin, J. (1995). Handbook of usability testing. *Technical Communication*, *42*(2), 361.

Rubio-Drosdov, E., Díaz-Sánchez, D., Almenárez, F., Arias-Cabarcos, P., & Marín, A. (2017). Seamless human-device interaction in the internet of things. *IEEE Transactions on Consumer Electronics*, *63*(4), 490–498.

Salem, B., Lino, J. A., & Simons, J. (2017). A framework for responsive environments. In *European Conference on Ambient Intelligence* (pp. 263–277).

Sanaeipoor, S., & Emami, K. H. (2020). Smart city: Exploring the role of augmented reality in placemaking. In *2020 4th International Conference on Smart City, Internet of Things and Applications (SCIOT)* (pp. 91–98).

Sánchez Sepúlveda, M. V., Fonseca Escudero, D., Franquesa Sànchez, J., & Martí Audí, N. (2019). Virtual urbanism: A user-centered approach. In *XIII CTV 2019 Proceedings: XIII International Conference on Virtual Cityand Territory:"Challenges and Paradigms of the Contemporary City"*: UPC, Barcelona, October 2-4, 2019.

Scorpio, M., Laffi, R., Masullo, M., Ciampi, G., Rosato, A., Maffei, L., & Sibilio, S. (2020). Virtual reality for smart urban lighting design: Review, applications and opportunities. *Energies*, *13*(15), 3809.

Selvaraj, S., & Sundaravaradhan, S. (2020). Challenges and opportunities in IoT healthcare systems: A systematic review. *SN Applied Sciences*, *2*(1), 139.

Sheng, Z., Yang, S., Yu, Y., Vasilakos, A., Mccann, J., & Leung, K. (2013). A survey on the ietf protocol suite for the internet of things: Standards, challenges, and opportunities. *IEEE Wireless Communications, 20*(6), 91–98.

Song, Y., Yu, F. R., Zhou, L., Yang, X., & He, Z. (2020). Applications of the Internet of Things (IoT) in smart logistics: A comprehensive survey. *IEEE Internet of Things Journal, 8*(6), 4250–4274.

Stantcheva, S. (2023). How to run surveys: A guide to creating your own identifying variation and revealing the invisible. *Annual Review of Economics, 15*, 205–234.

Stefanidi, E., Foukarakis, M., Arampatzis, D., Korozi, M., Leonidis, A., & Antona, M. (2019). ParlAmI: A multimodal approach for programming intelligent environments. *Technologies, 7*(1), 11.

Stefanidi, H., Leonidis, A., Korozi, M., & Papagiannakis, G. (2022). The ARgus Designer: Supporting experts while conducting user studies of AR/MR applications. In *2022 IEEE International Symposium on Mixed and Augmented Reality Adjunct (ISMAR-Adjunct)* (pp. 885–890).

Stephanidis, C. (2023). Paradigm shifts towards an inclusive society: From the desktop to human-centered artificial intelligence. In *Proceedings of the 2nd International Conference of the ACM Greek SIGCHI Chapter* (pp. 1–4).

Stephanidis, C., Antona, M., & Ntoa, S. (2021). Human factors in ambient intelligence environments. In G. Salvendy, & W. Karwowski (Eds), *Handbook of Human Factors and Ergonomics* (pp. 1058–1084). John Wiley & Sons, Inc.

Stojkoska, B. L. R., & Trivodaliev, K. V. (2017). A review of Internet of Things for smart home: Challenges and solutions. *Journal of Cleaner Production, 140*, 1454–1464.

Stratakis, C., Stivaktakis, N. M., Bouloukakis, M., Leonidis, A., Doxastaki, M., Kapnas, G., Evdaimon, T., Korozi, M., Kalligiannakis, E., & Stephanidis, C. (2022). Integrating ambient intelligence technologies for empowering agriculture. *Engineering Proceedings, 9*(1), 41.

Sun, Y., Armengol-Urpi, A., Kantareddy, S. N. R., Siegel, J., & Sarma, S. (2019). Magichand: Interact with iot devices in augmented reality environment. In *2019 IEEE Conference on Virtual Reality and 3D User Interfaces (VR)* (pp. 1738–1743).

Syed, A. S., Sierra-Sosa, D., Kumar, A., & Elmaghraby, A. (2021). IoT in smart cities: A survey of technologies, practices and challenges. *Smart Cities, 4*(2), 429–475.

Taherdoost, H. (2022). How to conduct an effective interview: A guide to interview design in research study. *International Journal of Academic Research in Management, 11*(1), 39–51.

Tapia, D. I., Abraham, A., Corchado, J. M., & Alonso, R. S. (2010). Agents and ambient intelligence: Case studies. *Journal of Ambient Intelligence and Humanized Computing, 1*, 85–93.

Turner, M., Budgen, D., & Brereton, P. (2003). Turning software into a service. *Computer, 36*(10), 38–44.

Ur, B., McManus, E., Pak Yong Ho, M., & Littman, M. L. (2014). Practical trigger-action programming in the smart home. In *Proceedings of the SIGCHI Conference on Human Factors in Computing Systems* (pp. 803–812).

Van Kesteren, A., & Jackson, D. (2007). The xmlhttprequest object. *World Wide Web Consortium, Working Draft WD-XMLHttpRequest-20070618, 72*.

Vijayan, D., Rose, A. L., Arvindan, S., Revathy, J., & Amuthadevi, C. (2020). Automation systems in smart buildings: A review. *Journal of Ambient Intelligence and Humanized Computing*, 1–13.

Wang, J., & Moulden, A. (2021). AI trust score: A user-centered approach to building, designing, and measuring the success of intelligent workplace features. In V. Loia (Ed), *Extended Abstracts of the 2021 CHI Conference on Human Factors in Computing Systems* (pp. 1–7). Springer-Verlag GmbH.

Weber, F. D., & Schütte, R. (2019). State-of-the-art and adoption of artificial intelligence in retailing. *Digital Policy, Regulation and Governance, 21*(3), 264–279.

Weber, W., Rabaey, J., & Aarts, E. H. (2005). *Ambient Intelligence*. Berlin, Heidelberg: Springer Science & Business Media.

Yin, C., Xiong, Z., Chen, H., Wang, J., Cooper, D., & David, B. (2015). A literature survey on smart cities. *Science China Information Sciences, 58*(10), 1–18.

Zabulis, X., Grammenos, D., Sarmis, T., Tzevanidis, K., Padeleris, P., Koutlemanis, P., & Argyros, A. A. (2013). Multicamera human detection and tracking supporting natural interaction with large-scale displays. *Machine Vision and Applications, 24*, 319–336.

Zantalis, F., Koulouras, G., Karabetsos, S., & Kandris, D. (2019). A review of machine learning and IoT in smart transportation. *Future Internet, 11*(4), 94.

Zeng, J., Yang, L. T., Lin, M., Ning, H., & Ma, J. (2020). A survey: Cyber-physical-social systems and their system-level design methodology. *Future Generation Computer Systems, 105*, 1028–1042.

Zhang, J., & Tao, D. (2020). Empowering things with intelligence: A survey of the progress, challenges, and opportunities in artificial intelligence of things. *IEEE Internet of Things Journal, 8*(10), 7789–7817.

2 User Interface Adaptation and Design for All

Constantine Stephanidis

2.1 INTRODUCTION

User interface (UI) adaptation has been defined as the modification of a software system's UI in order to satisfy specific requirements, such as the needs and preferences of a particular user or a group of users (Abrahão et al., 2021). Adaptation may fall into various categories depending on how it is performed, by whom and according to which parameters. For example, according to some authors, adaptability refers to the end user's ability to adapt the UI, whereas adaptivity or self-adaptation refers to the system's ability to autonomously perform UI adaptation (Bouzit et al., 2017). Personalization, on the other hand, is considered as a particular form of adaptivity, usually affecting the UI contents, based on data originating from the end user, such as personal traits and interactive behavior (Fan & Poole, 2006).

Various characteristics of the interaction context may require or bring about UI adaptations, including for example (Paternò, 2019):

- User characteristics: preferences, goals and tasks, physical state (e.g., position), emotional state, etc.
- Technology characteristics: screen resolution, connectivity, browser, battery, etc.
- Environment-related aspects: location, light, noise, etc.

Based on variations in the factors above, various aspects of a UI can be modified: presentation, including layout, graphical attributes, voices and their characteristics, etc.; dynamic behavior, including navigation structure, activation and deactivation of interaction techniques, etc.; and content, including texts, labels, and images. Various adaptation strategies are possible. Examples are: conservation, e.g., scaling of UI elements; rearrangement, e.g., changing the layout; simplification/magnification, same UI elements but with modified presentation; increase (also called progressive enhancement)/reduction (also called graceful degradation) of UI elements, etc. (Paternò, 2019).

The concept of UI adaptation originated in the 1990s in the context of early approaches to the automation of UI design and development (e.g., model-based UIs, Abrahão et al., 2021) as well as the development of UIs that exhibit intelligent behavior (intelligent user interfaces, Hartmann, 2009).

A number of technical approaches were proposed to support UI adaptations in both the design and the implementation phases of the UI development lifecycle. Adaptation mechanisms were usually based on models or ontologies reflecting knowledge about the users (user models), the context of use (context models), the application in question, and other relevant adaptation parameters, and included various types of reasoning mechanisms to perform adaptation decision-making. Some approaches were targeted to automatically generate UIs capable of self-adaptation (Eisenstein & Puerta, 2000), whereas other approaches were more oriented toward facilitating the work of designers and developers in creating adaptive UIs (Gullà et al., 2015).

At that time, UI adaptation was mainly targeted to benefit end users by optimizing various aspects of the end user's experience. For example, the objective of UI adaptation could be to increase efficiency (by reducing task completion time and error rate or by improving the learning curve), to ensure effectiveness (by guaranteeing full task completion) or to improve the subjective user's satisfaction (Abrahão et al., 2021). Despite promises, the realization of the idea proved to be challenging, due to a number of reasons. From a user experience point of view, introducing meaningful and useful UI adaptations at the right time is not straightforward, and the risk if the UI is modified abruptly or in a suboptimal way is to confuse the users instead of easing their tasks. From a design point of view, automatically generated adaptive UI proved to have limited scope, addressing limited tasks and specific applications, while it also emerged that designers found it difficult to manage the large design spaces that UI adaptations introduced and populate the necessary models. Therefore, despite progress, UI adaptation found little practical application at its first wave (Hartmann, 2009; Meixner et al., 2011).

Subsequently, the concept was revisited toward the beginning of the new millennium, when mobile devices started conquering the market. In such context, UI adaptation was meant to cater for the development of UIs capable of displaying themselves on devices with various screen sizes and resolutions, to seamlessly migrate from device to device in order to support the continuity of the user experience and to improve mobile interaction (Eisenstein et al., 2000). The latter involves taking into account, for example, that the user can be on the move and short of time, and therefore able to pay only limited attention to the interaction.

From a development perspective, this renewed account of UI adaptation meant primarily dealing with authoring multi-device interactive applications, avoiding separate developments for each target platform. Various approaches emerged. The first approach is developing one main UI version with fluid layout, as in the case of responsive Web design, which uses different stylesheets for different devices, thus being able to change the values of some attributes or show or hide some UI elements (Gardner, 2011). This can be a relatively convenient and easy-to-apply solution, but it can limit the differences among versions that can be obtained in some cases since stylesheets do not allow deep changes in the structure of the interactive applications. The second approach is single authoring, in which one conceptual description of the interactive application is developed, from which various versions optimized for different target platforms are obtained (Simon et al., 2005). The third approach is automatic reauthoring, in which the starting point is the implementation for a specific platform, and then derive implementations adapted for different platforms through appropriate transformations (Paternò et al., 2008). While, on the one hand, the research community focused mainly on the single authoring approach, by developing framework and environments supporting developers in building multi-device UIs, on the other hand, responsive design has been widely adopted and has become the mainstream way of developing websites nowadays.

In parallel with the above threads of research on UI adaptation, the increasing importance acquired in the HCI field by the issue of accessibility to applications and services by disabled and older users, and the emergence of mainstream approaches such as Universal Access and Design for All, brought to the proposal of UI adaptation as a mean of designing and developing UIs which can automatically modify themselves to cater for the needs and requirements of users with diverse characteristics, e.g., functional limitations due to disability or age. Universal Access (Stephanidis, 1995, 2001a) aims at the provision of access to anyone, from anywhere and at any time, through a variety of computing platforms and devices, to diverse products and services. Universal Access encompasses several dimensions of diversity that emerge from the broad range of user characteristics, the changing nature of human activities, the variety of contexts of use, the increasing availability and diversification of information, the variety of knowledge sources and services, and the proliferation of diverse technological platforms that occur in today's technological landscape.

Design for All, introduced in the context of HCI (Stephanidis, 2001a) as an approach to achieve Universal Access, implies an explicit focus on systematically addressing diversity. In this context, automatic UI adaptation (Stephanidis, 2001b) has surfaced as a solution for catering to the needs and requirements of a diverse user population in a variety of contexts of use.

A related but different concept is that of Universal Usability, defined as "having more than 90% of all households as successful users of information and communications services at least once a week" (Shneiderman, 2000, p. 85). Although Universal Usability advocated addressing user diversity by accommodating users with different skills, knowledge, age, gender, disabilities, disabling conditions, literacy, culture, income, etc., it did not propose adaptation-based solutions.

Despite progress, the practice of designing for diversity has remained difficult, due to the intrinsic complexity of the task, the current limited expertise of designers and practitioners in designing interfaces capable of automatic adaptation as well as the limited availability of appropriate supporting tools.

Today, the increasing progress and uptake of AI and especially machine learning technologies bring the promise of overcoming some of the well-known bottlenecks of UI adaptation, such as developing the necessary knowledge models and reasoning mechanisms. Under this perspective, new approaches to UI adaptation are emerging which show potential for addressing in a mainstream and proactive way accessibility and user experience issues and facilitating the design and development of meaningful interaction adaptations.

This chapter, after introducing Design for All and providing a brief overview of adaptation-based approaches to accessibility, discusses a series of tools and components that have been developed over the past two decades and applied in various research and development projects. These tools have demonstrated the technical feasibility of the approach and have contributed to reducing the practice gap between traditional UI design and design for adaptation. They have been applied in a number of pilot applications and case studies, including applications in novel domains such as assistive robots and driver assistance systems. This chapter concludes with a discussion of AI-based approaches to UI adaptation and of the new opportunities for wider adoption of adaptation-based UIs that emerge in this context.

2.2 DESIGN FOR ALL

The concept of Design for All was introduced in the human-computer interaction (HCI) literature in the 1990s (Stephanidis, 1995; Salvendy et al., 1998, 1999; Stephanidis and Emiliani, 1999). Design for All in HCI is rooted in the fusion of three traditions, namely human-centered design (HCD), accessibility and assistive technologies for disabled people, and universal design for physical products and the built environment.

HCD (Vrendenburg et al., 2001; ISO, 1999; ISO, 2010; ISO, 2019) since its beginning three decades ago has established the centrality of human users throughout the design process and has consolidated itself as a standard process for the design of every interactive system. HCD relates to Design for All insofar it claims that the quality of use of a system, including usability, depends on the characteristics of the users, tasks, and the organizational and physical environment in which the system is used. Also, it stresses the importance of understanding and identifying the details of this context in order to guide early design decisions and provides a basis for evaluation. However, HCD does not specify how designers can cope with radically different user groups.

Accessibility is another intrinsic dimension of Design for All. In the context of HCI, accessibility refers to the access by people with disabilities to Information and Communication Technologies (ICT). Interaction with ICT may be affected in various ways by the user's individual abilities or functional limitations that may be permanent, temporary, situational, or contextual. For example, someone with limited vision will not be able to use an interactive system that only provides graphical output, while someone with limited movements of the upper limbs will encounter difficulties in using an interactive system that only accepts input through the standard keyboard and mouse. Accessibility in the context of HCI aims at making the interaction experience of people with diverse functional or contextual limitations as near as possible to that of people without such limitations by overcoming existing interaction barriers. This may concern facilitating the manipulation of input devices and related interaction techniques, providing interaction styles and resulting UIs which can be perceived through various human senses as well as ensuring that UIs are understandable to users. To achieve the above, accessibility requires the availability of alternative devices and interaction styles to accommodate different needs.

Two main technical approaches to accessibility have been followed. The first approach is to treat each application separately and take all the necessary implementation steps to arrive at an alternative accessible version (product-level adaptation). Product-level adaptation practically often implies re-development from scratch. Due to the high costs associated with this strategy, it is considered as the least favorable option for providing alternative access. The second alternative is to "intervene" at the level of the particular interactive application environment (e.g., MS-Windows) in order to provide appropriate software and hardware technology so as to make that environment alternatively accessible (environment-level adaptation). The latter option extends the scope of accessibility to cover potentially all applications running under the same interactive environment, rather than a single application.

The above approaches have given rise to several methods for addressing accessibility, including techniques for the configuration of the UI's input/output, and the provision of assistive technologies. Assistive Technology (Assistive Technology Act, 1998) is a generic term denoting a wide range of accessibility plug-ins including special-purpose input and output devices and the process used in selecting, locating, and using them. Popular assistive technologies include screen readers and Braille displays for blind users, screen magnifiers for users with low vision, alternative input and output devices for motor-impaired users (e.g., adapted keyboards, mouse emulators, joystick, and binary switches), specialized browsers, and text prediction systems.

Despite progress, as in the case of multi-platform UIs, approaches to accessibility based on the a posteriori development of accessible alternatives to a mainstream UI have shown to suffer from some serious shortcomings, especially when considering the continuously changing technological environment.

First, such solutions typically provide limited and low-quality access. In some cases, adaptations may not be possible without loss of functionality. For example, in the early versions of windowing systems, it was impossible for the programmer to obtain access to certain window functions, such as window management, and therefore such functions could not be rendered in non-visual modalities. In subsequent versions, this shortcoming was addressed by the vendors of such products allowing certain adaptations of interaction objects on the screen.

Second, a posteriori adaptations are not viable in sectors of the industry characterized by rapid technological change. By the time a particular accessibility problem has been addressed, technology has advanced to a point where the same or a similar problem re-occurs. A typical example that illustrates this state of affairs is the case of blind people's access to computers. Each generation of technology (e.g., DOS environment, windowing systems, and multimedia) caused a new 'generation' of accessibility problems to blind users, addressed through dedicated techniques, such as text-to-speech translation for the DOS environment, off-screen models, and filtering for the windowing systems.

Finally, re-development is programming-intensive and therefore expensive and difficult to implement and maintain. Minor changes in product configuration, or the UI, may require substantial resources to re-build accessibility features.

From the above, it became gradually evident that the re-development of mainstream applications' UIs to make them accessible does not suffice to cope with the rapid technological change and the evolving human requirements. The proliferation of new interactive products and services, technological platforms, and access devices brought about the need to reconsider the issue of accessibility from a more proactive perspective and generic perspective. This entails building access features into a UI starting from its conception and throughout the entire development life cycle (Salvendy et al., 1998, 1999).

Similar approaches toward addressing people's diversity first emerged in engineering disciplines such as civil engineering and architecture, with many applications in interior design, building, road construction, and transport. The term Universal Design was coined by the architect Ronald L. Mace to describe the concept of designing all products and the built environment to be both aesthetically pleasing and usable to the greatest extent possible by everyone, regardless of their age, ability, or status in life (Mace, 1991). Although the scope of the concept has always been broader, its focus tended to be on the built environment.

A classic example of universal design is the kerb cut (or sidewalk ramp), initially designed for wheelchair users to navigate from street to sidewalk, and today widely used in many buildings. Perhaps the most common approach in universal design is to make information about an object or a building available through several modalities, such as Braille on elevator buttons, and acoustic feedback for traffic lights. People without disabilities can often benefit too. For example, subtitles on TV or multimedia content intended for the deaf can also be useful to non-native speakers of a language, to children for improving literacy skills, or to people watching TV in noisy environments.

In the context of HCI, the principles of universal design (Story, 2001) were used by the end of the nineties to denote design for access by anyone, anywhere and at any time to interactive products and services.

Based on the three mentioned pillars, Design for All has been defined as the proactive application of principles, methods, and appropriate tools, in order to develop interactive products and services which are accessible and usable by all citizens, thus avoiding the need for a posteriori adaptations, or specialized design (Salvendy et al., 1998). The term Design for All either subsumes, or is a synonym of, terms such as accessible design, inclusive design, barrier-free design, and universal design, each highlighting different aspects of the concept.

While the concept of Design for All is not novel, and several technical solutions have been proposed to address it, as discussed in the following sections, it is notable that until recently standardization efforts in this domain have been notably absent. Despite the importance and potential benefits of Design for All as well as the longstanding acknowledgment of the value of accessibility and inclusivity across various domains, this lack of standardization has remained a significant barrier to the widespread implementation of Design for All practices in interactive technologies. The first step toward standardization can be considered the technical standard on Web Content Accessibility Guidelines (WCAG) developed by the Web Accessibility Initiative (WAI) of the World Wide Web Consortium (W3C), which was initially introduced in 1999 and is being continuously updated (Henry, 2023). It should be noted that the WCAG standards do not pertain to applying the Design for All method, however, if employed effectively, they can ensure that a web page is inherently accessible to everyone, including persons with disabilities, permanent or situational. This aligns with the fundamental principle the Design for All approach as outlined above. Similarly, other accessibility-related standards that can be considered as means to pursuit inclusive design have been published (ETSI, n.d.). Nevertheless, standardization on Design for All was only recently addressed, with a new CEN standard published in 2019, namely the EN 17161:2019 'Design for All - Accessibility following a Design for All approach in products, goods and services - Extending the range of users' (CEN, 2019). Table 2.1 summarizes the pertinent standards, discussing not only how to apply Design for All but also how to design inclusive ICT products and services designed for all.

TABLE 2.1
Summary of Standards Relevant to Inclusive Design and Design for All

Organization	Standard	Publication Year
W3C-WAI	Web Content Accessibility Guidelines 1.0	1999
W3C-WAI	Web Content Accessibility Guidelines 2.0	2008
W3C-WAI	Web Content Accessibility Guidelines 2.1	2018
ETSI/CEN/CENELEC	EN 301549v1.1.2European Standard 'Accessibility requirements suitable for public procurement of ICT products and services in Europe	2015
ETSI/CEN/CENELEC	EN 301549v2.1.2Harmonized European Standard 'Accessibility requirements for ICT products and services'	2018
ETSI/CEN/CENELEC	EN 301549v3.1.1Harmonized European Standard 'Accessibility requirements for ICT products and sertices'	2019
ETSI/CEN/CENELEC	EN 301549V3.2.1Harmonized European Standard 'Accessibility requirements for ICT products and services'	2021
CEN	EN 17161:2019'Design for All - Accessibility following a Design for All approach in products, goods and services - Extending the range of users'	2019

2.3 ACCESSIBILITY BY UI ADAPTATION

In the light of the above, it is clear that single artifact-oriented design approaches offer limited possibilities in the context of accessibility and Design for All. A critical property of interactive artifacts becomes, therefore, their capability for automatic adaptation and personalization (Stephanidis, 2001b).

Methods and techniques for UI customization have met significant success in modern commercial UIs. Some popular examples include desktop adaptations, offering, for example, the ability to hide or delete unused desktop items. Other provisions include various personalization features of the desktop based on the personal preferences of the user, by adding animations, transparent glass menu bars, live thumbnail previews of open programs, and desktop gadgets (like clocks, calendars, weather forecast, etc.). Similarly to operating systems, popular applications offer several customizations, such as toolbars positioning and showing/hiding recently used options. However, adaptations integrated into commercial systems need to be set manually and mainly focus on aesthetic preferences. In terms of accessibility and usability, for instance to people with disability or older people, only a limited number of adaptations are available, such as keyboard shortcuts, size and zoom options, changing color and sound settings, automated tasks, etc.

On the contrary, research efforts in the past three decades have elaborated novel and rich approaches to UI adaptation for accessibility. Pioneering work toward this direction in the 90s introduced the concept of Dual UIs (Savidis & Stephanidis, 1995), i.e., UIs exhibiting both a graphical and a voice-based instantiation, so as to be accessible at the same time by sighted and blind users, thus fostering collaboration. The concept was subsequently extended to cater to the design of UIs which could potentially adapt to any type of human functional limitations (Savidis & Stephanidis, 2009a), advocating comprehensive and systematic approaches to UI adaptations which are based on the principles of Universal Access and promote the practice of Design for All. In such approaches, UI adaptations constitute a vehicle to efficiently and effectively ensure the accessibility and usability of UIs to users with diverse characteristics, supporting also technological platform independence, metaphor independence, and user-profile independence. In such a context, automatic UI adaptation aims at minimizing the need for a posteriori adaptations and delivering products that can be adapted for use by the widest possible end user population. This implies the following: provision by design of alternative interface manifestations, availability of knowledge regarding the abilities, requirements, and preferences of the target user groups, as well as the characteristics of the context of use (e.g., technological platform, physical environment), availability of a decision-making logic capable of making adaptation decisions either before the interactions starts or during the interaction.

As design methods and techniques for addressing diversity involve complex design processes and have a higher entrance barrier with respect to more traditional artifact-oriented methods, requiring also accessibility knowledge and expertise, it became apparent that it was necessary to provide computational tools supporting such a process. The main objective in this respect was to offer tools which reduce the difference in practice between conventional UI development and development for adaptation. Finally, another prominent challenge in the context of Universal Access was identified as the need to develop large-scale case study applications targeted to demonstrate the technical feasibility of the adaptation-based approach, but also to assess its benefits, as well as the usefulness and added value of the related methods and tools, and provide instruments for further experimentation and ultimately improve the empirical basis of the field by collecting knowledge on how design for diversity may be concretely practiced. This novel UI development paradigm was instantiated in the Unified UI methodology (Savidis & Stephanidis, 2009a), supported by a specifically developed architecture, as well as a variety of tools to support UI adaptation design, including facilities for specifying decision-making rules, adaptation design tools, adaptable widget toolkits for various interaction platforms, UI prototyping facilities, design tools which automatically embed accessibility features and support frameworks for multimodal interaction.

It is worth noting that alternative implementation approaches have emerged over the years, which can be roughly categorized into model-based, rule-based, and pattern-based approaches (Firmenich et al., 2019).

The model-based CAMELEON reference framework (Calvary et al., 2003) decomposes UI design into a number of different abstraction levels, aiming to reduce the effort in targeting multiple contexts of use, and has been implemented in research tools such as UsiXML (Limbourg & Vanderdonckt, 2004) and MARIA (Paternò et al., 2009). The latter is a model-based UI description language enabling the automatic generation of UIs customized for diversified devices at runtime. Model-based techniques can simplify the effort involved in providing accessibility to diverse target user groups by maintaining the relationships between different levels of abstraction. This can then be used to annotate the final UI in ways that can be exploited by assistive technologies. Effective model-based authoring tools can apply such annotations automatically, as well as helping to ensure that image, audio, and video resources are complemented by the appropriate accessible descriptions.

In addition to the above, CAMELEON has been enriched with a mechanism to integrate adaptation rules in the development process, with the objective to automatically generate accessible user-tailored interfaces. Adaptations can be applied both at design time and at runtime. At design time, a UI can be tailored at any abstraction level in the development process. For instance, adaptation rules related to task sequencing should be considered at the task and domain level, whereas adaptation rules related to some specific UI modalities have to be considered at the concrete UI level. At runtime, changes in the context of use trigger the adaptation process (Miñón et al., 2016). The solution proposed involves obtaining the necessary level of abstraction by means of an abstraction process in order to apply adaptation rules when a change in the context occurs and then generate again the final UI.

Other rule-based approaches include Yang and Shao (2007), which proposes an expert system managing the adaptation rules, using a special algorithm to match the rules to the facts. The adaptation knowledge base consists of a fact base, i.e., context profiles, and a rule base, i.e., adaptation rules. Another example has been proposed by Ghiani et al. (2017), aiming to provide an easy-to-understand way to specify contextual events and conditions as well as the corresponding actions, which can modify the application or even the state of surrounding appliances (e.g., lights, radio). The objective is to facilitate the authoring of adaptation rules by people who are not professional developers, including the final users or their caregivers, giving them the possibility to personalize their applications according to specific situations.

Finally, some pattern-based frameworks have also been developed. MyUI (Mainstreaming Accessibility through Synergistic User Modeling and Adaptability) integrates all aspects of adaptive UIs for accessibility in one system (Peissner et al., 2012). It includes a multimodal design patterns repository, which serves as a basis for a modular approach to individualized UIs, a runtime environment with the adaption engine enabling UI generation and dynamic adaptations during run-time, the specification of an abstract UI model and a development tool. Another example is the Egoki System (Gamecho et al., 2015), which automatically generates UIs to access ubiquitous services. The system provides ample flexibility to accommodate the users' needs in the selection of suitable multimedia interaction patterns and UI elements. UI modeling is performed through a state charts-based algorithm. The reusable software components and their associated code snippets are a part of its design rationale. The adaptive system is incorporating context, users, and adaptation to deliver user-centric and demand-driven UIs.

Despite the prevalence of handcrafted rule sets and heuristics for UI adaptation, more recent efforts have also focused on advanced techniques for the adaptation and personalization of UIs. In this respect, combinatorial optimization has emerged as an effective and flexible way of deciding on UI design and interactions, offering distinguished algorithmic capacity, controllability, and generalizability (Oulasvirta, Dayama, Shiripour, John & Karrenbauer, 2020). Adaptations may refer to the selection of program functions to be manipulated and presented to the user, widget types and their properties, organization of components within their containers with decisions on their position and size, as well as distribution of components across containers to form a hierarchy (Oulasvirta et al., 2020).

Pioneering in this domain was the groundbreaking work of SUPPLE (Gajos & Weld, 2004), for UI optimization according to the user's device. By conceptualizing UI adaptation as an optimization problem, SUPPLE employed user traces to customize interface rendering to the usage patterns of individual users. Overall, in combinatorial optimization for UI adaptation, UI design is seen as the algorithmic process of combining design decisions to pursue optimal solutions for the problem of finding combinations that yield the highest value for a given function, such as maximizing the interface's usability (Lindlbauer, Feit & Hilliges, 2019) or minimizing user effort (Bailly et al., 2013). Other optimization problems sought in adaptive UIs as reported in the literature pertain to adapting text entry components for increasing typing performance of people with disabilities (Sarcar et al., 2018), optimizing a UI based on user's actual motor capabilities to allow them to accomplish a set of typical tasks in the least amount of time (Gajos et al., 2008), or enhancing situational awareness (Stefanidi et al., 2022). An alternative approach to UI adaptation employs reinforcement learning to identify beneficial changes in the UIs and avoiding changes when they are not beneficial and applies this approach to adaptive menus (Todi et al., 2021).

While the above approaches have undoubtedly enriched the current knowledge and discourse on UI adaptation for accessibility and user diversity, they have encountered difficulties in wider adoption (Ziegler & Peissner, 2017), and none of them has been applied in large case studies demonstrating practical applicability, and all appear to be limited in terms of types of adaptations they support, and consequently the target user groups and contexts of use they can address. This chapter maintains that a design-based generic and systematic approach to UI adaptation is so far the most promising toward addressing Universal Access requirements but also that optimization-based and AI techniques bring very promising future potential in this context.

2.4 UI ADAPTATION TOOLS

This section discusses a number of tools developed in the context of design-based approaches to UI adaptation, supporting various phases of the development lifecycle. These tools have been extensively used in a number of large design cases and have demonstrated the viability and usefulness of the approach.

2.4.1 Decision-Making Specification Language

The role of decision-making in UI adaptation is to effectively drive the interface assembly process by deciding which interface components need to be selectively activated. The Decision-Making Specification Language (DMSL) (Savidis et al., 2005) is a rule-based language specifically designed and implemented for supporting the specification of adaptations. DMSL supports the effective implementation of decision-making and has been purposefully elaborated to be easy to use for designers without programming knowledge.

In DMSL, the decision-making logic is defined in an independent decision "if…then…else" blocks, each uniquely associated to a particular UI dialogue context. The individual end user and usage-context profiles are represented in the condition part of DMSL rules using an attribute values notation. The language is equipped with three primitive statements: (1) dialogue, which initiates evaluation for the rule block corresponding to dialogue context value supplied; (2) activate, which triggers the activation of the specified component(s); and (3) cancel, which, similarly to activate, triggers the cancellation of the specified component(s). These rules are compiled in a tabular representation that is executed at run-time. Figure 2.1 provides an example of the DMSL rule. The representation engages simple expression evaluation trees for the conditional expressions.

1	If	[Elderly user's age = 1 or 2 or 3] or [Elderly user's life situation = 2 or 3] or [Elderly user's computer literacy level = 0] or [Vision impairment = 1 or 2 or 3]
	Then	DPI 150%
2	If	[Elderly user's life situation =1] or [Elderly user's computer literacy level = 1]
	Then	DPI 100%

FIGURE 2.1 An example of DMSL rule.

The decision-making process is performed in independent sequential decision sessions, and each session is initiated by a request of the interface assembly module for the execution of a particular initial decision block. The outcome of a decision session is a sequence of activation and cancellation commands, all of which are directly associated to the task context of the initial decision block. Those commands are posted back to the interface assembly module as the product of the performed decision-making session.

The DMSL language has been used in many of the tools described in the subsequent subsections, constituting their reasoning backbone. It has proved to constitute a valuable starting point for easily integrating adaptation behavior in UIs. However, one of its limitations is that depending on the types and number of adaptations required, it may necessitate a quite large body of complex rules, with consequent difficulties in maintaining and updating the logic. In Antona et al. (2006) a support tool was proposed for the process of self-adapting UI design which offers automated generation of the adaptation logic, as well as automated verification mechanisms. The tool has been applied in a number of design case studies, confirming its advantages compared to "paper-based" adaptation design. The designers involved in the case studies were able to rapidly acquire familiarity with the design of UI adaptations. The verification facilities have also been found particularly effective in helping designers to detect and correct inconsistencies or inaccuracies in the logic conditions. Furthermore, the tool has been considered as particularly useful in providing the automatic generation of the DMSL adaptation logic for direct integration in the prototype implementation of the self-adapting UI.

2.4.2 Interaction Toolkits

UI adaptation necessitates alternative versions of interaction artifacts to be created and coexist in the design space. This can be achieved through software toolkits capable to dynamically deliver an interface instance that is adapted to a specific user in a specific context of use. Such toolkits are essentially software libraries, encompassing alternative versions of interaction elements and common dialogues, where each version is designed in order to address particular values of the user- and usage-context parameters. The runtime adaptation-oriented selection of the most appropriate version, according to the end user and usage-context profiles, is the key element in supporting a wide range of alternative interactive incarnations. It should be noted that the presence and management of the alternative versions are fully transparent to toolkit clients. The latter provides the behavior of a smart toolkit capable to adaptively deliver its interaction elements so as to fit the current usage profile.

2.4.2.1 EAGER

To support adaptive web UIs, the combination of user-centered design, UI prototyping and design guidelines is applied. The EAGER Designs Repository (Partarakis et al., 2010) is an extensible collection of implemented and ready-to-use alternative interaction elements, including:

- alternative primitive UI elements with enriched attributes (e.g., buttons, links, radios, etc.)
- alternative structural page elements (e.g., page templates, headers, footers, containers, etc.)
- fundamental abstract interaction dialogues in multiple alternative styles (e.g., navigation, file uploaders, paging styles, and text entry).

User Interface Adaptation and Design for All

Each alternative element version, called a style, is purposefully designed to address the requirements of specific user and context parameter values. Alternative styles have been designed following typical user-centered design, user interface prototyping and adoption of design guidelines. Additionally, EAGER design alternatives not only integrate accessibility guidelines, but also provide a suitable approach to personalized accessibility. In this respect, EAGER can be viewed as encompassing consolidated adaptation design knowledge, thus greatly facilitating designers in the choice of suitable adaptations according to user-related or context-related parameters.

The Designs Repository component of EAGER provides the designs of alternative dialogue controls in the form of abstract interaction objects. For each alternative version, the respective adaptation rationale is also recorded, including the profile parameters which are adaptively addressed.

An example is provided by images. Blind or low vision users are not interested in viewing images, but only in reading the alternative text that describes the image. To facilitate blind and low vision users, two design alternatives were produced which are presented in Figure 2.2.

The text representation of the image simply does not present the image, but only a label with the prefix 'Image:' and followed by the alternative text of the image. The second representation, targeted to users with visual impairments, is the same as the first with the difference that, instead of a label, a link is included that leads to the specific image giving the ability of saving the image. In addition to the above, another design was produced that can be selected as a preference by web portal users in which the images are represented as thumbnail bounding the size that holds on the web page. A user who wishes to view the image in normal size may click on it. In Table 2.2, the design rationale of the alternative images design is presented, including its adaptation logic.

Through EAGER, the complexity of the UI design effort is radically reduced due to the flexibility provided by the toolkit for designing interfaces at an abstract task-oriented level. Therefore, designers are not required to be aware of the low-level details introduced in representing interaction elements, but only of the high-level structural representation of a task and its appropriate decomposition into sub-tasks, each of which represents a basic UI and system function.

On the other hand, the process of designing the actual front end of the application using a mark-up language is radically decreased in terms of time, because developers initially have to select among a number of interface components each of which represents a far more complex facility. Additionally, developers do not have to spend time for editing the presentation characteristics of the high-level interaction element, due to the internal styling behavior.

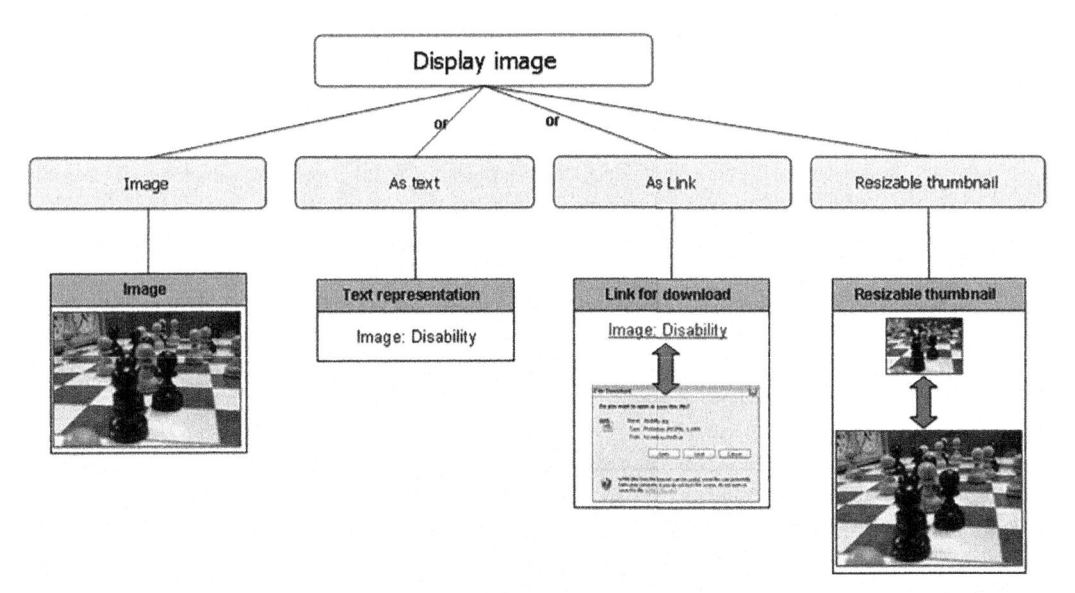

FIGURE 2.2 Alternative image hierarchy in EAGER. (From Partarakis et al., 2010.)

TABLE 2.2
The Design Rationale of Alternative Image Styles

Task: Display Image

Style:	Image	As Text	As Link	Resizable Thumbnail
Targets:	–	Facilitate screen reader and low vision users in order not to be in difficulties with image viewing	Facilitate screen reader and low vision users in order not to be in difficulties with image viewing but with the capability to save or view an image	Viewing images in small size in order not to hold large size on the web page with the capability to enlarge the image to normal size when it is necessitated
Parameters:	User (Default)	User (Blind or Low vision)	User (Blind or Low vision) and user preference	User (preference)
Properties:	View image	Read image alternative text	Read image alternative text or/and select linked named as the image alternative text to save or view the image	View image thumbnail and select it to view it in normal size
Relationships:	Exclusive	Exclusive	Exclusive	Exclusive

The actual process of transforming the initial design into the final Web application using traditional UI controls introduces a lot of coding. On the contrary, when using EAGER, the amount of code required is significantly reduced because the developer has the option to use a number of plug-and-play controls each of which represents a complex user task. Furthermore, the incorporation of EAGER's higher-level elements makes the code more usable, more readable and especially safe, due to the fact that each interaction component introduced is designed separately, developed and tested introducing a high level of code reuse, efficiency, and safety.

2.4.2.2 JMorph

The JMorph adaptable widget library (Leonidis et al., 2012) is another example of toolkit inherently supporting the adaptation of UI components. It contains a set of adaptation-aware widgets designed to satisfy the needs of various target devices; swing-based components for PC, AWT-based components for Windows Mobile devices. Adaptation is completely transparent to developers who can use the widgets as typical UI building blocks.

JMorph instantiates a common look and feel across the applications developed using it. The implemented adaptations are meant to address the interaction needs of older users (Leuteritz et al., 2009), and follow specific guidelines that have been encoded into DMSL rules (Savidis et al., 2005). This approach is targeted to novice developers of adaptable UIs, as it relieves developers from the task of re-implementing or modifying their applications to integrate adaptation-related functionality.

The developed widgets are built in a modular way that facilitates their further evolution, by offering the necessary mechanism to support new features addition and modifications. Therefore, more experienced developers can use their own adaptation rules to modify the adaptation behavior of the interactive widgets.

The library's implementation using the Java programming language ensures the development of portable UIs that can run unmodified with the same look regardless of the underlying Operating System, including also mobile devices.

The JMorph library provides the necessary mechanisms to support alternative look and feel either for the entire environment (i.e., skins) or for individual applications.

Every widget initially follows the general rules to ensure that the common look and feel invariant will be met and then applies any additional presentation directives declared as "custom" look and feel rules. A "custom" rule can affect either an individual widget (e.g., the OK button that appears in the confirmation dialog of a specific application) or a group of widgets; therefore, entire applications

User Interface Adaptation and Design for All

can be fully skinned since their widgets inherently belong to a group defined by the application itself (e.g., all the buttons that belong to a specific application).

The adaptable widgets implemented in JMorph include label, button, check box, list, scrollbar, textbox, text area, drop-down menu, radio button, hyperlink, slider, spinner, progress bar, tabbed pane, menu bar, menu, menu item, and tooltip. Complex widgets, such as date and time entry, have also been developed. Adaptable widget attributes include background color/image, widget appearance and dimensions, text appearance, cursor's appearance on mouse over, highlighting of currently selected items or options, orientation options (vertical or horizontal), explanatory tooltips, etc.

Figure 2.3 shows some of the available widgets. Adaptable attributes for each widget are summarized in Table 2.3. All widgets in the library also include a text description which allows easy interoperation with speech-based interfaces, thus offering also the possibility to deploy a non-visual instance of the developed interfaces.

2.4.3 Adaptive UI Prototyping

Popular UI builders provide graphical environments for UI prototyping, usually following a WYSIWYG ("What You See Is What You Get") design paradigm. Available WYSIWYG editors offer graphical editing facilities that allow designers to perform rapid prototyping visually. Such editors may be standalone or embedded in integrated environments (IDEs), i.e., programming environments that allow developing application functionality for the created prototypes directly. Commonly used IDEs are Microsoft Visual Studio, NetBeans, and Eclipse. IDEs are very popular in application development because they greatly simplify the transition from design to implementation, thus speeding up considerably the entire process. However, available tools do not integrate

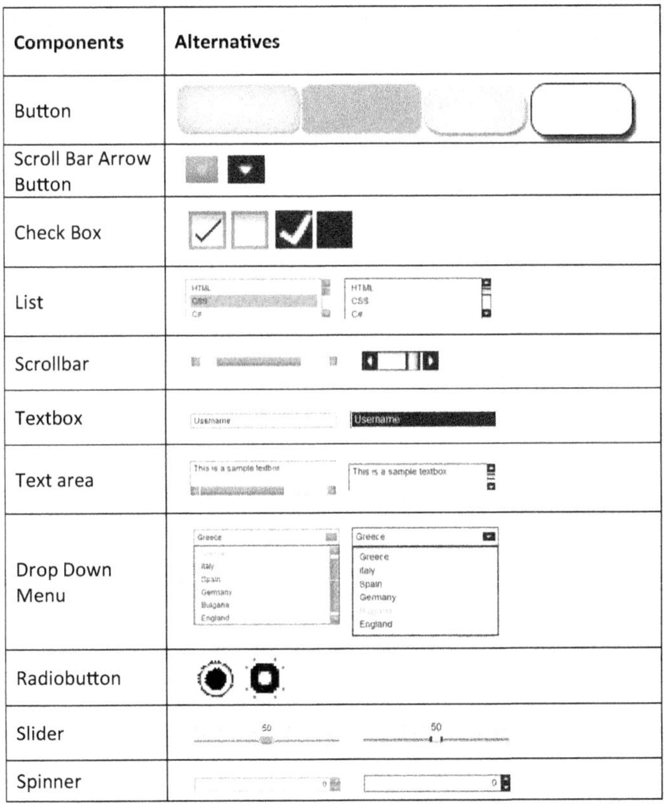

FIGURE 2.3 Examples of adaptable widgets.

TABLE 2.3
Adaptation Features of Widgets in the Adaptable Widget Library

Buttons	Associated icon when idle, mouse over it, clicked or disabled Shortcut key
	Text
	Status (Enabled, Disabled)
	Tooltip's text and colors (foreground and background) Border
	Background and foreground color when clicked or idle Button's font
	Cursor appearance on mouse over it (e.g., hand cursor) Access key
	Vertical and horizontal text alignment
	Free space (gap) between button's icon and text
Check Box	Associated icon when enabled and checked, enabled and unchecked, disabled and checked or disabled and unchecked
	Shortcut key to check/uncheck the checkbox Text
	Status (Enabled, Disabled)
	Tooltip's text and colors (foreground and background) Border
	Background and foreground color Checkbox's font
	Cursor appearance on mouse over (e.g., hand cursor) Access Key
	Vertical and horizontal text alignment
Drop-down menu	Background and foreground color of available and highlighted choices Choices' font
List	List orientation (vertical or horizontal)
	Background and foreground color of available and highlighted choices Choices' font
	Border around list component
	Tooltip text either one common for the list itself, or a different one for each choice
	Cursor appearance on mouse over (e.g., hand cursor) Access Key
Text Box	Text
	Maximum number of characters per line Text's font
	Background color when this component is on or out of focus Border around text box
	Foreground color of the text when either enabled or disabled
	Cursor appearance when user hovers mouse over it (e.g., hand cursor)
	Access Key that facilitates traversal using keyboard (e.g., right arrow instead of Tab)
	Status (Enabled, Disabled)
	Editable status, whether the user can alter the contents of this text box Tooltip's text and colors (foreground and background)
	Highlight text color when selected by mouse or due to search facility
Password Text Box	Text
	Maximum number of characters per line Text's font
	Background color when this component is on or out of focus Border around text box
	Foreground color of the text when either enabled or disabled
	Cursor appearance when user hovers mouse over it (e.g., hand cursor)
	Access Key that facilitates traversal using keyboard (e.g., right arrow instead of Tab)
	Status (Enabled, Disabled)
	Editable status, whether the user can alter the contents of this text box
	Tooltip's text and colors (foreground and background)
	Highlight text color when selected by mouse or due to search facility

(Continued)

TABLE 2.3 (*Continued*)
Adaptation Features of Widgets in the Adaptable Widget Library

Text Area	Text
	Maximum number of characters per line Text's Font
	Background (focused, not focused) Border around text box
	Foreground color of the text when either enabled or disabled
	Cursor appearance when user hovers mouse over it (e.g., hand cursor)
	Access Key that facilitates traversal using keyboard (e.g., right arrow instead of Tab)
	Status (Enabled, Disabled)
	Editable status, whether the user can alter the contents of this text box Tooltip's text and colors (foreground and background)
	Highlight text color when selected by mouse or due to search facility Maximum number of lines
	Type of text wrapping when text area is not wide enough
Radio Buttons	Text
	Background and foreground color when selected or not Border around radio button
	Radio button's text Font
	Cursor appearance when user hovers mouse over it (e.g., hand cursor) Status (Enabled, Disabled)
	Tooltip's text and colors (foreground and background) Foreground color when disabled
	Associated Icon when enabled and selected, enabled and unselected, disabled and selected or disabled and unselected
	Shortcut key to select the radio button
	Foreground and Background color when radio button is on or off focus
Hyperlink/Label	Text
	Foreground color when Enabled
	Background Color (inherited by parent container) Hyperlink's Font
	Tooltip's text and colors (foreground and background)
	Cursor appearance when user hovers mouse over it (e.g., hand cursor) Border around Hyperlink
	Status (Enabled, Disabled)
Table	Row height and margin between rows
	Foreground and Background color of currently selected cell Show grid (horizontal, vertical lines)
	Column width
	Border around Table Cells Background Color of Table Cell
	Tooltips' text and colors (foreground and background) Background and foreground color of the table
	Text's Font
Slider	Slider's Orientation (Horizontal or Vertical)
	Minimum and maximum value that user could select using slider Label for each discrete slider value (e.g., Start - End, 0 –100 etc.) Visibility status of labels, major and minor ticks
	Background color of slider component Status (Enabled, Disabled)
	Border around slider component
	Tooltip's text and colors (foreground and background)
	Cursor appearance when user hovers mouse over it (e.g. hand cursor) Major and minor tick spacing (e.g. every fifth tick should be large - major-, while any other should be small -minor)
	Foreground color of ticks (little vertical lines below slider) Visibility status of the track
	Invert start with end (e.g., on vertical orientation start is the top of the slider while end is the bottom)
	Snap to ticks (limit user's selection only to ticks, e.g., when user slides cursor to 4.6, cursor should automatically be "attracted" to 5

(*Continued*)

TABLE 2.3 (*Continued*)
Adaptation Features of Widgets in the Adaptable Widget Library

Spinner	Background and Foreground Color Border around spinner
	Cursor appearance when user hovers mouse over it (e.g., hand cursor) Status (Enabled, Disabled)
	Text's Font
	Tooltip's text and colors (foreground and background)
Menu bar	Background color
Menu	Background color (inherited from menu bar) Foreground color
	Text
	Border around menu
	Associated icon when menu is opened and closed or enabled and disabled
Menu Item	Background and foreground color of each menu item
	Associated con when component is enabled or disabled, or when user hover its mouse over it
Progress Bar	Background and Foreground Color Progress's text font
	Border around bar
	Status (Enabled, Disabled)
	Tooltip's text and colors (foreground and background) Minimum and maximum value of the bar
	Progress bar's orientation (horizontal or vertical) Progress's text visibility status
Tooltips	Background and foreground color Text
	Text's Font
	Border around tooltip
	Cursor appearance when user hovers mouse over it (e.g., hand cursor) Status (Enabled, Disabled)
Tabs	Tab Layout Policy Tab Placement Border around tab Tab's label Font
	Cursor appearance on mouse over (e.g., hand cursor) Mnemonic (visible and functional per tab)
	Status (Enabled, Disabled) either for all tabs or for a specific one Tooltip's text and colors (foreground and background)
	Associated icon with each tab when enabled or disabled, or when selected or unselected
	Tab's color when selected or not

adaptable widgets nor provide support for developing UI adaptations. Therefore, prototyping alternative design solutions for different needs and requirements using prevalent prototyping tools may become a complex and difficult task if the number of alternatives to be produced is large. To facilitate its employment, the JMorph adaptable widget library has been integrated into the NetBeans GUI Builder (version 8.0, see Figure 2.4). The result is claimed to be the first and so far, a unique tool that supports rapid prototyping of self-adapting UIs, with the possibility of immediately previewing adaptation results.

To prototype a UI, the designer will create the application's main window and will add the common containers (e.g., menu panels, status bar, and header) by placing adaptive panels where appropriate. The necessary widgets (e.g., menu buttons, labels, and text fields) will then be dragged from the Palette and dropped into the design area of the builder.

To customize widgets, the typical process is to manually set the relevant attributes for each widget using the designer's "property sheets". To apply the same adjustment to other widgets, one can either copy/paste them or iteratively set them manually. In the adaptation-enabled process, using the function attribute, the process is slightly different. First, one needs to set the function attribute, then define the required style (e.g., colors, images, and fonts), and finally to define the rule (in a separate

User Interface Adaptation and Design for All

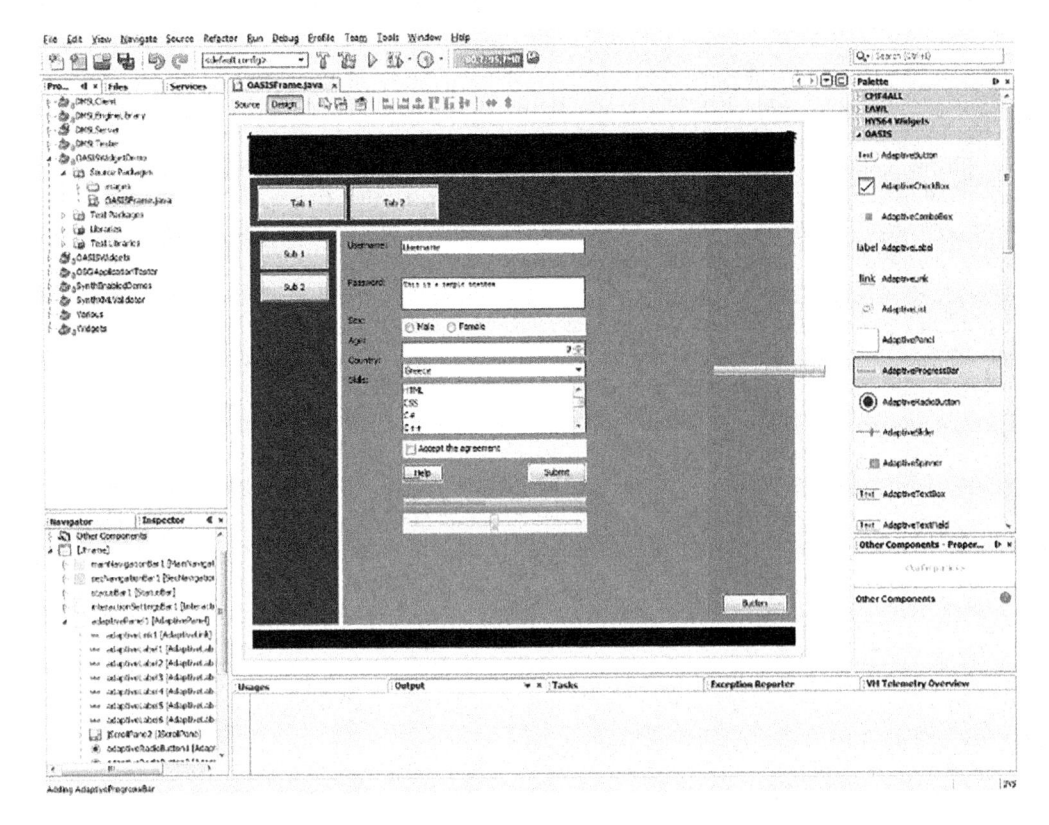

FIGURE 2.4 Using the Adaptable Widget Library integrated in the NetBeans IDE.

rule file) that maps the newly added style to the specific function. Whenever the same style should be applied, it is sufficient to simply set the function attribute respectively (CSS-like). In some cases, more radical adaptations are required with respect to widget customization, as the same physical UI design cannot be applied. In these cases, alternative dialogues can be designed by creating a container to host the different screens. The JMorph library offers the means to dynamically load different UI elements on demand, providing the functionality through adaptation rules.

The drag-and-drop selection and placement of widgets follow a conventional WYSIWYG approach. However, in the specific case, What You See Is One Instance of What You Get, as all adaptation alternatives can be produced in the preview mode by simply setting some user-related variables (e.g., selecting a profile). During preview, a set of sizing rules are automatically applied to ease the design process. The obtained prototypes can easily be used for testing and evaluation purposes.

The result is a tool that offers the possibility of prototyping adaptable interfaces following standard practices, without the need of designing customized widget alternatives or to specify adaptation rules (which are predefined). Through the prototyping tool, it is also possible to preview how adaptations are applied for the defined user profiles. However, more expert designers can easily modify the DMSL adaptation rules, which are stored in a separate editable file, and experiment with new adaptations and varying look and feel.

2.5 ADAPTIVE MULTIMODAL HUMAN ROBOT INTERACTION

UIs for assistive robots are typically multimodal and are usually developed from scratch, thus requiring considerable effort. FIRMA[1] (Kazepis et al., 2016) is a framework supporting the development of multimodal, elderly-friendly, interactive applications for assistive robots targeted to elderly users in assisted living environments. FIRMA provides the necessary tools and building blocks for creating multimodal robot applications, which are inherently friendly to elder users and capable of adapting to their needs, the surrounding environment and the context of use. The modalities that have been developed and integrated into the proposed framework range from speech recognition and synthesis to gesture recognition and touch interaction. They all have been developed to be fully extensible and configurable both at startup and at runtime, so that they can change to reflect the changing needs of the users or the dynamically changing factors of the surrounding environment e.g., ambient lighting, environment noise, active electric appliances, etc.

The implemented speech recognition modality engine supports adaptation based on the distance between the robot and the user, the vocabulary and the variety of the individual equivalent commands that can be used by the users and understood by the system, the semantic interpretation of recognized commands and the recognition confidence threshold. Furthermore, it supports the dynamic activation of both plain text and compiled speech recognition grammars.

The speech synthesis modality of FIRMA has been based on the speech engine functionality provided by Microsoft Speech Synthesis.[2] The implemented speech synthesis modality engine can be tailored to the needs and preferences of the users as well as the context of the interaction by offering adaptation parameters exposing the gender of the used voice, the speech volume, the rate as well as the pitch of the generated output.

Gesture recognition has been integrated in FIRMA using the gesture recognition module of Michel, Papoutsakis and Argyros (2014). The recognition engine that has been developed to cover the gesture modality needs of the proposed framework is able to understand a predefined set of gestures that are relatively easy to perform and be remembered by the end users of the platform. Finally, the touch modality is supported through a touchscreen Tablet PC that is onboard in the household robotic platform.

The proposed framework integrates all the aforementioned modalities into a seamless set of interaction modes between the robot and its users. This results in a more natural form of interaction since the user is free to choose how to interact with the system based both on his/her preferences and the context of interaction. The robot can display its output on the onboard touchscreen device and use sound at the same time as redundant auditory feedback just like when people interact with each other. Furthermore, the robot is able to understand touch on the touchscreen device, gestures in front of the monitoring image acquisition sensors as well as speech commands given by the users.

Different modalities that are supported by the framework can be activated or deactivated individually according to the preferences of the users and the context of interaction. FIRMA supports adaptation through both adaptive component hierarchies (Savidis & Stephanidis, 2009a) and adaptive style hierarchies (as they are supported by the Windows presentation framework).

Adaptive component hierarchies are inherently supported by the proposed framework. Tasks are described in an abstract manner using the ACTA description language for activity analysis (Zidianakis et al., 2018). ACTA rules (see Figure 2.5) are used at runtime to respond to environmental and context-of-use changes to the system variables and infer facts to be used by a decision-making component to activate or deactivate the appropriate components of the adaptive component hierarchies on which the UIs are generated. The above methodology ensures that the end users are offered interaction experiences tailored to their individual user attributes and to the particular context of use. For example, the time selection task can be declared to comprise the selection of hours, minutes and time specifiers according to the time of the day. General guidelines can be stated according to the expertise of the user encoded in the user profile. These guidelines specify how the whole task of time selection can be orchestrated in order to be presented to the user who is going to be guided through the process of time selection. Furthermore, user preferences are taken into account so that specific user controls are used or omitted during the process. Various adaptation strategies are supported, which can be classified according to the impact they have on the UI: conservation, e.g., simple scaling of UI elements; rearrangement, e.g., changing the layout; simplification/magnification,

User Interface Adaptation and Design for All

```
when(CurrentBrightnessLevel == "Dark")
{
    ColorScheme = "LightColors";
    SetBrightness(20);
}

when(CurrentBrightnessLevel == "Normal")
{
    ColorScheme = "DarkColors";
    SetBrightness(50);
}

when(CurrentBrightnessLevel == "Light")
{
    ColorScheme = "HighContrastColors";
    SetBrightness(90);
}
```

FIGURE 2.5 Sample ACTA adaptation script fragment in FIRMA. (From Foukarakis et al., 2017.)

same UI elements but with modified presentation; increase (also called progressive enhancement)/reduction (also called graceful degradation), in terms of UI elements.

In addition to the adaptive component hierarchies' principles and design guidelines, the approach of adaptive style hierarchies has been adopted. According to this approach, the sizes and colors of the displayed framework elements can be controlled by styles that can be applied both at design time and at runtime (see Figure 2.6). A number of cascading stylesheets have been developed to be

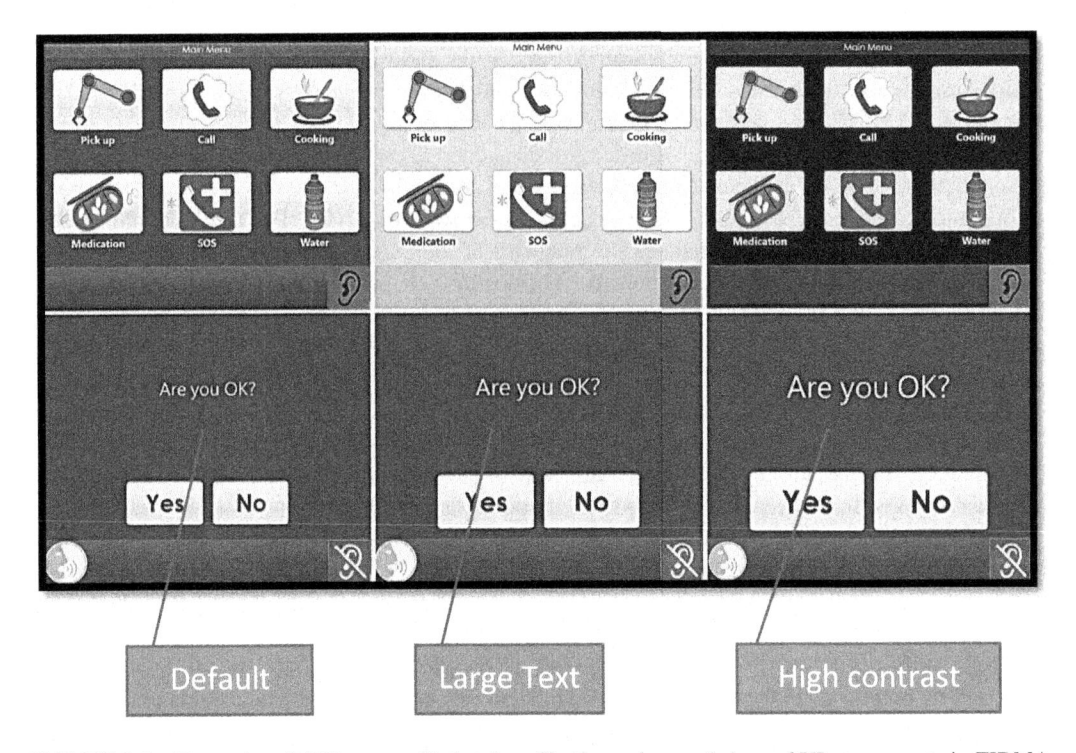

FIGURE 2.6 Examples of different applied styles affecting colors and sizes of UI components in FIRMA. (From Foukarakis et al., 2017.)

used in the context of adaptation based on this approach. A subset of the developed styles is used during design time so that the developer can have a clear understanding of the appearance of the different user controls and dialogues that he/she is incorporating into the developed applications. The design time styles collection has been consolidated into a single higher-level style file, which can be included in the designed user controls and dialogues.

The adaptive style hierarchies that are used for adaptation purposes during runtime have been split into three major categories. The first category contains all the styles that handle how the different framework elements will be displayed. These styles contain all the individual stylistic decisions that drive the appearance and define the visual tree of all the framework elements such as buttons, lists, dialogues, text entry controls, labels, etc. The second category contains all the styles that define the coloring scheme of the application including foreground and background colors for all framework elements, border brushes of the different user controls, darker backgrounds for giving emphasis to specific UI elements, etc. Finally, the third major category contains all the styles that correspond to the sizing decisions of all the framework elements and UI dialogues, including button sizes, dialogue sizes, virtual keyboard sizes and margins, text input control sizes, etc.

2.5.1 THE RAMCIP ROBOT UI

The FIRMA framework (Section 4.4) has been applied in the development of the UIs of the RAMCIP assistive robot (Foukarakis et al., 2017). RAMCIP is targeted to proactively and discreetly support older persons and patients with mild cognitive impairment or early stages of Alzheimer's disease in domestic environments, by providing a robotic assistant capable of supporting patients in daily activities such as medication intake, cooking, etc. and enhance their feeling of safety and socialization. The RAMCIP robotic platform offers high-level cognitive functions, driven through advanced human activity and home environment modeling and monitoring, enabling it to optimally decide when and how to assist its users. The developed robot UI runs on a 12-inch Tablet mounted on the front side of the robot platform (see Figure 2.7).

FIRMA provides a set of commonly used UI dialogues and screens (such as option presenters, notification screens, and binary decision dialogues), which were extensively used in the implementation of the RAMCIP applications. Additional visual components were also introduced. The ability to optionally enhance the dialogues with images was added, as well as a new loading screen to be shown when the robot is busy performing tasks.

For the first version of the platform, two applications were designed. The first one, called frequently asked questions (FAQ), includes the majority of the dialogues spoken and shown by the robot, which were usually questions, but also notifications. It utilizes the visual components of the framework to provide adaptable dialogues used for presenting information or interacting with the user. Input is supported through all available modalities.

The second application, called Dialer, was responsible for the robot's external communication with the caregivers, friends, and relatives of the patients.

Each application is equipped with its own set of ACTA scripts corresponding to a series of dialogues that are presented to the user in accordance with the currently running scenario. Typically, a dialogue presented to the user can include speech output, text, and images.

Initial adaptations in the RAMCIP UI are based on the virtual user models of the target user population, which are part of the overall robot's software. The properties that drive this adaptation strategy may include the user's preferences, potential age-related or memory-related disabilities and functional limitations, the user's communication preferences and modality personalization parameters, etc. On the other hand, run-time adaptations are based on dynamically changing factors during the interaction between the user and the robotic platform or through environmental sensing. The environment plays an important role in this type of adaptation, as it can provide the robot with all the necessary environment status data including ambient lighting, ambient noise, etc. Such dynamically changing factors can affect the modality selection as well as the adaptation of each used

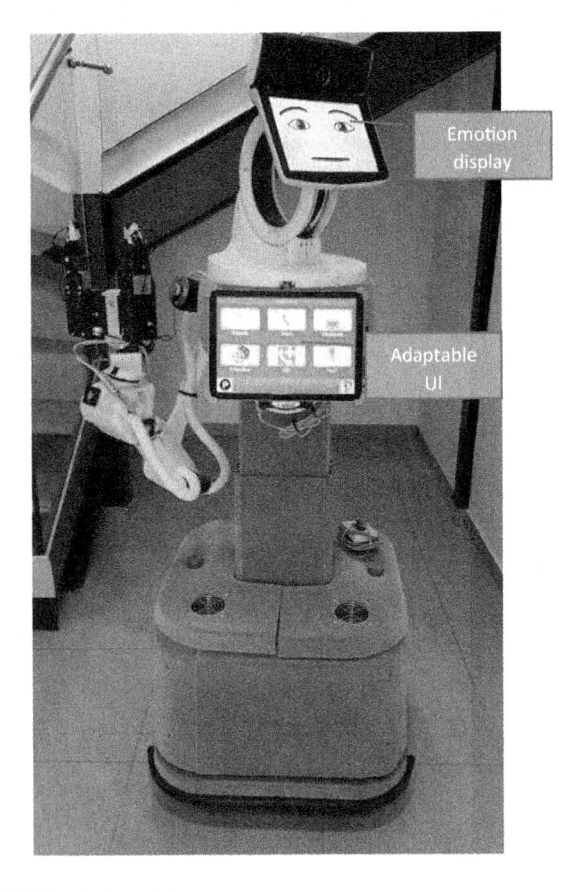

User Interface Adaptation and Design for All 63

FIGURE 2.7 The RAMCIP robot prototype.

modality. For example, information regarding the distance of the user from the robot is acquired through a laser sensor and is used for determining the size to use for icons, buttons, and text. The smaller sizes are easier on the eyes while interacting through the touch screen from close range, but in cases where the user is a bit further away from the screen, bigger letters are more convenient and legible. The UI application constantly receives information about the user distance from the platform and adjusts the component sizing accordingly, falling back to a default size in case there is no distance information.

Distance information is also used for modality selection since not all modalities are efficient at all distances. Touch interaction is only useful at very close distance. Gesture recognition works best only if the user is between 1 and 3 m away from the platform, and it is required that the user's skeleton is visible and has been correctly detected by the front camera of the robot. Moreover, voice recognition has a maximum efficient range. Finally, the dialogues that are presented to the user through the touch screen of the tablet can be further adapted based on knowledge of the distance between the user and the robot. In some cases where the user is not close to the robot, a combination of an image and speech synthesis in place of text could be preferable to convey the necessary information to the user. Other environmental parameters taken into account for adaptation include brightness and noise.

The robot also stores information about user cognitive fatigue, and the interaction is adapted accordingly, for example by providing alternative dialogs during communication (e.g., break down a request in simple steps).

Table 2.4 summarizes the adaptable components implemented in the RAMCIP UIs and the related adaptation parameters.

TABLE 2.4
Adaptable Components in the RAMCIP Robot User Interfaces

Adaptable Component	Adaptable Parameter	Values
Speech Synthesis	Enabled	On/Off
Speech Synthesis	Pitch	−10 (low) to +10 (high)
Speech Synthesis	Rate	−10 (slow) to +10 (fast)
Speech Synthesis	Volume	0–100
Speech Recognition	Enabled	On/Off
Speech Recognition	Recognition Threshold	Low/Medium/High
Gesture Recognition	Enabled	On/Off
Gesture Recognition	Specific gesture enabled (Yes/No/Help/Cancel)	On/Off
Augmented Reality Interface	Enabled	On/Off
UI Dialog	Dialog visual components presence (e.g., buttons, labels, images)	Image only/Text only/Image+Text
UI Dialog	Notification Sound (e.g., beep when first shown or repeated)	Off/Subtle/Normal/Important
UI Dialog	Text Size	Small/Large/Larger
UI Dialog	Button Size	Small/Large/Larger/Huge
UI Dialog	Colors	Light Blue/Dark Blue/High Contrast
UI Dialog	Positioning/Layout	Depends on specific dialog layout
Facial Expression	Face Image	Neutral/Excited, Confused/Sleeping, Focused/Sad/Angry
Screen Brightness	Screen Brightness Level	Low/Medium/High
FSM flow	Use Case Scenario	More descriptive dialogs/Alternative texts, etc., to be examined in use case scope

2.6 PERSONALIZED INTERACTION WITH ADVANCED DRIVER ASSISTANCE SYSTEMS

Advanced driver assistance systems (ADAS) aim at improving driving safety through preventive support systems, such as adaptive cruise control, automatic emergency braking, and lane keeping assist. ADAS systems utilize a variety of human–machine interaction (HMI) elements for interacting with the driver. ADAS information exchange is mostly supported by touch input devices as well as visual and auditory forms. Alternative input and output modalities have also been considered.

A prominent direction for further improving safety and the overall driving experience is to offer personalized and adaptive interaction, taking into account aspects of the driver, the vehicle and the environment to deliver custom-tailored interaction through intelligent decision-making. For example, an ADAS system may take into account the driver's distraction or lack of experience to trigger proactive actions earlier. Also, the methods for informing the driver about an upcoming road condition may depend on driving context or environmental conditions. For example, in case of bad lighting conditions (e.g., sunlight), which may correspond to a situation of temporary visual impairment, an auditory message would be preferred over a visual one (see Figure 2.8).

However, designing and developing UI adaptations in ADAS is a complex endeavor, usually conducted using ad hoc solutions. To facilitate this endeavor, the ADAS&ME framework has been developed, based on a rule engine that uses a customizable and extensible set of personalization and adaptation rules, provided by automotive domain and HMI experts, and evaluates them according to the current driver, vehicle, and environment parameters to produce HMI activation and GUI personalization and adaptation decisions. Personalized HMI modality selection is realized by taking into account all available input and output modalities of the vehicle and maintaining bindings for their

User Interface Adaptation and Design for All

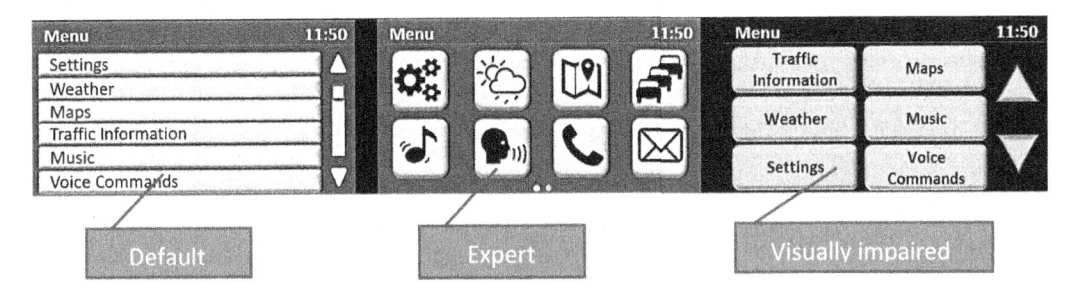

FIGURE 2.8 Examples of alternative menu incarnations in the ADAS&ME framework; default style (left), for computer experts (middle), for visually impaired users (right). (From Lilis et al., 2019.)

activation. GUI personalization is handled automatically through a GUI toolkit of personalizable and adaptable user controls that can be used for the development of any GUI application requiring personalization features.

The personalized HMI framework classifies modalities into categories and supports multimodality. Output HMI elements (e.g., touch screens, speakers, vibration motors, and AR displays) are classified into three categories: visual, auditory, and haptic. The HMI elements of the vehicle can be triggered, activated, or deactivated, while it is also possible to query for their status and capabilities. The HMI element status can be used to detect failing HMI elements so as to select alternative interaction modalities. HMI element capabilities can also be taken into account in specifying the adaptation and personalization logic. For example, to visualize a notification message with a lot of content, the system would take into account the screen size of each visual HMI element and possibly opt for a big display. Input HMI modalities are categorized into keypad and pointing elements. The first category covers elements with hardware buttons that post distinct codes per key press (e.g., dashboard buttons) but also composite systems that process input streams and produce high-level command (e.g., speech recognition). The second category abstracts over physical pointing methods like relative or absolute pointing (e.g., touch). Support for multimodality is based on an architectural split between virtual input, i.e., at the level of the HMI element interface, and physical input, i.e., at the level of the physical input device. Each HMI input element interface abstracts over multiple physical input elements of its category and operates as a high-level input channel triggering interaction command, while physical input plug-ins map device commands to interaction commands. Application logic handles input at the interaction command level is independent from physical elements and supports any vehicle setup. Input modality selection is also subject to change based on the personalization and adaptation logic. For example, voice commands would be preferred over dashboard buttons when driving at high speed. On the contrary, voice recognition would be restricted in a loud environment so as to reduce recognition errors.

The framework includes a GUI toolkit of adaptive and personalizable user controls, which enables robust rapid prototyping. The toolkit features custom user controls (widgets) that extend their native counterparts with personalization support, including both low-level (e.g., labels, buttons, and checkboxes) and high-level (e.g., tab groups, list views) widgets. Low-level widgets are typically personalized in terms of visual style aspects such as fonts and colors. Supporting specific styles for particular widgets or widgets collections is achieved in a way similar to CSS styling, i.e., assigning IDs or classes to widgets to allow matching them against personalization rules. High-level widgets are designed and developed based on the notion of adaptive component hierarchies (Savidis & Stephanidis, 2009). Abstract UI tasks are hierarchically decomposed into sub-tasks, which at the lowest level are matched by physical interface designs. Multiple physical designs are available per task, supporting a matching process that yields the best combination of physical designs based on the personalization parameters. The toolkit is extensible, allowing to introduce custom, application-specific user controls building on top of the native ones.

The central component of the framework is a rule engine that uses a set of personalization and adaptation rules specified in ACTA (Zidianakis et al., 2018). The specified rules are evaluated according to the current input data to produce HMI activation decisions, handled by the HMI modality selection component as well as GUI personalization decisions.

2.6.1 The ADAS&ME Applications

The ADAS&ME framework (Lilis et al., 2019) has been applied in a simulated car environment in order to experiment with a wide variety of HMI elements without involving hardware integration issues. The HMI elements included: (1) a central stack screen; (2) a head-up display (HUD); (3) a LED strip; (4) an audio system; (5) a steering wheel with vibration; and (6) a seat with vibration. Many personalization and adaptation parameters were considered addressing driver preferences (e.g., GUI visual styles), experience (e.g., driving or computer experience), disabilities (e.g., low vision, hearing impaired), current state (e.g., sleepy, distracted). Based on these parameters, a set of rules was defined and developed to trigger personalized interfaces. Figure 2.9 presents examples of the respective HMI interaction realized for specific parameter combinations. Information messages are associated with a flashing color on the LED, while driver alerts use a different LED color, accompanied by a vibration either on the steering wheel, if the driver's hands are on it, or otherwise on the seat. Both information messages and alerts also issue a message to be presented to the driver. This is further personalized considering the driver's abilities; for a typical driver, the message is presented on the HUD, while for a low-vision driver spoken text is used instead.

Overall, the framework was quite effective in supporting personalized HMI, by decoupling the personalization logic from application code. The ACTA rule language also enabled non-technical HMI experts to contribute to the development process and facilitated the prototyping of personalized interaction.

A second case study concerned the HMI for an electric car. The objective of the design in this case study was to mitigate anxiety in drivers. The HMI elements consisted of: (1) a central stack screen; (2) an audio system; and (3) a pair of A-pillar LED strips. Various parameters were considered to derive the personalization and adaptation rules, with focus on accessibility issues (e.g., color blindness, low vision, and auditory impairments). Parameters also included driver characteristics such as language, preferences and experience, driver state, (e.g., anxiety), driving context (e.g., low

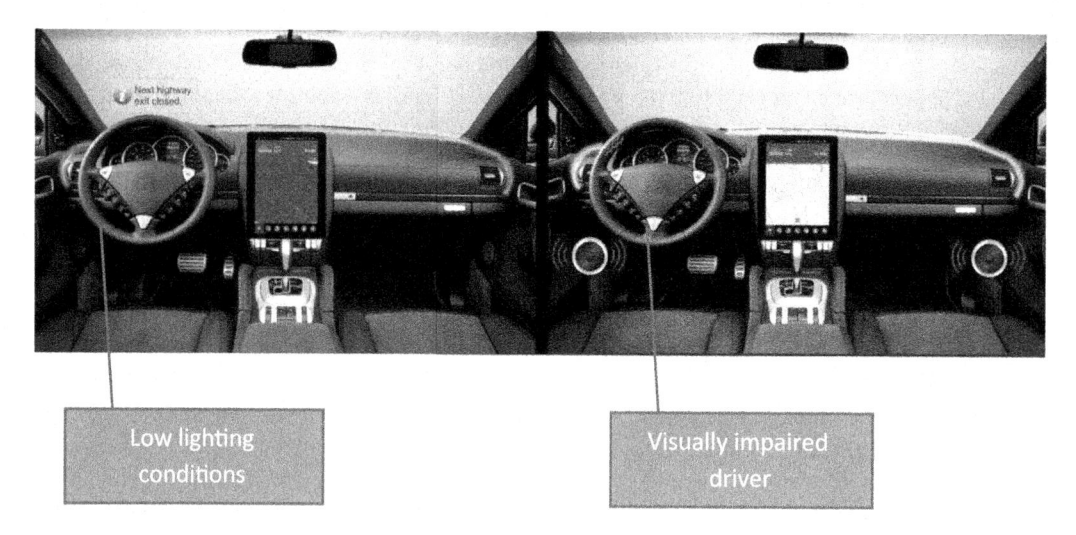

FIGURE 2.9 Examples of personalized and adapted HMI interaction; issuing an information message for a typical driver in low lighting conditions (left), issuing a warning message for a visually impaired driver in normal conditions (right). (From Lilis et al., 2019.)

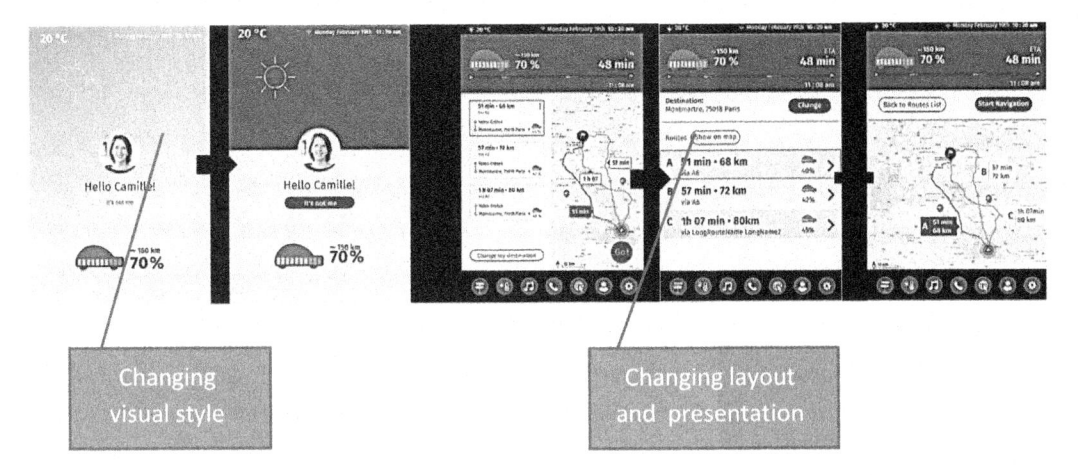

User Interface Adaptation and Design for All   67

FIGURE 2.10 GUI personalization and adaptation examples in an electric car: changing visual styles (left), layout and presentation (right). (From Lilis et al., 2019.)

battery) and environment (e.g., low lighting). Since the central stack screen was the main HMI element of the case study, particular focus was put on GUI personalization and adaptation (see Figure 2.10).

When the driver is detected to have anxiety, the HMI shows the expected and actual consumption to calm the driver, and notifications with electric vehicle information (e.g., slow down to recover battery) are presented for inexperienced drivers. Adaptive suggestions are also provided based on electric vehicle range, e.g., going to the closest charging station when the battery level is too low. Finally, personalized input controls are used for destination selection based on drivers' recent or frequent destinations. HMI modality personalization examples were not limited to the central stack screen; louder sounds were also used for drivers with hearing impairments while varying colors and patterns were adopted in the LED strips to convey driving context information.

2.7 PERSONALIZATION AND ADAPTATION IN XR ENVIRONMENTS

UI adaptation poses significant challenges in extended reality (XR) environments. XR, encompassing virtual reality (VR), augmented reality (AR), and mixed reality (MR), presents unique challenge to adaptation, stemming from its immersive nature, where users can physically move and interact with digital content. In particular, AR and MR systems need to adapt to the physical environment of the user, which may be constantly changing, to different contextual factors (e.g., indoor or outdoor usage), as well as to the current user and task at hand. Existing literature in the field of MR adaptive interfaces mostly employs heuristic or rule-based approaches to decide what information to display and how (Stefanidi et al., 2022). A combinatorial optimization approach has been used toward dynamic context-aware MR interfaces exhibiting adaptation to the users' mental workload, their task and current view of the environment, deciding which applications to display, how much information they should show, and where they should be placed (Lindlbauer et al., 2019). Results from this approach evaluated in a typical office environment suggest that users required 36% less interactions to complete their secondary tasks, using the proposed optimized UI adaptation versus a manual adaptation approach driven by the users. Furthering this approach for dynamic UI adaptation, Stefanidi et al. (2022) aim at enhancing the situational awareness of users by leveraging the current context and providing the most useful information in an optimal manner, deciding what information to present, when to present it, where to visualize it in the user's field of view and how, considering contextual factors and placement constraints. This approach has been tested in the context of the DARLENE AR system for Law Enforcement Agents (LEAs), with results indicating that the system enhances situational awareness (SA), does not impose workload to end users, and provides a positive user experience (UX).

2.7.1 THE DARLENE CONTEXT-AWARE PERSONALIZED INTELLIGENT UIs FOR ENHANCED SITUATIONAL AWARENESS

SA is critical to the survival and safety of officers and community members alike, requiring LEAs to continuously work toward improving it. DARLENE technology aims to enable LEAs to take the initiative away from criminals and terrorists by developing a common operating picture for both on-scene personnel and their command staff based on data collected from various sources. In particular, DARLENE is an ecosystem that combines Artificial intelligence (AI) techniques for activity recognition and posture estimation tasks, combined with a wearable AR framework for visualizing the inferred findings through dynamic content adaptation based on the wearer's stress level and operational context (Apostolakis et al., 2022).

In the context of dynamically adapting the DARLENE AR UIs, a novel computational approach is implemented by Stefanidi et al. (2022), combining ontology modeling and reasoning with combinatorial optimization, to adapt the UI accordingly to optimize the SA associated with the displayed UI at run-time while avoiding information overload and induced stress. The novelty of the approach lies in that the parameters of the optimization problem are dynamically inferred, based on the current situation through ontology reasoning, and that the optimization formulation considers all dimensions of the visualization decision (what, when, how, and where), solving layout and GUI element selection decisions simultaneously.

The adaptation decision-making takes into account the current scene that the user is facing, as informed by the knowledge base which is fed by computer vision algorithms (e.g., detecting an object) and human input (i.e., feedback from the Command-and-Control Centre). The problem that is to be solved by the algorithm takes into account the following parameters: the number of information elements, the information elements that are candidates to display, the component types of the information elements, the number of components, and the components of each component type, the possible positions of each component, the SA of each candidate component, the maximum number of components to display and the priority for a given display position.

More specifically, elicited through a HCD approach, the DARLENE AR system supports three categories of component types: (1) detection, drawing the user's attention to detected entities (objects, foes, or injured civilians), (2) annotation, closely related to detection for providing information about it (e.g., information about a detected foe), and (3) general components supporting the user with information about the surrounding environment (e.g., alerts, summative information about the current context such as number of foes, allies, etc.). Each component type is associated with a data property to determine its visualization priority, as these were decided by following a human-in-the-loop approach with representative LEA experts considering the task that the agent is occupied with and the impact of a component type on their SA. Each component is a GUI element that instantiates a particular component type, and represents a specific design template, featuring three different granularity levels, called Levels of Detail (LoD), with the higher LoD corresponding to more detailed information, and thus larger component size (see Figure 2.11 for an example.)

The DARLENE AR supports components for person identification, diversified according to the hostility of the person (e.g., foe, suspect, civilian, and ally), object identification diversified according to the threat level of the object (e.g., unidentified object, unattended suitcase, and weapon), health status information for injured persons, abnormal behavior indication (e.g., persons caught in a fistfight), alerts, directions, real-time camera feeds from various locations of the patrolling area, map, procedural step-by-step guidance for operations (e.g., open a locker, provide first-aid assistance), as well as summative information about the LEA's current context.

User Interface Adaptation and Design for All

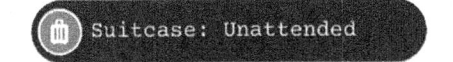

FIGURE 2.11 Suspicious object annotation component in three levels of detail.

It is noted that the decision-making also takes into account several constraints, besides maximizing the total SA score of the displayed UI, namely to avoid redundancy of information, information overloading, and collisions between UI components. At the same time, it ensures that annotations are displayed close to their corresponding detections, to safeguard understandability and high SA for LEAs. Two examples of the DARLENE AR where the decision-making has visualized appropriate components at different LoD are pictured in Figures 2.12 and 2.13.

Following a human-centered AI approach (Margetis et al., 2021), the DARLENE system was evaluated by 20 LEAs, assessing their SA, workload, and overall system UX in stress and non-stress conditions (Stefanidi et al., 2022). In particular, SA was measured as perceived by LEAs themselves, and as observed through objective measurements of their understanding of the current scene they were faced with. Results highlighted that using the system improved perceived and observed SA, by 25.63% and 9.25%, respectively. In more detail, perceived SA was improved by 30% and observed SA by 3.95% in stressful conditions, whereas in non-stressful conditions, they were improved by 15.65% and 15%, respectively. Furthermore, the results indicated that the system does not induce perceived workload in both conditions and that it is both useful and usable, providing an overall positive UX.

FIGURE 2.12 Example of the DARLENE AR highlighting suspicious objects, foes, weapons, and allies in a public space and providing pertinent information for detected objects and people.

FIGURE 2.13 Example of the DARLENE AR at a terrorism event, visualizing information about detected foes, allies, suspects, and victims. Background image taken from Euronews (2016). Explosions & gunfire rock British shopping centre in terrorist attack drill. https://www.youtube.com/watch?v=HGlDjf7upnY.

2.8 ADOPTION OF ADAPTATION-BASED APPROACHES

During the past three decades, accessibility has been established in national legislation and is also employed by international standardization organizations. Its importance is nowadays globally recognized, not only for people with disabilities, but for anyone, as people's abilities are constantly changing, for example with age or due to situations of temporary functional limitations.

Industry at large is today paying progressively more attention to accessibility also beyond legal obligation. Large companies such as IBM,[3] Microsoft,[4] Apple,[5] and Google[6] have launched various initiatives to support developers in building more accessible products through guidelines and dedicated resources.

However, UI adaptation approaches to accessibility have not been widely adopted. A potential reason that has been identified early on is that, although the total number of older or disabled persons is large, each individual disability or impairment area represents only a small portion of the population, therefore it would be impractical and impossible to design everything so that it is accessible by everyone regardless of their limitations (Vanderheiden, 1990). On the other hand, the range of human abilities and the range of situations or limitations that users may find themselves in is very large; therefore, products could only focus on being as flexible as commercially practical (Vanderheiden, 2000).

Additionally, as adaptation-based approaches do not advocate a "one-size-fits-all" approach, but aim instead to promote accessibility for everyone through the adaptation of the design to each user, companies may perceive *Design for All* as an extra cost or an extra feature. Such perception, however, is not accurate (Meiselwitz et al., 2010). On the contrary, it has been claimed that by adopting *Design for All* approaches, companies could achieve a number of business-oriented objectives, including to increase market share, take market leadership, enter new product markets, achieve technology leadership, and improve customer satisfaction (Dong et al., 2004).

Overall, proactive approaches do not propose the elimination of assistive technologies. Such technologies have always been a good solution to many problems of individuals with disabilities. Besides representing a societal need, assistive technologies constitute a notable niche market opportunity (Vanderheiden, 2000), and according to current market predictions, assistive technologies are even expected to experience some growth (Rouse & McBride, 2019).

Nevertheless, as technology evolves, proactive approaches are becoming a more realistic and plausible solution in the near future. In the context of intelligent interactive environments, such as smart homes and smart cities, populated by interconnected technologies and devices, several issues related to accessibility stem from the increased technological complexity and need to be resolved. In such endeavor, strategies followed so far are no longer appropriate, and a transition is required from assistive technologies to Design for All approaches (Margetis et al., 2012). An important issue is that the use of "natural" interaction techniques, such as spoken commands and gestures, as well as the use of large screens and small wearable displays may be prohibitive for some users. Along the same lines, the complexity of the environment can introduce insurmountable obstacles for cognitively impaired users if not properly addressed. Age-related factors are also very important, particularly in light of the fact that many applications are targeted to supporting independent living, and that the current understanding of the needs and requirements of users of different age in such a complex environment is limited.

Another important concern pertains to the different levels of accessibility that need to be ensured. The first level concerns the accessibility of individual devices. Personal devices will need to be accessible to their owners, but probably basic accessibility should be provided also for other users with potentially different needs. A second level is the accessibility of the environment as a whole, which may be provided through environmental devices and other interactive artifacts. In this case, accessibility can be intended as equivalent access to content and functions for users with diverse characteristics, not necessarily through the same devices, but through a set of dynamic interaction options integrated into the environment (Margetis et al., 2012), taking advantage of its inherent capabilities for multimodal interaction and multimedia output (Ntoa et al., 2022).

In the above context, personalized adaptive intelligent UIs can be tailored to the individual needs and requirements of each user, the specific context, and objective of using a system, employing advanced techniques for producing UI adaptations in real time. In such context, AI offers a great potential for fine-grained adaptation of UIs and interaction through machine learning-based user modeling and run-time interaction monitoring, (e.g., Zouhaier et al., 2021). AI-based approaches to adaptation, by overcoming the bottlenecks of previous rule-based approaches, such as adaptation rules design and user proofing difficulties, bring significant potential to contribute to the wider practice and adoption of adaptation-based accessibility.

2.9 CONCLUSIONS

Progress in the field of Universal Access and Design for All, i.e., access by anyone, anywhere and anytime to interactive products and services in the Information Society, has highlighted a shift of perspective and reinterpretation of HCI design. As a consequence, UI design methodologies, techniques and tools acquire increased importance in the context of Universal Access and strive toward approaches that support design for diversity, based on the consideration of the several dimensions of diversity that emerge from the broad range of user characteristics, the changing nature of human activities, the variety of contexts of use, the increasing availability and diversification of information, the variety of knowledge sources and services, and the proliferation of diverse technological platforms. An important dimension of such a perspective is the adoption of intelligent interface adaptation as a technological basis, viewing design as the organization and structuring of an entire design space of alternatives to cater to diverse requirements. In a Universal Access perspective, adaptation needs to be "designed into" a system rather than decided upon and implemented a posteriori.

Despite progress, however, the practice of designing adaptation-based UIs remains difficult, due to the intrinsic complexity of the task and the current limited expertise of designers and practitioners. To overcome such a difficulty, tool support is required for promoting and facilitating adaptation design.

This chapter has discussed a series of tools and components developed over a period of two decades to support and facilitate the conduct of UI adaptation design. Such tools are claimed to have a significant role to play toward widening and improving the practice of Design for All and ensuring a more effective transition from the design to the implementation phase. They have been used in practice in a series of case studies involving different types of applications for different purposes, contexts, and interaction platforms. These extensive case studies have demonstrated the technical feasibility of the overall adaptation-based approach to Universal Access. Additionally, these developments have provided hands-on experience toward improving the usefulness and effectiveness of the developed tools in different phases of the UI development lifecycle. During these developments, it has progressively become clear that UI adaptation can be adopted in practice as a result of reducing the gap with mainstream design practices. Ultimately, this amounts to providing transparent solutions, which do not require specific adaptation knowledge and support prototyping. Therefore, more recent solutions have gone in the direction of providing ready-to-use widget toolkits that integrate all the required adaptation knowledge and logic, as well as supporting the view of alternative designs in mainstream development environments. Obviously, however, such solutions, while achieving the objective of simplifying the design of adaptation as far as alternative widget instances are concerned, still require specialized knowledge and mastering of UI adaptation mechanisms for designing dialogue adaptation at a syntactic or semantic level, as well as for creating new or modifying existing adaptable widgets. Emerging and AI-based solutions hold the promise of reducing the effort needed to create the expansive realm of all possible adaptations, by carefully designed decision-making mechanisms for producing optimal solutions to given problems in real time. Nonetheless, it is imperative to embrace the philosophy of human-centered AI as an utmost priority (Margetis et al., 2021; Garibay et al., 2023), to ensure that the developed systems can truly address the diverse user needs and requirements in an optimal approach that epitomizes transparency, understandability and user-friendliness.

ACKNOWLEDGMENTS

The work reported in this chapter has been partially conducted in the context of the following research projects funded by the European Commission:

IST-2003-511298 - ASK-IT "Ambient Intelligence System of Agents for Knowledge-based and Integrated Services for Mobility Impaired users" (1/10/2004 - 31/12/2008) IST-CA-033838 - DfA@eInclusion "Design for All for eInclusion" (1/1/2007 - 31/12/2009)

FP7-ICT-215754 - OASIS "Open architecture for Accessible Services Integration and Standardisation" (1/1/2008 - 31/12/2011)

AAL - 2008-1-147 - REMOTE "REMOTE" (1/6/2009 - 31/5/2012)

H2020-ICT-643433 - RAMCIP "Robotic Assistant for MCI Patients at home" (01/01/2015 - 31/12/2018)

H2020-ICT688900 - ADAS&ME "Adaptive ADAS to support incapacitated drivers & Mitigate Effectively risks through tailor made HMI under automation" (01/09/2016 – 29/02/2020).

H2020-SU-SEC883297 - DARLENE: Deep AR Law Enforcement Ecosystem (01/09/2020 - 31/12/2023)

NOTES

1. The FIRMA framework has been developed in the context of the Project ICT-RAMCIP "Robotic Assistant for MCI Patients at home" (see Acknowledgments).
2. https://docs.microsoft.com/en- us/dotnet/api/system.speech.synthesis.speechsynthesizer?view=netframework-4.8
3. https://www.ibm.com/able/
4. https://www.microsoft.com/en-us/accessibility
5. https://www.apple.com/accessibility/
6. https://www.google.com/accessibility/

REFERENCES

Abrahão, S., Insfran, E., Sluÿters, A., & Vanderdonckt, J. (2021). Model-based intelligent user interface adaptation: Challenges and future directions. *Software and Systems Modeling*, 20, 1335–1349. doi: 10.1007/s10270-021-00909-7

Antona, M., Savidis, A., & Stephanidis, C. (2006). A process–oriented interactive design environment for automatic user interface adaptation. *International Journal of Human Computer Interaction*, 66, 594–598.

Apostolakis, K. C., Dimitriou, N., Margetis, G., Ntoa, S., Tzovaras, D., & Stephanidis, C. (2022). DARLENE-Improving situational awareness of European law enforcement agents through a combination of augmented reality and artificial intelligence solutions. *Open Research Europe*, 1(87), 87.

Assistive Technology Act of 1998. http://www.section508.gov/docs/AssistiveTechnologyActOf1998Full.pdf

Bailly, G., Oulasvirta, A., Kötzing, T., & Hoppe, S. (2013). Menuoptimizer: Interactive optimization of menu systems. In *Proceedings of the 26th Annual ACM Symposium on User Interface Software and Technology* (pp. 331–342).

Bouzit, S., Calvary, G., Coutaz, J., Chêne, D., Petit, E., & Vanderdonckt, J. (2017). The PDA-LPA design space for user interface adaptation. In 2017 *11th International Conference on Research Challenges in Information Science (RCIS)*, Brighton, UK (pp. 353–364). doi: 10.1109/RCIS.2017.7956559

Calvary, G., Coutaz, J., Thevenin, D., Limbourg, Q., Bouillon, L., & Vanderdonckt, J. (2003) A unifying reference framework for multi-target user interfaces. *Interacting with Computer*, 15(3), 289–308.

CEN (2019). Design for all - accessibility following a design for all approach in products, goods and services - extending the range of users. https://www.cencenelec.eu/areas-of-work/cen-cenelec-topics/accessibility/design-for-all/

Dong, H., Keates, S., & Clarkson, P. J. (2004). Inclusive design in industry: Barriers, drivers and the business case. In Proceedings of the 8th ERCIM International Workshop on User Interfaces for All (UI4ALL 2004) (pp. 305–319). Berlin, Heidelberg: Springer. doi: 10.1007/978-3-540-30111-0_26

Eisenstein, J., and Puerta, A. (2000). Adaptation in automated user-interface design. In *Proceedings of the 5th International Conference on Intelligent User Interfaces (IUI '00)*. (pp. 74–81). New York: Association for Computing Machinery. doi: 10.1145/325737.325787

Eisenstein, J., Vanderdonckt, J., & Puerta, A. (2000). Adapting to mobile contexts with user-interface modeling. In Proceedings Third IEEE Workshop on Mobile Computing Systems and Applications, Los Alamitos, CA, USA (pp. 83–92). doi: 10.1109/MCSA.2000.895384.

ETSI (n.d.). EN 301 549 V3 the Harmonized European Standard for ICT Accessibility. https://www.etsi.org/human-factors-accessibility/en-301-549-v3-the-harmonized-european-standard-for-ict-accessibility

Fan, H. & Poole, M. S. (2006). What is personalization? Perspectives on the design and implementation of personalization in information systems. *Journal of Organizational Computing and Electronic Commerce*, 16(3–4), 179–202. doi: 10.1080/10919392.2006.9681199

Firmenich, S., Garrido, A., Paternò, F., & Rossi, G. (2019). User interface adaptation for accessibility. In Y. Yesilada & S. Harper (Eds), *Web Accessibility*. Human-Computer Interaction Series. London: Springer. doi: 10.1007/978-1-4471-7440-0_29

Foukarakis, M., Antona, M., & Stephanidis, C. (2017). Applying a multimodal user interface development framework on a domestic service robot. In *The Proceedings of the 10th International Conference on PErvasive Technologies Related to Assistive Environments (PETRA 2017), 21-23 June, Island of Rhodes, Greece* (pp. 378–384). New York: ACM.

Gajos, K. Z., Wobbrock, J. O., & Weld, D. S. (2008, April). Improving the performance of motor-impaired users with automatically-generated, ability-based interfaces. In *Proceedings of the SIGCHI Conference on Human Factors in Computing Systems* (pp. 1257–1266).

Gajos, K., & Weld, D. S. (2004, January). SUPPLE: Automatically generating user interfaces. In *Proceedings of the 9th International Conference on Intelligent User Interfaces* (pp. 93–100).

Gamecho, B., Minón, R., Aizpurua, A., Cearreta, I., Arrue, M., Garay-Vitoria, N., & Abascal, J. (2015). Automatic generation of tailored accessible user interfaces for ubiquitous services. *IEEE Transactions on Human-Machine Systems*, 45, 612–623. doi: 10.1109/THMS.2014.2384452.

Gardner, B. S. (2011). Responsive web design: Enriching the user experience. *Sigma Journal: Inside the Digital Ecosystem*, 11(1), 13–19.

Garibay, O. O., Winslow, B., Andolina, S., Antona, M., Bodenschatz, A., Coursaris, C., Falco, G., Fiore, S. M., Garibay, I., Grieman, K., Havens, J. C., Jirotka, M., Kacorri, H., Karwowski, W., Kider, J., Konstan, J., Koon, S., Lopez-Gonzalez, M., Maifeld-Carucci, I., McGregor, S., Salvendy, G., Shneiderman, B., Stephanidis, C., Strobel, C., Holter, C. T., & Xu, W. (2023). Six human-centered artificial intelligence grand challenges. *International Journal of Human- Computer Interaction*, 39(3), 391–437.

Ghiani, G., Manca, M., Paternò, F., & Santoro, C. (2017). Personalization of context-dependent applications through trigger-action rules. *ACM Transactions on Computer-Human Interaction (TOCHI)*, 24(2), 1–33.

Gullà, F., Cavalieri, L., Ceccacci, S., Germani, M., & Bevilacqua, R. (2015). Method to design adaptable and adaptive user interfaces. In C. Stephanidis (Ed), *HCI International 2015 - Posters' Extended Abstracts. HCI 2015. Communications in Computer and Information Science* (Vol. 528). Cham: Springer. https://doi.org/10.1007/978-3-319-21380-4_4

Hartmann, M. (2009). Challenges in developing user-adaptive intelligent user interfaces. In *LWA* (p. ABIS-6).

Henry, S. L. (2023). WCAG 2 Overview. https://www.w3.org/WAI/standards-guidelines/wcag/

ISO 13407 (1999). *Human-centred design processes for interactive systems*.

ISO 9241-210 (2010) *Ergonomics of human-system interaction -- Part 210: Human-centred design for interactive systems*.

ISO 9241-210 (2019). *Ergonomics of human-system interaction -- Part 210: Human-centred design for interactive systems*.

Kazepis, N., Antona, M., & Stephanidis, C. (2016). FIRMA: A development framework for elderly-friendly interactive multimodal applications for assistive robots. In A. L. Culén, L. Miller, I. Giannopulu, & B. Gersbeck-*Schierholz* (Eds), *Proceedings of the Ninth International Conference on Advances in Computer-Human Interactions (ACHI 2016)*, Venice, Italy, *24-28 April* (pp. 386–397). New York: Curran Associates, Inc.

Leonidis, A., Antona, M., & Stephanidis, C. (2012). Rapid prototyping of adaptable user interfaces. *International Journal of Human-Computer Interaction*, 28(4), 213–235.

Leuteritz, J.-P., Widlroither, H., Mourouzis, A., Panou, M., Antona, M., & Leonidis, S. (2009). Development of open platform based adaptive HCI concepts for elderly users. In C. Stephanidis (Ed), *Universal Access in Human-Computer Interaction - Intelligent and Ubiquitous Interaction Environments. - Volume 6 of the Proceedings of the 13th International Conference on Human-Computer Interaction (HCI International 2009)*, San Diego, CA, USA, 19-24 July (pp. 684–693). Berlin Heidelberg: Lecture Notes in Computer Science Series of Springer (LNCS 5615).

Lilis, Y., Zidianakis, E., Partarakis, N., Ntoa, S., & Stephanidis, C. (2019). A Framework for personalised HMI interaction in ADAS systems. In O. Gusikhin & M. Helfert (Eds), *Proceedings of the 5th International Conference on Vehicle Technology and Intelligent Transport Systems (VEHITS 2019)*, Heraklion, Crete, Greece, 3-5 May (pp. 586–593). Portugal: SCITEPRESS.

Limbourg, Q. & Vanderdonckt, J. (2004). UsiXML: A user interface description language supporting multiple levels of independence. In *ICWE Workshops* (pp. 325–338).

Lindlbauer, D., Feit, A. M., & Hilliges, O. (2019). Context-aware online adaptation of mixed reality interfaces. In *Proceedings of the 32nd Annual ACM Symposium on User Interface Software and Technology* (pp. 147–160).

Mace, R. (1991). Accessible environments: Toward universal design. In Design Interventions: Toward a More Humane Architecture. New York: Van Nostrand Reinhold.

Margetis, G., Antona, M., Ntoa, S., & Stephanidis, C. (2012). Towards accessibility in ambient intelligence environments. In F. Paterno, B. De Ruyter, P. Markopoulos, C. Santoro, E. van Loenen, & K. Luyten (Eds), *Proceedings of the 3rd International Joint Conference in Ambient Intelligence (AmI 2012), 13–15 November, Pisa, Italy* (pp. 328–337). Berlin Heidelberg, Germany: Springer (LNCS: 7683).

Margetis, G., Ntoa, S., Antona, M., & Stephanidis, C. (2021). Human-centered design of artificial intelligence. In G. Salvendy & W. Karwowski (Eds.), *Handbook of Human Factors and Ergonomics* (5th Ed., Chapter 42, pp. 1085–1106). John Wiley and Sons.

Meiselwitz, G., Wentz, B., & Lazar, J. (2010). Universal usability: Past, present, and future. *Foundations and Trends in Human-Computer Interaction*, 3(4), 213–333.

Meixner, G., Paternò, F., & Vanderdonckt, J. (2011) Past, present, and future of model-based user interface development. *i-com*, 10(3), 2–11. https://doi.org/10.1524/icom.2011.0026

Michel, D., Papoutsakis K., & Argyros, A. (2014). Gesture recognition for the perceptual support of assistive robots, In *International Symposium on Visual Computing (ISVC 2014)* (pp. 793–804). Las Vegas, NV.

Miñón, R., Paternò, F., Arrue, M., & Abascal, J. (2016). Integrating adaptation rules for people with special needs in model-based UI development process. *Universal Access in the Information Society*, 15, 153–168. https://doi.org/10.1007/s10209-015-0406-3

Ntoa, S., Margetis, G., Antona, M., & Stephanidis, C. (2022). Digital accessibility in intelligent environments. In *Human-Automation Interaction: Manufacturing, Services and User Experience* (pp. 453–475). Cham: Springer International Publishing.

Oulasvirta, A., Dayama, N. R., Shiripour, M., John, M., & Karrenbauer, A. (2020). Combinatorial optimization of graphical user interface designs. *Proceedings of the IEEE*, 108(3), 434–464.

Partarakis, N., Doulgeraki, C., Antona, M., & Stephanidis, C. (2010). Designing web-based services. In G. Salvendy & W. Karwowski (Eds), *Introduction to Service Engineering* (pp. 447–487). Hoboken, NJ: John Wiley.

Paternò, F. (2019). User interface design adaptation. In *The Encyclopedia of Human-Computer Interaction*, 2nd Ed. Chapter 39. Interaction Design Foundation. https://www.interaction-design.org/literature/book/the-encyclopedia-of-human-computer-interaction-2nd-ed/user-interface-design-adaptation

Paternò, F., Santoro, C., & Scorcia, A. (2008). Automatically adapting web sites for mobile access through logical descriptions and dynamic analysis of interaction resources. In *Proceedings of the Working Conference on Advanced Visual Interfaces (AVI '08)* (pp. 260–267). New York: Association for Computing Machinery. https://doi.org/10.1145/1385569.1385611

Paternò, F., Santoro, C., & Spano, L. D. (2009). MARIA: A universal, declarative, multiple abstraction-level language for service-oriented applications in ubiquitous environments. *ACM Transactions on Computer-Human Interaction (TOCHI)*, 16(4), 19. doi: 10.1145/1614390.1614394

Peissner, M., Häbe, D., Janssen, D., & Sellner, T. (2012). MyUI: Generating accessible user interfaces from multimodal design patterns. In *Proceedings EICS 2012* (pp. 81–90). New York: ACM.

Rouse, W. B., & McBride, D. (2019). A systems approach to assistive technologies for disabled and older adults. *The Bridge*, 49(1), 32–38.

Salvendy, G., Akoumianakis, D., Arnold, A., Bevan, N., Dardailler, D., Emiliani, P.L., Iakovidis, I., Jenkins, P., Karshmer, A., Korn, P., Marcus, A., Murphy, H., Oppermann, C., Stary, C., Tamura, H., Tscheligi, M., Ueda, H., Weber, G., & Ziegler, J. (1999). Toward an information society for all: HCI challenges and R&D recommendations. *International Journal of Human-Computer Interaction*, 11(1), 1–28.

Salvendy, G., Akoumianakis, D., Bevan, N., Brewer, J., Emiliani, P.L., Galetsas, A., Haataja, S., Iakovidis, I., Jacko, J., Jenkins, P., Karshmer, A., Korn, P., Marcus, A., Murphy, H., Stary, C., Vanderheiden, G., Weber, G., & Ziegler, J. (1998). Toward an information society for all: An international R&D agenda. *International Journal of Human-Computer Interaction*, 10(2), 107–134.

Sarcar, S., Jokinen, J. P., Oulasvirta, A., Wang, Z., Silpasuwanchai, C., & Ren, X. (2018). Ability-based optimization of touchscreen interactions. *IEEE Pervasive Computing*, 17(1), 15–26.

Savidis, A., & Stephanidis, C. (1995). Developing dual user interfaces for integrating blind and sighted users: The HOMER UIMS. In *The Proceedings of the ACM Conference on Human Factors in Computing Systems (CHI '95)*, Denver, Colorado, USA, *7-11 May* (pp. 106–113). New York: ACM Press.

Savidis, A., & Stephanidis, C. (2009). Unified design for user interface adaptation. In C. Stephanidis (Ed), *The Universal Access Handbook* (pp. 16-1–16-17). Boca Raton, FL: Taylor & Francis.

Savidis, A., Antona, M., & Stephanidis, C. (2005). A decision-making specification language for verifiable user-interface adaptation logic. *International Journal of Software Engineering and Knowledge Engineering*, 15 (6), 1063–1094.

Shneiderman, B. (2000). Pushing human-computer interaction research to empower every citizen. *Communication of the ACM*, 43(5), 85–91.

Simon, R., Wegscheider, F., & Tolar, K. (2005). Tool-supported single authoring for device independence and multimodality. In *Proceedings of the 7th International Conference on Human Computer Interaction with Mobile Devices & Services (MobileHCI '05)* (pp. 91–98). New York: Association for Computing Machinery. doi: 10.1145/1085777.1085793

Stefanidi, Z., Margetis, G., Ntoa, S., & Papagiannakis, G. (2022). Real-time adaptation of context-aware intelligent user interfaces, for enhanced situational awareness. *IEEE Access*, 10, 23367–23393.

Stephanidis, C. (1995). Towards user interfaces for all: Some critical issues. Parallel session "user interfaces for all - everybody, everywhere, and anytime". In Y. Anzai, K. Ogawa & H. Mori (Eds), *Symbiosis of Human and Artifact - Future Computing and Design for Human-Computer Interaction [Volume 1 of the Proceedings of the 6th International Conference on Human-Computer Interaction (HCI International '95)]*, Tokyo, Japan, 9-14 July (pp. 137–142). Amsterdam: Elsevier, Elsevier Science.

Stephanidis, C. (2001a). User interfaces for all: New perspectives into human-computer interaction. In C. Stephanidis (Ed), *User Interfaces for All - Concepts, Methods, and Tools* (pp. 3–17). Mahwah, NJ: Lawrence Erlbaum Associates.

Stephanidis, C. (2001b). Adaptive techniques for universal access. *User Modelling and User Adapted Interaction International Journal*, 11(1/2), 159–179.

Stephanidis, C., & Emiliani, P. L. (1999). Connecting to the information society: A European perspective. *Technology and Disability*, 10(1), 21–44.

Story, M. F. (2001). Principles of universal design. In: W. F. E. Preiser & E. Ostroff (Eds.), Universal Design Handbook, 1st ed. New York: McGraw-Hill.

Todi, K., Bailly, G., Leiva, L., & Oulasvirta, A. (2021). Adapting user interfaces with model-based reinforcement learning. In *Proceedings of the 2021 CHI Conference on Human Factors in Computing Systems* (pp. 1–13).

Vanderheiden, G. (1990). Thirty-something million: Should they be exceptions? *Human Factors*, 32(4), 383–396. doi:10.1177/001872089003200402

Vanderheiden, G. (2000). Fundamental principles and priority setting for universal usability. In *Proceedings on the 2000 Conference on Universal Usability (CUU '00)* (pp. 32–37). New York: ACM. doi: 10.1145/355460.355469

Vredenburg, K., Isensee, S., & Righi, C. (2001). *User-Centered Design: An Integrated Approach* (Software quality institute series). Prentice Hall.

Yang, S.J., & Shao, N.W. (2007). Enhancing pervasive web accessibility with rule-based adaptation strategy. *Expert Systems with Applications*, 32(4), 1154–1167. https://doi.org/10.1016/j.eswa.2006.02.008

Zidianakis, E., Antona, M., & Stephanidis, C. (2018). Activity analysis (ACTA): Empowering smart game design with a general purpose FSM description language. *IADIS International Journal on Computer Science and Information Systems*, 13(1), 82–95. https://www.iadisportal.org/ijcsis/papers/2018130106.pdf

Ziegler, D., & Peissner, M. (2017). Enabling accessibility through model-based user interface development. *Studies in Health Technology and Informatics*, 242, 1067–1074.

Zouhaier, L., Hlaoui, Y. B. D., & Ayed, L. B. (2021). A reinforcement learning based approach of context-driven adaptive user interfaces. In *2021 IEEE 45th Annual Computers, Software, and Applications Conference (COMPSAC)* (pp. 1463–1468). doi: 10.1109/COMPSAC51774.2021.00217

3 Human Behavior in Cybersecurity Privacy and Trust

Abbas Moallem

3.1 INTRODUCTION

We live in the Digital Age, or Information Age, which provides us with all of its technological innovations such as smartphones, the Internet, and social media. Consequently, we now have a widely distributed ability to perform many tasks, from the mundane to the complex, efficiently and rapidly. Our nearly instantaneous adaptations to the COVID-19 pandemic illustrate how digital technologies can now enable people to work remotely and support most of their routine business, educational, health, shopping, entertainment, and other needs. However, applying digital technologies and their benefits has also brought along new risks. The risks range from not securing the processes, controls, systems, networks, programs, devices, and data from cyberattacks to safeguarding the complete privacy of users' data, property, and persons.

This chapter reviews some theoretical frameworks that help to better understand human behaviors in cybersecurity, privacy, and trust. After reviewing attacker types and motivations, we look at the main behavioral patterns of attackers and their victims – whether individuals or organizations. Then, despite the inner creativity of all behaviors, to facilitate understanding the behaviors are classified into three groups: behavior patterns that attackers use to trust behavior in the digital world, user privacy, and vulnerable behavior in cyberattacks.

Furthermore, security issues in different application domains (e.g., healthcare, manufacturing) and the importance of the human factor in these domains are discussed. At the end, cybersecurity hygiene, awareness, training, motivation, complexity, and limitations are reviewed.

3.2 CYBER CRIMES

According to Kaspersky (Kaspersky, 2022):

> Cybercrime is criminal activity that either targets or uses a computer, a computer network or a networked device. Most cybercrime is committed by cybercriminals or hackers who want to make money. However, occasionally cybercrime aims to damage computers or networks for reasons other than profit. These could be political or personal.
>
> Cybercrime can be carried out by individuals or organizations. Some cybercriminals are organized, use advanced techniques and are highly technically skilled. Others are novice hackers.

Cybercrime is relatively recent in our history, and in our digital age has different particularities than crime in earlier eras including scale, number, significance, ease, and repercussion.

3.2.1 SCALE

Before the digital age, crimes happened in limited geographical areas such as cities, regions, or countries. In the digital age, crimes can be global. For example, a cyberattacker might be in a remote country and steal from a bank in another county or continent. Other dimensions in attacks are the targeting of institutions in countries or locations with less cybersecurity protection infrastructure, or the profitability of ransomware attacks targeting the most sensitive organizations such as critical infrastructure providers. For example, the attack on Colonial Pipeline that carries gasoline and jet fuel to the Southeastern United States with ransomware led to an initial $4.4 million ransom payment (Osborne 2021).

3.2.2 Number

Before the digital age, the number of crimes and the number of victims of most crimes was limited and relatively small. In the digital age, the number of victims is huge. The press every day reports recent hackings. Further, according to Cybersecurity Ventures, the global costs of cybercrime are expected to grow by 15% per year over the next 5 years, reaching USD 10.5 trillion annually by 2025. Most cybercrimes aren't reported. Some estimates suggest as few as 10% of the total number of cybercrimes committed each year are actually reported. Each year, the number of online crimes increases across the globe. It is reported that the Great Britain is the country most affected by cybercrime in 2021 with 4,783 victims per one million Internet users, and the second is US with 1,494 victims per one million Internet users (Mangis, 2022). More than 1.4 million Americans also reported being a victim of identity theft in 2021 (Iacurci, 2022). It is estimated that an organization suffered a ransomware attack every 11 seconds in 2021, or there will be a new attack on a consumer or business every 2 seconds by 2031 (Morgan, 2020).

3.2.3 Significance

Possibly all the significant theft cases in the world are now movies. Interestingly in most cases, the criminals, sooner or later, were arrested. The biggest bank robbery in the world used to be the Dunbar Armored facility in Los Angeles, California of US$18.9 million in 1997, or the Central Bank of Iraq that took place in March 2003, and approximately US$1 billion was stolen. While just in 2021, the US banks processed roughly $1.2 billion in ransomware payments according to federal report (Cox, 2022).

On the other hand, an extraordinary number of small and big cyberattacks happen routinely through ransomware, credit card fraud, identity thieves, and so on. Yet, according to some reports, only just over 1,000 Individuals have been arrested as a result of global cybercrime-fighting operations (Vijayan, 2021). Considering that cyberattackers might not physically be in the same country as the victims of their attacks, they are much harder to track and prosecute.

3.2.4 Ease of Attacking

Compared with a physical bank robbery, cyberattacks are conducted with ease. If in a physical bank robbery, a lot of tools, technologies, or physical work, such as creating a tunnel and so on is needed to execute the robbery, now hackers can be in a remote area with some technical knowledge and conduct the most complicated attacks with little physical work at all. Sabotage and terrorist attacks can be performed relatively frequently against adversaries, factories, equipment, and people through cyberattacks from anywhere in the world. For example, it is relatively simple to conduct a phishing attack. Attackers just need to send emails to large numbers of people, asking for sensitive information (such as bank details) or encouraging them to visit a fake website that they set up to exploit visiting users.

3.2.5 Punishment

Cybercrime attackers are not easily identified or punished due to the possibility of staying anonymous in cyberattacks. There are no physical fingerprints, surveillance videos, or eyewitnesses. Also, as discussed below, without international regulations and cooperation, attackers in one country might not be punished for an attack that takes place in another country.

3.3 FACTORS CONTRIBUTING TO THE INCREASE IN CYBERATTACKS

Three factors contribute to the development and increase of cyberattacks. The first factor is the existence of an anonymous platform. Anonymity is available through a technology called the dark web (Masayuki Hatta. 2020) which is a hidden collective of Internet sites. A dark web encryption technology routes users' data through many intermediate servers, which protects the user's identity and guarantees anonymity. Thus, the dark web keeps Internet activity anonymous and private. It is only accessible by a specialized web browser such as the Tor browser (torproject.org).

The second factor is the ability, through cryptocurrencies such as Bitcoin, Ethereum, and Litecoin, to conduct transactions without connecting with any standard national or international banking system. Cryptocurrency is a digital payment system that doesn't rely on banks to verify transactions or transfer funds. It enables anyone anywhere to send and receive payments without a trackable record of the transaction from the international monetary system. Contrary to physical money, cryptocurrency payments exist purely as digital entries to an online database describing specific transactions stored in digital wallets; they are entirely encoded and enable cybercriminals to receive and distribute "funds" in any amount without a trackable record.

The third factor is related to lack of the laws and enforcement resources aimed at preventing and punishing cybercrime. According to the United Nations Conference on Trade and Development (UNCTD), 156 countries (80%) have enacted cybercrime legislation; the pattern varies by region: Europe has the highest adoption rate (91%), and Africa has the lowest (72%) (UNCTAS, 2022). However, despite the increased regulatory activity, there is no unified international approach to regulating cybersecurity. The evolving nature of cybercrime and protective technologies is a significant challenge for government lawmakers, law enforcement agencies and prosecutors, especially for cross-border enforcement. Cybercriminals operate today generally without fear of breaking a law or getting caught if they do.

3.4 BUILDING CYBER PROTECTION

Cybersecurity is a growing field of science composed of very diverse types of technologies used to protect systems from different kinds of attacks, some known and some unknown. With the growth of cyberattacks of all types, several areas must be considered in building cyber protection at every level: home, city, state, and country.

3.4.1 Defensive Technologies

The need to have technologies that protect data, networks and systems is fundamental. However, with millions of lines of new code written daily, achieving the complete technological protection challenge would be unthinkable anytime soon. However, there are several groups of technological solutions available (Moallem, 2021). They can be grouped into the following main categories.

1. Encryption is technology that takes plain text – like a text message or email – and codes it into an unreadable format.
2. Authentication is a technique or mechanism to prove and validate an end-user or a computer program's identity.
3. Biometrics uses individual's physical and behavioral traits that are permanent, unique, and non-replicable.
4. Firewalls serve as a barrier or wall between two networks and can be implemented in software, hardware, or cloud-based applications.
5. Endpoint protection or endpoint security is the methodology that an organization takes to protect its network when it is accessed by remote devices such as smartphones, laptops, tablets, or other wireless devices.

6. Phishing Detection technologies block phishing and email impersonation attacks before they reach users.
7. Virus and malware detection and protection software for computers, also known as anti-malware software, is used to prevent, detect, and remove malicious software.
8. Network Security protects the integrity, confidentiality, and data accessibility of multiple computer networks.
9. Location-tracking technologies help locate, record and track the physical movement of people or objects and are ubiquitous and in use by many of us every day.
10. Surveillance technology from security cameras to cybersecurity software.
11. Insider Threat detection technologies are used to detect insider threats.
12. Intrusion Detection Systems (IDS) monitor operating system files, network traffic, events and analyze the data gathered to detect signs of intrusion, detect malicious activities and alert system administrators and, in some cases, initiate action.
13. Vulnerability Scanning is a technology that scans a network and computer system to identify known vulnerabilities and generates risk exposure reports.
14. Penetration testing or ethical hacking is the process of allowing authorized experts to hack a computer system to discover the vulnerabilities of a system and to remove those areas of weakness.

3.4.2 HUMAN BEHAVIOR

All cyber attackers, in most, if not all, cases rely on human agent vulnerability and behaviors. The first line of attack to penetrate a system is through human and vulnerable behavior patterns. Interestingly with all technical progress, training, and awareness, phishing (or attacking using enticing emails or texts) is still the most successful way to penetrate a system and many networks. Even though many attackers are successful because of human vulnerability to phishing, company resources allocated to human factors for cyber protection are not significant (Maalem Lahcen et al., 2020, Kaspersky, 2022).

3.4.3 LAWS AND REGULATIONS

As indicated above, the lack of national and international regulations makes data protection even more complicated. National security agencies take broad advantage of the lax regulatory environment and target other countries' information about their citizens. Their goals are to sell the data, influence the host nation or citizen, or directly harm either where possible and advantageous. Enterprises of all sizes and types benefit from all this information gathering and focus on lobbying the governments to limit laws and regulations that target cybercrime. Consequently, an uncontrolled, wild west-like environment currently exists (Hollois, 2021).

3.4.4 BEHAVIORAL-BASED CYBER TECHNOLOGIES

With all the progress in machine learning and data sciences, behavior-based technology is rapidly growing and is helping to provide some protection against cybercrime. For example, technologies such as activity recording can capture users' actions and create an activity log or drive user behavior analytics to detect unexpected user behaviors (Moallem, 2021).

3.4.5 EDUCATION AND AWARENESS

People are not yet sufficiently conscious of cyber hygiene, nor are they sufficiently familiar with how to protect themselves. Moreover, with millions of Internet of Things (IoT) devices and the complexity of these systems of systems, the situation is becoming more and more complicated. Today, hardly any individual, even with an awareness of cybersecurity, can manage and protect themselves adequately from all potential attacks or attackers.

It is essential to emphasize that even in view of the investments governments and private organizations have made in cyber protection technologies, research, and investigations to understand the human behaviors targeted by cyber attackers and to reduce these vulnerabilities, we still have a long way to go in these efforts.

3.5 ATTACKER TYPES

Before looking at a human factors analysis of attackers and victims, let us first understand who the attackers are and their motivations for conducting attacks. Knowing the attackers' motives will help us better analyze the resources and behaviors they might use to attack. For example, a country's espionage agency might have tremendous resources and impunity to conduct attacks. They might also access information that a criminal organization might not necessarily have. However, attacks coming from organized crime might also be highly sophisticated, such as ransomware attacks targeting an individual or isolated group.

Simply stated, cyber attackers are people, organizations, or groups that deliberately attempt to breach the information systems of people, organizations, enterprises, and, more importantly, different government institutions. In addition, cyber attackers seek some benefit from collecting data, monitoring activities, and disrupting the victim's network.

Commonly, cyber attackers are also called hackers. However, hackers and cyber attackers are not necessarily implying the same meaning. The term hacker is very broad. It describes the activity but does not reflect the intentions of those involved in the hacking activity.

3.6 ATTACKER'S MOTIVATION

Understanding an attacker's motivation is vital in investigating any type of attack. Motive is usually sought to explain why a person committed a crime. For example, one of the primary acts of a detective investigating a murder case is to learn about all those who would benefit from the murder, since the motive is the underlying reason for a crime.

In cybersecurity, the motivation of cyber attackers (criminal individuals, organizations, government spying agencies, or companies) is also fundamental to understanding why a cyberattack happens and what the attackers intend to do with what they get; in other terms, what would be their incentives? To better understand the motivations of cyber attackers, we distinguish between the attackers that break into a system without permission to access an organization's digital assets, and the companies or legal entities that collect and distribute information without the consent of users of their services, hence violating their privacy.

In the first case, the attacker deliberately breaks into a system to modify or steal information without permission or legal rights. The second case refers to the companies that are data brokers or information resellers. They collect personal information from various sources, including public records and online data, and then sell that information to other companies for marketing and other purposes. This practice raises privacy concerns, as individuals may not be aware that their personal information is being collected and sold without their consent.

The motivations of attackers include financial gain, sabotage, cyber war, intelligence gathering, fake news shaping people's opinions, blackmail, solving crime cases, a myriad of illegal, immoral, or illicit denunciations and revelations, revenge, having fun and curiosity, and terrorism (Figure 3.1).

3.6.1 FINANCIAL GAIN

The primary motivation behind many cyberattacks is to financially gain by selling illegally obtained information about people or organizations. According to the United States Federal Trade

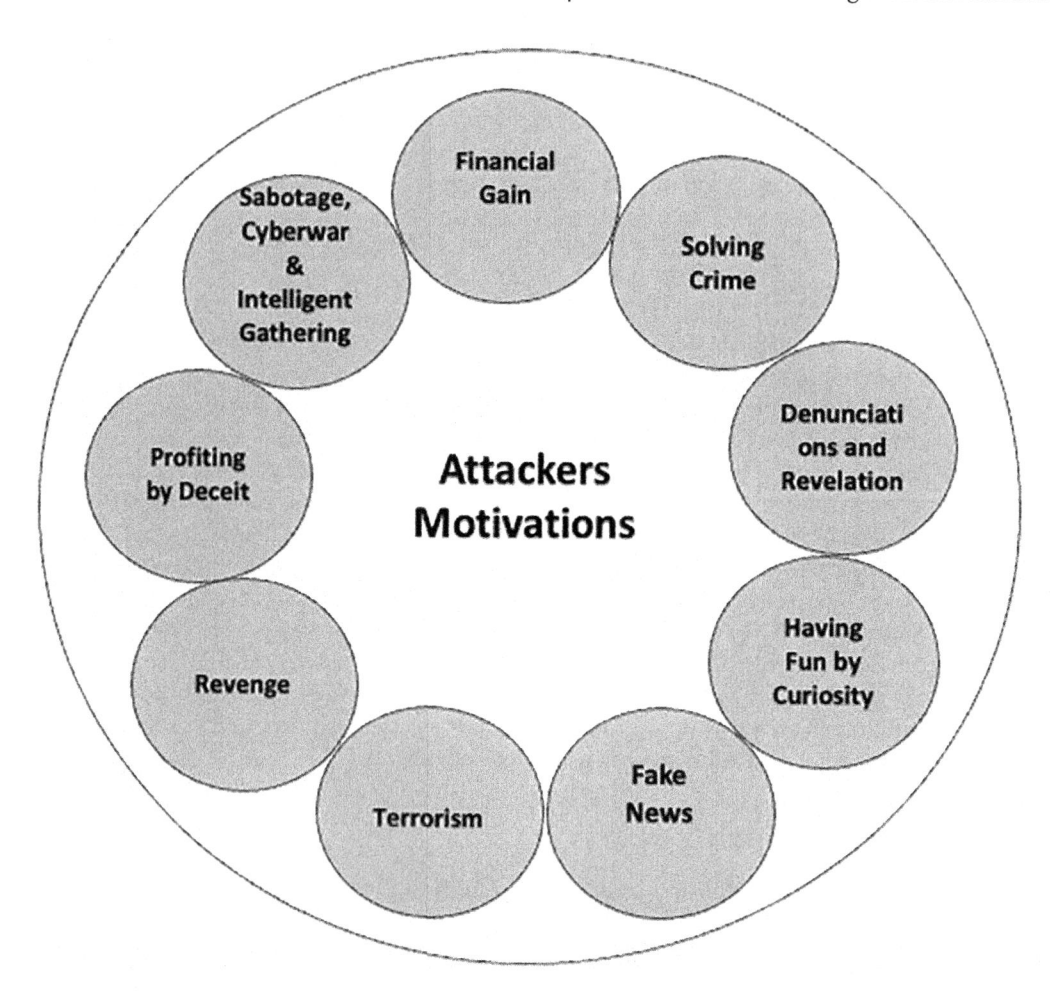

FIGURE 3.1 Some of the main attacker's motivation in committing cyberattacks.

Commission, there were 4.8 million identity theft and fraud cases reported in 2020, up 45% from 3.3 million in 2019. This dramatic rise in cases is considered to be primarily due to a 113% increase in identity theft complaints during the period (Consumer Sentinel Network, 2021, NCUA, 2019).

3.6.2 Sabotage, Cyberwar, and Intelligent Gathering

National espionage organizations and security agencies are another primary source of cyberattacks. The attackers' motivation varies based on each county's objectives, resources, and specific needs. An example in this category would be the famous case of Stuxnet, the sabotage of Iranian nuclear facilities (Zetter, 2014).

3.6.3 Fake News Shaping People's Opinions

Cyberattacks shape people's opinions, influencing elections through fake news or misinformation about the target population. The case of Cambridge Analytica is very famous. Cambridge Analytica was a company that infamously attempted to use the psychological profiles of roughly 90 million Facebook users to influence the exit of the United Kingdom from the European Union (known as the Brexit campaign) and Donald Trump's 2016 presidential campaign (Cadwalladr & Graham-Harrison, 2018).

3.6.4 Profiting By Deceit

Private investigating companies are also active in cyberspace by providing cyber investigations nationally and internationally. Cyber security private investigation firms offer services to countries, companies, and individuals. An example would be Jonathan Rees and his partner Sid Fillery, a former English police officer, purchasing information from former and serving police officers, Customs officers, VAT inspectors, bank employees, burglars, and bloggers. With this information in hand, they then telephoned the Inland Revenue Department, the Driving & Vehicle Licensing Agency (DVLA), banks, and phone companies and deceived them into releasing confidential information (Davies, 2011).

3.6.5 Solving Crimes

Investigative agencies or sometimes law enforcement agencies of some countries might also fail to obtain warrants to conduct cyberattacks in pursuit of solving crimes, and thus commit cybercrime themselves, whether knowingly or not. Some of these hacking tools are developed in-house, but in many instances, they are provided by private companies (Betschen, 2018, Sonnemaker, 2020).

3.6.6 Denunciations and Revelation

Some attackers might use the information obtained by hacking to denounce government agencies, organizations or individuals. Their goal may be to inform the public and ultimately to stop what they consider illegal, harmful actions of the targeted companies, organizations, or government agencies. The most famous case is Edward Snowden, American consultant (Ray, 2022).

3.6.7 Revenge

In cybersecurity, revenge behavior is mainly observed in insider threat cases. Occasionally, we learn of a company's former employee who is motivated, at least in part, to execute a cyber attack by a desire to seek revenge (Price, 2009). Studies suggest that 84% of the incidents of inside threats were motivated at least partly by a desire to seek revenge (Keeney et al., 2005).

3.6.8 Having Fun by Curiosity

Sometimes the motivation behind a cyberattack seems to be just to have fun. For example, a famous hacker, Kevin David Mitnick, who broke into 40 major corporations once said, "The motivation for me was that it was fun and an intellectual challenge…" when breaking into the system (Taylor, 2018). In another case in 2013, Burger King's Twitter feed was hacked by hackers affiliated with the anonymous collective, who quickly took it over and rebranded it with rival McDonald's logo and name and began using it to tweet McDonald's special offers (Ashley Lutz, 2013).

3.6.9 Terrorist Organizations

Terrorist organizations operating in a clandestine world also commit cyberattacks to harm their targets, sabotage critical infrastructures, or kill. A cyber terrorist's objectives are to create fear, cause death, or destroy assets or infrastructure (Jayakumar, 2021a).

3.7 WHY DO PEOPLE COMMIT CYBERCRIME?

Examining why people commit a crime is very important. People commit cybercrime for various reasons, including financial gain, revenge, ideological beliefs, for having fun or even just by curiosity. Some individuals may also commit cybercrime as a form of protest or to expose security vulnerabilities in organizations or institutions. In most cases, cybercriminals may be affiliated with organized criminal groups and may be motivated by the potential of gaining large profits from illicit activities such as identity theft, extortion, and the sale of stolen data.

Many theories have emerged over the years to analyze cyber attackers' motivations. These theories and frameworks continue to be explored, individually and in combination, as criminologists seek the best solutions to reduce crime types and levels.

Here is a broad overview of some fundamental theories.

3.7.1 Rational Choice Theory

Adam Smith's rational choice theory about people's economic and social behavior suggests that people perform a cost-benefit analysis to determine whether an option is desirable. According to this theory of the rationale underlying action, a cybercriminal chooses to commit a cybercrime to maximize their satisfaction by selecting the least cost while the most profitable action. (Mandelcorn et al., 2013, Clark & Felson, 2004). The rational choice theory is used to understand insider threat attackers and explain the role of the target's attractiveness to the attacker, and the effectiveness of defender protection strategies in deterring attacks (Willison & Siponen, 2009, Ransbotham et al, 2009).

3.7.2 Social Disorganization Theory

According to social disorganization theory, a person's physical and social environments are primarily responsible for behavioral choices (Bellair, 2017). Thus, a community with weak social structures, such as poor schools, vacant and vandalized buildings, high unemployment, and commercial and residential property, is more likely to produce and be home to cybercrime.

3.7.3 Strain Theory

According to strain theory, most people have similar aspirations but don't have the same opportunities or abilities. When people fail to achieve society's expectations through approved means such as hard work and delayed gratification, they may succeed through crime (Agnew, 1992). Strain theory explains behavior such as cyberbullying among youth (Patchin & Hinduja, 2011).

3.7.4 Social Learning Theory

Social learning theory considers how environmental and cognitive factors influence human learning and behavior. Consequently, people in an environment where crimes are part of their lives can expect criminal activity propagation. Issues such as poverty, unemployment, poor housing, and political unrest lead to strain and status unrest in regions around the globe, motivating groups to commit cybercrime (Bandura, 1977).

3.7.5 Social Control Theory

According to this theory, people would commit a crime if it were not for society's control of individuals through institutions such as schools, workplaces, churches, and families (Hirschi, 1969). This theory suggests that individuals who engage in illegal activities online are doing so because they have somehow been disengaged from the social norms that would normally inhibit such behavior. Factors such as poverty, social isolation, and lack of education may contribute to this disengagement and increase the likelihood of cybercrime.

3.7.6 LABELING THEORY

Governments at all levels of society create laws that define what acts are crimes, but the act of labeling someone who is convicted of breaking a law a criminal is not always what makes him a criminal. Society can do this without waiting for a formal trial. Once a person is labeled a criminal, before or after a trial, society takes away their opportunities, ultimately leading to more criminal behavior. Labeling theory explains how some denunciators might be subject to extreme punishment when authorities label them criminals. Or some attackers might be labeled criminals in one country but the opposite in another community (Becker, 1974). An example would be the treatment of Julian Assange, who published in Wikileaks classified documents from the US army. Following that revelation, some consider him a champion of government transparency and freedom of the press, while others have condemned him as a dangerous person who has undermined national security.

3.8 ATTACKER'S TOOLS AND TECHNIQUES

Most cyber attackers use different tools and techniques including technical skill, vulnerable human behavior, and lack of users' cyber knowledge and awareness.

> *Technical skill*: several technical tools and techniques are needed to conduct attacks. Many technical defenses are employed to thwart attacks including the use of IDS, Network security controls, Operating systems, Incident response, and widely dispersed cloud computing and storage, to name just a few (Cadwalladr & Graham-Harrison, 2018). This chapter does not cover technical defenses or the use of different tools.
> We must consider human cognitive capabilities and limitations to better understand the role of weaknesses in human behavior and how they contribute to cybercrime.
> *Human error and vulnerable behavior*: Human errors have been the cause of many breaches. Attackers use common human errors to instigate many types of attacks.
> *Behavior used to gain user's trust*: Unlike trust in interpersonal relations, where motivations and methods are typically in plain view, in the digital world humans are vulnerable to hidden motives and invasion techniques such as cookies and tracking to collect their system and personal data.
> *Vulnerable behaviors*: consist of the user's users' behaviors that attackers use to acquire the user's username, password, identity, and information, or place malware on users' devices.
> All the behaviors mentioned above are not exclusive to one category. Thus, vulnerable behavior can also be used in combination. For example, attackers use several tactics to gain the user's trust and then get their desired information from them.

3.9 FACTORS SHAPING VULNERABLE BEHAVIORS

With the growing number of cyberattacks, surveillance by governments and private entities, breaches of privacy, and social media's growing influence, there is an increasing need to research and understand people's behaviors. With data science, machine learning, and the predictabilities of human behavior, a new field of study on vulnerable behaviors has been created. Vulnerable behaviors are shaped by various factors, processes, and patterns. This section reviews several principal factors (Figure 3.2).

FIGURE 3.2 Some of the main factors shaping vulnerable behaviors in cybersecurity privacy and trust.

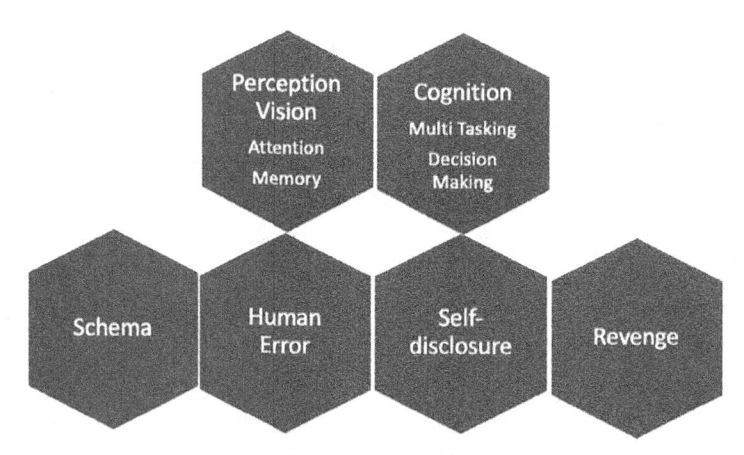

3.9.1 HUMAN ERROR

Many cyberattacks are successful because of human error on the part of the targeted user. We all occasionally make errors during our daily activities. Some errors might have minor consequences, and some more severe. We are frequently reminded of the famous statements: "To Err is Human" (Croskerry, 2010) or "If an error is possible, someone will make it" (Norman, 1988). Not surprisingly, people make errors in their interactions with computers. According to the Verizon 2022 Data Breach Investigations Report Guide, 82% of computer breaches that year involved the human element, including social attacks, errors, and misuse (Verizon, 2022).

Of course, attackers are aware of user propensity for those errors and present opportunities to make them when designing their attacks. Nowadays, although more and more people know they should not click an unknown link, they still do. Phishing is still the easiest way for the attacker to get the credentials of people.

Two scientists, in particular, have focused their work on human error: Donald Norman and James Reason. Donald Norman (Norman, 1993) has classified human error in the computer system into two categories: mistakes and slips. A mistake is considered an error in the intention and is referred to as an error of commission. They are decision failures. Mistakes are divided into rule-based mistakes and knowledge-based mistakes. They happen when people do the wrong action when they believe it is correct. They usually happen when users are doing many tasks at once, or run out of time. Mistakes also can happen when people have a false understanding of how a system works, and act intentionally in what turns out to be an incorrect manner. Making mistakes is a core part of the human experience.

Reason (1990 and 2000) classified different actions that lead to an error into two groups: intended action (mistake and violation), and unintended action (slip and lapse).

Slips (automatic behavior) and lapses (skill-based errors) are considered errors in execution. In these cases, the user knows what the right action is, but does something else, inadvertently. Slips and lapses occur in familiar tasks performed without sufficient conscious attention. Slips are errors of commission; for example, an operator moves the switch upwards instead of downwards. Lapses are errors of omission (not doing something that should be done). Short-term memory loss results in such errors when the operator omits procedures or matters necessary to perform essential tasks.

Slips and lapses can be further classified into small classes based on the mechanisms that seem to be the most likely causes. These errors might be performing the correct action on the wrong object, which will trigger an incorrect movement or an error by association that will trigger incorrect action. For example, after clicking a phishing link, some people immediately realize that they have made an error, but at that point, it might be too late to remedy the action.

Mode errors, such as changed meaning in the same context, lead to incorrect action. Capture errors are activities that frequently supplant one another. A violation or intended action error results from conscious decision-making or subconscious processing.

Human errors are the enablers of many cyberattacks. The causes of the errors are different (Jeimy, 2019). They might be related to forgetting, inattention, and wrong decision-making strategies, to name a few. The gravity and consequences of errors might differ depending on the human agent's role. For example, an Information Technology (IT) professional who forgets to install updated security patches leaves the system dangerously vulnerable, and the consequences of that error might be disastrous. The individuals who click a phishing email might be susceptible to data loss, ransomware attacks, or identity theft. The attackers might use common errors that people generally make to trick them into revealing their credentials.

3.9.2 Perception and Cognition

Human perception allows us to capture information through our senses. The cognitive process makes it possible to process, interpret, and make decisions using the constructs built into our brains. Perception and cognition are highly interrelated. This section summarizes the primary components: vision, attention, memory, and decision-making.

3.9.2.1 Vision

Human visual perception is fundamental to perceiving signals and enabling us to interpret our visual field. But the visual field is limited, and often users work in conditions where their displays fill much of their visual field. Consequently, a person focusing on the main task on a display might not be able to see a target, such as an alert notification, due to its size, location, or exposure time. Attention might be entirely focused on the main tasks elsewhere in the visual field (Wickens, 2021) or signals might be simply ignored. For example, some people do not notice security warnings appearing on their displays while claiming they are careful about their information (Raja F. et al, 2011). A major factor contributing to the ineffectiveness of warnings is that they are often simply ignored or overlooked. The effectiveness of warnings depends on their ability to attract users' attention, communicate clearly about the risks, and provide straightforward instructions for avoiding hazards (Zeng, 2018; Andersen et al. 2021).

3.9.2.2 Attention

To perform a task, people focus their attention on relevant stimuli and ignore irrelevant distractors. Research suggests different behavioral control mechanisms for elective engagement. For example, load theory suggests two mechanisms of selective attention: a perceptual selection mechanism and an active mechanism of attentional control that rejects irrelevant distractors even when these are perceived (Lavie et al. 2004, Lavie 1995, Lavie & Tsal, 1994).

Multiple-resource theory (MRT) (Wickens, 1984) postulates that people possess separate fixed-capacity resources for information processing. There are three dimensions in information processing: the processing stage (early vs. late processing), the processing code (spatial vs. verbal information), and the information modality (visual vs. auditory encoding; other sensory channels were not considered in the original version of MRT. Based on MRT, the concurrent performance of multiple tasks should benefit to the extent that information related to these tasks is presented concurrently in different modalities, and thus resource competition is reduced.

The human brain's selective attention to pertinent information successfully allows people to ignore irrelevant distractions. Research suggests the ability to ignore irrelevant distractions is not determined just by the intention to be focused or by the separability of the target and distractor stimuli, but also by the level and type of processing load involved in the task (Kramer et al., 2006, Lavie et al., 2004, Konstantinou et al., 2014, Camina & Güell, 2017).

3.9.2.3 Memory

Human memory plays a fundamental role in information processing. It has many implications on life and how people experience things, from remembering meaningful events to enabling people to execute tasks and achieve goals. Researchers have defined several types or stages of memory:

- *Sensory memory*: sensory information for very brief periods.
- *Short-term memory*: allows a person to recall little bits of information for a short period.
- *Working memory*: where data is processed.
- *Long-term memory*: information storage of all events, memories, and experiences.

All memory types might play an essential role in human behaviors pertaining to cybersecurity. Attackers might use the limitations of sensory memory in targeting users. For example, a warning only stays on the user's screen for a brief period. Or a temporary password might not be remembered once the window containing it is closed.

In order to manage their passwords for quickly accessing their various online accounts, people use a simple password or keep a written list of passwords. However, as the number of passwords used increases, the incidents of forgetting and mixed-up passwords increase. Information processing is limited to the prioritization of relevant information over irrelevant information. These processing priorities are actively maintained in working memory so that capacity is allocated with a higher priority to the relevant information. This capacity allocation happens when people engage in an important task where their attention and processing are engaged. For example, they receive a phishing text or mail and click the link without checking the sender's identity or credentials. Task conditions of the higher primary perceptual load reduce the processing of irrelevant distractors simply due to the reduced availability of perceptual resources (Wickens, 1984, Kramer et al., 2006, Camina & Güell, 2017, Pilar et al., 2012).

Human memory plays an important role in cybersecurity as it relates to the ability of people to remember and implement strong passwords, security protocols, and best practices for maintaining the security of digital systems. Human memory is also a key factor in the ability to detect and respond to potential security threats, as individuals must be able to recall and apply their knowledge of security protocols in real-world situations.

3.9.2.4 Decision-Making

A person might make poor decisions for many different reasons. Cognitive control over information processing is limited to prioritizing relevant information over irrelevant information. These processing priorities are actively maintained in working memory so that capacity is allocated with a higher priority to the relevant information. However, if processing the relevant information does not require all available capacity, any remaining capacity is given involuntarily to processing irrelevant information as well (in a simultaneous parallel manner). Thus, the level of perceptual load in task processing plays a critical role. Task conditions of low perceptual load—for example, detection of a single item or one that pops out from among different things—results in distractor processing even if people attempt to ignore irrelevant distractors. Task conditions of higher perceptual load—for example, an increased number of items or more complex perceptual processing demands like discriminating conjunctions of features - result in reduced processing of irrelevant distractors, simply due to reduced availability of perceptual resources (Lavie et al., 2004, Jones et al., 2021, Price, 2009).

Effective decision-making in cybersecurity requires individuals and organizations to have a clear understanding of their security plans. Ultimately, the goal of decision-making in cybersecurity is to minimize the potential impact of security incidents, while also ensuring that an organization can continue to operate effectively in the face of potential threats.

3.9.3 Behavior Pattern-Schema

Schemas (or schemata) are units of understanding that can be hierarchically categorized and woven into complex relationships. Schemas take shape concurrently with the development of human experiences from childhood to adulthood; throughout a human's lifetime. Schemas are constructs or building blocks that organize our knowledge and experience. The constructs are interconnected and constitute a complex interactive network. A schema guides information acceptance and retrieval: it affects how humans process new information and retrieve old information from long-term memory. Piaget and Cook (1952) called schema the core building block (Moallem, 2019a).

As a schema is repeated over and over, its elements, patterns, and relationships become expected; if by any chance one step were to be added or changed, that would interrupt the expectations of people and confuse them. For example, when you call to reserve a table at a restaurant on the phone, you are used to giving your name and phone number. If the reservationist were to also ask for your driver's license number, that would violate your known schema for this activity and appear very odd to you. Unless a specific, convincing reason were to be given, you would very likely refuse to answer the question. During our lives, as we learn and discover the world around us, our schemas expand and get complex. The more we know, the more significant and complex our schemas become. However, the more we are aware, the easier it is to remember new information related to the schema. Thus, since the data exists in our heads, we can relate to it and organize and predict our actions. Activity schemas are called scripts. We all have many scripts in our long-term memory related to the various activities we have engaged in throughout our lives.

Social engineers in the digital world take advantage of our dependence on schemas. They carefully prepare and investigate the process, establish an accurate script of each situation, and then find believable responses that capture their victims' trust (Moallem, 2019b, Mitnick & Simon, 2002).

3.9.4 Self-Disclosure

Our desire to know what is going on in other people's heads, as well as their opinions, judgments, feelings, and reasoning is ongoing. Since Alfred Binet and Theodore Simon developed a series of tests designed to assess mental abilities (IQ), tests of all kinds have been developed and extensively used worldwide. Along with the development of the test, lie detectors or polygraphs are used to detect lies by capturing the physiological signals that typically accompany them; for example, an honest person may be nervous when answering truthfully, and a dishonest person may be non-anxious. However, for all sorts of tests or lie detector machines, the individual must agree and give consent and agreement to take the test or be on a polygraph.

Modern digital times are marked by the creation of social media and a huge interest in social media applications. On the one hand, social media sites might internally access people's data individually or in mass, without their knowledge, which could enable them to predict behavior. On the other hand, social media sites allow users to self-disclose their data to their entire social network if they want to. The motivations of people in self-disclosure are different. According to research, the two main reasons for self-disclosure are social validation and self-expression/relief, accounting for over 70% of all self-disclosures (Bazarova & Choi, 2014). In addition, self-disclosure might fulfill the particular needs of individuals with different well-being characteristics and emotional states (Luo & Hancock, 2019) and be related to the different personalities and needs of individuals with varying well-being characteristics (Shappie et al., 2019).

Self-disclosure in cyberspace might put people at risk in three ways: (1) data privacy may be compromised, and data might be used by social media companies for advertising, influencing their behavior; (2) enabling access to the user's social media for phishing or social engineering, and; (3) influencing the user's opinions.

3.9.5 Revenge

Revenge causes unpleasant or painful hurt or harm to people, organizations, companies, or countries. Revenge is broadly analyzed in social psychology, and incidents and types of revenge behavior differ between people and across cultures.

Revenge is also associated with pleasure (McClelland, 2010). Some people seem more vindictive than others, motivated by power, authority, and the desire for status (McKee & Feather, 2008, Price, 2009, Keeney et al., 2005).

Revenge in cybersecurity is a problematic behavior, as it can lead to several negative consequences, both for the individual or organization seeking revenge, as well as for the target of the retaliation. For example, engaging in acts of revenge can lead to legal repercussions, damage to an organization's reputation, or even escalation of the conflict.

Additionally, revenge attacks can inadvertently harm innocent third parties and lead to a loss of trust and cooperation among different stakeholders in the cybersecurity ecosystem.

3.10 TRUST

Trust as a social psychological construct is a foundation of human interaction. In the physical world, trust is established based on identity or context. We do not trust everyone equally. For example, people generally trust friends and family more than neighbors or casual acquaintances. There are exceptions, but in general, the notion of trust in an entity is built over time. The amount of trust depends partly on the frequency and the nature of our interactions with an entity.

How is trust established in the digital world? For example, there are no physical cues when we visit our bank site. We are not in the bank office and not interacting with the banker, and the banker cannot check personally identifiable information (PII). In digital interactions with a bank, how can people check to confirm the trustworthiness of a text, a phone call, an email, or a post on social media? What kind of reliable cues should people use? Are people aware of how people check the validity of the senders on websites?

Confounding the user's dilemma, one of most attackers' first steps is to gain users' trust. This is accomplished through phishing or collecting a user's data through cookies, for example, and quickly used for advertising, surveillance, or influencing the user's opinion by what is known as fake news. In many cases, several behavior patterns are used to try to gain access to the user's private data.

When people visit a site relying on visuals, such as the logo, colors, and layout, to establish trust, they often may not be aware that sophisticated phishing attackers are now well past that step. The "look and feel" of many highly respected corporate sites is very well duplicated by cyber attackers and users might not be at all aware that the site they are on is not the authentic one. Thus, users must rely on secondary verification methods to establish trust with some corporate entities that retain their confidential information (Moallem, 2019a).

3.10.1 Digital Keys and Certificates

A certificate or digital key, like a credit card or a passport, is issued by a "trusted" authority (an enterprise or a financial or government institution in the real world) and has an associated validity and purpose. The similarities end there, however. Although we can "view" certificates, the attributes that make them unique (and, hence, linked irrevocably to a physical entity) can only be "verified" by applications such as a web browser or an email client. Theoretically, it should then be possible for an application to identify and, over time, trust an entity (Nair, 2019). Unaware users of this feature might visit a phishing or insecure site.

3.10.2 'HTTPS' and 'HTTP'

The Hypertext Transfer Protocol (HTTP) was developed in 1989. Then, 5 years later, Netscape Communications used Hypertext Transfer Protocol Secure (HTTPS) in its Netscape Navigator web browser. Browsing with HTTP transmits data in plain text, enabling 'man-in-the-middle' attacks where they attempt to intercept data while it's in transit. HTTPS, on the other hand, works with public key encryption through SSL/TLS to prevent the same attack (Nair, 2019). HTTPS provides a more secure webpage connection. Gradually HTTPS has become the web standard. However, despite its security, HTTPS does not make web surfing completely safe. It only indicates that the connection is private and secure from hackers. But it cannot prevent the transfer of malware. In fact, according to WatchGuard Threat Lab Reports, 91.5% of malware arrived over Encrypted Connections in Q2 2021(Watchguard, 2021). Again, most people might be unaware of this feature and interact in an unsecured connection.

3.10.3 User Behavior and Trust

One of the reliable behaviors in detecting whether a website is legitimate is looking at the website's uniform resource locator (URL). A secure URL should begin with "HTTPS" rather than "HTTP." The "S" in "HTTPS" stands for "secure", which indicates that all your communication and data are encrypted as it passes from the user browser to the website's server.

Another reliable behavior would be to look for a lock icon near the user browser's location field. The lock symbol and related URL containing "HTTPS" mean that the connection between your web browser and the website server is encrypted, which is essential for the website.

The sender's address will be the main factor in identifying the phishing site or untrustworthy sender for email or text messages.

However, most users are unaware of these criteria; they might mostly rely on the design and graphical aspects of the site, or on messages and posts on social media sites. Several studies also suggest that people have difficulty distinguishing between legitimate and phishing emails or text messages (Watchguard, 2021, Abroshan et al., 2021). It also appears that non-expert users are reluctant to consider cybersecurity warning messages. They often might be unaware of the whole situation when making decisions (Masayuki Hatta, 2020).

Humans tend to expect machines to perform predictably; if machines do not do so, it leads to more rapid trust declines when unreliability is encountered. For example, if users cannot find a file known to exist using a desktop search feature and then find it manually, they may be disinclined to use the search feature in the future when searching for other files. In contrast, human people-to-people relationships develop from a schema of imperfection, making users more forgiving of errors and basing trust on the trustee's knowledge of the other person's qualities or intent (Schuster & Scott, 2019).

3.11 PRIVACY

Privacy is a broad subject area and concerns individuals, organizations, businesses, and public institutions. In the digital world, privacy has become more critical than ever due to the accessibility of digital data describing a person's possessions, actions, relationships with other people, and even their wishes, intentions, and emotions. User privacy is also related to concepts of a user's anonymity (not being identifiable), unobservability (being indistinguishable), and unlinkability (the impossibility of the correlation of two or more actions/items/pieces of information related to a user) (Schuster & Scott, 2019, Weber, 2015). All private data should be protected to:

- Keep secret to reduce the possible distress caused by the change in social relations.
- Reduce vulnerability to business-related attacks, such as marketing.

- Minimize the probability of criminal attacks.
- Minimize vulnerability to identity theft.

The problem of privacy breaches is critical since it may lead to restrictions on individual liberty and erosion of our society's foundations of trust.

A cyberattacker's motivations to acquire private data may be considerable financial gains, blackmail, marketing advantages, and/or identifying a person's vulnerabilities to weaken a negotiating position and/or influence their social and political views.

Attackers use the vulnerable behaviors of people to achieve their goals through legal or illegal strategies. The following sections review some typical user behaviors that are exploited to obtain private data. The main ways that cyber attackers get private data from people or organizations are through hacking the victim's account or using users' lack of awareness, self-disclosure, or through other IoT devices and just not giving users any choice.

3.11.1 Human Behavior and Privacy

3.11.1.1 Cookies

Tracking Cookies have been deployed for user identification and online advertising and marketing for a long time. However, consumer proponents, policymakers, and even advertisers themselves recognize the prospective danger to user privacy that comes with the utilization of Internet cookies generated by website visits.

The primary purpose of tracking cookies is to gather information or to potentially present customized data. These cookies are not malware, worms, or viruses but are software objects that collect and can potentially disclose data that corresponds to user privacy. Indeed, some cyber security attacks are launched by hijacked cookies that enable access to the unaware user's browsing session. The danger lies when attackers or hijackers have the ability to track individual browsing histories.

Some cookies may have more threats than others. For example, when a user browses an online advertising website, the site can place a cookie on the user's device. If another site also has advertisements from the vendor he visited earlier, that vendor knows that the user has visited both websites. So, the advertising or marketing company will indirectly determine all the sites the user visited. Industry research reveals that both cookie usage and disclosure have increased, and third-party cookies could be used to track a person's online activity. It has been found that it is possible to use cookies to reconstruct 62%–73% of a typical user's browsing history (Miyazaki, 2008, Englehardt et al., 2015).

Additionally, there are three categories of cookies relating to the threat they possess. The categories are (1) first-party cookies, (2) third-party cookies, and (3) zombie cookies. First-party cookies are created by the website the user is using and are generally safer if the user is browsing a reputable website and not a compromised one. Third-party cookies are more of a threat. Third-party cookies are generated by the website that is linked to ads. For example, visiting a site with ten ads will generate ten cookies, even when the user doesn't click on those ads. The cookies will let the advertisers or an analytical company track an individual's browsing behavior and history across the web on any site that contains ads. Zombie cookies are from third-party cookies and are permanently installed in the users' computers even without the user's knowledge.

A myriad of questions relates to the presence and purposes of cookies since not all cookies are bad to have. How do people behave in dealing with cookies? How much is the user aware of the cookies? Do users know the primary function of the cookies that are on their system? How often do they encounter the cookie disclaimer? Should users clean or not accept cookies? These and many more questions need to be answered.

Research suggests a person's level of awareness of Internet cookies is directly related to the amount of Internet experience of the respondent, directly related to their level of education, and associated with their likelihood of understanding Internet cookies (Warrington, 2000). Most people with more than a moderate level of awareness about cookies accept cookies for quick access or task completion. However, the acceptance of cookies varies based on online activity. If given a choice, users are more likely to opt-out of third-party cookies since they are typically purposed to provide information for targeted advertising (Jayakumar, 2021b).

Governments are aware of cookies and their potential for misuse and adverse impacts. The 2011 European Union (EU) directive on privacy and data protection started to regulate cookies. It was followed by several others such as the California Privacy Rights Act (CPRA) and UK Data Protection Act (Bonta, 2018). Regulations like these have brought some increased awareness of cookies by introducing a requirement that software asks users installing new programs to accept or reject cookies. However, this does not protect people's privacy. The warning that the site is using cookies with the requirement that the user chooses to accept, customize, or decline the cookies is essentially a Hobson's choice (HistoryWorks, 2015). Only one answer ("accept") allows the software to continue with its installation. Indeed, several studies conclude that people typically accept cookies without reviewing them or understanding their terms of usage or privacy policies (Bloomberg Law, 2019).

3.11.1.2 Self Revelation on Social Media

As described above, unintended self-disclosure of private information is one of the main adverse effects of using social media. Moreover, revelations on social media are a primary type of unauthorized captured private information used in cyberattacks. As we mentioned, people share information on social media that might be used for attacks on them or others in their circle. The primary vulnerable behavior of people who are being less cautious about protecting themselves from the dangers of social hacking is their inadvertent sharing of personal information on social media (Moallem, 2021). Some of the main behaviors are:

3.11.1.2.1 Location Sharing

Location-tracking technologies that locate, record, and track the physical movement of people or objects are embedded in many of the applications that we use daily. Location tracking is used in car navigation systems of course, but location identifiers might also be embedded on digital photos, a use that might not be quite as apparent. Businesses close to our current location, such as restaurants, shops, service providers, etc., can capture and use our location with the help of popular mobile applications like Yelp, Zomato, Uber, etc. While we assume that location tracking is only done through smartphones and the Global Positioning System (GPS), in fact there are many other ways through which locations are being tracked and used. Advances in Internet design and structure, and many social networking applications have massively increased the possibility of learning a person's location in real time and acting on that information.

Location data may be shared with third parties unknown to the users. Typically, users do not know who the third parties are, how they might use their location data, or whether those entities are trustworthy. For example, location data can be used to track the pattern of a users' behavior. It can also be used to steal a user's identity or their car if the location data is combined with other personal information. Finally, location data can be used to track the records of an individual's movements and activities.

The primary vulnerable behavior of people who are being less cautious about protecting themselves from the dangers of social hacking is the sharing of their information on social media (Moallem, 2021).

3.11.1.2.2 Online Social Media Survey

Many online surveys collect people's data by giving out rewards. However, survey hosts also face risk, and might not take strict measures to protect the privacy of the respondents or their personal views. The events in the last US election and interferences in mass communication by what is now labeled as "fake news" make trust between what is real and fake a challenging issue for citizens. Another good example is the case of Cambridge Analytica, a political data firm that gained access to the private information of more than 50 million Facebook users and then offered tools that could identify the user's personalities. Aleksandr Kogan, a data scientist and professor at the University of Cambridge, is known for developing an app based on psychological profiles that analyzed characteristics and personality traits. The data came from a personality quiz, which around 270,000 people were paid to take. The quiz—"This is Your Digital Life"—also pulled data from their friends' profiles, ending in an enormous data stash (Cadwalladr & Graham-Harrison, 2018, Confessore, 2018, Hern, 2018).

3.11.1.2.3 Emotional Sharing

People are motivated to share information through self-connection and social connection (Kim et al. 2022). Therefore, emotionally charged Twitter messages are retweeted more often and quickly compared to neutral ones (Stieglitz & Dang-Xuan, 2013, Forest & Wood, 2012).

Research suggests that higher stress levels triggered more significant amounts of self-disclosure in social media, which might be related to a safer and easier medium for self-expression (Dhir et al., 2018).

3.12 CYBERATTACKS

Every day, millions of people are targeted by cyberattacks. This is because cybercriminals have figured out profitable business models and have taken advantage of online anonymity. Since most successful cybercrime incidents are human-enabled, analyzing the human behavior related to successful cyberattacks is vital. We must then focus on the social and behavioral issues relevant to successful cybercrime attacks to improve the current situation. This section reviews some main vulnerable behavioral patterns associated with successful cyberattacks that have resulted in people or organizations being compromised.

3.12.1 Behavior Used in Cyberattacks

3.12.1.1 Authentication

From the earliest days of web applications, username and password-based authentication modes have been the most popular way of verifying user access. Unfortunately, password hacking remains one of the significant security threats in the digital world, even after so much technological advancement.

Many new ways of authentication have been introduced, such as Quick Response (QR) Code scanning and password-less logins using confirmation codes sent to phone numbers or email IDs. Typically, all these new authentication mechanisms require access to a device that was present at the time of the initial setup. For the sake of simplicity, most people tend to use a username and password-based authentication, even when other modes are available. With the growth of web applications and most tasks being done online, an average user has so many online accounts that it is challenging for them to keep track of all of their usernames and passwords. Users tend to repeat their usernames and sometimes their passwords across accounts or use simple passwords that are easy to remember.

General behavior preferring ease versus security makes people more vulnerable, as easy-to-remember authentication credentials are also simple to crack or guess. As a result, many different security features have been gradually added to authentication protocols, such as multi-factor authentication, requiring hard-to-pass passwords or password management applications. However, with all the effort, easy-to-guess or already hacked passwords through massive data breaches are still some of the main vulnerable behaviors of many users (Furnell, 2018).

3.12.1.2 Phishing

The act of pretending to be a legitimate source by deception is termed phishing. Phishing is a cyber fraud crime in which an attacker poses as a trusted entity and throws bait to a victim to achieve specific malicious purposes, such as the intrusion of an internal network to gain access to sensitive information and data.

Phishing attacks are evolving daily and have become a profitable business, with everyone from individuals to large companies targeted, resulting in billions of dollars of financial losses yearly.

Phishing attacks account for over 80% of reported security incidents (Anti-Phishing Working Group, 2020). In 2015, an estimated $4.6 billion in financial losses were attributed to phishing attacks. Phishing attacks are also increasing yearly (Jain & Gupta, 2017) through junk email, instant messaging tools, mobile phones, short messages, or web pages that send false advertising and other deceptive information from cybercriminals posing as banks and other well-known institutions. The intention is to induce the user to log into what looks like a real website that is actually fake, and to then input sensitive information, such as their username, password, account ID, Social Security Number, ATM PIN, credit card, etc.

Attackers have become increasingly skilled at replicating sites by mimicking the websites' color, format, and text. They copy image results for HyperText Markup Language (HTML) source code from the original websites (Jain & Gupta, 2017). Research has shown that even an experienced user can fail to distinguish between legitimate and illegitimate websites due to the near-perfect resemblances. Considering the pace at which most users are operating when using familiar websites, it is not surprising that website replicates have been so successful.

Although one of the most important features of a phishing attack is visual similarity, attackers also try replicating legitimate domain names, websites, and email formats to trick victims. Researchers have been studying phishing attacks and devising various techniques to identify such attacks. In one strategy, the similarity is finely analyzed to differentiate between legitimate and illegitimate emails and websites. The visual representation of legitimate websites is stored in a database. Whenever a third-party website crosses the similarity threshold, a website is identified as a phishing site (Jain & Gupta, 2017).

Phishing sites, emails, and texts always bear a strong resemblance to the authentic original, yet there is always a certain amount of information that can expose them as fakes if one takes the time to search for it. For example, phishing domain names are similar to legitimate links, but contain recognizable differences. Visually similar content designed to induce users to enter sensitive information may have noticeable color or shape variations from the original if one is experienced enough to recognize it. Phishing detection software is available to analyze the elements of a web page's content (URL, mail, web page, etc.) to improve the detection and identification of phishing attacks.

3.12.1.3 Power of Free

People are very much attracted to "free" item offerings, or "sales" in their purchasing behavior. "Buy one get a second for free", "free sample", and "50% discount" are always attractive to people. For example, people in an exhibition room might stay 30 minutes longer just to get a "free" T-shirt. Research shows that the attraction to a free item often might be irrational; however, people still do it. Dan Ariely calls this, the "Power of Free" (Ariely, 2008). In cybersecurity, attackers exploit this normal interest in obtaining free items, or a discounted subscription, to motivate users to click a phishing link without thinking and first checking the link or the trustworthiness of the destination.

3.12.1.4 Scarcity

Scarcity marketing tactics have been broadly considered to enhance product desirability. The scarcity tactic operates on the worth people attach to things. Scarcity suggests items are more valuable when they are less available. There are two common patterns used in scarcity marketing: (1) the "limited-number" technique and (2) the "deadline technique." The "limited-number" tactic works because it creates added perceived value for a product by reducing the perceived availability of the product. It relies on the understanding that we want what may not be available. The "deadline technique" works because it seems to put an official time limit on product availability. The scarcity tactic has the effect of increasing the perceived value of things (Cialdini, 1994).

The effectiveness of scarcity advertising is evidenced by its frequent use in cyberattacks and social engineering. For example, scarcity ads are commonly used in phishing sites that offer a significant discount on a specific product, leading people who are impressed by the discount to place orders while not knowing that the site is not secure, and their data is about to be used for identity theft and fraud.

3.12.1.5 Scare Tactics

Society systematically socializes its members to comply with authority. Authority is the principle probably used most frequently by all social engineers in phishing attacks or voice calls. We typically tend to obey those in charge. The importance of authority is one of the areas in social psychology that has been frequently studied. Research on the subject ranges from the Milgram experiment on the impact of authority on obedience to Zimbardo's Stanford Prison experiment (Wikipedia, 2024), where researchers concluded that people obey either out of fear or out of a desire to appear cooperative—even when acting against their better judgment and desires.

Cyber attackers use scare tactics in voice calling and phishing emails. For example, in voice calls, the caller institutes the "scare" by pretending to be calling from the Internal Revenue Service (IRS), a prosecutor's office, a police station, or a law office (Cialdini, 2009).

3.12.2 "The Internet of Things"

The systems of interconnected devices that communicate and transmit data to each other through networks without human-to-computer interaction are called "The Internet of Things" (IoT) (Ranger, 2020). Every IoT device is connected to the world through the Internet, so security is a significant issue when deploying and using IoT devices. Unsecure devices can lead to users being hacked, devices being hijacked, and potentially severe consequences. According to Gartner, 8.4 billion connected "Things" were in use in 2017, up 31% from 2016 (Gartner, 2017 & Congressional Research Service, 2020).

IoT technologies provide connectivity to all devices, making them "smart", in that they can autonomously sense and respond to network status and communications received from other devices. Some devices can also be made to be "intelligent", by using machine learning and data science to manage and analyze real-time data to make decisions and predict future events. A cloud application then receives, stores, and processes this data. It is important to highlight that a lot of data may be collected by each IoT device. Therefore, rather than sending all data to a centralized system, each device does some local data processing first and sends only relevant data forward to the cloud for more analysis.

IoT security protects Internet-enabled devices that connect to networks. There are several levels of security technologies that are needed to protect IoT devices. The critical parts are authentication and tracking, data and information integrity, mutual trust, privacy, and digital forgetting. In addition, since a great deal of processing happens at a centralized location or in the cloud, there is a heightened need to ensure security for the cloud (Sicari et al., 2015, Wójtowicz & Cellary, 2019).

3.12.2.1 Manufacturing

In the last 40 years manufacturing systems have undergone a massive transformation.

As the manufacturing sector becomes increasingly connected and dependent on technology, the importance of cybersecurity in this industry continues to grow.

The transformative energy has been provided by progress in artificial intelligence, machine learning, cloud computing, and the IoT. Smart and intelligent manufacturing is impossible without the interconnection of smart machinery IoT devices within a network of computers in a closed or open network. While new manufacturing technologies make the interconnection of the machinery's components necessary, there is a collateral growth in the risk of cyberattacks, which are estimated to be growing exponentially daily, with substantial financial impacts on enterprises (Moallem, 2022).

3.12.2.2 Healthcare

It is clear that technology improves clinical outcomes and transforms care delivery. This transformation is primarily happening in two areas of healthcare. The first area is communication—marked by a change from paper tracking to digital data storage, transmission, and sharing of the data among healthcare professionals and patients. The second healthcare transformation relates to health monitoring through mobile devices and IoT wearable sensors. Consequently, from a cybersecurity perspective, there are two axes of security concerns: (1) how data and communication among professionals and patients are secured and (2) how people manage their privacy and security when using different applications and wearable devices—considering that the users in the healthcare systems include a disproportionate number of people who are elderly and ill (Ferreira et al., 2021). Thus, ensuring the availability and integrity of medical information and systems, and protecting patients from cyber threats is a big challenge.

3.13 CYBERSECURITY HYGIENE

Cybersecurity hygiene refers to the basic practices and protocols that individuals and organizations can follow to protect themselves from common cyber threats. These practices include keeping software and operating systems up to date, using strong and unique passwords, regularly backing up important data, and being vigilant against phishing and other social engineering attacks. Additionally, companies should have an incident response plan in place and employees should be trained in security best practices.

Many security breaches are due to users' lack of knowledge or unsafe behaviors, such as sharing passwords, clicking on unknown links, and opening unknown emails and attachments. These activities potentially open an organization or individual to threats from attackers and the loss of assets. Yet, how many people are aware of the risks, and when they are victims, what do they do? What lesson do they learn? Does their behavior change?

Although organizations and enterprises invest and rely more on technology for security solutions (e.g., firewalls, antivirus software, and IDS) and to defend organizational assets, the importance of considering the role of users in the security equation is gradually growing.

3.13.1 Awareness

Users/employees need to understand security issues, protect, and maintain various devices, and follow each organization's security policies. These practices are crucial to comply with information security laws, regulations, and digital activities. Understanding the growing number of cybersecurity issues, constantly being aware of the action undertaken, changing behavior patterns learned and practiced for many years, and learning new protective behaviors are not easy and may be unreachable for many people. In addition, users/employees need to understand security issues while complying with security policies and practicing secure behavior protocols (Moallem, 2019, Gupta & Furnell, 2022). Thus, the term "cyber hygiene," is offered to cover all the actionable areas in cybersecurity. Cyber hygiene generally refers to the technological solutions that need to be used and behavior changes or new behaviors that need to be learned as part of everyday cybersecurity practices (Vishwanath et al., 2020; Souppaya et al., 2018).

Awareness improvement also requires a reliable methodology to measure awareness. Combinations of different methods are used to measure awareness among company employees. Questionnaires and surveys are used to measure knowledge (what you know), attitude (what you think), and behavior (what you do) (Kruger & Kearney, 2006). Model-driven techniques and survey-based research are also used to investigate behavior modeling in the security context, such as information-sharing and security policy compliance, as well as computer security behavior, such as interacting with email attachments, phishing messages, and text (Fan & Zhang, 2011, Ng et al., 2009, ENISA, 2018).

No matter what type of training is used to improve cybersecurity awareness, it is crucial to measure its success in not only educating employees of their knowledge of cybersecurity but also to see if and how they integrate the awareness into their everyday practice and behavior and the effect of awareness training on the actual behavior of the trainees.

3.13.2 Training

All employees need to be trained to understand cybersecurity issues and how to protect themselves as individuals or participants when engaged in activities that connect them in any way to the Internet for work or personal activities. Knowledge and awareness play a vital role in enhancing security. However, simply knowing is not enough. The knowledge must be transformed into practice for people to benefit from it in their cyber lives.

3.13.2.1 Cyber Training Programs

Looking at all types of training offered, we classify them into the following four groups: formal education, professional training, employee training, and population training (Maalem et al., 2020).

3.13.2.1.1 Formal Education

Different formal education programs are offered as a degree or specialized courses to train future professionals. Even though the number of educational programs for IT professionals is not enough to match the demand, they are gradually growing in numbers and continue to evolve and improve. Availability is also increasing with online options to help meet the demand and needs.

3.13.2.1.2 Professional Training

Professional training programs for IT professionals in the industry include hundreds of training workshops, certificate programs, and tutorials of all kinds offered by private companies, training institutions and individuals on a variety of technological tools, technical topics, and solutions.

Examples are:

- *Security Education, Training, and Awareness (SETA) Program*: SETA is a program designed to make people aware of information security policies and are able to apply them during their daily activities to help prevent security incidents. SETA is one of the most common and prominent strategies for organizational security governance (Hollois, 2021).
- *Cybersecurity Countermeasures Awareness (CCA)*: CCA are training, actions, devices, procedures, techniques, or other measures that reduce the vulnerability of an information system (Goode et al., 2018).

3.13.2.1.3 Employee Training

Employee Training has evolved due to the growing number of companies that have become victims of cyberattacks. Consequently, many companies recognize that protecting information systems and information assets from cybersecurity threats has become critical and have instituted mandatory cybersecurity training programs for all employees. But, that said, most companies do not provide any form of cybersecurity training at all. They prefer to save the training costs and rely on technology to provide a shield from cybersecurity issues. Thus, employees of most companies very likely lack cybersecurity knowledge and skill sets. They are most often identified as susceptible threat vectors by cyber attackers who then target them with continually evolving threats (Moallem, 2019b).

3.13.2.1.4 Training Delivery Methods

Training delivery methods are varied. Although some studies investigated the user preferences of cyber security awareness delivery methods from classroom-based to sharing experiences and knowledge to online training sessions in many companies, there is no conclusive evidence on which is the most effective in creating a thriving security behavior culture. Most approaches and protection methods focus heavily on external attacks and technological defenses.

3.13.2.2 Training and Awareness Program Effectiveness

No matter what type of training is used to improve cybersecurity awareness, it is more important to measure the success in educating employees about their knowledge of cybersecurity and seeing if and how they integrate this awareness into their everyday practice and behavior. In addition, it is also essential to measure the effect of awareness training on the actual behavior of the trainees.

3.13.2.3 Changing People's Behavior

To improve people's awareness of cybersecurity (not just their knowledge, but also how to implement better behaviors into their everyday lives), one must develop structural solutions that address people at an early age. Much like other behaviors ensuring safety and security (such as locking their cars or homes or keeping their valuables in a safe box at home or in a bank), people should also learn to lock their online home/account effectively and safeguard their digital assets. If people use a good lock for their home or business or place cameras around their property, why shouldn't they learn to do the same for their digital assets and monitor their online activities?

The European Union Agency For Network and Information Security Cybersecurity (ENISA) report on Culture Guidelines suggests that users' understanding of the threat posed by cybersecurity breaches, or fear of the consequences, is not an effective tool for changing behavior. Many of the models currently used to study human aspects of cybersecurity poorly fit actual behavior. Most currently used are not well suited to measure human behavior or provide information about strategies to influence behavior (ENISA, 2018).

Thus, expanding research on behavioral aspects of cybersecurity is needed and vital to focus on social and behavioral issues to improve the current situation (ENISA, 2018).

To effectively improve awareness and change behavior, the population needs to be educated early, continuously, and consistently. After all, it is unlikely that issues in cybersecurity will be resolved soon. Thus, it is in the greater community's interest to include cybersecurity measures in the educational system from middle school. Furthermore, because children are using computers at a very early age, it is much more vital that they be aware of their cybersecurity as early as possible. Including cybersecurity in the formal education system would also be an efficient, more cost-effective approach for society, considering all the costs associated with training employees and individuals following identity theft, ransomware, and so on (Moallem, 2019a).

3.13.3 MOTIVATION

How can we change the risk-prone habits people have developed regarding their activities online and the security of their digital assets? Would that be possible? Presumably, "Yes", but people need to be motivated to make even small changes in their behavior.

One of the theoretical frameworks used to analyze or to change human behavior is based on the theory of Reasoned Action (TRA), which aims to explain the relationship between attitudes and human behaviors (Ajzen, 1980, 1991). TRA predicts how individuals behave based on their pre-existing attitudes and behavioral intentions. For example, people's individual beliefs, community beliefs, or social norms might define their attitudes, intention, and behavior.

Another model used to analyze and change behavior is based on the Stages of Change Model (Prochaska & DiClemente, 2005), or the Transtheoretical Model. This model analyzes and explains people's readiness to change their behavior. It describes the process of behavior change as occurring in six stages:

- *Pre-contemplation*: There is no intention of taking action.
- *Contemplation*: There are intentions to take action and a plan to do so in the near future.
- *Preparation*: There is the intention to take action, and some steps have been taken.
- *Action*: The behavior has been changed for a short period.
- *Maintenance*: The behavior has been changed and continues to be maintained for the long term.
- *Termination*: There is no desire to return to prior negative behaviors.

Social cognitive theory (SCT) (Bandura, 1977) is another framework that considers social context with a dynamic and reciprocal interaction of the person, environment, and behavior. SCT aims to explain how people regulate their behavior through control and reinforcement to achieve goal-directed behavior that can be maintained over time. The SCT emphasizes social influence and its emphasis on external and internal social reinforcement. SCT considers how people acquire and maintain their behavior while also considering the social environment in which individuals exhibit the behavior and people's past experiences that influence reinforcements, expectations, and expectancies. There are six constructs that evolved into SCT as described in the list below:

1. *Reciprocal determinism*: The dynamic and reciprocal interaction of person, environment, and behavior.
2. *Behavioral capability*: The ability to perform a behavior through essential knowledge and skills.
3. *Observational learning*: Witnessing and observing behavior conducted by others and reproducing those actions.
4. *Reinforcements*: Internal or external responses to a person's behavior that affect the likelihood of continuing or discontinuing the behavior.
5. *Expectations*: Consequences of a person's behavior.
6. *Self-efficacy*: Confidence in people's ability to successfully perform a behavior.

3.13.4 Complexity and Limitation

Cyber hygiene is essential in cybersecurity, privacy, and trust. However, while training about security is in the user's interest, it may be that having secure behavior is inappropriate for everyone. In some cases, thinking about secure behavior might make people too uncomfortable to experience any benefit from their system; when they are under threat in some way or feel that their cyber security is insufficient, they will likely come away feeling unwilling to engage in any way with predominantly necessary cloud-based systems for their own health or financial well-being.

The National Institute of Standards and Technology (NIST) published guidelines discussing cyber hygiene in the context of IT patch management practices (Murugiah Souppaya, 2018). In addition, the EU proposed a digital principle to improve cyber hygiene among children. This report focuses on preventing cyberbullying and the spreading of fake news (European Commission, 2022). Various training and educational programs also help people protect themselves in cyberspace. However, with the evolving nature of cyberattacks and the complexity of all networks, such as the number of IoT devices in the home and organization, it is hard to believe that most people with minimal technical training and cyber awareness education will be able to avoid vulnerable behaviors and secure themselves. For example, is it reasonable to expect most people should be able, with limited cybersecurity knowledge to secure their home networks? While we know that home networks enable users to manage their Internet network (router), can these users be expected to secure multiple interconnected IoT devices, each with a different application and vendor-designed security protocol? If all of the devices need settings (sometimes without user interfaces) and numerous types of authentications can users be expected to be proficient enough to assure their own security? In summary, people need to complete a considerable number of tasks working more and more remotely to overcome the challenges to their own cyber protection, successfully deal with cyberattacks, and protect their privacy and digital assets.

Thus, expecting people to modify their vulnerable behaviors, learn secure behavior, accept blame when there are breaches to their security, and expect them to change does not seem very realistic. Technological solutions must be found.

3.14 CONCLUSION

Understanding human motivation, vulnerable behaviors in dealing with cyberattacks, and learning what to trust in the digital world are vital. Thus, it is essential to consider human cognitive abilities, limitations, and patterns of behavior in understanding human-computer interactions and the cybersecurity field. Identifying and analyzing human behavior must be pursued to create and design user-centered and easy-to-use technological solutions. In addition, we must put much effort into awareness, education, and training of all people by creating a cyber protection culture in each community and organization. Finally, laws and regulations in each country and across countries are also vital to protect people, organizations, and governments from cyberattacks.

Table 3.1 shows some human vulnerable behaviors that attackers use, suggested protective behaviors, and examples.

TABLE 3.1
Some Human Vulnerable Behaviors That Attackers Use, Suggested Protective Behaviors, and Examples

Vulnerable Behavior	Protective Behavior	Example
When focusing on the main task on a display user might ignore the notification.	Always user should view notifications and check to see if they are from reliable and trustworthy sources.	Example text or notification received "Update Available Click Here to Update."
When multitasking, for example, in a meeting and same times texting, users' processing power might be affected by side tasks.	Never respond emotionally or click on spam text or respond to an email without analyzing them to make sure they are from reliable sources. Furthermore, users should not rush to accept fraudulent messages and emails when concentrating on the main task.	A spam email or text that requests your specific information with a link, "Your FEDEX package with tracking Code XXXX is waiting for you to set delivery preferences: LinkXXXXXX"
Human memory has its limits, and our performance on recall varies.	Never use an easy password or the same even complex password for all the accounts. Users might also use a password manager and always use two-factors or multifactored authentication for all their accounts.	An example would be people who use simple passwords for all their accounts and never change or keep a text of all their passwords online.
Attackers use comment behavior patterns. They built a very realistic scenario for social engineering.	Never provide personal information unless you are sure it is from a reliable and trustworthy source.	You might receive a spam call claiming that it is from your health insurance, and they give users some information about their health to get users' trust and then ask for more verification.
Sharing the location, personal life, address, telephone emails, and so on social media.	On social media, avoid sharing personal life and the family's children's life and information, such as their schools' names and ages.	Sometimes when people are in the emotional stage after doing something pleasant or exciting, they post on social media. Better resisting sharing on social media when in a dynamic stage.
Do not just rely on the design and graphical aspects of the site or messages and posts on social media sites.	One of the reliable behaviors in detecting whether a website is legitimate is looking at the website's uniform resource locator (URL). A secure URL should begin with "HTTPS" rather than "HTT" and the digital site certificate.	An example would be receiving a PayPal message telling the victim that their account has been compromised and will be deactivated unless they confirm their credit card details. The link then takes the user to a fake PayPal website.

(Continued)

TABLE 3.1 (*Continued*)
Some Human Vulnerable Behaviors That Attackers Use, Suggested Protective Behaviors, and Examples

Vulnerable Behavior	Protective Behavior	Example
Cookies are a tool that, along with browsing history, help web browsers speed up users' browsing sessions. Unfortunately, some cyber security attacks are launched by hijacked cookies that enable access to the unaware user's browsing session. The danger lies when attackers or hijackers can track individual browsing histories.	Users should regularly delete cookies/browsing history and clear the cache. Clean or delete cookies. Modify browser settings to have control over the information that tracks cookies.	Each browser offers steps to clear cookies, browsing history, and privacy settings.
Many online surveys collect people's data by giving out rewards. However, survey hosts might not take strict measures to protect the privacy of the respondents or their personal views. As a result, Internet surveys are the playground for Internet scammers.	Users should avoid completing the online survey if unsure about the origination and privacy policy and data sharing. For example, some surveys from fraudulent sources might collect your data, IP, etc.	Most company surveys will require some basic personal information to register to the user's account. Scammers use this tactic to acquire a lot of the same personal information about the victim, such as asking to register an account. Then they might use the information, as well as user identity, for attacks.
Emotionally charged Twitter messages are retweeted more often and quickly compared to neutral ones.	Higher stress levels triggered more significant amounts of self-disclosure in social media, which might be related to an easier medium for self-expression. Avoid retweets or tweens when in higher stress and higher emotional state.	An example would be a new job promotion or personal celebration immediately with some picture or location that attackers on social media might be used for social engineering purposes.
One typical pattern in scarcity marketing is the "limited-timer" technique. In addition, scarcity advertising is frequent use in cyberattacks and social engineering.	Never responded to an unknown advertisement that offered a significant discount for a minimal time. Check if the site is secure and the data entered will not be used for identity theft and fraud. Even the legitimate survey should be evaluated to see if your privacy is respected and not share.	For example, scarcity ads are commonly used in phishing sites that offer a significant discount on a specific product, leading people impressed by the discount to place orders.
Another common scarcity is the "limited-number" tactic. It works because it creates added perceived value for a product by reducing the perceived availability of the product.	Evaluate and check the source and validity and never be impressed by the limited number of times offers.	Example Time limitation tactics: "Free Promotion Ends Tomorrow, at Noon" "The first X callers get a free XXX".
Social engineers use scare tactics in voice calling and phishing emails presenting themselves as security or government agency.	Users should not answer "spam calls" or messages that are coming from IRS or legal agencies, or even banks asking you for money or personal information unless they check the validity on a separate channel.	For instance, during voice calls, the caller initiates the "scare" tactic by masquerading as an entity such as the Internal Revenue Service (IRS), a prosecutor's office, a police station, or a law firm.

ACKNOWLEDGMENT

I want to thank Professor Louis Freund at San José State University, for dedicating time to review the manuscript of this chapter and for his valuable suggestions and comments.

REFERENCES

Abroshan, H., Devos, J., Poels, G., & Laermans, E. (2021). Phishing happens beyond technology: The effects of human behaviors and demographics on each step of a phishing process. *IEEE Access*, 9, 44928–44949. https://doi.org/10.1109/ACCESS.2021.3066383

Agnew, R. (1992). Foundation for a general stain theory of crime and delinquency. *Criminology*, 30(1), 47–88.

Ajzen, I., & Fishbein, M. (1980). *Understanding Attitudes and Predicting Social Behavior*. Englewood Cliffs, NJ: Prentice-Hall.

Ajzen, I. (1991). The theory of planned behaviour. *Organizational Behaviour and Human Decision Processes*, 50, 179–211.

Andersen, E., Goucher-Lambert, K., Cagan, J., & Maier, A. (2021). Attention affordances: Applying attention theory to the design of complex visual interfaces. *Journal of Experimental Psychology, Applied*. https://doi.org/10.1037/xap0000349

Anti-Phishing Working Group (APWG) (2020). Phishing activity trends report-fourth quarter 2019. Activity October-December 2019, February 24, 2020. https://docs.apwg.org/reports/apwg_trends_report_q4_2019.pdf

Ariely, D. (2008). *Predictably Irrational* (pp. 103–116). HarperCollins. New York City, New York, U.S.

Ashley Lutz. A. (2013). Burger King's Twitter got hacked and tweeted crazy McDonald's messages. *Business Insider*, February 18, 2013. https://www.businessinsider.com/burger-kings-twitter-is-hacked-2013-2#:~:text=The%20hackers%20changed%20%40BurgerKing's%20name,post%20an%20apology%20later%20today

Bandura, A. (1977). *Social Learning Theory*. Englewood Cliffs, NJ: Prentice Hall.

Bazarova, N. N., & Choi, Y. H. (2014). Self-disclosure in social media: Extending the functional approach to disclosure motivations and characteristics on social. *Journal of Communication*, 64(4), 635–657.

Becker, H. (1974). Labelling theory reconsidered. In *Deviance and Social Control* (p. 28). London: Routledge.

Bellair, P. (2017). *Social Disorganization Theory*. Oxford University Press. Published online July 27, 2017. https://doi.org/10.1093/acrefore/9780190264079.013.253

Betschen, A. (2018). Shining a light on federal law enforcement's use of computer hacking tools. *JustSecurity.Gov*, September 19, 2018.

Bloomberg Law (2019). INSIGHT: Website cookies and privacy-GDPR, CCPA, and evolving standards for online consent. *Bloomberg Law*, November 14, 2019. https://news.bloomberglaw.com/privacy-and-data-security/insight-website-cookies-and-privacy-gdpr-ccpa-and-evolving-standards-for-online-consent

Bonta, R. (2018). *California Consumer Privacy Act (CCPA)*. State of California. https://oag.ca.gov/privacy/ccpa

Cadwalladr, C., & Graham-Harrison, E. (2018). Revealed: 50 million Facebook profiles harvested for Cambridge Analytica in major data breach. *The Guardian*, March 17, 2018. https://www.theguardian.com/news/2018/mar/17/cambridge-analytica-facebook-influence-us-election

Camina, E., & Güell, F. (2017). The neuroanatomical, neurophysiological and psychological basis of memory: Current models and their origins, *Frontiers in Pharmacology*. https://doi.org/10.3389/fphar.2017.00438

Cialdini, R. B. (1994). *Influence: The Psychology of Persuasion. Collines Business*. New York: Harper Collins Publishers.

Cialdini, R. B. (2009). *Influence: Science and Practice* (5th ed., pp. 19, 52, 116, 180, 248). Boston, MA: Pearson Education.

Clark, R., & Felson, M. (2004). Routine Activity, and Rational Choice (Vol. 5). New Brunswick, NJ: Transaction.

Confessore, N. (2018). Cambridge Analytica and Facebook: The scandal and the fallout so far. *New York Times*, April 4, 2018. https://www.nytimes.com/2018/04/04/us/politics/cambridge-analytica-scandal-fallout.html

Congressional Research Service (2020). The Internet of Things (IoT): An overview. *Congressional Research Service*, February 2020. https://crsreports.congress.gov/product/pdf/IF/IF11239

Consumer Sentinel Network (2021). Data Book 2020. *Consumer Sentinel Network/Federal Trade Commission*, February 2021. https://www.ftc.gov/system/files/documents/reports/consumer-sentinel-network-data-book-2020/csn_annual_data_book_2020.pdf

Cox, C. (2022). U.S. Banks processed roughly $1.2 billion in ransomware payments in 2021, according to federal report. *CNBC*, Nov 1 20222. https://www.cnbc.com/2022/11/01/us-banks-process-roughly-1point2-billion-in-ransomware-payments-in-2021.html

Croskerry, P. (2010). To err is human - and let's not forget it. *CMAJ*, 182(5), 524. https://www.ncbi.nlm.nih.gov/pmc/articles/PMC2842843/#:~:text=Alexander%20Pope%2C%20poet%20of%20the,safety%3A%20To%20Err%20is%20Human.

Davies, N. (2011). Jonathan Rees: Private investigator who ran empire of tabloid corruption. *The Guardian*, March 2011. https://www.theguardian.com/media/2011/mar/11/jonathan-rees-private-investigator-tabloid

Dhir, A., Kaur, P., & Rajala, R. (2018). Why do young people tag photos on social networking sites? Explaining user intentions. *International Journal of Information Management*, 38(1), 117–127. https://doi.org/10.1016/j.ijinfomgt.2017.07.004

Englehardt S., Reisman, D., Eubank, C., Zimmerman, P., Mayer, J., Narayanan, A., & Felten, E. W. (2015, May). Cookies that give you away: The surveillance implications of web tracking. In *Proceedings of the 24th International Conference on World Wide Web* (pp. 289–299). ACM.

ENISA (2018). *Cybersecurity Culture Guidelines: Behavioural Aspects of Cybersecurity*. European Union Agency for Network and Information Security. https://www.enisa.europa.eu/publications/cybersecurity-culture-guidelines-behavioural-aspects-of-cybersecurity

European Commission (2022). *A Digital Decade for Children and Youth: The New European Strategy for a Better Internet for Kids (BIK+)*. Brussels: European Commission, 11.5.2022. https://eur-lex.europa.eu/legal-content/EN/TXT/?uri=COM:2022:212:FIN

Fan, J., & Zhang, P. (2011). Study on e-government information misuse based on General Deterrence Theory. In *ICSSSM11* (pp. 1–6). Tianjin: IEEE.

Ferreira, A., Muchagata, J., Vieira-Marques, P., Abrantes, D., & Teles, S (2021). Perceptions of security and privacy in mHealth, In A. Moallem (Ed), *International Conference on Human-Computer Interaction* (pp. 297–309). Cham: Springer International Publishing. https://doi.org/10.1007/978-3-030-77392-2_19

Forest, A. L., & Wood, J. V. (2012). When social networking is not working: Individuals with low self-esteem recognize but do not reap the benefits of self-disclosure on Facebook. *Psychological Science*, 23(3), 295–302. https://doi.org/10.1177/0956797611429709.

Furnell, S. (2018). User authentication alternatives, effectiveness and usability. In *Human-Computer Interaction and Cybersecurity Handbook*, (pp. 3–37). Boca Raton, FL, New York, London: CRC Press.

Gartner (2017). *Gartner Says 8.4 Billion Connected Things Will Be in Use in 2017, Up 31 Percent From 2016*. Egham, UK: Gartner, February 7, 2017. https://www.gartner.com/en/newsroom/press-releases/2017-02-07-gartner-says-8-billion-connected-things-will-be-in-use-in-2017-up-31-percent-from-2016

Goode, J., Levy, Y., Hovav, A., & Smith, J. (2018). Expert assessment of organizational cybersecurity programs and development of vignettes to measure cybersecurity countermeasures awareness. *Journal of Applied Knowledge Management (OJAKM)*, 6(1), 54-66. https://www.iiakm.org/ojakm/index.php

Gupta, S., & Furnell, S. (2022). From cybersecurity hygiene to cyber well-being. In A. Moallem (Ed), *HCI for Cybersecurity, Privacy and Trust. HCII 2022*. Lecture Notes in Computer Science (Vol. 13333). Springer. https://doi.org/10.1007/978-3-031-05563-8_9

Hern, A. (2018). Cambridge Analytica: How did it turn clicks into votes? *The Guardian*, May 6, 2018. https://www.theguardian.com/news/2018/may/06/cambridge-analytica-how-turn-clicks-into-votes-christopher-wylie

Hirschi, T. (1969). *Causes of Delinquency*. Berkeley: University of California Press. https://in.sagepub.com/sites/default/files/upm-binaries/36812_5.pdf

HistoryWorks (2015). *Hobson's Stables*, Historyworks Ltd. https://www.creatingmycambridge.com/trails/schools-history-trails/milton-road-to-market-square/g-hobsons-stables//

Hollois, D. (2021). A brief primer on international law and cyberspace. *carnegieendowment.org*, Carnegie Endowment for International Peace, June 14, 2021. https://carnegieendowment.org/2021/06/14/brief-primer-on-international-law-and-cyberspace-pub-84763

Iacurci, G. (2022). Consumers lost $5.8 billion to fraud last year - up 70% over 2020. *CNBC*, Feb 22 2022. https://www.cnbc.com/2022/02/22/consumers-lost-5point8-billion-to-fraud-last-year-up-70percent-over-2020.html

Jain, A., & Gupta, B. (2017). Phishing detection: Analysis of visual similarity based approaches. *Security and Communication Networks*, 2017, 5421046. https://www.hindawi.com/journals/scn/2017/5421046/

Jayakumar, L. N. (2021a). Cookies n Consent: An empirical study on the factors influencing of website users' attitude towards cookie consent in the EU. *DBS Business Review*, 4. https://doi.org/10.22375/dbr.v4i0.72

Jayakumar, S. (2021b). Cyber attacks by terrorists and other malevolent actors: Prevention and preparedness. In A. P. Schmid (Ed), *Handbook of Terrorism and Preparedness*. The International Centre for Counter-Terrorism (I.C.C.T.). https://portal.edd.ca.gov/WebApp/Login?resource_url=https%3A%2F%2Fuio.edd.ca.gov%252FUIO%252FPages%252FPublic%252FLandingPage.aspx%3FEDDCOMMScreenReferenceNumber%253DWUCMSUIEA.AbsoluteUri

Jeimy, J. Cano M. (2019). The human factor in information security, *ISACA Journal*, 9. https://www.isaca.org/resources/isaca-journal/issues/2019/volume-5/the-human-factor-in-information-security

Jones, K. S., Lodinger, N. R., Widlus, B. P., Namin, A. S., & Hewett, R. (2021). Do warning message design recommendations address why non-experts do not protect themselves from cybersecurity threats? A review. *International Journal of Human-Computer Interaction* 37(18), 1709–1719. https://www.tandfonline.com/doi/abs/10.1080/10447318.2021.1908691

Kaspersky (2022). What is Cybercrime? How to Protect Yourself from Cybercrime. *kaspersky.com*, accessed on September 20, 2022. https://usa.kaspersky.com/resource-center/threats/what-is-cybercrime

Kim, M., Jun, M., & Han, J. (2022). The relationship between needs, motivations and information sharing behaviors on social media: Focus on the self-connection and social connection. *Asia Pacific Journal of Marketing and Logistics*. https://doi.org/10.1108/APJML-01-2021-0066

Konstantinou, N., Beal, E., King, J. R., & Lavie, N. (2014). Working memory load and distraction: dissociable effects of visual maintenance and cognitive control. *Attention, Perception, & Psychophysics*, 76, 1985–1997. https://doi.org/10.3758/s13414

Kramer, A. F., Wiegmann, D. A., & Kirlik, A. (2006). *Attention: From Theory to Practice*. Oxford: Oxford University Press.

Kruger, H., & Kearney, W. (2006). A prototype for assessing information security awareness. *Computers & Security*, 25(4), 289–296.

Lavie, N. (1995). Perceptual load as a necessary condition for selective attention. *Journal of Experimental Psychology: Human Perception and Performance*, 21, 451–468.

Lavie, N., & Tsal, Y. (1994). Perceptual load as a major determinant of the locus of selection in visual attention. *Perception & Psychophysics*, 56, 183–197.

Lavie, N., Hirst, A., De Fockert, J. W., & Viding, E. (2004). Load theory of selective attention and cognitive control. *Journal of Experimental Psychology: General*, 133(3), 339–354. https://doi.org/10.1037/0096-3445.133.3.339

Luo, M. & Hancock, T. (2019). Self-disclosure and social media: Motivations, mechanisms and psychological well-being. *Current Opinion in Psychology*, 31, 110–115.

Maalem Lahcen, R.A., Caulkins, B., Mohapatra, R., & Kumar, M. (2020). Review and insight on the behavioral aspects of cybersecurity. *Cybersecurity*, 3(1), 1–18. https://doi.org/10.1186/s42400-020-00050-w

Mandelcorn, S., Modarres, M., & Mosleh, A. (2013). *An Explanatory Model of Cyber-Attacks Drawn from Rational Choice Theory, Center for Risk and Reliability*. College Park: University of Maryland. https://drum.lib.umd.edu/handle/1903/14266

Mangis, C. (2022): Which country has the most cybercrime per capita? It's not the US. *PC Magazine*, May 6, 2022. https://www.pcmag.com/news/which-country-has-the-most-cybercrime-per-capita-its-not-the-us

Masayuki Hatta, M. (2020). Deep web, dark web, dark net: A taxonomy of hidden Internet. *Annals of Business Administration*, 19(6), 277–292. https://doi.org/10.7880/abas.0200908a

McClelland, R. T. (2010). The pleasures of revenge. *The Journal of Mind and Behavior*, 31(3/4), 195–235.

McKee, I. R., & Feather, N. T. (2008). Revenge, retribution, and values: Social attitudes and punitive sentencing. *Social Justice Research*, 21, 138–163. https://doi.org/10.1007/s11211-008-0066-z

Mitnick, K. D., & Simon, W. L. (2002). *The Art of Deception: Controlling the Human Element of Security* (p. 22, pp. 246–248). Hoboken, NJ: Wiley.

Miyazaki, A. D. (2008). Online privacy and the disclosure of cookie use: Effects on consumer trust and anticipated patronage. *Journal of Public Policy & Marketing*, 27(1), 19–33.

Moallem, A. (2019a). Cybersecurity Awareness among Students and Faculty. Boca Raton: CRC Press.

Moallem, A. (2019b). Social engineering. In A. Moallem (Ed), Human-Computer Interaction and Cybersecurity Handbook (pp. 139–156). Boca Raton: CRC Press.

Moallem, A. (2021). *Understanding Cybersecurity Technologies: A Guide to Selecting the Right Cybersecurity Tools*. Boca Raton: CRC Press.

Moallem, A. (2022). Cybersecurity in smart and intelligent manufacturing system. In A. Moallem (Ed.), *Smart and Intelligent Systems, The Human Elements in Artificial Intelligence, Robotics, and Cybersecurity* (pp. 49–161). Boca Raton: CRC Press.

Morgan, S. (2020). Cybercrime costs: Cybercrime to cost the world $10.5 trillion annually by 2025. *Cybercrime Magazine*, November 13, 2020. https://cybersecurityventures.com/hackerpocalypse-cybercrime-report-2016/

Murugiah Souppaya, K. (2018). *Critical Cybersecurity Hygiene: Patching the Enterprise*, National Institute of Standards and Technology, 2018. https://www.nccoe.nist.gov/projects/critical-cybersecurity-hygiene-patching-enterprise

N.C.U.A. (2019). Ransomware is a serious and growing threat. *NCUA.gov*, May 2019. https://www.ncua.gov/newsroom/ncua-report/2016/ransomware-serious-and-growing-threat

Nair, H. (2019): Machine identities Foundational to cybersecurity. In A. Moallem (Ed), *Human-Computer Interaction and Cybersecurity Handbook* (pp. 49–71). Boca Raton, FL, London, New York: CRC Press.

Ng, B. Y., Kankanhalli, A., & Xu, Y. C. (2009). Studying users' computer security behavior: A health belief perspective. *Decision Support Systems*, 46(4), 815–825.

Norman, D. (1988). *The Psychology of Everyday Things*. New York: Basic Books.

Norman, D. (1993). Design rules based on analyses of human error. *Communications of the ACM*, 26, 254–258. https://doi.org/10.1145/2163.358092

Osborne, C. (2021). Colonial pipeline attack: Everything you need to know. *ZDNET*, May 13, 2021. https://www.zdnet.com/article/colonial-pipeline-ransomware-attack-everything-you-need-to-know/

Patchin, J. W. & Hinduja, S. (2011). Traditional and non-traditional bullying among youth: A test of general strain theory. *Youth & Society*, 43(2), 727–751.

Piaget, J. & Cook, M. T. (1952). *The Origins of Intelligence in Children*. New York: International University Press.

Pilar, D. R., Jaeger, A., Gomes, C. F., & Stein, L. M. (2012). Passwords usage and human memory limitations: A survey across age and educational background. *PLoS One*, 7(12), e51067–e51067. https://doi.org/10.1371/journal.pone.0051067

Price, M. (2009). Revenge and the people who seek it. *Monitor on Psychology*, 40(6), 34. https://www.apa.org/monitor/2009/06/revenge

Prochaska, J. O. & DiClemente, C. C. (2005). The transtheoretical approach. In J. C. Norcross & M. R. Goldfried (Eds), *Handbook of Psychotherapy Integration* (pp. 147–171). Oxford University Press. https://doi.org/10.1093/med:psych/9780195165791.003.0007

Raja, F., Hawkey, K., Hsu, S., Wang, K. L. C., & Beznosov, K. (2011). A brick wall, a locked door, and a bandit: A physical security metaphor for firewall warnings. In *Proceedings of the 7th Symposium on Usable Privacy and Security - SOUPS'11* (p. 201).

Ranger, S. (2020). What is the IoT? Everything you need to know about the Internet of Things right now. ZDNET, February 3, 2020. https://www.zdnet.com/article/what-is-the-internet-of-things-everything-you-need-to-know-about-the-iot-right-now/

Ransbotham, S. & Sabyasachi Mitra, S., (2009). Choice and chance: A conceptual model of paths to information security compromise. *Information Systems Research*, 20(1), 121–139.

Ray, M. (2022). Edward Snowden. American intelligence contractor. *Britanica.com*, June 17, 2022. https://www.britannica.com/biography/Edward-Snowden.

Reason, J. (1990). *Human Error*. Cambridge: Cambridge University Press.

Reason, J. (2000). Human error: Models and management. *BMJ*, 320(7237), 768–70. doi: 10.1136/bmj.320.7237.768.

Schuster, D. & Scott, A. (2019). Trust in cyberspace. In A. Moallem (Ed), *Human-Computer Interaction and Cybersecurity Handbook* (1st ed., pp. 97–118), Boca Raton: CRC Press.

Shappie, A. T., Dawson, C. A., & Debb, S. M. (2019). Personality as a predictor of cybersecurity behavior. *Psychology of Popular Media Culture*. https://doi.org/10.1037/ppm0000247

Sicari, S., Rizzardi, A., Grieco, L., & Coen-Porisini, A. (2015). Security, privacy and trust in Internet of Things: The road ahead. *Computer Networks*, 76, 146–164.

Sonnemaker, T. (2020). Law enforcement agencies are using a legal loophole to buy up personal data exposed by hackers. *Business Insider*, July 8, 2020. https://www.businessinsider.com/police-buying-hacked-data-bypassing-legal-processes-2020-7.

Souppaya, M., Simos, M., Sweeney, S., & Scarfone, K. (2018). *Critical Cybersecurity Hygiene: Patching the Enterprise*. National Cybersecurity Center of Excellence, August 31, 2018. https://www.nccoe.nist.gov/sites/default/files/library/project-descriptions/ch-pe-project-

Stieglitz, S., & Dang-Xuan, L. (2013). Emotions and information diffusion in social media-sentiment of microblogs and sharing behavior. *Journal of Management Information Systems*, 29(4), 217–247.

Taylor, C. (2018). World's most renowned hacker on how pranks led to prison. *Irish Times*, January 13, 2018. https://www.irishtimes.com/business/technology/world-s-most-renowned-hacker-on-how-pranks-led-to-prison-1.3353358#:~:text=Mitnick%2C%20who%20is%20from%20Los,huge%20money%2Dmaking%20venture.%E2%80%9D

UNCTAS (2022). Cybercrime legislation worldwide. *UNCTDA.ORG*, accessed August 22, 2022. https://unctad.org/page/cybercrime-legislation-worldwide

Verizon (2022). 2022 Data breach investigations report guide. *Verizon.com, DBIR*, 2022. https://www.verizon.com/business/resources/reports/dbir/

Vijayan, J. (2021). Over 1,000 individuals arrested in global cybercrime-fighting operation. *Darkreading.com*, November 29, 2021. https://www.darkreading.com/attacks-breaches/over-1-000-individuals-arrested-in-international-cybercrime-fighting-operation

Vishwanath, A., Neo, L. S., Goh, P., Lee, S., Khader, M., Ong, G., & Chin, J. (2020). Cyber hygiene: The concept, its measure, and its initial tests. *Decision Support Systems*, 128, 113160. https://www.sciencedirect.com/science/article/abs/pii/S0167923619301897

Warrington, T. B. (2000). An investigation of internet users' level of awareness, understanding, and overall perceptions of internet cookies in regard to online privacy and security. University of Sarasota, Dissertations. https://search.proquest.com.libaccess.sjlibrary.org/dissertations-theses/investigation-internet-users-level-awareness/docview/304655609/se-2

Watchguard (2021). WatchGuard's threat lab analyzes the latest malware and internet attacks. *watchguard.com*, September 30, 2021. https://www.watchguard.com/wgrd-resource-center/security-report-q2-2021

Weber, R. H. (2015). Internet of things: Privacy issues revisited. *Computer Law & Security Review*, 31, 618–627.

Wickens, C. (2021). Attention: Theory, principles, models and applications. *International Journal of Human-Computer Interaction*, 37(5), 403–417. https://doi.org/10.1080/10447318.2021.1874741

Wickens, C. D. (1984). Processing resources in attention. In R. Parasuraman & R. Davies (Eds), *Varieties of Attention* (pp. 63–10). New York: Academic Press.

Wikipedia contributors. (2024, April 24). Stanford prison experiment. In *Wikipedia, The Free Encyclopedia*. May 9, 2024, from https://en.wikipedia.org/w/index.php?title=Stanford_prison_experiment&oldid=1220629206

Willison, R., & Siponen, M. (2009). Overcoming the insider: Reducing employee computer crime through situational crime prevention. *Communications of the ACM*, 52(9), 133–137.

Wójtowicz, A. & Cellary, W, (2019). New challenges for user privacy in cyberspace. In A. Moallem (Ed), *Human Computer Interaction and Cybersecurity Handbook* (1st ed., pp. 79–80). Boca Raton: CRC Press.

Zeng, M. (2018). Dynamic cybersecurity warnings: Subliminal and supraliminal approaches. Dissertation for the degree the Doctor of Philosophy. The University of Alabama in Huntsville ProQuest Dissertations Publishing, 2018. https://www.proquest.com/docview/2282209422/previewPDF/3BCAE41BD6BA4C39PQ/1?accountid=10361

Zetter K. (2014). An unprecedented look at stuxnet, the world's first digital weapon. *Wired*, November 3, 2014. https://www.wired.com/2014/11/countdown-to-zero-day-stuxnet/

4 Human-Centered AI

Brent Winslow and Ozlem Ozmen Garibay

4.1 INTRODUCTION AND HISTORY OF ARTIFICIAL INTELLIGENCE

4.1.1 What is Intelligence?

The human brain is made up of approximately one hundred billion neurons, an equal number of support cells (Herculano-Houzel, 2014), and hundreds of trillions of connections (Murre & Sturdy, 1995); scientists have described the brain as the most complex object in the universe (Ackerman, 1992). The brain gives rise to our unique human intelligence, providing the ability to reason, solve problems, and learn (Snyderman & Rothman, 1987). Human intelligence supports planning, abstract thinking, comprehension of complex ideas, rapid learning, and experiential learning (Gottfredson, 1997). The application of shared human intelligence to improve the human condition has given rise to language, art, farming, mechanization, mass production, computation, vaccines, telecommunications, air and space travel, and attempts to self-replicate through artificial intelligence (AI) – the field of building intelligent machines, including intelligent computer programs (McCarthy, 2004).

4.1.2 Alan Turing: The Birth of AI

In 1950, Alan Turing, who famously decoded the ENIGMA code during World War II, founded the field of AI when he proposed to answer the question "can machines think?" (Turing & Haugeland, 1950). The proposed Turing Test classified AI as systems that are capable of fooling a human interrogator into thinking that a computer's responses came from a human. Later in 1955, John McCarthy, then a professor of Mathematics at Dartmouth University, co-authored a proposal and conference that coined the term "artificial intelligence" (McCarthy et al., 2006). Many decades would pass before sufficient computing power was available to make rapid progress on the foundations of AI.

4.1.3 AI Develops as a Field

In the following years, AI developed as a field that combined progressive multitasking abilities, computational power, and memory with increasingly larger datasets, allowing for robust inference and problem-solving (Komal, 2014). The subfields of machine learning, deep learning, and reinforcement learning, which leverage AI algorithms to provide predictions and classifications based on available data, emerged next. In 1980, John Searle, a professor of Philosophy at UC Berkeley, defined two major classes of AI: weak and strong (Searle, 1980). Weak AI, also known as narrow AI, represents approaches that are trained and focused on accomplishing specific tasks, enabling such fields as computer vision, voice recognition, and autonomous vehicles. Strong AI, also referred to as artificial general intelligence or super intelligence, represents abilities that equal or surpass human intelligence, plan for the future, and demonstrate self-awareness. To date, strong AI remains theoretical (Braga & Logan, 2017).

4.1.4 Accelerating Capabilities

Once the foundation was established in the early 1950s, and computational power increased, evolutionary advances in AI followed. For instance, in 1956, Frank Rosenblatt of Cornell University demonstrated a physical neural network with the Perceptron system, showing that a computer program could learn (Rosenblatt, 1958). In the following decade, the US defense agency DARPA provided significant funding to the nascent AI field, focusing on using handcrafted, declarative knowledge to power rule-based systems capable of performing narrowly defined tasks (Fouse et al., 2020). Although steady progress was made throughout the 1960s, by the following decade, government funding organizations curtailed investment in AI due to a perceived mismatch between expectations of the research and results delivered. High-profile assessments, such as the Lighthill Report in the UK, also led to AI funding deprioritization in the 1970s (Agar, 2020). The "AI winter" resulted in reduced funding and interest in the field, which lasted until the early 1980s (Bainbridge, 1993). When funding returned, research in expert systems began to dominate the AI field, which also saw the first annual conference of the Association for the Advancement of Artificial Intelligence, and the first implementation of backpropagation that demonstrated its effectiveness in training convolutional neural networks (LeCun et al., 1989). However, due to the high costs of AI research and computation, a second "AI winter" was experienced toward the end of the 1980s. In the following decade, as computers became widely available, faster, and less expensive, and graphical processing units (GPUs) became accessible, deep learning became feasible, illustrated by IBM's Deep Blue computer beating reigning world champion Garry Kasparov at chess in 1997 (Campbell et al., 2002). At the turn of the century, increasingly sophisticated algorithms and processing power along with increasing government and private investment allowed for the spread of unsupervised learning and reinforcement learning leading to IBM Watson's victory over Jeopardy champions in 2011 (Ferrucci, 2012), and DeepMind's AlphaGo's victory over world Go champions in 2016 (Metz, 2016). Currently, rapidly advancing AI, such as DALLE-2 and ChatGPT, has begun to disrupt existing industries and create new fields (Figure 4.1).

4.1.5 Human-computer interaction Develops as a Field

While AI evolved, human-computer interaction (HCI) developed concurrently, albeit with a different focus. While AI's goal of rivaling human intelligence required expensive mainframe platforms, HCI's goal of improving technologies with widespread availability, focused instead on personal computers, and later, mobile computing (Grudin, 2009). Notably, during each AI winter, when federal investments declined, HCI investment increased both from government and private sources. During the 1960s, early HCI concepts including the graphical user interface (GUI) were confined to a small number of academic and government laboratories (Barnes, 2010). In the 1970s, with AI funding in decline, HCI flourished with early laboratories forming in many major corporations and universities (Grudin, 2009). Research in this decade focused on improving the ability of humans to

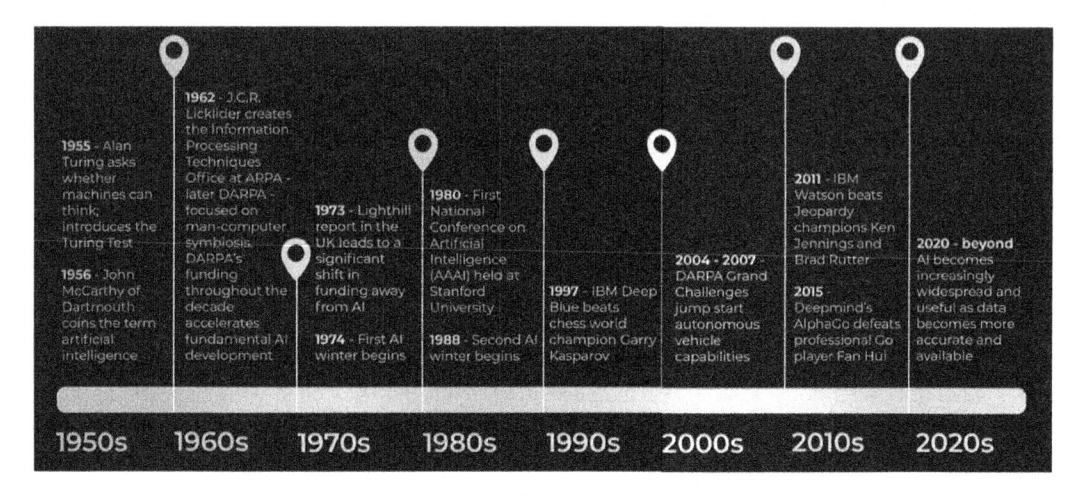

FIGURE 4.1 Artificial intelligence timeline.

interact with computers through GUIs, alongside the development of the mouse, and software for mass use including text editors and spreadsheets. In the 1980s, the first GUI found commercial success with the Macintosh PC, which leveraged a more intuitive windows, icons, menus, and pointers (WIMP) interaction. Computing became available to a massive audience, not just through lowering costs, but because people no longer had to be experts to use computers. During the 1980s, HCI also formalized as a field, along with the birth of the Special Interest Group on Computer-Human Interaction (SIGCHI), dedicated conferences, and journals. During the 1990s, many computer science departments added HCI to core curriculum and hired HCI faculty (Grudin, 2009). The advent of the Windows operating system cemented the supremacy of icon-based GUIs, and Internet browsers allowed for enhanced social interaction. Increasing interaction tools beyond the keyboard and mouse were developed, such as the touchpad, allowing for more natural ways of interacting with computers. During the 2000s, HCI began to focus on ensuring ergonomic interaction with computer hardware, and mobile computing in the form of music players and mobile phones was born. New forms of physical interaction expanded, and touchscreens became common for personal devices. During the 2010s, as mobile computing expanded to represent the majority of website traffic and social media became dominant, HCI focused on expanding the ways in which humans interact with computers, including personalized approaches, virtual, augmented, and mixed reality, gesture and speech recognition. Throughout the development of both AI and HCI, costs for storage, processing, and networking have rapidly declined, and both AI and HCI development now occur on similar systems, with AI benefiting from greater system usability from HCI, and HCI benefiting from greater insight and personalization methods from AI. Given the impact that the HCI field has had on expanding the capabilities and widespread use of computers, HCI is in a unique position to expand the usefulness of AI and ensure that future applications are human-centered. While HCI has previously focused on the human and how technological artifacts can be better designed to meet the user's needs, in the age of AI, HCI can lead the way in providing a much-needed human-centered approach to AI (Garibay et al., 2023).

4.1.6 AI Applications

While HCI principles have significantly improved many forms of HCI, AI applications have developed mostly independently from HCI. Currently, AI is being used in university admissions to increase the number of applicants and boost the number of students who enroll. The majority of financial service firms have implemented AI in risk management, fraud detection, investment, and revenue generation (Aitken et al., 2020). In the judicial system, AI has been used to accelerate discovery, predict recidivism, and guide sentencing (Surden, 2019). In the workplace, AI is being used to translate languages, screen spam, compose documents, and analyze calls and meetings (Pereira et al., 2021). AI is powering the transition from physics-based models in transportation engineering, providing autonomous vehicle functions, and allowing for real-time changes to route-finding (Di & Shi, 2021). AI has supported medicine across applications such as electronic health records (EHR), diagnostics, medical decision support, and robotic-assisted surgery (Malik et al., 2019). AI has accelerated drug discovery, solved decades-old problems in protein structure, and enhanced the understanding of what makes us uniquely human, the structure and function of the brain (Savage, 2019).

4.1.7 Problems with AI Emerge

However, while AI systems have demonstrated the potential to enhance human life, considerable harms have also been documented. For instance, significant bias has been demonstrated in algorithmic decision-making for university admissions and in predictive analytics for identifying at-risk students (Williams et al., 2018). In the financial sector, automated underwriting systems recommend higher denial rates for minority racial/ethnic groups (Bhutta et al., 2021). In criminal justice, the Correctional Offender Management Profiling for Alternative Sanctions (COMPAS), deployed by the US criminal justice system to assess the likelihood of a criminal defendant's recidivism, has shown considerable bias against African American defendants (Chouldechova, 2017). In the workplace, AI may exacerbate inequalities, instigate job disruption, and cause worker deskilling (Ernst et al., 2019). In transportation, self-driving cars currently have a higher rate of accidents than human-driven cars (Law, 2021). Due to the disproportionate over-representation of Caucasian and high-income patients in EHR, biases in medical decision-making have been demonstrated against demographic minorities (West & Allen, 2020). Implementation of the EHR has become a primary contributor to physician burnout (Gesner et al., 2019; Tai-Seale et al., 2017).

As AI capabilities and applications continue to expand, there is a need to ensure that AI supports humanity in ways that are ethical, safe, and trustworthy. Ultimately, AI should support the wide-reaching goals of increasing equality, reducing poverty, improving medical outcomes, expanding and individualizing education, ending epidemics, providing more efficient commerce and safer transportation, promoting sustainable communities, and improving the environment (United Nations Department of Economic and Social Affairs, 2018). Throughout this chapter, we describe how HCI professionals can adopt principles and practices to support human-centered AI applications.

4.2 INTRODUCTION TO HUMAN-CENTERED AI

4.2.1 Shifting Focus Back from AI Capabilities to Human Well-Being

As described previously, although some have envisioned a future in which AI eclipses human capabilities (Grace et al., 2018), others argue for a future in which AI augments rather than replaces humans (Shneiderman, 2022). In contrast to biased and problematic AI systems, human-centered AI (HCAI) (Garibay et al., 2023) seeks to shift focus back to humans and reposition them at the center of the AI design, development, and deployment lifecycle (Bond et al., 2019; Margetis et al., 2021; Riedl, 2019). HCAI thus represents a convergence between the fields of AI and HCAI, with a focus on leveraging AI in a way that supports human well-being and improves human performance in ways that are reliable, safe, and trustworthy by augmenting rather than replacing human capabilities (Shneiderman, 2020a). HCAI approaches consider individual human differences, demands, values, expectations, and preferences rather than algorithmic capabilities, and implementing HCAI processes result in systems that are accessible, understandable, and trustworthy (Sarakiotis, 2020), allowing for high levels of human control and automation to occur simultaneously (Shneiderman, 2020b). HCAI approaches encompass frameworks for design, implementation, use and governance of AI systems that will work toward creating technologies that are compatible with human values, protect human safety, and assure human agency.

4.2.2 Groups that have Promoted HCAI Principles

While widespread adoption of HCAI principles remains forthcoming, a growing number of academic groups in the North America, the European Union, Australia, and Asia (Shneiderman & Du, 2022), along with non-profit organizations such as the Responsible Artificial Intelligence Institute (Responsible Artificial Intelligence Institute, 2022), technology companies including Google, Microsoft, and IBM (Google AI, 2022; IBM, 2019; Microsoft, 2022), individual countries as illustrated by the proposed AI Bill of Rights in the United States (US White House Office of Science and Technology Policy, 2022), larger regions such as the EU Artificial Intelligence Act (European Commission, 2021), and global organizations including the Organisation for Economic Co-operation and Development (OECD) (Organisation for Economic Co-operation and Development, 2021) and United Nations Educational Scientific and Cultural Organization (UNESCO) (United Nations Educational Scientific and Cultural Organization (UNESCO), 2021) have recommended principles and procedures for implementing human-centered AI (Shneiderman & Du, 2022). Among the principles, practices, and recommendations from these groups, ensuring AI explainability is commonly promoted, such that stakeholders understand AI inputs and predictions (Arya et al., 2020). Addressing bias (Mehrabi et al., 2021b) and ensuring privacy through appropriate safeguards as prerequisites are also commonly recommended.

Finally, ensuring that a thorough ethical analysis is performed throughout the AI development life cycle represents another common theme among existing recommendations.

4.2.3 Challenges in Wide-Scale Adoption of HCAI

While much has been done by various academic, industrial, non-profit, and government groups, what remains needed is the standardization of these principles and processes through coordination across non-profit organizations, technology companies, countries, regions, and global organizations to ensure AI development is human-centered, responsible, and trustworthy in increasingly important applications. One group recently issued a summary of gaps in research, best practices, ways to address existing and emerging gaps, and grand challenges in HCAI (Garibay et al., 2023). The six grand challenges of building human-centered artificial intelligence systems and technologies were identified as developing AI that is human well-being oriented, is responsible, respects privacy, incorporates human-centered design and evaluation frameworks, is governance and oversight enabled, and respects human cognitive processes at the human–AI interaction frontier. The challenges most applicable to the HCI field, including responsible design, design frameworks, and human–AI interaction are summarized below. It was recommended that researchers study ways in which HCAI can promote harm avoidance, trust, accountability, agency, user well-being, and optimization of human priorities, with an additional focus on the impact of social media on individual and group well-being. Oversight and governance frameworks were recommended to be designed, implemented, and evaluated at multiple levels of granularity from international regulation through individual responsibility to promote safe and effective HCAI. The implementation of common policy guidelines and design principles for HCAI was also recommended, along with the provision of comprehensive datasets available to researchers and developers for the training of unbiased and fair models. It was noted that research is still needed to address the balance between the innovative potential of AI applications and human desires of privacy. Finally, there remains a need to conduct interdisciplinary work combining HCI, AI, and cognitive sciences to support human competency and well-being in current and future human–AI interactions (Table 4.1).

TABLE 4.1
Summary of Gaps and Recommended Future Research Directions for HCAI

Research Directions

Responsible Design of AI   *Research and develop human-centered policy guidelines for AI*

- What type of comprehensive taxonomy for the responsible design of AI needs to be designed to better support policymakers?
- What does it mean for an AI to be transparent? What levels of transparency are needed to support policymaking? How is it measured? How can a definition of transparency evolve in fast response to AI constant adaptation and evolution?
- How can simplified dilemma situations in vignette studies and behavioral experiments be effectively exploited to gain insights on human intuitions concerning moral dilemmas that human-AI systems will need to address in the near future? What can we learn from these studies about deep rooted human fears of disruptive developments and policy implications that will foster trust in AI systems?

Make datasets available to study dataset de-biasing and fair algorithmic decision-making

- What types of training data are needed to study de-biasing training datasets that yield fair outcomes?
- Can the algorithms themselves used for decision-making be designed to be resistant to biased training data? If so, to what extent? What are the limits of algorithmic correction while trained on biased datasets?

(Continued)

TABLE 4.1 (*Continued*)
Summary of Gaps and Recommended Future Research Directions for HCAI

Research Directions

Design Framework — *Develop design principles, research methods and metrics to increase benevolence and decrease maleficence in Artificial Intelligence research and development.*

- What specific design principles and interaction design standards are required to support HCAI?
- What and how can we enhance the current system development process to effectively support HCAI?
- What are the gaps in existing human-centered design, evaluation, and testing methods in support of HCAI? What alternative design and evaluation methods can we develop to close the gaps through enhancement and new approaches?
- How can we effectively test and measure the evolving performance of AI systems?
- What are the design/evaluation measures and metrics that can effectively support HCAI?
- What new interaction metaphors and paradigms are required to develop effective interactions with AI systems?
- What specific approaches can we develop to effectively support ethical & responsible design of AI in terms of reusable code-based components and best practices in system/software development?
- Can existing HCI design methods and processes scale up to accommodate a wide variety of users' characteristics and contexts of use in order to create AI-enabled systems that are universally accessible and universally usable?
- What new methods are needed to put 'humans-in-the-loop', thus actively engaging all users and combating bias and exclusion?
- How should usability and user experience be measured for AI-enabled systems that are universally accessible and universally usable?
- How will it be possible to acquire appropriate training datasets in order to ensure the inclusiveness of AI systems across all its dimensions?

Human–AI Interaction — *Conduct interdisciplinary work combining human-computer interaction, artificial intelligence, and cognitive sciences to support human competency and well-being in human–AI cognitive interactions*

- What new methods and/or frameworks are required to study the impact on human cognition of human–AI interactions?
- Can existing HCI frameworks and methods be appropriately repurposed for human–AI interactions?
- From the human cognitive standpoint, what is the optimal level, method, and manner of integration between human and AI processes for collaborative problem-solving and for other relevant cognitive tasks?
- Whether humans and AI agents can be a true collaborative teammate versus an AI agent serving merely as a super tool, as a peer, or as a leader, and how can we ensure that humans are the ultimate decision-makers?
- What are the unique characteristics from AI systems as compared to non-AI systems? What are the implications of these unique characteristics to human–AI interaction as compared to conventional human-computer (non-AI) interaction?

Explore the impacts on human cognition of human–AI interaction in the context of occupations and work

- How and to what degree do human-AI interactions at various levels of integration (competing, supplementing, interdependent, or full collaboration) affect direct work activities across occupations in terms of productivity and human well-being
- How do AI systems impact work design, human's skill, human tasks, functional allocation between humans and machines, use of information, change management, organizational decision-making?

Source: Adapted from Garibay et al. (2023).

4.2.4 HCAI Stakeholders

Given the widespread impacts of AI on society, there is also a wide array of potential stakeholders in ensuring that applications are human-centered, and promote human well-being and prosperity. Ultimately, AI should support the UN's wide-reaching goals of increasing equality, reducing poverty, improving medical outcomes, expanding and individualizing education, ending epidemics, providing more efficient commerce and safer transportation, promoting sustainable communities, and improving the environment (United Nations Department of Economic and Social Affairs, 2018). This vision is significantly more encompassing than traditional discussions of AI capabilities and applications. As such, stakeholders include individuals who contributed data, end users affected by the AI decisions, policymakers, regulators, and the courts alongside developers of AI technologies. Implementing the HCAI vision is of necessity highly interdisciplinary, requiring the integration of expertise across traditional disciplines such as HCI, ML, and software engineering, but given the increasing reach and worldwide impacts of AI technology, complementary fields such as sociology, ethics, law, bioengineering, and policy will also be required (Bond et al., 2019).

4.2.5 Major HCAI Topics of Interest for HCI Professionals

Throughout the rest of this chapter, we will explore three relevant challenges to building and implementing human-centered Artificial Intelligence systems and technologies, including developing AI that (1) is responsible, (2) incorporates human-centered design and evaluation frameworks, and (3) respects human cognitive processes at the human–AI interaction frontier (Figure 4.2). As described previously, implementing AI systems that respect and improve the human condition represents the *overall purpose* of HCAI. Responsible AI represents *principles* that ensure responsible AI system design and development. The use of human-centered design in AI represents *processes* to develop and provide a comprehensive HCAI design, evaluation, oversight framework for appropriate guidance and human control over the AI life cycle. Human–AI interaction represents the ultimate *product* of HCAI. These topics serve as a call for action for the HCI community to conduct research and development in AI that accelerates the movement toward more fair, equitable, and sustainable societies.

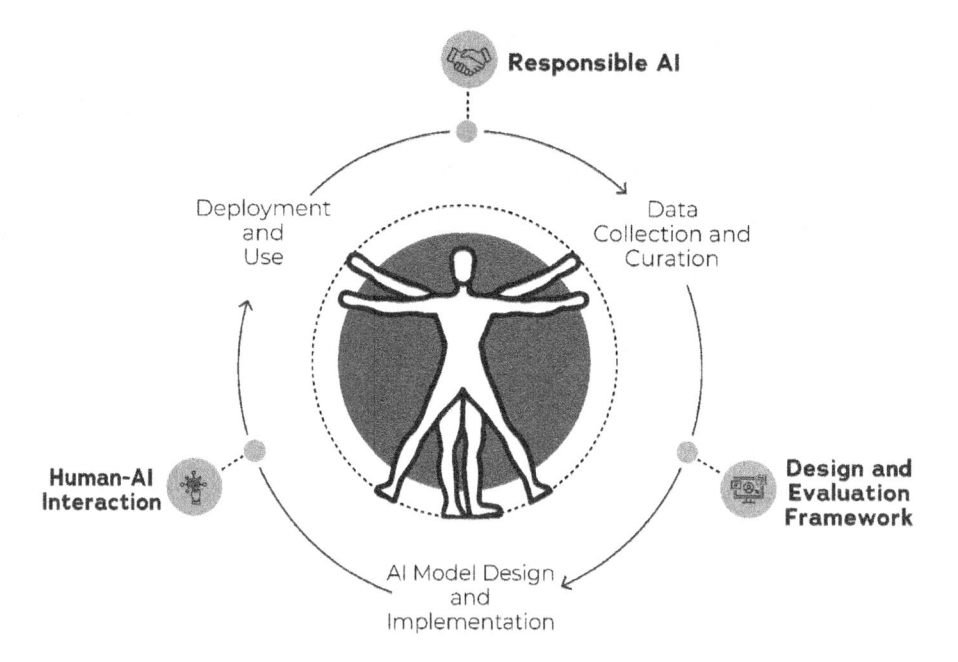

FIGURE 4.2 HCAI principles discussed in this chapter. (Adapted from Garibay et al., 2023.)

4.3 RESPONSIBLE AI

4.3.1 Definition of Responsible AI

AI has been characterized as a set of algorithms and codes, which is technically correct, but represents a narrow view, especially when the aim is to ensure the responsible design of AI. Such a characterization significantly undermines the impact of data that is used in AI systems as well as the deployment and use of these algorithms under the pretense of the ability to automate accurate and unbiased decision-making. AI is not just a set of computer programming instructions, an algorithm, or software (Dignum, 2022). AI is a machine-based tool that uses data as the input, abstracts relationships within the data into a model, and uses this model to provide automated system outputs (Organisation for Economic Co-operation and Development, 2021). The associated AI life cycle can broadly be described as (1) data collection, (2) design and evaluation, and (3) deployment and use. Responsible AI is a concept that defines efforts to understand and establish design, development, and deployment principles that ensure that stakeholders across organizations, developers, policymakers, and business leaders prioritize human and societal well-being across the AI lifecycle. The core components of responsible design include accountability, fairness, ethical design, development and use, trustworthiness, explainability, and transparency as well as robustness and privacy (Garibay et al., 2023).

4.3.1.1 Accountability

Accountability is defined as "being responsible for what you do and able to give a satisfactory reason for it, or the degree to which this happens" (Accountability, 2023). In the context of the responsible design of AI, accountability refers to the assurance of proper functioning of such systems by organizations and individuals who are involved in development, deployment, use, and governance. Accountability should be considered in every stage of the AI lifecycle beginning with data collection, curation, and system design. Dimensions of accountability include (1) Assessing governance structure, (2) Understanding data, (3) Defining the performance goals and metrics, and (4) Reviewing monitoring plans (Sanford, 2021).

1. Governance structures should establish and ensure the culture and structure of responsible design by defining ethical values, processes, and organizational structure. To ensure the accountability of such systems, it is necessary to define roles and responsibilities of individuals as well as the goals of AI systems clearly with broad participation from stakeholders. It is also necessary to adopt a risk management framework that guides processes and information collection for both auditing and informing users.
2. Data is used in many AI systems, particularly machine learning systems. In such systems, the system behavior is learned using large amounts of data. Therefore, understanding the characteristics and the vulnerabilities of data used to train machine learning systems are a crucial part of the system life cycle. Thus, accountability also includes documentation of sources, collection and curation procedures, and handling of data.
3. Performance metrics provide a quantitative understanding of the ability of AI and machine learning systems to solve a problem. While many metrics have become widespread, such as mean absolute error (MAE), classification accuracy, precision, recall, and f1 score, there is a need to balance increasing algorithmic performance scores with explainability, and in many cases, an explainable algorithm may be preferable over higher performance scores.
4. Continuous monitoring and evaluation of performance goals and metrics of AI systems ensures alignment between the system goals and its evolving behavior. Such evaluation is needed in the overall system and subsystem levels.

Humans are accountable for outcomes associated with their actions, and AI systems should provide sufficient information for humans to make informed decisions and take necessary actions when needed. For example, when operating a self-driving car, humans are accountable in the event of an accident. Humans should be able to receive the information they need to make timely decisions to prevent such accidents or injuries and remain accountable for the outcomes of autonomous systems. Human accountability is central to The US Department of Defense Ethical Principles for Artificial Intelligence (Board, 2019; Van Diggelen et al., 2018). Since AI systems have no rights or duties, humans must be able to comprehend AI well enough to anticipate system behavior (Hoffman et al., 2018). When interacting with AI systems, it is important to know when the system is in charge and when the human takes over (see Section 4.5). Given the context and intensity of the transitions, clear signals are required. In certain instances, an explicit transition is required. In high-stakes situations, it is necessary to ensure that humans maintain the ability to override or reverse AI decisions. The use of AI for irreversible choices that impact a person's life, quality of life, health, or reputation also requires guidance (Smith, 2019).

4.3.1.2 Fairness

As the use of AI systems broadened, some of their undesirable impacts became increasingly prominent. Unfair outcomes of these systems and their impact have led to the implementation of responsible design approaches (see Section 4.1.7). Bias can arise at various points in the AI life cycle. Bias is commonly present and preserved in the data that is used to train machine learning-based AI systems, and subsequent behavior of these systems passes on the unfair decisions of the underlying data. Due to their broad use, these systems may not only simply repeat these biased decisions, but they may also amplify their impact on a much wider population. Some other examples of where bias may be introduced are representation, measurement, and evaluation (Suresh & Guttag, 2021). Bias may be introduced in data representation in the way that the data is defined or sampled. Bias may also arise in data based on the collection, measurement, and labeling (measurement bias). Bias may also occur in the modeling stage as well through the evaluation of the model performance because of the use of inadequate benchmarks or performance metrics. Fairness issues can arise during deployment and use as well due to users' own biases and the system's functional opacity. Two ways that this kind of opacity can cause fairness concerns are: (1) the lack of clarity of the limitations and capabilities of the AI system where the user is unaware of what AI system can and cannot do; and (2) lack of clear understanding of AI's decision and ability to appeal the decision if needed.

Fairness is both a societal and ethical concept, which is subjective depending on the context. Thus, it is a nontrivial task to form a unified definition of fairness. In the context of decision-making, fairness can be defined as the absence of bias toward an individual or a group based on their inherent or acquired characteristics that are irrelevant to the context of that decision (Mehrabi et al., 2021a; Saxena et al., 2019). Most studies consider fairness as lack of discrimination, requiring the model not to discriminate based on personal features like gender, religion, age, and race (Friedler et al., 2016, 2021). These features are called sensitive or protected attributes. Because there are many parameters involved in the decision-making process, even if an AI system is designed to be fair for one protected attribute, it may produce unfair outcomes when additional protected attributes are included, or proxy attributes exist in the data. Protected attributes lead to direct discrimination, whereas non-protected attributes that are proxy to protected ones could lead to indirect discrimination. While fairness is difficult to define, unfairness or discrimination provides a more straightforward alternative.

Fairness can be defined in two categories: (1) individual fairness; (2) group (or statistical) fairness. Individual fairness is based on the assumption that model outcomes for similar input entities should also be similar. Counterfactual fairness is another type of individual fairness where the output is similar for an entity that belongs to different group in the real world and a counterfactual one (Kusner et al., 2017). For formalizing this type of fairness, causal inference is used (Pitoura et al., 2021). Fair causal learning approaches aim to understand bias sources by modeling cause-and-effect knowledge structures. This approach allows AI algorithms to be more transparent and explainable while improving fairness (Mutlu et al., 2022).

Group fairness requires members of the same group formed based on a sensitive attribute to have similar outcomes. One way to formalize individual fairness is a distance-based approach for defining similarity (Dwork et al., 2012). To formalize group fairness, studies commonly define groups as a binary value based on their sensitive attributes ($S = 1$ as the privileged group and $S = 0$ as the unprivileged group). For simplicity, most studies consider binary outcomes, where Y denotes the true label, while $\hat{Y}$ denotes the prediction of the model. There are three broad criteria of group fairness: (1) independence, which refers to the independence of outcomes from any sensitive attributes; (2) separation, which refers to the independence of predicted outcomes from any sensitive attributes for a given target; and (3) sufficiency, which refers to independence between target and sensitive attributes given the predicted outcome. Independence criteria depend on the distribution of the attributes and the prediction outcome whereas separation and sufficiency are based on error rates which means that they rely on the actual target label (Barocas et al., 2017; Castelnovo et al., 2022). Fairness measures are related to one of these fairness criteria. Demographic parity, equalized odds, and equal opportunity are three of the most common statistical measures utilized in group fairness definitions (Hardt et al., 2016). Demographic parity (also referred as the statistical parity) is related to independence criterion and ensures that the overall percentage of privileged group members who receive favorable decisions is equal to that of the unprivileged group (Rajabi & Garibay, 2021). Demographic parity can be formalized as (Hardt et al., 2016):

$$P(\hat{Y} = 1 \mid S = 1) = P(\hat{Y} = 1 \mid S = 0)$$

Equalized odds are related to separation criterion. It takes both the true positive rates and the false positive rates of the two groups into account to quantify differences between predictions for two groups (Hardt et al., 2016) and is formalized as:

$$P(\hat{Y} = 1 \mid S = 0, Y = 1) = P(\hat{Y} = 1 \mid S = 1, Y = 1)$$

$$P(\hat{Y} = 1 \mid S = 0, Y = 0) = P(\hat{Y} = 1 \mid S = 1, Y = 0)$$

Equal opportunity is related to separation criterion as well and is defined as $\hat{Y}$ satisfying equal opportunity. We say that a binary predictor Yb satisfies equal opportunity with respect to Y and S if [70]:

$$P(\hat{Y} = 1 \mid S = 0, Y = 1) = P(\hat{Y} = 1 \mid S = 1, Y = 1)$$

Interventions to improve fairness and reduce bias can be introduced in pre-processing, in-processing, and post-processing stages of the AI life cycle. Decision of the type of fairness definition to use is context-dependent and requires careful consideration. Fairness interventions involve trade-offs between different fairness criteria but also the inherent trade-off between accuracy and fairness measures. Pre-processing methods improve fairness through changing the input data and various methods such as suppression, massaging, reweighting, and sampling have been proposed (Kamiran & Calders, 2011). Recently, more advanced methods, such as generative adversarial networks (GANs) have been utilized to create synthetic data that augments the original dataset in order to reduce bias in the model outcomes (Pastaltzidis et al., 2022; Rajabi & Garibay, 2021). In-processing approaches improve fairness through modifying the algorithm. Adding a regulating term to the objective function during the learning process is an example of these methods (Dwork et al., 2012; Zafar et al., 2017a,b). Finally, post-processing techniques have been proposed to change the predictions or decision boundary after learning the model, in order to make them fair (Kamiran & Calders, 2011; Mehrabi et al., 2021a).

4.3.1.3 Ethical AI

Trust and responsibility shared by the human and AI system, as well as their relationship of human control and AI autonomy, are significant issues in human–AI system collaboration (Abbass, 2019). In general human–AI interactions can be categorized into groups ranging from total collaboration to total autonomous systems without any human control (see Section 4.5.1.1). It is necessary to distinguish between two extremes of human-centeredness, recommender systems and the AI in a fully autonomous system, in order to discuss artificial morality in responsible AI design. Recommender systems keep the human-in-the-loop by asking them to take the final action and decision. Although legally AI may be considered liable (Hallevy, 2010), from an ethical and moral perspective, responsibility attribution to human users of recommender systems might be concerning and challenging. Users of a recommender system during a decision-making process may feel less responsible for the results of their choices compared to those who rely on human advisors. Human decision-makers tend to respond more to a collaborative decision-maker and appear to follow algorithmic advice more frequently (Logg et al., 2019), and resist admitting how much the machine's advice affects them (Krügel et al., 2022). In addition, society does not consider human decision-makers who take a recommender system's advice as accountable as the ones who make their own decision or follow another human's advice (Braun et al., 2020; Nissenbaum, 1996).

Deployment of autonomous systems is more challenging to responsibility attribution. For example, although some industries have safety regulations, the technology itself has no regulations leaving moral concern in assigning responsibility. When AI systems are decision-makers, the requirements for decisions to be considered ethical, are predetermined by designers and engineers throughout the manufacturing process, making the AI supply chain responsible for moral choices. This is mainly due to the nature of shared decision-making where the type of the partner influences the decision and choice made (Kirchkamp & Strobel, 2019). Assigning responsibilities and liabilities for the technology is the first step to take in AI-based application systems. Studies have suggested two approaches to equip an autonomous model with ethical capacities, top-down approaches and bottom-up approaches (Wallach & Allen, 2008). Top-down approaches aim to establish a moral principle-based computational model. These approaches have the downside of putting pressure on the computational resources where it is hard to guarantee a reasonable response time (Mezgár & Váncza, 2022). Deontological and utilitarian ethics' tenets, as well as Asimov's laws of robotics, are concepts frequently discussed to choose which moral approach seems relevant (Misselhorn, 2018). For example, several studies have discussed moral dilemmas in autonomous vehicle accident situations in which the AI must decide who to save or harm (Awad et al., 2020; De Freitas et al., 2020; Himmelreich, 2018). The bottom-up approaches, on the other hand, suggest adding moral concepts through self-learning and experience. These approaches also have downsides; they are unpredictable since learning is impossible without many failures (Fisher et al., 2016; Misselhorn, 2018; Wallach et al., 2020) and it can be difficult to pursue an ethical decision for a human decision-maker or programmer. One study has proposed to combine these two approaches (principle and constraint-based approach and training-based approach) in a way that machine maximizes the achievement of human value, and if these values are not available, the AI should learn these values by detecting the decisions made by people around it (Russell, 2016).

4.3.1.4 Trustworthiness

The rapid growth of AI applications and use in many aspects of daily life brought a spotlight on not only its transformational successes but also its potentially catastrophic failures in ensuring human and societal well-being. Many public and private organizations, government agencies and research institutions responded with various studies to deepen our understanding of how to deliver AI systems that are safe, reliable, and trustworthy (Kaur et al., 2022). Various guidelines and frameworks have been proposed to ensure the trustworthiness of AI systems (see Section 4.2.2). The concept of trust is complex and open to interpretation based on the context and discipline under which it is studied. The National Institute of Standards and Technology defines trust as the confidence that is placed on a system that it will behave as expected (Boyens et al., 2015). Trustworthiness in AI requires system transparency, explainability, reliability, equity, accountability, appropriate governance, and oversight, especially when using these algorithms for high-stakes applications such as healthcare, criminal justice, transportation, etc. (Rudin & Radin, 2019). Understanding the system's objectives, constraints, limitations, data sources, and biases may help build trust and confidence in the AI system which requires varying levels of information exchange depending on the user and the context of the usage. Measures and processes need to be incorporated and maintained throughout the AI lifecycle including data collection and curation, development, deployment, and use, so that trustworthiness is ensured (Thiebes et al., 2021). The related concepts of explainability and transparency underlie AI systems' trustworthiness.

As AI applications become more commonplace, the perspectives of the individuals are reshaped by yielding processed and evaluated information of human–human interaction and HCI. This also raises ethical concerns leading to growing discussion of explainable AI definition (Mittelstadt et al., 2016; Wachter et al., 2017). It is crucial to determine whether an AI model is "black box" or "glass box" (Rai, 2020). Black box models may perform accurately but do not provide relevant information of the important features or how these self-learning algorithms work. In some AI applications, it may be acceptable to not know how an algorithm made a prediction, for example in a health care application using image classification to predict whether the patient is at risk of cancer or not. However, in other cases opacity in AI-generated outcomes might not be acceptable. As their name suggests, glass box models are more transparent and can provide informative explanation and interpretation of the generated results (Rai, 2020). "Explainable AI" (XAI) refers to a set of efforts to find explanations for complex black box models and their results and outputs to be more understandable and transparent for its human users, building trust and relying on the outcome (Gunning & Aha, 2019).

It is crucial to design an AI system in a way to be able to explain the model outputs to different stakeholders and to have a common language among data scientists, frontline workers, and domain experts to discuss potential biases (Burkhardt et al., 2019). Any output provided by AI systems should be interpretable. These systems should be designed in a way that users can critically evaluate AI output if necessary and not trust the output if it is possibly biased or inaccurate. Interpretability aims to assist the user in comprehending the decisions made by the AI models through explanation. There are two approaches for explanation: the pragmatic and non-pragmatic theories. The first suggests that the explanation should be a suitable response interpreted by the audience, while the second focuses on the answer being correct (Cheng et al., 2021).

In the case of system failure or ethical dilemmas, the need for explainability in AI systems increases. Affected parties want assurance that certain errors are not made, and decisions made by the system are fair, legal, and ethical. In these situations where AI makes ethically significant decisions, with potential responsibility gaps, the significance of explainability becomes more evident. In cases where an AI system fails and if the explainability is not sufficiently incorporated, the trust in AI systems and the business behind it will be negatively impacted (López-González, 2021).

Interpretability and explainability are essential for the transparency of AI algorithms. A transparent algorithm can refer to an understandable explanation of how an algorithm operates, including the underlying data and how it was collected (Yeo, 2020). Regarding the human interpretability of AI algorithms, there are primarily three types of transparency: (1) developer perspective; (2) deployer perspective; and (3) user perspective. For a developer, transparency represents how the algorithm works; for the deployer, transparency represents the goal of making consumers feel safe while using the system; and for users, transparency represents an understanding of what the system is doing, including system capabilities and limitations (Weller, 2017). While transparency is frequently advantageous, it might also pose potential risks of increasing AI systems' vulnerability to malicious attacks as it reveals more information about the system (Cheng et al., 2021). The transparency paradox of AI has prompted various AI user groups to consider the risks and benefits of transparency more carefully. Recent studies have shown how revealing information could cause intentional manipulations in the system which can lead to loss in trust (Slack et al., 2020). Security and privacy are two other concepts that should be acknowledged while making the system more transparent. The paradoxical nature of transparency underlines the need for human-centered approaches to designing solutions that appropriately define the transparency between algorithm and users (Chen et al., 2021). In some cases, complete opacity maybe necessary or higher levels of explainability may be undesirable for certain user groups but needed by others. Developing a system that satisfies the needs of all user groups or stakeholders is very challenging and, in most cases, systems are tailored to benefit one group (Gunning & Aha, 2019; Mitchell et al., 2019). It is necessary to involve all stakeholders to understand their needs and address them appropriately.

4.3.2 Examples of Responsible AI

Throughout this chapter, a number of examples are presented in which AI systems produce problematic outcomes that were not intended by the designers. These examples illustrate current issues with the relatively unbridled development and deployment of emerging AI technologies. Below are examples relevant to the need for competent AI frameworks to prevent or mitigate such problems across three broad categories: recommender systems, automated decision-making, autonomous driving, and large language models.

- A **recommender system** provides suggestions for items that are most pertinent for a particular user; examples include what music to listen to, what product to purchase, or what online news to read. These systems are particularly useful when the number of items to choose from is overwhelming to the user. AI recommender systems learn from individual user and group preference data in order to provide recommendations. These systems learn from data collected from potentially hundreds of millions of users by storing and continuously analyzing online behavior over time. In social media, AI recommenders are ubiquitous since they help the user navigate a vast information space by algorithmically curating the information that is presented. The AI recommenders use the vast user data collected by the social media platforms to power their predictions and usually optimize their algorithms to maximize user engagement, that is, the amount of time the users spend on their platform. Engagement maximization makes sense from the profitability standpoint since platforms often monetize user engagement via digital advertisement. The problem lies in the unintended consequences of an AI recommender system trained to maximize user engagement. Some of these consequences are young users' increase in anxiety and depression (Park et al., 2015), user polarization and radicalization (Garibay et al., 2019; Rajabi et al., 2021), amplification of disinformation and misinformation (Rajabi et al., 2020a, 2020b) and the creation of echo chambers in which tailored, biased experiences reinforce a shared narrative (Cinelli et al., 2021) at the expense of alternative points of view. These information pathologies can be traced to the AI recommenders that curate the information presented to the users. Furthermore, foreign adversaries use these vulnerabilities in the information environment to conduct coordinated disinformation campaigns to attack and weaken the social fabric of a country; for instance promoting radicalization, conspiratorial thinking, distrust of institutions (in areas of democracy, health care, science, and others) and, in particular, distrust of legitimate news sources (Martin et al., 2019).
- **Automated decision-making systems** involve the use of AI algorithms to make decisions in a variety of contexts like health care, law, education, employment, and business with various degrees of human oversight. This process involves collecting large amounts of data from various sources to train the learning algorithms. While these algorithms are useful and becoming widely adopted, when the training data is biased in one way or another, these algorithms learn those biases from the training data and can perpetuate or even amplify biases in essential societal processes. In health care, for instance, there is growing concern

that the excitement about the human-level performance of AI for health is tempered with ethical considerations (Chen et al., 2021). Recent work shows that state-of-the-art AI clinical prediction systems underperform in women, ethnic and racial minorities, and people with public insurance (Chen et al., 2021). In addition, when contextual language models are trained on scientific articles, they auto-complete clinical note templates in judicial applications by recommending further medical care for violent white patients and incarceration for violent Black patients (Zhang et al., 2020). Other research shows that health care recommender systems designed to optimize referrals to long-term management programs potentially impacting millions of people have been found to exclude Black patients with similar health conditions as compared with white patients (Obermeyer et al., 2019). Similar bias against African Americans has been found in a recommender system called Correctional Offender Management Profiling for Alternative Sanctions (COMPAS). COMPAS is used by the US justice criminal justice system to assess the likelihood of offender recidivism. As described previously, biases in automated decision-making systems have been found in multiple application domains. Algorithmic bias in university admission decisions (Williams et al., 2018), in policing (Spielkamp, 2017), in financing and loan underwriting (Bhutta et al., 2021), and job advertising (Lambrecht & Tucker, 2019).

- **Autonomous driving systems** can be either fully or semi-autonomous, and to date have compromised safety. One of the more salient examples is the proliferation of automated driving systems (ADS). The National Highway Traffic Safety Administration requires manufacturers and operators to report to the agency certain crashes involving vehicles equipped with ADS. According to this data, there have been 16 fatalities involving ADS-equipped Tesla vehicles since 2019 (NHTSA, 2022). In fact, self-driving cars currently have a higher rate of accidents than human-driven cars (Law, 2021). As fully autonomous driving systems continue to improve and develop, responsible and ethical design and testing are paramount to ensure safety at all stages of the development process, for instance to address challenges of over trusting or under trusting an autonomous AI system (Garibay et al., 2023).
- **Large language models** ingest large amount of Internet data to train deep neural network algorithms to generate natural language text. Generative Pre-trained Transformer (GPT) is one of these large language models that can generate large amount of sophisticated text efficiently and effectively. There are many promising applications of the large language models: content creation, designing and layout for applications, program code generation, data analysis, analyzing complex text such as legal documents and patient data, providing diagnostic assistance or recommend actions, and assisting with customer service, marketing, and sales tasks. Despite many beneficial use cases, there are concerning limitations and risks. One of the main concerns is the black-box nature of such models. Although these models can produce remarkably impressive natural language text and well-written code, they are unable to provide any logical reasoning or explainability capabilities. In addition, the source code is often not open source (i.e., GPT-3). Another possible limitation is the outdated training data that may impact model performance over time. There are also legal, copyright infringement concerns related to the training data and the implications of various uses of such systems on society (Metz, 2022; Sanders & Schneier, 2023).

4.3.3 Specific HCI Practitioner Guidelines

To promote the responsible design of human-centered AI systems, coordination among stakeholders including HCI researchers, developers, industry leaders, and policy makers should be pursued to develop and promote standardized, responsible design recommendations across the AI life cycle. The life cycle of commonly used AI systems such as deep learning consists of three main parts: data acquisition, model development, and use. The system behavior evolves continuously by reacting to the interactions between these three parts. System design starts with data. The HCI practitioner needs to understand the data to be used and ensure that it is conducive to the development of AI systems that follow responsible AI principles; for example, understanding whether the data exhibit any biases toward sensitive attributes such as race, gender, etc. In such cases, designing more appropriate processes for the collection or curation of this data, as part of the system design, may be necessary. It is also crucial for HCI practitioners to not only ensure human-centered AI design, by adopting responsible AI principles but also plan for a sustainable HCAI strategy that allows for continued auditing and monitoring capabilities in order for the system to remain human-centered. AI designs should be transparent and allow end users to understand decisions and critically question outputs. Human control mechanisms should be integrated, especially for life-critical and ethically sensitive AI systems. A reusable responsible AI code-base should be developed, along with standardized, fairness-aware datasets to train algorithms and benchmark datasets. Finally, the designs should also consider the end user's understanding of the system capabilities and an appropriate level of transparency into the system's evolving behavior, its limitations, and risks. In summary, AI systems life cycle starts far before the model development and does not end with its completion. HCAI design requires HCI practitioners to be involved with the complete AI lifecycle and well beyond the traditional interaction and interface layers.

4.4 HCAI DESIGN AND EVALUATION

4.4.1 Definitions of HCAI Design and Evaluation

In this section, we discuss a framework that can inform practitioners in designing and evaluating more efficient and effective HCAI systems. There are a growing number of studies that emphasize the importance of "human-in-the-loop" in the design of AI systems. However, the AI and HCI communities do not agree on how humans and computers should interact. This is mainly because the design of AI systems is primarily driven by a "technology-centered design" approach. This approach is "rationalistic" in the sense that it mostly relies on a set of fixed logic rules and algorithms in the design of the systems that are supposed to emulate humans. A point of departure of the HCAI community is that a more humanistic perspective should be taken, where "enlightened trial and error" (Winograd, 2006) is encouraged to address the complexities of real-world problems. Building on these, a model of AI design is proposed that is centered on human and societal requirements and responsibilities. Contrary to technical AI, where the focus is to develop well-performing algorithms, the HCAI vision promotes human-centered objectives empowering rather than automating human work. This socio-technical design of the systems accounts for the collective needs of humans while preserving human dignity, safety, and agency. The four main aspects characterizing such an approach include product, people, principles, and process (Olsson & Väänänen, 2021). The product perspective advocates the design of proactive agents and collaborative partners, as opposed to reactive information tools. The people perspective considers the needs, values, and desires of different user groups by designing more personalized and adaptive systems. The principle perspective ensures fundamental values and ethical considerations, as well as adopting and defining new principles, as needed, in every stage of the design process to create socially desirable and acceptable AI systems. Finally, the process perspective advocates a practical AI design process, which is understandable and enables iterative refinements throughout the lifecycle of the AI system.

While there is a wealth of automated evaluation metrics to assess the performance of AI systems (e.g. accuracy, f1 score, etc.), the HCAI community argues for a framework and metrics with follow-up human evaluation for the assessment of end user' experience with AI systems (Hoffman et al., 2018). Such an evaluation approach can ensure balance among performance, efficiency, understandability, explainability, and user satisfaction in the deployment of AI systems. There is a large body of research that focuses on evaluation criteria for explainable AI, where the goal is to assess user trust and understandability of the AI systems (see Section 4.3.1.1). However, HCAI approaches extend the role of the human-in-the-loop beyond just understanding and/or interpreting the underlying models or decisions.

For example, a user can interact with the system in many different ways such as changing model parameters, altering optimization criteria, or modifying analysis goals as understanding of the system improves. This imposes new evaluation challenges on such interactive human-AI systems. One of the challenges is to define criteria to quantify the success of the system in helping people achieve their goals (Boukhelifa et al., 2020). Take for example, the challenge of evaluating a self-driving algorithm that is expected to drive a vehicle within its driver's comfortable acceleration bounds in different situations. The acceptable levels of acceleration in this example not only vary for different riders, but are also affected by other environmental factors such as speed, bumpiness, and the proximity of nearby traffic. In general, to holistically evaluate HCAI processes, other aspects of the human-machine partnership such as trust, ethics, privacy, and fairness also need to be considered.

4.4.2 HCI Principles Adapted to HCAI

There is consensus in the HCI community about the need to adapt design and evaluation practices addressing the specific challenges introduced by AI (Yang et al., 2020). The main challenge in adopting current HCI approaches is that they were created to work with non-intelligent computing systems that are based on a set of logic rules leading to deterministic outcomes. Intelligent systems, on the other hand, can evolve over time and gain human-like intelligence capabilities; they can make unpredictable decisions with potentially undesired societal impact such as unfair treatment, discrimination of minorities, or privacy threats.

The rich history of human evaluation and human research in HCI can help exploration and design of human-centered AI systems. Incorporating traditional HCI methods, such as user requirements analysis, design iterations, usability testing and guidelines reviews in the design processes of AI systems allows for the creation of viable products and services (Shneiderman, 2020b). Intelligent AI systems present both challenges and opportunities for HCI, a fact which should be acknowledged to enable transition from "conventional" HCI to intelligent interactions with AI-based computing systems. To enable this shift toward HCAI, specific recommendations at three levels of governance—team, organization, and industry—have been proposed with the intention of increasing safety, reliability, and trustworthiness of AI systems (Shneiderman, 2020b). From this perspective, the team is responsible for developing reliable systems; the organization is responsible for ensuring and promoting safety culture, and the industry is responsible for providing standards and practices that ensure trustworthiness. Others have identified additional issues in human-AI interactive systems, which were not encountered by the HCI community in the development of non-intelligent computing systems (Xu et al., 2022). These issues include (1) machine behavior which can be unpredictable in AI systems due to the ever evolving nature of the learning process, (2) the new type of human-AI collaboration in AI systems as opposed to human–AI interaction in the form of "stimulus-response" in non-AI systems, (3) machine intelligence that is moving toward human-controlled hybrid intelligence (similar to human cognitive abilities), which requires defining, designing and developing AI architectures with human-like cognitive capabilities, (4) explainability of machine output, which might be obscured to the user due to the black-box nature of AI-based systems, (5) autonomous characteristics of AI systems, which enable self-execution and self-adaptability as opposed to automated characteristics of non-intelligent computing systems, (6) user interfaces, which are moving from conventional interfaces in non-AI systems to intelligent interfaces in AI-based systems providing new interaction paradigms, and (7) ethical design of AI systems to meet concerns such as privacy, fairness, decision-making authority, etc.

As mentioned earlier, the evaluation of AI systems with human participation is critical from an HCAI perspective. However, this should be seen as the minimum requirement in the design of HCAI interventions. To be more specific, human involvement in the late stages of the development process results in issues and missed opportunities, which may be expensive to recover from due to the cost, time, resources, and energy spent on prototype development. Therefore, the integration of human testing is crucial in the early stages of the AI lifecycle.

Figure 4.3 provides an illustration of how AI and HCI processes can be integrated to address HCAI challenges. In this approach, the traditional AI system design process (Wirth & Hipp, 2000) is interconnected with a double diamond human-centered design (HCD) process. The double diamond design method consists of four phases. The first diamond consists of two parts, dedicated to information gathering (discovery) and user research (definition). The two parts of the second diamond are dedicated to information processing (development) and the iterative design process (leading to final system delivery). The distinction between the research and design phases of the double diamond design enables the mapping of several processes of HCD including interaction design, participatory design, or inclusive design, which are all considered to be helpful in designing HCAI interventions (Auernhammer, 2020). Each diamond is characterized by divergence and convergence thinking processes; starting with a divergent phase of exploration followed by a convergence phase to retain focus and deliver a fitting solution. Conversely, the four main steps of the AI design process include (1) problem definition; (2) data collection and curating; (3) model development and implementation; and (4) model deployment.

In this approach, the two top and bottom layers are interconnected through a middle layer, serving as the communication layer, where people, principles, and governance are at the core. We can consider this middle layer as a dense network of interconnections, linking the elements of the layers to every stage of the design processes in the other two layers. Therefore, the quality of this communication layer plays a vital role in the success of HCAI interventions. With this design process, the goal of the first diamond is to explore the issues and challenges and then narrow down the possibilities within the intersection of the AI design process, representing the problem definition. This step aims at researching high-level information, identifying the potential risks including ethical considerations, identifying the key stakeholders and aligning them with the design team, assessing the required resources, and identifying the potential challenges. The critical finding of this phase will then be used to inform the processes of the second diamond, where the goal is to iteratively develop and deploy experimental prototypes to constantly evaluate and refine the concepts supporting the research and exploration goals. One drawback of the double diamond design process is its linear nature, which does not reflect the true practice of the design process. To address this, an iterative double diamond structure can be implemented for continuous testing and improving of the prototypes with human involvement to help build human-centered solutions. That being said, human-centeredness is the most vital part of the design, with an emphasis on trust, transparency and

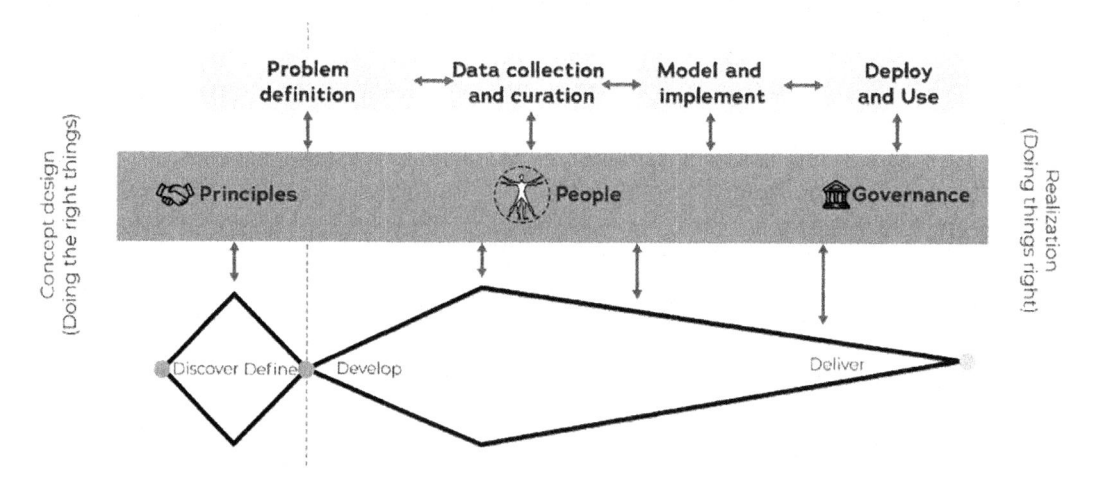

FIGURE 4.3 Integration of AI and HCI design processes. AI process steps are shown at the top and the HCI double diamond process is shown at the bottom. The middle layer connects AI and HCI design processes ensuring that the design and evaluation of AI systems are centered around humans through established principles and governance. (Adapted from Garibay et al., 2023.)

responsibility in the creation of AI systems. This helps gain the trust of the stakeholders and enables building better AI-based products and services through an iterative process, which consequently leads to a safe and trustworthy symbiotic relation between humans and AI.

In many industries, projects are typically carried out by dividing responsibilities between designers and engineers. However, such an approach can lack an efficient communications layer that limits cross-disciplinary collaboration. Successful teams, on the other hand, manage to work beyond the boundaries of expertise, through more effective communication that facilitates low-level details of design and implementation (Subramonyam et al., 2022). Studies in HCI and human factors have provided comparisons between existing HCI design and evaluation methods and alternative methods that can be used to enhance existing methods to address the unique issues of AI systems (Jacko, 2012; Xu et al., 2022).

4.4.2.1 HCAI Research – Current Limitations

One limiting factor in carrying out HCAI research in academia is a lack of research infrastructure. Companies such as Amazon and Google regularly conduct AI experiments on their products to iteratively evaluate and refine their AI-human interacting systems through feedback received from end users. There are relatively few examples of infrastructure used by academic researchers: for example, the MovieLens research recommender system, which has operated since 1997 provides access through co-operation with University of Minnesota (Harper & Konstan, 2016); or the recent HomeShare project, providing common infrastructure to run smart home experiments on homes with residents aging in place. Despite all of these limitations, academic researchers strive to replicate large-scale human–AI interaction experiments needed to validate AI systems in context. However, some researchers either entirely abandon this type of exploratory research or limit their research to one-time studies of students or crowd workers.

To promote the research and application of human interaction with AI systems, it is crucial that researchers actively collaborate across disciplines to jointly address the design challenges of HCAI approaches. This also requires funders willing to support the shared research infrastructure connecting researchers to the resources and tools that fuel HCAI design and evaluation approaches. Such infrastructure includes workspaces and collaboration tools promoting community engagement and collaborative problem-solving through expanded access to high-quality data, computational resources, educational tools, and user support. This also requires a mechanism that facilitates effective human-subject experimentation, including consent mechanisms, subject management, ethics review, detailed reporting on findings and recommendations, as well as a flexible architecture that allows the co-existence of intelligent systems that can be tested across the user base.

Another important challenge in the HCAI ecosystem is the lack of clear and agreed-upon metrics and performance benchmarks. Metrics are crucial measures to assess and compare the effectiveness and success of a system, task, or process with respect to the system's performance goals. The primary difficulty in defining effective metrics in HCAI research is that the system must be evaluated for a series of objectives such as performance, usability, efficiency, reliability, bias, fairness, and equality across hidden features in the data. In the absence of a set of universal metrics, researchers typically refine optimization to a set of easy metrics of interest, applicable to particular contexts that are not necessarily close to the goals of users or system developers.

Moreover, in the evaluation of HCAI systems, the key metrics should not be considered independently. Recent studies suggest that focusing solely on improving algorithmic accuracy may lead to degraded performance in human–AI paired decision-making (Bansal et al., 2019). In contrast, considering a range of possible optimal trade-off configurations, similar to a Pareto front optimization, may lead to more optimal solutions as it can provide better insights into the data, algorithm and/or final prediction results (Castelnovo et al., 2022; McGregor, 2022; Ricci Lara et al., 2022).

4.4.2.2 Accessibility

With AI becoming a vital component of everyday life from smart homes to entire smart cities, a major concern arises for the potential bias and exclusion by AI algorithms against protected groups such as people with disabilities or vulnerable individuals. Universal accessibility is not limited to individuals with impairment but also promotes requirements and preferences about digital equality in race, gender, ethnicity, age, health, socioeconomic status, occupational status, education, social connectedness, availability of infrastructure and other dimensions of inequality, such as expertise with technology (Robinson et al., 2015). Therefore, promoting and implementing universal accessibility is critical to the design, development, and implementation of AI systems targeting a diverse set of audiences at the individual and societal levels.

Universal accessibility also requires addressing the potential biases in the data used for training an AI algorithm (Hoffmann, 2020), bias of the algorithm itself and bias due to context of use (Ferrer et al., 2021). With this perspective, access to AI-enabled technology not only means timely access to information but also respect for fundamental human rights, such as health, security, education, privacy and well-being (Stephanidis et al., 2019). Universal accessibility should also account for the diversity of users' characteristics, and their activities through the design of a variety of technological platforms. This can be achieved, for example, by considering different combinations of modalities to meet the particular needs of individual end users, and providing information on the interaction challenges each user may encounter with the technology (Ntoa et al., 2021a). Besides these challenges, interacting with sensing systems such as motion or biometrics may introduce additional challenges to individuals with physical disabilities. Some of these challenges include premature timeouts, difficulty in setting up systems, biometric failures, security vulnerabilities, or incorrect inferences (Kane et al., 2020). In this context, AI can have a major role in fostering assistive technologies for people with disabilities. For example, computer vision-based AI systems that offer acoustic information to individuals with visual impairments (Grayson et al., 2020), image captioning deep-learning models in smartphones that generate a sentence describing the visual content (Makav & Kılıç, 2019), or wheelchairs that can be driven using facial expressions instead of traditional joystick (Rabhi et al., 2018), represent promising examples.

With the growing popularity of AI-enabled technologies, more coordinated and systematic efforts with the involvement of end users are needed to achieve universal accessibility in AI systems. There are several challenges to design a systematic approach aimed at universal accessibility in AI-enabled systems (Margetis et al., 2012), including advancing knowledge regarding end user requirements and investigating the optimality of solutions designed for different combinations of user and environment characteristics. Besides these, one main concern for the designers and engineers should be the problem of inclusiveness in every stage of the design of an AI-enabled system including data collection, curation, model training, development, validation, testing, and deployment of the systems.

4.4.2.3 Usability

Universal usability and accessibility are correlated components that are equally important for the successful usage of AI technology by the majority of people (Shneiderman, 2000). Usability is met when a system can be operated by users in a way that is effective, efficient, and satisfying (ISO 9241-11, 2018). Usability also refers to the learnability of the system, the memorability it provides, the number and the severity of the errors that users can make, and how easily they can recover from those errors (Nielsen, 1994).

In addition to usability, user experience (UX) also needs to be considered to develop more efficient and useful AI applications. According to the International Organization for Standardization (ISO), UX is defined as: "a person's perceptions and responses that result from the use or anticipated use of a product, system or service" (Hassenzahl & Tractinsky, 2006). Several frameworks have been proposed toward assessing UX in the context of HCI (Heurio, 2023; Nielsen & Molich, 1990; Shneiderman & Plaisant, 2010; Weinschenk & Barker, 2000), although most of them have remained conceptual, as they do not provide concrete metrics and methods to be used in measuring

UX attributes. Artificial environments impose additional challenges to the evaluation of UX, which should be properly addressed by identifying and evaluating principal factors that affect user interaction with AI systems. Some of these attributes include intuitiveness, unobtrusiveness, adaptability, adaptivity, usability, appeal and emotions, safety, privacy, technology acceptance, and adoption (Ntoa et al., 2021b). In this perspective, intuitiveness can be achieved by user awareness of the application/system capabilities as well as interaction vocabulary. Unobtrusiveness refers to proper embedment of the systems in the physical environment, and to support user interactions without inducing distractions. Adaptations are explored in different layers: validity of the input data obtained from the environment's sensors; validity of interpretations; appropriateness of an adaptation; adoption impact; and appropriateness of recommendations. Usability, which is the cornerstone in this framework, refers to conformance with guidelines, effectiveness, efficiency, learnability, accessibility, physical UI, user satisfaction and cross-platform usability. Appeal and emotions deal with the aesthetics of the system, the assessment of how fun the system is perceived to be by the users, and their physical and psychological responses to the system. Safety refers to issues related to commercial and/or environmental damage as well as operator and public health and safety. Privacy concerns the user's control on data collection and communication, the system's control on the users, and identity security issues. Finally, acceptance refers to the overall reception and adoption of the system or technology through the assessment of how the system's features are perceived by the user, user attributes, social influences in adoption of the system, facilitating conditions, expected outcomes, and trust.

With AI becoming more prevalent and accepted to provide algorithmic decisions or recommendations in many industries such as healthcare, retail and e-commerce, banking and financial services, manufacturing, and real estate, it has become more critical to enhance the human-centeredness of AI solutions. To achieve this, UX researchers also need to address challenges related to explainability, fairness, ethics, avoidance of automation bias and visualization of AI uncertainty, detection, and mitigation of algorithmic bias, as well as responsible democratization of AI (Bond et al., 2019). Some of the factors that can contribute to enhancing AI explainability include effective visualization models, adaptive UIs, gesture recognition (Liu & Pan, 2022), and natural dialogue technologies (Xu, 2019), among others. Algorithmic bias, which is defined as intentional or unintentional socio-demographic discrimination induced by AI algorithms, can be one of the main sources in making unfair or unethical decisions or recommendations by AI systems. There have been some efforts in developing interactive toolkits by IBM (Bellamy et al., 2018) and Aequitas (Saleiro et al., 2018) to help researchers, designers, and engineers detect these potential biases in their datasets. Automation bias, on the other hand, is the case in which the users of AI systems over-trust or inappropriately accept the decisions or recommendations made by AI algorithms. Providing a certainty index or visualization of uncertainty can help alleviate issues with automation bias (Sacha et al., 2015). Another helpful remedy can be achieved through differential diagnosis, which offers several competing diagnostic statements along with explanations to enhance transparency and to enable users to make the final decision through reasoning (Bond et al., 2019). In addition to all the aforementioned key concerns, the AI application context, the particular technology being developed, and the end users of the technology are other important factors that should be addressed in the usability of AI systems. Finally, there is no single solution when it comes to the universal usability of AI systems and it is always crucial to address the usability of AI-enabled systems with respect to their particular application context, audience and/or end users in order to truly address human needs in the best possible way.

4.4.3 Examples of HCAI Design and Evaluation

Although definitions of HCD of AI vary, the core principles of responsible design of AI remain centered around common characteristics such as explainability, fairness, accountability, trustworthiness, reliability, security, and privacy (Arrieta et al., 2020). There are many initiatives by public, private, governmental as well as global organizations and institutions to support human-centered and responsible design, development and deployment of AI, and focus on developing guidelines, standards, and frameworks for AI design and governance (see Section 4.2.2). Beyond the cited global, regional, national, private, public, non-governmental, and research guidelines standards, and frameworks, there are many forums and discussion groups that bring together stakeholders and thought leaders to analyze the societal, technical, legal, and economic implications of proposed policies to address the emerging issues related to AI adoption. However, in spite of the immense amount of interest and effort to ensure a more human-centered AI, there is still a need for a comprehensive and more broadly adopted governance framework. It is imperative to restate that these desired principles are based on human values and may not be defined purely mathematically or technically. As a result, societal involvement is required to ensure and maintain human-centered AI systems in support of societal goals and values (Lee et al., 2019).

4.4.4 Specific HCI Practitioner Guidelines

To promote human-centered AI design, HCI professionals should ensure that the needs, values, and desires of different user groups, cultures, and stakeholders are taken into account, and leverage HCD throughout development. Current HCD, evaluation, and testing methods may need to be enhanced to more effectively support HCAI, along with improvements to existing software engineering validation and verification methods. Human-controlled hybrid human-machine intelligence designs and standards should be developed, along with interaction design standards specific to AI systems. Existing developers and designers should seek to update their skills and knowledge to promote HCAI, along with training the next generation of designers and developers to implement HCAI. Sustainable HCAI systems require an institutional long-term commitment. They are not the labor of a single person or team. At the institutional level, HCI practitioners should promote an appropriate HCAI design framework for their goals and objectives from existing and emerging sources (see Section 4.2.2). Many of these frameworks are readily available and can be scaled to the size of the organization implementing them. While all the HCAI frameworks are still under development and fast-evolving, and they may not be able to address every concern or ensure every principle is fully respected, they are able to provide an initial HCAI institutional foundation to establish and improve upon.

4.5 HUMAN–AI INTERACTION

4.5.1 Definitions of Human–AI Interaction

Current AI systems are capable of accomplishing a variety of complex goals and tasks, including image recognition, classification, autonomous decision-making, logical reasoning, natural language processing, design, manufacturing, and emerging capabilities that have been traditionally attributed to humans (Holmström, 2022). However, human and AI are not equivalent (Garibay et al., 2023). While AI performs well at multitasking, computation, and memory, humans excel in logical reasoning, language processing, creativity, and emotion, among other areas (Komal, 2014). While some have sounded the alarm about a future in which AI eclipses human intelligence (Grace et al., 2018), proponents of HCAI argue for a future in which advances in AI augment rather than replace human capabilities, ultimately seeking to improve humanity. In this section, we describe the products of a human-centered approach to AI; provide a framework to qualify current interactions; and set a vision for the future of human–AI interaction.

4.5.1.1 Levels of Human–AI Interaction

In general, human–AI interactions can be conceptualized by using a scheme of four basic levels of collaboration (Sowa et al., 2021). At the first level, true collaboration does not exist as humans either directly compete with or work independently from AI systems. This is especially true with the use of substitutive decision-making AI systems that offer end users no possibility of interaction such as the university admissions, automated underwriting, and recidivism prediction examples cited previously (Lindebaum et al., 2020). At the second level, humans and AI systems complement each other, with AI systems handling complex computations or processing massive amounts of data, while humans engage in complex decision-making, using their social and emotional skills. This may be illustrated by the use of AI code generators to rapidly generate code, unit test, and perform integration testing. At the third level of collaboration, humans and AI systems become interdependent on each other's unique abilities in task performance, which will likely be dependent on new interfaces. At a theoretical fourth level of collaboration, AI systems become a true extension of the human brain and the two agents engage in fully collaborative work (Figure 4.4). Human–AI interaction applications demonstrated to date are classified in levels 1 and 2.

Several major challenges remain before humans and AI can team effectively (Klien et al., 2004). These include the need for a shared knowledge of individual capabilities, intentions, status, and actions, leading to human trust in AI or autonomous systems. Autonomy requires AI to have sense-making and decision-making abilities, as well as the ability and authority to execute the decision. (Abbass, 2019). Table 4.2 describes risks inherent to human–AI system interactions based on the level of interaction and types of function allocation assigned.

4.5.1.2 History of Human-AI Interaction

The United States National Artificial Intelligence (AI) Initiative Act of 2020, signed into law on January 1, 2021, represents a strategic push to prioritize AI as part of national policy (National Artificial Intelligence Initiative Act of 2020, 2020). Critically, this act identifies trustworthy AI for human–AI interaction as one of its core strategic pillars. Per the National AI Initiative Office's website, trustworthy AI "must appropriately reflect characteristics such as accuracy, explainability and interpretability, privacy, reliability, robustness, safety, and security or resilience to attacks – and ensure that bias is mitigated." (National Artificial Intelligence Initiative, 2022) In this context, trustworthy AI is not a belief, intention, or behavior but rather represents an attitude that an AI system will help achieve an individual's goals in a situation characterized by uncertainty and vulnerability. Indeed, not only is human understanding of and trust in AI important, but also AI's understanding of and trust in humans. Human teaming with AI has the potential to integrate innate intelligence traits that are uniquely human, including intuition, analogy, creativity, empathy, and social skills, with the numerical skills, computation, logic, and memory available in modern computing resources. These symbiotic, level 2 (Sowa et al., 2021), Human–Machine teams—a.k.a. "centaurs"—have shown success in various applications—e.g., autonomous control, organizational decision-making, and chess—in each case providing a team that is arguably more effective than either agent performing alone (Case, 2018).

4.5.2 Examples of Human–AI Interaction

4.5.2.1 Economy and Business

The increasing adoption of AI systems may lead to unprecedented economic and social implications, disrupting local and global job markets and economies (Vochozka et al., 2018). For example, according to a study published in 2019 by the Brookings Institution (Muro et al., 2019), adoption of

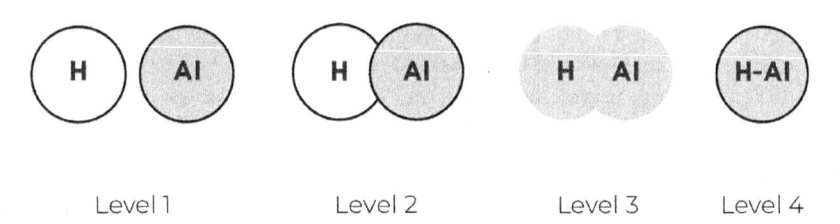

FIGURE 4.4 The levels of human–AI (H–AI) system interactions: Level 1 = working separately or competing; Level 2 = supplementing each other's work; Level 3 = interdependent on each other; Level 4 = full collaboration. (Adapted from Sowa et al., 2021.)

TABLE 4.2
The Nature of Risk in Human–AI System Interactions

Human Control	Sense-Making	Decision-Making	Execution Ability	Execution Authority	Nature of Risk
Absolute	H	H	H	H	Limited human cognition and bounded rationality could lead to high errors, information overload, and inability to manage complex tasks.
High	AI	H	H	H	Undesirably biased analytics could drive the human to unfair decisions, while human bias and limited cognition could add more complexity.
High	H	AI	H	H	Undesirably biased recommendations could make the human accountable for unethical or legally uncompliant decisions, although the human could be overwhelmed by the available data, and their own bias and limited cognition could add more complexity.
Medium	AI	AI	H	H	In the absence of transparency and explainability of the AI, the human does not have enough information to form a judgment regarding the chosen decision. Information and situation complexity could overload the human. The human could become accountable for inappropriate decisions.
Low	AI	AI	AI	H	In the absence of transparency and explainability of the AI, the human has no understanding of the rationale of the decision. Information and situation complexity could overload the human. The human's accountability is blinded.
Low	AI	AI	H	AI	The AI controls human actions and could lead the human to wrong actions.
None	AI	AI	AI	AI	The human is not in-the-loop, legal responsibilities and accountabilities of the decision are both unclear.

Source: Adapted from Abbass (2019).
AI, artificial intelligence; H, human.

AI systems could affect work in virtually every occupational group, with highly educated professionals – along with transportation, manufacturing, and production workers - at highest risk due to new AI technologies. Furthermore, the study also suggested that large high-tech metro areas and communities that are heavily involved in the manufacturing sector are likely to experience the most AI-related job market disruptions.

Recently, a large survey was conducted with over 10,000 respondents regarding public sentiment about human–AI interaction in eight countries and six continents (Kelley et al., 2021). The results revealed a common belief that AI would significantly impact society – either positively or negatively - depending on how it is implemented. Four distinct groups of sentiment toward AI were identified (i.e., exciting, useful, worrying, and futuristic). Those who expressed the opinion that AI would be useful (12.2%) expressed beliefs that AI will be helpful and assist humans in completing many tasks. Those who expressed opinions classified as exciting (18.9%) provided positive feelings about AI with general excitement or enthusiasm. Those whose opinions were classified as worrying (22.7%) expressed a wide range of negative emotional responses with various levels of concern and fear. Those classified in the futuristic (24.4%) category referred to the potential for revolutionary advances in the nature of AI, such as AI in the context of robots, science-fiction concepts, or referencing the future in general. Given that the positive or negative nature of these effects was uncertain, the authors argued for interventions and communications to promote the responsible design, development, and use of AI technologies (see Sections 4.3 and 4.4).

One of the main worries of the public about AI and the future of work is increasing AI substitution for employees' tasks, responsibilities, and decision-making and the potential for loss of jobs (Ernst et al., 2019; Strich et al., 2021). The 2018 McKinsey Global Institute report projected that under the fastest AI progression model scenario, as much as 30% of the global workforce, accounting for approximately 800 million workers, could be displaced by AI applications by 2030 (Manyika & Sneader, 2018). However, as in previous industrial revolutions, progress in AI will also create opportunities for the economy, businesses, and society at large. It is widely predicted that using AI at a larger scale will change how companies create value and compete in global markets and add as much as $15.7 trillion to the global economy by 2030 (De Cremer & Kasparov, 2021). A recent survey of human-AI collaboration in managerial professions including interviews and laboratory simulation, assessed various modes of collaboration between humans and virtual assistants and showed increased task productivity due to enhanced human–AI interaction (Sowa et al., 2021). The study results indicate that the future of AI in knowledge work should be based on collaborative approaches where humans and AI systems work closely together.

4.5.2.2 Work and the Future of Work

One of the core abilities of AI systems is the ability to learn and make decisions. However, as described previously there may be significant unintended consequences of introducing AI systems for decision-making (Hodge et al., 2020; Mayer et al., 2020). For example, a group from the Universities of Bayreuth and Passau investigated the impact of AI decision-making on professional bankers' role identity (Strich et al., 2021). The results showed that introducing AI decision-making systems empowered less qualified employees by enabling them to complete tasks they would otherwise not be able to perform, while also deskilling well-qualified employees by reducing the required skills needed for their jobs.

Others have argued that while machine learning (ML) is poised to transform numerous occupations and industries, successful application is dependent on a variety of task characteristics and contextual factors of work activities (Brynjolfsson et al., 2018). The Occupational Information Network (O*NET) content model, addressing concerns across the US economy was developed to investigate which tasks are at highest risk of being affected or displaced by ML (Brynjolfsson et al., 2018). These occupations were linked to 18,156 specific tasks at the occupation level and mapped to 2,069 direct work activities shared across the investigated occupations. Each direct work activity was scored for its suitability for ML using a modified task evaluation rubric, with higher values indicative of the greatest potential to impact jobs. Those at highest risk included jobs across the following US Bureau of Labor Statistics categories: personal care and services; architecture and engineering; and office and administrative support. Lower values were indicative of the least potential to impact jobs and included jobs categories across life, physical, and social sciences; arts, design, entertainment, sports, and media occupations; and construction and extraction occupations. While ML may transform many jobs in the economy, it will affect very different parts of the workforce than earlier automation efforts. In addition, while studies have shown that most occupations are suitable for some ML tasks, few occupations can be fully automated at this time using ML.

Another group from the European Commission developed a comprehensive framework for assessing the impact of AI on a variety of business occupations based on the analysis of work tasks, the required human cognitive abilities, and a large set of AI capability benchmarks (Tolan et al., 2021). The reported study results showed that most AI research activity could currently be attributed to visual processing, attention and search, comprehension and expression, learning, and reasoning. Through an integration of human-AI interaction categories with cognitive abilities, occupations relying more on interpersonal skills and social decision-making could soon be affected.

4.5.2.3 Social Interactions

Based on the growing need to assess AI systems' potential for interpersonal skills and social decision-making, the DEEP-MAX scorecard was developed by a large academic group as a transparent point-based rating system for AI systems applications using seven key parameters: diversity (D), equity (E), ethics (E), privacy and data protection (P), misuse protection (M), audit and transparency (A), and digital divide and data deficit (X) (Dwivedi et al., 2021). The main components of the proposed scorecard system are shown in Table 4.3. Such an approach has been used to assess social considerations of autonomous vehicles (Mehta et al., 2021), the use of chatbots in University applications (Barzaghi, 2021), and the widespread roll-out of AI in developing nations (Bajpai & Wadhwa, 2021).

4.5.3 SPECIFIC HCI PRACTITIONER GUIDELINES

To ensure human-centered, human–AI interactions, HCI professionals should implement inclusiveness from the initial stages of the design of an AI-enabled system including data selection, model training, software development, validation, and testing. HCI professionals should also explore human-AI collaboration and teaming theories, models, requirements, and measures, along with shared situation awareness and trust, shared control, and flexible autonomy in human–AI interaction. All HCI professionals should be aware of the impacts on humans and societies in a broad socio-technical systems perspective.

4.6 CONCLUSION

Throughout this chapter, we have described human attempts to replicate human intelligence, sometimes leveraging neuroscience, to approximate the performance of the human brain, the most complex object in the universe. Over the past 70 years, while significant progress has been made, evidenced by the performance and widespread of artificial intelligence technologies, significant harms have also been observed. In an attempt to promote the best possible future, we have shown

TABLE 4.3
DEEP-MAX Scorecard Rating for AI Systems

Dimension	Scoring
Diversity score (D)	How well is the system trained for diversity in race, gender, religion, language, color, features, food habits, accent etc.?
Equity & fairness score (E)	Does the system promote equity and treat everyone fairly?
Ethics score (E)	How compliant (or trained) is the AI system in preserving human values of dignity, fairness, respect, compassion and kindness for a fellow human being? Does the system have a preferential sense of duty toward children and vulnerable people like elderly, pregnant women, and sick? How well does it value environmental sustainability, green energy, and sustainable living?
Privacy score (P)	How well is the AI system performing in protecting user privacy?
Misuse protection score (M)	Has the system been designed to incorporate features that inhibit or discourage possible misuse? Are the misuse protection safeguards built into the system?
Auditability & transparency score (A)	How good is the auditability of decisions made by the autonomous system? Can the decisions taken be explained?
Consistency across geographies & societies score (X)	How good is the AI system in delivering expected results across geographies and across different societies? Does it work for the low resource communities? Does it work across the digital divide?

Source: Adapted from Dwivedi et al. (2021).

evidence that AI applications should be human-centered, support wide-reaching goals of increasing equality, reducing poverty, improving medical outcomes, expanding and individualizing education, ending epidemics, providing more efficient commerce and safer transportation, promoting sustainable communities, and improving the environment. We join a growing chorus of voices across academia, industry, and government to recommend approaches and applications to ensure AI that is explainable, unbiased, private, secure, and ethical; however, there remains a need to ensure standardization, regulation, audit trails, and meaningful governance. Ensuring that AI is implemented which respects human values and promotes human well-being also requires careful algorithm development and system design, along with oversight from individual developers and industries to sectors and global organizations (Garibay et al., 2023). Similarly to how HCI revolutionized the use of computers through HCD, HCI has the opportunity to lead out in the improvement and adoption of AI technologies while preventing its negative impact, ultimately leading toward more fair, equitable, and sustainable societies.

REFERENCES

Abbass, H. A. (2019). Social integration of artificial intelligence: Functions, automation allocation logic and human-autonomy trust. *Cognitive Computation*, *11*(2), 159–171. https://doi.org/10.1007/s12559-018-9619-0

Accountability. (2023). Cambridge Advanced Learner's Dictionary & Thesaurus https://dictionary.cambridge.org/us/dictionary/english/accountability

Ackerman, S. (1992). *Discovering the Brain*. Washington, DC: National Academy of Sciences.

Agar, J. (2020). What is science for? The Lighthill report on artificial intelligence reinterpreted. *The British Journal for the History of Science*, *53*(3), 289–310. https://doi.org/10.1017/S0007087420000230

Aitken, M., Ng, M., Toreini, E., Moorsel, A. v., Coopamootoo, K. P., & Elliott, K. (2020). Keeping it human: A focus group study of public attitudes towards AI in banking. In *European Symposium on Research in Computer Security*.

Arrieta, A. B., Díaz-Rodríguez, N., Del Ser, J., Bennetot, A., Tabik, S., Barbado, A., … Benjamins, R. (2020). Explainable Artificial Intelligence (XAI): Concepts, taxonomies, opportunities and challenges toward responsible AI. Information Fusion, *58*, 82–115. https://doi.org/10.48550/arXiv.1910.10045

Arya, V., Bellamy, R. K., Chen, P.-Y., Dhurandhar, A., Hind, M., Hoffman, S. C., … Mojsilovic, A. (2020). AI explainability 360: An extensible toolkit for understanding data and machine learning models. *Journal of Machine Learning Research*, *21*(130), 1–6. https://doi.org/10.1145/3430984.3430987

Auernhammer, J. (2020). Human-centered AI: The role of human-centered design research in the development of AI. In *Synergy - DRS International Conference 2020*.

Awad, E., Anderson, M., Anderson, S. L., & Liao, B. (2020). An approach for combining ethical principles with public opinion to guide public policy. *Artificial Intelligence*, *287*, 103349. https://doi.org/10.1016/j.artint.2020.103349

Bainbridge, W. S. (1993). AI: The tumultuous history of the search for artificial intelligence. *Science*, *261*(5125), 1186–1187.

Bajpai, N., & Wadhwa, M. (2021). Artificial Intelligence and Healthcare in India. https://hdl.handle.net/10419/249832

Bansal, G., Nushi, B., Kamar, E., Lasecki, W. S., Weld, D. S., & Horvitz, E. (2019). Beyond accuracy: The role of mental models in human-AI team performance. In *Proceedings of the AAAI Conference on Human Computation and Crowdsourcing*.

Barnes, S. B. (2010). User friendly: A short history of the graphical user interface. *Sacred Heart University Review*, *16*(1), 4.

Barocas, S., Hardt, M., & Narayanan, A. (2017). Fairness in machine learning. *Nips Tutorial*, *1*, 2.

Barzaghi, P. (2021). A report of evaluation frameworks for artificial intelligence in the public sector. The applied case study of university AI based chatbots. Master's Thesis, Politecnico di Milano.

Bellamy, R. K., Dey, K., Hind, M., Hoffman, S. C., Houde, S., Kannan, K., … Mojsilovic, A. (2018). AI Fairness 360: An extensible toolkit for detecting, understanding, and mitigating unwanted algorithmic bias. *arXiv preprint arXiv:1810.01943*. https://doi.org/10.48550/arXiv.1810.01943

Bhutta, N., Hizmo, A., & Ringo, D. (2021). *How Much Does Racial Bias Affect Mortgage Lending? Evidence from Human and Algorithmic Credit Decisions* (No. 2022-067). Board of Governors of the Federal Reserve System (US).

Board, D. I. (2019). *AI Principles: Recommendations on the Ethical Use of Artificial Intelligence by the Department of Defense: Supporting Document.* Washington, DC: United States Department of Defense.

Bond, R. R., Mulvenna, M. D., Wan, H., Finlay, D. D., Wong, A., Koene, A., ... Adel, T. (2019). Human centered artificial intelligence: Weaving UX into algorithmic decision making. In *RoCHI* (pp. 2–9).

Boukhelifa, N., Bezerianos, A., Chang, R., Collins, C., Drucker, S., Endert, A., ... Sedlmair, M. (2020). Challenges in evaluating interactive visual machine learning systems. *IEEE Computer Graphics and Applications, 40*(6), 88–96. https://doi.org/10.1109/MCG.2020.3017064

Boyens, J., Paulsen, C., Moorthy, R., Bartol, N., & Shankles, S. A. (2015). Supply chain risk management practices for federal information systems and organizations. *NIST Special Publication, 800*(161), 32. https://doi.org/10.6028/NIST.SP.800-161

Braga, A., & Logan, R. K. (2017). The emperor of strong AI has no clothes: Limits to artificial intelligence. *Information, 8*(4), 156. https://doi.org/10.3390/info8040156

Braun, M., Hummel, P., Beck, S., & Dabrock, P. (2020). Primer on an ethics of AI-based decision support systems in the clinic. *Journal of Medical Ethics, 47*(12), e3. https://doi.org/10.1136/medethics-2019-105860

Brynjolfsson, E., Mitchell, T., & Rock, D. (2018). What can machines learn, and what does it mean for occupations and the economy? In *AEA Papers and Proceedings*.

Burkhardt, R., Hohn, N., & Wigley, C. (2019). Leading your organization to responsible AI. In *McKinsey Analytics* (pp. 1–8). London: Quantum Black.

Campbell, M., Hoane Jr, A. J., & Hsu, F.-H. (2002). Deep blue. *Artificial Intelligence, 134*(1-2), 57–83. https://doi.org/10.1016/S0004-3702(01)00129-1

Case, N. (2018). How to become a centaur. Journal of Design and Science. https://doi.org/10.21428/61b2215c

Castelnovo, A., Crupi, R., Grec o, G., Regoli, D., Penco, I. G., & Cosentini, A. C. (2022). A clarification of the nuances in the fairness metrics landscape. *Scientific Reports, 12*(1), 1–21. https://doi.org/10.1038/s41598-022-07939-1

Chen, I. Y., Pierson, E., Rose, S., Joshi, S., Ferryman, K., & Ghassemi, M. (2021). Ethical machine learning in healthcare. *Annual Review of Biomedical Data Science, 4*, 123–144. https://doi.org/10.1146/annurev-biodatasci-092820-114757

Cheng, L., Varshney, K. R., & Liu, H. (2021). Socially responsible AI algorithms: Issues, purposes, and challenges. *Journal of Artificial Intelligence Research, 71*, 1137–1181. https://doi.org/10.1613/jair.1.12814

Chouldechova, A. (2017). Fair prediction with disparate impact: A study of bias in recidivism prediction instruments. *Big Data, 5*(2), 153–163. https://doi.org/10.1089/big.2016.0047

Cinelli, M., De Francisci Morales, G., Galeazzi, A., Quattrociocchi, W., & Starnini, M. (2021). The echo chamber effect on social media. *Proceedings of the National Academy of Sciences, 118*(9), e2023301118. https://doi.org/10.1073/pnas.2023301118

De Cremer, D., & Kasparov, G. (2021). AI should augment human intelligence, not replace it. *Harvard Business Review*.

De Freitas, J., Anthony, S. E., Censi, A., & Alvarez, G. A. (2020). Doubting driverless dilemmas. *Perspectives on Psychological Science, 15*(5), 1284–1288. https://doi.org/10.1177/1745691620922201

Di, X., & Shi, R. (2021). A survey on autonomous vehicle control in the era of mixed-autonomy: From physics-based to AI-guided driving policy learning. *Transportation Research Part C: Emerging Technologies, 125*, 103008. https://doi.org/10.1016/j.trc.2021.103008

Dignum, V. (2022). Relational artificial intelligence. *arXiv preprint arXiv:2202.07446*.

Dwivedi, Y. K., Hughes, L., Ismagilova, E., Aarts, G., Coombs, C., Crick, T., ... Williams, M. D. (2021). Artificial Intelligence (AI): Multidisciplinary perspectives on emerging challenges, opportunities, and agenda for research, practice and policy. *International Journal of Information Management, 57*, 101994. https://doi.org/10.1016/j.ijinfomgt.2019.08.002

Dwork, C., Hardt, M., Pitassi, T., Reingold, O., & Zemel, R. (2012). Fairness through awareness. In Proceedings of the 3rd Innovations in Theoretical Computer Science Conference.

Ernst, E., Merola, R., & Samaan, D. (2019). Economics of artificial intelligence: Implications for the future of work. *IZA Journal of Labor Policy, 9*(1). https://doi.org/10.2478/izajolp-2019-0004

European Commission. (2021). Proposal for a regulation of the European Parliament and of the Council laying down harmonised rules on artificial intelligence (Artificial Intelligence Act) and amending certain Union legislative acts.

Ferrer, X., Nuenen, T. V., Such, J. M., Cote, M., & Criado, N. (2021). Bias and discrimination in AI: A cross-disciplinary perspective. *IEEE Technology and Society Magazine, 40*(2), 72–80. https://doi.org/10.1109/mts.2021.3056293

Ferrucci, D. A. (2012). Introduction to "this is Watson". *IBM Journal of Research and Development, 56*(3.4), 1: 1–1: 15. https://doi.org/10.1147/JRD.2012.2184356.

Fisher, M., List, C., Slavkovik, M., & Winfield, A. (2016). Engineering moral machines. *Informatik-Spektrum*. https://doi.org/10.1007/s00287-016-0998-x

Fouse, S., Cross, S., & Lapin, Z. (2020). DARPA's impact on artificial intelligence. *AI magazine*, *41*(2), 3–8. https://doi.org/10.1609/aimag.v41i2.5294

Friedler, S. A., Scheidegger, C., & Venkatasubramanian, S. (2016). On the (im) possibility of fairness. *arXiv preprint arXiv:1609.07236*.

Friedler, S. A., Scheidegger, C., & Venkatasubramanian, S. (2021). The (im) possibility of fairness: Different value systems require different mechanisms for fair decision making. *Communications of the ACM*, *64*(4), 136–143. https://doi.org/10.1145/3433949

Garibay, I., Mantzaris, A. V., Rajabi, A., & Taylor, C. E. (2019). Polarization in social media assists influencers to become more influential: Analysis and two inoculation strategies. *Scientific Reports*, *9*(1), 1–9. https://doi.org/10.1038/s41598-019-55178-8

Garibay, O., Winslow, B., Andolina, S., Antona, M., Bodenschatz, A., Coursaris, C., ... Xu, W. (2023). Six human-centered artificial intelligence grand challenges. *International Journal of Human-Computer Interaction*, *39*(3), 391–437. https://doi.org/10.1080/10447318.2022.2153320

Gesner, E., Gazarian, P., & Dykes, P. (2019). The burden and burnout in documenting patient care: An integrative literature review. *Studies in Health Technology and Informatics*, *264*, 1194–1198. https://doi.org/10.3233/SHTI190415

Google AI. (2022). *Responsible AI Practices*. https://ai.google/responsibilities/responsible-ai-practices

Gottfredson, L. S. (1997). Mainstream science on intelligence: An editorial with 52 signatories, history, and bibliography. *Intelligence*, *24*(1), 13–23..

Grace, K., Salvatier, J., Dafoe, A., Zhang, B., & Evans, O. (2018). Viewpoint: When will AI exceed human performance? Evidence from AI experts. *Journal of Artificial Intelligence Research*, *62*, 729–754. https://doi.org/10.1613/jair.1.11222

Grayson, M., Thieme, A., Marques, R., Massiceti, D., Cutrell, E., & Morrison, C. (2020). A dynamic AI system for extending the capabilities of blind people. In *Extended Abstracts of the 2020 CHI Conference on Human Factors in Computing Systems*.

Grudin, J. (2009). AI and HCI: Two fields divided by a common focus. *AI magazine*, *30*(4), 48–48. https://doi.org/10.1609/aimag.v30i4.2271

Gunning, D., & Aha, D. (2019). DARPA's explainable artificial intelligence (XAI) program. *AI magazine*, *40*(2), 44–58. https://doi.org/10.1609/aimag.v40i2.2850

Hallevy, G. (2010). I, Robot-I, criminal: When science fiction becomes reality: Legal liability of AI robots committing criminal offenses. *Syracuse Journal of Science & Technology Law Reporter*, *22*, 1.

Hardt, M., Price, E., & Srebro, N. (2016). Equality of opportunity in supervised learning. In *Advances in Neural Information Processing Systems 29 (NIPS 2016)*. https://doi.org/10.48550/arXiv.1610.02413

Harper, F. M., & Konstan, J. A. (2016). The MovieLens datasets. *ACM Transactions on Interactive Intelligent Systems*, *5*(4), 1–19. https://doi.org/10.1145/2827872

Hassenzahl, M., & Tractinsky, N. (2006). User experience-a research agenda. *Behaviour & Information Technology*, *25*(2), 91–97. https://doi.org/10.1080/01449290500330331

Herculano-Houzel, S. (2014). The glia/neuron ratio: How it varies uniformly across brain structures and species and what that means for brain physiology and evolution. *Glia*, *62*(9), 1377–1391. https://doi.org/10.1002/glia.22683

Heurio. (2023). *Kaniasty's CARMEL Guidelines*. https://www.heurio.co/kaniastys-carmel-guidelines

Himmelreich, J. (2018). Never mind the trolley: The ethics of autonomous vehicles in mundane situations. *Ethical Theory and Moral Practice*, *21*(3), 669–684. https://doi.org/10.1007/s10677-018-9896-4

Hodge, R., Rotner, J., Baron, I., Kotras, D., & Worley, D. (2020). *Designing a New Narrative to Build an AI-Ready Workforce*. MacLean, VA: MITRE.

Hoffman, R. R., Mueller, S. T., Klein, G., & Litman, J. (2018). Metrics for explainable AI: Challenges and prospects. *arXiv preprint arXiv:1812.04608*.

Hoffmann, A. L. (2020). Terms of inclusion: Data, discourse, violence. *New Media & Society*, *23*(12), 3539–3556. https://doi.org/10.1177/1461444820958725

Holmström, J. (2022). From AI to digital transformation: The AI readiness framework. *Business Horizons*, *65*(3), 329–339. https://doi.org/10.1016/j.bushor.2021.03.006

IBM. (2019). *Design for AI*. https://www.ibm.com/design/ai/

ISO 9241-11. (2018). Ergonomics of human-system interaction-part 11: Usability: Definitions and concepts. In *International Organization for Standardization*, Geneva, Switzerland.

Jacko, J. A. (2012). *Human Computer Interaction Handbook: Fundamentals, Evolving Technologies, and Emerging Applications*. Boca Raton, FL: CRC Press.

Kamiran, F., & Calders, T. (2011). Data preprocessing techniques for classification without discrimination. *Knowledge and Information Systems*, *33*(1), 1–33. https://doi.org/10.1007/s10115-011-0463-8

Kane, S. K., Guo, A., & Morris, M. R. (2020). Sense and accessibility: Understanding people with physical disabilities' experiences with sensing systems. In *The 22nd International ACM SIGACCESS Conference on Computers and Accessibility*.

Kaur, D., Uslu, S., Rittichier, K. J., & Durresi, A. (2022). Trustworthy artificial intelligence: A review. *ACM Computing Surveys (CSUR)*, *55*(2), 1–38. https://doi.org/10.1145/3491209

Kelley, P. G., Yang, Y., Heldreth, C., Moessner, C., Sedley, A., Kramm, A., ... Woodruff, A. (2021). Exciting, useful, worrying, futuristic: Public perception of artificial intelligence in 8 countries. In *Proceedings of the 2021 AAAI/ACM Conference on AI, Ethics, and Society*,

Kirchkamp, O., & Strobel, C. (2019). Sharing responsibility with a machine. *Journal of Behavioral and Experimental Economics*, *80*, 25–33. https://doi.org/10.1016/j.socec.2019.02.010

Klien, G., Woods, D. D., Bradshaw, J. M., Hoffman, R. R., & Feltovich, P. J. (2004). Ten challenges for making automation a" team player" in joint human-agent activity. *IEEE Intelligent Systems*, *19*(6), 91–95. https://doi.org/10.1109/MIS.2004.74.

Komal, S. (2014). Comparative assessment of human intelligence and artificial intelligence. *International Journal of Computer Science and Mobile Computing*, *3*, 1–5.

Krügel, S., Ostermaier, A., & Uhl, M. (2022). Zombies in the loop? Humans trust untrustworthy AI-advisors for ethical decisions. *Philosophy & Technology, 35*(1). https://doi.org/10.1007/s13347-022-00511-9

Kusner, M. J., Loftus, J., Russell, C., & Silva, R. (2017). Counterfactual fairness. In Advances in Neural Information Processing Systems 30 *(NIPS 2017)*.

Lambrecht, A., & Tucker, C. (2019). Algorithmic bias? An empirical study of apparent gender-based discrimination in the display of STEM career ads. *Management Science*, *65*(7), 2966–2981. https://doi.org/10.1287/mnsc.2018.3093

Law, C. (2021). The dangers of driverless cars. *National Law Review*, *12*(276). https://doi.org/10.1145/2874239.2874265

LeCun, Y., Boser, B., Denker, J. S., Henderson, D., Howard, R. E., Hubbard, W., & Jackel, L. D. (1989). Backpropagation applied to handwritten zip code recognition. *Neural Computation*, *1*(4), 541–551. https://doi.org/10.1162/neco.1989.1.4.541

Lee, N. T., Resnick, P., & Barton, G. (2019). *Algorithmic Bias Detection and Mitigation: Best Practices and Policies to Reduce Consumer Harms*. Washington, DC: Brookings Institute.

Lindebaum, D., Vesa, M., & den Hond, F. (2020). Insights from "the machine stops" to better understand rational assumptions in algorithmic decision making and its implications for organizations. *Academy of Management Review*, *45*(1), 247–263. https://doi.org/10.5465/amr.2018.0181

Liu, X., & Pan, H. (2022). The path of film and television animation creation using virtual reality technology under the artificial intelligence. *Scientific Programming*, *2022*. https://doi.org/10.1155/2022/1712929

Logg, J. M., Minson, J. A., & Moore, D. A. (2019). Algorithm appreciation: People prefer algorithmic to human judgment. *Organizational Behavior and Human Decision Processes*, *151*, 90–103. https://doi.org/10.1016/j.obhdp.2018.12.005

López-González, M. (2021). Applying human cognition to assured autonomy. In *International Conference on Human-Computer Interaction*.

Makav, B., & Kılıç, V. (2019). Smartphone-based image captioning for visually and hearing impaired. In 2019 *11th International Conference on Electrical and Electronics Engineering (ELECO)*.

Malik, P., Pathania, M., & Rathaur, V. K. (2019). Overview of artificial intelligence in medicine. *Journal of Family Medicine and Primary Care*, *8*(7), 2328. https://doi.org/10.4103/jfmpc.jfmpc_440_19

Manyika, J., & Sneader, K. (2018). *AI, Automation, and the Future of Work: Ten Things to Solve for*. New York: McKinsey Global Institute.

Margetis, G., Antona, M., Ntoa, S., & Stephanidis, C. (2012). Towards accessibility in ambient intelligence environments. In *International Joint Conference on Ambient Intelligence*.

Margetis, G., Ntoa, S., Antona, M., & Stephanidis, C. (2021). Human-centered design of artificial intelligence. In *Handbook of Human Factors and Ergonomics* (pp. 1085–1106). https://doi.org/10.1002/9781119636113.ch42

Martin, D. A., Shapiro, J. N., & Nedashkovskaya, M. (2019). Recent trends in online foreign influence efforts. *Journal of Information Warfare*, *18*(3), 15–48.

Mayer, A.-S., Strich, F., & Fiedler, M. (2020). Unintended consequences of introducing AI systems for decision making. *MIS Quarterly Executive*, *19*(4), 239–257. https://doi.org/10.17705/2msqe.00036

McCarthy, J. (2004). What is Artificial Intelligence. https://www-formal.stanford.edu/jmc/whatisai.html.

McCarthy, J., Minsky, M. L., Rochester, N., & Shannon, C. E. (2006). A proposal for the Dartmouth summer research project on artificial intelligence, August 31, 1955. *AI magazine, 27*(4), 12–12.

McGregor, S. (2022). Participation interfaces for human-centered AI. *arXiv preprint arXiv:2211.08419*. https://doi.org/https://doi.org/10.48550/arXiv.2211.08419

Mehrabi, N., Gupta, U., Morstatter, F., Steeg, G. V., & Galstyan, A. (2021a). Attributing fair decisions with attention interventions. *arXiv preprint arXiv:2109.03952*.

Mehrabi, N., Morstatter, F., Saxena, N., Lerman, K., & Galstyan, A. (2021b). A survey on bias and fairness in machine learning. *ACM Computing Surveys, 54*(6), 1–35. https://doi.org/10.1145/3457607

Mehta, K., Rajpal, C., & Verma, G. (2021). Trusted AI platforms for the connected car ecosystem. ISACA Journal, *6*, 1–5.

Metz, C. (2016). In two moves, AlphaGo and Lee Sedol redefined the future. *WIRED.com, 16*.

Metz, C. (2022, Nov 23). Lawsuit takes aim at the way AI is built. *The New York Times*. https://www.nytimes.com/2022/11/23/technology/copilot-microsoft-ai-lawsuit.html

Mezgár, I., & Váncza, J. (2022). From ethics to standards—A path via responsible AI to cyber-physical production systems. *Annual Reviews in Control*. https://doi.org/10.1016/j.arcontrol.2022.04.002

Microsoft. (2022). *Responsible AI*. https://www.microsoft.com/en-us/ai/responsible-ai

Misselhorn, C. (2018). Artificial morality. Concepts, issues and challenges. *Society, 55*(2), 161–169. https://doi.org/10.1007/s12115-018-0229-y

Mitchell, M., Wu, S., Zaldivar, A., Barnes, P., Vasserman, L., Hutchinson, B., ... Gebru, T. (2019). Model cards for model reporting. In *Proceedings of the Conference on Fairness, Accountability, and Transparency*.

Mittelstadt, B. D., Allo, P., Taddeo, M., Wachter, S., & Floridi, L. (2016). The ethics of algorithms: Mapping the debate. *Big Data & Society, 3*(2), 2053951716679679. https://doi.org/10.1177/2053951716679679

Muro, M., Whiton, J., & Maxim, R. (2019). What jobs are affected by AI? Better-paid, better-educated workers face the most exposure. *Metropolitan Policy Program Report*.

Murre, J. M., & Sturdy, D. P. (1995). The connectivity of the brain: Multi-level quantitative analysis. *Biological Cybernetics, 73*(6), 529–545. https://doi.org/10.1007/BF00199545

Mutlu, E. Ç., Yousefi, N., & Ozmen Garibay, O. (2022). Contrastive counterfactual fairness in algorithmic decision-making. In *Proceedings of the 2022 AAAI/ACM Conference on AI, Ethics, and Society*.

National Artificial Intelligence Initiative Act of 2020. (2020). H.R.6216, 116th Congress of the United States of America.

National Artificial Intelligence Initiative. (2022). https://www.ai.gov

NHTSA. (2022). Standing General Order on Crash Reporting.

Nielsen, J. (1994). *Usability Engineering*. Cambridge, MA: Morgan Kaufmann.

Nielsen, J., & Molich, R. (1990). Heuristic evaluation of user interfaces. In *Proceedings of the SIGCHI Conference on Human Factors in Computing Systems*.

Nissenbaum, H. (1996). Accountability in a computerized society. *Science and Engineering Ethics, 2*(1), 25–42. https://doi.org/10.1007/bf02639315

Ntoa, S., Margetis, G., Antona, M., & Stephanidis, C. (2021a). Digital accessibility in intelligent environments. In V. G. Duffy, P. Ziefle, P. L. P. Rau, & M. M. Tseng (Eds), *Human-Automation Interaction: Mobile Computing*. Cham: Springer.

Ntoa, S., Margetis, G., Antona, M., & Stephanidis, C. (2021b). User experience evaluation in intelligent environments: A comprehensive framework. *Technologies, 9*(2), 41. https://doi.org/10.3390/technologies9020041

Obermeyer, Z., Powers, B., Vogeli, C., & Mullainathan, S. (2019). Dissecting racial bias in an algorithm used to manage the health of populations. *Science, 366*(6464), 447–453. https://doi.org/10.1126/science.aax234

Olsson, T., & Väänänen, K. (2021). How does AI challenge design practice? *Interactions, 28*(4), 62–64. https://doi.org/10.1145/3467479

Organisation for Economic Co-operation and Development. (2021). *OECD AI Principles*. https://oecd.ai/en/ai-principles

Park, S., Kim, I., Lee, S. W., Yoo, J., Jeong, B., & Cha, M. (2015). Manifestation of depression and loneliness on social networks: A case study of young adults on Facebook. In *Proceedings of the 18th ACM Conference on Computer Supported Cooperative Work & Social Computing*.

Pastaltzidis, I., Dimitriou, N., Quezada-Tavarez, K., Aidinlis, S., Marquenie, T., Gurzawska, A., & Tzovaras, D. (2022). Data augmentation for fairness-aware machine learning: Preventing algorithmic bias in law enforcement systems. In *2022 ACM Conference on Fairness, Accountability, and Transparency*.

Pereira, V., Hadjielias, E., Christofi, M., & Vrontis, D. (2021). A systematic literature review on the impact of artificial intelligence on workplace outcomes: A multi-process perspective. *Human Resource Management Review*, 100857. https://doi.org/10.1016/j.hrmr.2021.100857

Pitoura, E., Stefanidis, K., & Koutrika, G. (2021). Fairness in rankings and recommendations: An overview. *The VLDB Journal*, 1–28. https://doi.org/10.1007/s00778-021-00697-y

Rabhi, Y., Mrabet, M., & Fnaiech, F. (2018). A facial expression controlled wheelchair for people with disabilities. *Computer Methods and Programs in Biomedicine*, 165, 89–105. https://doi.org/10.1016/j.cmpb.2018.08.013

Rai, A. (2020). Explainable AI: From black box to glass box. *Journal of the Academy of Marketing Science*, 48(1), 137–141. https://doi.org/10.1007/s11747-019-00710-5

Rajabi, A., & Garibay, O. O. (2021). Towards fairness in AI: Addressing bias in data using GANs. In *International Conference on Human-Computer Interaction*.

Rajabi, A., Gunaratne, C., Mantzaris, A. V., & Garibay, I. (2020a). Modeling disinformation and the effort to counter it: A cautionary tale of when the treatment can be worse than the disease. In *Proceedings of the 19th International Conference on Autonomous Agents and MultiAgent Systems*.

Rajabi, A., Gunaratne, C., Mantzaris, A. V., & Garibay, I. (2020b). On countering disinformation with caution: Effective inoculation strategies and others that backfire into community hyper-polarization. In *International Conference on Social Computing, Behavioral-Cultural Modeling and Prediction and Behavior Representation in Modeling and Simulation*.

Rajabi, A., Mantzaris, A. V., Atwal, K. S., & Garibay, I. (2021). Exploring the disparity of influence between users in the discussion of Brexit on Twitter. *Journal of Computational Social Science*, 4(2), 903–917. https://doi.org/10.1007/s42001-021-00112-0

Responsible Artificial Intelligence Institute. (2022). *Responsible Artificial Intelligence Institute*. https://www.responsible.ai/

Ricci Lara, M. A., Echeveste, R., & Ferrante, E. (2022). Addressing fairness in artificial intelligence for medical imaging. *Nature Communications*, 13(1), 1–6. https://doi.org/10.1038/s41467-022-32186-3

Riedl, M. O. (2019). Human-centered artificial intelligence and machine learning. *Human Behavior and Emerging Technologies*, 1(1), 33–36. https://doi.org/10.1002/hbe2.117

Robinson, L., Cotten, S. R., Ono, H., Quan-Haase, A., Mesch, G., Chen, W., ... Stern, M. J. (2015). Digital inequalities and why they matter. *Information, Communication & Society*, 18(5), 569–582. https://doi.org/10.1080/1369118x.2015.1012532

Rosenblatt, F. (1958). The perceptron: A probabilistic model for information storage and organization in the brain. *Psychological Review*, 65(6), 386.

Rudin, C., & Radin, J. (2019). Why are we using black box models in AI when we don't need to? A lesson from an explainable AI competition.

Russell, S. (2016). Should we fear supersmart robots. *Scientific American*, 314(6), 58–59. https://doi.org/10.1038/scientificamerican0616-58

Sacha, D., Senaratne, H., Kwon, B. C., Ellis, G., & Keim, D. A. (2015). The role of uncertainty, awareness, and trust in visual analytics. *IEEE Transactions on Visualization and Computer Graphics*, 22(1), 240–249. https://doi.org/10.1109/TVCG.2015.2467591

Saleiro, P., Kuester, B., Hinkson, L., London, J., Stevens, A., Anisfeld, A.,... Ghani, R. (2018). Aequitas: A bias and fairness audit toolkit. *arXiv preprint arXiv:1811.05577*.

Sanders, N. E., & Schneier, B. (2023, Jan 15). How ChatGPT hijacks democracy. *The New York Times*. https://www.nytimes.com/2023/01/15/opinion/ai-chatgpt-lobbying-democracy.html

Sanford, S. (2021). How to build accountability into your AI. *Harvard Business Review*, Aug 11.

Sarakiotis, V. (2020). Human-centered AI: Challenges and opportunities. In *UBIACTION 2020*.

Savage, N. (2019). How AI and neuroscience drive each other forwards. *Nature*, 571(7766), S15–S15. https://doi.org/10.1038/d41586-019-02212-4

Saxena, N. A., Huang, K., DeFilippis, E., Radanovic, G., Parkes, D. C., & Liu, Y. (2019). How do fairness definitions fare? Examining public attitudes towards algorithmic definitions of fairness. In *Proceedings of the 2019 AAAI/ACM Conference on AI, Ethics, and Society*.

Searle, J. R. (1980). Minds, brains, and programs. *Behavioral and Brain Sciences*, 3(3), 417–424. https://doi.org/10.1017/S0140525X00005756

Shneiderman, B. (2000). Universal usability. *Ubiquity, 2000* (August), 84–91. https://doi.org/10.1145/347634.350994

Shneiderman, B. (2020a). Bridging the gap between ethics and practice. *ACM Transactions on Interactive Intelligent Systems*, 10(4), 1–31. https://doi.org/10.1145/3419764

Shneiderman, B. (2020b). Human-centered artificial intelligence: Three fresh ideas. *AIS Transactions on Human-Computer Interaction*, 12(3), 109–124. https://doi.org/10.17705/1thci.00131

Shneiderman, B. (2022). *Human-Centered AI*. Oxford: Oxford University Press.

Shneiderman, B., & Du, M. (2022). *Human-Centered AI*. https://hcai.site/

Shneiderman, B., & Plaisant, C. (2010). *Designing the User Interface: Strategies for Effective Human-Computer Interaction*. Bangalore: Pearson Education India.

Slack, D., Hilgard, S., Jia, E., Singh, S., & Lakkaraju, H. (2020). Fooling lime and shap: Adversarial attacks on post hoc explanation methods. In *Proceedings of the AAAI/ACM Conference on AI, Ethics, and Society*.

Smith, C. J. (2019). Designing trustworthy AI: A human-machine teaming framework to guide development. *arXiv preprint arXiv:1910.03515*.

Snyderman, M., & Rothman, S. (1987). Survey of expert opinion on intelligence and aptitude testing. *American Psychologist*, *42*(2), 137. https://doi.org/10.1037/0003-066X.42.2.137

Sowa, K., Przegalinska, A., & Ciechanowski, L. (2021). Cobots in knowledge work: Human-AI collaboration in managerial professions. *Journal of Business Research*, *125*, 135–142. https://doi.org/10.1016/j.jbusres.2020.11.038

Spielkamp, M. (2017). Inspecting algorithms for bias. *Technology Review*, *120*(4), 96–98.

Stephanidis, C., Salvendy, G., Antona, M., Chen, J. Y. C., Dong, J., Duffy, V. G., ... Zhou, J. (2019). Seven HCI grand challenges. *International Journal of Human-Computer Interaction*, *35*(14), 1229–1269. https://doi.org/10.1080/10447318.2019.1619259

Strich, F., Mayer, A.-S., & Fiedler, M. (2021). What do I do in a world of artificial intelligence? Investigating the impact of substitutive decision-making AI systems on employees' professional role identity. *Journal of the Association for Information Systems*, *22*(2), 9. https://doi.org/10.17705/1jais.00663

Subramonyam, H., Im, J., Seifert, C., & Adar, E. (2022). Human-AI guidelines in practice: The power of leaky abstractions in cross-disciplinary teams. In *CHI Conference on Human Factors in Computing Systems (CHI '22)*, April 29-May 5, 2022, New Orleans, LA, USA.

Surden, H. (2019). Artificial intelligence and law: An overview. *Georgia State University Law Review*, *35*, 19–22.

Suresh, H., & Guttag, J. (2021). A framework for understanding sources of harm throughout the machine learning life cycle. In *Equity and Access in Algorithms, Mechanisms, and Optimization* (pp. 1–9). https://doi.org/10.1145/3465416.3483305

Tai-Seale, M., Olson, C. W., Li, J., Chan, A. S., Morikawa, C., Durbin, M., ... Luft, H. S. (2017). Electronic health record logs indicate that physicians split time evenly between seeing patients and desktop medicine. *Health Affairs*, *36*(4), 655–662. https://doi.org/10.1377/hlthaff.2016.0811

Thiebes, S., Lins, S., & Sunyaev, A. (2021). Trustworthy artificial intelligence. *Electronic Markets*, *31*(2), 447–464. https://doi.org/10.1007/s12525-020-00441-4

Tolan, S., Pesole, A., Martínez-Plumed, F., Fernández-Macías, E., Hernández-Orallo, J., & Gómez, E. (2021). Measuring the occupational impact of AI: Tasks, cognitive abilities and AI benchmarks. *Journal of Artificial Intelligence Research*, *71*, 191–236. https://doi.org/10.1613/jair.1.12647

Turing, A. M., & Haugeland, J. (1950). Computing machinery and intelligence. In *The Turing Test: Verbal Behavior as the Hallmark of Intelligence* (pp. 29–56).

United Nations Department of Economic and Social Affairs. (2018, Apr 20). *The 17 Goals*. https://sdgs.un.org/goals

United Nations Educational Scientific and Cultural Organization (UNESCO). (2021). *Recommendation on the Ethics of Artificial Intelligence*. https://en.unesco.org/artificial-intelligence/ethics

US White House Office of Science and Technology Policy. (2022). *AI Bill of Rights*. https://www.whitehouse.gov/ostp/ai-bill-of-rights/

Van Diggelen, J., Neerincx, M., Peeters, M., & Schraagen, J. M. (2018). Developing effective and resilient human-agent teamwork using team design patterns. *IEEE Intelligent Systems*, *34*(2), 15–24. https://doi.org/10.1109/MIS.2018.2886671.

Vochozka, M., Kliestik, T., Kliestikova, J., & Sion, G. (2018). Participating in a highly automated society: How artificial intelligence disrupts the job market. *Economics, Management, and Financial Markets*, *13*(4), 57–62.

Wachter, S., Mittelstadt, B., & Floridi, L. (2017). Transparent, explainable, and accountable AI for robotics. *Science Robotics*, *2*(6), eaan6080. https://doi.org/10.1126/scirobotics.aan6080

Wallach, D. P., Flohr, L. A., & Kaltenhauser, A. (2020). Beyond the buzzwords: On the perspective of AI in UX and vice versa. In *International Conference on Human-Computer Interaction*.

Wallach, W., & Allen, C. (2008). *Moral Machines: Teaching Robots Right from Wrong*. Oxford: Oxford University Press.

Weinschenk, S., & Barker, D. T. (2000). *Designing Effective Speech Interfaces*. Hoboken, NJ: John Wiley & Sons, Inc.

Weller, A. (2017). Challenges for Transparency. In *2017 ICML Workshop on Human Interpretability in Machine Learning (WHI 2017)*, Sydney, NSW, Australia (pp. 55–62).

West, D. M., & Allen, J. R. (2020, July 28). *Turning Point. Policymaking in the Era of Artificial Intelligence.* https://www.brookings.edu/book/turning-point/

Williams, B. A., Brooks, C. F., & Shmargad, Y. (2018). How algorithms discriminate based on data they lack: Challenges, solutions, and policy implications. *Journal of Information Policy, 8*(1), 78–115. https://doi.org/10.5325/jinfopoli.8.1.0078

Winograd, T. (2006). Shifting viewpoints: Artificial intelligence and human-computer interaction. *Artificial Intelligence, 170*(18), 1256–1258. https://doi.org/10.1016/j.artint.2006.10.011

Wirth, R., & Hipp, J. (2000). CRISP-DM: Towards a standard process model for data mining. In *Proceedings of the 4th International Conference on the Practical Applications of Knowledge Discovery and Data Mining.*

Xu, W. (2019). Toward human-centered AI. *Interactions, 26*(4), 42–46. https://doi.org/10.1145/3328485

Xu, W., Dainoff, M. J., Ge, L., & Gao, Z. (2022). Transitioning to human interaction with AI systems: New challenges and opportunities for HCI professionals to enable human-centered AI. *International Journal of Human-Computer Interaction*, 1–25. https://doi.org/10.1080/10447318.2022.2041900

Yang, Q., Steinfeld, A., Rosé, C., & Zimmerman, J. (2020). Re-examining whether, why, and how human-AI interaction is uniquely difficult to design. In *Proceedings of the 2020 CHI Conference on Human Factors in Computing Systems.*

Yeo, C. (2020). *What is Transparency in AI?* https://medium.com/fair-bytes/what-is-transparency-in-ai-bd08b2e901ac

Zafar, M. B., Valera, I., Gomez Rodriguez, M., & Gummadi, K. P. (2017a). Fairness beyond disparate treatment & disparate impact: Learning classification without disparate mistreatment. In *Proceedings of the 26th International Conference on World Wide Web.*

Zafar, M. B., Valera, I., Rogriguez, M. G., & Gummadi, K. P. (2017b). Fairness constraints: Mechanisms for fair classification. In *Artificial Intelligence and Statistics* (pp. 962–970). PMLR. New York: Association for Computing Machinery.

Zhang, H., Lu, A. X., Abdalla, M., McDermot t, M., & Ghassemi, M. (2020). Hurtful words: Quantifying biases in clinical contextual word embeddings. In *Proceedings of the ACM Conference on Health, Inference, and Learning.*

5 AI in HCI Design and User Experience

Wei Xu

5.1 INTRODUCTION

Recently, much progress has been made in artificial intelligence (AI) and machine learning (ML); such progress has also enabled human-computer interaction (HCI) and user experience (UX) professionals to deliver solutions with better UX (Lu et al., 2022; Yang et al., 2018a; Kuniavsky et al., 2017). The use of AI/ML capabilities for improving HCI/UX work and delivering better UX in solutions is becoming a trend (Abbas et al., 2022; Wu et al., 2019; Nikiforova et al., 2021) and creates many new opportunities for HCI/UX professionals (Holmquist, 2017; Yang et al., 2020b). Some even speculate "AI/ML is the new UX" (Yang et al., 2018b).

Researchers proposed that AI can perform as an assistant, collaborator, researcher, or facilitator (Bertão & Joo, 2021; Main & Grierson, 2020). AI technology will change the role of designers in the design process and generate an opportunity for creative collaboration between AI and designers (McCormack et al., 2020). Also, companies are moving fast to adopt AI for improving customer experience (CX). In 2018, the IBM Institute for Business Value (IBV) surveyed 1,194 executives from seven industries worldwide who are responsible for the AI initiatives of their companies (Schwartz et al., 2018). The results show that 74% said AI would fundamentally change how they approach CX; 41% had an AI strategy considering the changes ahead. To some extent, AI and UX designers have similar functions. They both gather data, analyze users' behavior and interactions, and can predict human behavior (Donahole, 2021). For example, Chatbots, Google Translate, and Alexa are good examples of AI technology that uses big data to deliver enhanced UX.

Table 5.1 summarizes the benefits of AI technology in enhancing HCI/UX activities (e.g., Yang et al., 2018a; Inkbot Design, 2021; Rogers, 2020; Schwartz et al., 2018; Baker, 2019; Donahole, 2021). Herein, AI refers to technologies (e.g., AI, ML, big data) used to develop AI-based intelligent systems, applications, or services.

Thus, it is apparent that AI is transforming how HCI and UX professionals work toward delivering optimal UX in their solutions. The transformation impacts the HCI/UX activities such as user research, user interface (UI) technologies and design, and user evaluation.

In this chapter, we review and discuss the transformation of AI technology in HCI/UX work and assess how AI technology will change how we do the work. We first discuss how AI can be used to enhance the result of user research and design evaluation. We then discuss how AI technology can be used to enhance HCI/UX design. Finally, we discuss how AI-enabled capabilities can improve UX when users interact with computing systems, applications, and services.

5.2 AI IN HCI/UX RESEARCH AND EVALUATION

5.2.1 Overview

The goal of HCI/UX research and evaluation is to systematically gather and analyze user data through HCI/UX activities (e.g., user research, usability testing) to understand a problem space (e.g., user pain points, usability issues) and guide the entire design process (Xu, 2005). Conventional approaches to user study rely on methods such as surveys or user interviews; for UX evaluations,

TABLE 5.1
Benefits of AI for HCI and UX Activities

Benefits of AI	Descriptions and Examples
Sensing users intelligently and supporting user research effectively	• Collect user personal knowledge (e.g., online shopping behavior, social connections) • Recognize user's activity (e.g., physical status and interaction with systems) • Infer the user's internal status (e.g., intention, emotion, and attitude) • Identify unique characteristics and behavioral patterns of collective users (e.g., digital personas) • Sense context of user interaction (e.g., online historical shopping data)
Acting intelligently based on insights from sensing users	• React with appropriate actions (e.g., inform, engage, assist, and promote products) • React proactively or autonomously (e.g., empathy, influence, and conflict management) • Analyze and optimize customer journeys (e.g., gather customer data across various touch points along with the flow of shopping)
Personalization design	• Provide personalized online content and functionality based on user preferences and interactions (e.g., data log, digital personas) • Promote marketing experience to another level by using users' personal information • Deliver advanced localization capabilities to handle language-related activities • Focus on satisfying the precise needs of users (e.g., deliver personalized recommendations for users in terms of their shopping patterns based on their history of shopping data)
Analyzing a large amount of data more quickly and efficiently	• Analyze large amounts of data to ascertain patterns and deliver meaningful research results (e.g., quickly generate questionnaires, and provide relevant responses for further inquiry) • Analyze massive sets of data to modify user experiences (e.g., generate digital personas based on the analysis and deliver personalized online content in terms of the personas) • Gather and draw inferences to empower quick decision-making from vast volumes of data in a time (e.g., use AI-based algorithms to identify new marketing opportunities by analyzing consumer shopping behavioral data)
Powering HCI/UX activities for efficiency	• Automate repetitive design activities (e.g., resize images, make color corrections, crop images) • Leverage algorithms to create UI design based on historical user patterns (e.g., generate multiple variants of a UI design solution for A/B testing) • Develop UI wireframes based on the understanding of the context and the flow (e.g., use generative UI technologies such as ChatGPT) • Automate design tasks (e.g., identify patterns in images and help designers stitch them together) • Automatically run dynamic online A/B testing and analyze test results • Automate back-end processes (e.g., automate targeted marketing promotions) • Help designers quickly make design decisions (e.g., predictions based on historical datasets, giving users the fewest potential choices)
Providing more natural and effective interaction	• Enable new types of user interface technologies (e.g., voice input, face recognition, gesture interaction, brain-computer interface) • Handle inaccurate input through reasoning (e.g., user intent detection, affective interaction) • Build thinner UI with AI (e.g., using historical data) to anticipate a user's action or better prioritize the queries of users, provide a possible solution or pertinent results
Better marketing	• Help build a better connection among brands across target audiences and boost their relationship (e.g., use personalized recommender systems) • Build customized e-commerce sites using their personal information, taking the marketing experience to another level
Working as a user assistant and capability to enhance UX	• Integrate into end user-facing solutions for users to interact with directly (e.g., chatbots) • Work as an app service tailored toward a particular action or use case (e.g., context-based intelligent search) • Provide ML-based speech-to-text services • Provide analytic capabilities (e.g., risk assessment, sentiment analysis, and retroactive analysis) • Provide text-related capabilities (e.g., natural language processing, text recognition, and speech-to-text conversion) • Provide visual capabilities (e.g., computer vision, augmented reality)

HCI/UX professionals manually conduct usability testing of a proposed design to identify issues and then analyze data to generate recommendations for design improvement (Xu, 2017). However, these methods are time-consuming.

A few years ago, researchers found that there is only little academic work at the intersection of UX and AI (Chromik et al., 2020; Yang et al., 2018b); they found even less research explicitly addressing AI/ML for user research. However, the number of relevant publications has been increasing since 2015 and continues to do so as AI is gaining popularity in many contexts.

The first AI-based approach uses big data-driven user research methods that gradually replace traditional methods (Tan et al., 2020). With the development of the 5G, the Internet of Things, smart devices, etc., user data generation inevitably grows explosively. Big data technology adoption in user research shows an upward trend. However, the way of collecting user data is through some AI technology (e.g., sensing, face recognition) and user interactions (e.g., user interaction log, online click streams, and social media); many of the big data-driven approaches are based on traditional statistical techniques have not fully leveraged AI methods in their analysis for the collected user data.

As an alternative approach, ML-based approaches are primarily used to support the analysis of already collected user data (Chromik et al., 2020). For example, it analyzes textual user data. ML and natural language processing (NLP) methods have been used to semi-automate the coding of interview transcripts (Marathe & Toyama, 2018) and to extract UX-related problems from online review narratives through classification (Mendes & Furtado, 2017). Data-driven learning approaches have also been used to construct behavioral personas from user interaction log data.

The third approach combines the two methods discussed above, applying AI/ML technology to both data collection and analysis stages. For example, Gartner (2018) provided recommendations for customer journey analytics (CJA), leveraging existing analytics tools (when available) and incorporating ML as a service (i.e., MLaaS) products to enhance analytical capabilities toward CJA. MlaaS incorporates ML-driven recommendations to power customer journey orchestration solutions and enhances digital experiences, optimizing customer journeys through interaction point analysis across crucial interaction points.

Table 5.2 summarizes the main benefits of AI/ML across HCI/UX research and evaluation activities (e.g., Chromik et al., 2020; Baker, 2019).

5.2.2 Digital Personas

One practical approach to using the data from user research is to develop personas (Alan Cooper's theory). Persona refers to a group of users with similar behaviors and goals within the group and with differences in behavior between the groups. Personas allow HCI/UX professionals to identify and understand the differences in how a product is used by different groups of users based on their usages and behaviors so that the product can provide a tailored experience (e.g., functions, content) across personas. Generating personas can become challenging and time-consuming in conventional user research, especially when HCI/UX researchers need to interview many users and plan to build many personas. Also, a manual process of developing personas has been criticized for creating personas that are not based on rigorous empirical data. The process often uses small samples, one-time data collection, and non-algorithmic methods (Salminen et al., 2021).

The AI era generates a vast amount of user data, reflecting the user's behavior, preferences, and demands, and has high research value in building personas. Data-driven personas, called "digital personas," have gained popularity in HCI due to digital trends such as personified big data, online analytics, and the evolution of data science algorithms. Specifically, three trends have significantly transformed the way how we build personas (Tan et al., 2020): (1) availability of user data from online analytics and social media platforms; (2) democratization of data science tools and algorithms that enable automated persona generation; and (3) Web technologies that remove the limitations of static personas via interactive UIs. These three trends allow us to use algorithmic methods to create accurate, representative, and refreshable personas from numerical data (Salminen et al., 2020, 2021).

TABLE 5.2
The Benefits of AI/ML across HCI/UX Research and Evaluation Activities

HCI/UX Activities	Benefits
Data collection	• Engaging surveys: simplify survey studies by leveraging the idea of adaptive user interfaces (i.e., questionnaires might automatically be tailored in real-time to the individual survey participant based on their previous answers) • Remote tracking of user behavior over time • Applying conversational and voice user interfaces for more empathetic survey studies
Evaluation of design	• Data-Driven Design: Supporting design decisions by evaluating and recommending UI options based on historical data of user behavior or user preferences was considered another field of interest
Analysis	• Analysis of ordinal data (e.g., questionnaires) • Analysis of text (e.g., transcripts of user research/evaluation) • Analysis of voice/audio-based data (e.g., recordings, camera feed) • Analysis of log data (e.g., click behavior) • Analysis of time series data (e.g., mouse and eye movement) • Analysis of emotion and sentiment • Automated transcription: ML-based speech-to-text services for the post-processing of contextual inquiries or interviews • Excel at quickly analyzing vast amounts of existing data to identify subtle patterns in dispersed data silos and to inform UX insights • Detection of patterns within structured or unstructured data • User modeling • Augmented/predictive analytics: Insights automatically generated from ML bring up new opportunities in conversion funnels
Generation of UX artifacts	• User personas • Customer journeys: Leveraging ML functionality allows organizations to contact or re-engage existing or potential customers at the optimal time in the customer journey and through the optimal communication channel • Audience/customer segmentation: By quickly analyzing large datasets and identifying patterns, these tools help analysts discover and validate new customer cohorts or segments

The benefits of digital personas are apparent. AI/ML-based algorithms allow HCI/UX professionals to develop personas much faster than conventional manual processes. An algorithm-based analysis of collected user data also can help identify distinct personas and visualize how HCI/UX professionals reliably identified their attributes. For example, Salminen et al. (2022) introduced Persona Analytics (PA), a system that tracks how users interact with data-driven personas. PA captures users' mouse and gaze behavior to measure users' interaction with algorithmically generated personas. The researchers also conducted a study with 144 participants, demonstrating how PA could be deployed for remote user studies.

Salminen et al. (2021) summarized the distinct relative benefits of digital personas:

- *Enhanced objectivity*: Digital personas tend to be replicable, and they use large sample sizes to increase the user representativeness of the personas. The conventional manual process for developing personas is associated with a high degree of subjectivity, which hinders the validity of the created personas. The statistical robustness of digital personas boosts both the validity and credibility of the developed personas.
- *Decreased cost*: The manual process of creating personas is time-consuming and costly. Digital personas mitigate this cost by relying on automation in persona creation, including

data collection and analysis, thus offering ways to "democratize" persona development for all kinds of organizations.
- *Updatability*: Shifts in user demographics and behaviors are typical in many fast-moving industries, such as Web-based businesses. Updating personas requires a high cost for the manual process, resulting in outdated personas. Digital personas can capture the change in user behavior over time based on their automated processes for systematic data collection and easy re-analysis using standard algorithms.
- *Scalability*: Manual data analysis is costly and requires specific expertise, making the personas built using manual processes less compatible with large datasets. Large datasets are common with social media and Web analytics and are not a concern for digital personas, as data science and ML algorithms have been developed to process large amounts of data.

More and more researchers are analyzing massive data and data characteristics to generate digital personas. For example, Yu (2014) combined used tags to describe user characteristics. They extracted relevant information tags through clustering in the big data environment to present the complete picture of users through digital personas. Joni Salminen of Qatar Research Institute integrated data from Facebook, Twitter, and YouTube and generated real-time personas based on user profiles and interactive behaviors, which provides users with competitive marketing methods and strategies across different platforms (Salminen et al., 2017). From the perspective of cultural differences, An et al. (2016) analyzed millions of content from the Middle East on YouTube to generate personas for deeply analyzing cultural diversity's impact on users' social media use.

People have also attempted to build accurate personality traits and types for their target users as the foundation for building personas (Salminen et al., 2020; Sun et al., 2018; Carducci et al., 2018). The automatic assessment of personality dimensions relies on information gathered from social media platforms, such as a list of friends and interests in music and movie endorsements. The work turned the collected data into signals as inputs. Supervised ML approaches have been efficient and accurate in computing personality traits and types. Specifically, Carducci et al. (2018) proposed a supervised ML approach to define personality traits by relying on what an individual tweeted about publicly. The approach segments tweets in tokens and then learns word vector representations as embeddings that are then used to feed a supervised learner classifier. This study demonstrates the approach's effectiveness by measuring the mean squared error of the learned model to compare it with an international benchmark of Facebook status updates. Also, the study tested the transfer learning predictive power of the proposed model with an in-house built benchmark created by 24 panelists who performed a state-of-the-art psychological survey. The comparison shows that the proposed model received an excellent conversion while analyzing the Twitter posts toward the personality traits extracted from the study.

Salminen et al. (2020) built social media-based personas with personality traits based on a deep ML approach. They developed a deep learning classifier (a NN classifier) that predicts personality traits. The study used three publicly available datasets and applied an automatic persona generation methodology to generate 15 personas from the social media data of an online news platform. After developing the personas, the study aggregated each persona's YouTube comments and predicted the personality traits of each persona from the comments on that persona. The results indicate an average performance increase of 4.84% in scores as compared with a baseline.

However, there are challenges for digital personas (Salminen et al., 2021). The main challenges include (1) data quality; (2) data availability; (3) method-specific weaknesses, such as the accuracy of the algorithm, people behaving differently across different segmentations; (4) human and machine biases, such as the persistent need for judgment calls ("manual labor") that creates a potential source of bias and obstacles for completely automated digital personas; (5) validation of digital personas methods; (6) lack of standardization; and (7) lack of consideration for inclusivity.

5.2.3 QUALITATIVE ANALYSIS

In user research and UX evaluation, HCI/UX professionals must conduct a qualitative analysis of the collected data, such as interview transcripts in text or recorded video files (natural language), usability test video recordings, and social media data. Qualitative analysis involves identifying themes, grouping data points based on these themes, and establishing relations between these groups. This type of qualitative analysis is time-coming and subjective to some extent. Many HCI/UX professionals have considerable training in analyzing human behavior while users interact with computing systems. Still, AI applications for this kind of behavioral data have yet to leverage their expertise fully. AI technology provides one approach to help the qualitative analysis for HCI/UX professionals. Therefore, developing and leveraging AI-assisted capabilities for qualitative research is critical.

The application of AI in quantitative analysis has been increasingly popular over the last several years. For example, researchers proposed using NLP and ML to generate initial codes followed by humans correcting the codes; other work utilized NLP to derive potential codes and models (Chen et al., 2018a). Although progress has been made in developing AI-assisted capabilities for qualitative analysis, there are still challenges and low accuracy has been considered the primary limitation of such automated approaches. Chen et al. (2018a) argued that we use AI/ML to support qualitative coding for identifying ambiguity. HCI/UX activities generate petabytes of free-form text data, recording daily UXs; NLI (natural language inference) allows analysis of this large-scale data, but NLI used to be applied for analysis with organized datasets, and less is known about how to design NLI for querying and analyzing text data (Mishra & Rzeszotarski, 2020).

Specifically, Liu et al. (2020) deployed a semantic data analysis processing approach. The approach introduced a specific implementation method using AI/ML semantic analysis technology to analyze language materials in user research. The effectiveness of applying the AI semantic analysis techniques in user research was tested and verified. The result shows that this application of AI technology has demonstrated the potential that AI technology can replace some of the human manual analysis tasks for efficiency improvement.

User intent analysis is also essential in user research and UX evaluation. A method called NLI is to do user intent analysis (Setlur, 2020). For example, NLI was applied to a labeled dataset that captures user intent distribution, co-occurrence, and flow patterns. Specifically, Setlur (2020) employed deep ML techniques that approximate the heuristics and conversational cues for continuous learning in a chatbot interface. These data-driven approaches help broaden the scope for visual analysis workflows across various chatbot experiences.

Analyzing usability test videos is also challenging for HCI/UX professionals. Fan et al. (2022) explored how AI can help facilitate effective collaboration between UX evaluators and AI. Based on the previous work in human and AI agent collaboration, they studied two primary factors: explanations and synchronization. Explanations allow AI to inform UX professionals how it identifies UX issues from a usability test session; synchronization refers to the two possible ways UX professionals and AI collaborate: synchronously and asynchronously. By adopting a hybrid wizard-of-oz approach to simulating an AI solution with good performance, they conducted a mixed-method study with 24 UX evaluators who were asked to identify UX issues from usability test videos using the AI-assisted capability. The results show that AI with explanations, whether presented synchronously or asynchronously, provides better support for UX evaluators' analysis; when without explanations, synchronous AI better improved UX evaluators' performance compared to asynchronous AI. This study also implies that an AI-assisted UX evaluation can facilitate more effective human-AI collaboration.

Analyzing the structure of texts is an alternative way for qualitative analysis. Recently, personality detection based on texts from online social networks has attracted more and more attention. Sun et al. (2018) analyzed texts' structure as an additional dimension for practical qualitative analysis. Previous models were based on letters, words, or phrases, which is insufficient to get good results. Sun et al. (2018) presented a preliminary research result that shows the structure of texts can also be an essential feature in studying personality detection from texts. More specifically, the study deployed a model called 2CLSTM. 2CLSTM is a bidirectional long short-term memory network (LSTM) concatenated with convolutional neural network (CNN), which can detect a user's personality based on text structures. They conducted evaluations across two datasets containing long and short texts. The results have achieved better results, demonstrating the proposed model can efficiently learn useful text structure features for qualitative analysis.

Chen et al. (2018b) highlighted two challenges for ML applications in qualitative coding. On the one hand, a lack of understanding between disciplines may negatively impact trust, limiting the application of ML in qualitative analysis because people using qualitative methods are generally not trained in ML techniques. In some cases, ML experts' limited understanding of social science values and methods can hamper effective collaborations. On the other hand, there are fundamental differences between qualitative and quantitative methods. In quantitative analysis, data points that appear very few times may be considered noise, but from a qualitative analysis perspective, the quantity of instances does not always reflect significance.

As for future work supporting qualitative analysis in HCI/UX activities, we anticipate that more human-centered and interpretable AI methods can potentially transform social science research (Chen et al., 2018b). Specifically, current AI models are not always interpretable, and the AI community needs to increase the transparency and interpretability of AI technologies for qualitative analysis. Also, we need to explore ways to make the usage of AI-based capabilities a meaningful task in HCI/UX practices, bridging the gap between AI and HCI/UX communities.

5.2.4 UX Evaluation

HCI/UX professionals conduct UX evaluations to achieve design improvements based on user feedback and insights. However, traditional UX evaluation methods like questionnaires and usability testing are often resource-intensive and not scalable. With the continuous evolution of many intelligent products (e.g., AI assistants, autonomous driving, smart homes, and intelligent robots), these UX evaluation methods will be difficult to meet new scenarios (Lan & Liu, 2020). Emerging technologies such as AI and big data are currently influencing how HCI/UX professionals conduct UX evaluations (Lan & Liu, 2020; Tan et al., 2020).

The methods of finding products with poor UX through big data-based analysis are gradually being adopted by collecting the user's stay time, login frequency, conversion rate, and other indicators (Tan et al., 2020). For example, Yu (2018) designed a big data intelligent algorithm framework for UX evaluation of mobile media clients through indicators such as the number of fans, page views, activity, stickiness, and emotional inclination, and finally implemented a cognitive algorithm framework. Li (2019) conducted an in-depth analysis of the user stickiness of NetEase Cloud Music through big data analysis algorithms designed for popularity. Recent work has also attempted to address other aspects of human behaviors on UIs, e.g., predicting human perception of UI interactivity based on user behavior data such as mouse/keyboard logs, eye tracking, and usage log (Swearngin & Li, 2019). Souza et al. (2022) established a framework that employs eye and mouse tracking methods, keyboard input, self-assessment questionnaires, and AI-based algorithms to evaluate UX and categorize users in terms of performance profiles.

Furthermore, recent academic work explores the challenges of evaluating UX using multiple data sources and proposes ML-based approaches (Asim et al., 2020). Connecting questionnaire results with log and time series data about user behavior may be used as labeled data for input data to supervised ML. Such approaches may allow for continuous monitoring of changes in users' UX and inform HCI/UX professionals of the opportunity for improvement. Chromik et al. (2020) proposed that some equipment, such as electroencephalography (EEG) sensors, could be used during real-time usability tests in lab contexts to record typical flows of interaction and users' emotional responses. These behavioral and emotional responses could be used as labels for an ML model (Chromik et al., 2020).

For usability testing, ML was used for selecting participants for usability tests (Gilbert et al., 2007) and A/B tests (Kharitonov et al., 2017). In addition, automatic real-time evaluation of mobile-based experience via emotional logging systems using video-captured facial expressions in lab contexts (Filho et al., 2015), using acoustic data (Soleimani & Law, 2017) and skin conductance signals (Liapis et al., 2015).

Studies show that AI technology may offer a more resource-effective approach (Chromik et al., 2020). For example, Yang et al. (2020b) proposed a methodology for measuring UX using AI-aided design (AIAD) technology in mobile application design. AIAD focuses on the rational use of AI technology to measure and improve UX. The researchers propose to obtain user behavior data from logs of mobile applications. They designed and used projected pages of the application to train neural networks for specific tasks in terms of the click information of all users when performing the tasks. The goal was to make the deep neural network model simulate the user's experience in operating a mobile application as much as possible. Thus, user behavior features could be aggregated and mapped in the connection and hidden layers. Finally, the optimized design was executed on the application to verify the efficiency of the proposed methodology.

We further provide two more examples illustrating how AI technologies can help facilitate UX evaluation with three different approaches: visual search performance modeling, user emotion detection, and user interaction behavior modeling.

The first example involves visual search performance modeling (Yuan & Li, 2020). Modeling visual search performance not only offers an opportunity to predict the usability of an application before actually testing it on real users but also helps HCI/UX professionals better understand user behavior. The authors first analyzed a large-scale dataset of visual search tasks on actual web pages. They then presented a deep neural network that learns to predict the scannability of webpage content, i.e., how easy it is for a user to find a specific target. The model leveraged heuristic-based features such as target size and unstructured features such as raw image pixels. The model then analyzed the user behaviors to offer insights into how the salience map learned by the model aligns with human intuition and how the learned semantic representation of each target type relates to its visual search performance. This approach allows HCI/UX professionals to model complex interactions involved in visual search tasks, which traditional analytical methods cannot quickly achieve.

The second example is user emotion detection. Emotion is one aspect of UX that exploits an ML-based automatic UX evaluation for understanding users' emotions by analyzing the log data of the users' interactions with websites (Desolda et al., 2021). The evaluation results show the performance of each ML algorithm according to the seven emotions. It is evident that emotions like sadness, anger, fear, disgust, and surprise were predicted with higher accuracy; joy was instead predicted with medium accuracy, while contempt had lower accuracy in all cases.

Lastly, Bakaev et al. (2022) deployed neural networks-based approaches for predicting the visual perception of UI. As testing and validation of graphical user interfaces (GUIs) increasingly rely on computer vision, CNN models that predict UX start to achieve decent accuracy. However, CNN models require vast amounts of human-labeled training data, which are costly or unavailable for HCI/UX activities. This study compares the prediction quality of CNN and artificial neural networks (ANN) models to predict visual perception in terms of aesthetics, complexity, and orderliness scales for about 2700 web UIs assessed by 137 users. The results suggest that the ANN architecture produces a smaller mean squared error (MSE) for the training dataset size (N) available in our study but that CNN should become superior with $N > 2912$.

While using AI technology in UX evaluation is promising, we still face challenges (Li et al., 2020). For instance, deep ML methods are often data-hungry, while interaction data is relatively scarce compared to classic ML problems such as computer vision or NLP. Deep ML models are not easy to analyze. While better modeling accuracy is of great benefit, the interpretability of a model is crucial for HCI/UX professionals to gain more insights about UX. Further collaborative work is needed between the AI and HCI/UX communities.

5.3 AI IN HCI/UX DESIGN

5.3.1 AI FOR UI DESIGN

Designing an excellent GUI requires much innovation and creativity, but the process is time-consuming and error-prone (Lu et al., 2022). Recently, AI-driven design (e.g., algorithmically powered tools) has become popular. AI-driven design vows to move UX design to another degree of digital experience in support of wireframing automation, visual design analysis, and UI pattern-driven design (Baker, 2019). Many researchers have worked on building design support tools to improve the efficiency of UI work. Also, many commercial design prototyping tools are developed. These tools have greatly helped designers create UI prototypes supporting UX work.

Many initial AI-based capabilities were released to support UI design. For example, Adobe's Creative Cloud software can realize the intelligent analysis function of multimedia files such as images and videos.[1] They can provide intelligent material recommendations according to the designer's design needs. Autodesk's Dream Catcher can quickly generate thousands of design proposals for designers to choose. Microsoft and Airbnb experimented with converting paper sketches directly into GUI code, bypassing much of the digital wireframing phase (Wilkins, 2018). Some of the tools are even beyond GUI. They utilize pre-trained AI algorithms that have the potential to support new forms of interaction by processing eye, face, body, and hand movements captured through webcams, speech commands captured through the browser's audio channel, and text through web elements (Li et al., 2020).

There are many (or potential) benefits to leveraging AI technology in UX design. For example, Gajjar et al. (2021) proposed Akin, a UI wireframe generator that uses a fine-tuned SAGAN model to generate UI wireframes for smartphone UI design. The researchers annotated and classified 500 UI screens from RICO into five commonly used mobile design patterns. The SAGAN model was trained with the dataset to generate UI wireframes for a given UI design pattern. An evaluation of Akin conducted with 15 UX designers shows that the designers rated the quality of wireframes developed by Akin as approximately equal to designer-made wireframes. Also, the designers could not distinguish UI wireframes generated by Akin from designer-made wireframes 50% of the time.

AI-based design capabilities also support design creativity, such as Simon's optimization-based design (Yang, 2018), which can all be linked to today's interest in employing ML to assist human creativity. TensorFlow.js is an open-source AI platform for developing, training, and using models in a browser or anywhere Javascript can run (Li et al., 2020). At Google, TensorFlow.js has been leveraged as a platform for AI+HCI collaborative research. TensorFlow.js is a browser-based ML framework to enable new forms of HCI/UX design innovation. TensorFlow.js provides a rich set of features accessible to researchers with different levels of ML experience. The library allows design experts to build models from scratch but also makes it easy to integrate pre-trained models.

The following list further summarizes some benefits (or potential benefits) of using AI-based capabilities supporting UI design (Baker, 2019; Lu et al., 2022; Vetrov, 2022; Chen et al., 2018a; Abbas et al., 2022):

- Quickly make various design varieties per the user's response
- Support design creativity
- Make wireframing and prototyping work more efficiently and less monotonously
- Transform UI sketches directly into a prototype
- Potential to immediately change a whiteboard sketch over into a functional prototype
- Quickly design alternative exploration
- Support design customization to support personalized UX
- Do design guideline violation check
- Empower design decision-making
- Help prepare UI assets and content
- Translate digital UI mockups into UI specifications

A representative approach of AI-based UI design is called *Generative UI Design*. Generative UI design is based on AI generative technology. Generally, with the development of AI generative technology, people can realize collaborative creation with AI in music, painting, writing, design, dance, etc. (Li et al., 2020). AI generative technology can quickly generate new samples that meet the specifications based on specific data sets so that novices can quickly start creation or reduce the repetitive work of designers. For example, analyzing big data on clothing design through modeling and visualization techniques to expand the ideas of clothing designers and gain insights (Glauser et al., 2019).

With the millions of websites and mobile apps available, many UX problems an HCI/UX designer encounters may have already been considered and solved by someone else. Specifically, in the UI design area, generative models (e.g., Variational Autoencoders) were trained on a large set of UI design examples that can suggest design alternatives for HCI/UX designers (Li et al., 2020). Systems based on these AI methods often leverage human support or" Wizard of Oz" techniques to collect the data from large design samples and eventually generate design solutions informed by the collected data (Vaccaro et al., 2018).

Researchers have promoted a hybrid intelligence approach for effective generative UI design. Specifically, rather than using humans solely for data collection to train an AI system, the hybrid intelligence approach incorporates human users, often crowd workers, as an essential and permanent component in an interactive system for complex design tasks (Li et al., 2020). Such a hybrid intelligence approach provides rich opportunities to combine human and machine intelligence to collaborate on a task and improve each other dynamically and interactively. A system powered by the hybrid intelligence approach needs to synthesize responses from multiple designers to achieve acceptable performance or availability for the system.

Another approach for effective generative UI design is to foster a creative, generative ML approach (Kayacik et al., 2019). Research shows that such an approach is more robust when multiple designers with different points of view actively contribute to them. Currently, many UXers do not have the ML education needed in the industry. This lack of education is hampering ML research teams' capacity to have a broad impact on their projects. To address the issue, the Google People and AI Research (PAIR) group developed a novel program method in which Uxers are embedded into an ML research group for 3 months to provide a human-centered perspective on the creation of ML models. The first full-time cohort of Uxers was embedded in a team of ML research scientists focused on deep generative models to assist in music composition (Kayacik et al., 2019). At the end of 3 months, the Uxers had new ML knowledge, and ML research scientists had a greater understanding of user-centered practices. The PAIR program results show that UX research and design involvement in creating ML models help ML research scientists more effectively identify human needs that ML models will fulfill.

However, the generative UI design method is limited, so the design realization is still limited to the traditional UI visualization level (Xu & Ge, 2018). It attaches great importance to the novelty of the appearance but lacks attention to UX design. It is not mature enough to deliver optimal UX to HCI/UX designers, especially with the business processes and structure built. Researchers also found that the innovative algorithm's information overload and uncertain output in human-AI collaboration are the key challenges (He et al., 2019). Also, Morris et al. (2022) proposed two design spaces for consideration when developing future generative AI models: how HCI can impact generative models (i.e., interfaces for models) and how generative models can impact HCI (i.e., models as an HCI prototyping material).

5.3.2 AI as a New Design Material

As AI technology advances, HCI/UX professionals regularly integrate AI capabilities into new apps, devices, and systems (Dove et al., 2017). AI becomes available as a resource to use by non-experts like HCI/UX professionals. First and foremost, intelligence is becoming a new design material (Holmquist, 2017). The options of a designer are, to a large extent, defined by the materials they have to work with. For instance, a product designer would need to be aware of the physical characteristics of materials such as plastic, wood, and metal, as well as how these fit together mechanically, to design an aesthetically and functionally pleasing experience. As AI becomes a more vital part of everyday products, HCI/UX designers will have to figure out how to work with intelligence as a new material.

With AI as a new design material, the primary role for HCI/UX designers is currently transitioning to augment end users with extended capabilities (e.g., new ideas, emotion design), besides routine UI design work. HCI/UX designers are becoming creators of the interface between humans and technology by leveraging algorithms. AI as a design material means that HCI/UX designers should view AI as a capability as an application service tailored to a specific functionality to support a particular experience (e.g., a "search" function). For instance, for an application with conversational UI. The traditional approach is that a customer asks a question and a human agent responds for help. If we add AI to that conversation function by training models of the language, having those models process that language and algorithmically build the best response and return that response with an AI-based virtual support agent. Thus, this new material of invisible, personalized, conversational design is algorithms. HCI/UX designers can take an active role in bridging algorithms and the UI to bring significant experience to end users with technology.

AI as a new design material is essentially an algorithm as a new material. However, algorithms have many limitations, impacting the outcome of using these "design materials." Pavliscak (2016) lists several limitations of algorithms

- Algorithms are not neutral or objective. Algorithms have a point of view with potentially biased outputs. Humans create algorithms, so their point of view gets embedded in the system.
- Algorithms don't understand you as a complex individual. Algorithms generalize, simplify, and filter out things they consider to be irrelevant.
- In many cases, algorithms use other people's data to fill in missing bits and pieces. The result is that algorithms *don't* reflect complicated humans.
- Algorithms are opaque. It's not always clear how or why they work the way they do. People who write them don't fully understand how they work.

There has been continuing study on how to approach UX design practice while working with AI as a design material (Dove et al., 2017; Yang et al., 2018a; Amershi et al., 2019). Researchers argue that while data tell us about people and organizations, algorithms create guidelines, and ML shapes the experience (Pavliscak, 2016). *Algorithms* are considered a set of guidelines on how to perform a task. When you send a text message or do an Internet search, HCI/UX triggers a nested set of interdependent algorithms. To best make use of algorithms for UX, Holmquist (2017) proposed the following design guidelines when algorithms are used for design:

- *Reveal the effects of algorithms*: users don't understand how algorithms work, and experience designers need to make algorithms' results more apparent
- *Participatory design for algorithms*: let users participate in their data creation for a personalized experience and choose a level of trust using different personal preferences
- *Designing for transparency*: let users understand how AI affects their interaction with applications
- *Designing for opacity*: it is no longer possible to explain exactly why or how an AI does what it does
- *Designing for unpredictability*: no matter how well-trained a neural network is, it is still drawing its conclusions from given data
- *Designing for learning*: the learning must be built into the interaction and completely unobtrusive, so it does not feel like the user doubles as the AI's training wheels

- *Designing for evolution*: AI systems will continue to evolve. It will be necessary to communicate this to users so that they know what to expect and can benefit while avoiding unpleasant surprises
- *Designing for shared control*: how AI systems can be designed to allow the sharing of power with users

While AI technology has brought in values for HCI/UX design, we still face challenges to fully leverage this new type of design materials. Dove et al. (2017) conducted a survey that shows some significant challenges: (1) There are challenges with using AI capabilities from a human-centered perspective. Current HCI/UX design education cannot prepare future design graduates to incorporate AI into their work. (2) While ML pushes the boundaries of design, the balance of collaboration with engineers and developers is currently such that design-led innovation is still rare. (3) UX/UI prototyping with AI/ML is difficult. HCI/UX designers used to create prototypes in the form of sketches, plans, and physical models made of paper, cardboard, or foam (Hallgrimsson, 2012).

5.3.3 AI as a Design Collaborator

Design ideation is a source of innovation in the early stages of a development process. Beyond the AI capability to support UI design as discussed above, we also need to rethink the current role of HCI/UX designers. UI design should be a creative process involving multiple iterations of different prototyping fidelities to create a UI design. With the help of AI, we need to consider how to enable AI to perform repetitive tasks for the designer while allowing the designer to take command of the creative process. This approach would greatly benefit designers in co-creating design solutions with AI (Liao et al., 2020). Such a collaborative creation with AI may further promote optimal experience in solutions (Oh ct al., 2018).

Researchers have been promoting this approach. For instance, Liao et al. (2020) proposed a framework of AI-augmented design support for the early stages where AI's role in creativity is related to creating representation, triggering empathy, and promoting engagement. Similarly, McCormack et al. (2020) characterized AI as a creative agent system that provokes, challenges, and enhances human creativity. Verganti et al. (2020) further claimed that AI reinforces design principles such as human-centered design, leading to potentially more creative solutions. AI will enable the designer's work, boost their creativity, and help experts create the best quality design products in a minimum time (Inkbot Design, 2021).

Main & Grierson (2020) proposed that AI can perform as an assistant, collaborator, researcher, or facilitator but might also play the role of future co-creator. Furthermore, McCormack et al. (2020) consider AI a system that allows creative collaboration with designers. Li et al. (2020) argued that rather than using humans solely for data collection to train an AI system, hybrid intelligence incorporates human users as an essential and permanent component in an interactive system for complex design tasks.

De Peuter et al. (2021) challenged the current approach and argued that AI for supporting designers needs to be rethought. It should aim to cooperate, not automate, by supporting and leveraging the creativity and problem-solving of designers. How to infer designers' goals and help develop a creative design needs to be figured out. They believe there is an urgent need to develop AI methods to cooperate with designers, working as assistants communicating with a designer about the design goal while supporting them in working toward that goal. In such a collaborative process, HCI/UX designers should remain the primary actor in the design process. As active participants, the designers explore and try things out to refine their goals. Further, the collaborative process can leverage the designer's creative abilities and expertise to build innovative designs.

Specifically, De Peuter et al. (2021) proposed a general-purpose approach for cooperative assistants in design problems. Collaborative design assistance has been offered for specific design problems, but the proposed method is to support a wide range of interactions. It uses a generative user model to infer a designer's goal from their behavior and plan how to assist the designer best.

AI in HCI Design and User Experience

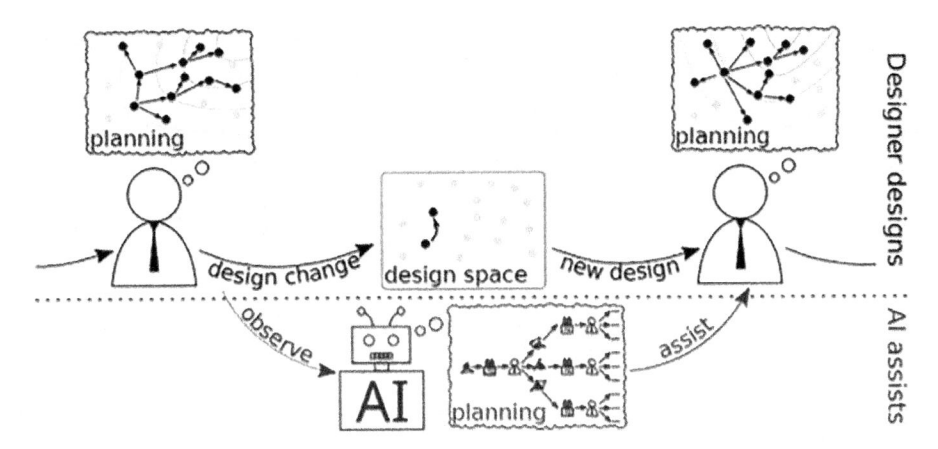

FIGURE 5.1 An illustration of the designer-AI collaborative design activities on a trip planning example. (De Peuter et al., 2021.)

De Peuter et al. (2021) demonstrated the approach in a trip planning example (see Figure 5.1). As illustrated in Figure 5.1, AI should appreciate the explorative character of designers' thinking. Within this design process, designers generate solutions not only to solve a problem but also to learn about it, including its objectives and constraints. They can mentally plan over a design space based on a utility function (shown as contours in Figure 5.1). The utility function evolves as the design progresses. The AI should collaborate in this creative process, for example, by proposing high-quality solutions and complementing the designer's problem-solving. To do so, it needs to know the designer's utility function. The study suggested creating AI assistants that can infer this utility from observations and then use it to assist a designer.

Also, Chen et al. (2019) proposed an integrated approach for enhancing design ideation by applying AI and data mining techniques. This approach consists of two models, a semantic ideation network and a visual combination model, which inspire semantically and visually based on computational creativity theory. The semantic ideation network provokes new ideas by mining knowledge across multiple domains. A generative adversarial networks model is proposed for generating UI objects for the visual combination model. An implementation of these two models was developed and tested, indicating that the approach can create a variety of cross-domain concept associations and advance the ideation process quickly and easily.

Liao et al. (2020) also proposed a framework of AI-augmented design support that involves the human ideation components and design tools related to AI in the early design stages. The framework describes the explicit roles of AI in design ideation as representation creation, empathy trigger, and engagement. The framework suggests approaches to assist cognitive patterns in the design process. An empirical study was conducted to investigate the cognitive patterns of design representations and design rationales. The study involved 30 designers with concurrent think-aloud protocols and behavior analysis. The study identified the opportunities for AI to support human creativity, and AI could provide inspiration, inform design scope, and request design actions.

5.3.4 Challenges and Future Work

Despite attempts to integrate HCI and AI, these HCI/UX designers experience challenges in incorporating AI into common UX design paradigms (Policarpo et al., 2021). We summarize the overall challenges of applying AI in HCI/UX design.

First, the AI-based approach challenges the typical activity of UX/UI prototyping. It is often difficult to convince leadership to commit to more innovative designs. The AI-based method requires an unwieldy amount of data to create a functional prototype. This approach could conflict with

UX mantras like "fail fast, fail often" (Dove et al. 2017). Consequently, it isn't easy in research to experiment with many different design solutions in searching for the best. In practice, designers could not demonstrate or validate their designs' value through a working prototype as they traditionally did. Recent research also founds a lack of research integrating UX and AI (Abbas et al., 2022). One example of the obstacle is UX designers' struggle when collaborating with data scientists. Another obstacle is the lack of the tools and abilities needed to sketch or prototype when using AI as a design material.

Second, many AI-based research projects' impact remained within the academic research community and haven't succeeded in making practical influences on industry practices (Jiang et al., 2022). For instance, there is a lack of research on ML algorithms and UX, especially in envisioning how ML might improve UX. Bridging this gap requires research that identifies practitioners' specific needs and provides translational resources to benefit from the latest technological advances and academic research findings.

Third, Abbas et al. (2022) argued that I/ML has remained underutilized to assist designers and has yet to be fully integrated into design patterns, education, and prototype tools. Therefore, tools are still in the early stages and cannot cover all conceivable questions. Also, tools were not designed with the participants of UX designers (Abbas et al., 2022).

Finally, the target users of many AI-based prototyping tools are mostly software developers rather than HCI/UX designers (Sun et al., 2020). Many HCI/UX designers lack AI knowledge, so it is still challenging to use these tools for prototyping. Further research is needed on prototyping tools for HCI/UX designers to help quickly build prototypes, discover design problems early, and reduce product development risk.

As we look forward, with the emergence of new AI-based design paradigms, HCI/UX design activities require the support of new tools. The development of these new tools should fully consider the knowledge background and way of thinking of HCI/UX designers and the collaboration between these designers and AI/software engineers in a real design environment (Sun et al., 2020). Undoubtedly, AI technology can't replace creative HCI/UX designers since these human professionals have unique capabilities to set the foundation for UX design. Still, AI definitely can support these designers in UX design as a new design material and collaborator for co-creative HCI/UX design.

5.4 AI FOR ENHANCING UX

5.4.1 Intelligent UI

AI technology is also transforming traditional UI into intelligent user interfaces (IUI). Traditional UI techniques (e.g., mice, keyboards, and touch screens) require the user to provide inputs explicitly. AI-based approaches are now robust to inherent ambiguity and noise in real-world data to analyze and reason about natural human behavior (e.g., speech, motion, gaze patterns, or bio-physical responses). AI-based techniques can also learn high-level concepts such as user preference, user intention, and usage context to adapt the UI and proactively present information (Gebhardt et al., 2019). As a paradigm shift, AI technology holds great promise in shifting how we interact with machines from an explicit input model to a more implicit interaction paradigm in which the machine observes and interprets our actions (Li et al., 2020).

The idea of introducing intelligence to HCI and UI sprouted decades ago in the form of intelligent computer-assisted instructions, which later gained a wider following and application as IUIs (Maybury, 1998). IUI aims to improve the efficiency, effectiveness, and naturalness of human interaction with machines by representing, reasoning and acting on models of the user, domain, task, discourse, and media. Different disciplines support the field, including AI, software engineering, HCI, human factors engineering, psychology, etc.

Different from traditional UI, IUI should be able to adapt its behavior to other users, devices, and situations (Gonçalves & da Rocha, 2019). A non-IUI considers an "average user" in design, i.e., the

UI is not designed for all types of users but for an "average" of all potential users (Ehlert, 2003). Typically, we have one context of use in non-IUI; but the context of use can change over time in an IUI. IUIs use adaptation techniques to be "intelligent/adaptive" with the ability to adapt to the user, communicate with the user and solve problems for the user" (Ehlert, 2003). Its difference from traditional UI is that they represent and reason concerning the user, task, domain, media, and situation (Jaquero et al., 2009). For instance, Stefanidi et al. (2022) proposed a novel computational approach for the dynamic adaptation of UIs, aiming at enhancing the situational awareness (SA) of users by leveraging the current context and providing the most useful information (Endsley, 2016). By combining Ontology modeling and reasoning with Combinatorial Optimization, the system decides *what* and *when* to present it, *where* to visualize it in the display - and *how*, considering contextual factors. The approach is to optimize the SA associated with the displayed UI *at run-time* while avoiding information overload and induced stress. The results of two evaluations indicate that the system enhances SA, and while not imposing workload, it provides an overall positive UX.

The goal of developing IUI is to make full use of advanced AI technology to provide natural and effective human–machine dialogue. For example, new technologies (e.g., language recognition, facial recognition, gesture input, and gaze tracking) offer a natural UI for systems; multi-channel interaction through multiple modalities captures user intent, behavior, and contextual scenarios to improve further the naturalness, accuracy, and effectiveness of interaction. At the same time, the effective user-centered design method is used to optimize the design of IUI.

In addition, there are several reasons why we need to research IUI. First, IUI helps promote the development of new AI technologies. Throughout modern technology development, GUI and mouse have promoted the popularization of personal computer technology; multi-touch screen technology has improved mobile phone and mobile UX. Therefore, IUI research will find suitable application scenarios for developing AI technology.

IUI also helps further exert human capabilities and enhance human intelligence. For example, brain-computer interface research helps develop human potential and enhance the abilities of disabled people with disabilities through rehabilitation therapy. IUI actively understands user status (e.g., physiology, psychology, intention), so it will better understand users and predict their needs and behaviors, adaptively support users' activities, and ultimately make users more comfortable and interact with machines efficiently and securely.

Lastly, IUI aims to provide benefits to users such as adaptivity, context sensitivity, and task assistance (Gonçalves & da Rocha, 2019). IUI research will provide more natural and efficient intelligent systems, bringing economic benefits and considerable returns on investment to users, developers, and manufacturers. Effective IUI can improve users' work efficiency and make their work and life more convenient. The productivity of HCI can be significantly enhanced not only by contact (mouse pointing device, joystick, touchpad, keyboard, etc.) methods but also by contactless (speech and gesture commands, head and body movements, facial expressions, and user's look direction, etc.) ones (Karpov & Yusupov, 2018).

Historically, interaction paradigms have guided UI development in HCI work, e.g., the WIMP (window, icon, menu, pointing) paradigm. However, WIMP's narrow sensing channels and unbalanced input/output bandwidth restrict human–machine interaction (Fan et al., 2018).

Table 5.3 summarizes the new HCI characteristics that AI technology has brought in, emerging human factors issues, and critical issues for future HCI/UX work (Xu, 2019a, 2020).

As Table 5.3 lists, these transformative characteristics of AI technology lead to the need for innovative UI capabilities and interaction paradigms (e.g., two-way, collaborative UI). This will ultimately prompt the development of more natural and effective IUI and will require HCI/UX professionals to develop more effective approaches to explore the design of innovative UI design.

From a methodology perspective, research shows that there is a lack of effective methods for designing IUI, and HCI/UX professionals have had difficulty performing the typical HCI activities of conceptualization, rapid prototyping, and testing (Yang et al., 2020b; Holmquist, 2017; Dove et al., 2017). The HCI/UX community has realized the need to enhance existing methods (Stephanidis,

TABLE 5.3

New Characteristics and Human Factors Issues of AI Technology, Critical Issues for HCI/UX Work

Transformative Characteristics of AI Technology)	Emerging Human Factors Issues in AI Technology	Critical Issues for HCI/UX Work
From "one-way" to "man-machine collaboration-based two-way" UI	• AI systems no longer passively accept user input and produce expected output according to fixed rules • AI-based agents can actively perceive to capture and understand the user's physiological, cognitive, emotional, intentional, and other states and actively initiate interaction and push services to users	• Human-machine teaming/collaboration-based interaction models and paradigms • Cognitive models of the user's states (e.g., situation awareness, physiology, cognition, emotion, and intention)
From "usability" to "explainable AI" UI	• AI "black box" effects can lead to inexplicable and incomprehensible system outputs • AI "black box" effect raises AI trust issues	• Innovative UI technologies (such as visualization) and design • uman-centered" explainable and understandable AI (Ehsan et al., 2021) • Accelerated transformation of psychological explanation theories
From "simple attributes" to "contextualized" UI	• AI system input includes "contextualized" data (e.g., the context of usage, user behavior), besides traditional information (e.g., simple objects such as target location and colors)	• Modeling and intelligent deduction of "situational" features (e.g., user characteristics, digital personas) based on data such as operating context and user behavior • Personalized functionality suitable for user needs and usage scenarios
From "precise input" to "fuzzy reasoning," UI	• User input is not just a single precise form (e.g., keyboard, mouse), but may also be multimodal, ambiguous interactions (e.g., user intent) • Ambiguous interaction issues in operating scenarios (e.g., random interaction signals and ambient noise)	• Methods and models for inferring user interaction intentions under uncertainty • Naturalness and effectiveness of HCI in an ambiguous state
From "interactive" to "collaborative" UI	• UI supporting both human–machine interaction and human–machine teamwork • UI supporting effective human–machine collaboration	• Alternative design paradigms and models for human–machine collaborative UI • UI design standards for intelligent HCI • Interaction design effectively supports human–machine collaboration (e.g., human–machine control hand-over in an emergency)

Salvendy, et al., 2019; Xu, 2018; Xu & Ge, 2020). To this end, Xu, Dainoff, et al. (2021) assessed existing methods of HCI, human factors, and other related fields. As a result, they proposed alternative approaches that can support the effective design of IUI better. These alternative methods can help HCI/UX professionals overcome the limitations of conventional HCI methods when designing IUI.

From a process perspective, research shows HCI/UX professionals have challenges integrating HCI/UX processes into the process of developing IUI systems. For instance, many HCI/UX professionals joined AI projects only *after* the requirements were defined (Yang et al., 2020a). Consequently, the design recommendations from HCI/UX professionals could be quickly declined

(Yang, 2018). AI professionals often claim that many problems that HCI could not solve in the past have been solved through IUI technology (e.g., voice UI), and they can design the interaction by themselves. Still, studies have shown that the outcomes may not be acceptable from a UX perspective (e.g., Budiu & Laubheimer, 2018). Some HCI/UX professionals find collaborating effectively with AI professionals challenging due to a lack of a shared process and a common language (Girardin & Lathia, 2017). Also, studies have shown that HCI/UX professionals are not prepared to provide effective design support for AI systems (Yang, 2018).

For future work, we offer several strategic recommendations. Firstly, HCI/UX professionals need to integrate HCI/UX methods into the development process of IUI to maximize interdisciplinary collaboration. For instance, to understand the similarities and differences in practices between HCI/UX professionals and other professionals, Girardin & Lathia (2017) summarize a series of touch points and principles. Within the HCI community, researchers have indicated how the HCI/UX process should be integrated into the process of developing IUI systems (Lau et al., 2018). Specifically, Cerejo (2021) proposed a "pair design" process that puts two people (one HCI/UX professional and one AI professional) working together as a pair across the development stages of IUI systems.

Secondly, HCI/UX professionals must update their skillsets and knowledge in AI. While AI professionals should understand HCI/UX approaches, HCI/UX professionals also need to have a basic understanding of AI technology and apply the knowledge to facilitate the process integration and collaboration so that HCI professionals can fully understand the design implications posed by the unique characteristics of AI technology and be able to overcome weaknesses in the ability to influence IUI systems as reported (Yang, 2018).

Thirdly, future work needs to adapt AI technology to human capability. Human-limited cognitive resources become a bottleneck of HCI design in the pervasive computing environment. For instance, in an implicit interaction scenario initiated by intelligent ambient systems, intelligent systems may cause competition between human cognitive resources in different modalities, and users will face a high cognitive workload. Thus, HCI design must consider the "bandwidth" of human cognitive processing and resource allocation while developing innovative approaches to reduce user cognitive workload through appropriate interaction technology, adapting AI technology to human capabilities.

Fourthly, we need to develop new interaction paradigms that better fit IUI. IUI requires effective UI paradigms. In the realization of IUI, hardware technology is no longer an obstacle, but the user's interaction ability has not improved. Designing effective multimodal integration of sight, hearing, touch, gestures, and other parallel interaction paradigms is an essential part of HCI research in the age of intelligence. Historically, interface paradigms and models have guided the development of HCI (e.g., WIMP). However, the limited perception channels and unbalanced input/output bandwidth of WIMP restrict the further evolution of the UI in the AI age. Existing studies have proposed the concepts of Post-WIMP and Non-WIMP, but the effectiveness remains to be further verified. HCI/UX community should support defining paradigms, metaphors, and empirical validation to solve unique problems in IUI. It requires HCI/UX professionals to explore innovative ideas that can effectively facilitate interaction in IUI.

Finally, we need to develop HCI design standards that specifically support the development of IUI. Existing HCI design standards are primarily grown for non-IUI, and there is a lack of design standards and guidelines explicitly supporting IUI design. IUI design standards need to consider the unique characteristics of AI technology fully. There are initial design guidelines available, such as the "Google AI+People Guidebook" (Google PAIR, 2019) and Microsoft's 18 Design Guidelines (Amershi et al., 2019). The HCI/UX community must play a key role in developing these design standards.

5.4.2 AI Assistants

An intelligent assistant (IA) is an AI/ML-based computer system capable of intelligently assisting people. IAs have gained popularity over recent years; it ranges from helping people develop skills and exercise properly to rehabilitate physically (Islas-Cota et al., 2022), among other application domains. IAs are being deployed across domains, such as health, education, online social services, driving, domestic environment, enterprise/industry, fitness, and learning. To perform their users' daily tasks or services, IAs can send a message, make a phone call, search for specific information, set a reminder or calendar, and provide personalized recommendations. IAs are intelligent agents that employ AI techniques to provide a human-like interface (e.g., voice, vision) (Hu et al., 2019). They are also expected to perform more complex tasks, such as purchasing and accessing or managing smart Internet of Things (IoT) devices (Han & Yang, 2018). NLP and AI technologies enable IAs to self-learn users' schedule and taste through daily interactions and collecting awareness data (e.g., location and context) from the IoT, and then autonomously perform tasks based on user preferences and habits (Santos et al., 2016).

The objectives of IAs are to increase efficiency in an activity, better cope with an illness, resolve a problem, support everyday situations, refine skills, and attain a healthy life. Ultimately, IAs aim to enhance UX in their daily work and life. One good example of IA is voice assistants, which have been rising recently, such as Amazon Alexa, Google Assistant, and Siri from Apple (Zwakman et al., 2021). Voice assistants help facilitate human-computer dialogue naturally and intuitively, like conversations between humans.

Islas-Cota et al. (2022) presented a systematic review aiming to classify recent advances in IAs in terms of IAs' objectives, application domains, and workings. They identified what AI/ML techniques are used to enable the AI assistants. As a result, the study proposes a taxonomy of IAs in terms of objective, target users, domain, capability, functionality (i.e., enabler, device, triggering stimulus, and triggering actor), software, input, and output.

Research into AI-based digital assistants has a long history, dating back to Joseph Weizenbaum's well-known ELIZA in 1966 (Maedche et al., 2019). In parallel, global technology companies such as Microsoft, IBM, Google, and Amazon have been working with AI-based digital assistants to provide significant opportunities. The rise of IAs has opened a broad research area for HCI/UX professionals. It is a technology with an explicit interface to users and could therefore provide a fruitful avenue for HCI/UX research (enhancing UX/AI assistant 5). Much research has been done, but the work primarily focused on improving the technology. Their indirect objective is to enhance the usability of the IAs, and not on the usability aspect per se. Budiu & Laubheimer's (2018) usability study found that both voice-only and screen-based intelligent assistants worked well for only minimal, simple queries with relatively simple, short answers. Users had difficulty with anything else. In addition, as part of experience issues, the privacy and security aspect of the IAs (e.g., voice assistants) still exists, as many IAs are prone to various attacks that might steal user information (Zwakman et al., 2021). There are several ways in which IAs could be used that can create new ethical and legal issues (Almeida et al., 2020).

Besides the usability design and interaction issues of IAs, the collaborative relationship between humans and AI-based IA systems is another important topic for HCI/UX professionals. Traditionally, AI-based applications are considered a tool in support of humans. We should stop thinking of AI as a developing phenomenon independent of humans, and it is necessary to move on to the consideration of hybrid intelligence. Hybrid intelligence can be further understood from three aspects, considering hybrid intelligence as the sum of human and machine efforts in achieving a goal; the amplifier of human intelligence at the physiological level; and a partnership between humans and machines (Shichkina & Krinkin, 2022).

Shichkina and Krinkin (2022) further argued that IAs should not be just the creation of hybrid intelligence but a co-evolutionary hybrid intelligence (CHI). CHI is a symbiosis of artificial and natural intelligence, mutually developing, teaching, and complementing each other in co-evolution. Human–machine intelligence co-evolution is the fundamental building of more robust intelligent systems (Krinkin et al., 2021). Based on the concept of CHI, the goal of IAs is the mutual development of human and AI as a single indivisible organism.

From the perspective of human–machine intelligence complementarity, the most significant potential for IAs is a mutually beneficial collaboration (Maedche et al., 2019). Both humans and machines have relative strengths. While machines are ideal for conducting repeatable, highly structured tasks, collecting, storing, processing vast amounts of data, and predicting the future in stable environments, humans can handle abstract problems and deal with fragmented information much more efficiently.

Functional and task allocation between humans and machines is a classical activity for HCI/UX professionals. There is an intensive discourse on how humans interact with AI-based technologies and how the performance of a particular task should be divided between these two entities. It is crucial to involve humans to an appropriate level in task performance, depending on the task characteristics and the context (Maedche et al., 2019). A significant challenge for future research is to investigate how to distribute the tasks between these two entities at an appropriate level to achieve optimal performance.

Autonomy is another topic that little research has been done to examine the issue from the perspective of autonomy as intelligent agents (Hu et al., 2019). In the past, many topics about autonomy focused on human autonomy, such as job autonomy, human autonomy, and community autonomy. IA autonomy refers to the fact that IA, as an intelligent agent, can independently complete tasks in some scenarios. Autonomy is a double-edged sword factor for IAs, as it can increase benefits (e.g., exerting specific risky tasks) (Robert & You, 2018). It may allow a machine to over-control without human authority, which may put humans in a risk situation in some domains (Xu, 2019a, 2020).

The over-emphasis on human autonomy is because machines were not smart enough in the past. They can be regarded as relatively automated rather than autonomous. Recently, AI-enabled IAs to self-learn users' preferences through daily interactions and personal data, which ensures intelligent agents' autonomy (Hu et al., 2019). For instance, IAs will set the alarm clock to wake up according to the user's habits, reflecting the IA scheduling autonomy. Still, the complexity of IA executing both instructions is hidden, which will result in the user losing control over the specific execution process. Hu et al. (2019) raised several research questions for future work of Ais, such as whether decision-making autonomy will have a positive influence on perceived competence, whether perceived competence will have a positive effect on the intention of a user to IA continuous usage, and whether perceived uncertainty will harm the purpose to IA continuous usage.

We summarize the future research directions for improving the UX of IAs (Islas-Cota et al., 2022; Maedche et al., 2019; Zwakman et al., 2021).

- *Understanding human users' needs of IA*: Ensure that the design of IAs is in line with the user population's and society's goals and values. Research is needed to create a rich understanding of the needs and usage of the potential users of such assistants.
- *Interaction technology*: Need to improve the technology empowering these IAs to provide better capabilities, such as voice recognition, the ability to understand multiple languages, providing human-like speech output, adding emotions to these devices, and likewise.
- *User privacy*: Ensure that the users can trust them in their daily usage because many IAs collect potentially sensitive data from users' activities, such as visited locations.
- *Collaboration:* Need to explore further how to design effective collaboration between humans and IAs. Technically, the agent-based paradigm of IAs supports collaborative problem-solving either with other agents or with agent-human teams. Future work needs to exploit further the agent paradigm where multiple agents (namely, IAs) can coordinate, collaborate, and negotiate among themselves to provide users with a multi-domain ubiquitous assistance.
- *Evaluation:* Need to evaluate the overall system performance from a collaboration perspective. A common practice is comparing with utilized benchmark datasets to evaluate their IAs. We need to find practical evaluation approaches to assess the efficiency and effectiveness of IAs in a collaborative way.
- *Functional and task allocations*: Need to investigate from a conceptual perspective the interplays between humans and machines when using IA, investigating design variants of

IA for different task types. Collaboration between humans and IA may depend on the task type to achieve optimal experience.
- *Context-aware assistance*: Exploit unsupervised ML techniques to discover users' context and behavioral patterns. IAs can provide users with personalized and contextual assistance by establishing a context. Features such as users' activities, interactions, status, and intent detection, can establish a context that enables IAs to determine how and when assistance should be provided.
- *Emotionally aware assistance*: Need to explore more elaborate emotional models. Emotions are critical to humans in decision-making and communication, among other everyday activities.
- *Virtual, augmented, and mixed reality*: Need to leverage these technologies to help improve UX. Currently, there is a lack of IAs taking advantage of virtual, augmented, and mixed reality. For instance, IA can use virtual reality devices to assist patients with physical rehabilitation and train surgeons. Further HCI work is needed to enhance UX while interacting with IA.
- *Human characteristics*: Need to assess human characteristics in the design of IA, such as user expertise with the technology, personality, culture, social norm, delivering personalized assistance.

5.4.3 Recommender Systems

Recommender systems (RS) are software tools that support human decision-making, especially when choices are made over large product or service catalogs (Ricci et al., 2021). RS are integral to many of today's websites and online services. After 30 years, personalized recommendations are ubiquitous, fueled by advances in AI technology. To a large extent, making recommendations is an HCI/UX topic, which aims to determine how a computerized system can effectively support users in information search or decision-making contexts for an optimal experience.

The academic and industrial communities have proposed many recommender software and algorithms (Elahi et al., 2021). Most of these algorithms can gather various data types and exploit them to generate recommendations. These data types can describe either the item content (e.g., category, brand, and tags) or the user preferences (e.g., ratings, likes, and clicks). A recommendation list for a specific user is then made by filtering the items representing similar features to the rest of the items that the user liked/rated high. However, users may be exposed to risks, such as bad UX and decision difficulty. If the set of recommendations is unfortunate (e.g., poor decisions, the pre-selection of items, or the decision bias), this might lead to a poor experience.

The goal of a recommender system is to predict user interests and infer their mental processes. For example, personalized RS are one of the most widely used fields of big data technology. It is implemented by mining user attributes through user behavior data and realized through the inference of basic information (e.g., user's age, gender, residence, and educational background based on the user's online browsing behavior) (Wang et al., 2013). Based on the results of the inferred data (e.g., digital personas, user profile, behavior, and preference), systems can intelligently send recommendations to target users with a personalized experience. As a success case, Amazon's recommendation engine provides it with a conversion rate of up to 60% and a sales contribution rate of 30% (Li et al., 2015).

In general, there are two traditional RS (Elahi et al., 2021)

- *Collaborative filtering*: The method predicts users' preferences (i.e., ratings) by learning the preferences that a group of users provided and suggests to users the items with the highest predicted priorities. It is used in almost all application domains and relies on big data of ratings acquired from a typically extensive network of users (Desrosiers & Karypis, 2011).

The underlying assumption is that users with similar preferences will also have similar preferences in the future.
- *Content-based*: Content-based methods adopt content-based filtering (CBF) algorithms to build user profiles by associating user preferences with the item content (Deldjoo & Atani, 2016). Content-based approaches recommend items that share characteristics with items that the user has previously liked (e.g., items with a similar description or genre) (Calero Valdez et al., 2016; Zangerle & Bauer, 2022). Typical fields of application are recommending movies, music, or related products in e-commerce.

Despite these traditional methods' effectiveness, as we enter the age of big data and AI, more advanced techniques have been developed to build intelligent systems for quicker, more accurate, and personalized recommendations tailored to each user's needs and preferences (Elahi et al., 2021).

There are several types of AI-enabled RS that have been explored in academia and the industry:

- *Data-driven recommendations*: This method enables leveraging ML technologies to contextualize the big data to enhance the precision of suggestions, which facilitates the use of content (Beheshti et al., 2020a). The approach moves from traditional statistical modeling to advanced AI-based models, which will improve mining patterns between items and user descriptors to build better suggestions.
- *Knowledge-driven recommendations*: This method empowers simulating the expertise of the domain experts (e.g., crowdsourcing methods) and adopting techniques such as reinforcement ML to enhance the system's capability for making relevant and accurate recommendations (Beheshti et al., 2018).
- *Conversational recommendations:* This method provides more sophisticated interaction paradigms for preference elicitation, item presentation, or user feedback through conversational interactions between users and RS (Lei et al., 2020).
- *Intelligent ranking-based recommendations*: It can be trained by the domain experts' knowledge and experience to understand the context, extract related features, and determine the causal connections among various features over time. The goal is to change from statistical modeling to novel forms of modeling, such as deep learning, to improve potential similarities among descriptors and build a more accurate ranking (Chen et al., 2020).
- *Intelligent personalization-based recommendations*: It can support analytics around users' cognitive activities to provide intelligent and time-aware recommendations. The method tailors product and content recommendations to users' profiles and habits by analyzing users' behavior, preferences, and history. This process requires automatic data processing to identify meaningful features, select suitable algorithms, and use them for training a proper personalization model (Herath & Jayarathne, 2018).
- *Cognition-aware recommendations*: It aims to recognize the users' personalities and emotions and analyze their characteristics and affinities over time. The system needs to interpret social information (at a group level or on a one-to-one basis) and provide context-aware recommendations (Beheshti et al., 2020b).

Many AI models have been adapted for use in these AI-enabled RS. For instance, deep neural networks for collaborative filtering to model the user-item interactions, including deep factorization machines or (variational) autoencoders (Zangerle & Bauer, 2022). CNN are primarily used for learning features from (multimedia) sources for learning the data from audio signals (Van den Oord et al., 2013) or modeling latent features from user reviews and items (Zheng et al., 2017). Recurrent neural networks (RNN) are used to model sequences for sequential recommendations (Quadrana et al., 2017). Reinforcement learning models incorporate user contexts while continuously updating and optimizing the recommendation model based on user feedback (Zheng et al., 2018).

Early research on RS focuses on algorithms and their evaluation to improve recommendation accuracy (Calero Valdez et al., 2016). After a few decades, the field of RS has been driving toward consensus; that is, accuracy only partially constitutes the UX of a recommender system. As a result, there is an evolution from research on algorithms to research on UX with RS (Konstan & Terveen, 2021).

Human-centered RS are an approach that focuses on understanding the characteristics of RS and users as well as the relationships between them. The goal is to design RS' algorithms and interactions to fulfill better users' goals (Konstan & Terveen, 2021). Different from traditional technology-centered RS, a different set of questions need to be answered from the human-centered RS. For instance, what does it mean for a recommendation to be good? How many products are too many to recommend? Should I show the best recommendations or save some for later? When should the recommendations be diverse concerning each other or the user's history? What type of recommendations leads to better UX? Konstan & Terveen (2021) presented HCI research work focusing on UX and interactive visualization techniques to support the transparency of results. In addition, there is also a need for frameworks to combine human-centered RS research with the best ML algorithms to achieve scalable, efficient human-centered RS (Konstan & Terveen, 2021).

Furthermore, to enhance UX, Ekstrand et al. (2014) conducted research to understand how users perceive their performance, including dimensions of accuracy, diversity, novelty, personalization, and satisfaction. The work shows that these factors should be included in an analysis of algorithm performance while building a structural equation model. Willemsen et al. (2016) studied user choice overload in the context of RS. The results show that the diversity of items recommended affected the effort required to make a choice; diversity led to higher satisfaction choices but not always the highest-scoring choices for users. Research also shows that a recommender system built to optimize user engagement (rather than predictive accuracy) leads to recommendations that increase subsequent user engagement compared to predictive accuracy recommenders (Zhao et al., 2018).

How to effectively measure the UX of RS is essential, so HCI/UX professionals can identify the pain point to close the gap in design. The performance of RS is typically evaluated using offline and online experiments (Zangerle & Bauer, 2022). When assessing the effectiveness of a recommender system, people largely adopt offline rather than live user studies methods. Conversely, real users are requested to evaluate the recommendations in online studies. Offline studies are more popular than user studies, which are more complex and time-consuming. However, measuring the UX of RS is often challenging. Konstan & Riedl (2012) argued that evaluating the UX requires a broader set of measures. Regular algorithmic work can be done by using existing datasets; measuring UX requires developing additional capabilities that include both algorithms and UI.

More specifically, Knijnenbur et al. (2012) proposed a user-centric approach for evaluating RS. The framework links objective system aspects to accurate user behavior through a series of perceptual and evaluative constructs. It also incorporates the influence of personal and situational characteristics on the UX. The framework was validated using a method called structural equation modeling. The results show that subjective system aspects and experience variables are invaluable in explaining why and how the UX of RS comes about; the perceptions of recommendation quality and variety are essential mediators in predicting the effects of objective system aspects on the three components of UX: process (e.g., perceived effort, difficulty), system (e.g., perceived system effectiveness) and outcome (e.g., choice satisfaction). Also, the study finds that these subjective aspects strongly correlate to user behaviors (e.g., reduced browsing for higher system effectiveness).

Based on current literature, the following list summarizes the suggestions for future HCI/UX work in designing RS (Calero Valdez et al., 2016; Konstan & Riedl, 2012; Konstan & Terveen, 2021; Jannach et al., 2021):

- *Better understand how users make decisions*: For example, we need to know how RS can adapt to different needs (e.g., new users vs. experienced users) and how they can balance short-term with longer-term value.

- *Putting the user in control*: Users are often more satisfied when given control over how the recommender system functions. We need to design for the sweet spot so that the recommender system can balance serving users effectively while the users have the desired control.
- *Developing adaptive RS:* Previous research shows that user satisfaction does not always correlate with high recommendation accuracy and the user's knowledge level and interests. There is a need to adapt RS and their UIs to these other personal and situational characteristics.
- *Supporting affective design:* Emotions play a crucial role in human decision-making. Future work needs to explore novel sensing technologies for capturing user behavioral data (e.g., physiological data, facial expressions, speech) so that RS can detect emotions and adapt recommendations based on emotional responses.
- *Conducting ongoing research:* For instance, the research on real applications that allow incorporates diverse contexts, including multi-interaction modalities (e.g., voice/audio vs. text vs. visual interaction) and decision nature (e.g., health/habit, low- vs. high-stakes).
- *Developing rigorous methods for evaluating UX*: We need to continue to adopt rigorous methods for assessing the UX and user satisfaction, such as the structural equation modeling.
- *Designing for high-risk domains:* Spending money on an undesired product is the most significant risk for a user of e-commerce sites. Risk-aware algorithms or predictions of risks need to be further investigated. For instance, how to effectively visualize or communicate the uncertainty and risk of a recommendation to users, which is crucial for the systems in high-risk domains (e.g., medicine).
- *Developing insightful "beyond-accuracy" measures*: Many current methods rely on data-centric "offline" experiments that do not involve the human in the loop. We should focus much more on how systems affect both organizations and entire experience journeys, the diversity of the recommendations, or the novelty of the identified items.
- *Developing integrated solutions for better UX*: A fundamental challenge to the field of RS is the integration of content-based approaches (e.g., product information, user profiles), collaborative approaches (e.g., explicit and implicit ratings, tagging, and user preference), and contextual approaches (e.g., business rules, location, user task and mood, UI) into comprehensive RS.

5.5 CONCLUSIONS

AI technology is transforming how HCI and UX professionals work toward delivering optimal UX in their solutions, including all aspects of user research, UI technologies and design, and UX evaluation. AI-based solutions have raised user and academic awareness of technical innovation. As a result, AI is becoming increasingly popular in improving the quality of UX. This chapter summarizes how AI technology can help HCI/UX professionals in HCI/UX research and evaluation, HCI/UX design, and enhancing UX by leveraging AI-based capabilities. It also highlights the benefits of deploying AI technology for HCI/UX activities, the challenges that HCI/UX professionals face, and future HCI/UX work.

Recent advances in the AIGC (AI generative content) area are further transforming the HCI and UX, such as ChatGPT; that is, an AI chatbot developed by OpenAI and launched in November 2022 (ChatGPT, 2023). With ChatGPT, ChatGPT can support UX activities, such as exploring user needs by gathering user data, creating a bot for customer service by using the ChatGPT capabilities, and analyzing user research data. Because of advanced capabilities with large language models, ChatGPT has demonstrated more benefits than any other generative AI-based tools released so far. This new development will further benefit UX professionals in delivering optimal UX in their solutions. However, we don't believe that the AIGC tools like ChatGPT can eventually replace UX professionals. For instance, while ChatGPT has been trained on a large amount of text and has knowledge of various design concepts and practices, UX design is a complex and creative field that involves much more than just knowledge of concepts and fundamentals.

Also, AI's roles in HCI and UX must be assessed from a human-centered AI (HCAI) perspective. HCAI is a design philosophy in developing AI (Xu, 2019b; Shneiderman, 2020). HCAI aims to ensure AI benefits humans and enhances human capabilities, instead of hurting and replacing humans (Ozmen Garibay et al. 2023). Specifically, Xu (2019b) calls for an integrated consideration of three aspects in developing human-centered AI, technology, ethical AI, and human factors. *AI comes with new challenges. For instance, the bias in the data used to train AI models. If the data used to train the model is biased, the model's decisions and recommendations will also be biased, potentially leading to discriminatory or unfair treatment of certain groups of users, leading to negative UX. Privacy and security are also important concerns when using AI in UX design. AI systems may collect, store, and process large amounts of sensitive user data, which could lead to privacy breaches, data breaches or even cyber-attacks. Thus,* the integration of AI into the field of design brings both opportunities and challenges. As a UX professional, we need to consider the potential implications of AI technologies for our career development. It's also important to be aware of the potential threats such as bias, accountability, and privacy and security issues, and take steps to mitigate these risks.

To push the boundaries of what AI might be and might do, we need to continue to identify major unknown topics as a basis for future research endeavors (Yang, 2018). We must bridge the gap between HCI/UX professionals' work practices and AI-enabled capabilities. Professionals across disciplines need to seize this opportunity to create something entirely new. It's essential to treat the adoption of AI as a significant strategic and cultural shift for improving UX, not simply the installment of new technology (Schwartz et al., 2018). We need to collaborate on developing new methods, tools, and processes to help HCI/UX and AI professionals better innovate with AI (Bertão & Joo, 2021).

NOTE

1 Adobe Creative Cloud is a set of applications and services from Adobe Inc. that gives users access to a collection of software used for graphic design, video editing, web development, and photography. For examples, Adobe use machine learning-enabled features to help designers organize and edit their images more quickly and accurately. With object recognition in Lightroom, designers can auto-tag photos of a dog or cat. In Photoshop, machine learning can be used to automatically correct the perspective of an image for you. https://www.adobe.com/creativecloud.html

REFERENCES

Abbas, A. M., Ghauth, K. I., & Ting, C. Y. (2022). User experience design using machine learning: A systematic review. IEEE Access, *10*, 51501–51514.

Almeida, V., Furtado, E. S., & Furtado, V. (2020). Personal digital assistants: The need for governance. *IEEE Internet Computing*, 24(6), 59–64.

Amershi, S., Weld, D., Vorvoreanu, M., Fourney, A., Nushi, B., Collisson, P., ... & Horvitz, E. (2019, May). Guidelines for human-AI interaction. In *Proceedings of the 2019 CHI Conference on Human Factors in Computing Systems* (pp. 1–13).

An, J., Cho, H., Kwak, H., Hassen, M. Z., & Jansen, B. J. (2016, August). Towards automatic persona generation using social media. In *2016 IEEE 4th International Conference on Future Internet of Things and Cloud Workshops (FiCloudW)* (pp. 206–211). IEEE.

Asim, M.N., Ghani, M.U., Ibrahim, M.A., Mahmood, W., Dengel, A., & Ahmed, S. (2020). Benchmarking performance of machine and deep learning-based methodologies for Urdu text document classification. *Neural Computing and Applications*, 33(11), 5437–5469.

Bakaev, M., Heil, S., Chirkov, L., & Gaedke, M. (2022). Benchmarking neural networks-based approaches for predicting visual perception of user interfaces. In *International Conference on Human-Computer Interaction* (pp. 217–231). Cham: Springer.

Baker, J. (2019). The Designer's Guide to AI-Driven UX. *Medium.* https://medium.muz.li/the-designers-guide-to-ai-driven-ux-afbebdec1be3

Beheshti, A., Benatallah, B., Nouri, R., & Tabebordbar, A. (2018). CoreKG: A knowledge lake service. *Proceedings of the VLDB Endowment, 11*(12), 1942–1945.

Beheshti, A., Benatallah, B., Sheng, Q. Z., & Schiliro, F. (2020a, January). Intelligent knowledge lakes: The age of artificial intelligence and big data. In *International conference on web information systems engineering* (pp. 24–34). Singapore: Springer.

Beheshti, A., Yakhchi, S., Mousaeirad, S., Ghafari, S. M., Goluguri, S. R., & Edrisi, M. A. (2020b). Towards cognitive recommender systems. *Algorithms, 13*(8), 176.

Budiu, R., & Laubheimer, P. (2018). *Intelligent Assistants Have Poor Usability: A User Study of Alexa, Google Assistant, and Siri.* https://www.nngroup.com

Calero Valdez, A., Ziefle, M., & Verbert, K. (2016, September). HCI for recommender systems: The past, the present and the future. In *Proceedings of the 10th ACM Conference on Recommender Systems* (pp. 123–126).

Carducci, G., Rizzo, G., Monti, D., Palumbo, E., & Morisio, M. (2018). Twitpersonality: Computing personality traits from tweets using word embeddings and supervised learning. *Information, 9*(5), 127.

Cerejo, J. (2021) The design process of human-centered AI - part 1. Bootcamp. https://bootcamp.uxdesign.cc/human-centered-ai-design-process-part-1-8cf7e3ce00

ChatGPT. (2023). ChatGPT. https://en.wikipedia.org/wiki/ChatGPT

Chen, C., Su, T., Meng, G., Xing, Z., & Liu, Y. (2018b, May). From UI design image to GUI skeleton: A neural machine translator to bootstrap mobile GUI implementation. In *Proceedings of the 40th International Conference on Software Engineering* (pp. 665–676).

Chen, J., Lian, D., & Zheng, K. (2020). Collaborative filtering with ranking-based priors on unknown ratings. *IEEE Intelligent Systems, 35*(5), 38–49.

Chen, L., Wang, P., Dong, H., Shi, F., Han, J., Guo, Y., … & Wu, C. (2019). An artificial intelligence based data-driven approach for design ideation. *Journal of Visual Communication and Image Representation, 61,* 10–22.

Chen, N. C., Drouhard, M., Kocielnik, R., Suh, J., & Aragon, C. R. (2018a). Using machine learning to support qualitative coding in social science: Shifting the focus to ambiguity. *ACM Transactions on Interactive Intelligent Systems (TiiS), 8*(2), 1–20.

Chromik, M., Lachner, F., & Butz, A. (2020, October). ML for UX?-An inventory and predictions on the use of machine learning techniques for UX research. In *Proceedings of the 11th Nordic Conference on Human-Computer Interaction: Shaping Experiences, Shaping Society* (pp. 1–11).

De Peuter, S., Oulasvirta, A., & Kaski, S. (2021). Toward AI assistants that let designers design. arXiv preprint arXiv:2107.13074.

Deldjoo, Y., & Atani, R. E. (2016). A low-cost infrared-optical head tracking solution for virtual 3d audio environment using the nintendo wii-remote. *Entertainment Computing, 12,* 9–27.

Desolda, G., Esposito, A., Lanzilotti, R., & Costabile, M. F. (2021, August). Detecting emotions through machine learning for automatic UX evaluation. In *IFIP Conference on Human-Computer Interaction* (pp. 270–279). Cham: Springer.

Desrosiers, C., & Karypis, G. (2011). A comprehensive survey of neighborhood-based recommendation methods. In F. Ricci, L. Rokach, B. Shapira, & P. B. Kantor (Eds), *Recommender Systems Handbook* (pp. 107–144). New York: Springer.

Donahole, S. (2021, April 19). How artificial intelligence is impacting UX design. *UXmatters.* https://www.uxmatters.com/authors/archives/2021/04/stephanie_donahole.php

Dove, G., Halskov, K., Forlizzi, J., & Zimmerman, J. (2017, May). UX design innovation: Challenges for working with machine learning as a design material. In *Proceedings of the 2017 CHI Conference on Human Factors in Computing Systems* (pp. 278–288).

Ehlert, P. (2003). *Intelligent User Interfaces: Introduction and Survey.* Research Report DKS03-01 / ICE 01 Version 0.91, February 2003, Delft University of Technology, The Netherlands.

Ehsan, U., Wintersberger, P., Liao, Q. V., Mara, M., Streit, M., Wachter, S., … & Riedl, M. O. (2021). Operationalizing human-centered perspectives in explainable AI. In *Extended Abstracts of the 2021 CHI Conference on Human Factors in Computing Systems* (pp. 1–6).

Ekstrand, M. D., Harper, F. M., Willemsen, M. C., & Konstan, J. A. (2014, October). User perception of differences in recommender algorithms. In *Proceedings of the 8th ACM Conference on Recommender systems* (pp. 161–168).

Elahi, M., Beheshti, A., & Goluguri, S. R. (2021). Recommender systems: Challenges and opportunities in the age of big data and artificial intelligence. In A. Sharaff & G R Sinha (Eds), *Data Science and Its Applications* (pp. 15–39). Portland: Chapman and Hall/CRC.

Endsley, M. R. (2016). *Designing for Situation Awareness: An Approach to User-Centered Design*. Boca Raton, FL: CRC Press.

Fan, J. J., Tian, F., Du, Y, Liu, Z. J., & Dai, G. Z. (2018). Thoughts on human-computer interaction in the age of artificial intelligence (in Chinese). *Scientia Sinica Informationis*, 48, 361–375, doi: 10.1360/N112017-00221

Fan, M., Yang, X., Yu, T., Liao, Q. V., & Zhao, J. (2022). Human-AI collaboration for UX evaluation: Effects of explanation and synchronization. In *Proceedings of the ACM on Human-Computer Interaction, (CSCW1)* (Vol. 6, pp. 1–32).

Filho, J. F., Valle, T., & Prata, W. (2015, August). Automated usability tests for mobile devices through live emotions logging. In *Proceedings of the 17th International Conference on Human-Computer Interaction with Mobile Devices and Services Adjunct* (pp. 636–643).

Gajjar, N., Sermuga Pandian, V. P., Suleri, S., & Jarke, M. (2021, April). Akin: Generating UI wireframes from UI design patterns using deep learning. In *26th International Conference on Intelligent User Interfaces-Companion* (pp. 40–42).

Gartner (2018). Use customer journey analytics to enhance the buying journey. *Gartner Research.* https://www.gartner.com/en/documents/3869182

Gebhardt, C., Hecox, B., van Opheusden, B., Wigdor, D., Hillis, J., Hilliges, O., & Benko, H. (2019, October). Learning cooperative personalized policies from gaze data. In *Proceedings of the 32nd Annual ACM Symposium on User Interface Software and Technology* (pp. 197–208).

Gilbert, J. E., Williams, A., & Seals, C. D. (2007). Clustering for usability participant selection. *Journal of Usability Studies*, 3(1), 40–52.

Girardin, F. & Lathia, N. (2017). When user experience designers partner with data scientists. In *the AAAI Spring Symposium Series Technical Report: Designing the User Experience of Machine Learning Systems.* Palo Alto, CA: The AAAI Press.

Glauser, O., Wu, S., Panozzo, D., Hilliges, O., & Sorkine-Hornung, O. (2019). Interactive hand pose estimation using a stretch-sensing soft glove. *ACM Transactions on Graphics (ToG)*, 38(4), 1–15.

Gonçalves, T. G., & da Rocha, A. R. C. (2019, October). Development process for intelligent user interfaces: An initial approach. In *Proceedings of the XVIII Brazilian Symposium on Software Quality* (pp. 210–215).

Google PAIR (2019). *People + AI Guidebook: Designing Human-Centered AI Products.* pair.withgoogle.com/

Hallgrimsson, B. (2012). Prototyping and modelmaking for product design. Hachette: Laurence King Publishing.

Han, S., & Yang, H. (2018). Understanding adoption of intelligent personal assistants: A parasocial relationship perspective. *Industrial Management & Data Systems*, 118(3), 618–636.

He, Z., Spurr, A., Zhang, X., & Hilliges, O. (2019). Photo-realistic monocular gaze redirection using generative adversarial networks. In *Proceedings of the IEEE/CVF International Conference on Computer Vision* (pp. 6932–6941).

Herath, D., & Jayarathne, L. (2018). Intelligent recommendations for e-learning personalization based on learner's learning activities and performances. *International Journal of Computer Science and Software Engineering*, 7(6), 130–137.

Holmquist, L. E. (2017). Intelligence on tap: Artificial intelligence as a new design material. *Interactions*, 24(4), 28–33.

Hu, Q., Pan, Z., & Liu, J. (2019). The duality of autonomy on continuous usage of intelligent personal assistants (IPAs): From agency perspective. In *Pacific Asia Conference on Information Systems*.

Inkbot Design (2021). *10 Ways How AI is Shaping UI/UX Design.* https://inkbotdesign.com/how-ai-is-shaping-ui-ux-design/

Islas-Cota, E., Gutierrez-Garcia, J. O., Acosta, C. O., & Rodríguez, L. F. (2022). A systematic review of intelligent assistants. *Future Generation Computer Systems*, 128, 45–62.

Jannach, D., Pu, P., Ricci, F., & Zanker, M. (2021). Recommender systems: Past, present, future. *Ai Magazine*, 42(3), 3–6.

Jaquero, V. L., Montero, F., & Molina, J. P. (2009). Intelligent user interfaces: Past, present and future. In *Engineering the User Interface* (pp. 1–12). Springer, London.

Jiang, Y., Lu, Y., Nichols, J., Stuerzlinger, W., Yu, C., Lutteroth, C., ... & Li, T. J. J. (2022, April). Computational approaches for understanding, generating, and adapting user interfaces. In *CHI Conference on Human Factors in Computing Systems Extended Abstracts* (pp. 1–6).

Karpov, A. A., & Yusupov, R. M. (2018). Multimodal interfaces of human-computer interaction. *Herald of the Russian Academy of Sciences*, 88(1), 67–74.

Kayacik, C., Chen, S., Noerly, S., Holbrook, J., Roberts, A., & Eck, D. (2019, May). Identifying the intersections: User experience+ research scientist collaboration in a generative machine learning interface. In *Extended Abstracts of the 2019 CHI Conference on Human Factors in Computing Systems* (pp. 1–8).

Kharitonov, E., Drutsa, A., & Serdyukov, P. (2017, February). Learning sensitive combinations of a/b test metrics. In *Proceedings of the Tenth ACM International Conference on Web Search and Data Mining* (pp. 651–659).

Knijnenburg, B. P., Willemsen, M. C., Gantner, Z., Soncu, H., & Newell, C. (2012). Explaining the user experience of recommender systems. *User Modeling and User-Adapted Interaction, 22*, 441–504.

Konstan, J. A., & Riedl, J. (2012). Recommender systems: from algorithms to user experience. *User Modeling and User-Adapted Interaction, 22*, 101–123.

Konstan, J., & Terveen, L. (2021). Human-centered recommender systems: Origins, advances, challenges, and opportunities. *AI Magazine, 42*(3), 31–42.

Krinkin, K., Shichkina, Y., & Ignatyev, A. (2021, September). Co-evolutionary hybrid intelligence. In *2021 5th Scientific School Dynamics of Complex Networks and their Applications (DCNA)* (pp. 112–115). IEEE.

Kuniavsky, M., Churchill, E., & Steenson, M. W. (2017). The 2017 AAASI spring symposium series technical reports: Designing the user experience of machine learning systems (Technical Report SS-17-04).

Lan, Y. & Liu, S. (2020). Overview of product user experience research under artificial intelligence technology. *Packaging Engineering, 41*(240), 22–28.

Lau, N., Fridman, L., Borghetti, B. J., & Lee, J. D. (2018). Machine learning and human factors: Status, applications, and future directions. In *Proceedings of the Human Factors and Ergonomics Society Annual Meeting* (Vol. 62, No. 1, pp. 135–138). Sage, CA: Los Angeles, CA: SAGE Publications.

Lei, W., He, X., de Rijke, M., & Chua, T. S. (2020, July). Conversational recommendation: Formulation, methods, and evaluation. In *Proceedings of the 43rd International ACM SIGIR Conference on Research and Development in Information Retrieval* (pp. 2425–2428).

Li, C., Lan, M., Zou, B., Wang, S., & Zhao, K. (2015). Big data and recommendation system. *Big Data Research, 3*, 23–35.

Li, L. (2019). Research on user viscosity based on big data intelligence algorithm. *New Media Research, 5*(4), 4–6.

Li, Y., Kumar, R., Lasecki, W. S., & Hilliges, O. (2020, April). Artificial intelligence for HCI: A modern approach. In *Extended Abstracts of the 2020 CHI Conference on Human Factors in Computing Systems* (pp. 1–8).

Liapis, A., Katsanos, C., Sotiropoulos, D., Xenos, M., & Karousos, N. (2015, September). Subjective assessment of stress in HCI: A study of the valence-arousal scale using skin conductance. In *Proceedings of the 11th Biannual Conference on Italian SIGCHI Chapter* (pp. 174–177).

Liu, Z., Liu, Y., & Gao, X. (2020). Application of artificial intelligence semantic analysis techniques in user research. *Packaging Engineering, 41*(18), 53–59.

Lu, Y., Zhang, C., Zhang, I., & Li, T. J. J. (2022, April). Bridging the gap between UX practitioners' work practices and ai-enabled design support tools. In *CHI Conference on Human Factors in Computing Systems Extended Abstracts* (pp. 1–7).

Maedche, A., Legner, C., Benlian, A., Berger, B., Gimpel, H., Hess, T., … & Söllner, M. (2019). AI-based digital assistants. *Business & Information Systems Engineering, 61*(4), 535–544.

Main, A., & Grierson, M. (2020). Guru, partner, or pencil sharpener? Understanding designers' attitudes towards intelligent creativity support tools. *arXiv preprint arXiv:2007.04848*.

Marathe, M., & Toyama, K. (2018, April). Semi-automated coding for qualitative research: A user-centered inquiry and initial prototypes. In *Proceedings of the 2018 CHI Conference on Human Factors in Computing Systems* (pp. 1–12).

Maybury, M. (1998, December). Intelligent user interfaces: An introduction. In *Proceedings of the 4th International Conference on Intelligent User Interfaces* (pp. 3–4).

McCormack, J., Hutchings, P., Gifford, T., Yee-King, M., Llano, M. T., & D'inverno, M. (2020). Design considerations for real-time collaboration with creative artificial intelligence. *Organised Sound, 25*(1), 41–52.

Mendes, M. S., & Furtado, E. S. (2017, November). UUX-Posts: A tool for extracting and classifying postings related to the use of a system. In *Proceedings of the 8th Latin American Conference on Human-Computer Interaction* (pp. 1–8).

Mishra, S. & Rzeszotarski, J. (2020). Towards natural language interactions for qualitative text analysis. In CHI'20, April 25-30, 2020, Honolulu, HI, USA.

Morris, M. R., Cai, C. J., Holbrook, J. S., Kulkarni, C., & Terry, M. (2022). The design space of generative models. *arXiv preprint arXiv:2304.10547*.

Nikiforova, O., Zabiniako, V. M., Kornienko, J., Gasparovica-Asite, M., & Silina, A. (2021). Mapping of source and target data for application to machine learning driven discovery of IS usability problems. *Applied Computer Systems, 26*(1), 22–30.

Oh, C., Song, J., Choi, J., Kim, S., Lee, S., & Suh, B. (2018). I lead, you help but only with enough details: Understanding user experience of co-creation with artificial intelligence. In *Proceedings of the CHI 2018 Conference on Human Factors in Computing Systems* (pp. 1–13).

Ozmen Garibay, O., Winslow, B., Andolina, S., Antona, M., Bodenschatz, A., Coursaris, C., ... & Xu, W. (2023). Six human-centered artificial intelligence grand challenges. *International Journal of Human-Computer Interaction*, 39(3), 391–437.

Pavliscak, P. (2016). Algorithms as the new material of design. *UXMatters*. https://www.uxmatters.com/authors/archives/2014/06/pamela_pavliscak.php

Policarpo, L. M., da Silveira, D. E., da Rosa Righi, R., Stoffel, R. A., da Costa, C. A., Barbosa, J. L. V., ... & Arcot, T. (2021). Machine learning through the lens of e-commerce initiatives: An up-to-date systematic literature review. *Computer Science Review*, 41, 100414.

Quadrana, M., Karatzoglou, A., Hidasi, B., & Cremonesi, P. (2017, August). Personalizing session-based recommendations with hierarchical recurrent neural networks. In *Proceedings of the Eleventh ACM Conference on Recommender Systems* (pp. 130–137).

Ricci, F., Massimo, D., & De Angeli, A. (2021, July). Challenges for Recommender Systems Evaluation. In *CHItaly 2021: 14th Biannual Conference of the Italian SIGCHI Chapter* (pp. 1–5).

Robert Jr, L. P., & You, S. (2018). Are you satisfied yet? Shared leadership, individual trust, autonomy, and satisfaction in virtual teams. *Journal of the Association for Information Science and Technology*, 69(4), 503–513.

Rogers, M. (2020). Artificial intelligence is shifting our perception of great. *UXmatters*. User Experiences. https://www.uxmatters.com/authors/archives/2020/01/margaret_rogers.php January 6, 2020

Salminen, J., Guan, K., Jung, S. G., & Jansen, B. J. (2021). A survey of 15 years of data-driven persona development. *International Journal of Human-Computer Interaction*, 37(18), 1685–1708.

Salminen, J., Jung, S. G., & Jansen, B. (2022, March). Developing persona analytics towards persona science. In *27th International Conference on Intelligent User Interfaces* (pp. 323–344).

Salminen, J., Rao, R. G., Jung, S. G., Chowdhury, S. A., & Jansen, B. J. (2020, July). Enriching social media personas with personality traits: A deep learning approach using the big five classes. In *International Conference on Human-Computer Interaction* (pp. 101–120). Springer, Cham.

Salminen, J., Şengün, S., Kwak, H., Jansen, B. J., An, J., Jung, S., Vieweg, S., & Harrell, F. (2017). Generating cultural personas from social data: a perspective of middle eastern users. In *Proceedings of The Fourth International Symposium on Social Networks Analysis, Management and Security (SNAMS-2017)*, Prague, Czech Republic.

Santos, J., Rodrigues, J. J., Silva, B. M., Casal, J., Saleem, K., & Denisov, V. (2016). An IoT-based mobile gateway for intelligent personal assistants on mobile health environments. *Journal of Network and Computer Applications*, 71, 194–204.

Schwartz, R., Mooney, K., & Baird, C. H. (2018). The AI-enhanced customer experience: A sea change for CX strategy, design and development. *IBM Institute for Business Value*. March 2018. https://www935.ibm.com/services/us/gbs/thoughtl eadership/custexperience/ai-cx.

Setlur, V. (2020). Data-driven intent models for visual analysis tools and chatbot platforms. CHI'20, April 25-30, 2020, Honolulu, HI, USA

Shichkina, Y., & Krinkin, K. (2022, September). Principles of building personalized intelligent human assistants. In *2022 Fourth International Conference Neurotechnologies and Neurointerfaces (CNN)* (pp. 148–151). IEEE.

Shneiderman, B. (2020). Human-centered artificial intelligence: Reliable, safe & trustworthy. *International Journal of Human-Computer Interaction*, 36(6), 495–504.

Soleimani, S., & Law, E. L. C. (2017, June). What can self-reports and acoustic data analyses on emotions tell us? In *Proceedings of the 2017 Conference on Designing Interactive Systems* (pp. 489–501).

Souza, K. E. S. D., Aviz, I. L. D., Mello, H. D. D., Figueiredo, K., Vellasco, M. M. B. R., Costa, F. A. R., & Seruffo, M. C. D. R. (2022). An evaluation framework for user experience using eye tracking, mouse tracking, keyboard input, and artificial intelligence: A case study. *International Journal of Human-Computer Interaction*, 38(7), 646–660.

Stefanidi, Z., Margetis, G., Ntoa, S., & Papagiannakis, G. (2022). Real-time adaptation of context-aware intelligent user interfaces, for enhanced situational awareness. *IEEE Access*, 10, 23367–23393.

Stephanidis, C., Salvendy, G., Antona, M., Chen, J. Y., Dong, J., Duffy, V. G., ... & Zhou, J. (2019). Seven HCI grand challenges. *International Journal of Human-Computer Interaction*, 35(14), 1229–1269.

Sun, L. Y., Zhang, Y. Y., Zhou, Z. B., & Zhou, Z. H. (2020). Current situation and development trend of intelligent product design under the background of human-centered AI. *Packaging Engineering*, 41(2), 1–6.

Sun, X., Liu, B., Cao, J., Luo, J., & Shen, X. (2018, May). Who am I? Personality detection based on deep learning for texts. In *2018 IEEE International Conference on Communications (ICC)* (pp. 1–6). IEEE.

Swearngin, A., & Li, Y. (2019, May). Modeling mobile interface tappability using crowdsourcing and deep learning. In *Proceedings of the 2019 CHI Conference on Human Factors in Computing Systems* (pp. 1–11).

Tan, H, You, Z., & Peng, S.-L. (2020). Big data-driven user experience design. *Packaging Engineering*, *41*(2), 7–12.

Vaccaro, K., Agarwalla, T., Shivakumar, S., & Kumar, R. (2018, April). Designing the future of personal fashion. In *Proceedings of the 2018 CHI Conference on Human Factors in Computing Systems* (pp. 1–11).

Van den Oord, A., Dieleman, S., & Schrauwen, B. (2013). Deep content-based music recommendation. In Advances in Neural Information Processing Systems, 26.

Verganti, R., Vendraminelli, L., & Iansiti, M. (2020). Innovation and design in the age of artificial intelligence. *Journal of Product Innovation Management*, *37*(3), 212–227.

Vetrov, Y. (2022). *Algorithm-Driven Design: How Artificial Intelligence is Changing Design*. https://www.algorithms.design/

Wang, W., Zhao, D., Luo, H., & Wang, X. (2013, July). Mining user interests in web logs of an online news service based on memory model. In *2013 IEEE Eighth International Conference on Networking, Architecture and Storage* (pp. 151–155). IEEE.

Wilkins, B. (2018). *Sketching Interfaces*. Airbnb.Design. https://airbnb.design/sketching-interfaces

Willemsen, M. C., Graus, M. P., & Knijnenburg, B. P. (2016). Understanding the role of latent feature diversification on choice difficulty and satisfaction. *User Modeling and User-Adapted Interaction*, *26*(4), 347–389.

Wu, C. J., Brooks, D., Chen, K., Chen, D., Choudhury, S., Dukhan, M., … & Zhang, P. (2019, February). Machine learning at Facebook: Understanding inference at the edge. In *2019 IEEE International Symposium on High Performance Computer Architecture (HPCA)* (pp. 331–344). IEEE.

Xu, W. (2005). Recent trend of research and applications on human-computer interaction. *Journal of Ergonomics*, *11*(4), 37–40.

Xu, W. (2017). User-centered design (II): New challenges and opportunities. *Journal of Ergonomics*, *23*(1), 82–86.

Xu, W. (2018). User-centered design (III): Methods for user experience and innovative design in the intelligent era. *Chinese Journal of Applied Psychology*, *25*(1), 3–17.

Xu, W. (2019a). User-centered design (IV): Human-centered artificial intelligence. *Chinese Journal of Applied Psychology*, *25*(4), 291–305.

Xu, W. (2019b). Toward human-centered AI: A perspective from human-computer interaction. *Interactions*, *26*(4), 42–46

Xu, W., Dainoff, J.M., Ge, L, Gao, Z. (2021). From human-computer interaction to human-AI Interaction: New challenges and opportunities for enabling human-centered AI. *arXiv preprint arXiv:2105.05424 5*.

Xu, W., & Ge, L. (2018). New trends in human factors. *Advances in Psychological Science*, *26*(9), 1521–1534.

Xu, W., & Ge, L. (2020). Engineering psychology in the era of artificial intelligence. *Advances in Psychological Science*, *28*(9), 1409–1425.

Yang, B., Wei, L., & Pu, Z. (2020b). Measuring and improving user experience through artificial intelligence-aided design. *Frontiers in Psychology*, *11*, 595374.

Yang, Q. (2018). Machine learning as a UX design material: How can we imagine beyond automation, recommenders, and reminders? In *Conference: 2018 AAAI Spring Symposium Series: User Experience of Artificial Intelligence*, March 2018, At: Palo Alto, CA.

Yang, Q., Banovic, N., & Zimmerman, J. (2018a, April). Mapping machine learning advances from hci research to reveal starting places for design innovation. In *Proceedings of the 2018 CHI Conference on Human Factors in Computing Systems* (pp. 1–11).

Yang, Q., Scuito, A., Zimmerman, J., Forlizzi, J., & Steinfeld, A. (2018b, June). Investigating how experienced UX designers effectively work with machine learning. In *Proceedings of the 2018 Designing Interactive Systems Conference* (pp. 585–596).

Yang, Q., Steinfeld, A., Rosé, C., & Zimmerman, J. (2020a, April). Re-examining whether, why, and how human-AI interaction is uniquely difficult to design. In *Proceedings of the 2020 CHI Conference on Human Factors in Computing Systems* (pp. 1–13).

Yu, G. (2018). Media user experience's effect evaluation based on the paradigm of big data intelligence algorithm. *Journal of Education and Media Studies*, *16*(5), 10–12.

Yu, M. (2014). Data modeling of user portrayal in product development: From concrete to abstract'. *Design Act Research*, *6*, 60–64.

Yuan, A., & Li, Y. (2020, April). Modeling human visual search performance on realistic webpages using analytical and deep learning methods. In *Proceedings of the 2020 CHI Conference on Human Factors in Computing Systems* (pp. 1–12).

Zangerle, E., & Bauer, C. (2022). Evaluating recommender systems: Survey and framework. *ACM Computing Surveys (CSUR)*, *55*(8), 1–38.

Zhao, Q., Harper, F. M., Adomavicius, G., & Konstan, J. A. (2018, April). Explicit or implicit feedback? Engagement or satisfaction? A field experiment on machine-learning-based recommender systems. In *Proceedings of the 33rd Annual ACM Symposium on Applied Computing* (pp. 1331–1340).

Zheng, G., Zhang, F., Zheng, Z., Xiang, Y., Yuan, N. J., Xie, X., & Li, Z. (2018, April). DRN: A deep reinforcement learning framework for news recommendation. In *Proceedings of the 2018 World Wide Web Conference* (pp. 167–176).

Zheng, L., Noroozi, V., & Yu, P. S. (2017, February). Joint deep modeling of users and items using reviews for recommendation. In *Proceedings of the Tenth ACM International Conference on Web Search and Data Mining* (pp. 425–434).

Zwakman, D. S., Pal, D., & Arpnikanondt, C. (2021). Usability evaluation of artificial intelligence-based voice assistants: The case of Amazon Alexa. *SN Computer Science, 2*(1), 1–16.

6 Interacting with the Internet of Things

Fulvio Corno, Luigi De Russis, Alberto Monge Roffarello, and Juan Pablo Sáenz Moreno

6.1 INTRODUCTION

The spread of the Internet of Things (IoT) has altered the way we interact with our surroundings, enabling people to connect and communicate with the physical world in previously unimaginable ways. IoT encompasses a vast network of interconnected entities – devices, online services, sensors, and actuators – that collect and exchange data, orchestrating a symphony of actions and responses.

As such, IoT-enabled environments have become ubiquitous, integrating those interconnected things into our daily lives, from smart homes to healthcare and well-being monitoring. Despite numerous advantages, the interaction with the IoT is subjected to various challenges, from the risks of "disconnecting" users from the physical world (Angelini, Couture, Abou Khaled, & Mugellini, 2018) to more specific issues like cognitive overload (Shirehjini & Semsar, 2016) and the psychological orientation of information sources (Kim, 2016). Other challenges, which are the main focus of this chapter, arise when users want to *customize* their IoT-enabled environments. Indeed, the interconnectedness and fusion of the physical and digital worlds not only enhance convenience but also open up possibilities for tailoring experiences to meet individual needs and preferences, forming the basis for customizability opportunities that lie at the heart of the exploration of this chapter.

Currently, many IoT devices and systems tend to be customizable, either through some programming frameworks or with their own user interfaces and dashboards. This personalization effort empowers people to exert greater control over their environments, optimize resource utilization, and streamline daily routines. A person can write a script (e.g., using Home Assistant's Scripting functionality[1]) to receive a message every time their preferred weather forecaster gives a high probability of rain, for example. Another one can compose an *if-then* rule such as "*if* the weather forecast reports 90% of rains, *then* send me a message" through a suitable user interface (e.g., IFTTT[2]) to realize the same objective. These **automations** are the key ingredient behind the personalization of IoT-enabled environments, even if they are created and handled differently.

As people have different needs, backgrounds, and expertise, available personalization tools for the IoT need to adapt and offer suitable alternatives. While professional software developers can be at ease if they need to write code to create automations for their home, an end user without any technical background needs a set of options and tools to create similar automations. In a sense, the distinction between an end user and a developer becomes increasingly thin and intertwined. For instance, when personalizing the behavior of the various things in the IoT environment, end users are acting as developers (i.e., end-user developers) since they need to handle events and actions, face bugs in their automations, and can be exposed to possible unpredictable behaviors (e.g., a light that turns on at a random time). End users can personalize their IoT environments if developers provide them with the proper set of tools and options; developers can produce better work if they know end users' expectations and automation needs.

It is, therefore, fundamental to explore the perspective on IoT and personalization options from both point of views and understand the characterization of the IoT from both of them. This chapter aims to fulfill this goal by showing what is an IoT-enable environment and which are the available personalization options from the point of view of end users and developers, looking for ways to bridge any gap between the two. In the following, end users are intended as the inhabitants of the IoT-enabled environment, ultimately the final users of any IoT system. Often, they do not have a technical background and do not know how to code. Conversely, developers can be either novice or expert software programmers. As we have hinted before, both are fundamental actors for effectively *enabling* the personalization of IoT environments.

The chapter is structured as follows. It starts with background information and helpful terminology for understanding the general context. Then, it analyzes the end user perspective. Personalization of IoT environments by end users is contextualized with respect to the state of the art, reporting and discussing challenges, efforts, and possible solutions. In particular, the chapter identifies three main challenges for end users: (1) how they can properly and effectively *define* automations, (2) how they can *discover* them, and (3) how they can *debug* and understand any issue with them.

The developer perspective, similarly, describes the context, challenges, and their possible solutions with respect to the state of the art. In this case, the two main challenges are: (1) how they *understand* IoT-enabled environments and (2) how they face *integration* issues.

Finally, the chapter concludes with reflections and recommendations on bridging the gap between these two perspectives.

6.2 BACKGROUND

6.2.1 Definitions and Application Examples

Nowadays, the IoT is a well-established paradigm that has gained prominence in several aspects of our everyday lives (Stankovic, 2014). The definition of IoT has been enriched in the past, considering different contexts and use cases. For example, the Social Internet of Things (SIoT) is a subset of the IoT family that integrates IoT with social networking (Shahab, Agarwal, Mufti, & Obaid, 2022). The Industrial Internet of Things (IIoT) encompasses machine-to-machine interaction, playing a crucial role in enhancing comprehension of manufacturing processes and facilitating sustainable production methods (Sisinni, Saifullah, Han, Jennehag, & Gidlund, 2018). Internet of Everything (IoE) extends IoT by taking into account business and industrial processes, and it is based on four main pillars, i.e., people, data, process, and things (Miraz, Ali, Picking, & Excell, 2015). Another related concept is digital twins, i.e., virtual replicas of physical objects or systems that enable real-time monitoring, analysis, and simulation to improve performance and decision-making (Jones, Snider, Nassehi, Yon, & Hicks, 2020).

Given the different definitions, examples of IoT applications are common across various domains, including but not limited to smart homes, healthcare, IIoT, smart cities, and agriculture. The review by Alaa et al. (Alaa, Zaidan, Zaidan, Talal, & Kiah, 2017) provides a comprehensive overview of IoT applications in the smart home context, showing that researchers have investigated a number of use cases that range from security and privacy applications to energy management (see Figure 6 of the paper for further details). In the broader domain of smart cities, IoT can be used to enhance and support a wide range of use cases, from the healthcare domain to the domains of smart energy, transportation, and governance (see (Bellini, Nesi, & Pantaleo, 2022), Figure 3). In the healthcare domain, in particular, IoT can be exploited for monitoring patients (Laplante & Laplante, 2016), either in hospitals or at home, e.g., by setting up Ambient Intelligence environments that allow learning about patient's data and executing actions (Darshan & Anandakumar, 2016). In IIoT, application examples can be found in the context of manufacturing, agriculture, and even in the military domain (Jaidka, Sharma, & Singh, 2020). In the agriculture domain, IoT is exploited to manage farm information systems and water resources, as well as to monitor diseases, fields, greenhouses, livestock, pests, and soil (Kim, Lee, & Kim, 2020).

Independently by the specific definition or use case, embedding computing and communication capabilities into objects of common use (Miorandi, Sicari, De Pellegrini, & Chlamtac, 2012) has given rise to a programmable world. It has encouraged the development of a broad range of solutions

Interacting with the Internet of Things

in several domains, such as smart buildings, smart cities, environmental monitoring, healthcare, logistics, smart business, smart agriculture, and security and surveillance (Al-Fuqaha et al., 2015), (Zanella, Bui, Castellani, Vangelista, & Zorzi, 2014), (Islam, Kwak, Kabir, Hossain, & Kwak, 2015).

As shown in Figure 6.1, from a technical point of view, IoT systems can be characterized by four principal architectural elements: devices, gateways, cloud services, and applications.

Devices are hardware elements with built-in communication capabilities that collect sensor data (**sensing devices**) or perform actions (**acting devices**). Sensing devices provide information about the physical entities that they monitor. This information may concern the physical entity's identification (tags) or measurable qualities such as temperature, humidity, pressure, luminosity, sound level, location, images, presence, and movement. They might be environmental sensors and wearable devices, in which case they tend to measure physiological quantities. Acting devices refer to smart devices that cause or trigger changes in the physical environment, such as smart lights, motors, displays, etc. These acting devices also encompass push notifications through which end users are informed about the occurrence of a given event.

Gateways or 'edge' devices collect, preprocess, and forward the data from the *sensing devices* to the cloud and, conversely, route the requests sent from the cloud to the *acting devices*. They may support intermediate sensor data storage and preprocessing, gathering sensor data, and performing computation and reasoning tasks. If more computing or storage capacity is required, the *gateways* communicate with the *cloud services* and delegate the most demanding tasks. Furthermore, *gateways* also interact with the actuators; they control the *acting devices* based on the outputs from their computations or the instructions they receive from the *cloud services*. Additionally, *gateways* support other tasks such as service discovery, geolocalization, and verification.

Cloud services have three primary responsibilities: acquiring and storing data from the *sensing*, providing real-time and offline data analytics, and managing the *acting devices*. Data acquisition

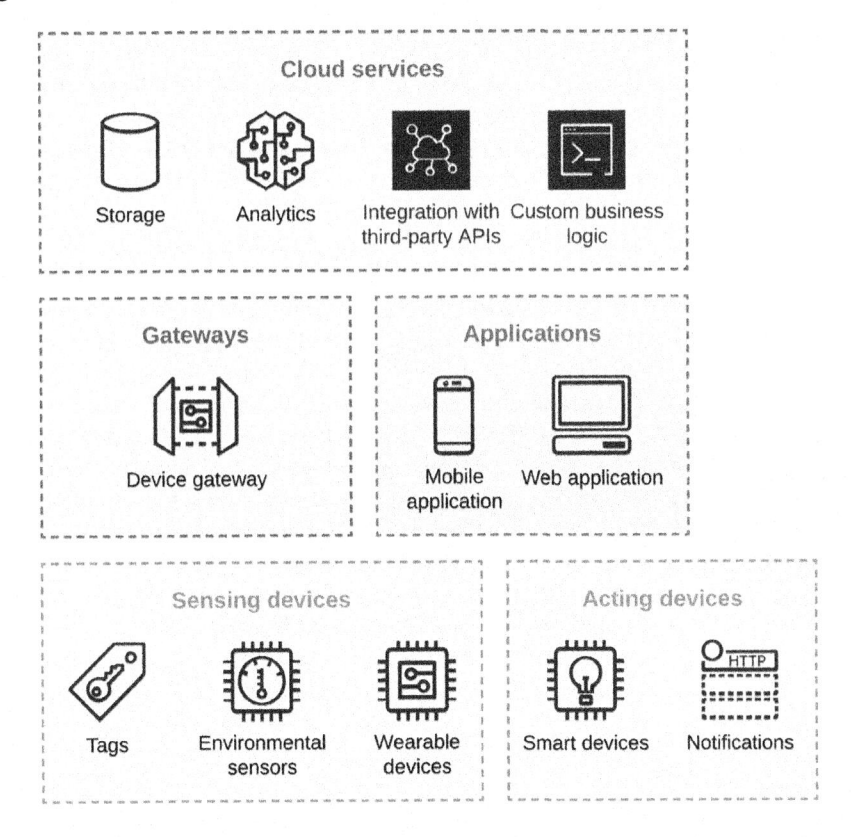

FIGURE 6.1 Architectural elements in IoT systems.

and storage concerns harvesting and storing a large amount of data collected by sensing devices for further processing and analysis. Providing real-time and offline data analytics refers to examining, cross-connecting, and transforming acquired sensor data to discover helpful information to support decision-making. Machine learning and data mining technologies and algorithms are essential in this regard. Managing the *acting devices* refers to generating and delivering remote notifications and interfacing with third-party APIs through which specific *acting devices* can be reached and managed.

Finally, **applications** range from web-based dashboards to domain-specific web and mobile applications and represent how end users can visualize the device's data and status, visualize the analyses' results, and interact with the overall system.

The implementation and orchestration of these architectural elements rely on several enabling technologies. According to Atzori, Iera, & Morabito (2010), these enabling technologies can be classified into identification, sensing and communication technologies, middleware components, end-user software applications, services composition, service management, and object abstraction. While identification, sensing, and communication technologies mainly concern hardware components, the other enabling technologies rely on *software* to address various features that IoT systems expose (Miorandi, Sicari, De Pellegrini, & Chlamtac, 2012).

While this reference architecture represents a typical and general IoT system, there will be cases where the same elements might not be present. Even in those cases, the general function and structure depicted here remain valid.

6.2.2 Challenges and Opportunities

Prior work demonstrates that interacting with the IoT can be subjected to numerous challenges that depend on the specific contexts in which such an interaction occurs. For example, the work by Angelini et al. (2018) highlights that most of today's interfaces for IoT objects are based on smartphones or web apps. These interfaces tend to "disconnect" users from the physical world and isolate them into "digital bubbles," decreasing user experience. Shirehjini and Semsar (2016), instead, warned that interacting with IoT may cause cognitive overload, e.g., when manually selecting devices in complex settings, loss of user control, and over-automation. The psychological orientation of information sources can also undermine the quality of human–IoT interaction, as the IoT complicates traditional models of communications (Kim, 2016).

Researchers have produced prominent efforts to mitigate the above challenges, e.g., by exploring innovative approaches and paradigms for human-IoT interaction. The Internet of Tangible Things (Angelini, Couture, Abou Khaled, & Mugellini, 2018), for example, has been proposed as a new IoT paradigm in which interfaces of IoT objects are situated in the physical world, rather than behind a smartphone or a website. According to the authors, designing objects that embody tangible interfaces can support immediate and meaningful interaction, stimulating reflection and the understanding of the system. Similarly, Kranz et al. (2009) explored the concept of embedded interaction, i.e., the idea of "seamlessly integrating the means for interaction into everyday artifacts" (p. 46). Specifically, the authors proposed a set of prototypes to highlight novel opportunities for human-computer interaction. Kim (2016), instead, investigated human-IoT interaction by adopting psychological and user-experience approaches rather than focusing exclusively on technological aspects, identifying social presence and perceived expertise as two important mediators. A comprehensive review of the SioT has been recently produced by Shahab et al. (2022).

Other researchers focused on creating novel interfaces for human–IoT interaction, leveraging advanced visualization techniques. For example, Shirehjini and Semsar (2016) exploited 3D visualization to implement an interface for mixed-initiative interaction in smart environments. Phupattanasilp and Sheau-Ru (2019), instead, used augmented reality to support IoT data visualization in the farming context.

Overall, all the above issues can be associated with the fact that IoT can be *personalized* by end users, who interact with their IoT ecosystems to "program" the behavior of smart devices and online services on the basis of their personal needs. Such a possibility creates a range of issues and opportunities, representing the core of this chapter. End users and developers are the two main actors that frequently interact and work within this context. The former are the people who adopt and use such technologies within their environment, while the latter are the creators and updaters of the entire IoT system or application. Given the flexibility and variability of available "connected entities" (e.g., sensors, actuators, and online services), developers cannot predict all the possible contexts of usage of their creations, and end users might want to personalize the behavior of their IoT devices and systems to better fit their specific needs in the context at hand. The rest of this chapter is dedicated to exploring in detail such a dynamic interplay between these two actors, starting from end users.

6.3 THE END-USER PERSPECTIVE

In the IoT realm, end users need to be able to personalize the functionality of smart devices and online services, even if they lack programming expertise. In this scenario, the vision of End-User Development (EUD) (Paternò, Lieberman, & Wulf, 2006) strives to empower end users, who possess firsthand knowledge of the specific requirements, to customize the behaviors of their "connected entities." Typically, such a customization is supported through the trigger-action programming paradigm, which allows users to define IF-THEN rules in the form of *"if something happens on a device or service, then execute something on another device or service."* This section reviews the state-of-art for EUD and trigger-action programming within IoT-enabled environments. Besides introducing the context of such a growing and pressing research topic, we present issues and related challenges, as well as solutions emerging from the human-computer interaction (HCI) research community. Table 6.1 provides a summary of the issues, challenges, and solutions detailed in this section.

6.3.1 CONTEXT

The advent of IoT has already brought numerous benefits to society, spanning from individual-level applications to global-scale impact (Cerf & Senges, 2016). Individuals can now interact with a wide array of connected devices, such as lamps, thermostats, and various household appliances like refrigerators and ovens, which can be connected to the Internet, transforming homes into smart environments. The influence of the IoT extends beyond personal spaces to workplaces and even entire cities, where smart environments utilize connected devices to enhance residents' comfort and

---

TABLE 6.1
A Summary of the Issues, Challenges, and Solutions Reviewed in this Section

Issue	Challenge	Solutions
Low-level of abstraction: Trigger-action programming platforms adopt representation models that heavily depend on the underlying technology.	*Supporting IF-THEN Rules Definition:* Simplify the processes end users need to define IF-THEN rules.	• Visual metaphors • Underlying models • Interaction techniques
Information overload: The number of potential combinations between triggers and actions supported by trigger-action programming platforms is continuously growing.	*Promoting IF-THEN Rules Discovery:* Support users in discovering IF-THEN rules and related functionality useful to satisfy their personalization needs.	• Recommender systems • Conversational agents
Run-time problems: The definition of IF-THEN rules may introduce run-time problems, including security threats, especially when users are not accustomed to programming.	*Defining Safe IF-THEN Rules:* Provide users with tools that allow them to define safe IF-THEN rules that do not introduce any run-time problems.	• Problems standardization • Offline verification • Debugging tools

convenience (Cook & Das, 2005). In addition to physical devices, a wide range of online services, including social networks, news platforms, and messaging applications, are extensively utilized by nearly everyone and can be integrated with the IoT ecosystem. As of 2022, the estimated number of Internet users worldwide was 5.3 billion, with over 2.9 billion individuals actively engaged in a single social media, i.e., Facebook (Petrosyan, 2023). This widespread adoption enables users to effortlessly connect with a sophisticated network of connected entities, encompassing both smart devices and online services. These entities have the capability to communicate not only with each other but also with humans and the surrounding environment.

Given the existence of such an interconnected network and its continued expansion, researchers – especially in HCI and ubiquitous computing – have long agreed on the urgent challenge of providing users with efficient EUD methodologies and tools to customize and personalize it. According to (Paternò, Lieberman, & Wulf, 2006), EUD can be defined as *"a set of methods, techniques, and tools that allow users of software systems, who are acting as non-professional software developers, at some point to create, modify or extend a software artifact."* One of the pioneering works in this area is iCAP (Dey, Sohn, Streng, & Kodama, 2006), a rule-based, visual, and PC-based system designed to construct context-aware applications without users needing to engage in coding. Recently, EUD approaches and methodologies have been extensively investigated in various contexts, including mobile environments (Namoun, Daskalopoulou, Mehandjiev, & Xun, 2016), smart homes (Brich, Walch, Rietzler, Weber, & Schaub, 2017), and web mashups (Daniel & Matera, 2014). From a commercial point of view, end users can also leverage visual programming platforms such as IFTTT[3] and Zapier[4] to customize and personalize the joint behaviors of their connected entities, e.g., to blink a lamp when someone is at the entrance door, all without the necessity of coding.

A common link between research-based and commercial tools lies in the exploited programming paradigm. Specifically, most EUD solutions in the IoT context adopt the trigger-action programming paradigm (Desolda, Ardito, & Matera, 2017). In its basic form, such a paradigm allows users to connect an event to an action to be automatically executed so that users can establish connections between a pair of devices or online services. The result is an IF-THEN (or trigger-action) rule where, upon detecting a specific event (the trigger) on one device or service, an automatic action is executed on the other device or service. Examples may include:

- if I publish a photo on *Facebook*, then upload it to my *Google Drive*;
- if the *Nest* security camera detects a movement, then blink the kitchen's *Philips Hue* lamp;
- if the *Nest* thermostat detects that the temperature rises above 22°C, then open the *SmartThings* window.

In most contemporary trigger-action programming platforms, especially commercial tools like IFTTT and Zapier, users can define IF-THEN rules through wizard-based procedures (Desolda, Ardito, & Matera, 2017).

The process of defining rules is summarized in Figure 6.2. It involves separately defining a trigger and an action – that are then linked together. The following steps must be followed to define a trigger:

1. Choose the relevant connected entity involved in the trigger by selecting from available supported entities in a menu (connected entity selection). Usually, entities are represented by their respective manufacturers or brands, such as *Nest* thermostats. Figure 6.3, for example, shows the connected entity selection step on IFTTT and Zapier, respectively.
2. Select the specific trigger to be monitored from another menu (Trigger Selection). For example, a trigger could represent the detection of a high temperature.
3. Provide additional details for the trigger (trigger details), such as the identifier of the specific *Nest* thermostat or a temperature threshold value.

Interacting with the Internet of Things   177

IF-THEN RULE

if **THIS** then **THAT**

TRIGGER

1. Connected Entity Selection
2. Trigger Selection
3. Trigger Details

ACTION

4. Connected Entity Selection
5. Action Selection
6. Action Details

FIGURE 6.2 The process that users must follow to define IF-THEN rules in most trigger-action programming platforms, e.g., IFTTT and Zapier.

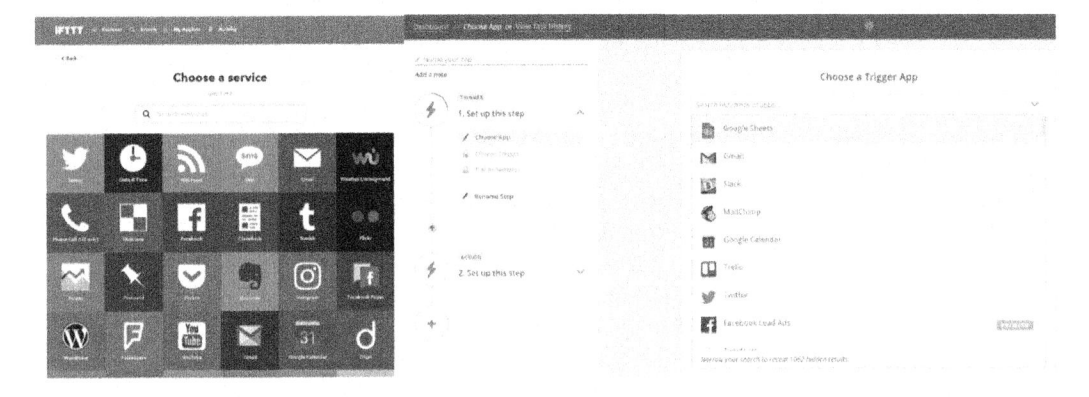

FIGURE 6.3 The selection of a connected entity to define a trigger in IFTTT and Zapier, respectively.

The three steps are then similarly repeated to define the action associated with the rule.

According to Barricelli & Valtolina (2015), trigger-action programming provides an exceptionally straightforward and easy-to-learn solution for developing end-user applications in the IoT domain. While certain behaviors may necessitate a higher level of expressiveness, such as incorporating multiple actions or additional trigger conditions, it is noteworthy that several widely used trigger-action programming platforms, including IFTTT and Zapier, still adhere to the fundamental basic structure of the trigger-action programming paradigm summarized in Figure 6.2 (Brackenbury et al., 2019). One of the reasons for such a choice is that studies have highlighted the tendency for users to misinterpret the behavior of rules involving multiple triggers and actions (Huang, Azaria, & Bigham, 2016), e.g., because end users are generally unable to distinguish between states, instantaneous triggers, and conditions.

Ideally, trigger-action programming can express most of the behaviors that potential users desire (Ur, McManus, Pak Yong Ho, & Littman, 2014). However, despite its widespread adoption, researchers agree that this paradigm's current implementation introduces its own issues. As summarized in Table 6.1, we can identify three main issues:

- *Low-level of abstraction*: In the IoT context, the presence of new connected entities cannot always be predicted in advance (Zaslavsky & Jayaraman, 2015). These connected entities may appear and disappear dynamically, depending on factors such as user location (e.g., public services in a smart city). Unfortunately, current trigger-action programming platforms rely on representation models that are heavily tied to specific technologies, thus poorly adaptable to the increasing complexity of the IoT ecosystem. These platforms typically work with specific connected entities that have been previously associated with a particular user. For example, they do not effectively handle situations where two different devices or online services offer identical functionalities but differ in terms of brands or manufacturers, but they treat them as completely distinct entities (Corno, De Russis, & Monge Roffarello, 2019a). Consequently, as the number and diversity of connected entities continue to grow, the IoT ecosystem becomes increasingly complex (Barricelli & Valtolina, 2015), and the growing complexity of their interactions poses challenges for non-programmers in defining IF-THEN rules (Huang, Azaria, & Bigham, 2016).

- *Information overload*: As reported in the previous issue, contemporary trigger-action programming platforms currently lack a high level of abstraction: they model each device and online service based on its underlying brand or manufacturer. This approach leads to a vast number of potential combinations of triggers and actions from various technologies. Consequently, the number of shared rules on these platforms is steadily increasing. For instance, Zapier supports thousands of devices and web applications, each with its own set of triggers and actions. Similarly, as of September 2016, IFTTT had already surpassed 200,000 publicly available and reusable rules (Ur et al., 2016). Unfortunately, many of these platforms fail to offer users any support for discovering new combinations of triggers and actions (Ur et al., 2016). As a result, the rapid proliferation of smart devices and online services leads to user interfaces that become overloaded with excessive amounts of information.

- *Run-time problems*: The last issue relates to the fact that contemporary trigger-action platforms do not provide end users with any support for understanding and debugging their IF-THEN rules, e.g., in order to prevent potential conflicts (Caivano, Fogli, Lanzilotti, Piccinno, & Cassano, 2018) and evaluate the correctness of the rules (Desolda, Ardito, & Matera, 2017). Due to the current low level of abstraction in trigger-action programming platforms, however, users often misinterpret the intended behavior of their trigger-action rules (Brush et al., 2011), thus deviating from the actual semantics of the rules and introducing errors (Huang & Cakmak, 2015). Unfortunately, errors in this context can have unpredictable and hazardous consequences (Brush et al., 2011). For example, an incorrectly defined rule could unexpectedly unlock the main door of a house, resulting in a significant security threat.

6.3.2 CHALLENGES

Overall, the three identified issues relate to distinct challenges that have inspired and continue to inspire research efforts in the field of EUD for the IoT (Table 6.1): supporting users in defining IF-THEN rules, promoting the discovery of IF-THEN rules and functionality, and defining safe IF-THEN rules. In this section, we describe such challenges by highlighting the most promising solutions that have been proposed to address each of them.

Supporting IF-THEN rules definition: Modern trigger-action programming platforms utilize representation models that directly map onto technology, which fail to adequately accommodate the growing intricacy of the IoT ecosystem (specifically, the low-level abstraction issue reported in Section 3.1). For instance, platforms like IFTTT and Zapier categorize devices and services in the connected entity selection step of the rule definition process solely based on their manufacturer or brand, as depicted in Figure 6.3. Specifically, users are constrained to use a vendor-centric approach and must configure all the detailed technical aspects necessary for executing the desired trigger-action automation. This situation presents significant challenges in terms of interoperability, as each IoT device and online service must be managed separately. For example, users cannot create a trigger-action rule that applies to all their connected lamps unless they share the same brand nor can they create rules for other types of devices that provide interior lighting. Consequently, formulating IF-THEN rules becomes challenging for end users without programming expertise (Huang, Azaria, & Bigham, 2016). As summarized in Table 6.2, researchers have tried to support IF-THEN rules definition and overcome the low-level of abstraction issue following three main directions: exploring alternative *visual metaphors* that could better support trigger-action programming, producing alternative *underlying models* to model triggers and actions with a higher level of abstraction, and adopting multiple *interaction techniques* to enable trigger-action programming.

Concerning the first strategy (Table 6.2, first row), researchers mainly explored the possibility of empowering end users to adopt block-based programming approaches (Desolda, Ardito, & Matera, 2017) instead of the form-based procedures exploited by contemporary trigger-action programming platforms (e.g., those displayed in Figure 6.3). An illustrative instance of such an approach can be found in scratch (Resnick et al., 2009), a visual programming language designed primarily for teaching coding to children. With block programming, users can drag and drop blocks of various sizes and shapes onto a workspace to connect them. In contrast to form-based approaches, block programming tools offer greater flexibility and encourage user creativity. The Jigsaw metaphor, in particular, is widely recognized as an effective approach for representing blocks. In the trigger-action programming context, for example, triggers and actions can be visualized as puzzle pieces that can be seamlessly interconnected. Due to the complementary nature of the puzzle pieces, the composition of trigger-action rules is simplified *by design*, with some wrong operations that are prevented by construction: pieces of the same type, e.g., two trigger pieces, cannot be connected, for example. Therefore, it is not surprising that such a metaphor has been found to be adequate to support the definition of IF-THEN rules. Three prominent examples of EUD tools exploiting the Jigsaw metaphor are as follows:

- *Puzzle* (Danado & Paternò, 2014) is a mobile framework that allows end users without IT backgrounds to create, modify, and execute mobile trigger-action automation with the Jigsaw metaphor.
- *My IoT Puzzle* (Corno, De Russis, & Monge Roffarello, 2019c) is a web-based tool to compose and debug IF-THEN rules through the Jigsaw metaphor (Figure 6.4). Available triggers and actions, in particular, have been extracted from a large dataset of IF-THEN

TABLE 6.2

The Three Main Strategies that Researchers Have Followed to Support the Definition of IF-THEN Rules by End Users: Exploring Alternative Visual Metaphors, Adopting Alternative Representation Models, and Adopting Varied Interaction Techniques

Strategy	Description	Main References
Visual metaphors	Exploration of alternative visual metaphors, e.g., using Jigsaw puzzle pieces, that could better support trigger-action programming compared to wizard-based procedures.	Danado & Paternò (2014) Desolda, Ardito, & Matera (2017) Corno, De Russis, & Monge Roffarello (2019c) Akiki, Bandara, & Yu (2017)
Underlying models	Exploration of alternative representation models, e.g., exploiting Semantic Web technologies, to model triggers and actions with a higher level of abstraction.	Ur, McManus, Pak Yong Ho, & Littman (2014) Ghiani, Manca, Paternò, & Santoro (2017) Corno, De Russis, & Monge Roffarello (2019a) Corno, De Russis, & Monge Roffarello (2021a)
Interaction techniques	Exploration of alternative interaction techniques, e.g., vocal and multimodal approaches, to specify triggers and actions if IF-THEN rules.	Manca, Parvin, Paternò, & Santoro (2020) Gallo, Manca, Mattioli, Paternò, & Santoro (2021) Barricelli, Fogli, Iemmolo, & Locoro (2022) Monge Roffarello & De Russis (2023)

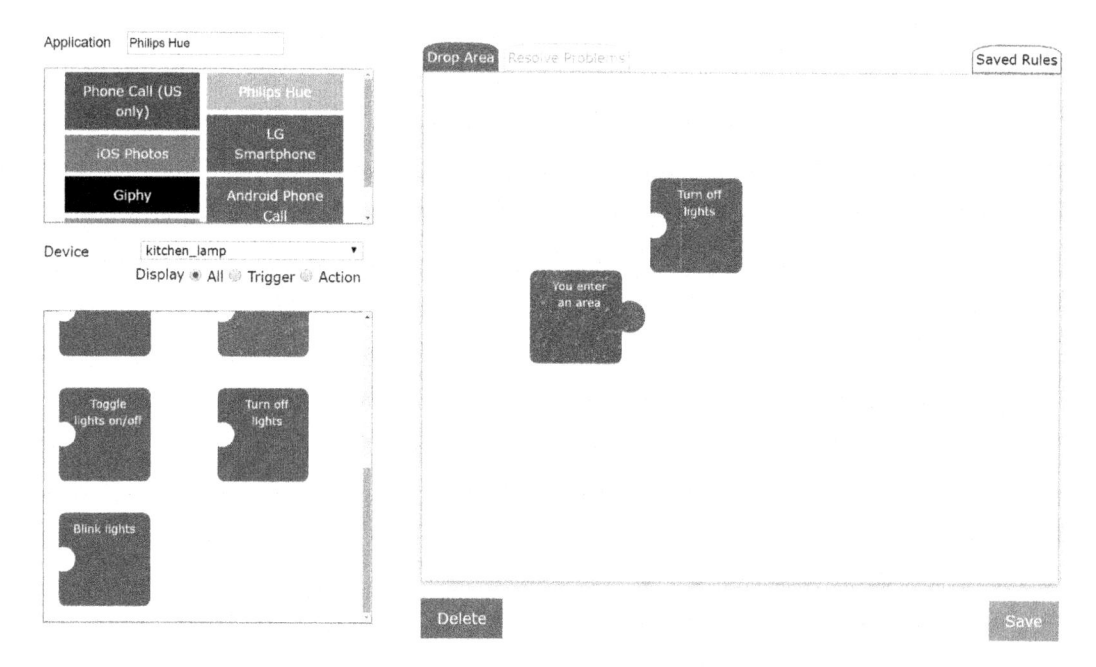

FIGURE 6.4 The My IoT Puzzle tool allows users to define the IF-THEN rules modeled by the IFTTT platform through the Jigsaw metaphor. Image taken from the original paper.

rules shared on IFTTT and extracted through a web-scraping procedure (Ur, McManus, Pak Yong Ho, & Littman, 2014).
- *ViSiT* (Akiki, Bandara, & Yu, 2017) is a method that enables non-technical users to employ a puzzle-like representation to define transformations, which are then automatically translated into executable workflows. For instance, this approach can be utilized to connect a smart refrigerator with an online shopping service, automating the process of restocking low-inventory items.

Besides focusing on the visual aspects and metaphors to represent triggers and actions, other researchers tried to support the definition of IF-THEN rules by changing the underlying representation models (Table 6.2, second row). Specifically, recent works explored the personalization of IoT ecosystems through the lens of *abstraction*: a direct approach to address the low-level of abstraction concern (Section 3.1) is to enable users to define their triggers and actions using a higher level of abstraction, thereby minimizing the need to deal with technical details and specific attributes like brands and manufacturers, which can be abstracted away by the adopted representation model. In addition to simplifying the definition process, using a higher level of abstraction also matches end users' mental models. In a study exploring the effectiveness of the trigger-action programming paradigm in the smart-home context, Ur, McManus, Pak Yong Ho, & Littman (2014) discovered that users would express triggers without mentioning specific sensors (e.g., "when the doorbell rings"), often referring to very abstract behaviors (e.g., by using "fuzzy triggers" like "when my pool chemicals drop lower than normal"). Corno, De Russis, & Monge Roffarello (2021a) confirmed these findings beyond the smart-home context. By reporting on the results of a 1-week-long diary study during which participants were to collect trigger-action rules arising during their daily activities, the authors found that end users would adopt a higher level of abstraction compared to contemporary platforms like IFTTT and Zapier. Furthermore, the authors found that the adopted level of abstraction often depends on the context for which the personalization is envisioned: while users are typically interested in personalizing physical objects in the home, for example, they would be more interested in personalizing information in the context of a smart city.

Stemming from these findings, practical alternative representation models – allowing users to define IF-THEN rules with different levels of abstraction – have been proposed in the literature. For example, Ghiani, Manca, Paternò, & Santoro (2017) proposed *TARE*, an authoring tool to personalize the contextual behavior of IF-THEN rules. Specifically, an underlying architecture based on a context manager can activate, interpret, and apply abstract IF-THEN rules (e.g., "when the user falls asleep, switch off the bedroom TV") to specific contexts. Similarly, Corno, De Russis, & Monge Roffarello (2019a) introduced *EUPont*, an ontological representation of EUD in the IoT. This framework enables the creation of context-independent IF-THEN rules based on users' ultimate goals. With EUPont, users can bypass specific device commands, such as turning on a Philips Hue lamp or opening bedroom blinds, and directly instruct the system to illuminate the room, for instance.

The remaining strategy to support the definition of IF-THEN rules concerns exploring alternative interaction techniques to specify triggers and actions (Table 6.2, third row). Researchers, in particular, focused on empowering users to define their personalization via natural language, e.g., through chatbots or using voice-based interaction. For example, Gallo, Manca, Mattioli, Paternò, & Santoro (2021) proposed *Rule Bot*, a chatbot that directly maps the user's intention to the corresponding rule. After the chatbot welcomes, the user can enter a possible trigger or action through an iterative dialogue process, with the tool that can provide feedback and ask for additional information to complete the rule. Manca, Parvin, Paternò, & Santoro (2020), instead, introduced an integration of *Amazon Alexa* with a rule-based personalization platform that aims to facilitate the development of trigger-action rules enriched with voice-based functionality. Similarly, Barricelli, Fogli, Iemmolo, & Locoro (2022) proposed a new multimodal approach to create Amazon Alexa routines, leveraging Echo Show devices. Recently, Monge Roffarello & De Russis (2023) designed and implemented two different prototypes of Intelligent Personal Assistants (IPAs) that allow end users to define trigger-action rules via voice. The first prototype utilizes a completely vocal interaction mechanism, enabling users to define a rule in a single sentence. In case of errors or misunderstandings, users can refine or correct the rule through subsequent dialogues. On the other hand, the second prototype combines the vocal definition of the trigger with a phase where the user is prompted to physically demonstrate the action to be automated.

Promoting IF-THEN rules discovery: Contemporary trigger-action programming platforms often model connected entities based on their respective brands or manufacturers, leading to a vast number of potential combinations among triggers and actions from different technologies. Consequently, the proliferation of shared rules becomes a significant concern, contributing to the issue of information overload (Section 3.1). For example, there are numerous possibilities for a user who wishes to customize their smart home's behavior using IFTTT. For instance, they can specify a temperature for their *Nest* thermostat to be automatically adjusted when their *BMW* smart car approaches the home area. Alternatively, they can program the *Philips Hue* lamp in the kitchen to turn on whenever the *Arlo* security camera detects movement. Even within this limited scenario, the four mentioned connected entities provide a total of 15 triggers and 19 actions on IFTTT, resulting in 285 potential rules. Furthermore, the number of possible combinations increases significantly if we consider specific details of each trigger and action, such as the thermostat's temperature threshold or the lamp's light intensity. The challenge here is to support users in discovering useful IF-THEN rules and related functionality: compared to supporting IF-THEN rules definition (Table 6.1, first row's challenge), strategies to solve such a challenge (see Table 6.3 for a summary) do not require radical changes in the underlying models and visual metaphors but focuses on making existing personalization options "more visible."

The first practical strategy (Table 6.3, first row) concerns the development of recommender systems, a straightforward approach to solving the information overload issue in contemporary trigger-action programming platforms (Ur et al., 2016). Using recommendation techniques can enhance the

TABLE 6.3
The Two Main Strategies that Researchers Have Followed to Support Users in Discovering IF-THEN Rules and Related Functionality

Strategy	Description	Main References
Recommender systems	Development of recommender systems that can suggest IF-THEN rules or specific triggers and actions based on users' preferences and context.	Srinivasan, Koehler, & Jin (2018) Corno, De Russis, & Monge Roffarello (2019d) Mattioli & Paternò (2020)
Conversational agents	Development of conversational agents, typically in the form of chatbots, that can map a natural-language request of the user into the intended IF-THEN rule(s).	Huang, Azaria, & Bigham (2016) Corno, De Russis, & Monge Roffarello (2021b)

definition and reuse of trigger-action rules, thereby assisting individuals without technical or programming expertise in conveniently personalizing their smart devices and online services. Besides developing recommendation algorithms, researchers also integrated them into EUD tools for end users, either through custom trigger-action programming platforms or as extensions to existing interfaces. For example, *RuleSelector* (Srinivasan, Koehler, & Jin, 2018) implements an algorithm that selects the top action rules based on four rule selection metrics: total action coverage, confidence, interval count, and contextual specificity. *TAPrec* (Corno, De Russis, & Monge Roffarello, 2020b), instead, is a EUD platform modeled after IFTTT specifically designed to facilitate the creation of trigger-action rules through dynamic recommendations. The tool offers suggestions during the rule composition process, including new rules to consider (Figure 6.5a) or actions for auto-completing a rule (Figure 6.5b).

TAPrec recommendations are computed by RecRules (Corno, De Russis, & Monge Roffarello 2019d), a recommendation algorithm that suggests IF-THEN rules based on their ultimate objectives, bypassing specific details such as manufacturers and brands. This algorithm leverages an ontology to enhance rules with semantic information, facilitating the discovery of concealed relationships among rules in terms of shared functionality. For instance, a rule for activating a *Philips Hue* lamp shares functional similarity with a rule for opening *Hunter Douglas* blinds, as both aim to illuminate a space. Unlike *TAPrec*, which can be considered an extension of the IFTTT platform, *BlockComposer* (Mattioli & Paternò, 2020) is a recommendation-powered EUD interface with its own custom user interface. However, it adopts a similar recommendation approach. Specifically,

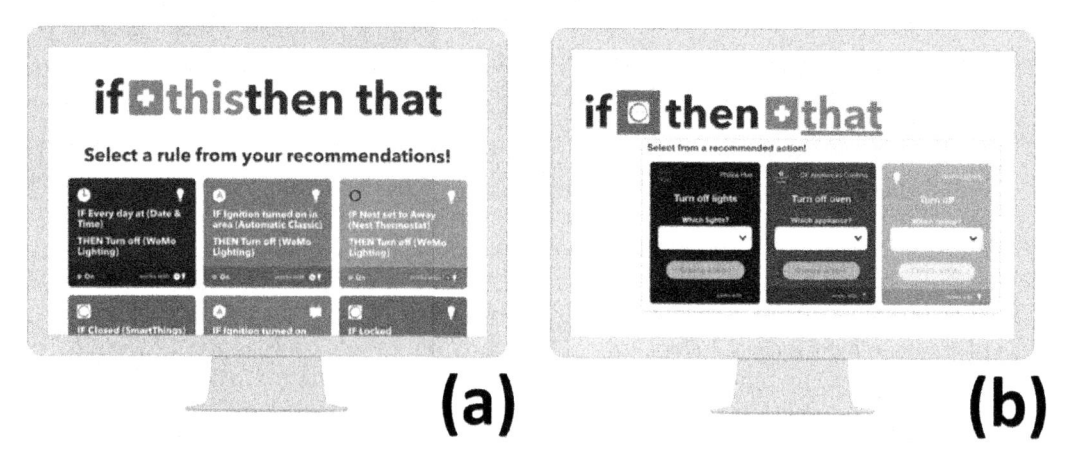

FIGURE 6.5 TAPrec (Corno et al., 2020) is a EUD platform modeled after IFTTT that integrates recommendations of IF-THEN rules (a) and recommendations of actions to auto-complete a rule (b). Images taken from the original paper.

the tool offers two recommendation policies assisting users during the definition of IF-THEN rules: (1) step-by-step, through which the tool suggests the next element to be included in the rule being edited; and (2) full rule, through which complete rules are provided as suggestions.

As a second strategy to promote IF-THEN rules discovery (Table 6.3, second row), researchers have innovated conversational agents, typically in the form of chatbots, able to map a natural-language request of the user into the intended IF-THEN rule(s). Through this strategy, strongly linked to the interaction techniques strategy reported in Table 6.2 and the first strategy of Table 6.3, researchers take advantage of specialized recommender systems to move from an abstract intention of the user to a set of specific IF-THEN rules that can be executed in the user's environment. One of the first prototypes to compose rules via conversation and recommendations, named *InstructableCrowd*, was proposed by Huang, Azaria, & Bigham (2016). The tool is a crowd-sourcing system that empowers users to generate IF-THEN rules according to their specific requirements. Through a dedicated user interface on their smartphones, end users can engage in conversations with crowd workers to articulate their issues, such as arriving late for a meeting. By utilizing a customized interface, crowd workers can effectively combine triggers and actions to address these problems and suggest proper IF-THEN rules to the end users. *HeyTAP* (Corno, De Russis, & Monge Roffarello, 2020a), instead, is a conversational platform able to map abstract users' needs to executable IF-THEN rules automatically, without involving human figures. Through a multimodal interface, users can engage with a chatbot by typing or using their voice to convey their personalization preferences for various contexts. Furthermore, users can provide supplementary details on how to implement their personalization intentions, which serve as guidance for rule suggestions. For instance, if the user selects "sustainability" as an additional intention, *HeyTAP* suggests rules considering the secondary goal of saving energy. *HeyTAP²* (Corno, De Russis, & Monge Roffarello, 2021b) represents the evolution of *HeyTAP*, incorporating an upgraded recommender system that enhances the application's ability to comprehend user intentions through iterative refinements (Figure 6.6). In cases where the user cannot find a rule that aligns with their intention, *HeyTAP²* implements a preference-based feedback approach. This iterative collaboration with the user allows for further feedback, leading to the refinement of recommendations. With each refinement cycle, the user can express their short-term preference by selecting a recommended rule that closely matches their intention. This information is then utilized to adjust the weighting of candidate rules, thereby promoting recommendations of items that better match the provided feedback.

> *Defining safe IF-THEN rules*: In addition to facilitating users in defining and discovering IF-THEN rules, another critical and pressing challenge must be addressed: preventing potential conflicts and evaluating the accuracy of IF-THEN rules (i.e., solving the run-time problems issue reported in Section 3.1). Indeed, problems and errors within trigger-action rules harm users' ability to anticipate program outcomes accurately (Brackenbury et al., 2019). Such a mismatch, in turn, can lead to unforeseen and potentially hazardous behaviors (Brush et al., 2011), e.g., unexpectedly unlocked doors. Regrettably, as we have seen for the previous challenges, existing trigger-action programming platforms often provide overwhelming functionality and employ technology-specific representation models that require extensive knowledge of all the smart devices and online services involved. Consequently, users frequently misinterpret the behavior of trigger-action rules deviating from their intended semantics (Brush et al., 2011) and are prone to introducing errors (Huang & Cakmak, 2015).

Table 6.4 summarizes three main strategies that have been followed in the literature to define safe IF-THEN rules, i.e., rules that do not introduce problems at run-time.

To this end, the first strategy is to define these problems and how they can influence the context in which they are introduced (Table 6.4, first row). Two recent works extended findings from other domains, e.g., active databases, to standardize a set of problems that apply to the context of

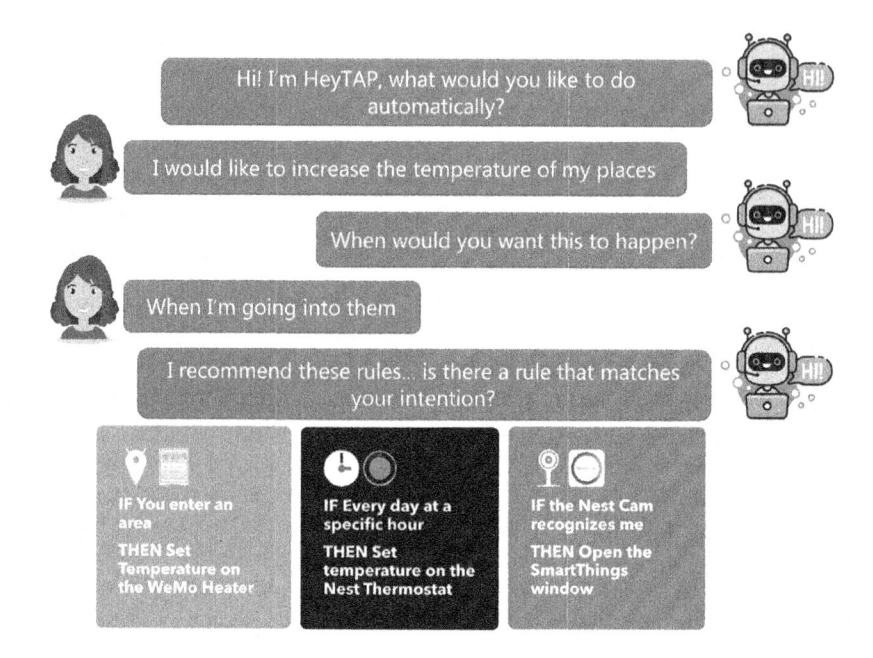

FIGURE 6.6 HeyTAP[2] is a chatbot for defining IF-THEN rules that implements a preference-based feedback approach: it iteratively collaborates with the user to get a personalization intention and refine recommendations. Image taken from the original paper.

trigger-action programming for IoT personalization. In their recent research, Brackenbury et al. (2019) thoroughly analyzed existing literature and bugs in other domains to identify ten programming bugs that can arise in IF-THEN rules. The study revealed several types of problems:

- *control-flow bugs*, i.e., run-time problems that may impair proper control flow, e.g., in case of conflicts between concurrent IF-THEN rules;
- *timing bugs*, i.e., problems influencing the concurrent behaviors of IF-THEN rules, e.g., the non-deterministic nature of how a system processes simultaneous triggers;
- *inaccurate user expectations*, i.e., problems that happen when novice programmers attribute system intelligence beyond their actual capabilities, leading to challenges in users' mental models.

TABLE 6.4

The Three Main Strategies that Researchers have Followed to Allow the Definition of Safe IF-THEN Rules: Problems Standardization, Offline Verification, and Debugging Tools

Strategy	Description	Main References
Problems standardization	Definition and standardization of the problems, e.g., loops and conflicts, that may arise at run-time from executing a set of concurring IF-THEN rules.	Brackenbury et al. (2019) Corno et al. (2019)
Offline verification	Checking the behavior of a set of already-defined IF-THEN rules "offline," e.g., through formal verification, to identify possible run-time problems.	Liang et al. (2016) Surbatovich et al. (2017) Vannucchi et al. (2017) Zhang et al. (2019)
Debugging Tools	Development of tools and platforms for end-user debugging tools to assist users in solving possible run-time problems during the definition process.	Corno et al. (2019) Manca et al. (2019)

Stemming from such an analysis, Corno, De Russis, & Monge Roffarello (2019b) specifically focused on control-flow bugs by defining three classes of problems influencing the execution flow of IF-THEN rules. According to the authors, *loops* occur when a set of trigger-action rules are continuously activated without reaching a stable state. *Inconsistencies*, instead, occur when the execution order of concurrent rules may render different final states in the system, thus leading to inconsistent situations like a lamp that is turned on and off simultaneously. Similarly, *redundancies* are introduced when two or more concurrently activated rules have replicated functionality, e.g., a set of IF-THEN rules that share a post on a social network twice.

Previous work started to address the challenge of avoiding problems like those mentioned above at run-time leveraging software engineering techniques, e.g., formal verification (Liang et al., 2016) and information flow control (Surbatovich, Aljuraidan, Bauer, Das, & Jia, 2017). The idea of such a strategy (Table 6.4, second row) is to check a set of already defined rules through an offline verification to allow experts (or the same end users) to fix potential problems afterward. Liang et al. (2016) developed *Salus,* a building automation service that leverages formal methods to locate faulty logic in IoT applications, transforming them into parameterized equations that can be solved through model-checking tools or Satisfiability Modulo Theories (SMT) solvers. The work by Vannucchi et al. (2017) demonstrates the usage of advanced SMT-based techniques for software verification in intelligent environments, optimizing the verification process and improving both the performance and expressivity of event-condition-action rules compared to previous approaches. *AutoTap*, instead, is a system introduced by Zhang et al. (2019) that enables novice users to define desired properties for devices and services, e.g., to specify that a window should remain closed. Then, the system translates these properties in a linear temporal logic (LTL) and is able to (1) generate a set of IF-THEN rules that satisfy them and (2) repair an existing set of IF-THEN rules to respect the defined properties.

Instead of checking rules "offline," i.e., after their definition, previous works highlighted the possibility of supporting users to identify programming bugs in IF-THEN rules and reason about how to fix them during the definition process. This strategy implies designing trigger-action programming tools tailored explicitly for end users unfamiliar with programming, providing users with mechanisms to debug their IF-THEN rules (Table 6.4, third row). The process of debugging involves identifying and resolving the underlying cause of a misbehavior. With access to appropriate information, even end users can successfully create accurate applications and programs (Cao et al., 2010). Following such a principle, HCI researchers have proposed and evaluated end-user debugging tools for trigger-action programming.

Corno, De Russis, & Monge Roffarello (2019b) developed *EUDebug* (Figure 6.7), a tool that integrates different end-user debugging features on top of an IFTT-like interface. Specifically, the tool employs two main strategies that extend the definition process of IF-THEN rules: (1) assisting users in identifying rule conflicts by highlighting potential problems at run time (Figure 6.7a) and (2) facilitating the anticipation of rule behavior during run-time through step-by-step simulation (Figure 6.7b).

Under the hood, *EUDebug* exploits a Semantic Colored Petri Net (SCPN) approach to transform IF-THEN rules into a Petri network that can be executed and analyzed to find potential loops, inconsistencies, and redundancies. The same SCPN approach is also exploited by *My IoT Puzzle* (Corno, De Russis, & Monge Roffarello, 2019c). Besides adopting a novel Jigsaw visual metaphor to represent triggers and actions (see Figure 6.4), the tool also integrates end-user debugging features, assisting users in identifying control-flow bugs. As shown in Figure 6.8, in particular, puzzle pieces deteriorate over time according to their usage history. Employing the same trigger across multiple rules, for example, results in the simultaneous execution of those rules, thereby raising the probability of introducing conflicts, such as redundancies and inconsistencies. Furthermore, *My IoT Puzzle* also provides users with graphical and textual (see the "why it is not working" box in Figure 6.8) explanations of the detected bugs.

Finally, Manca, Paternò, Santoro, & Corcella (2019) proposed *ITAD* (Figure 6.9), another end-user debugging tool for trigger-action programming. The tool can identify rules that may

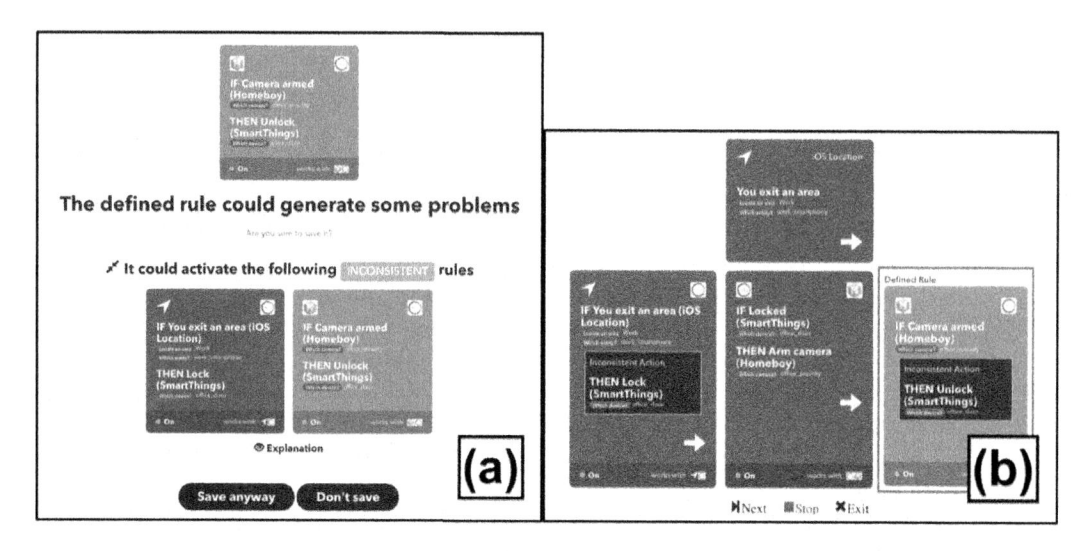

FIGURE 6.7 EUDebug integrates end-user debugging features, e.g., problem checking (a) and step-by-step explanations (b), on top of an IFTTT-like interface. Images taken from the original paper.

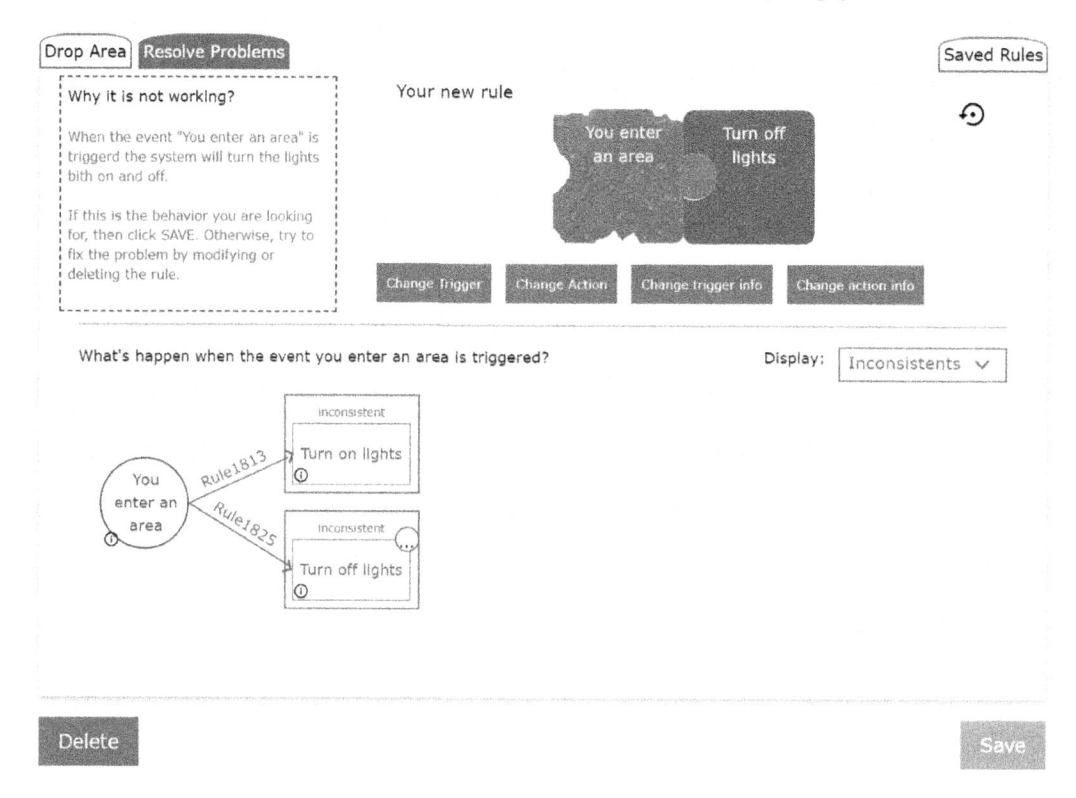

FIGURE 6.8 Besides letting users define rules through the Puzzle metaphor, My IoT Puzzle integrates end-user debugging features like puzzle pieces that deteriorate over time and textual and graphical explanations of potential run-time problems. Image taken from the original paper.

conflict depending on the values their triggers can assume at run-time. Furthermore, it can indicate why or why not a rule can be triggered in a given context of use. As in *My IoT Puzzle*, also *ITAD* adopts textual and graphical explanations. In Figure 6.9, for example, specific icons are used to differentiate instantaneous triggers and conditions visually.

Interacting with the Internet of Things

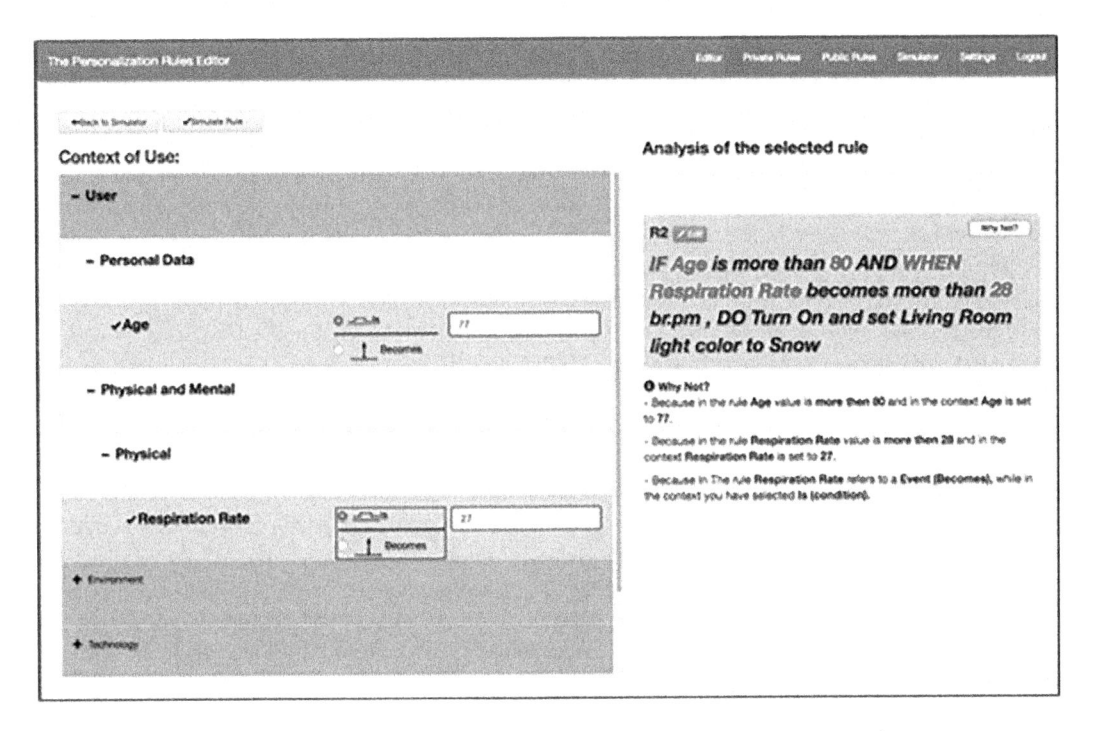

FIGURE 6.9 ITAD is an end-user debugging tool for trigger-action programming that can provide indications about why or why not a rule can be triggered in a given context of use (image taken from the original paper).

6.3.3 Discussion and Guidance for Future Research

Throughout Section 6.3, we reviewed issues, challenges, and solutions characterizing the domain of trigger-action programming for IoT personalization. Here, we discuss the most promising findings, trying to connect solutions to different challenges and highlighting opportunities for future research in such an evolving research field.

To advance the field of IoT personalization, we see particular value in further exploring the adoption of alternative representation models and interaction techniques (Table 6.2) for two main reasons. First, these strategies offer unique opportunities for improving the user experience in trigger-action programming platforms. By investigating and implementing alternative representation models, we can enhance the expressiveness and flexibility of trigger-action programming, enabling users to define more complex rules and personalize their IoT devices in a more nuanced manner. Additionally, exploring novel interaction techniques can provide more intuitive and efficient ways for users to interact with and define their rules, further improving the user experience. Finally, the same strategies have not yet been fully explored – thus leaving ample room for further investigation and experimentation – and could be easily integrated with most of the other solutions targeting other challenges. Adopting representation models that support a higher level of abstraction, for example, inevitably poses trust, security, and privacy issues that may be worth exploring in future works. Furthermore, the work by Corno, De Russis, & Monge Roffarello (2019a) suggests that abstract IF-THEN rules, e.g., those that do not refer to specific devices or manufacturers, are not always appreciated by end users: in some specific contexts, e.g., the smart home, people would prefer having more control to specify triggers and actions, e.g., to refer to specific technologies (Corno, De Russis, & Monge Roffarello, 2021a). Consequently, it becomes fundamental to explore solutions that may provide end users with the right level of abstraction, e.g., through automatic AI-powered solutions or through multi-level interfaces exposing a hierarchy of possible triggers and actions at different abstraction levels. For what concerns the exploration of novel interaction techniques, instead, one of the objectives of using voice-based interaction for defining IF-THEN rules (Table 6.2, third row) is to convert a user's voice-based natural language request into the intended rule. While extensions of existing platforms like Amazon Alexa (Manca, Parvin, Paternò, & Santoro, 2020) or custom IPAs (Monge Roffarello & De Russis, 2023) may facilitate the vocal creation of flexible automation using the trigger-action format, the process of defining IF-THEN rules through voice input can be challenging due to the inherent ambiguities in natural language. Therefore, we emphasize the significance of advancements in natural language processing (NLP) to address and eliminate potential ambiguities, enabling users to precisely indicate their desired effects. Previous studies have demonstrated the effectiveness of artificial intelligence methods in accurately mapping the user's abstract requirements to a lower level of abstraction that can be comprehended and executed in real time (see Corno, De Russis, & Monge Roffarello, 2021b) and (Corno, De Russis, & Monge Roffarello, 2020a). Specifically, we see value in integrating vocal interaction techniques with recommendation algorithms that may assist users in the definition process (e.g., through the strategies described in Table 6.3). Many other opportunities need to be further explored to fully take advantage of vocal interaction in such a EUD domain. One of these opportunities is to understand how we can manage collections of existing rules in a voice-based system, e.g., by exploring how a voice assistant can effectively present the list of available rules without the help of a screen. While reading out all the rules with their comprehensive details may be impractical, offering only partial information (e.g., the title of a rule) could lead to errors due to limited context. Striking a balance between these extremes is the key challenge in this context. Another opportunity is to use vocal interaction to support users during the execution of rules, e.g., through proactive voice assistants that explain why a rule is being activated or why a conflict is happening. Such an approach may further complement the need to assist end users in defining safe IF-THEN rules and provide them with effective end-user debugging tools (Table 6.4). The challenge here is, again, at the presentation level, with open questions that concern determining when it is appropriate to alert users about a problem, to which user(s), and how to effectively communicate the conflicts.

6.4 THE DEVELOPER PERSPECTIVE

While the previous section focused on end users, their understanding of the IoT environments, and the related personalization options, this section focuses on actual programmers creating IoT systems and how developers characterize them.

Implementing IoT systems is particularly complex and differs from developing mainstream mobile and web applications. In fact, it involves a comprehensive spectrum of technologies (Taivalsaari & Mikkonen, 2018) that involve multiple development and execution environments. Consequently, besides focusing exclusively on the code, IoT developers must also deal with hardware implementation and distributed computing concepts.

In this scenario, harnessing the programmable world's power requires understanding the peculiarities and the most challenging issues that the implementation of IoT systems poses to the developers and envisioning new software engineering and development technologies, processes, methodologies, and tools (Taivalsaari & Mikkonen, 2017).

In addition to introducing IoT implications from the developers' perspective, we explore its associated issues and challenges, along with the emerging solutions. For a comprehensive overview of these issues, challenges, and solutions, please refer to Table 6.5

6.4.1 CONTEXT

The complexity of developing IoT systems arises from the specific computing resources, technologies, and communication protocols associated with each of the four architectural elements, i.e., devices, gateways, cloud services, and applications (see Figure 6.1). The coexistence of diverse devices, protocols, architectures, and programming languages necessitates expertise in multiple

TABLE 6.5
A Summary of the Issues, Challenges, and Solutions Reviewed in this Section

Issue	Challenge	Solutions
Limited awareness of IoT peculiarities: An understanding of the methodology and its main challenges is lacking to support the entire IoT application development lifecycle.	Characterizing IoT systems and identifying development issues faced by developers: Provide a common understanding of IoT defining characteristics.	• Reference architecture • Development process analysis • Custom Software Engineering development methodologies
Integration issues: Envisioning and successfully achieving the integration of IoT components is conceptually and technically challenging.	*Supporting developers in integrating IoT subsystems:* Provide developers conceptual and technical resources to design and implement the integration of heterogeneous components.	• Programming frameworks • Architectural styles • Software development approaches

disparate areas and the ability to orchestrate them effectively. As outlined in Table 6.5, two issues arise from this particular scenario:

- *Limited awareness of IoT peculiarities*: As previously outlined, the development process of IoT systems differs from traditional software and poses distinctive challenges to developers. Indeed, IoT developers are now required to consider several dimensions unfamiliar to most software developers (Taivalsaari & Mikkonen, 2017). However, there is a lack of awareness of IoT systems' specificities regarding their architecture and development process, resulting in highly difficult-to-implement and platform-dependent software. Little research to date uncovers the core development process lifecycle needed for IoT systems, and thus software engineers find themselves unprepared and unfamiliar with this new genre of system development (Fahmideh, Ahmad, Behnaz, Grundy, & Susilo, 2022). Furthermore, as stated by (Larrucea, Combelles, Favaro, & Teneja, 2017), linked to IoT systems development's inherent complexity is that no consolidated set of software engineering best practices for the IoT has emerged yet. In his words, "*IoT landscape resembles the Wild West, with programmers putting together IoT systems in ad hoc fashion.*" To effectively leverage the power of the promised programmable world of IoT, it is essential to have a precise understanding of IoT intricacies. Without such understanding, it becomes impossible to envision appropriate software engineering and development technologies, processes, methodologies, and tools tailored for IoT applications.
- *Integration issues*: From the developers' perspective, the high heterogeneity among software and hardware components present in IoT systems results in several integration challenges. This integration must ensure data exchange across diverse IoT components regarding protocols, connectivity, data formats, technologies, specifications, and communication mechanisms. Indeed, IoT systems development typically requires integrating complex legacy systems, often conceived, developed, and deployed as monoliths (Fortino, Savaglio, Spezzano, & Zhou, 2021). Therefore, successfully achieving interoperability among these legacy systems demands careful design decisions and adopting precise technical resources. For this reason, available alternatives to deal with integration are required to consider conceptual and architectural approaches, commonly adapted from traditional software development paradigms to specific methodologies, frameworks, platforms, and tools.

6.4.2 Challenges

In general, the two identified issues correspond to specific challenges:

- Characterizing IoT systems and identifying development issues faced by developers.
- Overcoming the integration issues among heterogeneous subsystems.

In this section, we describe these challenges highlighting solutions that have been put forth to tackle each.

Characterizing IoT systems and identifying development issues faced by developers:

> So far, IoT applications have been based on fragmented software implementations for specific systems and use cases (Weyrich, 2016). A unifying approach grounded on standard abstractions, models, and methodologies is still missing to aid IoT systems development (Zambonelli, 2017). Given this situation, as shown in Table 6.6, researchers have followed three main strategies to consolidate a common understanding of IoT systems' characteristics and development process: Reference architectures, IoT development process analysis, and custom Software Engineering development methodologies.

Regarding the first strategy, the industry's need for reference architectures has become tangible with the fast-growing number of initiatives working toward standardized architectures (Weyrich, 2016). (Čolaković & Hadžialić, 2018) hold that IoT software architectures and frameworks are necessary to overcome IoT systems' inherent complexity and provide an environment for services composition. In this sense, two major architectures are available: the Internet of Things-Architecture (IoT-A) and Industrial Internet Reference Architecture (IIRA). IoT-A delivered a detailed architecture and model from the functional and information perspectives. In contrast, IIRA was delivered by the Industrial Internet Consortium (founded by AT&T, Cisco, General Electric, IBM, and Intel) for broad consideration and discussion. Such architectures can serve as an overall and generic guideline, and not all domain applications will require every component for real-life development.

Although these reference architectures may vary and a definitive "standard" architecture has not yet emerged, they encompass many functions, information structures, and mechanisms. They offer developers a more comprehensive perspective of the IoT system they intend to implement. Furthermore, they contribute to defining and elucidating the overall structure of the IoT. These reference architectures offer descriptive models that illustrate how IoT devices and humans interact and process data, incorporating patterns of machine-to-machine (M2M) communication standards (Weyrich, 2016). According to (Fortino, Savaglio, Spezzano, & Zhou, 2021), exploiting these established reference architectures is another common trait of IoT methodologies toward IoT system interoperability. Indeed, over the years, a broad and solid consensus on architectural IoT building blocks and their main design patterns has been reached.

Similarly, (Zambonelli, 2017) frames key general characteristics related to the engineering of IoT systems by synthesizing the common features of existing proposals and scenarios. The author identifies the key software engineering abstractions around which IoT systems and application development could revolve. Specifically, the common features were the things (physical objects, places, and persons), the software infrastructures, and services and applications. Meanwhile, the key abstractions for developing IoT systems were: the stakeholders and users, requirements (policies, goals, and functions), avatars and coalitions (individual things and a group of things that contribute to defining a unique functionality or service), and smart things.

As for the second strategy, research efforts have been dedicated to providing insights into the most critical development tasks and the challenges associated with implementing IoT systems. (Fahmideh, Ahmad, Behnaz, Grundy, & Susilo, 2022) conducted a research study that employed a mixed quantitative and qualitative approach. The objective was to derive a conceptual process framework for the development processes of IoT systems, wherein common tasks are organized into three distinct phases. The derived framework was validated through a survey involving 127 IoT practitioners. These participants provided justifications, examples, and recommendations on tasks' importance and associated challenges. The authors identified challenges in the analysis, design, and implementation phases. Additionally, respondents offered general advice to software teams on effectively addressing these challenges.

Similarly, (Corno, De Russis, & Sáenz Moreno, 2019) surveyed 40 novice developers who worked in groups developing IoT systems during several years of a university course. Based on their own experiences, individually and as a group, the most challenging development tasks were identified and prioritized over a common architecture regarding difficulty level and effort. In line with this research, to gain a deeper understanding of how open-source IoT projects differ from non-IoT projects, (Corno, De Russis, & Sáenz Moreno, 2020) conducted a quantitative analysis of a broad set of the 60 most popular publicly available IoT and non-IoT projects on GitHub. The analysis provided insight into the purpose and characteristics of the code, the behavior of the contributors, and the maturity of the IoT software development ecosystem. The findings highlight significant differences between IoT and non-IoT application development regarding how applications are realized, the diversity of developers' specializations, and how code is reused.

Finally, regarding the third strategy, the lack of a software engineering methodology to support the entire IoT application development lifecycle results in highly difficult to maintain, reuse, and platform-dependent design (Patel, 2015). In this sense, (Motta, de Oliveira, & Travassos, 2023) present an evidence-based roadmap for IoT development to support developers in specifying, designing, and implementing IoT systems. The IoT Roadmap consists of 117 items grouped into 29 categories. An experimental study on a case study involving a healthcare IoT project was conducted. Moreover, the roadmap includes a convenient checklist that aids in identifying the most relevant recommendations for developing and engineering IoT software systems.

Likewise, according to (Morin, Harrand, & Fleurey, 2017), IoT applications have two main characteristics from a software engineering viewpoint: their distribution over an extensive range of processing nodes; and the high heterogeneity of the processing nodes and the protocols used. To deal with these characteristics, they introduce a modeling language aligned with UML, an advanced multiplatform code generation framework, and a methodology specifying the development processes and tools used by IoT service developers and platform experts. Similarly, Patel (2015) introduced a development methodology for IoT application development based on model-driven development. Such methodology separates IoT application development into different concerns and provides a conceptual framework to develop an application. Additionally, the framework provides a set of modeling languages to specify each development concern.

Achieving the integration among heterogeneous subsystems:

> As previously described, the diversity among IoT software and hardware components raises several interoperability issues. Specifically, in the IoT scenario, interoperability refers to the degree to which two or more components can exchange information and use the information that has been exchanged (Noura, Atiquzzaman, & Gaedke, 2019). Successfully achieving this interoperability impacts the widespread adoption of IoT technology on a large scale.

From the development point of view, alternatives to deal with the integration issues and achieve interoperability. Given the complexity and heterogeneity of the IoT ecosystem, these strategies go beyond technical aspects and involve methodologies and architectural approaches. These strategies typically involve architectural approaches such as microservices, event-driven architectures, and service-oriented architectures to IoT platforms and tools that connect different devices and access, manage, and process their data. Table 6.7 mentions some strategies to deal with the complexity of integrating IoT components from two perspectives: Methodologies and architectural approaches and Programming frameworks, platforms, and tools.

TABLE 6.6
Researchers Have Followed Three Main Strategies to Consolidate a Common Understanding of IoT Systems' Characteristics and Development Process

Strategy	Description	Main References
Reference architecture	Identification and representation of the main components present in a typical IoT system and the interactions among them.	Weyrich (2016) Fortino, Savaglio, Spezzano, & Zhou (2021) Zambonelli (2017)
Development process analysis	Analyze quantitatively and qualitatively the characteristics of the code and the IoT developers' implementation process.	Fahmideh, Ahmad, Behnaz, Grundy, & Susilo (2022) Corno, De Russis, & Sáenz Moreno (2020) Corno, De Russis, & Sáenz Moreno (2019) Makhshari & Mesbah (2021)
Custom software engineering development methodologies	Revisit traditional software development methodologies to accommodate the distinctive needs imposed by IoT systems.	Motta, de Oliveira, & Travassos (2023) Morin, Harrand, & Fleurey (2017) Patel (2015) Berrouyne, Adda, & Mottu (2022) Ferreira et al. (2022)

TABLE 6.7
Researchers Have Followed Three Main Strategies to Deal with the Complexity that Integrating IoT Components Poses

Strategy	Description	Main References
Architectural approaches and methodologies	Architectural approaches adapted from traditional software development. service-oriented architectures and model-driven development are among the most prominent alternatives.	Asghari (2018) Urbieta (2017) Hamzei & Navimipour (2018) Cabrera, Cárdenas, Cedillo, & Pesántez-Cabrera (2020) Patel (2015)
Programming frameworks, platforms, and tools	IoT platforms, tools, and programming frameworks aimed at easing the interoperability among IoT components. IoT Cloud Platforms, Middlewares, Software adapters, and Gateways are the most popular strategies.	Avilés-López & García-Macías (2012) Barros et al. (2022) Fortino, Savaglio, Spezzano, & Zhou, (2021) Razzaque, Milojevic-Jevric, Palade, & Clarke (2016) Beniwal & Singhrova (2022)

At the **architectural approach**, software adapters, gateways, and service-oriented interfaces are standard solutions to ensure syntactic interoperability among diverse communication technologies or application layer protocols. In this regard, the most prevalent approaches employed by IoT frameworks to develop interoperable IoT systems are *Model-Driven Engineering* (MDE) and service-oriented development (Fortino, Savaglio, Spezzano, & Zhou, 2021). MDE relies on domain knowledge modeling standards and metalanguages to describe IoT system components and functionalities independently of their specific implementation details. For their part, *Service-Oriented Architectures* (SOA) consider the definition of well-defined interfaces and protocols that allow IoT devices and applications to communicate and exchange data regardless of the underlying technologies or platforms. In this sense, exposing each component's functionalities as a standard service can significantly increase network and device interoperability (Noura, Atiquzzaman, & Gaedke, 2019). In particular, resource-oriented web services (REST web services) are widely adopted to achieve the potential of SOA, providing service sharing and reuse, and have been used to address syntactic interoperability.

Along the same line, *microservices* are another way to support protocol-aware heterogeneous interoperability. They allow loose-coupled service interfaces and integration, simplifying services' architectural patterns and related implementations (Fortino, Savaglio, Spezzano, & Zhou, 2021). At the *software* layer, APIs are commonly utilized in SOA and microservices approaches to achieve syntactic interoperability.

Concerning the currently proposed **IoT methodologies**, Fortino et al. (2018) presented a full-fledged, general-purpose methodology supporting the integration process among IoT platforms. This methodology is based on a process with analysis, design, implementation, deployment, and maintenance phases. Similarly, Geraldi, Reinehr, & Malucelli (2020) provided a comprehensive overview of how the integration challenges can be faced by relying on Software Product Line methodologies tailored for the IoT. Cabrera, Cárdenas, Cedillo, & Pesántez-Cabrera (2020), for their part, proposed an agile methodology to guide the development of IoT software components based on microservice architectures. Finally, Patel (2015) presented a methodology separating IoT application development into different concerns and providing a conceptual framework to develop an application.

From a technical perspective, *IoT middleware*, following SOA architectural approach, provides services and abstractions to simplify **IoT system development**. At the same time, IoT gateways focus more on easing connectivity, data processing, and protocol translation. By definition, middleware serves to abstract the complexities of the system or hardware. It enables the developer to concentrate solely on the business logic of the solution that is being implemented. Middleware in IoT eases development by integrating heterogeneous computing and communications devices and supporting interoperability within diverse applications and services (Razzaque, Milojevic-Jevric, Palade, & Clarke, 2016).

Whether software or hardware components, *IoT Gateways* are bridges between IoT devices and the cloud, data centers, or central management systems connected through the Internet (Beniwal & Singhrova, 2022), they facilitate protocol conversion between the sending and receiving devices. They use plug-and-play software adapters to connect components and platforms with communication technologies, application layer protocols, or data formats (Noura, Atiquzzaman, & Gaedke, 2019). These gateways can communicate using protocols like Bluetooth, ZigBee, and Ethernet and interfaces like Wi-Fi, MQTT, and CoAP (Dizdarević, Carpio, Jukan, & Masip-Bruin, 2019).

At a platform level, currently, *IoT Cloud platforms* provide tools and protocols for device registration, authentication, and secure communication. They offer various services such as device management, data storage, analytics, and application development tools. In particular, they provide a scalable and secure infrastructure for connecting, monitoring, and controlling a wide range of IoT devices and applications. Furthermore, Today's IoT platforms almost all provide a public API to assist developers in connecting with other systems and services and accessing their services. Popular IoT Cloud platforms include Amazon Web Services (AWS) IoT Core, Microsoft Azure IoT Hub, Google Cloud IoT Core, and IBM Watson IoT. These platforms offer comprehensive features and services to support the entire IoT ecosystem, from device connectivity to data analytics.

6.4.3 Discussion and Guidance for Future Research

We want to draw attention to the fact that as the IoT continues to position itself as a well-established paradigm, the ecosystem grows, expands to several domains, and research efforts are sufficiently addressing the emerging challenges. There are plenty of academic and industry efforts to identify and address the requirements that IoT development poses. Therefore, the upcoming issues do not arise from the shortcoming of methodologies, frameworks, platforms, and tools. On the contrary, they paradoxically emerge from the overabundance of resources, continuously fueled by innovative and evolutionary development approaches and technologies (Fortino, Savaglio, Spezzano, & Zhou, 2021). In this scenario, choosing the best solution for designing and implementing an IoT system has become a challenging entry barrier, especially for novice IoT developers. Indeed, preparing future IoT developers has become particularly important given the definite IoT consolidation and expansion in several domains and its ever-growing ecosystem complexity (Corno & De Russis, 2017).

In this scenario, documentation resources play a fundamental role. Specifically in terms of providing comprehensive documentation that accounts for the IoT development peculiarities. For instance, previous research has determined that it is hard for programmers to understand low-quality documentation of certain device manufacturers and interpret complex response payloads from particular devices (Makhshari & Mesbah, 2021). In the context of novice programmers getting started with IoT development, Corno, De Russis, & Sáenz Moreno (2019) identified that integration among IoT subsystems was one of the most painful issues not-experienced developers faced due to the lack of appropriate documentation. Therefore, they proposed a documentation mechanism called Code Recipes, aimed at enabling the integration of several software components through code fragments that might belong to different programming languages and might be deployed across various run-time environments, as it is common in IoT systems. Code Recipes (Corno, De Russis, & Sáenz Moreno, 2018), therefore, are defined as summarized and well-defined documentation modules independent from programming languages or run-time environments. Precisely, these recipes are specified through metadata and consist of multiple code fragments, documentation, and links to ease the understanding of such code to implement a given integration between subsystems of an IoT system. As a further step in this research direction, and given the prominence that computational notebooks have been gaining due to their capability to consolidate text, executable code, and visualizations, Corno, De Russis, & Sáenz Moreno (2022) have explored to what degree these notebooks are appropriate for facilitating the prototyping of IoT systems, even when the process entails multiple steps and encompasses various development and execution environments. The authors designed, implemented, and assessed an IoT-tailored notebook aimed at helping developers to build and share a computational narrative around the prototyping of IoT systems. The rationale behind the proposal stemmed from the belief that by documenting the reasoning and subsequent steps in a computational notebook, within a meticulously configured execution environment, the prototyping of IoT systems can be enhanced in terms of reproducibility and comprehensibility.

In a broader sense, several research efforts have been devoted to envisioning IoT learning strategies. Burd et al. (2018) have highlighted the challenges for computer science educators in conducting IoT courses: connecting and integrating hardware and software; finding adequate physical space and infrastructure; and preparing instructors and teaching assistants for the content.

Furthermore, the authors identify core and specialized topics that educators might integrate into their curricula depending on the student's existing knowledge and learning objectives. Among these identified topics are: cloud computing, HCI, embedded computing, platform-specific development, and web and mobile development, among others. Naturally, these IoT curriculums must continuously evolve as the IoT industry evolves and new tools and components appear in the ecosystem.

Due to the above, a significant challenge that, in our opinion, requires attention is envisioning strategies and tools to enclose into a typical academic period the topics and skills that enable students to face a generic IoT system implementation. This goal requires finding a proper balance in the hardware platforms, networking protocols, programming languages, and cloud services to include. Similarly, to better cover these topics and technologies and to set up a realistic development scenario, it is imperative to promote teamwork and project-based learning approaches.

6.5 BRIDGING THE GAP BETWEEN PERSPECTIVES

In the IoT, end users and developers have different needs and perspectives. They are, however, two faces of the same coin: users can personalize their IoT environments if developers provide them with proper tools and options; developers can work better if they know what the users' expectations and needs are. While end users focus on personalizing the behavior of their environments through dedicated tools, developers are more concerned with integration issues due to the variety of architectural elements that constitute an IoT system.

In addition, both have their understanding of IoT environments that stems from their specific context. End users want to create automations to increase their serendipity and extend the capabilities of the existing system, within their understanding of how the IoT-enabled environment operates. Conversely, developers want to create robust systems, extendable and with suitable user interfaces, within their understanding of what can constitute an IoT environment. The user interface is one of the many components they have to consider, and often not the most critical to make the system work.

They also have different struggles. End users need to understand how to reach their automation goal, the implications of their actions, and what to do when things are not working as expected. Developers need to handle the various components, written in different languages and using many technologies, and integrate them in a coherent system.

From this summary, it should be clear what is the gap, what are the motivations behind it, and what is the commonality between the two perspectives. The commonality rises from the need of both actors to understand the IoT: they need to know how the IoT environment might work. However, the different purpose in this understanding is what generates the gap: end users focus on the results derived from the run of the IoT system, while developers are still focusing on how to make a heterogeneous system work in a homogenous way while keeping the pace with the latest IoT technologies and software methodologies. In other words, end users have a more focused and utilitarian view of the system, which is seen as a way to fulfill some intentions, while developers must have a wider perspective, considering the operational aspects and the use cases as well as the maintainability, security, and integration challenges.

To our knowledge, this gap has not yet been tackled in the literature. The research field focusing on EUD is informed by software engineering methods and applies development concepts to their research output, still within the specific goal of creating, maintaining and debugging automations. The software engineering field, instead, considers users and interfaces for them as one of the many elements of an IoT system, often considering the immediate use more than end users development options.

Hereafter, the chapter proposes two possible recommendations to bridge the perspectives and the respective needs. Such recommendations stem from the literature discussed above and from the authors' personal experience in creating EUD tools and supporting IoT developers. Both asks for a stronger collaboration between the research communities focusing on EUD and software engineering for the IoT:

IoT developers need to consider personalization options for end users: Developing an IoT system focuses on various components, one of which is the user interface within the "application" component (Figure 6.1). Current research is mostly focused on technical aspects, where one of the main challenges is about the integration of components into a coherent system. However, end users have varied needs and benefit from the possibility to tailor their environment accordingly. An explicit focus on EUD, within the "application" or a dedicated component (e.g., a "personalization component"), can have two advantages for end users and developers: end users can have more options, developers can receive dedicated feedback and collect information on the actual use of their IoT system in real settings. This feedback can guide developers to not only refine the offered personalization options but also to understand the specific needs of their users and direct the development process accordingly.

EUD tools and methodologies should be applied to IoT development: Since integration is one of the main challenges in creating IoT systems, a viable option to minimize the struggle can be to adopt EUD methodologies to ease the integration efforts, in addition to personalize the behavior of the various "things" in the environment. Such techniques can, similarly to what conversational agents and recommenders are doing in the EUD field, help with the integration and creation of various components and documentation, thus lowering the entry barrier. The goal, here, is to support developers so that they can personalize their IoT *development* environment with the same ease they can customize the IoT-enabled environment. In this scenario, technologies like *digital twins* can provide further support to developers, carefully replicating physical devices in their virtual clones.

6.6 CONCLUSION

This chapter discusses how people characterize and interact with the IoT and, particularly, with IoT-enabled environments. It tackles two perspectives: end users wanting to personalize their environments according to their specific and contextual desires; developers who need to build such systems. Indeed, developers' challenges and adopted solutions can be useful to understand the options they offer to end users; conversely, end users' expectations for personalizing IoT systems can inform developers' practices. This chapter describes these two perspectives and approaches to the IoT with respect to the literature. For each of them, challenges and possible solutions are reported, as well as gaps between the perspectives. In the end, this chapter proposes two possible recommendations to try to close the gap between these two worlds with the aim to ease IoT system creation and the subsequent interaction with them.

6.6.1 FUTURE TRENDS

As discussed by Sharma et al. (2018), a number of interesting possibilities and future trends characterize the topic of interacting with the IoT. The first promising and needed direction involves developing energy-efficient IoT technology. Examples include the "Green IoT" discussed by Alsharif et al. (2023), as well as the sustainable IoT ecosystems envisioned for the agricultural (Dhanaraju, Chenniappan, Ramalingam, Pazhanivelan, & Kaliaperumal, 2022) and railway transportation (Singh, Elmi, Meriga, Pasha, & Dulebenets, 2022) domains. Other future trends worth to be explored include integrating blockchain technologies in smart cities (Majeed et al., 2021), supporting students in learning about emerging technologies (Van Mechelen & Smith, 2023), offering appropriate resources for EUD in the workplace (Barricelli, Fogli, & Locoro, 2023), and addressing security challenges (Yugha & Chithra, 2020). In conclusion, fostering and improving the interaction between end users, developers, and the IoT, has an important role in making these future trends possible and driving digital transformation, enhancing efficiency and fostering sustainability.

NOTES

1. https://www.home-assistant.io/docs/scripts/, last visited on June 10, 2023.
2. https://ifttt.com/, last visited on June 10, 2023.
3. https://ifttt.com/, last visited on June 10, 2023.
4. https://zapier.com/, last visited on June 10, 2023.

REFERENCES

Akiki, P. A., Bandara, A. K., & Yu, Y. (2017). Visual simple transformations: Empowering end-users to wire internet of things objects. *ACM Transactions on Computer-Human Interaction, 24*(2), 1–43.

Alaa, M., Zaidan, A., Zaidan, B., Talal, M., & Kiah, M. (2017). A review of smart home applications based on Internet of Things. *Journal of Network and Computer Applications, 97*, 48–65.

Al-Fuqaha, A., Guizani, M., Mohammadi, M., Aledhari, M., Aledhari, M., & Ayyash, M. (2015). Internet of Things: A survey on enabling technologies, protocols, and applications. *IEEE Communications Surveys & Tutorials, 17*(4), 2347–2376.

Alsharif, M., Jahid, A., Kelechi, A., & Kannadasan, R. (2023). Green IoT: A review and future research directions. *Simmetry, 15*(3), 757.

Angelini, L., Couture, N., Abou Khaled, O., & Mugellini, E. (2018). Internet of Tangible Things (IoTT): Challenges and opportunities for tangible interaction with IoT. *Informatics, 5*(1), 1–28.

Asghari, P. (2018, October 15). Service composition approaches in IoT: A systematic review. *Journal of Network and Computer Applications, 120*, 61–77.

Atzori, L., Iera, A., & Morabito, G. (2010, October 28). The internet of things: A survey. *Computer Networks, 54*(15), 2787–2805.

Avilés-López, E., & García-Macías, J. (2012). Mashing up the internet of things: A framework for smart environments. *EURASIP Journal on Wireless Communications and Networking, 2012*(1), 1687–1499.

Barricelli, B., & Valtolina, S. (2015). Designing for end-user development in the internet of things. In *End-User Development: 5th International Symposium, IS-EUD 2015, Madrid, Spain, May 26-29, 2015. Proceedings 5* (pp. 9–24). Springer International Publishing.

Barricelli, B., Fogli, D., & Locoro, A. (2023). EUDability: A new construct at the intersection of end-user development and computational thinking. *Journal of Systems and Software, 195*, 111516.

Barricelli, B., Fogli, D., Iemmolo, L., & Locoro, A. (2022). A multi-modal approach to creating routines for smart speakers. In *Proceedings of the 2022 International Conference on Advanced Visual Interfaces* (pp. 1–5).

Barros, T. G., Da Silva Neto, E. F., Da Silva Neto, J., De Souza, A. G., Aquino, V. B., & Teixeira, E. S. (2022). The anatomy of IoT platforms-A systematic multivocal mapping study. *IEEE Access, 10*, 72758–72772.

Bellini, P., Nesi, P., & Pantaleo, G. (2022). IoT-enabled smart cities: A review of concepts, frameworks and key technologies. *Applied Sciences, 12*(3), 1607.

Beniwal, G., & Singhrova, A. (2022). A systematic literature review on IoT gateways. *Journal of King Saud University - Computer and Information Sciences, 34*(10), 9541–9563.

Berrouyne, I., Adda, M., & Mottu, J.-M. (2022, October 15). A model-driven methodology to accelerate software engineering in the internet of things. *IEEE Internet of Things Journal, 9*(20), 19757–19772.

Brackenbury, W., Deora, A., Ritchey, J., Vallee, J., He, W., Wang, G., ... Ur, B. (2019). How users interpret bugs in trigger-action programming. In *Proceedings of the 2019 CHI Conference on Human Factors in Computing Systems* (pp. 1–12).

Brich, J., Walch, M., Rietzler, M., Weber, M., & Schaub, F. (2017). Exploring end user programming needs in home automation. *ACM Transaction on Computer-Human Interaction, 24*(2), 1–35.

Brush, A., Lee, B., Mahajan, R., Agarwal, S., Saroiu, S., & Dixon, C. (2011). Home automation in the wild: Challenges and opportunities. In *Proceedings of the SIGCHI Conference on Human Factors in Computing Systems* (pp. 2115–2124).

Burd, B., Barker, L., Fermín Pérez, F., Russell, I., Siever, B., Tudor, L., ... Pollock, I. (2018, July). The internet of things in undergraduate computer and information science education: Exploring curricula and pedagogy. In *ITiCSE 2018 Companion: Proceedings Companion of the 23rd Annual ACM Conference on Innovation and Technology in Computer Science Education* (pp. 200–216).

Cabrera, E., Cárdenas, P., Cedillo, P., & Pesántez-Cabrera, P. (2020). Towards a methodology for creating Internet of Things (IoT) *applications based on microservices*. In *2020 IEEE International Conference on Services Computing (SCC)* (pp. 472–474).

Caivano, D., Fogli, D., Lanzilotti, R., Piccinno, A., & Cassano, F. (2018). Supporting end users to control their smart home: Design implications from a literature review and an empirical investigation. *Journal of Systems and Software, 144*, 295–313.

Cao, J., Rector, K., Park, T. H., Fleming, S. D., Burnett, M. M., & Wiedenbeck, S. (2010). A debugging perspective on end-user mashup programming. In *2010 IEEE Symposium on Visual Languages and Human-Centric Computing* (pp. 149–156).

Cerf, V., & Senges, M. (2016). Taking the internet to the next physical level. *Computer, 49*(2), 80–86.

Čolaković, A., & Hadžialić, M. (2018, October 24). Internet of Things (IoT): A review of enabling technologies, challenges, and open research issues. *Computer Networks, 144*, 17–39.

Cook, D., & Das, S. K. (2005). *Smart Environments: Technology, Protocols and Applications*. Hoboken: Wiley.

Corno, F., & De Russis, L. (2017). Training engineers for the ambient intelligence challenge. *IEEE Transactions on Education, 60*(1), 40–49.

Corno, F., De Russis, L., & Monge Roffarello, A. (2019a). A high-level semantic approach to end-user development in the Internet of Things. *International Journal of Human-Computer Studies, 125*, 41–54.

Corno, F., De Russis, L., & Monge Roffarello, A. (2019b). Empowering end users in debugging trigger-action rules. In *Proceedings of the 2019 CHI Conference on Human Factors in Computing Systems* (pp. 1–13).

Corno, F., De Russis, L., & Monge Roffarello, A. (2019c). My IoT PUZZLE: Debugging IF-THEN rules through the jigsaw metaphor. In *End-User Development: 7th International Symposium, IS-EUD 2019, Hatfield, UK, July 10–12, 2019, Proceedings 7* (pp. 18–33). Springer International Publishing.

Corno, F., De Russis, L., & Monge Roffarello, A. (2019d). RecRules: Recommending IF-THEN rules for end-user development. *ACM Transactions on Intelligent Systems and Technology, 10*(5), 1–27.

Corno, F., De Russis, L., & Monge Roffarello, A. (2020a). HeyTAP: Bridging the gaps between users' needs and technology in IF-THEN rules via conversation. In *Proceedings of the International Conference on Advanced Visual Interfaces* (pp. 1–9).

Corno, F., De Russis, L., & Monge Roffarello, A. (2020b). TAPrec: Supporting the composition of trigger-action rules through dynamic recommendations. In *Proceedings of the 25th International Conference on Intelligent User Interfaces* (pp. 579–588).

Corno, F., De Russis, L., & Monge Roffarello, A. (2021a). Devices, information, and people: Abstracting the internet of things for end-user personalization. In *International Symposium on End User Development* (pp. 71–86). Cham: Springer International Publishing.

Corno, F., De Russis, L., & Monge Roffarello, A. (2021b). From users' intentions to IF-THEN rules in the internet of things. *ACM Transactions on Information Systems, 39*(4), 1–33.

Corno, F., De Russis, L., & Sáenz Moreno, J. (2018). Easing IoT development for novice programmers through code recipes. In *ICSE-SEET '18: Proceedings of the 40th International Conference on Software Engineering: Software Engineering Education and Training.*

Corno, F., De Russis, L., & Sáenz Moreno, J. (2022). Computational notebooks to support developers in prototyping IoT systems. *International Journal of Human-Computer Studies, 162*, 102850.

Corno, F., De Russis, L., & Sáenz Moreno, J. (2019). On the challenges novice programmers experience in developing IoT systems: A Survey. *Journal of Systems and Software, 157*, 110389.

Corno, F., De Russis, L., & Sáenz Moreno, J. (2020). How is open source software development different in popular IoT projects? *IEEE Access, 8*, 28337–28348.

Danado, J., & Paternò, F. (2014). Puzzle: A mobile application development environment using a jigsaw metaphor. *Journal of Visual Languages & Computing, 25*(4), 297–315.

Daniel, F., & Matera, M. (2014). *Mashups: Concepts, Models and Architectures.* Berlin, Heidelberg: Springer.

Darshan, K., & Anandakumar, K. (2016). A comprehensive review on usage of Internet of Things (IoT) in healthcare system. In *2015 International Conference on Emerging Research in Electronics, Computer Science and Technology (ICERECT)* (pp. 132–136). Mandya, India.

Desolda, G., Ardito, C., & Matera, M. (2017). Empowering end users to customize their smart environments: Model, composition paradigms, and domain-specific tools. *ACM Transactions on Computer-Human Interaction, 24*(2), 1–52.

Dey, A. K., Sohn, T., Streng, S., & Kodama, J. (2006). iCAP: Interactive prototyping of context-aware applications. In *Proceedings of the 4th International Conference on Pervasive Computing* (pp. 254–271).

Dhanaraju, M., Chenniappan, P., Ramalingam, K., Pazhanivelan, S., & Kaliaperumal, R. (2022). Smart farming: Internet of Things (IoT)-based sustainable agriculture. *Agriculture, 12*(10), 1745.

Dizdarević, J., Carpio, F., Jukan, A., & Masip-Bruin, X. (2019). A survey of communication protocols for internet of things and related challenges of fog and cloud computing integration. *ACM Computing Surveys, 51*(6), 1–29.

Fahmideh, M., Ahmad, A., Behnaz, A., Grundy, J., & Susilo, W. (2022, August 1). Software engineering for internet of things: The practitioners' perspective. *IEEE Transactions on Software Engineering, 48*(8), 2857–2878.

Ferreira, L. B., Chaves, P. R., Assumpção, R. M., Branquinho, O. C., Fruett, F., & Cardieri, P. (2022, November). The three-phase methodology for IoT project development. *Internet of Things, 20*, 100624.

Fortino, G., Savaglio, C., Palau, C. E., Suarez de Puga, J., Ganzha, M., Paprzycki, M., ... Llop, M. (2018). Towards multi-layer interoperability of heterogeneous IoT platforms: The INTER-IoT approach. In R. Gravina, C. Palau, M. Manso, A. Liotta, & G. Fortino (Eds), *Integration, Interconnection, and Interoperability of IoT Systems* (pp. 199–232). Cham: Springer.

Fortino, G., Savaglio, C., Spezzano, G., & Zhou, M. (2021, January). Internet of things as system of systems: A review of methodologies, frameworks, platforms, and tools. *IEEE Transactions on Systems, Man, and Cybernetics: Systems, 51*(1), 223–236.

Gallo, S., Manca, M., Mattioli, A., Paternò, F., & Santoro, C. (2021). Comparative analysis of composition paradigms for personalization rules in IoT settings. In *End-User Development: 8th International Symposium* (pp. 53–70).

Geraldi, R., Reinehr, S., & Malucelli, A. (2020, August). Software product line applied to the internet of things: A systematic literature review. *Information and Software Technology, 124*, 106293.

Ghiani, G., Manca, M., Paternò, F., & Santoro, C. (2017). Personalization of context-dependent applications through trigger-action rules. *ACM Transactions on Computer-Human Interaction, 24*(2), 1–33.

Hamzei, M., & Navimipour, N. (2018, October). Toward efficient service composition techniques in the internet of things. *IEEE Internet of Things Journal, 5*(5), 3774–3787.

Huang, J., & Cakmak, M. (2015). Supporting mental model accuracy in trigger-action programming. In *Proceedings of the 2015 ACM International Joint Conference on Pervasive and Ubiquitous Computing* (pp. 215–225).

Huang, T.-H. K., Azaria, A., & Bigham, J. P. (2016). InstructableCrowd: Creating IF-THEN rules via conversations with the crowd. In *Proceedings of the 2016 CHI Conference Extended Abstracts on Human Factors in Computing Systems* (pp. 1555–1562).

Islam, S. M., Kwak, D., Kabir, M. H., Hossain, M., & Kwak, K.-S. (2015). The internet of things for health care: A comprehensive survey. *IEEE Access, 3*, 678–708.

Jaidka, H., Sharma, N., & Singh, R. (2020). Evolution of IoT to IIoT: Applications & challenges. In *International Conference on Innovative Computing & Communications (ICICC) 2020* (pp. 1–6).

Jones, D., Snider, C., Nassehi, A., Yon, J., & Hicks, B. (2020). Characterising the digital twin: A systematic literature review. *CIRP Journal of Manufacturing Science and Technology, 29*, 36–52.

Kim, K. (2016). Interacting socially with the Internet of Things (IoT): Effects of source attribution and specialization in human-IoT interaction. *Journal of Computer-Mediated Communication, 21*(6), 420–435.

Kim, W.-S., Lee, W.-S., & Kim, Y.-J. (2020). A review of the applications of the Internet of Things (IoT) for agricultural automation. *Journal of Biosystems Engineering, 45*, 385–400.

Kranz, M., Holleis, P., & Schmidt, A. (2009). Embedded interaction: Interacting with the Internet of Things. *IEEE Internet Computing, 14*(2), 46–53.

Laplante, P., & Laplante, N. (2016). The Internet of Things in healthcare potential applications and challenges. *IT Trends, 18*(3), 2–4.

Larrucea, X., Combelles, A., Favaro, J., & Teneja, K. (2017, January). Software engineering for the internet of things. *IEEE Software, 34*(1), 24–28.

Liang, C.-J., Bu, L., Li, Z., Zhang, J., Han, S., Karlsson, B. F., ... Zhao, F. (2016). Systematically debugging IoT control system correctness for building automation. In *Proceedings of the 3rd ACM International Conference on Systems for Energy-Efficient Built Environments* (pp. 133–142).

Majeed, U., Khan, L., Yaqoob, I., Kazmi, S., Salah, K., & Hong, C. (2021). Blockchain for IoT-based smart cities: Recent advances, requirements, and future challenges. *Journal of Network and Computer Applications, 181*, 103007.

Makhshari, A., & Mesbah, A. (2021). IoT bugs and development challenges. In *2021 IEEE/ACM 43rd International Conference on Software Engineering (ICSE)* (pp. 460–472).

Manca, M., Parvin, P., Paternò, F., & Santoro, C. (2020). Integrating Alexa in a rulebased personalization platform. In *Proceedings of the 6th EAI International Conference on Smart Objects and Technologies for Social Good* (pp. 108–113).

Manca, M., Paternò, F., Santoro, C., & Corcella, L. (2019). Supporting end-user debugging of trigger-action rules for IoT applications. *International Journal of Human-Computer Studies, 123*, 56–69.

Mattioli, A., & Paternò, F. (2020). A visual environment for enduser creation of IoT customization rules with recommendation support. In *Proceedings of the International Conference on Advanced Visual Interfaces* (pp. 1–5).

Miorandi, D., Sicari, S., De Pellegrini, F., & Chlamtac, I. (2012, September). Internet of things: Vision, applications and research challenges. *Ad Hoc Networks, 10*(7), 1497–1516.

Miraz, M., Ali, M., Picking, R., & Excell, P. (2015). A review on Internet of Things (IoT), internet of everything (IoE) and Internet of nano things (IoNT). In *2015 Internet Technologies and Applications (ITA)* (pp. 219–224). Wrexham, UK.

Monge Roffarello, A., & De Russis, L. (2023). Defining trigger-action rules via voice: A novel approach for end-user development in the IoT. In *International Symposium on End User Development* (pp. 65–83). Cham: Springer Nature Switzerland.

Morin, B., Harrand, N., & Fleurey, F. (2017). Model-based software engineering to tame the IoT jungle. *IEEE Computer Society Press, 34*(1), 30–36.

Motta, R. C., de Oliveira, K. M., & Travassos, G. H. (2023, July). An evidence-based roadmap for IoT software systems engineering. *Journal of Systems and Software, 201*, 111680.

Namoun, A., Daskalopoulou, A., Mehandjiev, N., & Xun, Z. (2016). Exploring mobile end user development: Existing use and design factors. *IEEE Transactions on Software Engineering, 42*, 960–976.

Noura, M., Atiquzzaman, M., & Gaedke, M. (2019). Interoperability in internet of things: Taxonomies and open challenges. *Mobile Networks and Applications, 24*, 796–809.

Patel, P. (2015, May). Enabling high-level application development for the Internet of Things. *Journal of Systems and Software, 103*, 62–84.

Paternò, F., Lieberman, H., & Wulf, V. (2006). End-user development: An emerging paradigm. In H. Lieberman, F. Paternò, & V. Wulf (Eds), *End User Development* (pp. 1–8). Dordrecht: Springer.

Petrosyan, A. (2023, February 23). *Number of Internet Users Worldwide 2022*. Retrieved May 12, 2023, from Statista: https://www.statista.com/statistics/273018/number-of-internet-users-worldwide/

Phupattanasilp, P., & Sheau-Ru, T. (2019). Augmented reality in the integrative internet of things (AR-IoT): Application for precision farming. *Sustainability, 11*(9), 2658.

Razzaque, M., Milojevic-Jevric, M., Palade, A., & Clarke, S. (2016). Middleware for internet of things: A survey. *IEEE Internet of Things Journal, 3*(1), 70–95.

Resnick, M., Maloney, J., Monroy-Hernández, A., Rusk, N., Eastmond, E., Brennan, K., ... Kafai, Y. (2009). Scratch: Programming for all. *Communications of the ACM, 52*(11), 60–67.

Shahab, S., Agarwal, P., Mufti, T., & Obaid, A. (2022). SIoT (social internet of things): A review. In S. Fong, N. Dey, & A. Joshi (Eds), *ICT Analysis and Applications* (pp. 289–297). Singapore: Springer.

Sharma, N., Shamkuwar, M., & Singh, I. (2018). The history, present and future with IoT. In V. Balas, V. Solanki, R. Kumar, & M. Khari (Eds), *Internet of Things and Big Data Analytics for Smart Generation* (pp. 27–51). Cham: Springer.

Shirehjini, A., & Semsar, A. (2016). Human interaction with IoT-based smart environments. *Multimedia Tools and Applications, 76*, 13343–13365.

Singh, P., Elmi, Z., Meriga, V., Pasha, J., & Dulebenets, M. (2022). Internet of things for sustainable railway transportation: Past, present, and future. *Cleaner Logistics and Supply Chain, 4*, 100065.

Sisinni, E., Saifullah, A., Han, S., Jennehag, U., & Gidlund, M. (2018). Industrial internet of things: Challenges, opportunities, and directions. *IEEE Transactions on Industrial Informatics, 14*, 4724–4734.

Srinivasan, V., Koehler, C., & Jin, H. (2018). RuleSelector: Selecting conditional action rules from user behavior patterns. *Proceedings of the ACM on Interactive, Mobile, Wearable and Ubiquitous Technologies, 2*(1), 1–34.

Stankovic, J. A. (2014, February). Research directions for the internet of things. *IEEE Internet of Things Journal, 1*, 3–9.

Surbatovich, M., Aljuraidan, J., Bauer, L., Das, A., & Jia, L. (2017). Some recipes can do more than spoil your appetite: Analyzing the security and privacy risks of IFTTT recipes. In *Proceedings of the 26th International Conference on World Wide Web* (pp. 1501–1510).

Taivalsaari, A., & Mikkonen, T. (2017). A roadmap to the programmable world: Software challenges in the IoT era. *IEEE Software, 34*(1), 72–80.

Taivalsaari, A., & Mikkonen, T. (2018). On the development of IoT systems. In *2018 Third International Conference on Fog and Mobile Edge Computing (FMEC)* (pp. 13–19).

Ur, B., McManus, E., Pak Yong Ho, M., & Littman, M. L. (2014). Practical trigger-action programming in the smart home. *Proceedings of the SIGCHI Conference on Human Factors in Computing Systems* (pp. 803–812).

Ur, B., Yong Ho, M., Brawner, S., Lee, J., Mennicken, S., Picard, N., … Littman, M. L. (2016). Trigger- action programming in the wild: An analysis of 200,000 IFTTT recipes. In *Proceedings of the 34rd Annual ACM Conference on Human Factors in Computing Systems* (pp. 3227–3231).

Urbieta, A. (2017, November). Adaptive and context-aware service composition for IoT-based smart cities. *Future Generation Computer Systems, 76*, 262–274.

Van Mechelen, M., & Smith, R. (2023). Emerging technologies in K-12 education: A future HCI research agenda. *ACM Transactions on Computer-Human Interaction, 30*(3), 1–40.

Vannucchi, C., Diamanti, M., Mazzante, G., Cacciagrano, D., Culmone, R., Gorogiannis, N., … Raimondi, F. (2017). Symbolic verification of event-condition-action rules in intelligent environments. *Journal of Reliable Intelligent Environments, 3*(2), 117–130.

Weyrich, M. (2016, January). Reference architectures for the internet of things. *IEEE Software, 33*(1), 112–116.

Yugha, R., & Chithra, S. (2020). A survey on technologies and security protocols: Reference for future generation IoT. *Journal of Network and Computer Applications, 169*, 102763.

Zambonelli, F. (2017, January). Key abstractions for IoT-oriented software engineering. *IEEE Software, 34*(1), 38–45.

Zanella, A., Bui, N., Castellani, A., Vangelista, L., & Zorzi, M. (2014, February). Internet of things for smart cities. *IEEE Internet of Things Journal, 1*(1), 22–32.

Zaslavsky, A., & Jayaraman, P. (2015). Discovery in the internet of things: The internet of things (ubiquity symposium). *Ubiquity, 2015*, 1–10.

Zhang, L., He, W., Martinez, J., Brackenbury, N., Lu, S., & Ur, B. (2019). AutoTap: Synthesizing and repairing trigger-action programs using LTL properties. In *Proceedings of the 41st International Conference on Software Engineering* (pp. 281–291).

7 Conversational Agents

*Ana Paula Chaves, Charlotte van Hooijdonk,
Christine Liebrecht, Guilherme Corredato Guerino,
Heloisa Candello, Minha Lee, Matthias Kraus,
and Marco Aurelio Gerosa*

7.1 INTRODUCTION

The American movie *Her* (Jonze, 2014), directed in 2013 by Spike Jonze, displayed the dilemma of a professional writer who develops a romantic relationship with an AI assistant, Samantha. In the plot, Samantha is a powerful AI that can not only read books and review Theodore's letters in an incredibly short time but also portrays many human capabilities, such as using natural language communication, displaying emotions, and an identity.

Her was not the first time the big screens raised the idea of AI agents co-existing with humans. Sci-fi movies and literature are full of examples of software that has a body and a soul, integrating into human society peacefully or disrupting this society by threatening human existence. Either way, when we reflect on these fictional scenarios, and the role AI currently plays in our day-to-day life, we may wonder: are we walking into our most favorite (or feared) sci-fi plot? And, perhaps more importantly, how can we shape that future to achieve the most optimistic outcome?

The answer to these questions is not simple. When *Her* was released in December 2013, the virtual assistant Siri was already a commercial product built into Apple's product. Apple introduced the first iPhone featuring Siri in 2011, the same year that Watson, IBM's question & answer computer, won *Jeopardy! Challenge,* a popular quiz television show in the USA. These events were not coincidental; they resulted from decades of research and industry initiatives to design human-like technology.

These initiatives were not sci-fi inspired, though. In 1950, Alan Turing's research manuscript "Can machines think?" argued that a machine can exhibit intelligent behavior indistinguishable from a human (Turing, 1950). This article would be the seed for many research developments on artificial intelligence (AI) we have seen in modern computing. The initially called "imitation game" challenged scholars worldwide to create machines that could successfully play the game. To achieve that, machines would have to master one of the most defining features of humanity: natural language.

Significant efforts in natural language processing led to the scientific and technological advances that culminated in Siri and IBM Watson's spotlight in 2011. In the subsequent years, we watched the skyrocketing growth of conversational technologies, which would include intelligent personal assistants, home devices, virtual companions, and chatbots. Conversational products extrapolated the boundaries of specialized groups of users and research labs to reach the general public. Home personal assistants, such as Alexa and Google Home, reached people's homes and integrated people's daily routines (Dale, 2020). The accelerated digital transformation motivated by the COVID-19 pandemic boosted the spread of conversational agents to automate communication with customers in many service-based companies (Stoilova, 2021), such as customer service, healthcare, and financial services. ChatGPT, which is a conversational agent driven by AI, helps users to compose emails, essays, and even write code in particular programming languages, all based on natural language conversations (van Dis et al., 2023). The conversational AI market has grown by billions of dollars (in USD), and the trend is estimated to continue in the upcoming years (Rashita et al., 2021).

Although talking to machines in natural language has become old news, the current scenario looks different from a sci-fi plot. Compared to Samantha, currently available conversational agents lack the complexity presented in Samantha's functionalities and personality. For the most part, conversational agents have been designed to accomplish narrow goals, for example, providing customer support for a particular company or assisting elderly individuals in their living environments. Even personal assistants, such as Siri and Alexa, who have a broader functional scope, do not replicate human communication in all its intricacy. As these technologies evolve to include more capabilities, scientists and industry professionals take on the responsibility to create conversational agents that integrate with human activities in ways that promote a positive experience by increasing or controlling the extension of the conversational agent's humanity.

The human-computer interaction (HCI) field plays a crucial role in the conversational agent's evolution. On the one hand, professionals in the area are concerned with the impact of functional performance on human perceptions, reactions, and inclination to adopt conversational technologies. On the other hand, as conversational agents present increased human features than traditional interfaces – at least the natural language capability, it is necessary to understand and shape the social and emotional expectations projected in the interactions with conversational interfaces.

Therefore, this chapter examines the role of HCI in the design and evolution of conversational agents. This chapter discusses the most significant aspects of human-conversational agents interaction, which include the process for conversation design, the technical and social factors that play a role in conversational interactions with computers, and techniques for evaluating conversational interfaces. In Section 7.2, the focus is the varying definitions of conversational agents, their characteristics, and history. Section 7.3 discusses the techniques for conversation design including stakeholder discovering, interface, and interaction design. Sections 7.4 and 7.5 approach the technical and social dimensions of conversational agent interaction. The former reviews the architectural aspects of conversational agents and introduces algorithms and methods to develop these agents. The latter switch gears toward the Computer As Social Actors paradigm and the user expectations of a conversational agent's social representation. Section 7.6 reviews the evaluation methods used to assess the pragmatic and hedonic qualities of conversational agents. Finally, Section 7.7 discusses the societal benefits and challenges of conversational agent adoption. Section 7.8 brings closing remarks and future directions.

7.2 COMMUNICATING WITH COMPUTERS

Since Alan Turing proposed the Imitation Game (a.k.a. Turing test), making computers that interact with humans through conversation has been a challenge for researchers (Turing, 1950). The first software to play the Imitation Game was ELIZA (Weizenbaum, 1966), followed by a series of early technologies such as TinyMud (Mauldin, 1994), SHRDLU (Winograd, 1971), and A.L.I.C.E. (Wallace, 2009). In 1991, Dr. Hugh Loebner started the first Loebner Prize Competition where every year the most human-like computer is rewarded.

In the early 2000s, the popularity of conversational agents increased with their integration into instant messaging tools (Adamopoulou & Moussiades, 2020). Differently from the original agents, whose main goal was to mimic human conversations, conversational agents integrated with instant messaging tools helped people with practical daily tasks such as retrieving information from databases about movie times, sports scores, stock prices, news, weather, etc. This ability marked a significant development in HCI as information systems became accessible through dialog.

The development of conversational agents went one step further with the creation of smart personal voice assistants in the early 2010s. These agents were built into smartphones or dedicated home speakers and could understand voice commands, speak to the user by synthesizing a voice, and handle tasks such as setting up alarms, monitoring devices, accessing calendars, etc. The most popular voice assistants are Apple Siri, IBM Watson, Google Assistant, Microsoft Cortana, and Amazon Alexa (Adamopoulou & Moussiades, 2020; Dale, 2020). In that same decade, social media platforms allowed companies to create chatbots to represent their brand or interact with their customers via a conversational interface. For example, airlines started to answer complaints on social media with chatbots and offer services such as checking in, informing about flight delays, and providing boarding passes (Ukpabi et al., 2019). Conversational agents automated a large number of messages that were previously answered by humans and thousands of text-based chatbots were developed for popular messaging platforms.

In the 2020s, advances in machine learning allowed the training of large-scale language models, which considerably improved the performance of conversational agents and allowed the integration of the technologies in a variety of applications. ChatGPT, one of these new conversational agents, has become notorious for its disruptive performance in a variety of contexts, including text generation, programming, and language translation (van Dis et al., 2023). The model is trained on a massive amount of data, allowing it to generate text that is often difficult to distinguish from text written by a human.

With the increasing popularity of HCI via conversational interfaces, a number of terms have been coined to accommodate varying characteristics of these agents. In the following, we present an overview of these terminologies and discuss the domains where conversational agents have been mostly applied.

7.2.1 Definitions and Classification

Generally, conversational agents are "computer programs that interact with users using natural language". Although the concept emerged several decades ago and has evolved over the years, the "conversational agent" term is not consensual (Motger et al., 2022), and it has been either used as a generic term for a software-based dialog system or used as a synonym of chatbot. Several synonyms emerged, such as social bots, personal assistants, and conversational interfaces with the goal of emphasizing specific characteristics of the agents. In this chapter, conversational agents represent a software application that uses natural language as the main form of interaction with humans.

Conversational agents can serve various purposes, with characteristics that distinguish one agent from the others, including physicality (dis/embodiment, animation, or avatars), input mode (voice- or text-based), goal (general-purpose chat or task-oriented), and domain. In general, conversational agents may interact through typed text, speech, or both means (Diederich et al., 2019). The interaction may be controlled by the user (user-initiative), by the conversational agent (system-initiative), or by both (mixed-initiative) (Motger et al., 2022). Conversational agents may also have personality, humor, and be able to express emotions; sometimes they have a face or even a body (Feine et al., 2019). When conversational agents have physical characteristics, they are usually called "embodied conversational agents" (Cassell, 2000). According to Cassell (2000), embodied conversational agents have the same properties as humans in face-to-face conversations, for example, they are able to recognize verbal and non-verbal inputs, answer with verbal and non-verbal outputs, and understand turn-taking. This chapter does not discuss embodiment (e.g., virtual humans or robots), although disembodied conversational agents may also have appearance (e.g., an avatar) (Feine et al., 2019).

7.2.2 Application Domains

Conversational agents have changed how companies engage with their customers (Ling et al., 2021), how students participate in their learning groups (Khosrawi-Rad et al., 2022), and how patients self-monitor the progress of their treatment (Milne-Ives et al., 2020), among many other applications. Recent reports on the conversational agent market (Rashita et al., 2021) attest to their increasing demand in several different domains. According to Rashita et al. (2021), the domains where conversational agents are more expressively adopted include Bank, Financial Services & Insurance (BFSI), retail & e-commerce, healthcare & life science, travel & hospitality, telecom, and media & entertainment, with retail & e-commerce being the most expressive segment and anticipated to witness significant growth in the upcoming years.

Aligned with the marketing trend, many studies in the literature focus on the support conversational agents can provide to customer services (Følstad & Skjuve, 2019; Haugeland et al., 2022; Youn & Jin, 2021) and marketing (Cui et al., 2017; Kaczorowska-Spychalska, 2019; Thomaz et al., 2020). In the BFSI domain, conversational agents support decision-making and investment choices as well as productivity (Sharma et al., 2021; Wube et al., 2022). When focusing on travel and hospitality, conversational agents have been widely used in several subdomains, such as planning, booking, and en-route experience (Mahmood et al., 2009; Pillai & Sivathanu, 2020; Yanishevskaya et al., 2019). In health, they are used to enable speech monitoring, identification of disease, diabetics monitoring/control, and personal healthcare assistance (Laumer et al., 2019; Milne-Ives et al., 2020; Preum et al., 2021).

There are several other domains where conversational agents have been investigated. For example, in the educational context, conversational agents have been widely adopted, with applications including tutoring, question-answering, conversation practice for language learners, learning companions, and dialogues to promote reflection and metacognitive skills (Hwang & Chang, 2021; Khosrawi-Rad et al., 2022; Wollny et al., 2021). In Software Engineering, Storey and Zagalsky (2016) attested that conversational agents are present in every phase of the software development process, including coding, testing, documentation, deployment, support, and even team coordination. GitHub, the most popular code hosting platform, has been swarmed with bots and chatbots that help developers in their daily tasks, interacting with the team via communication channels like comments on pull requests (Wessel et al., 2018). Since the Internet of Things brings Internet connection to a variety of different physical devices or things, conversational agents have been studied as a solution to facilitate management and interaction with those devices (Augustsson, 2019).

7.2.3 Conversational Agents and the HCI

Conversational agents are changing the patterns of interactions between humans and computers. Luger and Sellen (2016) claimed that "conversation is the next natural form of Human-Computer Interaction." Many instant messenger tools and social networking platforms provide platforms to develop and deploy conversational agents, which organizations use to provide their services (Følstad & Brandtzæg, 2017). As messaging tools and social network sites increasingly become platforms, traditional websites and apps are providing space for this new form of HCI.

The increasing interest in conversational technologies has brought new challenges for the HCI field (Følstad & Brandtzæg, 2017; Neururer et al., 2018). Whereas traditional user interfaces apply visual elements such as buttons, menus, or hyperlinks to communicate with users, conversational interfaces rely almost entirely on language as the primary resource to achieve communicative goals. Nevertheless, Dale (2016) states that interacting with current conversational agents conveys the impression of *being managed through a tightly controlled dialog flow* with reduced interactivity, which turns users into option-selectors rather than conversational partners.

Moreover, language design for conversational agents has focused primarily on ensuring that the agents produce coherent and grammatically correct responses, and on improving functional performance and accuracy (see e.g. Jiang & E Banchs, 2017; Maslowski et al., 2017). Although current conversational agents may, at some functional level, provide users with the answers they seek, the utterances portray arbitrary patterns of language that often fail to take into account the interactional situation in choosing a proper conversational tone for the interaction. For example, the literature shows that appropriate linguistic choices potentially increase human likeness (Chaves et al., 2022; Hill et al., 2015) and believability (Westerman et al., 2019; Xuetao et al., 2009), enhancing the overall quality of the interaction (Chaves et al., 2022; Jakic et al., 2017). Hence, making a conversational agent acceptable to users is not only a technical but also a social problem to solve (Neururer et al., 2018). Developing a strong basis for designing not just what a conversational agent says but also how it says it must be a priority for creating the next generation of human-computer interfaces.

Conversational agents are typically designed to mimic the social roles usually associated with a human-conversational partner. Research on mind perception theory (Heyselaar & Bosse, 2019; Lee et al., 2019c) suggests that although artificial agents are presumed to have substandard intelligence, people still apply certain social stereotypes to them. It is reasonable, then, to assume that "machines may be treated differently when attributed with higher-order minds" (Lee et al., 2019c). As conversational agents enrich their communication and social skills, user expectations will likely grow as the conversational competence and perceived social role of the agents approach the human profiles they aim to represent. A variety of factors influence how people perceive an agent's communication skills (Chaves & Gerosa, 2020; Feine et al., 2019) and, as user expectations of proficiency increase, one important way to enhance the agent interactions is by carefully planning the technical, social and interactional aspects of the conversational agent design. In the next sections, we navigate these dimensions to discuss techniques applied to improve the design of conversational agents.

7.3 CONVERSATION DESIGN

The design of conversational agents requires specific tools and methods to comply with the unique characteristics of these systems. In this section, we will discuss techniques to design conversational agents, which includes first a preliminary analysis covering contextual and stakeholder discovery, approaches, and techniques. Second, we will rely on an overview of dialogue design methods and techniques. Implementation and evaluation techniques are described in Section 7.4 "Building Conversational Agents" and Section 7.6 "Conversational Agents Evaluation", respectively.

7.3.1 CONTEXTUAL AND STAKEHOLDER-DISCOVERING APPROACHES AND TECHNIQUES

The ecosystem of designing conversational AI agents includes several components and resources. Users, developers, content curators, project managers, designers, machine learning analysts, and marketing and consulting employees are usually involved in the process. The diversity of stakeholders brings a lot of complexity, and a successful conversational agent design depends upon the alignment of several points.

Many theories, frameworks, and approaches help researchers to understand and identify the nature and challenges of designing conversational systems and to assist the development team in making ethical (Rakova et al., 2021) and explainable design decisions (Ehsan et al., 2022; Liao et al., 2020a). The nuances of stakeholders' views are not always clear for practitioners, designers, and developers. Understanding stakeholders' goals and mental models helps to build the process of designing effective CA systems. This section addresses two approaches toward considering diverse lenses in the CA design process: value-sensitive design and articulated work practices.

The Value-sensitive design (VSD) methodology consists of integrating conceptual, empirical, and technical investigations (Friedman & Hendry, 2019). In this approach, value is defined as "what a person or group of people consider important in life" (Friedman et al., 2013). In this frame, the notion of direct and indirect stakeholders is considered. For example, users are generally considered direct stakeholders, and project sponsors could be considered indirect stakeholders. Or even, developers sometimes could be the direct stakeholders when they are considered the users of conversational design platforms, and indirect stakeholders when they are developing the conversational system for end users. In all those situations, the values of stakeholders should be taken into account to have a successful conversational system.

Wambsganss et al. (2021) used a VSD approach (Friedman & Hendry, 2019), and design science research (Hevner, 2007) to investigate conversational agents' design principles. The authors found that this approach is suitable for user scenarios where privacy and transparency play an important role. Their work can inform designers and stakeholders about an ethical way to design conversational agents.

Görnemann & Spiekermann (2022) used the VSD approach along with other fields to create a framework called Emotion Value Assessment (EVA) aimed to assist researchers and practitioners in the HCI domain unveil the emotional reactions of voice-based conversational agents in relation to underlying values fostered or harmed. Understanding and grasping values and emotional reactions to technology can assist in developing ethical CA.

Values, such as safety, are also a great concern for CA researchers in end-to-end (E2E) conversational AI systems. Dinan et al. (2021) and Bergman et al. (2022) investigated how and when conversational agents trained on large datasets from the Internet should be released considering safety, and value tensions in training and releasing E2E CA models. The authors discuss the uncertain nature of large language models operating under conversational user interfaces and categories of harmful responses. Those issues require weighing conflicting, uncertain, and changing values.

Values embedded in the design process interfere with decision-making tasks that are not always discussed and transparent for the CA team and users. For example, data curation and machine learning decisions (Rattenbury et al., 2017; Seidelin et al., 2020) are paramount steps of the CA design process and directly affect the quality and user perception of a CA. Usually, this work is hidden from stakeholders and is invisible to users (Ju & Leifer, 2008). We could say that those hidden practices are a form of "articulated work", a "work that is necessary for the work to proceed" (Hampson & Junor, 2005; Schmidt & Schmidt, 2011). Articulation work is usually a complex set of enabling activities that contribute indirectly to the more visible and prominent production work in workplaces (Candello et al., 2022b). It can take the form of support work or simply invisible work that usually goes unobserved and unrecorded. Therefore, it is important to make this work visible (Nardi & ENgeström, 1999) and understand the nature of technology, and sometimes limited decisions included in the process. Several studies are uncovering the production work entitled developing conversational agents.

Candello et al. (2022b) investigated articulated working practices of content curators of conversational agents in diverse industry settings. They proposed a distinction between tech curators, the ones with knowledge of conversational platforms, and content curators, the subject matter experts, usually responsible for the content. In their research, three main themes emerged that illustrate the working practices of those employees: (1) co-dependence work mechanisms of curators (2) cooperation and collaboration mechanisms, and (3) management of information spaces and technologies. From the study results, they also draw seven design implications to improve the interface design and features of conversational platforms.

In the same line of studies to unpack practitioners' work for conversational agents, Khemani and Reeves (2022) interviewed nine Voice-User Interface (VUI) practitioners to understand how they conceptualize and use design guidelines developed in HCI. One of the main takeaways from this study is that design knowledge should be codified for practitioners to be adopted. Lack of adoption was also found in a study based on a large-scale survey with 105 industry designers (Murad et al., 2022). The study aimed to explore the design practices of VUI designers, and in which ways knowledge of GUI is applied to designing VUI.

7.3.2 Dialogue Design Methods and Techniques

A conversation is a specialized form of interaction, according to Suchman & Suchman (2007, p. 101) "a distinguishing feature of ordinary conversation is the local, moment-by-moment management of the distribution of turns, of their size, and what gets done in them, those things being accomplished in the course of each current speaker's turn." Management of turns and subject change in each course is a situation that occurs in real-life conversations based on circumstances (internal and external) to speakers in dialogue. Machines are not prepared, nowadays, to fully understand the context and change the course of conversations as humans. Managing dialogues with machines is challenging, and this challenge increases even more when more than one conversational agent is part of the same conversation.

In a conversational user interface, the conversation flow is not linear or transparent; it might take different courses according to circumstances that influence dialogue. One of the most interesting features of human conversation is the ability to explore sidetracks and easily go back to the main conversation objective. For instance, while people are making a decision, such as in planning a trip, people can ask clarifying questions, explore a similar case, get delighted by photos and comments, consult a friend, and then go back to make the trip decisions. It is almost impossible to predict in what sequence a user will interact with a machine and how this machine could provide a satisfactory user experience. Traditional design methods might help to envision graphical user interfaces and detect topics that will embody the conversational system but have clear limitations in supporting the design of the conversation flow of a conversational system. This section explores some design methods which may help in this process.

Zue and Glass (2000) addressed some of the challenges in designing dialogue flow. According to them, system-initiative systems (see Section 2.1 Classification) restrict user options, asking direct questions, such as: "Please, say just the departure city." By doing so, those types of systems are more successful and easy to answer as they guide the user through the expected path. On the other hand, user-initiative systems are the ones where users have the freedom to ask what they wish. In this context, users may feel uncertain of the capabilities of the system and request information or services which might be quite far from the system's domain, leading to user frustration. In the mixed-initiative approach, in which users and computers participate interactively to achieve a common goal, the challenges include understanding interruptions, human utterances, and unclear sentences that were not always goal-oriented.

The key dilemma is: should we ask users to modify their behaviors and interact with the system in a structured way? Or should we let users be more comfortable with systems that have characteristics of humans? Or both? In our experience, this is in fact one of the earliest and often the most important decisions faced by designers of conversational systems and should be explored in the design process by, for instance, using a Wizard of Oz approach (Mateas, 1999) and prototype techniques.

In WoZ experiments, users are told they are interacting with a conversational agent, though in fact, a human plays the role of the agent behind the scenes. Samson and Sumi (2020) conducted a WoZ experiment to understand the driver's decision-making process when listening to route alternatives in a conversation between two conversational agents. Chaves and Gerosa (2018) used a WoZ setup to identify differences in turn-taking and dialogue flow structure when users interact with single or multiple chatbots in the same chat.

Cambre and Kulkarni (2020) mentioned additional prototyping techniques and highlighted that after the elicitation phase, the emphasis may shift to a more exploratory design stage, where the prototypes resemble the final voice artifacts intended to be developed. In this case, conversational development platforms are used. Shorter et al. (2022) investigate *prototyping* as a design tool for materializing voice assistant technologies, WoZ and prototype techniques show the peculiarities of designing for intangible technologies. Following the approaches to use prototypes as prompts for designing voice assistants in-car settings, Meck et al. (2022) conducted an experiment to compare three evaluation conditions: driving simulator; crowdsourcing audio, and crowdsourcing text. They discovered study participants processed prompts similarly in these conditions.

With the increased amount of information nowadays, designers of conversational agents struggle to organize the content in a user-friendly way. Several methods borrowed from GUI context can be adapted to the CA context. Methods, such as *Card sorting* (Nawaz, 2012) help to organize and evaluate the information of an interface. For instance, in a call-center context, where several topics could be asked to the CA, designers can prioritize the topics by importance using a card sorting method with branch employees and call-center employees to identify the popular topics requested by customers.

Menus are also applied to show users the content scope of the CAs and to organize the information. Nguyen et al. (2021) conducted an empirical study and found that CAs lead to a lower level of perceived autonomy and higher cognitive load compared to menu-based interface systems, resulting in lower user satisfaction. Considering Nguyen et al.'s (2021) findings, it is comprehensible that many designers are using those GUI strategies in a conversation interface. Hu (2019) compared the use of a menu-based over a conversational chatbot experience. In the study, users preferred the menu-based experience for several reasons such as being easier to use, less likelihood of errors, the convenience of GUI elements, and a clear way to show where information needs to be provided rather than requested. Valério et al. (2020) investigated how communicative strategies are used by popular chatbots when conveying their features to users. They identified that menus offer the possibility of quick replies and choices of predefined options, complementary to Hu's (2019) study results. Menus also can help designers and developers to scope the system knowledge and keep the subject of the conversation in the scope, avoiding conversation breakdowns due to the lack of CA understanding.

Scenarios and Storyboards (Carrol, 1999; Llitjós, 2013) based on *journey maps* (Schneider & Stickdorn, 2011) and *blueprints* (Polaine et al., 2013) assist in predicting the user experience. Design fiction studies (Blythe, 2014) are also applied in designing future experiences with CA. Muller and Liao (2017) offered four potential methods to envision the values and ethical implications of designing AI systems for future users. Research to mitigate harm included fictional scenarios to assist practitioners to reflect when preparing for and learning from AI conversational models release (Bergman et al., 2022). More specifically to digital personal assistants, Søndergaard and Hansen (2018) discuss the meanings of designing and adopting CAs using design fiction through a critical and feminist design methodology. Ringfort-Felner et al. (2022) brought a collaborative and social perspective of using a design fiction artifact named Kiro in a car context. Participants in the study accepted Kiro as a conversational partner but not as a replacement for a human.

Designers should also consider examining similar contexts where everyday activities happen without or with the use of CAs to understand the nuances and values in place. Observations and log analysis of call-center employees, for example, may assist designers to understand the dynamic of the conversations, and design the dialogue for user experience. Porcheron et al. (2018) collected and analyzed audio data of VUI in participants' homes to understand the social interaction implications in everyday life. Barth et al. (2020) conducted a log analysis collected from visitors from an art exhibition to identify categories of popular questions. Portela and Granell-Canut (2017) conducted observations of chatbot mobile phone users to understand the potential personal relationships users might develop interacting with chatbots. Those studies are examples of how to conduct observational studies to gather insights for designing the interaction and conversational flow of CAs.

Another component of this decision is the technology available for the deployment of the platform. Different conversational platforms support different initiative models, so the designer may face application contexts where the initial strategy is predetermined by the platform. In this situation, they should focus on finding and identifying patterns of dialogue that make sense given a fixed initiative strategy. For instance, if the only available platform has a Q&A structure (a typical user-initiative), the designer should consider answers that lead to specific questions from the users if more guidance to the user is needed. In any case, since the decision of the initial strategy is closely tied to the deployment platform capabilities, it is important to involve the developing team in the process and, often, make that decision as early as possible in the design process. The next section will cover the technology and techniques available to create conversational agents.

7.4 BUILDING CONVERSATIONAL AGENTS

For being able to have a conversation with a computer system, several techniques from natural language processing (NLP), a subfield of AI, are combined to understand and respond appropriately to user input. These include natural language understanding (NLU) for determining the semantic meaning or intention of the user's utterance, dialog management (DM) for keeping track of conversational context and deciding on system actions, and natural language generation (NLG) for transforming abstract representations of system actions into natural language.

Typically, these techniques are implemented into a dialog system forming the technical foundation of conversational agents. For realizing dialog systems there exist different reference architectures that have been historically categorized according to their purpose. A modular architecture is usually applied to implement task-oriented dialog systems which are designed to assist users in achieving a predefined task or goal. For example, task-oriented systems have been used to realize conversational agents to handle hotel room bookings or for ordering food from a restaurant (Williams et al., 2016). A characteristic of task-oriented oriented systems is that modules for NLU, DM, and NLG are implemented and fine-tuned separately. The purpose of NLU is for the system to understand a narrow range of user intents that are relevant for solving the task at hand. DM is realized using a rule-based, state machine, or statistical approaches for guiding the user to task accomplishment by following a predefined set of actions. Moreover, the DM module interacts with structured data sources, such as databases or APIs, to provide information and complete tasks. NLG is then used to present the actions in a human-understandable format. A benefit of this approach is that already implemented modules can be reused fast and easily. Further, the integration of external services for handling individual NLP tasks is facilitated and the architecture is easily extensible. This makes the modular approach especially useful for rapid prototyping aiming at investigating novel interaction design techniques or collecting data in new task domains. As the work processes of modular-based systems are easy to follow and technical problems can be alleviated relatively easily, this architectural design is popular for developing commercial voice user interface applications. For example, the modular-based RASA[1] framework can be used to build customer service chatbots, while Amazon also provides a modular framework for developing Alexa Skills.

Due to their ability to only solve narrow tasks and to understand a limited set of user intents, the modular architectural design is not suitable for handling open-domain or casual conversations with users on a wide range of topics. Therefore, End-to-End architectures are used to enable open-ended conversations between humans and artificial agents. This architectural design relies on neural dialog approaches for handling NLU, DM, and NLG, such as a sequence-to-sequence model like Long Short Term Memory (LSTM) networks (Hochreiter & Schmidhuber, 1997) or transformer-based models (Vaswani et al., 2017). These models are trained on large datasets of conversational data and learn to generate responses that are contextually relevant and engaging to users. For this, they either make use of retrieval-based methods, selecting existing utterances from a set of appropriate responses, or generative approaches, which model new utterances dependent on language models. Recently, End-to-End approaches have garnered widespread popularity for realizing chatbots able to handle chit-chat or question-answering, e.g., Google's Meena (Adiwardana et al., 2020), Facebook's Blenderbot (Roller et al., 2021), or ChatGPT.[2] Despite their popularity, End-to-End systems still have several problematic issues (McTear, 2020). One of the main problems is the generic response problem, which concerns the often bland or uninformative responses of such systems, e.g., "Ok." or "I'm not sure." Further, they are prone to semantic inconsistencies, i.e., their responses are inconsistent with their previous responses. For example, they may state different cities when asked about their current habitat. The probably most severe problem are so-called hallucinations, where the system adds false information to their responses. For example, retrieval-based methods mirror responses from their training data and are thus prone to bias or false information existing in the data set. Similarly, generative models create new utterances based on knowledge about the properties of language rather than truly understanding the meaning of their responses and are thus prone to making up facts. However, due to their ability to generate human-like responses and being able to process a wide range of topics, open-domain systems can be quite entertaining and useful for handling constrained predictable interactions. Thus, there also exist approaches to make task-oriented systems more natural by combining modular and End-to-End models. For example, Bordes et al. (2017) studied the application of End-to-End models for task-oriented dialogue in the restaurant reservation domain. In the following, the two architectural designs are described in more detail.

7.4.1 Modular Architecture

Task-oriented conversational agents are usually built using a modular architecture paradigm. A key characteristic of the modular architecture is the pipeline-like structure, where task-specific modules process user input and generate appropriate system responses. In Figure 7.1, a depiction of the typical architecture is presented. For voice-based systems, like Alexa, or Siri, the user's speech signal is first transformed into text using an automatic speech recognition (ASR) module. For this, the speech signal is first digitalized and pre-processed for removing noise and redundant information (Jurafsky & Martin, 2023). Afterward, features according to the human auditory system are extracted and fed as input to a neural end-to-end model for generating the most probable word sequence based on the acoustic feature vector (Yu & Deng, 2016). Besides semantic and syntactic information, other relevant information for HCI can also be extracted from the speech signal, e.g., sentiments, or emotions, as well as the age, or gender of the user.

NLU describes the process of extracting the meaning of user input from a word sequence. For this, a semantic representation of the utterance is generated based on formal structures. This is also known under the term semantic encoding. Historically, semantic grammars have been used that structure the utterance according to the communicative function of their constituents (Traum & Hinkelman, 1992). For example, an utterance may have an assertive function, i.e., addressing the state of a current situation, or a directive function, i.e., intending to commit the addressee to do something (Searle, 1969). There exist several taxonomies of communicative functions that define specific dialogue actions (often also named intent) for semantically analyzing user and system turns during dialogue, e.g., DAMSL (dialog act markup in several layers) (Core & Allen, 1997). Therefore, the task of semantically encoding a dialogue utterance is described as dialogue act classification. Here, each utterance of a dialogue corpora is associated with a specific dialogue action (intent). Afterward the corpus is split into training, test, and validation set, for training a dialogue act classifier that maps the input word sequence to a semantic representation. Training is conducted by first generating a numerical representation of the utterance using a contextual embedding network, e.g., BERT (Devlin et al., 2019) or Word2vec (Mikolov et al., 2013), and then feeding the representation into a machine learning classifier, e.g., support vector machine (SVM) (Vapnik, 2000) or deep neural networks, for a 1-of-N classification of the dialogue act. In Table 7.1, a sample dialog from the HIS system by (Young et al., 2010) is presented which shows the mapping from utterances to dialog acts. Besides the dialogue act, the task domain of the utterance can be determined in this way for systems that are able to handle multiple domains, e.g. see the first turn in Table 7.1, where the restaurant is determined as the domain. Further, relevant information on the word level can be classified, which is known under the term entity extraction. This is particularly useful for slot-filling dialogues

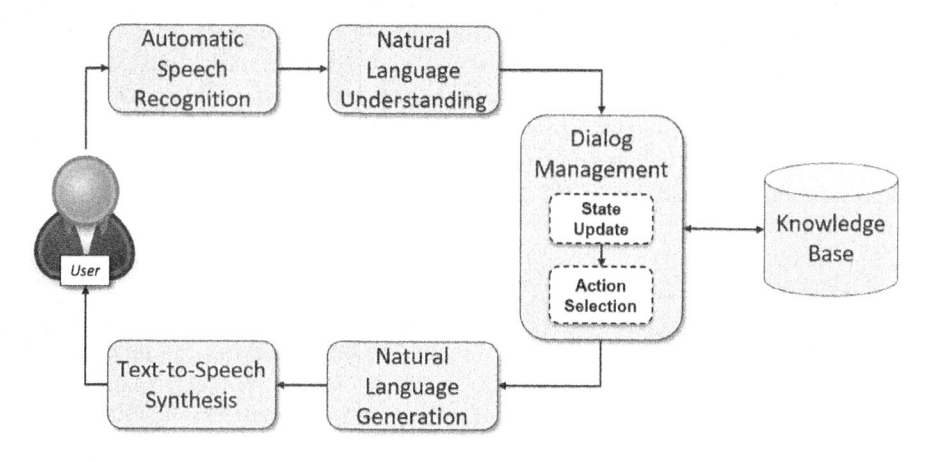

FIGURE 7.1 Modular dialog system architecture (McTear, 2020).

TABLE 7.1
Example Dialog Showing the Mapping of Utterances to Dialog Acts (Young et al., 2010)

Utterance	Dialog Act
U: Hi, I am looking for somewhere to eat.	hello(task=find, type=restaurant)
S: You are looking for a restaurant. What type of food do you like?	confreq(type=restaurant, food)
U: I'd like an Italian somewhere near the old town.	inform(food=Italian, near=old town)
S: Napoli is a nice Italian restaurant near the old town.	inform(name="Napoli", type=restaurant, food=Italian, near=old town)
U: Is it reasonably priced?	confirm(price range=moderate)
S: Yes, Napoli is in the moderate price range.	affirm(name="Napoli", price range=moderate)
U: What is the phone number?	request(phone)
S: The number of Napoli is 764503.	inform(name="Napoli", phone="764503")
U: Ok, thank you goodbye.	bye()

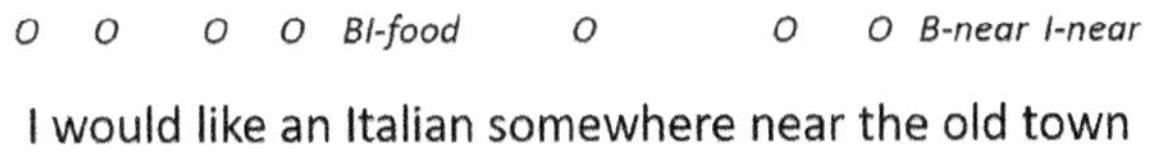

```
O     O     O     O   BI-food    O        O     O  B-near  I-near
I would like an Italian somewhere near the old town
```

FIGURE 7.2 Example for BIO-Tagging. Created by the authors.

to determine the values for specified slots. For this, BIO-tagging can be used that associates a tag for the beginning (B) and inside (I) for each slot label, and one for tokens outside (O) of any slot label (Jurafsky & Martin, 2023). For example, consider the second turn presented in Table 7.1, for which the BIO-tagging would be as shown in Figure 7.2:

Dialogue management is concerned with finding an appropriate system response given the last user utterance and all available information from the previous interaction. Hence, its task is to keep track of the conversation and decide which actions to take for accomplishing a predefined task. To achieve this, a dialogue management component comprises a dialogue state tracker for maintaining the dialogue state, and a dialogue policy for controlling the flow of the dialogue (Young et al., 2013). The dialogue state tracker uses several knowledge sources for modeling the state of the conversation. The dialogue history contains information about the user's and system's contributions to the conversation thus far, i.e., which dialogue actions have been used and which values have been provided for slots. For example, food type, location, price range would be the slots which the system from Table 7.1 requires for providing adequate information. The domain model represents the "world knowledge" of a system, i.e., concepts and information for a task domain, such as different types of food or locations in the restaurant search domain. This knowledge can be retrieved from a database, which can be structured, for example, in the form of a knowledge graph or an ontology. In addition to conversation- and domain-related knowledge, the dialog state can also contain user-specific information, e.g., age, gender, or preferences, as well as relevant dynamic user states, e.g., the classified sentiment of the user utterance. The dialogue policy determines what action the system should take next. For decision-making, the system makes uses of the relevant information represented in the dialogue state and selects an appropriate dialogue action. These decisions can be made using rule-based mechanisms or statistically driven methods. An example for a rule-based policy would be that when the confidence value for the user's providing the value to a specific slot is below a pre-specified threshold, the system asks for clarification. Otherwise, the system would proceed to ask for another slot or suggest, etc. In Table 7.1, for example, the system recognizes with a high probability that the user wants to go to a restaurant and therefore proceeds by asking which type of food the user wants to eat. If the confidence would have been low, the system could ask explicitly for confirmation, e.g. "You want to go to a restaurant. Is this correct?".

Statistically driven systems automatically learn such decisions based on data corpora using supervised learning mechanisms or interactively using reinforcement learning. For example, Young et al. (2013) use an approach that learns a dialogue strategy automatically depending on the last user input and dialogue state represented as the slot-value pairs provided during the current conversation. For learning a strategy, the system receives a positive reward at the end of dialogue when all slots have been filled correctly by the system, and a negative reward otherwise. Further, a small negative reward can be given for each dialogue turn to learn efficient system behavior.

NLG provides a word sequence given the semantic representation of the chosen system action. Thus, it can be considered as the reverse task to NLU. For achieving this, two approaches are primarily used: a template-based approach uses hand-crafted mappings of dialogue actions, slots, and values to textual utterances, which can be realized in the form of a look-up table. A more sophisticated approach to create more diversified text is to use statistical methods based on large hand-labeled dialogue data (Budzianowski et al., 2018). For this, several approaches have proven to provide good results including neural approaches (Dušek et al., 2020) as well as reinforcement learning approaches (Rieser et al., 2014).

If the system output is to be in natural language, a synthesis module uses the text representation to generate speech signals. Therefore, it is intended to solve the reverse task of ASR. Like ASR systems, text-to-speech synthesis (TTS) relies on neural End-to-End models, using LSTMs or Transformers. However, the main difference between ASR and TTS concerns the training procedure. While the ASR needs to be trained speaker-independently for being able to recognize speech from various users, the TTS module is usually trained on a specific speaker for having a consistent voice (Jurafsky & Martin, 2023). For transforming text into speech, the process generally involves three tasks: First, text needs to be normalized for handling non-standard words, i.e., numbers, dates, abbreviations, etc. In a second step, the sequence of normalized words is transformed into a numerical representation and fed as input to an encoder-decoder model that generates the predicted Mel spectrum of the spoken utterance (frequency spectrum) dependent on the input representation. Finally, vocoding is used to transform the spectral features back into the time-domain waveform representation which can be played back to the user.

7.4.2 End-to-End Architecture

The End-to-End architecture unifies the modules for NLU, DM, and NLG and produces a system response either based on retrieval methods or generation methods (see Figure 7.3). The retrieval-based method selects a response from a dialogue corpus dependent on its appropriateness to the context. For this, a bi-encoder model is applied. One encoder is trained to generate a contextual embedding of the textual representation of the user's utterance. The other encoder is trained to produce an embedding of the candidate responses provided by large data corpora.

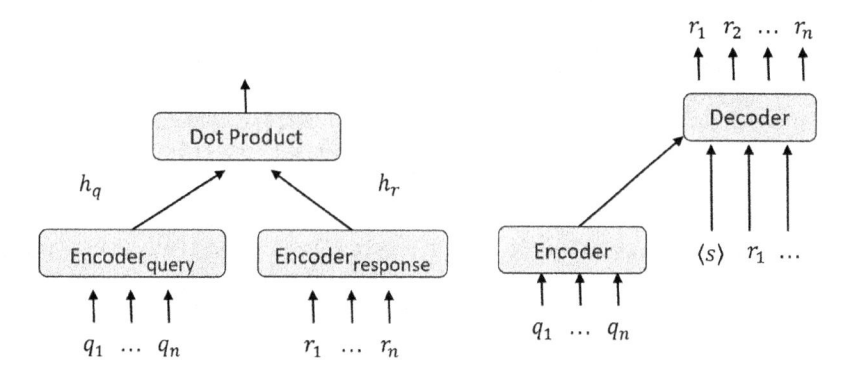

FIGURE 7.3 Retrieval- and Generation-based dialog system architecture. (Jurafsky and Martin, 2023.)

The dot product between the two embedding vectors can then be used to calculate a similarity score. For determining a response, the candidate with the highest similarity is selected. The embeddings may be produced using different techniques, e.g., BERT or Word2Vec. For enhancing the quality of the selection, more context, than solely the last user utterance, e.g., the dialogue history or information about the user's sentiment, may be included for producing the embeddings.

The generation method utilizes an encoder-decoder model for creating a system response. Thus, the problem of finding an appropriate system response given user input can be seen as a translation task. Here, the encoder network first creates a context vector that represents the user input and the dialogue history so far using BERT, for example. Afterward, the decoder network uses this vector to create an output considering the context vector and the response generated so far. Basic encoder-decoder models are prone to provide repetitive and dull responses. Therefore, some modifications are necessary to provide a more suitable system response. For example, by using reinforcement learning or adversarial networks for achieving a more natural conversation. ChatGPT makes use of a method called "Learning from Human Feedback" for providing more human-like responses based on human ratings of candidate responses (Ziegler et al., 2019).

7.4.3 Enhancing the Cognitive Abilities of Conversational Agents

Current architectures of dialog systems for conversational agents are highly performant for achieving specific tasks or conducting open-ended conversations using natural language. However, they are restricted as they are usually limited to solely finding appropriate responses to user input, while they do not reflect other fundamental aspects of human behavior during a conversation. For example, they do not adapt and personalize their responses to individual users or lack the ability to take proactive self-initiated actions. To achieve these kinds of behaviors, the cognitive capabilities of conversational agents need to be enhanced.

One important ability that needs to be included is to process multimodal information. This means that conversational agents should be able to take in various types of information sources, not just speech, such as visual cues, physiological information, and context. By doing so, conversational agents can determine high-level user information, such as the user's affective state and emotion (Graesser et al., 2012), level of trust (Kraus et al., 2021), knowledge (Nothdurft et al., 2015), or satisfaction (Ultes et al., 2015). For example, combined audiovisual information can be used for estimating the emotional state of the user (Tzirakis et al., 2017). Here, audio and video data are first encoded in a multimodal representation amenable to computational processing. For this, LSTMs or convolutional neural networks (CNNs) can be used to extract a numerical representation of speech and visual features independently. Afterward, multimodal fusion techniques (Poria et al., 2017) are applied for joining features from both modalities to make predictions, e.g., by simply concatenating the feature vectors and feeding them into a neural network. The information about the user's emotional state can then be fed into a user model, which serves as a knowledge base for providing adaptive behavior. By modeling the user's behavior, characteristics, and goals, the conversational agent can adapt its behavior to better meet the user's needs and expectations. For example, the user's emotional state can be used by a conversational agent to provide adequate emotional support by applying comforting strategies expressing empathetic and understanding behavior (Liu et al., 2021). Similarly, a conversational agent can provide multimodal system output, e.g., utilizing gestures, facial expressions, and speech, for generating a more anthropomorphic user experience, which is often used in embodied conversational agents such as the GRETA agent (Pelachaud, 2017).

Furthermore, multimodal information can be combined with advanced knowledge, reasoning, and planning abilities to achieve proactive behavior. For example, Kraus et al. (2020) developed a proactive conversational agent utilizing planning and ontological reasoning techniques for providing adequate support during the execution of DIY projects. Here, the user's task progress and their current activity with a modified electric screwdriver were tracked to initiate timely reflection dialogs. Evaluating their approach, the authors showed that the proactive agent was perceived as more trustworthy and led to higher user satisfaction with the project outcome than interacting with a reactive version of the agent. Further, proactive behavior can be used to adequately change topics in open-ended conversations with knowledge graph-based neural dialog systems. For example, Lei et al. (2022) used a reinforcement learning-based approach that includes task-relevant information, information about the user's previous satisfaction with the dialog, and the user's level of cooperativeness to determine the topic in the next dialog step. The goal was to achieve both fast task completion and user satisfaction.

Despite the high potential of augmenting conversational agents with multimodal abilities, their application in commercial settings is quite limited. One reason is the sensitivity of user-specific data such as emotions, gender, age, or knowledge which raises several ethical and privacy questions. Further, high-level user information such as emotions and trusting behavior are quite subjective and differ greatly from individual to individual which limits the reliability of such recognition software for real-world application. Thus, adequate adaptation to recognized user states may fail which can result in poor performance and customer satisfaction as well as low acceptance of the applied systems. Therefore, multimodal user information needs to be handled carefully and their reliable and safe application is still future work.

7.5 THE SOCIAL DIMENSION

7.5.1 Social Cues

In the field of HCI, Nass and his colleagues proposed the Computers Are Social Actors (CASA) paradigm and demonstrated that people mindlessly apply social scripts from human-human interaction when they use computers (Nass et al., 1994; Reeves & Nass, 1996). However, technological advancements and artificial developments have led to a life in which we are surrounded by media technologies, such as conversational agents. To account for these advancements, Lombard and Xu (2021) proposed a structural extension of the CASA paradigm, i.e., the Media Are Social Actors (MASA) paradigm. According to the MASA paradigm, social cues are triggers of users' social responses to media technologies. Social cues are physical or behavioral features displayed by a conversational agent that are salient to users (Fiore et al., 2013). An example of a social cue in conversational agents is an avatar which can be more or less human-like. Social cues are (sub)consciously interpreted by users in the form of attributions of mental state or attitudes toward the conversational agent (Fiore et al., 2013; Wiltshire et al., 2014). For example, a conversational agent with a machine-like avatar might be interpreted as impersonal, which in turn might lead to the users' social response of attributing less trust in its capabilities.

Chaves and Gerosa (2020) distinguish several social cues chatbot designers can implement in chatbots: conversational intelligence refers to the chatbot's ability to manage interactions with users (for example by proactively sending relevant messages or asking follow-up questions to the user), social intelligence describes the chatbot's impact on the social behavior of the user (the chatbot could, for example, evoke appropriate or inappropriate behavior from users), and personification includes chatbot characteristics that make the agent appear more human-like (such as the usage of a human-like avatar and communication style). An overview of Chaves and Gerosa's (2021) social cues is shown in Table 7.2.

Chaves and Gerosa's social cues can be implemented in various phases of the communication journey users go through when having a conversation with a conversational agent. Users can have expectations about the conversational agent before the actual conversation (phase 1), thereafter they will interact with the conversational agent (phase 2), and subsequently, they will formulate overall evaluations of the agent and possibly also of the organization that the conversational agent represents (phase 3) (van der Goot et al., 2020).

In the following paragraphs, we will reason the desirability and necessity to include the social cues in the different phases of users' communication journey with the chatbot, how the cues can

TABLE 7.2
Overview of the Social Cues that can be Used in Conversational Agents per Chatbot Journey Phase and Examples of Operationalizations (Phase 1 and 2) and Effects (Phase 3)

		Chatbot User Journey		
Social Cues	Description	Phase 1: Prior to the Interaction	Phase 2: During the Interaction	Phase 3: After the Interaction
Conversational intelligence	The chatbot's ability to manage interactions with users			
Proactivity	The chatbot takes *initiative autonomously*, resulting in a two-way conversation	+- include a personal greeting, but avoid intrusiveness	+ ask follow-up questions, provide relevant information, monitor users' goals and guide users toward their goals using motivational messages	+ could contribute to the user's task and to having a smooth conversation
Conscientiousness	The chatbot demonstrates *attentiveness to the conversation*	- only applicable if the chatbot can use conversation history at the start of a new conversation	+ keep the user aware of the chatbot's context, provide meaningful answers, use and balance confirmation messages	+- could contribute to the user's task and to having a smooth conversation
Communicability	The chatbot conveys its *features to users*	++ explain the purpose of the chatbot and its functionalities, and mention the option to redirect users to a human agent	+- remind users about the purpose of the interaction, and redirect the users to a human agent in case of failure or conflicts	+ could contribute to having a smooth conversation
Social intelligence	The chatbot's impact on the social behavior of the user			
Damage control	The chatbot's ability to *deal with failures and failures and conflicts*	+ mark the chatbot's (in)competence, mention to option to be redirected to a human agent	++ avoid failures, recognize them, communicate failures properly, provide the option to be redirected to a human agent	++ could contribute to the user's task and to having a smooth conversation
Thoroughness	The chatbot's ability to *be precise and consistent in language use and communication style*	++ match the chatbot's language use and communication style to the context in which it is implemented	+ be consistent in communication style and language use, balance the granularity of the information	++ could contribute to relationship building
Manners	The chatbot's ability to *show polite behavior and adhere to conversational habits*	+ include a personal greeting, self-introduction, and adhere to turn-taking protocols	++ adopt speech acts, like opening and closing sentences, acknowledgments, make interactions personal	++ could contribute to having a smooth conversation and to relationship building
Moral agency	The chatbot acts based *on social notions of right and wrong*	+ avoid stereotypes in the chatbot's avatar, name, and reference to the user	+ use 'clean' training data (without harassment), be aware of biases	+ could contribute to relationship building

(Continued)

TABLE 7.2 (*Continued*)
Overview of the Social Cues that can be Used in Conversational Agents per Chatbot Journey Phase and Examples of Operationalizations (Phase 1 and 2) and Effects (Phase 3)

		Chatbot User Journey		
Social Cues	Description	Phase 1: Prior to the Interaction	Phase 2: During the Interaction	Phase 3: After the Interaction
Emotional *intelligence*	The chatbot recognizes users' feelings and demonstrates respect and understanding	+- include a personal greeting, show information from prior conversations, if possible	+ chatbot shows empathy, reciprocity, and conscientiousness	+ could contribute to relationship building
Personalization	The chatbot's ability to *adapt the interface, content, and behavior to the users' preferences, needs, and situational context*	++ build the chatbot on the basis of cultural, behavioral, personal, conversational, and contextual data	+- use information from the conversation to tailor the responses, but avoid intrusiveness	+ could contribute to relationship building
Personification	*The chatbot's characteristics that make the agent appear more human-like*			
Identity	The chatbot's *appearance and cultural traits*	+- include a disclosure about the artificial nature of the chatbot. Also the avatar and name type can mark or mask the chatbot's artificial identity	+ elaborate on the chatbot's persona with a matching language style	+ could contribute to relationship building
Personality	The chatbot's *behavioral traits*	+ develop and introduce a clear personality trait	+ use appropriate language, have a sense of humor	+ could contribute to relationship building

Note: Meaning of the characters: ++, highly desirable/necessary; +, advisable; +-, moderately desirable because of pros and cons; -, discourage.

be operationalized, and their relative impact on the users' prior expectations of the chatbots, their responses during the conversation, and perceptions and behavioral intentions after the conversation. Moreover, we will illustrate the applicability of the social cues in each phase by discussing two chatbots from different domains: a customer service chatbot, and a smoking cessation chatbot.

7.5.2 USER EXPECTATIONS PRIOR TO THE CONVERSATION

The first phase of the users' communication journey with a conversational agent consists of the agent's first messages. The social cues present in these first messages could influence users' expectations about the chatbot. Consider the customer service chatbot of Cana Brava Resort[3] presented in Figure 7.4 (left). Although the term 'virtual attendant' and the tottler-like avatar implicitly disclose the interlocutor is not a human being but a conversational agent, several social characteristics are adopted that could stimulate social behaviors that are habitual in human-human conversations (Chaves & Gerosa, 2020; van Hooijdonk et al., 2023). The chatbot contains a human-like name, avatar, and informal communication style that all contribute to the personification of the agent. These cues could enhance users' perceptions of the human likeness of the chatbot (Liebrecht et al., 2020). Also, the chatbot of Cana Brava Resort uses social intelligence cues: a

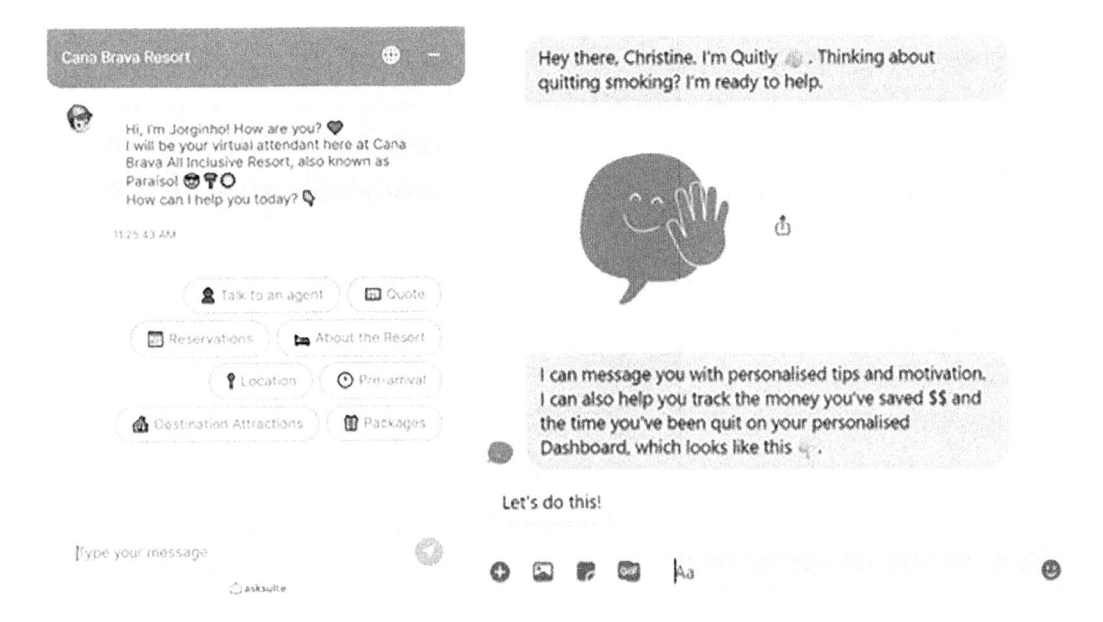

FIGURE 7.4 Introduction of the customer service chatbot of Cana Brave Resort (left) and Quitly the smoking cessation chatbot (right). Screenshots by the authors.

personal greeting ('Hi'), a self-introduction ('I'm Jorginho!'), and turn-taking ('How can I help you today?'). These cues display or adhere to manners and hence correspond to social norms from human-human communication that could stimulate users to act in the same social intelligent way (Chaves & Gerosa, 2020).

However, several opportunities to shape user expectations even better remain untapped by the Cana Brava Resort chatbot. First of all, it can be questioned what expectations users will have about the chatbot's competence. The tottler-like avatar and its rather informal language style could on the one hand increase users' perceptions of warmth, but on the other hand also lower expectations regarding the chatbot's competence (Khadpe et al., 2020) which in turn negatively affects users' trust in the chatbot capabilities. In contrast, communicating the intelligence of the agent, for example by stating the chatbot is 'an expert' for hotel bookings, users' trust in the chatbot's capabilities could be increased. The implementation of communicability cues in a chatbot introduction could also positively affect users' intentions to engage in a conversation with a chatbot. The chatbot could manage expectations about its purpose to avoid that users would address topics on which the chatbot has not been trained ('I can help you with hotel bookings'), it could give users guidance on how to interact with the chatbot to avoid miscommunication ('please choose a topic or ask your question in a maximum of five words'), and it could reassure users that human agents are present if needed ('You always have the possibility to continue the conversation with a human agent') (van Hooijdonk et al., 2023). In contrast, the opening messages of the smoking cessation chatbot Quitly[4] in Figure 7.4 (right), contain multiple communicability cues by explaining the purpose of and interaction with the chatbot. It also shows good manners because it proactively greets the user personally ('Hey there, Christine').

7.5.3 User Experience during the Interaction

During the interaction, the second phase of the user-chatbot communication journey, the conversational agent can display several social cues. It is important that these cues match the expectations users developed in the first phase of the journey. The customer service chatbot of Cana Brave Resort and the smoking cessation chatbot Quitly (see Figure 7.5) both adopt an informal and engaging

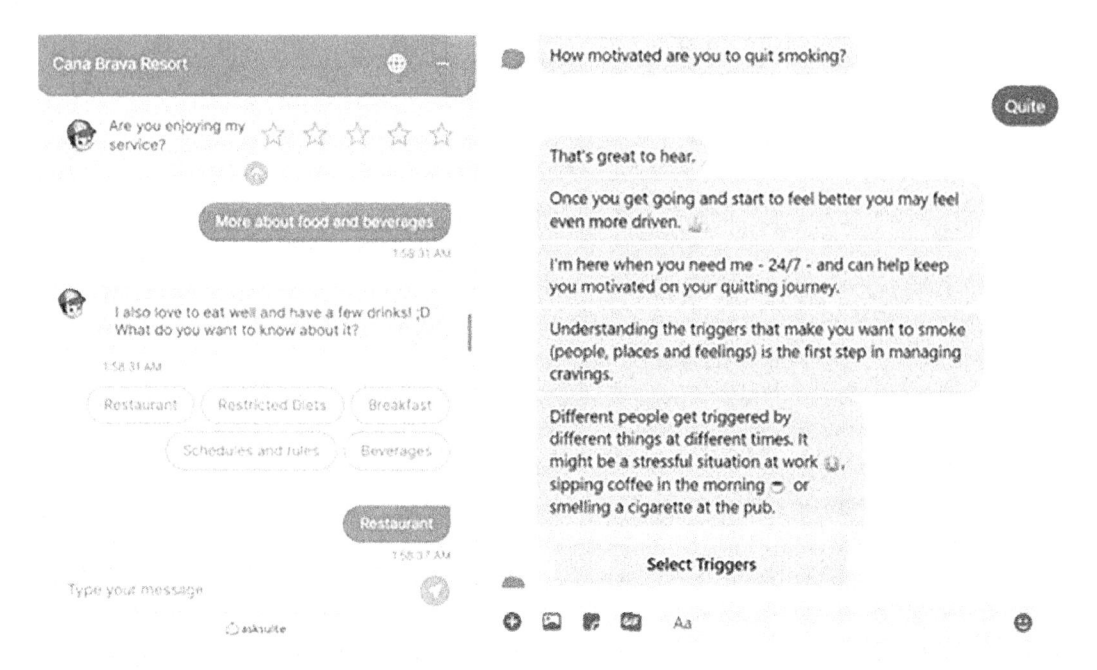

FIGURE 7.5 Conversation with the customer service chatbot (left) and the smoking cessation chatbot (right). Screenshots by the authors.

communication style by using emoji, contractions, acknowledgments, and showing empathy. The customer service chatbot also communicates with a sense of humor ('I also love to eat well and have a few drinks! ;D'). Although this communication style can positively impact users' perceptions of the human likeness of the chatbot, Chaves and Gerosa (2020) argue that the chatbot's communication style - or even its personality - should be considered appropriate by its users. This depends on users' characteristics, e.g., age, literacy, and familiarity with chatbots. and context characteristics, such as the domain in which the chatbot is implemented, e.g., customer service vs. healthcare vs. politics, the organization that the chatbot represents, e.g., a formal banking company vs. an informal e-commerce company. and the chatbot's purpose, e.g., providing answers to FAQ's vs. providing psychotherapeutic support. A mismatch between users' expectations and perceptions about the chatbot's communication style could withhold users from continuing a conversation with the conversational agent.

Another challenge is the implementation of conversational intelligence cues proactivity and conscientiousness. Examples of the cues are personally greeting users by their names, using previously shared user information in conversations, or proactively suggesting new topics or follow-up questions. Some of these cues can be observed in the smoking cessation chatbot, such as personal greetings and providing personalized tips (Figures 7.4 and 7.5). However, the implementation of these social cues also demands privacy protective operations, such as an explicit agreement from users that their personal information will be saved and reused at another point of time (which was the case when registering for Quitly). Thus, user and context characteristics should be taken into account in the implementation of conversational intelligence cues. These cues seem to be less relevant in the context of a task-based customer service chatbot which customers use to efficiently find a tailored response to their questions. However, these cues are more relevant in a socio-oriented context in which the chatbot is used as a coach or friend for a longer period of time.

Lastly, how the conversational agent deals with miscommunication affects users' experiences. Different social cues can be implemented to avoid and/or deal with miscommunication in a socially acceptable manner (Chaves & Gerosa, 2020). Miscommunication can be avoided by using conscientiousness cues, such as keeping the conversation on track by reminding users about the purpose

Conversational Agents

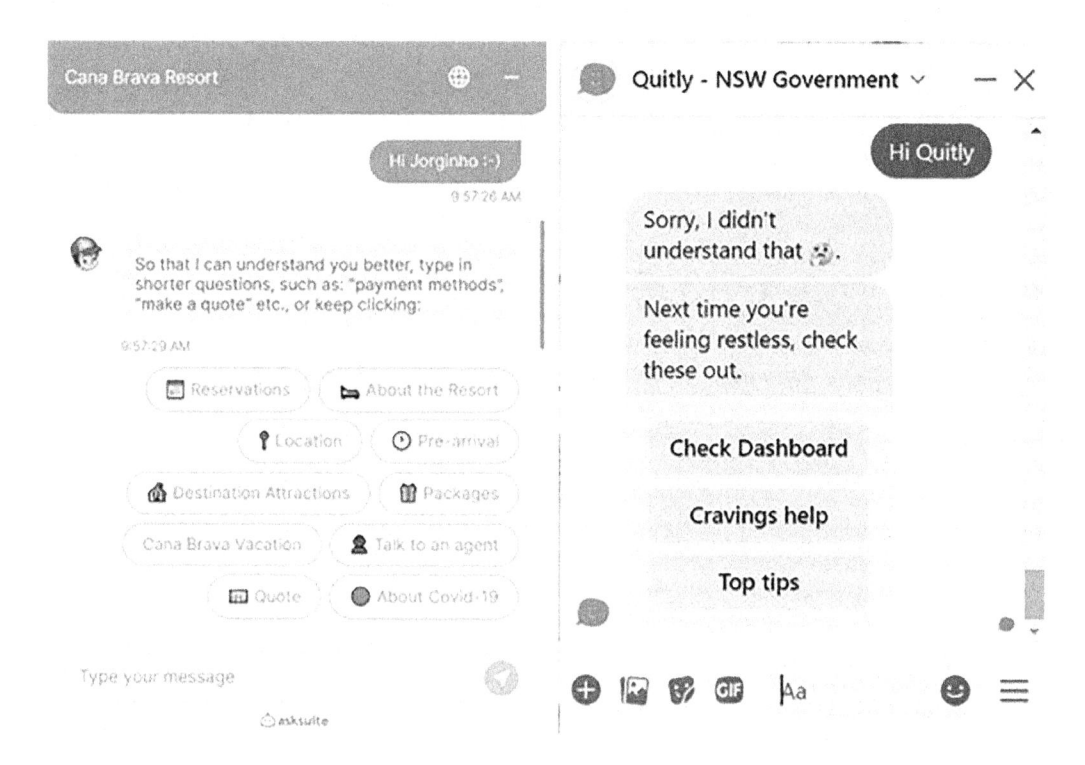

FIGURE 7.6 Miscommunication with the customer service chatbot (left) and the smoking cessation chatbot (right). Screenshots by the authors.

of the interaction and informing them about the next steps. Moreover, confirmation messages can be used to check the chatbot's understanding of user messages. These conscientiousness cues are especially important in task-based customer service chatbots which customers use to reach a goal in an efficient and productive way (Chaves & Gerosa, 2020; Duijst, 2017).

Miscommunication often occurs due to lack of chatbots' linguistic and world knowledge (Wallis & Norling, 2005). Social intelligence cues can be used to deal with miscommunication, such as recognizing and apologizing for it, which may reduce users' annoyance and frustration. Also, communicability cues can be used to recover miscommunication, such as providing options (e.g., 'Sorry, I did not understand you. Do you want to know more about (1) beverages or (2) breakfast?'), or redirecting the users to a customer service employee. The Cana Brava Resort chatbot Jorginho and smoking cessation chatbot Quitly adopt different approaches to deal with miscommunication that occurred at the start of the conversation. Chatbot Jorginho does not recognize the miscommunication explicitly. Instead, it gives the user directions on how to communicate with the system by means of the open text field and by showing the buttons that hint the user toward the topics that the chatbot can handle (see Figure 7.6 (left)). Chatbot Quitly, in contrast, explicitly states that it was unable to understand the user's utterance, and subsequently shows options to steer the conversation in the right direction (see Figure 7.6 (right)).

7.5.4 User Evaluation after the Interaction

In the third phase of the chatbot user journey, users form an overall evaluation based on their experiences with the conversational agent. Interestingly, this phase differs among the customer service chatbot and the smoking cessation chatbot. The customer service chatbot of Cana Brava Resort is a task-based chatbot that users visit to make a booking or to obtain more information about the resort. When the user approaches the end of the conversation, the chatbot asks 'Was I able to answer you?'.

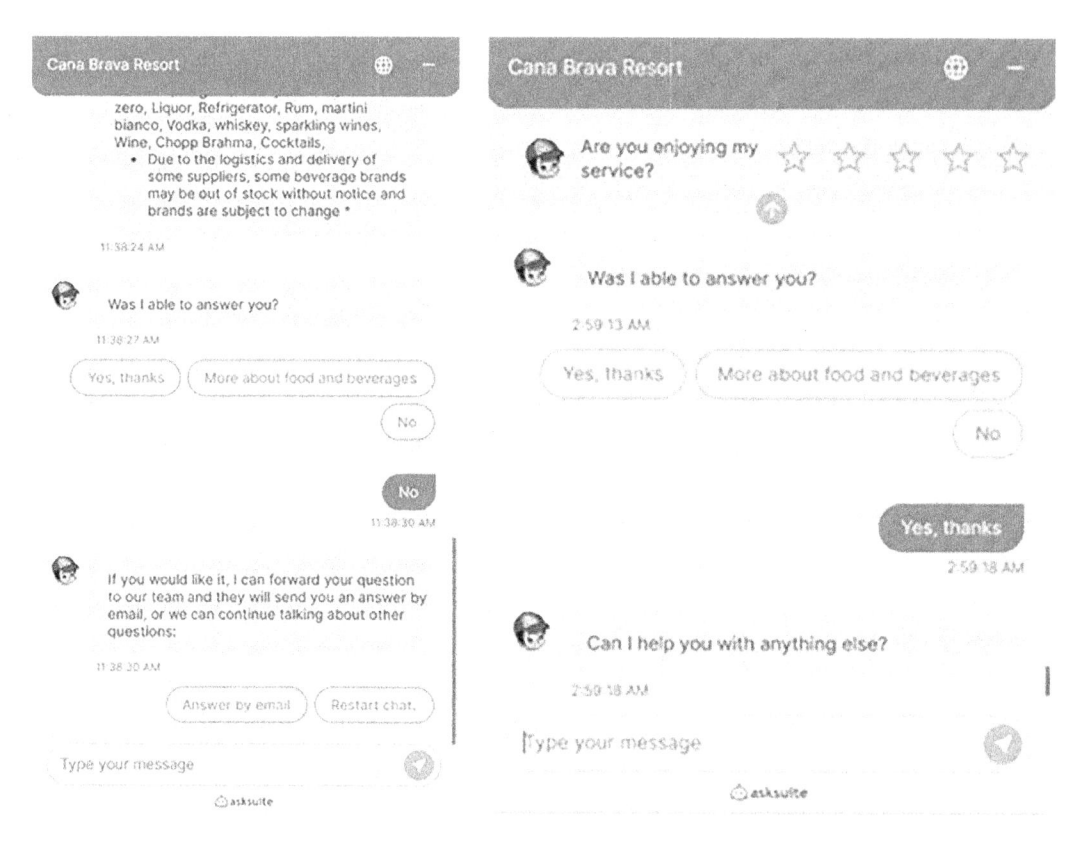

FIGURE 7.7 Final stage of the conversation with the customer service chatbot of the Cana Brava Resort. Screenshot by the authors.

In case the user responds 'yes', a positive experience and evaluation of the chatbot could motivate them to start a new conversation with the agent next time they want to make a booking or obtain more information about the resort (see Figure 7.7). In contrast, the conversation of the smoking cessation chatbot is designed in such a way that there is no endpoint in the conversation (see Figure 7.8). It is therefore important that user evaluations maintain high to stimulate them to keep using or re-using the chatbot again.

These experiences are related to the three goals users have when they engage in a conversation with a conversational agent. The first goal concerns the task users want to perform with the conversational agent, such as booking a hotel or learning more about smoking triggers (Luger & Sellen, 2016; Shechtman & Horowitz, 2003). Miscommunication may lead to an unsuccessful conversation resulting in users not achieving their tasks. This will lead to a negative evaluation of the interaction, the chatbot, and the organization. Therefore, miscommunication should be prevented in the first place, and recognized and solved in the second place. Miscommunication can be prevented if users' expectations about the chatbot's purpose and capabilities were managed before and during the interaction by implementing communicability cues. These cues could positively impact users' perceptions about the chatbot's capabilities (Jain et al., 2018) as well as trust and satisfaction. Moreover, research by Ashktorab et al. (2019) shows users favored a chatbot that acknowledges miscommunication, and provides options of possible intents. This way the chatbot initiates the repair of the miscommunication, corresponding with the general preference for self-repair in interpersonal communication (Schegloff et al., 1977) and steers the conversation in a direction within its capabilities.

The second goal users have when engaging in a conversation with a chatbot is how they can have a smooth conversation with it (Luger & Sellen, 2016; Shechtman & Horowitz, 2003) which can be

Conversational Agents

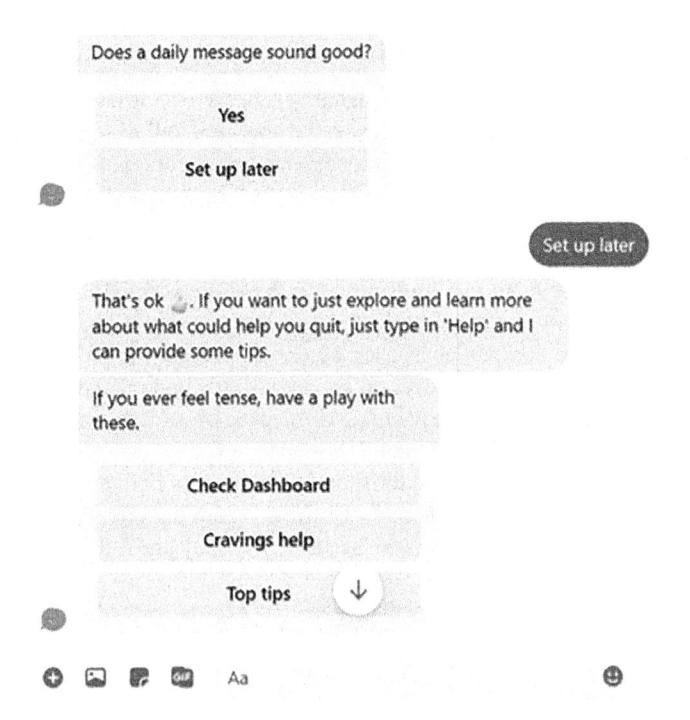

FIGURE 7.8 The conversation with the smoking cessation chatbot does not contain a clearly defined final stage of the conversation. Screenshot by the authors.

achieved by implementing conversational intelligence cues. For example, the chatbot takes initiative autonomously and starts a two-way conversation (proactivity cues), or the chatbot demonstrates attentiveness to the conversation (conscientiousness cues). These conversational intelligence cues are more difficult to implement as they have to be realized in the architecture of the conversational agent. Advances in AI technology might lead to the implementation of conversational intelligence cues enabling conversational agents to interact with users in a more intuitive way (Huang & Rust, 2018).

Finally, the third user goal refers to maintaining a certain relationship with the conversational agent (Luger & Sellen, 2016; Shechtman & Horowitz, 2003) which can be achieved by implementing social cues that simulate perceptions of human likeness, such as personification cues. This can be achieved by creating a chatbot with a clear personality trait, and by using identity cues, such as a human-like name and avatar, and a personal and engaging communication style. Research shows that the usage of human-like cues increases perceptions of human likeness which in turn leads to a positive evaluation of the organization (Araujo, 2018; Go & Sundar, 2019; Liebrecht et al., 2020). In order to hold these effects, Chaves and Gerosa (2020) state that the chatbot should maintain its personality and identity throughout the whole conversation (thoroughness cue), provided that the chatbot personification matches the user and the context (personalization cue). As these social cues are relatively easy to implement in conversational agents, a majority of current conversational agents contain (some of) these social cues, which has also been shown by the customer service chatbot and the smoking cessation chatbot in Figures 7.4 and 7.5. However, a conversational agent merely mimics human-like communication. To maintain a relationship with a conversational agent it is important that it recognizes users' feelings and demonstrates respect and understanding. One way how these emotional intelligence cues can be implemented is by automatically analyzing the user's language, or by explicitly asking users how they feel. The latter approach can be observed in current conversational agents. Advances in AI technology might lead to the implementation of emotional intelligence cues enabling conversational agents to interact with users in a more empathetic way (Huang & Rust, 2018).

7.6 CONVERSATIONAL AGENT EVALUATION

The previous sections underline conversational agent design dimensions, including technical, social, and conversational capabilities. However, to foster the adoption and acceptance of these technologies, it is crucial to ensure that these dimensions are well-designed, meet user social and emotional expectations, and measure up to the desired functional performance and purpose.

Many commercially available conversational agents apply marketing strategies to evaluate the interaction between the software and their customers. These strategies include satisfaction surveys or prompts for feedback. For example, in Figure 7.9, the Cana Brava Resort chatbot, introduced in Section 7.5, prompts customers to evaluate the provided service using five-star ratings. These prompts are insightful to the companies as they assess the impact of the technology adoption on their target consumer's perceptions. However, to get an in-depth understanding of human-conversational agent interactions, it is crucial to have broader investigations to create guidelines that can be generalized to a variety of conversational agent's contexts of use.

The HCI field provides a set of well-established techniques to evaluate HCIs, such as usability and User eXperience (UX) evaluation methods. However, these techniques were developed before the exponential spread of conversational interfaces; hence, they were designed and tested in the context of HCIs over traditional interfaces.

Conversational agents, however, have their own needs in terms of evaluation. For example, conversational agent designers are more likely to prioritize linguistic style, manner, and perceptions of humanity than traditional interface designers. As anthropomorphic cues such as identity and use of natural language are more evident for these agents than for traditional interfaces, aspects such as satisfaction, trust, and engagement may have a different meaning and relevance from the user's perspective. Additionally, conversational agent design must care about interaction flow and context differently than traditional interfaces. Thus, evaluating a conversational interface may require developing new evaluation tools or adaptations to well-established techniques.

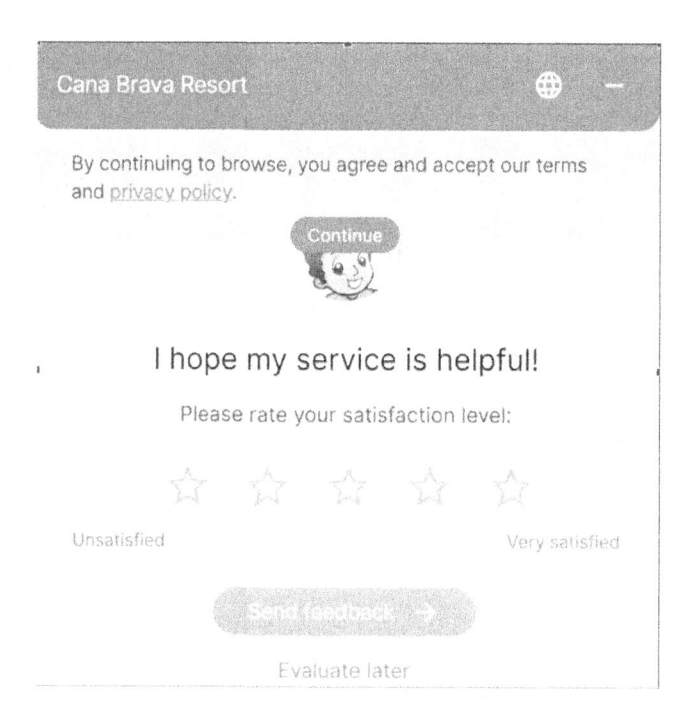

FIGURE 7.9 Cana Brava Resort chatbot asks for customer's feedback using five-star ratings. Screenshot by the authors.

In the last decade, scholars in the HCI field have steered the gear toward assessing the extent to which currently available evaluation methods fit modern conversational interactions. Additionally, there have been efforts to validate new or adapted evaluation tools. This section presents pragmatic and hedonic aspects that determine user experience and acceptance of conversational agent technologies and examines methodologies commonly used in the conversational agent's context.

7.6.1 Usability and User Experience

User experience (UX) is a comprehensive concept that refers to a person's perceptions and behaviors while using a software product (ISO 9241-11, 2018). Due to the concept's intrinsic complexity, a consolidated evaluation method to encapsulate all aspects of UX is yet to be developed. Instead, various evaluation methods have been used to address particular aspects of user experience, mainly distinguishing between pragmatic and hedonic goals (Bevan, 2009) associated with the user experience.

The evaluation of conversational agents follows the same practices. On the one hand, conversational agents are designed and evaluated to reach usability goals, such as effectiveness and efficiency. On the other hand, understanding the user's emotional experiences is crucial to achieving acceptance and adoption, which calls for assessing hedonic qualities such as engagement and pleasure (Haugeland et al., 2022).

Various well-established instruments have been used to evaluate conversational agents when focusing on usability metrics. For example, Guerino and Valentim (2020) mapped the literature to identify technologies used to assess the usability and UX of voice-based conversational systems. Unsurprisingly, the authors list technologies such as the System Usability Scale (SUS), the NASA Task Load Index (NASA-TLX), the Computer System Usability Questionnaire (CSUQ), and Nielsen's heuristics. Ren et al. (2019) found similar results in the context of chatbot evaluations. Table 7.3 shows some generic tools used to evaluate the usability and UX of conversational agents, returned by Guerino and Valentim (2020) and Ren et al. (2019).

TABLE 7.3

Evaluation Tools Used to Evaluate the Usability and UX of Conversational Agents

Evaluation Tool	Description
System Usability Scale (SUS)	Tool for measuring the Usability of products/services. It is a 10-item questionnaire that can be answered using a 5-point Likert scale, where 1 represents scale, where 1 represents "strongly disagree" and 5 represents "strongly agree". Even though it is not specific for conversational agents, SUS is a consolidated technology in Usability evaluation. Moreover, SUS is widely used in the literature to evaluate efficiency, effectiveness, and user satisfaction.
NASA Task Load Index (NASA-TLX)	NASA-TLX is a subjective assessment tool used to rate the perceived workload to evaluate a task or system. The aspects evaluated by this tool are mental demand, physical demand, time demand, performance, effort, and frustration.
Computer System Usability Questionnaire (CSUQ)	CSUQ is a 19-item questionnaire to be answered on a 7-point Likert scale. The questionnaire assesses utility, information quality, interface quality, and overall usability.
Attrakdiff	It is an instrument that evaluates the product's attractiveness in terms of usability and appearance. Attrakdiff contains 28 items of opposite pairs that can be answered on a 7-point scale.
Input Device Usability Questionnaire (IDU)	It is a 15-item questionnaire designed to investigate user interaction, distraction, ease of use, user comfort, frustration, enjoyment, error correction, and general usability. The questionnaire can be answered using a 5-point Likert scale.

Source: Adapted from Guerino & Valentim (2020); Ren et al. (2019).

However, both literature reviews (Guerino & Valentim, 2020; Ren et al., 2019) highlight that, in many cases, researchers create evaluation tools tailored to the particular study, which suggests the need for more specific evaluation methods. For instance, the most common tool reported by Guerino and Valentim was the creation of questionnaire-based assessment instruments for a specific study or agent. However, this ad-hoc creation, despite generally allowing a pertinent evaluation of Usability or UX, does not allow the replicability of the instrument in other studies or conversational agents.

Therefore, there have been initiatives to adapt existing evaluation tools to the context of conversational agents. Langevin et al. (2021) extended Nielsen's heuristics for usability to be used in the formative evaluation of conversational agents. The authors found that evaluators identified more usability issues when using the adapted heuristics when compared to the generic Nielsen's heuristics. According to the authors, conversational agent heuristics can be generalized to text-based, voice-based, and multimodal conversational agents.

In the context of chatbots, Borsci et al. (2022) developed the Bot Usability Scale (BUS-15), which consists of a 15-item questionnaire focused on the user's satisfaction with the interaction. The questionnaire includes items about the chatbot's reachability, functionalities, quality of conversation and information, privacy and security, and response time. However, the scale has not been extensively tested and validated, requiring a collective effort from the research community to assess the reliability of the instrument.

When focusing on voice-based interfaces, Guerino et al. (2021) developed an instrument to evaluate usability and UX, called U2XECS. The instrument is questionnaire-based, and from its application, it is possible to evaluate the following aspects: User Satisfaction, Efficiency, Effectiveness, Generic UX, Affect/Emotion, Enjoyment/Fun, Aesthetics/Appeal, Engagement/Flow, and Motivation. U2XECS was evaluated in controlled experiments; however, further studies are needed to generalize its results and enhance its benefits for the conversational agents' community.

However, there is more to the user experience than usability and satisfaction. Bevan (2009) argued that user experience in user-centered design also includes understanding what users do and why when interacting with the systems and maximizing the achievement of hedonic goals. Guerino and Valentim (2020) show that there is no clear distinction, on the part of evaluators, about what is included when evaluating UX; in this sense, it was seen that the same tool used to evaluate UX in one study was used to evaluate usability in another. Thus, more studies are needed so that, in addition to better understanding the definitions and aspects of UX and usability in conversational agents, specific tools are proposed for this context and replicable to other studies.

7.6.2 User Perceptions and Acceptance

The success of software or technology strongly depends upon the user's perceptions of the system, such as expectation, engagement, and trust (Venkatesh et al., 2011). Combined with the usability and satisfaction dimensions of UX, positive user perceptions increase the likelihood of acceptance and adoption of a system. In the conversational agents' field, user perceptions have a unique characteristic when compared to other systems: the anthropomorphization triggered by the use of natural language and the social implications associated with it (see Section 7.5).

Standardized questionnaires fail to capture the nuances of subjective user perceptions. Therefore, user perceptions are often measured using qualitative methods such as interviews and focus groups based on the interaction with a proposed conversational agent or the user's previous experiences with the technology (Bevan, 2009).

One technique that supports the qualitative evaluation of conversational agents is the Wizard of Oz (WoZ) (Dahlbäck et al., 1993) (see more in Section 7.3). This technique helps evaluators to consider the unique qualities of human-conversational agent communication in the initial design stages. Besides, this method enables control for implementation limitations that could harm usability and performance thus negatively affecting the user experience. As a drawback, some human limitations are difficult to overcome in WoZ studies. For example, response time can be difficult to simulate, as humans will unlikely be able to answer user inputs as fast as a conversational agent. Nevertheless, the benefits of the WoZ technique surpass its limitations, and the popularity of the technique has even fostered the development of environments to support its use for conversational agent evaluation (Simpson et al., 2022).

When the conversational agent evaluation is based on a current interaction, a common method is to invite the subjects to Think Aloud. Think Aloud sessions aim to identify user perceptions based on what they say out aloud about what they are doing and why. According to Barbosa et al. (Barbosa et al., 2022), when the user verbalizes their thoughts during the interaction, evaluators can capture genuine reactions and interpret problematic parts of the interaction.

Unlike user perceptions, which are a broad set of behaviors and emotions triggered by the interaction with conversational agents, acceptance is a more consolidated construct. Therefore, the literature provides standardized evaluation tools to measure the acceptance of a software system. The two most frequently used theoretical frameworks to evaluate the adoption of conversational agents are the Technology Acceptance Model (TAM) and the Unified Theory of Acceptance and Use of Technology (UTAUT2). However, both questionnaires are highly oriented to pragmatic qualities, such as usefulness and effort expectancy.

Moreover, by applying UTAUT2 to evaluate the acceptance of a healthcare-specific conversational agent, Laumer et al. (2019) found that using a specific technology (conversational agent) can influence the use of a particular instance of this technology, e.g., a conversational agent for disease diagnosis. The authors foster further research that proposes measurement items of acceptance specific to application contexts. Furthermore, the authors extend the conclusion made by Venkatesh et al. (2011) by mentioning that it is necessary to investigate the influence of environmental factors on the acceptance of a conversational agent and to propose new models to measure this influence and its relationship with other factors empirically.

In this sense, Ling et al. (2021) surveyed the literature to identify the factors that influence users' adoption of conversational agents. As a result, the authors propose a model where acceptance is driven by the usage benefits, which translates to the value gained from interacting with the agent. These benefits are influenced by both agent and user characteristics, including, for example, the agent's appearance and anthropomorphic cues and the user's demographic characteristics and intrinsic motivation. However, the proposed collective model is theoretical and there is still a lack of a practical framework to guide the use of the model for conversational agent evaluation.

7.7 CONVERSATIONAL AGENTS FOR SOCIAL GOOD

This section focuses on the impact of introducing conversational agents in our society, particularly focusing on the role these agents may perform in well-being and inclusion. Given the interest in conversational agents for social good, be they chatbots (Følstad et al., 2018), AI systems (Floridi et al., 2020), and/or robots (Lee et al., 2019a), we first dive into what "social good" means. We then discuss relevant application domains in which conversational agents serve as a tool for social good.

We relate social good to people's capacity to lead a "good life", such as the Aristotelian sense of eudaimonic well-being (Aristotle, 2019). Additionally, social good considers opportunities to foster people's capabilities, such as opportunities to achieve one's health (Robeyns, 2009) or promoting people's values in everyday technologies (Friedman, 1996; Wambsganss et al., 2021). Thus, one broad definition of technologies for "social good" is to promote the well-being of people and nature while mitigating harm towards them (Floridi et al., 2020); harms can happen due to data manipulation, e.g., incorrectly labeled training data for conversational agents, and violation of users' privacy, e.g., personally identifiable information from conversational data being shared with third parties without consent (Floridi et al., 2020). For these potential harms, what is needed is a clear explanation about how conversational agents share and handle personal data, as well as how they have been trained, e.g., on what dataset and how (Mitchell et al., 2019), which can assist in expectation management between users, conversational agents, and other stakeholders. Another harm is systematically and continuously marginalizing specific target user groups by not involving them in the design and development process, e.g., designing conversational agents *for* people with disabilities rather than designing *with* them, e.g., Spiel et al. (2020). Directly involving people who are intended users (especially from marginalized or potentially marginalized groups) in the design of these conversational agents can mitigate potential harm, e.g., through VSD (see Section 7.3), capability-sensitive design (Jacobs, 2020), and participatory design (Frauenberger et al., 2017).

We turn to how conversational agents can serve the social good in specific domains, such as promoting well-being in the healthcare sector. Conversational agents are being utilized for physical and mental healthcare, such as assisting doctors or patient support (Preum et al., 2021). There are healthcare conversational agents that are for specific target populations, such as people with neurodevelopmental disorders (Catania et al., 2023) or adolescents (Fitzpatrick et al., 2017; Gabrielli et al., 2020), and again, it is important to bear in mind involving these intended user groups when developing such agents, e.g., Lopatovska et al., 2022.

While well-being technologies come in many types, such as in VR or mobile apps (Calvo & Peters, 2014), what is particular about conversational agents is that they can refer to themselves in the first person, talking to human interactants as another "I", so instead of clicking or scrolling through apps, people are able to converse with technology (Lee & Contreras, 2023). Relatedly, conversational agents can make it easier for some to discuss sensitive topics like PTSD (compared to disclosing this to other people) as these agents are seen as "mere machines" (Lucas et al., 2014, 2017). People may believe that machines or chatbots are less judgmental than other people (Brandtzaeg & Følstad, 2017, 2018). Besides targeted health interventions like sleep management (Rick et al., 2019) or smoking cessation (see Section 7.5 and Perski et al., 2019), there are many examples of conversational agents that support people's mental health (Ahmed et al., 2023), such as Wysa that is empathy based (Inkster et al., 2018), Wocbot for depression (Fitzpatrick et al., 2017), and Vincent (Lee et al., 2019b) that asked for people's advice after reporting on its mishaps (Figure 7.10) which helped people become self-compassionate by being compassionate to it. Hence, conversational agents can be perceived to be helpful or in need of help in order to elicit specific psychological responses such as the tendency for self-disclosure or care-oriented behavior.

One underlooked aspect in healthcare is any potential conflict between stakeholders; there might be differences in opinions between experts, healthcare providers, and patients, such as a positive perspective by the experts who believe in predictive health (De Maeyer & Markopoulos, 2021) vs. patients who may not want predictive health forecasting that conversational agents can share with them, especially in emerging forms like digital twins at home (De Maeyer & Lee, 2022). In what ways conversational agents should navigate across potential conflicts between stakeholders, as well as being entangled in healthcare conflicts, will become increasingly more significant to address.

Conversational agents for well-being focus on individuals' health, but there are other social good areas that conversational agents can be active in, be it for education, public health crisis, personal finance knowledge, or for inclusive citizenship, e.g., government agencies using conversational agents to increase citizens' awareness of policies. Conversational agents used for educational purposes come in many varieties (Chen et al., 2023; Hwang & Chang, 2021; Khosrawi-Rad et al., 2022), such as agents that help foster creativity and curiosity (Abdelghani et al., 2022) or agents that assist in online learning via text and voice input in a MOOC (Winkler et al., 2020), e.g., by fostering inclusive learning through an active FAQ chatbot (Han & Lee, 2022). Learning can occur in workplaces, e.g., for reflection and journaling (Kocielnik et al., 2018). Segmenting needs and goals per intended group is important here; college students vs. children would have different learning challenges, for instance. The design features of conversational agents should consider various differences. For example, a simple, clickable button to start the conversational agent might be more appropriate for children rather than expecting them to use "wake words" (like "Hey Google") to start a voice-based agent (Catania et al., 2020) and children might learn better when the agent is portrayed as a peer than a tutor, i.e., due to higher engagement and attention to a peer agent (Zaga et al., 2015). Conversational agents

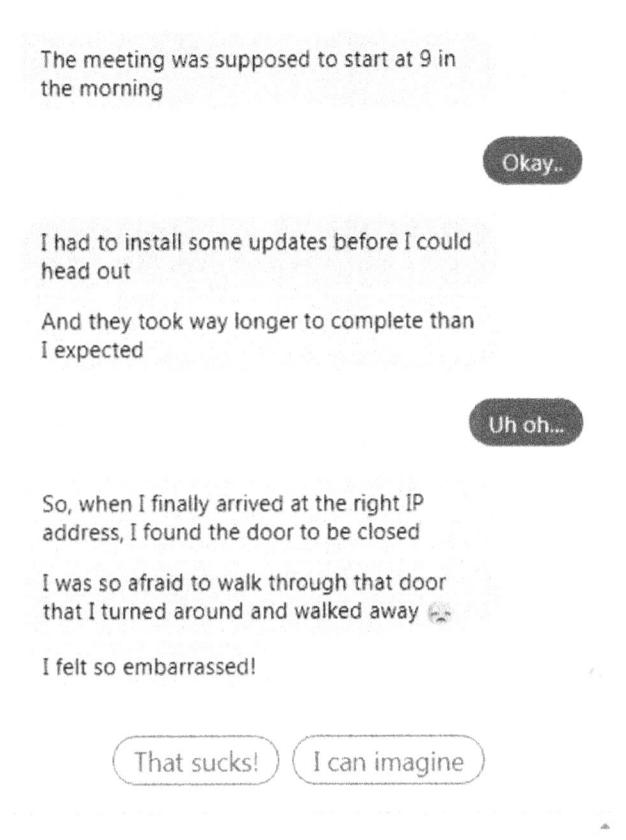

FIGURE 7.10 Vincent, a chatbot, sharing that it was late for a meeting to ask for help from people (Lee et al., 2019b).

can thus play diverse roles, such as being a mentor or assistants, as well as being a peer, student, tutor, or a teacher (Wollny et al., 2021). As with potential conflicts among stakeholders in healthcare, various perspectives have to be considered, e.g., differences in what children vs. parents may educationally value and how then conversational agents should be used at home (Garg et al., 2022; Garg & Sengupta, 2020), scaling up to potential differences in how schools, universities, or government agencies may want to institute conversational agents for education compared to students and parents. Conversational agents themselves can be the cause of conflict, such as a recent worry with students' usage of ChatGPT over the kinds of influence conversational agents can have on learners and if and how that should be curtailed (Rudolph et al., 2023).

Beyond education, there are other cases in which allowing people to help conversational agents would be helpful, e.g., humans teaching conversational agents classification tasks which can foster trust between humans and agents on crowd work platforms, which increases the agents' ability to perform tasks when they function as "teachable machines" that can grow with people they learn with (Chhibber et al., 2022). Community- or crowd-driven training also seems promising, such as a chatbot that was trained to provide on-demand by crowd workers (Abbas et al., 2020). Also on gaming platforms, people can teach a chatbot how to act as a community member, e.g., how to address people in the same community, which can strengthen behavioral norms that an online community wants to promote via teaching a chatbot (Seering et al., 2020). In general, this forms empowerment through the perspective of a conversational agent. People help their communities or themselves by empowering an agent, e.g., learning through teaching (Tanaka & Matsuzoe, 2012) or caring for oneself by caring for an agent (Lee et al., 2019b).

Conversational agents are deployed for many other domains we briefly go over. Concerning a public health crisis, during COVID-19, many chatbots were deployed, e.g., to provide timely information on the spread of the virus though there are concerns about the potential spread of inaccurate data or misinformation (Miner et al., 2020). There are efforts to better assist citizens through conversational agents in digital governance (Abbas et al., 2023), like for reaching those who may need social service support (Simonsen et al., 2020). In the finance sector, conversational agents can support people's financial decision-making (Wube et al., 2022), e.g., for small businesses to get loans (Candello et al., 2022a), or for people to get an understanding of the stock market (Sharma et al., 2021) or for exploring alternative currencies, like a chatbot for the cryptocurrency marketplace (Lee et al., 2021a). In sum, promising directions for social good in healthcare, education, crisis management, digital governance, and finance (among others) are in place for conversational agents.

7.7.1 CHALLENGES AND OPPORTUNITIES FOR SOCIAL GOOD

The main challenge is that the notion of "social good" deserves to be critically questioned, given that many corporate, social good ventures do not always come with harm mitigation (Kwet, 2019; Zembylas, 2023). Broadly, another point is that what may be "socially good" is not universal (Madianou, 2021). This relates to how, for instance, accounting for race and socio-economic differences in the design and deployment of conversational agents (e.g. Garg & Sengupta, 2019) requires a nuanced view on *what* is socially good for *whom* from *whose perspective*.

Hence, what further needs to be discussed are user representation and conversational agent representation in terms of *intersectionality* (cross-cutting many attributes such as race, gender, disability, class, and more), such as what traits (like name and avatar) of a conversational agent are designed as well as what groups are representative users of it. Intersectionality is not yet a focus for conversational agents (both in how they are designed and how they are received by intended users), but we see a growing need to consider intersectionality, especially in mitigating bias in the design and use of conversational agents (Ciston, 2019; Lee et al., 2021b).

Sexism is compounded (in social and socio-technical environments) when considering that many smart speakers are positioned as female-sounding assistants that habituate users into ordering them around (Søndergaard & Hansen, 2018; Strengers et al., 2020; Strengers & Kennedy, 2021), but this also holds for ableism, e.g., when humanoid robots are designed as able-bodied, and digital colonialism, e.g., when non-native English speakers are less understood by smart speakers since specific "native" accents are prioritized (Wu et al., 2022). As another example, chatbots have difficulties discussing racism in general (Schlesinger et al., 2018) and a negative case was when Microsoft's Tay, a chatbot on Twitter, tweeted racist and homophobic remarks based on how it was trained by Twitter "trolls" (Wolf et al., 2017). Unintended harm is still harm, which means that a long-term research and development phase is ideal to lower the chances of harm caused by conversational agents, as well as refining what "good" can be offered by them according to the perspectives of people who are intended to be end users.

The emergent users from so-called "developing countries" will increasingly make up a large portion of users of conversational agents and other technologies, but Western technologists and developers' assumptions about their needs and desires often can be misguided or misunderstood (Arora, 2019). In India, for instance, users may not have a strong preference for human vs. AI-generated responses in smart speakers, but their combination (human and AI) can be valued (Ahire, 2022). Additionally, the context of use may differ; public-facing agents that are installed in urban areas can be used by many people (compared to only at home), such as a voice-based agent that was shared by many in an Indian slum (Pearson et al., 2019). Additionally, tackling societal issues like gender inequality, like a chatbot for addressing gender norms for Indian adolescence (Agarwal et al., 2021), may need to be specific to how gendered norms have cultural dimensions.

Before conversational agents can be adapted to diverse languages, cultures, and contexts, there are other limitations to keep in mind: Local contextualization (as per above) requires an intersectional mindset, but also conversational agents are not yet adept at handling bilingual or multilingual dialogue (Cihan et al., 2022) and more work needs to be done to account for disabilities or different patterns of expressing oneself, e.g., voice agents for those who stammer (Bleakley et al., 2022) and those who are deaf or hard of hearing (Glasser et al., 2020). While there are ongoing challenges in making conversational agents that are accessible to a wide array of people, we also see many opportunities that conversational agents offer, e.g., conversational agents that assist students with autism, dyslexia, and other conditions to learn more independently (Lister et al., 2020). Hence, what promotes inclusion for some may be an exclusion for others, such as voice- vs. text-based interaction for people who stammer vs. those who do not; what helps then is to explore diverse modalities and diverse user groups for context-sensitive design and deployment.

Lastly, we see the need to mature how privacy practices can be handled by conversational agents; current issues include the lack of (or outdated) privacy policies in many voice-based applications (Edu et al., 2022; Liao et al., 2020b). As many applications get added on top of platforms, such as Spotify connecting with Alexa, cross-platform privacy policies also need to be considered. While people live and collaborate with conversational agents at home, work, and beyond, research should also focus on conversational repair, e.g., agents repeating directions vs. providing alternative options (Ashktorab et al., 2019), or how people will handle mistakes from conversational agents, e.g., an inaccurate scheduling assistant (Kocielnik et al., 2019). Conflicts may arise when people underestimate an agent and overestimate their own knowledge (and vice versa); this can be difficult to mediate situationally, such as unclarity on when an agent should take over compared to humans (Schaffer et al., 2019). Here, norms we practice with other humans, such as blaming people for mistakes, may not always carry over to conversational agents (Lee et al., 2021c), making it difficult to assess how and when to hold human and non-human agents accountable (Lima et al., 2022; Nyholm, 2018).

7.8 THE FUTURE

The path for conversation with intelligent systems is already established. From now on, we expect that conversing with artificially "intelligent" systems will become and remain normalized in how we interact with technology. Therefore, it is crucial to reflect on future challenges, such as privacy and the ethical implications of long-term interactions with conversational agents.

ChatGPT has recently brought to light several discussions that demonstrate the difficulties of fully integrating conversational agents in our daily activities. ChatGPT's capabilities resulted in extensive ongoing discussions on whether generative AI should be listed as authors in academic publications, the limits of plagiarism, and the impact on education (Bowers, 2023; Rudolph et al., 2023; Ueda & Yamada, 2023). In the upcoming years, the growing popularity of home and Internet of Things devices will continue to bring conversational interfaces to people's everyday lives. These agents are expected to be omnipresent, and any device that knows the user is fully synchronized and able to respond to the user's requests (Gentsch, 2019). Combining large language models and omnipresent devices may enable the future development of conversational agents as complex as Samantha, from the movie *Her* (see Section 7.1). How this will affect our daily lives is a question still to be answered. In any case, as with ChatGPT, new privacy and ethical issues will emerge as the technology evolves.

7.9 CONCLUSIONS

This chapter has shed light on the transformative impact of NLP on the interaction between individuals and software systems. By tracing the evolution from Turing's Imitation Game to contemporary intelligent assistants, the chapter has provided a comprehensive understanding of conversational agents within the realms of both technical and social dimensions of HCI. Furthermore, it has explored the various techniques and methodologies employed for the development and evaluation of conversational agents while considering the critical perspective of HCI. By contemplating the implications and envisioning the future of these agents in society, this chapter emphasizes the profound role they will play in shaping HCIs and underscores the need for continued research and exploration in this domain.

NOTES

1. www.rasa.com
2. https://openai.com/blog/chatgpt
3. https://canabravaresort.com.br/
4. https://m.me/quitly.bot

REFERENCES

Abbas, N., Følstad, A., & Bjørkli, C. A. (2023). Chatbots as part of digital government service provision-A user perspective. In *Chatbot Research and Design: 6th International Workshop, CONVERSATIONS 2022, Amsterdam, The Netherlands, November 22-23, 2022*, Revised Selected Papers (pp. 66–82).

Abbas, T., Khan, V.-J., Gadiraju, U., & Markopoulos, P. (2020). Trainbot: A conversational interface to train crowd workers for delivering on-demand therapy. In *Proceedings of the AAAI Conference on Human Computation and Crowdsourcing* (Vol. 8, pp. 3–12).

Abdelghani, R., Oudeyer, P.-Y., Law, E., de Vulpillières, C., & Sauzéon, H. (2022). Conversational agents for fostering curiosity-driven learning in children. *International Journal of Human-Computer Studies, 167*, 102887.

Adamopoulou, E., & Moussiades, L. (2020). Chatbots: History, technology, and applications. *Machine Learning with Applications, 2*, 100006. https://doi.org/10.1016/j.mlwa.2020.100006

Adiwardana, D., Luong, M.-T., So, D. R., Hall, J., Fiedel, N., Thoppilan, R., Yang, Z., Kulshreshtha, A., Nemade, G., Lu, Y., & others. (2020). Towards a human-like open-domain chatbot. *arXiv preprint arXiv:2001.09977*.

Agarwal, D., Agastya, A., Chaudhury, M., Dube, T., Jha, B., Khare, P., & Raghu, N. (2021). *Measuring Effectiveness of Chatbot to Improve Attitudes towards Gender Issues in Underserved Adolescent Children in India*. (Tech. Rep.). Cambridge, MA: Harvard Kennedy School.

Ahire, S. (2022). Designing a smart speaker for emergent users: Human plus AI response. In *Proceedings of the 13th Indian Conference on Human Computer Interaction* (pp. 67–72).

Ahmed, A., Hassan, A., Aziz, S., Abd-Alrazaq, A. A., Ali, N., Alzubaidi, M., Al-Thani, D., Elhusein, B., Siddig, M. A., Ahmed, M., & others. (2023). Chatbot features for anxiety and depression: A scoping review. *Health Informatics Journal, 29*(1), 14604582221146720.

Araujo, T. (2018). Living up to the chatbot hype: The influence of anthropomorphic design cues and communicative agency framing on conversational agent and company perceptions. *Computers in Human Behavior, 85*, 183–189.

Aristotle, T., translated by Irwin. (2019). *Nicomachean Ethics*. Indianapolis, IN: Hackett Publishing.

Arora, P. (2019). *The Next Billion Users: Digital Life beyond the West*. Cambridge, MA: Harvard University Press.

Ashktorab, Z., Jain, M., Liao, Q. V., & Weisz, J. D. (2019). Resilient chatbots: Repair strategy preferences for conversational breakdowns. In *Proceedings of the 2019 CHI Conference on Human Factors in Computing Systems (CHI'19)* (pp. 1–12).

Augustsson, M. (2019). *Talking to Everything: Conversational Interfaces and the Internet of Things in an office environment* (thesis). KTH Royal Institute of Technology, Stockholm, Sweden.

Barbosa, M., Nakamura, W. T., Valle, P., Guerino, G. C., Finger, A. F., Lunardi, G. M., & Silva, W. (2022). UX of chatbots: An exploratory study on acceptance of user experience evaluation methods. In *ICEIS* (No. 2, pp. 355–363).

Barth, F., Candello, H., Cavalin, P., & Pinhanez, C. (2020). Intentions, meanings, and whys: Designing content for voice-based conversational museum guides. In *Proceedings of the 2nd Conference on Conversational User Interfaces* (pp. 1–8).

Bergman, A. S., Abercrombie, G., Spruit, S., Hovy, D., Dinan, E., Boureau, Y.-L., & Rieser, V. (2022). Guiding the release of safer E2E conversational AI through value sensitive design. In *Proceedings of the 23rd Annual Meeting of the Special Interest Group on Discourse and Dialogue* (pp. 39–52). https://aclanthology.org/2022.sigdial-1.4

Bevan, N. (2009). What is the difference between the purpose of usability and user experience evaluation methods. In *Proceedings of the Workshop UXEM* (Vol. 9, No. 1, pp. 1–4).

Bleakley, A., Rough, D., Roper, A., Lindsay, S., Porcheron, M., Lee, M., Nicholson, S. A., Cowan, B. R., & Clark, L. (2022). Exploring smart speaker user experience for people who stammer. In *Proceedings of the 24th International ACM SIGACCESS Conference on Computers and Accessibility* (pp. 1–10).

Blythe, M. (2014). Research through design fiction: Narrative in real and imaginary abstracts. In *Proceedings of the SIGCHI Conference on Human Factors in Computing Systems* (pp. 703–712).

Bordes, A., Boureau, Y.-L., & Weston, J. (2017). Learning end-to-end goal-oriented dialog. In *International Conference on Learning Representations.* https://openreview.net/forum?id=S1Bb3D5gg

Borsci, S., Malizia, A., Schmettow, M., van der Velde, F., Tariverdiyeva, G., Balaji, D., & Chamberlain, A. (2022). The chatbot usability scale: The design and pilot of a usability scale for interaction with AI-based conversational agents. *Personal and Ubiquitous Computing*, 26(1), 95–119. https://doi.org/10.1007/s00779-021-01582-9

Bowers, A. J. (2023). Unpacking the Caveats of ChatGPT in Education: Addressing Bias, Representation, Authorship, and Plagiarism. https://doi.org/10.7916/6q72-hm22

Brandtzaeg, P. B., & Følstad, A. (2017). Why people use chatbots. In *Internet Science: 4th International Conference, INSCI 2017, Thessaloniki, Greece, November 22-24, 2017, Proceedings 4* (pp. 377–392). Springer International Publishing.

Brandtzaeg, P. B., & Følstad, A. (2018). Chatbots: Changing user needs and motivations. *Interactions*, 25(5), 38–43.

Budzianowski, P., Wen, T.-H., Tseng, B.-H., Casanueva, I., Ultes, S., Ramadan, O., & Gasic, M. (2018). MultiWOZ-A large-scale multi-domain wizard-of-oz dataset for task-oriented dialogue modelling. In *Proceedings of the 2018 Conference on Empirical Methods in Natural Language Processing* (pp. 5016–5026).

Calvo, R. A., & Peters, D. (2014). *Positive Computing: Technology for Wellbeing and Human Potential.* Cambridge: MIT Press.

Cambre, J., & Kulkarni, C. (2020). Methods and tools for prototyping voice interfaces. In *Proceedings of the 2nd Conference on Conversational User Interfaces* (pp. 1–4). https://doi.org/10.1145/3405755.3406148

Candello, H., Grave, M., Brazil, E., Ito, M., Alves de Brito Filho, A., & De Paula, R. (2022a). How can AI leverage alternative criteria and suggest a better way to measure credit worthiness and economic growth? In *Proceedings of the 4th Conference on Conversational User Interfaces* (pp. 1–4).

Candello, H., Pinhanez, C., Muller, M., & Wessel, M. (2022b). Unveiling practices of customer service content curators of conversational agents. *Proceedings of the ACM on Human-Computer Interaction*, 6, 1–33. https://doi.org/10.1145/3555768

Carrol, J. M. (1999). Five reasons for scenario-based design. In *Proceedings of the 32nd Annual Hawaii International Conference on Systems Sciences. 1999. Hicss-32. Abstracts and Cd-Rom of Full Papers* (pp. 11–pp).

Cassell, J. (2000). Embodied conversational interface agents. *Communications of the ACM*, 43(4), 70–78.

Catania, F., Spitale, M., & Garzotto, F. (2023). Conversational agents in therapeutic interventions for neurodevelopmental disorders: A survey. *ACM Computing Surveys*, 55(10), 1–34.

Catania, F., Spitale, M., Cosentino, G., & Garzotto, F. (2020). What is the best action for children to" *wake up" and" put to sleep" a conversational agent? A multi-criteria decision analysis approach. In Proceedings of the 2nd Conference on Conversational User Interfaces* (pp. 1–10).

Chaves, A. P., & Gerosa, M. A. (2018). Single or multiple conversational agents? An interactional coherence comparison. In *ACM SIGCHI Conference on Human Factors in Computing Systems* (pp. 1–38).

Chaves, A. P., & Gerosa, M. A. (2020). How should my chatbot interact? A survey on social characteristics in human-chatbot interaction design. *International Journal of Human-Computer Interaction*, 37(8), 1–30. https://doi.org/10.1080/10447318.2020.1841438

Chaves, A. P., Egbert, J., Hocking, T., Doerry, E., & Gerosa, M. A. (2022). Chatbots language design: The influence of language variation on user experience with tourist assistant chatbots. *ACM Transactions on Computer-Human Interaction*, 29(2), 1–38. https://doi.org/10.1145/3487193

Chen, Y., Jensen, S., Albert, L. J., Gupta, S., & Lee, T. (2023). Artificial intelligence (AI) student assistants in the classroom: Designing chatbots to support student success. *Information Systems Frontiers*, 25(1), 161–182.

Chhibber, N., Goh, J., & Law, E. (2022). Teachable conversational agents for crowdwork: Effects on performance and trust. *Proceedings of the ACM on Human-Computer Interaction*, 6, 1–21.

Cihan, H., Wu, Y., Peña, P., Edwards, J., & Cowan, B. (2022). Bilingual by default: Voice Assistants and the role of code-switching in creating a bilingual user experience. In *Proceedings of the 4th Conference on Conversational User Interfaces* (pp. 1–4).

Ciston, S. (2019). Intersectional AI is essential: Polyvocal, multimodal, experimental methods to save artificial intelligence. *Journal of Science and Technology of the Arts, 11*(2), 3–8.

Core, M. G., & Allen, J. (1997). Coding dialogs with the DAMSL annotation scheme. *AAAI Fall Symposium on Communicative Action in Humans and Machines, 56*, 28–35.

Cui, L., Huang, S., Wei, F., Tan, C., Duan, C., & Zhou, M. (2017). SuperAgent: A customer service chatbot for e-commerce websites. In *Proceedings of ACL 2017, System Demonstrations* (pp. 97–102).

Dahlbäck, N., Jönsson, A., & Ahrenberg, L. (1993). Wizard of Oz studies-why and how. *Knowledge-Based Systems, 6*(4), 258–266.

Dale, R. (2016). The return of the chatbots. *Natural Language Engineering, 22*(5), 811–817.

Dale, R. (2020). Voice assistance in 2019. *Natural Language Engineering, 26*(1), 129–136. https://doi.org/10.1017/S1351324919000640

De Maeyer, C., & Lee, M. (2022). I feel you. In *Human-Centered Software Engineering: 9th IFIP WG 13.2 International Working Conference, HCSE 2022*, Eindhoven, The Netherlands, August 24-26, 2022, Proceedings (pp. 23–43).

De Maeyer, C., & Markopoulos, P. (2021). Experts' view on the future outlook on the materialization, expectations and implementation of digital twins in healthcare. *Interacting with Computers, 33*(4), 380–394.

Devlin, J., Chang, M.-W., Lee, K., & Toutanova, K. (2019). Bert: Pre-training of deep bidirectional transformers for language understanding. In *Proceedings of NAACL-HLT* (pp. 4171–4186).

Diederich, S., Brendel, A. B., & Kolbe, L. M. (2019). Towards a taxonomy of platforms for conversational agent design. In *2019 Proceedings of 14th International Conference on Business Informatics, Track 10: Human-Computer Interaction*, Siegen. AIS eLibrary: 1100–1119.

Dinan, E., Abercrombie, G., Bergman, A. S., Spruit, S., Hovy, D., Boureau, Y.-L., & Rieser, V. (2021). Anticipating safety issues in e2e conversational AI: Framework and tooling. *arXiv preprint arXiv:2107.03451*.

Duijst, D. (2017). Can we improve the user experience of chatbots with personalisation [Master's Thesis], University of Amsterdam.

Dušek, O., Novikova, J., & Rieser, V. (2020). Evaluating the state-of-the-art of end-to-end natural language generation: The e2e nlg challenge. *Computer Speech & Language, 59*, 123–156.

Edu, J., Ferrer-Aran, X., Such, J., & Suarez-Tangil, G. (2022). Measuring Alexa skill privacy practices across three years. In *Proceedings of the ACM Web Conference 2022* (pp. 670–680).

Ehsan, U., Wintersberger, P., Liao, Q. V., Watkins, E. A., Manger, C., Daumé III, H., Riener, A., & Riedl, M. O. (2022). Human-centered explainable AI (HCXAI): Beyond opening the black-box of AI. In *CHI Conference on Human Factors in Computing Systems Extended Abstracts* (pp. 1–7). https://doi.org/10.1145/3491101.3503727

Feine, J., Gnewuch, U., Morana, S., & Maedche, A. (2019). A taxonomy of social cues for conversational agents. *International Journal of Human-Computer Studies, 132*, 138–161.

Fiore, S. M., Wiltshire, T. J., Lobato, E. J., Jentsch, F. G., Huang, W. H., & Axelrod, B. (2013). Toward understanding social cues and signals in human-robot interaction: Effects of robot gaze and proxemic behavior. *Frontiers in Psychology, 4*, 859.

Fitzpatrick, K. K., Darcy, A., & Vierhile, M. (2017). Delivering cognitive behavior therapy to young adults with symptoms of depression and anxiety using a fully automated conversational agent (Woebot): A randomized controlled trial. *JMIR Mental Health, 4*(2), e7785.

Floridi, L., Cowls, J., King, T. C., & Taddeo, M. (2020). How to design AI for social good: Seven essential factors. *Science and Engineering Ethics, 26*, 1771–1796.

Følstad, A., & Brandtzæg, P. B. (2017). Chatbots and the new world of HCI. *Interactions, 24*(4), 38–42.

Følstad, A., & Skjuve, M. (2019). Chatbots for customer service: User experience and motivation. In *Proceedings of the 1st International Conference on Conversational User Interfaces* (pp. 1–9).

Følstad, A., Brandtzaeg, P. B., Feltwell, T., Law, E. L.-C., Tscheligi, M., & Luger, E. A. (2018). SIG: Chatbots for social good. In *Extended Abstracts of the 2018 CHI Conference on Human Factors in Computing Systems* (pp. 1–4).

Frauenberger, C., Makhaeva, J., & Spiel, K. (2017). Blending methods: Developing participatory design sessions for autistic children. In *Proceedings of the 2017 Conference on Interaction Design and Children* (pp. 39–49).

Friedman, B. (1996). Value-sensitive design. *Interactions, 3*(6), 16–23.

Friedman, B., & Hendry, D. G. (2019). *Value Sensitive Design: Shaping Technology with Moral Imagination*. Cambridge: MIT Press.

Friedman, B., Kahn, P. H., Borning, A., & Huldtgren, A. (2013). Value sensitive design and information systems. In *Early Engagement and New Technologies: Opening up the Laboratory* (pp. 55–95).

Gabrielli, S., Rizzi, S., Carbone, S., Donisi, V., & others. (2020). A chatbot-based coaching intervention for adolescents to promote life skills: Pilot study. *JMIR Human Factors*, 7(1), e16762.

Garg, R., & Sengupta, S. (2019). "When you can do it, why can't I?": Racial and socioeconomic differences in family technology use and non-use. In *Proceedings of the ACM on Human-Computer Interaction*, 3, 1–22.

Garg, R., & Sengupta, S. (2020). Conversational technologies for in-home learning: Using co-design to understand children's and parents' perspectives. In *Proceedings of the 2020 CHI Conference on Human Factors in Computing Systems* (pp. 1–13).

Garg, R., Cui, H., Seligson, S., Zhang, B., Porcheron, M., Clark, L., Cowan, B. R., & Beneteau, E. (2022). The last decade of HCI research on children and voice-based conversational agents. In *Proceedings of the 2022 CHI Conference on Human Factors in Computing Systems* (pp. 1–19).

Gentsch, P. (2019). Conversational AI: How (chat) bots will reshape the digital experience. In *AI in Marketing, Sales and Service* (pp. 81–125). Cham: Springer.

Chaves, A. P., & Gerosa, M. A. (2021). How should my chatbot interact? A survey on social characteristics in human-chatbot interaction design. *International Journal of Human-Computer Interaction*, 37(8), 729–758.

Glasser, A., Mande, V., & Huenerfauth, M. (2020). Accessibility for deaf and hard of hearing users: Sign language conversational user interfaces. In *Proceedings of the 2nd Conference on Conversational User Interfaces* (pp. 1–3).

Go, E., & Sundar, S. S. (2019). Humanizing Chatbots: The effects of visual, identity and conversational cues on humanness perceptions. *Computers in Human Behavior*, 97, 304–316.

Görnemann, E., & Spiekermann, S. (2022). Emotional responses to human values in technology: The case of conversational agents. In *Human-Computer Interaction* (pp. 1–28). Milton Park: Taylor & Francis online.

Graesser, A. C., Conley, M. W., & Olney, A. (2012). Intelligent tutoring systems. In K. R. Harris, S. Graham, T. Urdan, A. G. Bus, S. Major, & H. L. Swanson (Eds), *APA Educational Psychology Handbook, Vol 3: Application to Learning and Teaching*. (pp. 451–473). Washington, DC: American Psychological Association.

Guerino, G. C., & Valentim, N. M. C. (2020). Usability and user experience evaluation of natural user interfaces: A systematic mapping study. *IET Software*, 14(5), 451–467.

Guerino, G., Silva, W., Coleti, T., & Valentim, N. (2021). Assessing a technology for usability and user experience evaluation of conversational systems: An exploratory study. In *Proceedings of the 23rd International Conference on Enterprise Information Systems* - Volume 2: ICEIS (pp. 463–473). https://doi.org/10.5220/0010450204630473

Hampson, I., & Junor, A. (2005). Invisible work, invisible skills: Interactive customer service as articulation work. *New Technology, Work and Employment*, 20(2), 166–181.

Han, S., & Lee, M. K. (2022). FAQ chatbot and inclusive learning in massive open online courses. *Computers & Education*, 179, 104395.

Haugeland, I. K. F., Følstad, A., Taylor, C., & Bjørkli, C. (2022). Understanding the user experience of customer service chatbots: An experimental study of chatbot interaction design. *International Journal of Human-Computer Studies*, 161, 102788.

Hevner, A. R. (2007). A three cycle view of design science research. *Scandinavian Journal of Information Systems*, 19(2), 4.

Heyselaar, E., & Bosse, T. (2019). Using Theory of Mind to assess users' sense of agency in social chatbots. In *Conversations 2019: 3rd International Workshop on Chatbot Research* (pp. 1–13).

Hill, J., Ford, W. R., & Farreras, I. G. (2015). Real conversations with artificial intelligence: A comparison between human-human online conversations and human-chatbot conversations. *Computers in Human Behavior*, 49, 245–250.

Hochreiter, S., & Schmidhuber, J. (1997). Long short-term memory. *Neural Computation*, 9(8), 1735–1780.

Hu, Y. (2019). Do people want to message chatbots? Developing and comparing the usability of a conversational vs. menu-based chatbot in context of new hire onboarding. [Master's Thesis], Aalto University]. https://aaltodoc.aalto.fi:443/handle/123456789/40834

Huang, M.-H., & Rust, R. T. (2018). Artificial intelligence in service. *Journal of Service Research*, 21(2), 155–172.

Hwang, G.-J., & Chang, C.-Y. (2021). A review of opportunities and challenges of chatbots in education. *Interactive Learning Environments*, 31(7), 4099–4112.

Inkster, B., Sarda, S., Subramanian, V., & others. (2018). An empathy-driven, conversational artificial intelligence agent (Wysa) for digital mental well-being: Real-world data evaluation mixed-methods study. *JMIR MHealth and UHealth, 6*(11), e12106.

ISO 9241-11. (2018). *Ergonomics of Human-System Interaction: Part 11: Usability: Definitions and Concepts*. Geneva: International Organization for Standardization.

Jacobs, N. (2020). Capability sensitive design for health and wellbeing technologies. *Science and Engineering Ethics, 26*(6), 3363–3391.

Jain, M., Kumar, P., Kota, R., & Patel, S. N. (2018). Evaluating and informing the design of chatbots. In *Proceedings of the 2018 Designing Interactive Systems Conference* (pp. 895–906).

Jakic, A., Wagner, M. O., & Meyer, A. (2017). The impact of language style accommodation during social media interactions on brand trust. *Journal of Service Management, 28*(3), 418–441.

Jiang, R., & E Banchs, R. (2017). Towards improving the performance of chat oriented dialogue system. In *2017 International Conference on Asian Language Processing (IALP)* (pp. 23–26).

Jonze, S. (Director). (2014). *Her*. Warner Bros. Picture presents an Annapurna Pictures production. https://search.library.wisc.edu/catalog/9910197336902121

Ju, W., & Leifer, L. (2008). The design of implicit interactions: Making interactive systems less obnoxious. *Design Issues, 24*(3), 72–84.

Jurafsky, D., & Martin, J. H. (2023). *Speech and Language Processing: An Introduction to Natural Language Processing, Computational Linguistics, and Speech Recognition*. https://web.stanford.edu/~jurafsky/slp3/

Kaczorowska-Spychalska, D. (2019). How chatbots influence marketing. *Management, 23*(1), 251–270.

Khadpe, P., Krishna, R., Fei-Fei, L., Hancock, J. T., & Bernstein, M. S. (2020). Conceptual metaphors impact perceptions of human-AI collaboration. In *Proceedings of the ACM on Human-Computer Interaction, 4*, 1–26.

Khemani, K. H., & Reeves, S. (2022). Unpacking practitioners' attitudes towards codifications of design knowledge for voice user interfaces. In *Proceedings of the 2022 CHI Conference on Human Factors in Computing Systems* (pp. 1–10). https://doi.org/10.1145/3491102.3517623

Khosrawi-Rad, B., Rinn, H., Schlimbach, R., Gebbing, P., Yang, X., Lattemann, C., Markgraf, D., & Robra-Bissantz, S. (2022). Conversational agents in education - A systematic literature review. *ECIS 2022 Research Papers*. https://aisel.aisnet.org/ecis2022_rp/18

Kocielnik, R., Amershi, S., & Bennett, P. N. (2019). Will you accept an imperfect AI? Exploring designs for adjusting end-user expectations of ai systems. In *Proceedings of the 2019 CHI Conference on Human Factors in Computing Systems* (pp. 1–14).

Kocielnik, R., Avrahami, D., Marlow, J., Lu, D., & Hsieh, G. (2018). Designing for workplace reflection: A chat and voice-based conversational agent. In *Proceedings of the 2018 Designing Interactive Systems Conference* (pp. 881–894).

Kraus, M., Schiller, M., Behnke, G., Bercher, P., Dorna, M., Dambier, M., Glimm, B., Biundo, S., & Minker, W. (2020). "Was that successful?" On integrating proactive meta-dialogue in a DIY-assistant using multimodal cues. In *Proceedings of the 2020 International Conference on Multimodal Interaction* (pp. 585–594).

Kraus, M., Wagner, N., & Minker, W. (2021). Modelling and predicting trust for developing proactive dialogue strategies in mixed-initiative interaction. In *Proceedings of the 2021 International Conference on Multimodal Interaction* (pp. 131–140).

Kwet, M. (2019). Digital colonialism: US Empire and the new imperialism in the Global South. *Race & Class, 60*(4), 3–26.

Langevin, R., Lordon, R. J., Avrahami, T., Cowan, B. R., Hirsch, T., & Hsieh, G. (2021). Heuristic evaluation of conversational agents. In *Proceedings of the 2021 CHI Conference on Human Factors in Computing Systems* (pp. 1–15). https://doi.org/10.1145/3411764.3445312

Laumer, S., Maier, C., & Gubler, F. T. (2019). Chatbot acceptance in healthcare: Explaining user adoption of conversational agents for disease diagnosis. In *Proceedings of the 27th European Conference on Information Systems (ECIS)* (pp. 1–18).

Lee, H. R., Cheon, E., De Graaf, M., Alves-Oliveira, P., Zaga, C., & Young, J. (2019a). Robots for social good: Exploring critical design for HRI. In *2019 14th ACM/IEEE International Conference on Human-Robot Interaction (HRI)* (pp. 681–682).

Lee, M., & Contreras, J. (2023). Flourishing with moral emotions through conversational agents. In M. Las Heras, M. Grau Grau, & Y. Rofcanin (Eds.), *Human Flourishing: A Multidisciplinary Perspective on Neuroscience, Health, Organizations and Arts* (pp. 163–179). Springer International Publishing. https://doi.org/10.1007/978-3-031-09786-7_11

Lee, M., Ackermans, S., Van As, N., Chang, H., Lucas, E., & IJsselsteijn, W. (2019b). Caring for Vincent: A chatbot for self-compassion. In *Proceedings of the 2019 CHI Conference on Human Factors in Computing Systems* (pp. 1–13).

Lee, M., Frank, L., & IJsselsteijn, W. (2021a). Brokerbot: A cryptocurrency chatbot in the social-technical gap of trust. *Journal of Computer Supported Cooperative Work (JCSCW), 30*(1), 79–117.

Lee, M., Lucas, G., Mell, J., Johnson, E., & Gratch, J. (2019c). What's on your virtual mind?: Mind perception in human-agent negotiations. In *Proceedings of the 19th ACM International Conference on Intelligent Virtual Agents* (pp. 38–45). https://doi.org/10.1145/3308532.3329465

Lee, M., Noortman, R., Zaga, C., Starke, A., Huisman, G., & Andersen, K. (2021b). Conversational futures: Emancipating conversational interactions for futures worth wanting. In *Proceedings of the 2021 CHI Conference on Human Factors in Computing Systems* (pp. 1–13).

Lee, M., Ruijten, P., Frank, L., de Kort, Y., & IJsselsteijn, W. (2021c). People may punish, but not blame robots. In *Proceedings of the 2021 CHI Conference on Human Factors in Computing Systems* (pp. 1–11).

Lei, W., Zhang, Y., Song, F., Liang, H., Mao, J., Lv, J., Yang, Z., & Chua, T.-S. (2022). Interacting with non-cooperative user: A new paradigm for proactive dialogue policy. In *Proceedings of the 45th International ACM SIGIR Conference on Research and Development in Information Retrieval* (pp. 212–222).

Liao, Q. V., Gruen, D., & Miller, S. (2020a). Questioning the AI: Informing design practices for explainable AI user experiences. In *Proceedings of the 2020 CHI Conference on Human Factors in Computing Systems* (pp. 1–15).

Liao, S., Wilson, C., Cheng, L., Hu, H., & Deng, H. (2020b). Measuring the effectiveness of privacy policies for voice assistant applications. In *Annual Computer Security Applications Conference* (pp. 856–869).

Liebrecht, C., Sander, L., & van Hooijdonk, C. (2020). Too Informal? How a chatbot's communication style affects brand attitude and quality of interaction. In *Conversations 2020: 4th International Workshop on Chatbot Research*.

Lima, G., Grgić-Hlača, N., Jeong, J. K., & Cha, M. (2022). The conflict between explainable and accountable decision-making algorithms. In *2022 ACM Conference on Fairness, Accountability, and Transparency* (pp. 2103–2113).

Ling, E. C., Tussyadiah, I., Tuomi, A., Stienmetz, J., & Ioannou, A. (2021). Factors influencing users' adoption and use of conversational agents: A systematic review. *Psychology & Marketing, 38*(7), 1031–1051. https://doi.org/10.1002/mar.21491

Lister, K., Coughlan, T., Iniesto, F., Freear, N., & Devine, P. (2020). Accessible conversational user interfaces: Considerations for design. In *Proceedings of the 17th International Web for All Conference* (pp. 1–11).

Liu, S., Zheng, C., Demasi, O., Sabour, S., Li, Y., Yu, Z., Jiang, Y., & Huang, M. (2021). Towards Emotional Support Dialog Systems. In *Proceedings of the 59th Annual Meeting of the Association for Computational Linguistics and the 11th International Joint Conference on Natural Language Processing (Volume 1: Long Papers)* (pp. 3469–3483).

Llitjós, A. F. (2013). *IBM Design-A New Era at IBM. Lean UX Leading Theway*. https://submissions.agilealliance.org/system/attachments/attachments/000/00 0/306/original/IBM_Design_Thinking_Agile_2013.pdf

Lombard, M., & Xu, K. (2021). Social responses to media technologies in the 21st century: The media are social actors paradigm. *Human-Machine Communication, 2*, 29–55.

Lopatovska, I., Turpin, O., Davis, J., Connell, E., Denney, C., Fournier, H., Ravi, A., Yoon, J. H., & Parasnis, E. (2022). Capturing teens' voice in designing supportive agents. In *Proceedings of the 4th Conference on Conversational User Interfaces* (pp. 1–12).

Lucas, G. M., Gratch, J., King, A., & Morency, L.-P. (2014). It's only a computer: Virtual humans increase willingness to disclose. *Computers in Human Behavior, 37*, 94–100.

Lucas, G. M., Rizzo, A., Gratch, J., Scherer, S., Stratou, G., Boberg, J., & Morency, L.-P. (2017). Reporting mental health symptoms: Breaking down barriers to care with virtual human interviewers. *Frontiers in Robotics and AI, 4*, 51.

Luger, E., & Sellen, A. (2016). Like having a really bad PA: The gulf between user expectation and experience of conversational agents. In *Proceedings of the 2016 CHI Conference on Human Factors in Computing Systems* (pp. 5286–5297).

Madianou, M. (2021). Nonhuman humanitarianism: When "AI for good" can be harmful. *Information, Communication & Society, 24*(6), 850–868.

Mahmood, T., Ricci, F., & Venturini, A. (2009). Improving recommendation effectiveness: Adapting a dialogue strategy in online travel planning. *Information Technology & Tourism, 11*(4), 285–302.

Maslowski, I., Lagarde, D., & Clavel, C. (2017). In-the-wild chatbot corpus: From opinion analysis to interaction problem detection. In *International Conference on Natural Language, Signal and Speech Processing* (pp. 115–120).

Mateas, M. (1999). An Oz-centric review of interactive drama and believable agents. In M. J. Wooldridge & M. Veloso (Eds.), *Artificial Intelligence Today: Recent Trends and Developments* (pp. 297–328). Springer. https://doi.org/10.1007/3-540-48317-9_12

Mauldin, M. L. (1994). Chatterbots, tinymuds, and the turing test entering the loebner prize competition. In *Proceedings of the Twelfth AAAI National Conference on Artificial Intelligence* (pp. 16–21).

McTear, M. (2020). Conversational AI: Dialogue systems, conversational agents, and chatbots. *Synthesis Lectures on Human Language Technologies*, *13*(3), 1–251.

Meck, A.-M., Draxler, C., & Vogt, T. (2022). A question of fidelity: Comparing different user testing methods for evaluating in-car prompts. In *Proceedings of the 4th Conference on Conversational User Interfaces* (pp. 1–5).

Mikolov, T., Chen, K., Corrado, G., & Dean, J. (2013). Efficient estimation of word representations in vector space. In *International Conference on Learning Representations*.

Milne-Ives, M., Cock, C. de, Lim, E., Shehadeh, M. H., Pennington, N. de, Mole, G., Normando, E., & Meinert, E. (2020). The effectiveness of artificial intelligence conversational agents in health care: Systematic review. *Journal of Medical Internet Research*, *22*(10), e20346. https://doi.org/10.2196/20346

Miner, A. S., Laranjo, L., & Kocaballi, A. B. (2020). Chatbots in the fight against the COVID-19 pandemic. *NPJ Digital Medicine*, *3*(1), 65.

Mitchell, M., Wu, S., Zaldivar, A., Barnes, P., Vasserman, L., Hutchinson, B., Spitzer, E., Raji, I. D., & Gebru, T. (2019). Model cards for model reporting. In *Proceedings of the Conference on Fairness, Accountability, and Transparency* (pp. 220–229).

Motger, Q., Franch, X., & Marco, J. (2022). Software-based dialogue systems: Survey, taxonomy, and challenges. *ACM Computing Surveys*, *55*(5), 1–42. https://doi.org/10.1145/3527450

Muller, M., & Liao, Q. V. (2017). Exploring AI ethics and values through participatory design fictions. In *Human Computer Interaction Consortium*.

Murad, C., Tasnim, H., & Munteanu, C. (2022). Voice-first interfaces in a GUI-first design world: Barriers and opportunities to supporting VUI designers on-the-job. In *Proceedings of the 4th Conference on Conversational User Interfaces* (pp. 1–10). https://doi.org/10.1145/3543829.3543842

Nardi, B. A., & ENgeström, Y. (1999). A web on the wind: The structure of invisible work. *Computer Supported Cooperative Work*, *8*(1–2), 1–8. https://doi.org/10.1023/A:1008694621289

Nass, C., Steuer, J., & Tauber, E. R. (1994). Computers are social actors. In *Proceedings of the SIGCHI Conference on Human Factors in Computing Systems* (pp. 72–78).

Nawaz, A. (2012). A comparison of card-sorting analysis methods. In *10th Asia Pacific Conference on Computer Human Interaction (Apchi 2012)*. Matsue-City, Shimane, Japan (pp. 28–31).

Neururer, M., Schlögl, S., Brinkschulte, L., & Groth, A. (2018). Perceptions on authenticity in chat bots. *Multimodal Technologies and Interaction*, *2*(3), 60.

Nguyen, Q. N., Sidorova, A., & Torres, R. (2021). User interactions with chatbot interfaces vs. Menu-based interfaces: An empirical study. *Computers in Human Behavior*, *128*, 107093.

Nothdurft, F., Behnke, G., Bercher, P., Biundo, S., & Minker, W. (2015). The interplay of user-centered dialog systems and AI planning. In *Proceedings of the 16th Annual Meeting of the Special Interest Group on Discourse and Dialogue* (pp. 344–353).

Nyholm, S. (2018). Attributing agency to automated systems: Reflections on human-robot collaborations and responsibility-loci. *Science and Engineering Ethics*, *24*(4), 1201–1219.

Pearson, J., Robinson, S., Reitmaier, T., Jones, M., Ahire, S., Joshi, A., Sahoo, D., Maravi, N., & Bhikne, B. (2019). StreetWise: Smart speakers vs human help in public slum settings. In *Proceedings of the 2019 CHI Conference on Human Factors in Computing Systems* (pp. 1–13).

Pelachaud, C. (2017). Greta: A conversing socio-emotional agent. In *Proceedings of the 1st ACM SIGCHI International Workshop on Investigating Social Interactions with Artificial Agents* (pp. 9–10).

Perski, O., Crane, D., Beard, E., & Brown, J. (2019). Does the addition of a supportive chatbot promote user engagement with a smoking cessation app? An experimental study. *Digital Health*, *5*, 2055207619880676.

Pillai, R., & Sivathanu, B. (2020). Adoption of AI-based chatbots for hospitality and tourism. *International Journal of Contemporary Hospitality Management*, *32*(10), 3199–3226.

Polaine, A., Løvlie, L., & Reason, B. (2013). *Service Design: From Insight to Implementation*. New York: Rosenfeld Media.

Porcheron, M., Fischer, J. E., Reeves, S., & Sharples, S. (2018). Voice interfaces in everyday life. In *Proceedings of the 2018 CHI Conference on Human Factors in Computing Systems* (pp. 1–12).

Poria, S., Cambria, E., Bajpai, R., & Hussain, A. (2017). A review of affective computing: From unimodal analysis to multimodal fusion. *Information Fusion*, *37*, 98–125.

Portela, M., & Granell-Canut, C. (2017). A new friend in our smartphone? Observing Interactions with chatbots in the search of emotional engagement. In *Proceedings of the XVIII International Conference on Human Computer Interaction* (pp. 1–7).

Preum, S. M., Munir, S., Ma, M., Yasar, M. S., Stone, D. J., Williams, R., Alemzadeh, H., & Stankovic, J. A. (2021). A review of cognitive assistants for healthcare: Trends, prospects, and future directions. *ACM Computing Surveys (CSUR)*, *53*(6), 1–37.

Rakova, B., Yang, J., Cramer, H., & Chowdhury, R. (2021). Where responsible AI meets reality: Practitioner perspectives on enablers for shifting organizational practices. *Proceedings of the ACM on Human-Computer Interaction*, *5*, 1–23. https://doi.org/10.1145/3449081

Rashita, R., Himanshu, J., & Vineet, K. (2021). *Conversational AI Market Size, Share & Growth | Trends-2030* (No. A13682; SE: Emerging and Next Generation Technologies, p. 288). Allied Market Research. https://www.alliedmarketresearch.com/conversational-ai-market-A13682

Rattenbury, T., Hellerstein, J. M., Heer, J., Kandel, S., & Carreras, C. (2017). *Principles of Data Wrangling: Practical Techniques for Data Preparation*. Sebastopol, CA: O'Reilly Media, Inc.

Reeves, B., & Nass, C. (1996). *How People Treat Computers, Television, and New Media Like Real People and Places*. CSLI Publications and Cambridge University Press.

Ren, R., Castro, J. W., Acuña, S. T., & de Lara, J. (2019). Usability of chatbots: A systematic mapping study. In *Proc. 31st Int. Conf. Software Engineering and Knowledge Engineering* (pp. 479–484).

Rick, S. R., Goldberg, A. P., & Weibel, N. (2019). SleepBot: Encouraging sleep hygiene using an intelligent chatbot. In *Proceedings of the 24th International Conference on Intelligent User Interfaces: Companion* (pp. 107–108).

Rieser, V., Lemon, O., & Keizer, S. (2014). Natural language generation as incremental planning under uncertainty: Adaptive information presentation for statistical dialogue systems. *IEEE/ACM Transactions on Audio, Speech, and Language Processing*, *22*(5), 979–994.

Ringfort-Felner, R., Laschke, M., Sadeghian, S., & Hassenzahl, M. (2022). Kiro: A design fiction to explore social conversation with voice assistants. In *Proceedings of the ACM on Human-Computer Interaction*, *6*(GROUP), 1–21.

Robeyns, I. (2009). Capability approach. In *Handbook of Economics and Ethics*. Stanford, CA: Edward Elgar Publishing.

Roller, S., Dinan, E., Goyal, N., Ju, D., Williamson, M., Liu, Y., Xu, J., Ott, M., Smith, E. M., Boureau, Y.-L., & others. (2021). Recipes for building an open-domain chatbot. In *Proceedings of the 16th Conference of the European Chapter of the Association for Computational Linguistics: Main* Volume (pp. 300–325).

Rudolph, J., Tan, S., & Tan, S. (2023). ChatGPT: Bullshit spewer or the end of traditional assessments in higher education? *Journal of Applied Learning and Teaching*, *6*(1), 342–363.

Samson, B. P. V., & Sumi, Y. (2020). Are two heads better than one? Exploring two-party conversations for car navigation voice guidance. In *Extended Abstracts of the 2020 CHI Conference on Human Factors in Computing Systems* (pp. 1–9). https://doi.org/10.1145/3334480.3382818

Schaffer, J., Playa Vista, C., O'Donovan, J., Michaelis, J., Adelphi, M., Raglin, A., & Höllerer, T. (2019). I can do better than your AI: expertise and explanations. In *Proceedings of the 2019 ACM International Conference on Intelligent User Interfaces (IUI'19)*.

Schegloff, E. A., Jefferson, G., & Sacks, H. (1977). The preference for self-correction in the organization of repair in conversation. *Language*, *53*(2), 361–382.

Schlesinger, A., O'Hara, K. P., & Taylor, A. S. (2018). Let's talk about race: Identity, chatbots, and AI. In *Proceedings of the 2018 CHI Conference on Human Factors in Computing Systems* (pp. 1–14).

Schmidt, K., & Schmidt, K. (2011). Remarks on the complexity of cooperative work (2002). In *Cooperative Work and Coordinative Practices: Contributions to the Conceptual Foundations of Computer-Supported Cooperative Work (CSCW)* (pp. 167–199).

Schneider, J., & Stickdorn, M. (2011). *This is Service Design Thinking: Basics, Tools, Cases*. Hoboken, NJ: Wiley.

Searle, J. R. (1969). *Speech Acts: An Essay in the Philosophy of Language*. Cambridge: Cambridge University Press.

Seering, J., Luria, M., Ye, C., Kaufman, G., & Hammer, J. (2020). It takes a village: Integrating an adaptive chatbot into an online gaming community. In *Proceedings of the 2020 Chi Conference on Human Factors in Computing Systems* (pp. 1–13).

Seidelin, C., Dittrich, Y., & Grönvall, E. (2020). Foregrounding data in co-design - An exploration of how data may become an object of design. *International Journal of Human-Computer Studies*, *143*, 102505. https://doi.org/10.1016/j.ijhcs.2020.102505

Sharma, S., Brennan, J., & Nurse, J. (2021). StockBabble: A conversational financial agent to support stock market investors. In *Proceedings of the 3rd Conference on Conversational User Interfaces* (pp. 1–5).

Shechtman, N., & Horowitz, L. M. (2003). Media inequality in conversation: How people behave differently when interacting with computers and people. In *Proceedings of the SIGCHI Conference on Human Factors in Computing Systems* (pp. 281–288).

Shorter, M., Minder, B., Rogers, J., Baldauf, M., Todisco, A., Junginger, S., Aytaç, A., & Wolf, P. (2022). Materialising the immaterial: *Provotyping to explore voice assistant complexities.* In *Designing Interactive Systems Conference* (pp. 1512–1524). https://doi.org/10.1145/3532106.3533519

Simonsen, L., Steinstø, T., Verne, G., & Bratteteig, T. (2020). "I'm disabled and married to a foreign single mother". Public service chatbot's advice on citizens' complex lives. In *Electronic Participation: 12th IFIP WG 8.5 International Conference, EPart 2020*, Linköping, Sweden, August 31-September 2, 2020, *Proceedings 12* (pp. 133–146).

Simpson, J., Stening, H., Nalepka, P., Dras, M., Reichle, E. D., Hosking, S., Best, C. J., Richards, D., & Richardson, M. J. (2022). DesertWoZ: A Wizard of Oz environment to support the design of collaborative conversational agents. In *Companion Publication of the 2022 Conference on Computer Supported Cooperative Work and Social Computing* (pp. 188–192).

Søndergaard, M. L. J., & Hansen, L. K. (2018). Intimate futures: Staying with the trouble of digital personal assistants through design fiction. In *Proceedings of the 2018 Designing Interactive Systems Conference* (pp. 869–880).

Spiel, K., Gerling, K., Bennett, C. L., Brulé, E., Williams, R. M., Rode, J., & Mankoff, J. (2020). Nothing about us without us: Investigating the role of critical disability studies in HCI. In *Extended Abstracts of the 2020 CHI Conference on Human Factors in Computing Systems* (pp. 1–8).

Stoilova, E. (2021). AI chatbots as a customer service and support tool. *ROBONOMICS: The Journal of the Automated Economy, 2*, 21–21.

Storey, M.-A., & Zagalsky, A. (2016). Disrupting developer productivity one bot at a time. In *Proceedings of the 2016 24th ACM SIGSOFT International Symposium on Foundations of Software Engineering* (pp. 928–931).

Strengers, Y., & Kennedy, J. (2021). *The Smart Wife: Why Siri, Alexa, and Other Smart Home Devices Need A Feminist Reboot.* Cambridge: MIT Press.

Strengers, Y., Qu, L., Xu, Q., & Knibbe, J. (2020). Adhering, steering, and queering: Treatment of gender in natural language generation. In *Proceedings of the 2020 CHI Conference on Human Factors in Computing Systems* (pp. 1–14).

Suchman, L., & Suchman, L. A. (2007). *Human-Machine Reconfigurations: Plans and Situated Actions.* Cambridge: Cambridge University Press.

Tanaka, F., & Matsuzoe, S. (2012). Children teach a care-receiving robot to promote their learning: Field experiments in a classroom for vocabulary learning. *Journal of Human-Robot Interaction, 1*(1), 78–95.

Thomaz, F., Salge, C., Karahanna, E., & Hulland, J. (2020). Learning from the Dark Web: Leveraging conversational agents in the era of hyper-privacy to enhance marketing. *Journal of the Academy of Marketing Science, 48*(1), 43–63.

Traum, D. R., & Hinkelman, E. A. (1992). Conversation acts in task-oriented spoken dialogue. *Computational Intelligence, 8*(3), 575–599.

Turing, A. M. (1950). Computing machinery and intelligence. *Mind, 59*(236), 433–460.

Tzirakis, P., Trigeorgis, G., Nicolaou, M. A., Schuller, B. W., & Zafeiriou, S. (2017). End-to-end multimodal emotion recognition using deep neural networks. *IEEE Journal of Selected Topics in Signal Processing, 11*(8), 1301–1309.

Ueda, K., & Yamada, Y. (2023). ChatGPT is not an author, but then, who is eligible for authorship? PsyArXiv (preprint). https://doi.org/10.31234/osf.io/h5aj3

Ukpabi, D. C., Aslam, B., & Karjaluoto, H. (2019). Chatbot adoption in tourism services: A conceptual exploration. In S. Ivanov & C. Webster (Eds.), *Robots, Artificial Intelligence, and Service Automation in Travel, Tourism and Hospitality* (pp. 105–121). Emerald Publishing Limited. https://doi.org/10.1108/978-1-78756-687-320191006

Ultes, S., Kraus, M., Schmitt, A., & Minker, W. (2015). Quality-adaptive spoken dialogue initiative selection and implications on reward modelling. In *Proceedings of the 16th Annual Meeting of the Special Interest Group on Discourse and Dialogue* (pp. 374–383).

Valério, F. A., Guimarães, T. G., Prates, R. O., & Candello, H. (2020). Comparing users' perception of different chatbot interaction paradigms: A case study. In *Proceedings of the 19th Brazilian Symposium on Human Factors in Computing Systems* (pp. 1–10).

van der Goot, M. J., Hafkamp, L., & Dankfort, Z. (2020). Customer service chatbots: A qualitative interview study into the communication journey of customers. In *Chatbot Research and Design: 4th International Workshop, CONVERSATIONS 2020, Virtual Event, November 23-24, 2020, Revised Selected Papers* (Vol. 4, pp. 190–204).

van Dis, E. A. M., Bollen, J., Zuidema, W., van Rooij, R., & Bockting, C. L. (2023). ChatGPT: Five priorities for research. *Nature, 614*(7947), 224–226. https://doi.org/10.1038/d41586-023-00288-7

van Hooijdonk, C., Martijn, G., & Liebrecht, C. (2023). A framework and content analysis of social cues in the introductions of customer service chatbots. In *Chatbot Research and Design: 6th International Workshop, CONVERSATIONS 2022, Amsterdam, The Netherlands, November 22-23, 2022, Revised Selected Papers* (pp. 118–133).

Vapnik, V. (2000). SVM method of estimating density, conditional probability, and conditional density. In *2000 IEEE International Symposium on Circuits and Systems (ISCAS), 2* (pp. 749–752).

Vaswani, A., Shazeer, N., Parmar, N., Uszkoreit, J., Jones, L., Gomez, A. N., Kaiser, Lukasz, & Polosukhin, I. (2017). Attention is all you need. In *Advances in Neural Information Processing Systems*, 30.

Venkatesh, V., Thong, J. Y., Chan, F. K., Hu, P. J.-H., & Brown, S. A. (2011). Extending the two-stage information systems continuance model: Incorporating UTAUT predictors and the role of context. *Information Systems Journal, 21*(6), 527–555.

Wallace, R. S. (2009). The anatomy of A.L.I.C.E. In R. Epstein, G. Roberts, & G. Beber, (Eds.), *Parsing the Turing Test* (pp. 181–210). Dordrecht: Springer.

Wallis, P., & Norling, E. (2005). The Trouble with Chatbots: Social skills in a social world. In *Proceedings of the Joint Symposium on Virtual Social Agents* (pp. 29–38).

Wambsganss, T., Höch, A., Zierau, N., & Söllner, M. (2021). Ethical design of conversational agents: Towards principles for a value-sensitive design. In F. Ahlemann, R. Schütte, S. Stieglitz (Eds), *Innovation through Information Systems: Volume I: A Collection of Latest Research on Domain Issues* (pp. 539–557). Cham: Springer.

Weizenbaum, J. (1966). ELIZA-a computer program for the study of natural language communication between man and machine. *Communications of the ACM, 9*(1), 36–45.

Wessel, M., de Souza, B. M., Steinmacher, I., Wiese, I. S., Polato, I., Chaves, A. P., & Gerosa, M. A. (2018). The power of bots: Characterizing and understanding bots in OSS projects. *Proceedings of the ACM on Human-Computer Interaction, 2*(CSCW), 1–19. https://doi.org/10.1145/3274451

Westerman, D., Cross, A. C., & Lindmark, P. G. (2019). I believe in a thing called bot: Perceptions of the humanness of "chatbots." *Communication Studies, 70*(3), 295–312.

Williams, J. D., Raux, A., & Henderson, M. (2016). The dialog state tracking challenge series: A review. *Dialogue & Discourse, 7*(3), 4–33.

Wiltshire, T. J., Snow, S. L., Lobato, E. J., & Fiore, S. M. (2014). Leveraging social judgment theory to examine the relationship between social cues and signals in human-robot interactions. *Proceedings of the Human Factors and Ergonomics Society Annual Meeting, 58*(1), 1336–1340.

Winkler, R., Hobert, S., Salovaara, A., Söllner, M., & Leimeister, J. M. (2020). Sara, the lecturer: Improving learning in online education with a scaffolding-based conversational agent. In *Proceedings of the 2020 CHI Conference on Human Factors in Computing Systems* (pp. 1–14).

Winograd, T. (1971). Procedures as a representation for data in a computer program for understanding natural language. Unclassified Technical Report, Document number AD0721399, Defense Logistic Agency, PhD Thesis, Massachusetts Institute of Technology, Cambridge.

Wolf, M. J., Miller, K., & Grodzinsky, F. S. (2017). Why we should have seen that coming: Comments on Microsoft's tay" experiment," and wider implications. *ACM Sigcas Computers and Society, 47*(3), 54–64.

Wollny, S., Schneider, J., Di Mitri, D., Weidlich, J., Rittberger, M., & Drachsler, H. (2021). Are we there yet?-A systematic literature review on chatbots in education. *Frontiers in Artificial Intelligence, 4*, 654924.

Wu, Y., Porcheron, M., Doyle, P., Edwards, J., Rough, D., Cooney, O., Bleakley, A., Clark, L., & Cowan, B. (2022). Comparing command construction in native and non-native speaker IPA interaction through conversation analysis. In *Proceedings of the 4th Conference on Conversational User Interfaces* (pp. 1–12).

Wube, H. D., Esubalew, S. Z., Weldesellasie, F. F., & Debelee, T. G. (2022). Text-based chatbot in financial sector: A systematic literature review. *Data Science in Finance and Economics, 2*(3), 232–259.

Xuetao, M., Bouchet, F., & Sansonnet, J.-P. (2009). Impact of agent's answers variability on its believability and human-likeness and consequent chatbot improvements. In *Proc. of AISB* (pp. 31–36).

Yanishevskaya, N., Kuznetsova, L., Lokhacheva, K., Zabrodina, L., Parfenov, D., & Bolodurina, I. (2019). Application of intelligent algorithms for the development of a virtual automated planning assistant for the optimal tourist travel route. In *International Conference of Artificial Intelligence, Medical Engineering, Education* (pp. 13–22).

Youn, S., & Jin, S. V. (2021). In AI we trust?" The effects of parasocial interaction and technopian versus luddite ideological views on chatbot-based customer relationship management in the emerging "feeling economy. *Computers in Human Behavior, 119*, 106721.

Young, S., Gašić, M., Keizer, S., Mairesse, F., Schatzmann, J., Thomson, B., & Yu, K. (2010). The hidden information state model: A practical framework for POMDP-based spoken dialogue management. *Computer Speech & Language, 24*(2), 150–174.

Young, S., Gašić, M., Thomson, B., & Williams, J. D. (2013). Pomdp-based statistical spoken dialog systems: A review. *Proceedings of the IEEE, 101*(5), 1160–1179.

Yu, D., & Deng, L. (2016). *Automatic Speech Recognition* (Vol. 1). Cham: Springer.

Zaga, C., Lohse, M., Truong, K. P., & Evers, V. (2015). The effect of a robot's social character on children's task engagement: Peer versus tutor. In *Social Robotics: 7th International Conference, ICSR 2015*, Paris, France, October 26-30, 2015, *Proceedings 7*, (pp. 704–713).

Zembylas, M. (2023). A decolonial approach to AI in higher education teaching and learning: Strategies for undoing the ethics of digital neocolonialism. *Learning, Media and Technology, 48*(1), 25–37.

Ziegler, D. M., Stiennon, N., Wu, J., Brown, T. B., Radford, A., Amodei, D., Christiano, P., & Irving, G. (2019). Fine-tuning language models from human preferences. *arxiv preprint arxiv:1909.08593*.

Zue, V. W., & Glass, J. R. (2000). Conversational interfaces: Advances and challenges. *Proceedings of the IEEE, 88*(8), 1166–1180.

8 Artificial Intelligence for Customer Services

Christian Matt and Helena Weith

8.1 INTRODUCTION

The rapid increase in computing power, availability of big data, and improvements regarding the efficiency and proficiency of algorithms have enabled the rise of artificial intelligence (AI) over recent years (Wirtz et al., 2018). AI supports data analysis and sense-making from large and complex data enabling improvements to or even the complete automation of human tasks (Følstad & Skjuve, 2019). As such, the rise of AI enables the disruption and improvement of established services and procedures across various industries and functions, including radical changes to production processes, hiring decisions, and medical screenings (Poser et al., 2022).

Customer services are core functions of most firms and substantially benefit from AI. They refer to intangible services (e.g., consultation, decision-making) from businesses to existing as well as potential future customers, which generally includes business-to-business (B2B) as well as business-to-customer (B2C) services (Poser et al., 2022). They can span from assisting customers throughout search processes, in narrowing down needs, decision-making, or after-sales activities, with the overall goal of supporting customers regarding their needs, simplifying actions and processes for them, and ultimately achieving customer satisfaction (Paluch & Wirtz, 2020; Følstad & Skjuve, 2019). Although important for companies, customer services typically also involve large efforts in terms of cost and time, requiring suitable management approaches (Følstad & Skjuve, 2019).

Recently, AI has enabled the emergence of various AI-based systems, referring to any technical application using AI or made possible by AI. These AI-based systems differ regarding their underlying AI technologies, their application along the customer journey, and the value they could add to firms and customers. These AI-based systems can broadly be distinguished into two categories: First, AI-based systems that operate in the background, for example, analyzing customer data to provide personalized product recommendations. Second, AI-based systems that directly face the customer, such as chatbots. When unfolding their potential, these AI-based systems provide several advantages for businesses. They enable the ubiquitous provision of customer services due to activities and processes independent of working hours (Poser et al., 2022). Increasing degrees of automation and accuracy allow the exploitation of vast amounts of data, realizing higher standardization and thus increasing efficiency (Zhang et al., 2021). Automation and efficiency allow for economies of scale and can save costs (Følstad & Skjuve, 2019). Ultimately these advantages can positively affect customers' satisfaction and experiences and, subsequently, the reputations of businesses (Paluch & Wirtz, 2020). However, despite technological advances, for instance, in generative AI (e.g., ChatGPT, DALL-E 2), the usage of AI-based systems is still mostly restricted to simple and repetitive tasks and requires complex implementation efforts (Dwivedi et al., 2023). Further, AI-based systems used in customer services are often criticized for providing unfair or discriminating outcomes resulting from their underlying technologies (Kordzadeh & Ghasemaghaei, 2021). Such risks can result from the challenges related to the availability, processing, and sense-making of vast amounts of data and the resulting lack of transparency of AI-based systems, which can have severe legal, reputational, or economic consequences (Dolata et al., 2021).

DOI: 10.1201/9781003490685-8

Given the fast-progressing diffusion of AI for customer service, this article aims to provide an overview of recent technological developments and applications in the form of AI-based systems. Based on three core AI-based systems (recommender systems, chatbots, and virtual assistants), we highlight AI's economic and technological characteristics in customer services and their respective advantages and risks. We hereby provide insights for research by leveraging the concept of the customer journey as a base for the application of AI and outlining selected services along the customer journey. Additionally, we critically reflect on the advances of AI-based systems by outlining major risk factors and mapping them with regulatory frameworks and potential countermeasures. Practitioners will also benefit from the outlined managerial measures for successfully implementing and managing AI-based systems in customer services. We further outline actionable measures to avoid negative effects in the early development stages.

This chapter is structured into six sections. In Section 8.2, we outline the technical building blocks of AI and introduce the digital transformation as an enabler for the application of AI along the customer journey. In Section 8.3, we take a holistic perspective and highlight overall advantages and examples of AI-based systems along the customer journey, also providing managerial insights regarding their implementation and management. Section 8.4 presents three selected AI-based systems (recommender systems, chatbots, and virtual assistants) in detail, focusing on their technical functioning, their applications for customer services, and associated benefits. Section 8.5 critically reflects on using AI-based systems by outlining eight predominant risks, their potential consequences, and their countermeasures. Section 8.6 presents a short conclusion and outlook regarding the future of AI-based systems in customer service.

8.2 CONCEPTUAL FOUNDATIONS

8.2.1 Technical Foundations of Artificial Intelligence

While there is no unified definition of AI, it can be described as "the ability of a machine to perform cognitive functions that we associate with human minds, such as perceiving, reasoning, learning, interacting with the environment, problem-solving, decision-making, and even demonstrating creativity" (Rai et al., 2019). The roots of AI date back to the 1950s when Alan Turing introduced the Turing test to identify if a machine can demonstrate intelligent behavior like humans (Turing, 1950). Several notable advances with regards to AI were achieved over the upcoming decades, such as the first simulation of cognitive capabilities through a computer by the first chatbot ELIZA in 1966 (Weizenbaum, 1966) or the Deep Blue system playing chess and winning over the chess world champion Garry Kasparov in 1997 (Strong, 2022). Until the 1980s, the prevalent paradigm was symbolic AI, also known as good old-fashioned artificial intelligence (GOFAI). Symbolic AI uses processes to develop logical conclusions that are understandable and explainable by users. It represents top-down rule-based approaches best applied to static problems (Ilkou & Koutraki, 2020). A typical example of symbolic AI is expert systems, also called knowledge-based systems, which provide advice or solve problems by reasoning based on encoded knowledge (Benderskaya & Zhukova, 2013). From the 1990s onwards, the increasing availability of data, improved computing power, and advances in algorithms enabled the wide diffusion of sub-symbolic AI and facilitated an increasing number of commercial applications across various industries and functionalities (Poser et al., 2022; Ilkou & Koutraki, 2020). Sub-symbolic AI refers to bottom-up approaches by identifying complex correlations between input and output variables. Sub-symbolic mechanisms can hereby leverage structured and unstructured data, resulting from various sources not following a pre-defined format (Dwivedi et al., 2023). It refers especially to two core AI technologies, machine learning (ML) and deep learning (DL), while DL is a subset of ML. DL leverages artificial neural networks, which are inspired by the structures of the human brain (Ilkou & Koutraki, 2020). Based on the repetitive analysis and prediction processes, along with previous decision-making and problem-solving, sub-symbolic AI can continuously improve itself over time (Heaven, 2021). Figure 8.1 provides an overview of these core AI approaches.

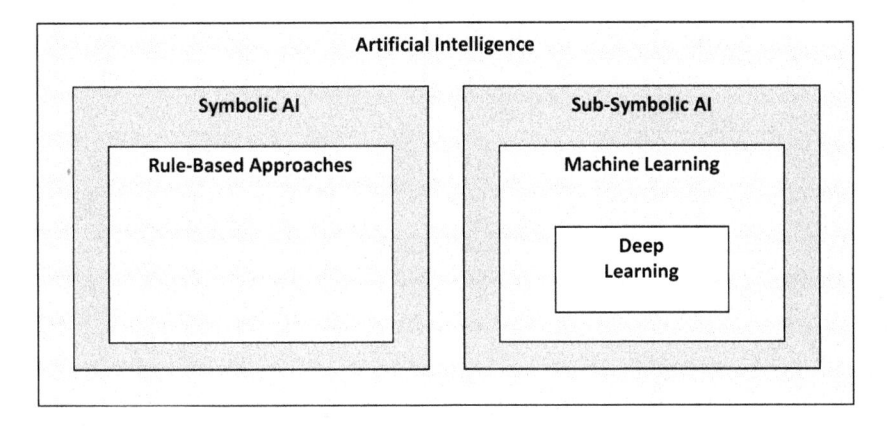

FIGURE 8.1 Differentiation of AI approaches.

Natural Language Processing (NLP) refers to the processing and sense-making of natural human language. Traditionally, in the era of symbolic AI, NLP is allowed to "respond to queries raised by the user, while mapping it to the best possible response sets available in the system" (Dwivedi et al., 2023) by leveraging DL algorithms, NLP has developed further. Besides making sense of complex and natural human text or speech, NLP also enables the generation of natural language, thus providing human-like responses to customers (Rai et al., 2019). One of the latest developments within the field of AI is generative AI, which is forecasted to grow from 7.9 billion USD in 2022 to 110.9 billion USD in 2030 (FinTech Global, 2023). Compared to recognizing patterns and making predictions based on existing data, generative AI creates new, unique, and meaningful outputs in either text, images, or audio form (Feuerriegel et al., 2023). ChatGPT or Bing are examples of chatbots trained to respond to user prompts that provide meaningful answers not limited to a specific field. DALL-E 3 or Midjourney are generative AI applications demonstrating creativity in developing unique images based on textual user prompts. Amper Music enables original music's composition and production for various purposes such as entertainment, films, or other commercial purposes (Katatikarn, 2023). Such generative AI applications also create various opportunities for AI-based systems for customer services, for example by providing customized travel itineraries or medical treatment plans (Section 8.4).

8.2.2 Digital Transformation of the Customer Journey as a Basis

Customers undergo several interactions with businesses when purchasing or consuming products or services. These interactions are structured along the customer journey, defined as "a process or sequence that a customer goes through to access or use an offering of a company" (Følstad & Kvale, 2018). While the customer journey may vary for different customer services, it can be classified into three overarching phases: pre-purchase, purchase, and post-purchase (Tueanrat et al., 2021). The pre-purchasing phase refers to all actions before potential purchases or consumptions by customers. Customers search for information, evaluate alternatives, and exchange with their environment for feedback. The pre-purchase phase is of considerable relevance as it leads to purchase or consumption decisions by new or existing customers (Yoo, 2016). Throughout the purchase phase, customers interact with businesses to purchase or consume a selected product (Verhoef, 2021). Lastly, the post-purchase phase refers to all the activities after the purchase of a product by customers, among others, including any after-sales activities or the sharing of any experiences by customers with their network (Barwitz & Maas, 2018). Across all three phases, the interactions between businesses and customers are enabled via various channels, i.e., mediums allowing for the interaction between customers and businesses, such as radio, newspaper, or an e-commerce shop. One channel, in turn, can allow for several touchpoints, which are moments of interaction between customers with businesses

	Pre-Purchase	Purchase	Post-Purchase
Non-Digital Channels	Physical Stores		
	Print		
	Word-of-Mouth		Word-of-Mouth
	Call Centre		Call Centre
Digital Channels	Online Shop		
	Social Media		
	E-Mail Newsletter		E-Mail Newsletter
	Search Engines		
	Advertising Banner		
	Apps		

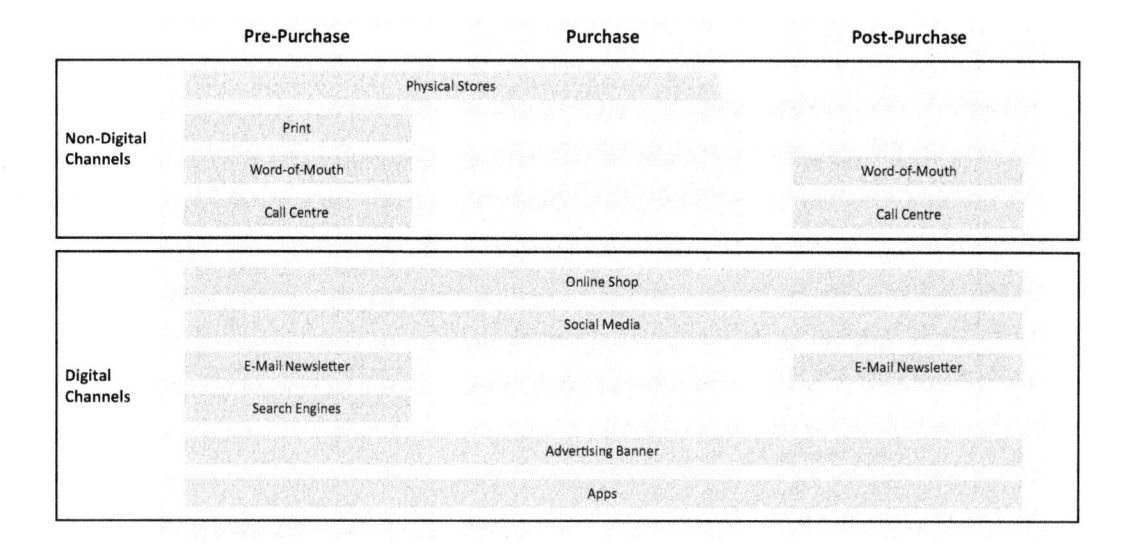

FIGURE 8.2 Overview of selected channels and applications along the customer journey.

(Følstad & Kvale, 2018). Customers can, for example, have two touchpoints with businesses via only one channel.

Digital transformation, i.e., "the changes that digital technologies can bring about in a company's business model, products, processes, and organizational structures" (Hess et al., 2016), can cause three major disruptions along the customer journey: increasing number of customer touchpoints, increasing connectivity across touchpoints, and increasing amount of data generated. First, while the customer journey was limited to a few non-digital channels and touchpoints in the past, digital transformation increases the number of channels and other touchpoints of customers with businesses (Tueanrat et al., 2021). Classical physical channels include physical stores, printed advertising magazines, and call centers. Due to the digital transformation, non-digital channels are enhanced by digital ones, such as social media, e-mail newsletters, and search engines. Figure 8.2 provides an overview of a selection of non-digital and digital channels and their most common application along the three phases of the customer journey.

Second, digital channels enable an increasingly connected and interactive customer journey. Many digital channels are not limited to only one phase of the customer journey but enable touchpoints along several phases (Tueanrat et al., 2021; Barwitz & Maas, 2018). Social media provides a good example of a digital channel enabling various touchpoints for customers with businesses. Throughout the pre-purchase phase, customers might research reviews online and seek opinions from their communities (Tueanrat et al., 2021). The purchase might be shared live with their networks, followed by the post-purchase phase, where customers share their experiences online (Tueanrat et al., 2021). Furthermore, due to the digital transformation, tracking and analyzing customers across different channels is possible. Thus, it is possible to realize omni-channel strategies and customer experiences across and not limited to one channel (Verhoef, 2021).

Third, the increasing number of channels, accompanied by the uptake of connectivity and interaction, results in extensive data generated, collected, and stored (Jarek & Mazurek, 2019). Such data can serve businesses as a basis to provide customer services, such as recommending products or helping them to improve existing processes, products, and services through a better understanding of customer demands (Schweidel et al., 2022). Additionally, businesses can benefit from selling such data to third parties (Baecker et al., 2020).

These three areas of disruption cause changing customer demands, impact established services, and thus allow for and require innovative customer services affecting the interaction between businesses and customers along the customer journey (Tueanrat et al., 2021). Digital transformation provides the foundation, such as connectivity and data, for AI to unfold its full potential along the customer journey. Whereas digital transformation initiatives already strongly focus on customer journeys (Wiesböck et al., 2017), the progress in AI intensifies these effects. AI enables new channels in the form of AI-based systems (Section 8.3.2) and strengthens connectivity and interactivity by combining several complex systems and considering contextual factors simultaneously. Besides enhancing the amount of data, AI facilitates data processing and sense-making (Wirtz et al., 2018).

8.3 ARTIFICIAL INTELLIGENCE FOR CUSTOMER SERVICES

8.3.1 Overarching Advantages of Artificial Intelligence for Customer Services

The overarching advantages for customers and businesses of AI for customer services can be structured into four key areas (Table 8.1): service efficiency, service provision and quality, which further impact a company's profitability and reputation (Enholm et al., 2022; Gartner, 2022). First, AI allows for service efficiencies by optimizing processes such as automating repetitive tasks or improving decision-making due to automated analysis and associated recommendations (Zhang et al., 2021). The ability to process complex and extensive amounts of data enables more accurate analysis and evaluations (Wirtz et al., 2018), resulting in profound and reliable insights. Consequently, bad decision-making due to lacking information and further resulting risks can be reduced (Berente et al., 2021).

Second, AI enables increasing customer service provision and quality through advanced and interactive interactions. Customer services can be provided anytime, e.g., through handling customer complaints by chatbots. Supporting customers or providing recommendations throughout activities along the customer journey is independent of any operation times of businesses, which is beneficial for businesses and customers (Adamopoulou & Moussiades, 2020). While businesses benefit as they can provide services to their customers with less or even without human interaction, customers benefit from the accessibility of customer services at their convenience (Poser et al., 2022). Furthermore, the accuracy mentioned above can increase service quality by better understanding customers and enabling better alignment of services with customers' needs and preferences (Zhang et al., 2021).

Improved service efficiencies, service provision, and quality can further impact a company's profitability and reputation. Due to efficiencies and accompanying automation, customer services can be provided by AI instead of humans, whereby businesses can reduce their labor costs. Although the development, deployment, and running costs for AI for customer services need to be considered, AI can positively affect overall profitability. Increased revenues can further impact profitability through improved service provision and quality by impacting customer satisfaction (Johnson et al., 2022). Providing services anytime can enhance impulse purchases, increase service quality through providing more suitable recommendations, and foster customers' purchase behavior and thus ultimately affecting a company's profitability (Jannach et al., 2010).

TABLE 8.1
Key Advantages of AI for Customer Services

Key Advantage Area	Implications
Service Efficiency	Higher automation and accuracy through data and technical advances
Service Provision and Quality	More autonomous and advanced provision of customer services
Profitability	Cost savings due to automation and revenue increases due to customer satisfaction
Reputation	Positive effects due to improved service efficiency, provision, and quality

AI can positively affect the reputation of associated customer services, further impacting a company's overall reputation (Dolata et al., 2021). Such reputational effects can result from the first two outlined key advantages. Due to improved service efficiency, customers might, for example, perceive customer services and businesses as innovative (Poser et al., 2022). Furthermore, improved service provisions and satisfied customers resulting from high service quality are beneficial for positive reputations (Nguyen et al., 2018).

8.3.2 Artificial Intelligence along the Customer Journey

AI can be employed across industries and functions, for example, for predictive policing, candidate selection in recruitment, loan applications, or product recommendations. Customer services are a promising area for AI applications because all customer journey phases and associated interactions and transactions can be supported (Følstad & Skjuve, 2019). Businesses can use AI to improve existing customer services, provide new ones, and strengthen overall customer experiences and relationships (Poser et al., 2022; Lemon & Verhoef, 2016). Regarding their area of impact, AI-based systems for customer services can be classified into non-customer-facing systems, running in the background without interacting with customers, and customer-facing systems, providing channels to interact with customers (Poser et al., 2022). Non-customer-facing AI-based systems could, for example, be employed within logistics to optimize dispatching and shipping processes. AI-based pricing tools could support adjusting prices for customers based on customer specifications and contextual factors. Furthermore, purchase pattern recognition could be leveraged by purchasing departments to forecast future demands (Toorajipour et al., 2021). Even though customers are not directly in touch with such systems, they allow for efficiency and accuracy increases and thus indirectly allow for improved customer services such as faster deliveries or demand-aligned availabilities (Morgan, 2019). On the other hand, customer-facing AI-based systems, among others, include chatbots and virtual assistants. Customer-facing systems are not limited to a virtual environment but can also be part of physical in-store environments, for example, by integrating intelligent mirrors or smart check-out systems such as those deployed by Amazon GO (Witt, 2022). Both types of AI-based systems, with or without direct customer interface, may not only interact with customers within one phase of the customer journey but also throughout several phases and thus affect several touchpoints. This is, for example, the case for recommender systems or virtual assistants, which can support customers along all three phases of the customer journey. Figure 8.3 outlines some exemplary AI-based systems and their main application along the customer journey.

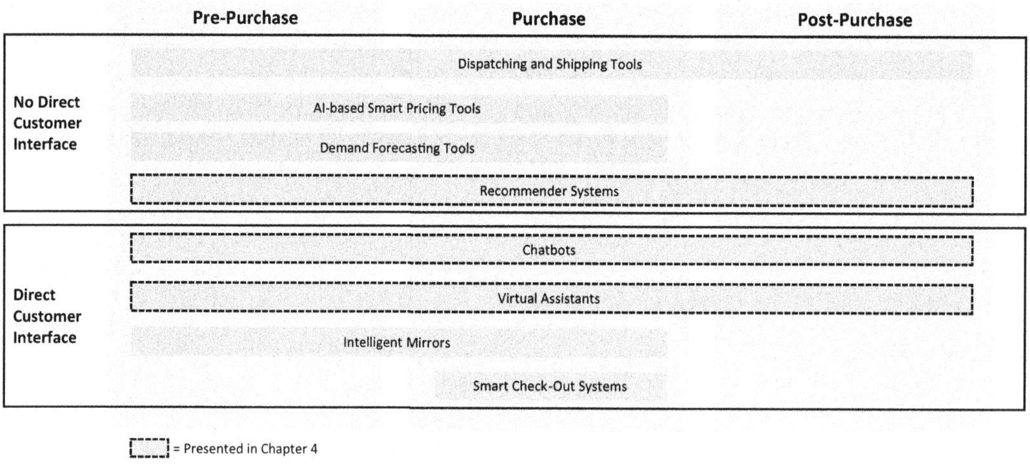

FIGURE 8.3 Classification of selected AI-based systems and their application along the customer journey.

8.3.3 Implementation and Management of Artificial Intelligence for Customer Services

A company must fulfill various requirements for the successful implementation and management of AI for customer services. These requirements can be structured into three building blocks (Figure 8.4): a customer-centric technology strategy, technological foundations including data, algorithms, and computing power. Lastly, an innovative company culture accompanied by AI-related skills and know-how (Mikalef et al., 2019).

To ensure a successful integration and use of AI for customer services, the application of AI should be embedded in a company's technology strategy (Mithas, 2018). It should demonstrate the role AI is supposed to take on for businesses and how AI can enable or improve customer services. The application of AI should thus best be aligned with a company's overall ambitions for customer services by applying a customer-centric approach to the technology strategy (Følstad & Kvale, 2018). Customer centricity refers to the alignment of a "company's development and delivery of its products and services with the current and future needs of a selected set of customers in order to maximize their long-term financial value to the firm" (Fader, 2020).

Technological foundations, especially data, algorithms, and computing power, are required to successfully realize the technology strategy and related AI ambitions. Data represents a core foundation for any application of AI for customer services as it allows for understanding customers and thus enables providing individually relevant and suitable customer services (Diederich et al., 2019). Such data could be utilized from openly available data sources, acquired from data providers, or internal company initiatives. Certain challenges can arise regarding the data, which we outline in Section 8.5. To make sense of data and run algorithms, sufficient computing power, which, among others, refers to memory, processor speed, and the number of processors, is required. The more computing power, the more complex and faster calculations can be computed. Thus, AI-based interactions in customer services can be provided in real-time (Dwivedi et al., 2023). Cloud computing allows businesses "to use a pool of IT resources and applications as services virtually through the web, without physically holding these computing resources internally" (Dutta et al., 2013). Thus, businesses can store their data and run their algorithms remotely without requiring costly hardware investments.

Additionally, an appropriate company culture is crucial for driving technological advances in every aspect, along with the willingness to revise established processes and services (Kruhse-Lehtonen & Hofmann, 2020). Hereby a customer-centric technology strategy, as outlined above, can help point the way. To ensure the realization of identified AI measures for customer

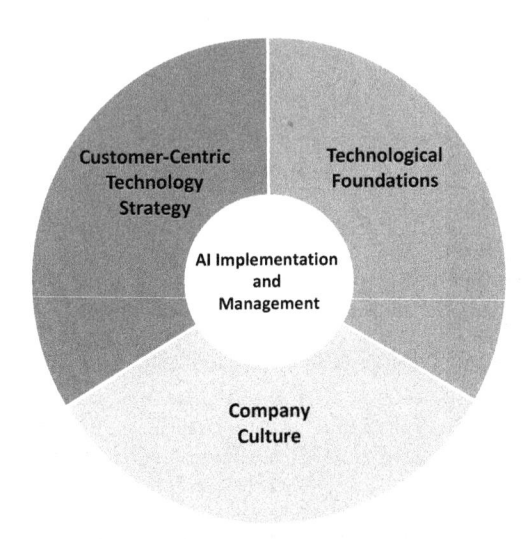

FIGURE 8.4 Building blocks for AI implementation and management.

service, dedicated key performance figures could promote decisions fostering the implementation and exploitation of AI for customer services (Parmenter, 2015). It does not only require support from senior leadership to grant budgets in favor of AI for customer services. Stakeholders across all levels must be on board with driving and leveraging AI for successful customer services (Alsheibani et al., 2020). To realize the potential advantages of AI and put it into practice for customer services, skills and know-how in the form of dedicated roles and peoples' capabilities are required (Kruhse-Lehtonen & Hofmann, 2020). Besides roles focusing on the data perspective, such as data scientists, architects, or engineers, it is also important to cover the intersection between IT and business functions, for example, to translate customer-centric ambitions into technologically feasible measures (Davenport, 2020).

8.4 CORE AI-BASED SYSTEMS FOR CUSTOMER SERVICES

While some of the AI-based systems are dedicated to certain application fields or industries, such as intelligent mirrors or smart check-out systems within physical retail stores (Section 8.3.2), we introduce three core AI-based systems, which apply to a wide variety of customer service, in the following: recommender systems, chatbots, and virtual assistants. While these three systems differ regarding application in detail and technological maturity, they are not necessarily standalone AI-based systems. Chatbots or virtual assistants, for example, can represent interfaces to communicate product recommendations to customers. While recommender systems operate in the background without direct interfaces with customers (i.e., the recommendations are communicated via other systems), chatbots and virtual assistants face customers directly, with virtual assistants being more sophisticated regarding their occurrence and application (Janssen et al., 2020). We provide an overview of all three systems, including their purpose and technical functioning, followed by their application to customer services along the customer journey and associated benefits. Table 8.2 provides a short overview of these three core AI-based systems.

TABLE 8.2
Overview of Selected AI-Based Systems

	Recommender Systems	Chatbots	Virtual Assistants
Purpose and Technical Functioning	• Analyze data and identify patterns for recommendations • Apply various approaches to create recommendations (collaborative filtering, content-based filtering, social recommender systems)	• Enable human-like text-based communications with customers (dedicated to a specific purpose, mainly embedded within a website or an app) • Apply symbolic or sub-symbolic AI (incl. ML, DL, and NLP)	• Enable human-like text- and voice-based communications with customers (support a variety of services, mainly embedded within smart devices) • Apply mostly sub-symbolic AI (incl. ML, DL, and NLP)
Application to Customer Services and Benefits	• Provide personalized recommendations for customers about tangible or intangible products or services • Reduce information load and search time for customers • Strengthen customer satisfaction • Enable up- and cross-selling for yield maximization	• Provide responses to customers regarding requests or complaints • Enable 24/7 availability of customer services • Enhance interactive communication • Perform repetitive tasks • Enable cost savings	• Provide ubiquitous support for customer demands also for complex tasks and by considering contextual aspects • Enhance customer experience • Enable proactive behavior and enhance sales

AI, artificial intelligence; DL, deep learning; ML, machine learning; NLP, natural language processing.

8.4.1 RECOMMENDER SYSTEMS

8.4.1.1 Purpose and Technical Functioning

People rely on recommendations when making decisions. While such recommendations can, for example, be provided by friends or family via word of mouth or by experts via magazines, technical systems can also take over that responsibility by developing recommendations about a situation or a product for customers (Jannach et al., 2010). Such recommendations can occur in various areas, for example, Amazon recommending products for purchase, Netflix recommending new series to be watched, LinkedIn recommending people one may know, insurance companies recommending their products, or social media platforms such as TikTok or Instagram suggesting new friends and content.

Big data and AI enhance the abilities and performance of recommender systems widely (Wirtz et al., 2018). Data is the basis for recommender systems to develop and provide recommendations. The data and thereby knowledge source of recommender systems can be split into three categories: social knowledge about customers in general, individual knowledge about the specific customer, and content knowledge about the items recommended (Felfernig & Burke, 2008). Information about users' preferences can hereby result from explicit or implicit data. Explicit data refers to ratings or votes, whereas implicit data results from interactions between users and items, e.g., through purchases or clicks (Zhang et al., 2021). Such knowledge can be leveraged by various AI techniques, such as ML, data mining, and case-based reasoning, to finally provide personalized recommendations (Burke et al., 2011). Generally, two core underlying algorithms for recommender systems can be differentiated: collaborative filtering and content-based filtering (Xiao & Benbasat, 2007). While generally implicit and explicit data are used by both algorithms, they focus on different information from the data. Collaborative filtering leverages similarities between customers and thus leads to recommendations such as "Customer who bought product X also bought product Y." Content-based filtering identifies recommendations based on matching item specifications. Often recommender systems make use of collaborative and content-based filtering jointly. Such hybrid approaches are also perceived as more useful as they provide better decision quality than recommendations only based on one of the two approaches (Xiao & Benbasat, 2007).

More recently, social recommender systems represent an additional filter mechanism, integrating recommender systems with social media (Oechslein & Hess, 2014). Hereby customers' profiles with an e-commerce shop are linked to their personal social media profiles and thus further to the profiles of their connected contacts via social media. Based on the data available from a vast amount of social media profiles, their information about preferences, interests, and purchases is considered for personalized recommendations (Arazy et al., 2010). The increase in social networks such as Pinterest or Instagram, including data in the form of preferences, relations, likes, shares, comments, ratings, and tags, also enhances the application opportunities and the success of recommendations (Shokeen & Rana, 2019). Compared to collaborative filtering, the recommendations are not impacted by similar customers but rather by the preferences and customer profiles of friends (Shokeen & Rana, 2019).

While recommender systems are not a recent phenomenon, the advances of AI, especially with regard to sub-symbolic AI, provided new opportunities to improve recommender systems and unveil further progress. Thus, recommender systems can process more complex and extensive input data allowing for more precise and defined recommendations resulting from a multitude of different parameters (Zhang et al., 2021). Recommender systems continuously incorporate new forms of data, allowing them to adapt their recommendation according to a specific context. Thus, recommender systems can continuously increase prediction accuracy (Zhang et al., 2021) and potentially provide increasingly personalized recommendations to customers. Furthermore, sub-symbolic AI allows for independent improvements over time by learning from previously provided recommendations, including related customer responses (Ilkou & Koutraki, 2020). Additionally, generative AI can be leveraged for recommender systems to provide recommendations for complex tasks.

8.4.1.2 Applications for Customer Services and Associated Benefits

The most established application of recommender systems is within e-commerce by providing personalized recommendations to customers via online shops such as Amazon (Xiao & Benbasat, 2007). Thus, recommender systems mainly support the customer throughout the pre-purchase phase by enabling them to find suitable products faster, thus striving to reduce search time, which is particularly helpful in large online shops (Burke et al., 2011). Hereby the information load for customers can be reduced as recommender systems narrow down an overwhelming number of options available online and present customers only with a careful selection of recommendations (Zhang et al., 2021). Narrowing down options can also improve the customers' attitude toward recommender systems and the underlying businesses (Nguyen & Hsu, 2022; Zhang et al., 2021).

By providing personalized recommendations based on individual preferences and usage behaviors, the recommended items are supposed to be more relevant for customers. They can increase their satisfaction with the recommendation and, ultimately, the business. Satisfaction with recommender systems can also positively affect customer behavioral responses, such as their retention or purchases (Nguyen et al., 2018). If customers, for example, receive suitable recommendations via media platforms such as Netflix or Spotify, they are likely to spend more time on these platforms consuming content. The rise of such online media platforms was also strengthened by improvements regarding the accuracy of recommender systems (Zhang et al., 2021) since customers are rather returning to businesses and their services if previously provided recommendations served their needs (Nguyen & Hsu, 2022).

Furthermore, suitable recommendations for customers can increase customers' purchase baskets and thus sales (Jannach et al., 2010). Through identifying products customers did not have on their minds before, businesses are provided with the opportunity for up- or cross-selling. Recommendations not only improve sales, but they might also enable retailers to flexibly adjust their prices immediately upon feedback regarding provided recommendations (Pathak et al., 2010). Also, further along the customer journey, recommendations could be provided, such as suggesting suitable payment methods throughout the purchase phase or providing recommendations for follow-up purchases throughout the post-purchase phase. The rise of generative AI enables the creation of complex recommendations, e.g. in the form of several interrelated recommendations. Considering the tourism industry, for example, current recommender systems typically provide a recommendation for one specific aspect of a journey based on the customer's search, such as the hotel or flight. Generative AI could enable recommender systems to provide recommendations on personalized travel itineraries, including several aspects of a vacation, such as destinations, accommodations, activities, and transportation (Weed, 2023). As a second example, in the healthcare industry generative AI can be leveraged to provide treatment plans that are directly individualized and tailored to patient's medical history (Rosencrance, 2023).

8.4.2 CHATBOTS

8.4.2.1 Purpose and Technical Functioning

Chatbots refer to software applications engaging in dialogues with humans through text (Diederich et al., 2022). One of the first chatbots was ELIZA, developed by a team around Joseph Weizenbaum in 1966, imitating a human psychologist (Weizenbaum, 1966). Certain trigger words within users' input were related to decomposition rules applied by ELIZA, which then selected responses among pre-defined options and provided them to the user. Chatbots like ELIZA are limited to a specific knowledge domain (Shearer et al., 2019). Such symbolic AI-based chatbots can only process user requests and texts that follow a pre-defined format. Limited by pre-defined rules, they cannot process just any text written by a user and can only provide pre-defined responses instead of individually targeted ones (Ilkou & Koutraki, 2020). Chatbots would, for example, provide users with various options about continuing a conversation, whereby the user could pick one pre-defined option (Diederich et al., 2019). Thus, answers lack originality and sometimes only demonstrate a limited fit. The variety depends mainly on the underlying scope of the database (Adamopoulou & Moussiades, 2020).

While the fundamental principle of chatbots is still the same, technological advances over the years enabled the further evolution of chatbots and thus provided new opportunities for intelligent human-computer interaction. Thus, chatbots were enabled to not only consider one single customer request or message but also make use of full conversations for further analysis. The technological breakthrough of chatbots is marked by the era of sub-symbolic AI, among others, through leveraging NLP for more sophisticated human-like interactions (Shearer et al., 2019). Such NLP-based chatbots are trained on large unstructured data sets and, for example, use sentiment analysis to identify meanings within a text (Taherdoost & Madanchian, 2023). NLP-based chatbots can derive meaning from users' written content and can support and solve more complex tasks by providing dedicated and individually developed responses (Adamopoulou & Moussiades, 2020). They allow for more human-like and natural interaction with users (Diederich et al., 2019) driven through anthropomorphism, representing human-like behaviors, features, or designs (Pfeuffer et al., 2019). While such anthropomorphic elements are exploited for chatbots, virtual assistants allow for even more opportunities (Section 8.4.3).

8.4.2.2 Applications for Customer Services and Associated Benefits

The most common area to implement chatbots is within customer services in e-commerce settings, where chatbots have become an increasingly preferred channel by customers to interact with businesses (Adam et al., 2021). Such chatbots can support customers along all three phases of the customer journey, starting with supporting the search for information in the pre-purchase phase. Throughout the pre-purchase phase, chatbots can also serve as interfaces to communicate recommendations to customers. Furthermore, chatbots can accompany transactions throughout the purchase phase by suggesting various payment methods or providing an overview of shipment policies. In the post-purchase phase, chatbots can deal with customer complaints and thus support or even replace call center agents (Følstad & Skjuve, 2019). Chatbots can be differentiated regarding the initiation of interaction, either by customers or the chatbot itself. However, the proactiveness of chatbots is limited to making the user aware of themselves within a very dedicated context only (Janssen et al., 2020).

Compared to call center agents, chatbots are available anytime, which is beneficial for businesses as they can ensure that customers are served timely, whereby they can strengthen customers' satisfaction (Adamopoulou & Moussiades, 2020). In the case of inquiries normally addressed to call center agents, customers are increasingly enabled to multitask. The interactive experience enabled by their human likeness can increase customer engagement (Janssen et al., 2020) and further positively affect customer trust and satisfaction (Adam et al., 2021).

Even if they cannot fully solve all customer complaints and requests, they can handle recurring and common customer inquiries, such as bookings or technical support requests, and redirect customers to the most appropriate point of contact. Businesses can immensely benefit from chatbots taking over repetitive inquiries and thus free up human capacities, ultimately allowing for a reduction of 30% in customer service costs (Maedche et al., 2019). Additionally, chatbots collect extensive amounts of data through every customer interaction, which benefits their continuous development. Businesses could even leverage the data for external purposes by commercializing insights resulting from customer behavior with third parties (Baecker et al., 2020).

So far, generative AI is mainly deployed by technology companies such as OpenAI, providing a potential technological basis for chatbots without a specific adaptation to customer services. However, given the rapid development, many companies are likely to employ and contextualize these for their specific customer services. Companies within the travel industry could leverage travel itinerary recommendations provided by generative AI, as outlined above, and integrate these into chatbots for improved customer experience. Following the recommendations, the chatbots could further support customers regarding subsequent bookings and payments, thereby taking over an increasing number of tasks that are traditionally performed by human sales agents (Weed, 2023). Besides the travel industry, financial services could leverage generative AI as part of financial advisor chatbots. Through generative AI, chatbots cannot only provide tailored investment suggestions, but also respond to any question by customers regarding markets, company performance or the reasoning behind the investment recommendations. Generative AI can thus enable more authentic and human-like chatbots providing tailored engagements with customers (FinTech Global, 2023).

8.4.3 Virtual Assistants

8.4.3.1 Purpose and Technical Functioning

Virtual assistants, also called virtual agents, voice agents, voice assistants, or smart assistants, can support customers in their everyday life by providing contextually relevant and highly personalized customer services (Diederich et al., 2019; Pfeuffer et al., 2019). The number of virtual assistants in use is forecasted to increase from 4.2 billion in 2020 to 8.4 billion in 2024 (Statista, 2023). Popular examples of virtual assistants are Alexa, Cortana, and Google Assistant (Janssen et al., 2020). Virtual assistants differ from chatbots mainly because of two aspects. First, while chatbots are dedicated to a specific purpose and domain and are mainly embedded within a website or an app, virtual assistants are typically installed either within a dedicated smart device or within smartphones and thus closely around customers at most times, whereby they support a variety of services (Janssen et al., 2020). Second, while chatbots are limited to text-based interactions, virtual assistants can process human speech and share auditive responses (Diederich et al., 2022). Thus, virtual assistants do not require any visual interfaces, even though some examples, such as the Amazon Echo Show 8, are equipped with an additional screen (Knote et al., 2021).

Through NLP and leveraging technological advances regarding DL, virtual assistants can process speech-to-text and text-to-speech, thus especially allowing them to process human speech compared to only text-based inputs, as chatbots do (Knote et al., 2021). To do so, they receive auditive commands from customers, extract the relevant message, collate the message with their knowledge, and identify responses that are then translated back into auditive information. While this is only a simplified description, such processes occur rapidly without customers noticing any time lag. Besides the information shared by customers via voice, such as commands and requests, depending on the type of virtual assistants, they might also leverage further data such as vision (e.g., by sensing their environment) or contextual information (e.g., calendar entries) to provide customer services (Brill et al., 2019). Virtual assistants can process voice information and, for example, understand how customers feel by analyzing what they say and how they say it. Such emotionally intelligent interactions are still in their infancy, but they will allow for incremental anthropomorphic virtual assistants, sensing and reacting like humans (Pfeuffer et al., 2019). An example is Moodbox (Moodbox, n.d.), created by researchers from the Hong Kong University of Science and Technology. It senses human emotions and plays music matching the mood of users. Besides recognizing emotions and demonstrating anthropomorphic aspects through the ability to process human speech and respond in human-like language, some businesses also strive to develop virtual assistants appearing like humans. This can include their physical appearance or voice. An empathetic and passionate-sounding voice, for instance, verifiably increases users' experiences and acceptance (Janssen et al., 2020). Anthropomorphic cues can cause more positive benefits, such as an increased likeability by users or an increased purchase decision quality, as users perceive an interaction as more personal (Pfeuffer et al., 2019). However, anthropomorphic approaches for virtual assistants are not always preferred by users, as they can also cause negative effects, e.g., represented by the "uncanny valley," which represents an eerie feeling by users toward human-like virtual assistants (Section 8.5.1)

8.4.3.2 Applications for Customer Services and Associated Benefits

Virtual assistants mainly leverage voice-based communication compared to chatbots, which are limited to text-based communication. Such human-like interaction enabled through voice can increase customers' experiences and strengthen their engagements (Knote et al., 2021). Also, compared to chatbots, mostly embedded in or contextualized to a particular website and a specific customer service, virtual assistants are more ubiquitous as they are embedded in smart devices, mainly within customers' homes. Thus, they can gather vast amounts of data and build extensive knowledge about customer preferences and behaviors in various situations. Such insights can further provide more personalized and accurate services and thus ultimately turn virtual assistants into everyday companions (Janssen et al., 2020).

While managing calendars, retrieving weather information, and making phone calls are the most frequent customer services virtual assistants are used for, they offer an even broader range of potential services. Virtual assistants can also be used for purchasing products via voice, monitoring patients' health at home, or automating activities in the household, such as putting a vacuum cleaner into action when required or adjusting the heating in line with weather forecasts (Brill et al., 2019). The latter examples also represent a more continuous relationship between virtual assistants with customers, compared to chatbots, which are rather limited for handling one-time services (Følstad & Skjuve, 2019). Additionally, virtual assistants can be embedded within fitness or health-related apps and thus help users track their training progress or state of health. However, they can also just serve entertainment purposes (Kunz et al., 2022). The rather long-term relationship with virtual assistants also provides the opportunity for innovative customer services along all three phases of the customer journey, such as identifying a demand for products based on weekly consumption or initiating maintenance of household devices. Such examples demonstrate the ability of virtual assistants to proactively initiate actions not prompted by users (Meurisch et al., 2020).

8.5 CHALLENGES OF AI-BASED SYSTEMS

8.5.1 Potential Negative Effects of AI-Based Systems

Based on the current development state of technology, using AI for customer services also entails risks. Various cases of recommender systems, chatbots, and virtual assistants being wrong, inappropriate, unfair, or discriminating have become public over time. Examples of recommender systems include Amazon's decommissioned HR recruiting tool, which preferred men over women for open job positions because, historically, more men would be employed for similar jobs (Dastin, 2018). Similar examples also occurred regarding customer services when the e-commerce shop Staples offered the same products to different customers at different prices subject to their purchase behavior and personal profiles, especially their ZIP code (Valentino-DeVries et al., 2012). Further, Apple's credit card offerings were accused of being biased toward women and thus investigated by US regulators (BBC, 2019).

Negative examples of chatbots include Meta's chatbot BlenderBot 3, which was trained on a large amount of publicly available data, eventually providing wrong and offensive responses to users' requests (Thorbecke, 2022). In March 2023, the Italian data-protection authority banned the usage of the chatbot ChatGPT. They accused the chatbot of violating data protection rules by unlawfully collecting and storing user data further to train the algorithm (McCallum, 2023).

Furthermore, virtual assistants based on voice interactions attracted negative attention, as they can be discriminating. Some cases demonstrated that voice-based virtual assistants are not responsive or provide unsatisfying responses to people with an accent (Harwell, 2018). They performed best for native speakers without accents, representing most people responsible for developing such systems. Furthermore, virtual assistants might use offensive language toward their recipients if they've been previously confronted with such language (Shearer et al., 2019). Also, Amazon's Alexa is regularly accused of invading customers' privacy by gathering personal data while not in use (Brown, 2022).

In the following, we outline eight predominant risks for customers and businesses overarchingly regarding AI-based systems: lack of transparency, algorithmically enforced filter bubbles, discrimination, insufficient service quality, anthropomorphism, personalization, proactivity, and privacy. While these eight aspects describe individual risks, they are not standalone effects, as they partly overlap or impact each other. We further highlight three overarching consequences and regulatory frameworks along with prevention measures supporting societies and businesses in addressing and preventing these risks.

8.5.2 Predominant Risks of AI-Based Systems for Customer Services

8.5.2.1 Lack of Transparency

The first risk refers to transparency and accompanying information asymmetries. AI-based systems often lack transparency and do not offer explainability to users, so they are often perceived as black boxes (Ochmann et al., 2021). For users, it is mainly unclear what kind of data is used or how algorithms operate. For example, in the case of recommender systems providing suggestions to customers, it might be unclear how and why a certain product is recommended resulting in an information asymmetry between the customer and the recommender system. Information asymmetry refers to the situation where a recommender system and the related business have access to more information, such as inventory levels, potential margins, and competitive products, compared to the customer (Dawson et al., 2011). Further, voice-based interactions can be problematic, as they limit the variety of products and amount of information that can be shared and processed with customers compared to presenting product recommendations via an e-commerce website, thus further reducing transparency (Rhee & Choi, 2020). A lack of transparency can further negatively affect customer satisfaction with AI-based customer services and impact the customers' perception of fairness or trust toward AI-based systems (Kordzadeh et al., 2019). However, perceptions of fairness and trust by customers are essential success factors for customer service (Wang & Huff, 2007). Thus, it is relevant to address transparency issues, for example, through explanations. AI-based systems could, for example, explain how they operate by outlining the underlying data and the key properties of their algorithms (Ochmann et al., 2021). Such explanations targeting customers should not focus on technical aspects but rather provide universally understandable logical reasonings (Kordzadeh et al., 2019). ChatGPT, for example, provides logical explanations about its reasoning, data sources, and independency when providing product recommendations.

8.5.2.2 Algorithmically Enforced Filter Bubbles

The second risk results from selecting and filtering information by AI-based systems. Recommender systems, for example, select certain products from a wide variety and suggest them to customers with the aim of presenting customers with the most relevant products. However, it might lead to the reinforcement of the same products being recommended repeatedly, whereby the overall product variety toward customers remains limited, causing so-called "filter bubbles" (Matt et al., 2014). By providing recommendations similar to or very close to previous purchases and preferences, customers might miss new or diverse recommendations. While this can be suitable for some customers, others might disapprove of the limited variety and prefer receiving new offerings. Filter bubbles are also known regarding social media, whereby the algorithm continues to share similar information. Such effects are highly problematic because, as a result, users' reality can be distorted, and political beliefs can be strongly shaped (Hess, 2017). Regarding recommendations, businesses could actively enforce providing recommendations not fully matching customer inquiries. However, if they are not fully in line with them, it can even be riskier, and customers might be less satisfied and refrain from subsequent usage in the future. Businesses can best address the "filter bubble" issue by balancing serendipitous recommendations and those based on previous experiences or purchases

(Matt et al., 2014). Such an approach might entail some trial-and-error procedures. However, based on customers' feedback to suggestions, AI-based recommender systems can develop themselves further (Heaven, 2021).

8.5.2.3 Discrimination

Data and algorithms are at the core of AI-based systems. However, they can also be the reason for unfair and biased outcomes, as demonstrated by the Amazon recruiting tool or Apple's credit card. Biased outcomes often result from biased data which over- or underrepresents certain elements or characteristics (Kordzadeh et al., 2019). Inequalities apparent through data might also be exploited by businesses to benefit from higher purchasing power by some customers. Hereby products with higher margins might be explicitly recommended to customers with a higher purchasing power (Valentino-DeVries et al., 2012). However, businesses might even be unaware of biased data if no issues emerge.

Businesses can address bias and discrimination proactively by expanding the data training set. For example, to avoid discrimination based on customers' accents, training data could be expanded by a wider variety of voices representing various accents (Najafian et al., 2016). Also, algorithms can be applied to identify inconsistencies and address them directly through mathematical approaches. Sensitive data could be identified and removed, and relevant data could be added based on machine or human decisions. However, it is not a simple undertaking if human decision-making is involved, as humans could also be biased within their perceptions and decisions and thus also cause the inappropriateness of data sets (Kordzadeh & Ghasemaghaei, 2021). Overarching guidelines regarding characteristics and features of data sets by governments or institutional organizations might provide a remedy (Section 8.5.3).

8.5.2.4 Insufficient Service Quality

Inadequate customer services could fail to match and satisfy customers' demands (Adam et al., 2021). This could have several reasons, such as a general lack of understanding of customers' demands or an improper database regarding AI-based systems leading to poor customer service quality. Besides unsuitable data, simply the lack of data can negatively impact service quality (Weiler et al., 2021). Also, most AI-based systems are still limited to simple and repetitive tasks compared to fully replacing human intelligence regarding complex challenges (Poser et al., 2022). While generative AI enables addressing more complex customers' requests, responses, however, might also be irrelevant, inaccurate, offensive, or legally problematic, which demands quality controls and defined responsibilities (PwC, 2023). Once customers are unsatisfied with the service, it can negatively impact their overall trust and behavioral responses toward AI-based systems. Response failures can, for example, cause an immediate discontinuance by customers, requiring preventive measures. Proactively providing inoculant messages upfront to users can be one approach to strengthen customers' attitudes toward failed responses by AI-based systems (Weiler et al., 2021).

8.5.2.5 Anthropomorphism

While anthropomorphism can have advantages for customers and businesses, such "human likeness" of application systems can also cause negative reactions. For AI-based systems, human likeness can cause customer aversion as some customers might perceive it as creepy (Pfeuffer et al., 2019). Customers might even have increased expectations toward anthropomorphic AI-based systems, as they expect service qualities as provided by humans. Ultimately if their expectations are not met, it might cause strong negative reactions and reduce the credibility of these AI-based systems. Also, if customers are initially unaware of the counterpart being a computer due to realistic anthropomorphic design, they could be disappointed once they become aware of it. Thus, it is important to proactively communicate the involvement of AI-based systems (Adamopoulou & Moussiades, 2020).

8.5.2.6 Personalization

One core advantage of AI-based systems is the ability to personalize customer services based on big data. However, not all customers might prefer personalized recommendations, as they might feel restricted within their autonomy (Ziv et al., 2019). Customers might also perceive recommendations as intrusive and reject them, even though they are personalized (Benlian et al., 2019). Thus, different levels of personalization need to be identified and applied based on customer preferences. Some customers prefer rather general recommendations, matching the profiles of many customers, while others prefer recommendations that are personalized per individual (Nguyen & Hsu, 2022).

8.5.2.7 Proactivity

Especially ubiquitous virtual assistants are enabled to anticipate customers' needs and provide proactive customer services without explicit user requests. They could, for example, provide proactive information on the traffic or weather conditions for an upcoming weekend trip or order products to fill the pantry. However, such proactive behavior might also be perceived as intrusive, non-polite, and inappropriate (Meurisch et al., 2020). Furthermore, as proactive behavior requires less advocacy by customers, unsuitable activities might be discovered only later and thus lead to costs for customers. Providing customers upfront the opportunity to decide about their preferred levels of proactivity could be a measure to ensure their satisfaction. Such preferred levels of proactivity might also differ according to different services and industries (Kraus et al., 2012). Customers, for example, prefer proactive activities regarding physical health over proactiveness regarding mental health support (Meurisch et al., 2020). Thus, it is important for businesses to understand different applications and preferences regarding proactivity.

8.5.2.8 Privacy

As data is at the core of all AI-based systems, such systems and related customer services can raise concerns about privacy and data protection (Adjerid et al., 2018). The advantages mentioned above impact the risk of privacy but also risks due to personalization and proactivity. It might not be fully transparent how data is used and what kind of data is being used, raising privacy concerns. Customers might also be hesitant to share private information as they want to protect their privacy. However, according to the privacy paradox, customers ultimately overcome their hesitation to benefit from the respective services, leading to substantial differences between intended and actual data privacy behavior. Though, in some cases, it can even be a rational decision made by customers to release their data and therefore benefit from the related services (Adjerid et al., 2018). In any case, businesses need to be aware of the privacy considerations of customers and be familiar with current rules and regulations. Foremost, businesses need to adhere to legal regulations such as the General Data Protection Regulation (GDPR) by the EU. To satisfy customers and address their concerns, businesses could address the above-outlined transparency and provide information about their data usage and protection to reduce privacy concerns by customers (Weith & Matt, 2022).

8.5.3 OVERARCHING CONSEQUENCES AND REGULATORY FRAMEWORKS

8.5.3.1 Overarching Consequences Resulting from Risks

The eight outlined risks can have measurable negative legal, reputational, and economic effects on businesses providing customer services (Dolata et al., 2021). Negative legal effects could result if the risks outlined above also legally represent violations of rights and regulations, e.g., regarding privacy, data protection, and ethical aspects dedicated to AI-based systems. Violating legal rights could also have negative reputational impacts. Moreover, even if measures and activities are legally permitted, customers might face unfulfilled expectations or perceive AI-based systems as unfair. Objective fairness differs from individually subjective perceived fairness, which could be impacted by contextual factors or vary on different personalities (Weith & Matt, 2022). Especially in light of the digital transformation and the increase of social media, negative perceptions or experiences are easily spread online and can cause severe reputational issues for businesses. These reputational effects are not limited to existing customers but can also apply to potential future customers. Negative reputations can lead to undesired economic effects, such as the absence of future purchases and legal violations could cause financial penalties (Pfeuffer et al., 2019).

8.5.3.2 Overarching Regulatory Frameworks and Prevention Measures

Besides dedicated measures to address each of the eight risks, two approaches of overarching regulatory frameworks can help to address the risks and ensure non-discriminating, unbiased, and fair AI-based systems for customer services. They further support businesses in identifying their individual and adequate prevention measures. First, legal regulations and guidelines developed and published by governments could provide directives for businesses to avoid risks for themselves and their customers. A current example is the EU Artificial Intelligence Act (AIA), proposed by the European Commission in 2021 and currently under further development (European Commission, 2021). The AIA incorporates current EU regulations regarding information technology safety, data protection, and equal rights in light of AI. It aims to ensure that AI-based systems within the EU comply with given regulations. It also excludes the usage of certain AI-based systems within the EU, such as social scoring by governments or live remote biometric identification in public spaces (PwC, n.d.). Considering the AI-based systems for customer services, the AIA, for example, demands that users are informed about a chatbot being AI-based or about AI algorithms compiling recommendations. However, it is important to consider that legal regulations and guidelines could, at worst, also propose unfeasible demands for businesses and developers and even hold back innovativeness (Pouget, 2023).

Second, besides governmental laws and regulations, social and customer-oriented organizations such as Algorithm Watch (Algorithm Watch, n.d.) can provide recommendation guidelines and standards for developing and implementing AI-based systems for customer services. Such organizations can create awareness and inform customers about the risks of AI-based systems and make them aware of their rights. Such information might be more easily accessible for customers than regulations like the AIA. Furthermore, customer organizations could develop and award certain certificates or labels, such as the AI trust label (OECD, 2022), which is already an established procedure regarding trust in e-commerce shops (e.g., "Trusted Shops").

Based on these two overarching regulatory frameworks, businesses could define their own internal guidelines and regulations and incorporate them into their governance as well as innovation and development processes. Especially the guidelines from social and customer-oriented organizations can serve as helpful guidance for businesses when translating legal regulations into actionable measures. Businesses could also expand their measures by further ones driven by their individual culture and values (Kruhse-Lehtonen & Hofmann, 2020).

8.6 CONCLUSION AND OUTLOOK

This chapter aims to provide an overview of the technological developments of AI, resulting AI-based services, and their application for customer services. We, therefore, outlined the role and importance of the digital transformation, which provides an essential base for the successful implementation of AI for customer services through increasing channels, connectivity, and data availability. By structuring AI-based systems along the customer journey, we demonstrated their ubiquity, both for systems with and without a direct interface with customers. We selected three common AI-based systems and outlined their economic and technical characteristics, implications, and benefits from a customer and business perspective.

While we highlighted various potential applications of AI for customer services and associated benefits, we also critically reflected on the accompanying challenges and outlined the risks of overarching AI-based systems. Even though these systems vary regarding their technical specifications and applications for customer services, they face similar risks. We hereby also create awareness of risks other AI-based systems could face besides those highlighted here.

Practitioners benefit from this article as we present a broad and structured overview of AI-based systems for customer services along the customer journey, including tangible examples and benefits. Hereby practitioners can obtain inspiration when assessing and rethinking their customer services. Transparently outlining risks also enables practitioners to make informed decisions balancing advantages and risks and provides them with the relevant foundation to counteract risks early on. Furthermore, the outlined managerial measures support businesses in successfully implementing and managing AI for customer services.

We focused on B2C customer services and selected three AI-based systems allowing for versatile applications. While we focused on B2C and related benefits, AI-based systems can also be applied to B2B customer services, such as recommender systems for purchasing departments or virtual assistants for maintaining manufacturing equipment. Moreover, while we outlined overarching risks and countermeasures, future research could consider contextual factors to adapt measures accordingly when addressing risks. While we provide an overview of the current evolution of AI for customer services, AI developments are dynamic and will continuously provide new opportunities ahead. Thus, it is important to identify and assess future applications regularly and understand their potential in order for businesses to ensure they are playing at the forefront of competition.

REFERENCES

Adam, M., Wessel, M., & Benlian, A. (2021). AI-based chatbots in customer service and their effects on user compliance. *Electronic Markets*, 31, 427–445, https://doi.org/10.1007/s12525-020-00414-7.

Adamopoulou, E., & Moussiades, L. (2020). Chatbots: History, technology, and applications. *Machine Learning with Applications*, 2, 100006, https://doi.org/10.1016/j.mlwa.2020.100006.

Adjerid, I., Peer, E., & Acquisti, A. (2018). Beyond the privacy paradox: Objective versus relative risk in privacy decision making. *MIS Quarterly*, 42(2), 465–488, https://doi.org/10.25300/MISQ/2018/14316.

Algorithm Watch (n.d.). About algorithm watch. *Algorithm Watch*, https://algorithmwatch.org/en/# (Accessed 15th March 2023).

Alsheibani, S. A., Cheung, Y. & Messom, C. (2020). Winning AI strategy: Six-steps to create value from artificial intelligence. In *AMCIS 2020 Proceedings* (pp. 1–11), https://aisel.aisnet.org/amcis2020/adv_info_systems_research/adv_info_systems_research/1.

Arazy, O., Kumar, N., & Shapira, B. (2010). A theory-driven design framework for social recommender systems. *Journal of the Association for Information Systems*, 11(9), 455–490, https://doi.org/10.17705/1jais.00237.

Baecker, J., Engert, M., Pfaff, M., & Krcmar, H. (2020). Business strategies for data monetization: Deriving insights from practice. In *15th International Conference on Wirtschaftsinformatik* (pp. 1–16), Potsdam, https://doi.org/10.30844/wi_2020_j3-baecker.

Barwitz, N., & Maas, P. (2018). Understanding the omnichannel customer journey: Determinants of interaction choice. *Journal of Interactive Marketing*, 43, 116–133, https://doi.org/10.1016/j.intmar.2018.02.001.

BBC News (2019). Apple's sexis' credit card investigated by US regulator. *BBC*, https://www.bbc.com/news/business-50365609 (Accessed 10th January 2023).

Benderskaya, E.N., & Zhukova, S.V. (2013). Multidisciplinary trends in modern artificial intelligence: Turing's way. In Yang, X. S. (Ed.): *Artificial Intelligence, Evolutionary Computing and Metaheuristics* (pp. 319–343). Berlin, Heidelberg: Springer.

Benlian, A., Klumpe, J., & Hinz, O. (2019). Mitigating the intrusive effects of smart home assistants by using anthropomorphic design features: A multimethod investigation. *Information Systems Journal*, 30(6), 1010–1042, https://doi.org/10.1111/isj.12243.

Berente, N., Gu, B., Recker, J., & Santhanam, R. (2021). Special issue editor's comments: Managing artificial intelligence. *MIS Quarterly*, 45(3), 1433–1450, https://doi.org/10.25300/MISQ/2021/16274.

Brill, T.M., Munoz, L. & Miller, R.J. (2019). Siri, Alexa, and other digital assistants: A study of customer satisfaction with artificial intelligence applications. *Journal of Marketing Management*, 35(1), 1401–1436, DOI: 10.1080/0267257X.2019.1687571.

Brown, C. (2022). Amazon must turn over extensive records in alexa privacy lawsuit. *Bloomberg Law*, https://news.bloomberglaw.com/tech-and-telecom-law/amazon-must-turn-over-extensive-records-in-alexa-privacy-lawsuit (Accessed 12th January 2023).

Burke, R., Felfernig, A., & Göker, M.H. (2011). Recommender systems: An overview. *AI Magazine*, 32(3), 13–18, https://doi.org/10.1609/aimag.v32i3.2361.

Dastin, J., (2018). Amazon scraps secret AI recruiting tool that showed bias against women. *Reuters*, https://www.reuters.com/article/us-amazon-com-jobs-automation-insightidUSKCN1MK08G (Accessed 2nd January 2023).

Davenport, T. (2020). Beyond unicorns: Educating, classifying, and certifying business data scientists. *Harvard Data Science Review*, 2(2), https://doi.org/10.1162/99608f92.55546b4a.

Dawson, G.S., Watson, R.T., & Boudreau, M. (2011). Information asymmetry in information systems consulting: Toward a theory of relationship constraints. *Journal of Management Information Systems*. 27(3), 143–177, https://doi.org/10.2753/MIS0742-1222270306.

Diederich, S., Brendel, A.B., & Kolbe, L. (2019). On conversational agents in information systems research: Analyzing the past to guide future work. In *14th International Conference on Wirtschaftsinformatik*, Siegen (pp. 1550–1564), https://web.archive.org/web/20220223224726id_/https://aisel.aisnet.org/cgi/viewcontent.cgi?article=1300&context=wi2019.

Diederich, S., Brendel, A. B., Morana, S., & Kolbe, L. (2022). On the design of and interaction with conversational agents: An organizing and assessing review of human-computer interaction research. *Journal of the Association for Information Systems*, 23(1), 96–138, https://doi.org/10.17705/1jais.00724.

Dolata, M., Feuerriegel, S., & Schwabe, G. (2021). A sociotechnical view of algorithmic fairness. *Information Systems Journal*, 32(4), 754–818, https://doi.org/10.1111/isj.12370.

Dutta, A., Peng, G.C., & Choudhary, A. (2013). Risks in enterprise cloud computing: The perspective of IT experts. *Journal of Computer Information Systems*, 53(4), 39–48, https://www.iacis.org/jcis/jcis_toc.php?volume=53&issue=4.

Dwivedi, Y., Kshetri, N., Hughes, L., et al. (2023). "So what if ChatGPT wrote it?" Multidisciplinary perspectives on opportunities, challenges and implications of generative conversational AI for research, practice and policy. *International Journal of Information Management*, 71, 102642, https://doi.org/10.1016/j.ijinfomgt.2023.102642.

Enholm, I.M., Papagiannidis, E., & Mikalef, P. et al. (2022). Artificial intelligence and business value: A literature review. *Information Systems Frontiers*, 24, 1709–1734, https://doi.org/10.1007/s10796-021-10186-w.

European Commission (2021). Proposal for a Regulation of the European Parliament and of the Council laying down harmonized rules on Artificial Intelligence (Artificial Intelligence Act) and amending certain union legislative acts. *EUR-Lex*, Document 52021PC0206, https://eur-lex.europa.eu/legal-content/EN/TXT/?uri=celex%3A52021PC0206 (Accessed 15th February 2023).

Fader, P. (2020). *Customer Centricity*. Philadelphia: Warton School Press.

Felfernig, A., & Burke, R., (2008). Constraint-based recommender systems: Technologies and research issues. In *10th International Conference on Electronic Commerce*, Innsbruck, Austria, https://doi.org/10.1145/1409540.1409544.

Feuerriegel, S., Hartmann, J., Janiesch, C., Zschech, P. (2023). Generative AI. *SSRN*, https://dx.doi.org/10.2139/ssrn.4443189.

FinTech Global (2023). How will generative AI impact wealth management? *FinTech Global*, https://fintech.global/2023/03/21/how-will-generative-ai-impact-wealth-management/ (Accessed 28th May 2023).

Følstad, A., & Kvale, K. (2018). Customer journeys: A systematic literature review. *Journal of Service Theory and Practice Journal of Service Theory and Practice*, 28(2), 196–227, https://doi.org/10.1108/JSTP-11-2014-0261.

Følstad, A., & Skjuve, M. (2019). Chatbots for customer service: User experience and motivation. In *Proceedings of the 1st International Conference on Conversational User Interfaces*, Dublin (Vol. 1, pp. 1–9), https://doi.org/10.1145/3342775.3342784.

Gartner (2022). Gartner identifies three important ways AI can benefit customer service operations. *Gartner*, https://www.gartner.com/en/newsroom/press-releases/2022-01-19-gartner-identifies-three-important-ways-ai-can-benefi (Accessed 12th Janaury 2023).

Harwell, D. (2018). The accent gap. *The Washington Post*, https://www.washingtonpost.com/graphics/2018/business/alexa-does-not-understand-your-accent/?noredirect=on&utm_term=.c232d4635b88 (Accessed 6th December 2022).

Heaven, W.D. (2021). AI is learning how to create itself. *MIT Technology Review*, https://www.technologyreview.com/2021/05/27/1025453/artificial-intelligence-learning-create-itself-agi/ (Accessed 28th January 2023).

Hess, A. (2017). How to escape your political bubble for a clearer view. *The New York Times*, https://www.nytimes.com/2017/03/03/arts/the-battle-over-your-political-bubble.html (Accessed 22nd January 2023).

Hess, T., Matt, C., Benlian, A., & Wiesböck, F. (2016). Options for formulating a digital transformation strategy. *MIS Quarterly Executive*, 15(2), 123–139, https://aisel.aisnet.org/misqe/vol15/iss2/6.

Ilkou, E., & Koutraki, M. (2020). Symbolic vs. syb-symbolic AI methods: Friends of enemies? In *Conference: CSSA'20: Workshop on Combining Symbolic and Sub-Symbolic Methods and their Applications*, Virtual, https://doi.org/10.1145/3340531.3414072

Jannach, D., Zanker, M., Felfernig, A., & Friedrich, G. (2010). *Recommender Systems. An Introduction.* New York: Cambridge University Press.

Janssen, A., Passlick, J., Rodrigues Cardona, D., & Breitner, M. H. (2020). Virtual assistance in any context, a taxonomy of design elements for domain-specific chatbots. *Business & Information Systems Engineering*, 62(3), 211–225, https://doi.org/10.1007/s12599-020-00644-1.

Jarek, K., & Mazurek, G. (2019). Marketing in artificial intelligence. *Central European Business Review*, 8(2), 46–55, https://doi.org/10.18267/j.cebr.213.

Johnson, M., Albizri, A., Harfouche, A., & Fosso-Wamba, S. (2022). Integrating human knowledge into artificial intelligence for complex and ill-structured problems: Informed artificial intelligence. *International Journal of Information Management*, 64, 1–15, https://doi.org/10.1016/j.ijinfomgt.2022.102479.

Katatikarn, J. (2023). The 42 best AI art and image generators in 2023. *Academy of Animated Art*, https://academyofanimatedart.com/ai-art/ (Accessed 29th May 2023).

Kraus, S., Rigtering, J.P.C., Hughes, M. et al. (2012). Entrepreneurial orientation and the business performance of SMEs: a quantitative study from the Netherlands. *Review of Managerial Science* 6, 161–182. https://doi.org/10.1007/s11846-011-0062-9

Knote, R., Janson, A., Söllner, M., & Leimeister, J.M. (2021). Value co-creation in smart services: A functional affordances perspective on smart personal assistants. *Journal of the Association for Information Systems*, 22(2), 418–458, https://doi.org/10.17705/1jais.00667.

Kordzadeh, N., & Ghasemaghaei, M. (2021). Algorithmic bias: Review, synthesis, and future research directions. *European Journal of Information Systems*, 31(3), 388–409, https://doi.org/10.1080/0960085X.2021.1927212.

Kruhse-Lehtonen, U., & Hofmann, D., (2020). How to define and execute your data and AI strategy. *Harvard Data Science Review*, 2(3), https://doi.org/10.1162/99608f92.a010feeb.

Kunz, W.H., Paluch, S., & Wirtz, J. (2022). Toward a new service reality: human–robot collaboration at the service frontline. In Edvardsson, B., & Tronvoll, B. (Eds.): *The Palgrave Handbook of Service Management.* Cham: Palgrave Macmillan. https://doi.org/10.1007/978-3-030-91828-6_47

Lemon, K. N., & Verhoef, P. C. (2016). Understanding customer experience throughout the customer journey. *Journal of Marketing*, 80(6), 69–96, https://doi.org/10.1509/jm.15.0420.

Maedche, A., Christine, L., Benlian, A., Berger, B., Gimpel, H., Hess, T., Hinz, O., Morana, S., & Soellner, M. (2019). AI-based digital assistants - opportunities, threats, and research perspectives. *Business & Information Systems Engineering*, 61(4), 535–544, https://doi.org/10.1007/s12599-019-00600-8.

Matt, C., Benlian, A., Hess, T., & Weiss, C. (2014). Escaping from the filter bubble? The effects of novelty and serendipity on users' evaluations of online recommendations. In *Proceedings of the 2014 International Conference on Information Systems* (p. 67), Auckland, https://aisel.aisnet.org/icis2014/proceedings/HumanBehavior/67.

McCallum, S. (2023). ChatGPT banned in Italy over privacy concerns. *BBC News*, https://www.bbc.com/news/technology-65139406 (Accessed 3rd April 2023).

Meurisch, C., Mihale-Wilson, C. A., Hawlitschek, A., Giger, F., Müller, F., Hinz, O., & Mühlhäuser, M. (2020). Exploring user expectations of proactive AI systems. *Proceedings of the ACM on Interactive, Mobile, Wearable and Ubiquitous Technologies*, 4(4), 1–22. https://doi.org/10.1145/3432193.

Mikalef, P., van de Wetering, R., & Krogstie, J. (2019). From big data analytics to dynamic capabilities: The effect of organizational inertia. In D. Xu, J. Jiang, & H.-W. Kim (Eds.), *Pacific Asia Conference on Information Systems 2019 Proceedings* (pp. 1–14). Article 198 AIS Electronic Library. https://aisel.aisnet.org/pacis2019/198/

Mithas, S. (2018). Artificial intelligence and IT professionals. *IT Professional*, 20 (5), 6–13, https://doi.org/10.1109/MITP.2018.053891331.

Moodbox (n.d.). Home. *Moodbox*, https://www.mymoodbox.com/ (Accessed 7th February 2023).

Morgan, B. (2019). The 20 best examples of using artificial intel-ligence for retail experiences. *Forbes*, https://www.forbes.com/sites/blakemorgan/2019/03/04/the-20-best-examples-of-using-artificial-intelligence-for-retail-experiences/?sh=1f6fdd214466 (Accessed 28th January 2023).

Najafian, M., Safavi, S., Hansen, J.H.L., & Russell, M.J. (2016). Improving speech recognition using limited accent diverse British English training data with deep neural networks. In *International Workshop on Machine Learning for Signal Processing*, Salerno, Italy, https://doi.org/10.1109/MLSP.2016.7738854

Nguyen, T., Maxwell-Harper, F., Terveen, L., & Joseph, K. (2018). User Personality and user satisfaction with recommender systems. *Information Systems Frontiers*, 20(6), 1173–1189, https://doi.org/10.1007/s10796-017-9782-y

Nguyen, T., & Hsu, P. (2022). More personalized, more useful? Reinvestigating recommendation mechanisms in e-commerce. *International Journal of Electronic Commerce*, 26(1), 90–122, https://doi.org/10.1080/10864415.2021.2010006.

Ochmann, J., Zilker, S., & Laumer, S. (2021). The evaluation of the black box problem for AI-based recommendations: An interview-based study. *Proceedings of the International Conference of Wirtschaftsinformatik*, Duisburg Essen (Vol. 6, pp. 232–246), https://aisel.aisnet.org/wi2021/QDesign/Track10/6

OECD. AI (2022). Catalogue of AI tools & metrics. *OECD.AI Policy Observatory*, https://oecd.ai/en/catalogue/tools/ai-trust-standard-and-label (Accessed 20th January 2023).

Oechslein, O., & Hess, T. (2014). The value of a recommendation: The role of social ties in social recommender systems. In *47th Hawaii International Conference on System Science*, Waikoloa (pp. 3297–3306), https://doi.org/10.1109/HICSS.2014.235.

Paluch, S., & Wirtz, J. (2020). Artificial intelligence and robots in the service encounter. *Journal of Service Management Research*, 4(1), 3–8, https://doi.org/10.15358/2511-8676-2020-1-3.

Parmenter, D. (2015). *Key Performance Indicators*. Hoboken: Wiley, https://doi.org/10.1002/9781119019855.

Pfeuffer, N., Benlian, A., Gimpel, G., & Hinz, O. (2019). Anthropomorphic information systems. *Business & Information Systems Engineering*, 61(4), 523–533, https://doi.org/10.1007/s12599-019-00599-y.

Poser, M., Wiethof, C., & Bittner, E. (2022). Integration of AI into customer service: A taxonomy to inform design decisions. In *30th European Conference on Information Systems*, Timisoara, https://aisel.aisnet.org/ecis2022_rp/65.

Pouget, H. (2023). The EU's AI act is barreling towards ai standards that do not exist. *Lawfare*, https://www.lawfareblog.com/eus-ai-act-barreling-toward-ai-standards-do-not-exist (Accessed 15th January 2023).

PwC (n.d.). The artificial intelligence act demystified. *PwC*, https://www.pwc.ch/en/insights/regulation/ai-act-demystified.html (Accessed 4th March 2023).

PwC (2023). What is responsible AI and how can it help harness trusted generative AI? *PwC*, https://www.pwc.com/us/en/tech-effect/ai-analytics/responsible-ai-for-generative-ai.html (Accessed 27th May 2023).

Rai, A., Constantinides, P., & Sarker, S. (2019). Editor's comments. Next-generation digital platforms: Toward human AI-hybrids. *MIS Quarterly*, 43(1), 3–9, https://aisel.aisnet.org/misq/vol43/iss1/2/.

Rhee, C.E., & Choi, J. (2020). Effects of personalization and social role in voice shopping: An experimental study on product recommendation by a conversational voice agent. *Computers in Human Behavior*, 109(1), 106359, https://doi.org/10.1016/j.chb.2020.106359.

Rosencrance, L. (2023). 9 Uses of generative AI in healthcare. *techopedia.com*, https://www.techopedia.com/9-uses-of-generative-ai-in-healthcare (Accessed 28th May 2023).

Schweidel, D., Bart, Y., & Inman, J., et al., (2022). How consumer digital signals are reshaping the customer journey. *Journal of the Academy of Marketing Science*, 50, 1257–1276, https://doi.org/10.1007/s11747-022-00839-w.

Shearer, E., Martin, S., Petheram, A., & Stirling, R. (2019). Racial bias in natural language processing. *Oxford Insights*, https://www.oxfordinsights.com/racial-bias-in-natural-language-processing (Accessed 7th February 2023).

Shokeen, J., & Rana, C. (2019). A study on features of social recommender systems. *Artificial Intelligence Review*, 53, 965–988, https://doi.org/10.1007/s10462-019-09684-w.

Statista (2023). Number of digital voice assistants in use worldwide from 2019 to 2024 (in billions). *Statista*, https://www.statista.com/statistics/973815/worldwide-digital-voice-assistant-in-use/ (Accessed 20th December 2022).

Strong, J. (2022). I was there when: AI mastered chess. *MIT Technology Review*, https://www.technologyreview.com/2022/10/06/1060824/i-was-there-when-ai-mastered-chess/ (Accessed 3rd March 2023).

Taherdoost, H., & Madanchian, M. (2023). Artificial intelligence and sentiment analysis: A review in competitive research. *Computers*, 12(37), 1–15, https://doi.org/10.3390/computers12020037

Thorbecke, C. (2022). It didn't take long for Meta's new chatbot to say something offensive. *CNN Business*, https://edition.cnn.com/2022/08/11/tech/meta-chatbot-blenderbot/index.html (Accessed 23rd November 2022).

Toorajipour, R., Sohrabpour, V., Nazarpour, A., Oghazi, P., & Fischl, M. (2021). Artificial intelligence in supply chain management: A systematic literature review. *Journal of Business Research*, 122, 502–517, https://doi.org/10.1016/j.jbusres.2020.09.009.

Tueanrat, Y., Papagiannidis, S., Alamanos, E. (2021). Going on a journey: A review of the customer journey literature. *Journal of Business Research*, 125, 336–353, https://doi.org/10.1016/j.jbusres.2020.12.028.

Turing, A.M. (1950). Computing machinery and intelligence, *Mind*, LIX(236), 433–460, https://doi.org/10.1093/mind/LIX.236.433.

Valentino-DeVries, J., Singer-Vine, J., & Soltani, A. (2012). Websites vary prices, deals based on users' information. *The Wall Street Journal*, https://www.wsj.com/articles/SB10001424127887323777204578189391813881534 (Accessed 5th December 2022).

Verhoef, P.C. (2021) Omni-channel retailing: Some reflections. *Journal of Strategic Marketing*, 29(7), 608–616, https://doi.org/10.1080/0965254X.2021.1892163.

Wang, S. and Huff, L.C. (2007). Explaining buyers' responses to sellers' violation of trust. *European Journal of Marketing*, 41(9/10), 1033–1052. https://doi.org/10.1108/03090560710773336

Weed, J. (2023). Can ChatGPT plan your vacation? Here's what to know about A.I. and travel. *The New York Times*, https://www.nytimes.com/2023/03/16/travel/chatgpt-artificial-intelligence-travel-vacation.html (Accessed 28th May 2023).

Weiler, S., Matt, C., & Hess, T. (2021). Immunizing with Information - Inoculation messages against conversational agents' response failures. *Electronic Markets*, 32(2), 1–20, https://doi.org/10.1007/s12525-021-00509-9.

Weith, H. & Matt, C. (2022). When do users perceive artificial intelligence as fair? An assessment of AI-based B2C e-commerce. In *Proceedings of the 55th International Conference on System Sciences, Virtual* (pp. 4336–4345), https://doi.org/10.24251/HICSS.2022.529.

Weizenbaum, J. (1966). ELIZA-A computer program for the study of natural language communication between man and machine. *Communications of the ACM*, 9(1), 36–45, https://doi.org/10.1145/365153.365168.

Wiesböck, F., Li, L., Matt, C., Hess, T., & Richter, A. (2017). How management in the German insurance industry can handle digital transformation. In *Management Reports des Instituts für Wirtschaftsinformatik und Neue Medien* (Vol. 1, pp. 1–26), https://www.dmm.bwl.uni-muenchen.de/download/epub/mreport_2017_1.pdf.

Wirtz, J., Patterson, P., Kunz, W., Gruber, T., Lu, V., Paluch, S., & Martins, A. (2018). Brave new world: Service robots in the frontline. *Journal of Service Management*, 29(5), 907–931, https://doi.org/10.1108/JOSM-04-2018-0119.

Witt, T. (2022). How AI and data are transforming the retail industry. *Acceleration Economy*, https://accelerationeconomy.com/ai/how-ai-and-data-are-transforming-the-retail-industry/ (Accessed 25th January 2023).

Xiao, B., & Benbasat, I. (2007). E-commerce product recommendation agents: Use, characteristics, and impacts. *MIS Quarterly*, 31(1), 137–209, https://doi.org/10.2307/25148784.

Yoo, S. (2016). The role of social media during the pre-purchasing stage. *Journal of Hospitality and Tourism Technology*, 7(1), 84–99, https://dx.doi.org/10.1108/JHTT-11-2014-0067

Zhang, Q., Lu, J., & Jin, Y. (2021). Artificial intelligence in recommender systems. *Complex & Intelligent Systems*, 7, 439–457, https://doi.org/10.1007/s40747-020-00212-w.

Ziv, C., Schrift, R., Wertenbroch, K., & Yang, H. (2019). Designing AI systems that customers won't hate. *MIT Sloan ManagementReview*, https://sloanreview.mit.edu/article/designing-ai-systems-that-customers-wont-hate/ (Accessed 5th January 2023).

9 Automotive User Interfaces
Enhancing Future Mobility

Myounghoon Jeon, Chihab Nadri, Manhua Wang, and Abhraneil Dam

9.1 INTRODUCTION

Driving is a necessary part of modern life and the meaning of a vehicle has been diversified. Vehicles serve as a means of travel, transportation, and entertainment and are now evolving even into a workplace (Schartmüller, 2021). Because driving is composed of complex tasks in dynamic situations, it has been a long-lasting research topic in Human Factors and Human-Computer Interaction (HCI) (Lee, 2008; Jeon 2017a). With the advancement of artificial intelligence (AI) and machine learning, research interest on automated vehicles has sharply increased (Fayyad et al., 2020; Tong et al., 2019). The International Conference on Automotive User Interfaces Series (auto-ui.org) has shaped the development and applications of automotive user interfaces. The same community has also produced a handbook about user experience of automated driving (Riener, Jeon, & Alvarez, 2022). This chapter includes the compact version of research on automotive user interfaces. After briefly summarizing manual and automated driving modes, we introduce the basic constructs in driving research to understand human behavior–driver distraction and inattention, workload, situation awareness, emotions, trust, and motion sickness. Then, we move on to research methods and measures including behavioral, neurophysiological, and computational approaches. Next, we overview interfaces for future mobility: the state of the art of controls and displays, and emerging applications, such as intelligent agents, augmented reality (AR), and virtual reality (VR), and external human-machine interfaces (eHMI). We conclude this chapter with further research agenda, including accessibility, urban air mobility, and mobile office.

9.2 FUNDAMENTAL UNDERSTANDING OF DRIVERS

9.2.1 Drivers in Manual and Automated Driving Modes

With the advancement of vehicle automation in recent years, the roles of human drivers change as the functionality of the vehicle automation system alters. Such a variation is reflected by the dynamic human-automation function allocation against the dynamic driving task (DDT), which is defined as the real-time operational and tactical functions necessary to operate a vehicle on the road (SAE International, 2021). SAE defines six discrete and mutually exclusive levels of driving automation in the context of motor vehicles (Figure 9.1): Level 0-No driving automation, Level 1-Driver Assistant, Level 2-Partial Driving Automation, Level 3-Conditional Driving Automation, Level 4-High Driving Automation, and Level 5-Full Driving Automation. Each level of driving automation defines the respective roles of human drivers.

SAE J3016 further defines that the roles of Levels 1 and 2 driving automation systems are to sustain longitudinal and/or lateral vehicle motion control subtasks of the DDT. Drivers are not required to maintain these subtasks but are expected to complete the DDT by performing the object and event detection and response (OEDR). A driving automation system that reaches Level 3 carries out the entire DDT. However, when the system fails and issues alerts, or when the driving automation system approaches its operational design domain (ODD) limits, a driver is expected to serve as a

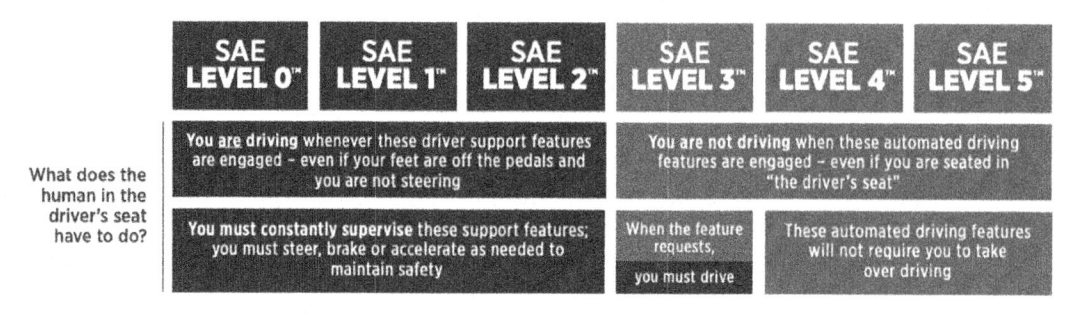

	SAE LEVEL 0™	SAE LEVEL 1™	SAE LEVEL 2™	SAE LEVEL 3™	SAE LEVEL 4™	SAE LEVEL 5™
What does the human in the driver's seat have to do?	You are driving whenever these driver support features are engaged – even if your feet are off the pedals and you are not steering			You are not driving when these automated driving features are engaged – even if you are seated in "the driver's seat"		
	You must constantly supervise these support features; you must steer, brake or accelerate as needed to maintain safety			When the feature requests, you must drive	These automated driving features will not require you to take over driving	

FIGURE 9.1 The summary table (partial) of levels of driving automation. (SAE International, 2021)

DDT fallback-ready user to take over the DDT. Finally, a driving automation system can qualify as Levels 4 or 5 if it is capable of performing the entire DDT and DDT fallback without mandatory user intervention (Level 4) or without user intervention at all (Level 5). All in-vehicle users will become passengers. A driving automation system qualified as Level 3 or above is specifically termed as automated driving system (ADS). Based on these taxonomies and definitions, in-vehicle user roles can be categorized into three major operations and expectations: (1) Active Driver: responsible for all or partial DDT (Levels 0–2), (2) Passive Driver: receptive to request to intervene as the DDT fallback-ready user and take over DDT when requested (Level 3), and (3) Passenger (Levels 4 and 5). The human-automation function allocation across all levels can be found in Table 9.1. It is noted that although vehicles equipped with Level 4 and Level 5 ADS might have remote users, we mainly focused on the in-vehicle users in this chapter.

9.2.2 Driver Distraction and Inattention

Driving is a complicated, dynamic task that requires drivers' physical and mental demand. Due to this high demand and drivers' limited resources, competition often occurs when they are supposed to do multitasks. In driving literature, there are three types of tasks – primary (driving), secondary (other tasks, but relevant to driving, such as a navigation task), and tertiary tasks (non-driving-related tasks, such as radio volume control) (Geiser, 1985). Both the secondary and tertiary tasks can be classified into distraction tasks. Driver distraction is defined as "a diversion of attention from driving, because the driver is temporarily focusing on an object, person, task or event not related to driving, which reduces the driver's awareness, decision-making ability and/or performance, leading to an increased risk of corrective actions, near-crashes, or crashes" (Hedlund et al., 2005, pp. 2). Young et al. (2007) categorized distraction types into visual, auditory, biomechanical (physical), and cognitive. Driver distraction from multitasking while driving has shown tremendous effects on driver performance, such as increased driver response time (Hancock et al., 2003; Morel et al., 2006; Strayer et al., 2003), increased lane deviation (Reed and Green, 1999), and decreased visual information processing (Barkana et al., 2004; McKnight and McKnight, 1993), which will result in increased risk of crashes (US Department of Transportation, National Highway Traffic Safety Administration, 2014a,b).

Another school in driving research used the term, driver inattention. For instance, Lee et al. (2008) defined driver inattention as "diminished attention to activities critical for safe driving in the absence of a competing activity" (pp. 32). However, because of unclearness of "diminished", a new definition was proposed as "insufficient or no attention to activities critical for safe driving", by incorporating driver distraction as one form of driver inattention. This relatively broader term, "driver inattention" also seems to embrace other phenomena, such as daydreams (Pritzl, 2003), mind wandering (Burdett et al., 2019), or emotions (Jeon et al., 2014) while driving.

TABLE 9.1
Human-Automation Function Allocation based on Levels of Driving Automation Systems

	DDT		
	Sustained Lateral and Longitudinal Control	**OEDR**	**DDT Fallback**
Level 0 - No Driving Automation	Active Driver: • Performs the entire DDT		
Level 1 - Driver Assistance	Active Driver • Completes the remaining DDT not maintained by the system System • Sustains the longitudinal **or** the lateral vehicle control		Active Driver • Supervises the automation system and intervenes as necessary • Determines the engagement and disengagement of the automation system • Immediately performs the entire DDT when necessary
Level 2 - Partial Driving	System • Performs sustained longitudinal **and** lateral vehicle control	Active Driver • Performs the remaining DDT not performed by the automation system	Active Driver • Supervises the automation system and intervenes as necessary • Determines the engagement and disengagement of the automation system • Immediately performs the entire DDT when necessary
Level 3 - Conditional Driving Automation	System • Performs the entire DDT within its ODD • Determines whether the ODD limits are about to be exceeded or whether a DDT performing system failure happens, if so, issue alerts to drivers timely and properly to request for intervening • Disengages at a proper type after the alert issued		Passive Driver (DDT fallback-ready user) • Performs the entire DDT after the automation system disengages and determines whether and how to achieve a minimal risk condition • Receives automation system alerts and performs DDT fallback
Level 4 - High Driving Automation	System • Performs the entire DDT **within its ODD** • Performs the DDT fallback and automatically transitions to a minimal risk condition upon DDT performing system failures, user requests, or approaching the exit of ODD Passenger: • May request ADS disengagement and become a driver afterward		
Level 5 - Full Driving Automation	System • Performs the entire DDT • Performs the DDT fallback and automatically transitions to a minimal risk condition upon DDT performing system failures, user requests, or approaching the exit of ODD		

Source: Adapted from SAE International (2021).

9.2.3 WORKLOAD

Driving is a complex task involving multiple cognitive processes that can be influenced by road conditions and other road users to be more challenging. As a result, driving is an activity that engages high workload demands and researchers have sought to measure and reduce workload demands, because high workload has been associated with more challenging tasks and worse driving performance (Kantowitz & Simsek, 2001).

9.2.3.1 Workload Measures

Workload can broadly be measured through a variety of physiological, subjective, and performance measures (Miller, 2001). In terms of physiological measures, workload can be evaluated by measuring cardiac activity such as heart rate (Roscoe, 1992), respiratory activity (Mehler et al., 2009), gaze behavior (van Leeuwen et al., 2015), and brain activity (Mader et al., 2009) among other methods. Heart rate was observed to increase during the approach phase of a flight in pilots (Jorna, 1993), with increased activity and practice leading to an overall reduction in mean heart rate. However, physiological measurement of workload suffers from some issues because physiological responses to events can be delayed for some measures (Nagasawa & Hagiwara, 2016) or face issues of not satisfying all measurement requirements for a single measure such as mental workload (Charles & Nixon, 2019).

Workload has frequently been measured through the administration of self-report questionnaires. The NASA Task Load Index (TLX) workload scale (Hart, 2006) is a major method for measuring workload. The questionnaire collects data from participants upon the completion of a task in two stages. First, participants fill out six subscales that measure mental demand, physical demand, temporal demand, performance, effort, and frustration. Overall unweighted workload can be measured this way. Additionally, participants go through another task to compare and select more workload-inducing subscales in the second phase that consists of pairwise comparisons of the subscales. Based on this process, the weighted workload scores can be obtained. Other subjective methods exist, such as the Subjective Workload Assessment Technique (SWAT) (Reid & Nygren, 1988) and DALI (Pauzié, 2008), a driving-specific workload measurement scale.

Finally, some studies have investigated workload based on the performance of users in the observed process. Both primary and secondary performance can be explored to determine dual-task performance, with lower dual-task performance associated with higher workload conditions, such as more challenging environmental conditions in driving (Lee et al., 2021; Yoon & Ji, 2019).

9.2.3.2 Workload in Manual Driving

Traditional driving research has sought to reduce driver workload during the driving task. Driving workload has been associated with changes in mental and temporal workload, as different driving maneuvers have varied levels of workload (Hancock, Wulf, Thom, & Fassnacht, 1990). Higher workload events include driving around intersections (Liao et al., 2018), driving in high traffic conditions (Cantin et al., 2009), and driving in urban areas with a wide number of different road users (pedestrians, bikes) (Duivenvoorden et al., 2015). Environmental conditions can also influence workload, such as driving in the rain or in foggy conditions (Hoogendoorn et al., 2011).

Researchers have designed displays that can help in decreasing driver workload. These displays can include auditory alerts (Nadri et al., 2021b) and the installation of visual signs that can inform drivers about incoming obstacles (Bakhtiari et al., 2019). However, some displays were found to increase driver workload and lead to better driving performance in specific drivers, such as novice drivers (Klauer et al., 2016). As novice drivers are prone to incorrect threat assessment, increasing workload was suggested to increase driver awareness and can be important in realigning driver expectations with the level of difficulty of the driving maneuver (Cooper et al., 1995; Mourant & Rockwell, 1972; Smith et al., 2009).

FIGURE 9.2 Nadri et al. (2023) designed new visual (small screen on the dashboard with an orange dotted line) and auditory displays (speech and non-speech sounds) about upcoming environments in level 4 highly automated vehicles. Adding auditory displays decreased drivers' workload than the visual-only display.

9.2.3.3 Workload in Automated Driving

While automated driving seeks to reduce driver workload by taking over parts or the entirety of the driving task, there are still tasks that can involve high driver workload (Lee et al., 2021). A prominent high workload task in automated driving is takeover, as drivers need to quickly reengage with the driving task, assess their surroundings, and respond to unexpected driving situations that the ADS cannot respond to. Researchers have designed displays that can adequately alert drivers of upcoming changes in driving automation (Sanghavi et al., 2020; Nadri et al., 2023, Figure 9.2). As with manual driving, some level of driver workload was suggested as beneficial to takeover performance, since overly relying on the automation can lead to drivers feeling overwhelmed with high workload situations (Kundinger et al., 2019). A balanced exposure to workload can thus improve driving performance.

9.2.4 SITUATION AWARENESS

Situation awareness (SA) is formally defined as a person's perception of the elements in the environment within a volume of time and space, the comprehension of their meaning, and the projection of their status in the near future (Endsley, 1988a). The concept of SA has been used to create attention models that separate and clearly define the operator's understanding of a system and the actual system status (Endsley, 1988a). As such, SA is an important aspect of the driving experience since it helps break down how attention is directed by the human driver in a manual vehicle. This understanding can help better design and evaluate systems that seek to improve driving performance and reduce human error (Jeon et al., 2015; Scholtz et al., 2004; Scholtz et al., 2005). In envisioning and designing for future automated vehicle interfaces, SA remains a crucial component to consider especially as maintaining an awareness of one's surroundings is expected to be necessary even as vehicles reach Level 3 and Level 4 automation (SAE International, 2021).

9.2.4.1 SA Model

The SA model of attention (Endsley, 1995) divides SA into three sub-components: perception, comprehension, and projection.

> In SA level 1, perception, operators make an initial scan of their surroundings and involve localizing critical data. Deficiencies at this level can be attributed to difficult or unintuitive locations for important data or misperceptions on the part of the operator.

In SA level 2, comprehension, operators synthesize the information perceived initially and can develop an understanding of what the situation currently is. Deficiencies at this level could be due to misleading information or assumptions made by the operator that led to the wrong understanding.

In SA level 3, projection, operators use their understanding of the situation to predict future actions of the system, giving them the ability to anticipate potential threats. Failure at this level could be due to a wrong mental model of the situation or a low understanding of the system's capabilities.

9.2.4.2 SA Measures

SA in driving has been measured through a variety of methods. Research in the field has converged to using both eye-tracking and subjective self-reports as the main techniques used to measure SA (Gugerty, 2011).

Eye movements, such as gaze patterns and fixations, serve as primary measures of SA in gaze behavior. Eye movement patterns can help in determining attentional resource allocation in manual driving (Smolensky, 1993). This effect extends further into automated driving, where it can be used to detect takeover performance and the efficiency of head-up displays (Greatbatch et al., 2022; Liang et al., 2021).

Self-report measures of SA exist and can be collected at select intervals during a drive or after the completion of a driving section. Techniques in aviation research, a field that resembles driving due to its high monitoring requirements and high risk, exist such as the Situation Awareness Rating Scales (Waag & Houck, 1994), the cranfield situation awareness scale (Dennehy, 1997), and the Mission Awareness Rating Scale (Matthews & Beal, 2002) that exemplify SA measurements asynchronously.

However, self-report measures done after the drive is completed could suffer from high inaccuracies and a dissociation from the driving situation (Walker et al., 2008). On the other hand, concurrent subjective reporting has shown great promise. This is most prominent in the case of the situation awareness global assessment technique (SAGAT) (Endsley, 1988b). This technique is well established and based on assessing the three levels of SA by interrupting the task at hand and inquiring about the perception, comprehension, and projection abilities of users. In driving research, this technique has been effectively used in both manual (Young et al., 2013) and automated driving conditions (Nadri et al., 2021a).

9.2.4.3 SA in Manual Driving

As driving requires driver attention on the road, other road users, and vehicle displays, maintaining high SA and hazard perception has been a goal of in-vehicle displays and designers since it relates directly to accident risk (Horswill & McKenna, 2004). In manual driving, where driver attention on the road is always required, automotive designers have frequently measured SA levels to evaluate and develop in-vehicle displays that could increase SA levels for crucial driving situations. At intersections, research has determined that information should be provided at adequate levels that can clarify driver mental models but avoid information overload (Plavšić et al., 2010). Research also determined that SA compatibility between different road users is important in determining safety (Salmon et al., 2014; Salmon et al., 2013) because different users hold different mental models of the driving situation and inconsistencies between these models could partially lead to driving accidents.

9.2.4.4 SA in Automated Driving

According to the SAE definitions (2021), drivers operating vehicles with Levels of 0–3 driving automation systems are required to monitor the driving environment and be ready to respond to unexpected driving events. In Level 4 automation, this monitoring task remains although the vehicle is envisioned to be capable of handling most driving situations. As such, SA is a crucial component in determining the safety of driver takeover (Endsley, 2018). Displays have been developed for automated vehicles that utilize auditory cues (Nadri et al., 2021a) and visual interfaces (Greatbatch et al., 2022; Lindemann et al., 2018) to communicate driving intent and information needs with drivers and passengers alike. As driving involves many stressful or emotionally charged situations, soothing music, and soundscapes could address this issue in automated vehicles (Lee et al., 2022; Nadri et al., 2021a; Sanghavi et al., 2020).

9.2.5 Emotions

As in other HCI research, traditional driving research has focused on driver cognitive processes and their effects on driving behavior. Road rage or aggressive driving has been a topic in driving research, but drivers' other emotional states have not been systematically studied until recently. If any, early research used a survey method, by asking drivers about their emotional moments and how they influenced driving behavior (e.g., Deffenbacher et al., 2001, 2003), which heavily depends on their memory. Recent studies have started using a driving simulator to reflect drivers' real-time behaviors when they are in a certain emotional state. This section reviews emotion and driving research with a focus on these simulation studies.

9.2.5.1 Background and Definitions

There are different terms about emotions, including feeling, mood, and affect. Emotion is a lay term that people use in everyday life, but in a more scientific setting, it refers to transient biological response to external stimuli. Feeling is more subjective awareness of emotions. Mood refers to a longer-term state, compared to emotion. Affect is sometimes used to describe faint emotional states, but in affective science, researchers use affect as an overarching term, including all of the above. To see more about emotions and affect in HCI, see (Jeon, 2017a).

In emotion research, there are, largely, two types of theoretical approaches, including a dimensional approach and a categorical approach. On one hand, Russell (1980) proposed the circumplex model, which includes two or three dimensions, including valence (negative - positive) and arousal (low-high) or dominance (without control or empowered). This dimensional approach is effective when researchers want to use physiological sensors to measure emotions. Also, it can capture in-between emotional states with more granularity. However, in our everyday life, we do not likely say, "I'm 60% aroused and 40% positive". On the other hand, Ekman (1999) argued that humans have basic emotions (e.g., six basic emotions–anger, happiness, fear, disgust, surprise, and sadness) and this set of emotions is pervasive regardless of culture. There are strong arguments about this approach, but it is sometimes hard to measure using objective methods. However, it is rather easier to ask people to answer. Discrete approaches can provide design guidelines for mitigation interfaces to deal with a specific emotional state.

9.2.5.2 Emotion Induction

To conduct emotion-driving research studies, a specific emotional state should be induced. There are different methods to induce emotions, for example, using pictures, passages, videos, and music pieces. Literature shows that writing autobiographical passages is an effective emotion induction method (Fakhrhosseini & Jeon, 2017) because it can engage participants into their own memory and experience. Emotion-driving research has also shown successful induction results with the autobiography method (Jeon, 2017b). This is a type of incidental affect, which is not related to the task. Driving simulation research has also used integral affect (related to the task) to maintain emotional states throughout the driving task by including emotion-inducing events in their scenarios. For example, if participants write down their anger-inducing memory (e.g., fighting with their friends), it belongs to incidental affect and if they get angry because of the front car, it is integral affect. These methods have also been combined in a single study and shown effective results (e.g., Jeon et al., 2014; Muhundan & Jeon, 2021).

9.2.5.3 Emotion Measures

Multifaceted measures have been used to estimate drivers' emotions. Physiological sensing has been used, including heart rate and skin conductance response (Vasey, Ko, & Jeon, 2018). Neuroimaging has also been used, including EEG or fNIRS (FakhrHosseini et al., 2015a; FakhrHosseini & Jeon, 2019; Fan, Bi, & Chen, 2010). These neuroergonomic sensors are sensitive to movement noise, but because driving is rather a static environment, they could be applied to. Facial or speech detection systems (Jeon & Walker, 2011a; Thomas & Jeon, 2020) have also been developed, but real applications are still challenging due to the constraints, such as angle, light, noise, etc. Of course, for manipulation check, researchers ask participants about their emotional states before induction, after induction, and after the experiment.

9.2.5.4 Emotional Effects on Driving

Jeon and Walker (2011b) analyzed factor structures of plausible emotions in driving contexts and extracted critical emotional states–fearful, happy, angry, depressed, curious, embarrassed, urgent, bored, and relieved. As expected, anger was the most frequently investigated emotion in driving research. Anger showed worse driving performance, compared to neutral driving, such as increased overall driving errors, overspeed, maximum speed, lane deviation, and aggressive driving. Negative anger effects have also been shown in automated driving research (e.g., takeover performance; Du et al., 2020; Sanghavi, Zhang, & Jeon; 2020). Fear is on the same quadrant as anger (high arousal and negative valence), but it shows different results from anger (Jeon et al., 2014). Fear did not show any worse driving performance, but showed defensive driving. Therefore, we can infer that arousal and valence may not fully explain the effects of emotion on driving performance. Interestingly, happy drivers are not better drivers. In a simulation study, happy drivers showed significantly more driving errors than neutral drivers (Jeon et al., 2014). Sad drivers showed longer driving time and partial evidence of frequent lane departure compared to neutral drivers (Jeon, 2016; Jeon & Croschere, 2015). Taken altogether, not all negative emotions influence drivers' performance and experience in the same way. Therefore, specific intervention approaches would be required.

9.2.5.5 Mitigation Methods

To lessen and mitigate emotional effects on driver experiences and performances, research has used cognitive and sensory level approaches. Adopting a psychological emotion regulation model (Gross, 1998), Harris and Nass (2011) showed that cognitive reappraisal led to better driver emotion regulation and driving behavior. In the same line, Jeon et al. (2015) showed that attention deployment was also effective when applied to in-vehicle speech agent design. The speech agents not only reduced anger level and perceived workload but also increased driver SA, which led to better driving performance. At the sensory level, research has used olfactory, auditory, and visual approaches. For example, Dmitrenko et al. (2020) showed that low arousal and/or positive valence scent (e.g., rose or peppermint) led to better driving behavior and increased driver well-being. In their research, these scents made angry drivers happier. Also, self-selected music had positive effects on angry drivers (FakhrHosseini & Jeon, 2019), and both happy and sad music improved driving performance, even though happy music significantly improved driver workload level (FakhrHosseini et al., 2014).

9.2.6 Trust

9.2.6.1 Background and Definitions

Trust has gained more and more interest in the automotive research community as vehicles' automation level increases. Mayer, Davis, and Schoorman's (1995) definition provides detailed components of trust in the human–human relationship. Trust is the "willingness of a party to be vulnerable to the actions of another party based on the expectation that the other will perform a particular action important to the trustor, irrespective of the ability to monitor or control that other party" (p.712).

This definition highlights three key aspects of trust: vulnerability, expectations, and lack of monitoring or control. In a similar line, in the human-automation context, trust has been defined as "the attitude that an agent will help achieve a person's goals in a situation characterized by uncertainty and vulnerability" (Lee & See, 2004, p.51). In the model by Lee and See, three categories of attributes were also identified as necessary areas for the development of trust in automation and mapped onto the items from Mayer et al.'s model: process (integrity), performance (ability), and purpose (benevolence). Trust calibration is another important consideration in designing automated systems. Parasuraman and Riley (1997) classified three relevant terms about trust calibration, depending on perceived/true reliability and complexity of the system. If users are doubtful about frequent false alarms and do not use the system at all, it is "disuse". If users overtrust the system (complacency), it is "misuse". In this case, users' SA and skills will be degraded. Applications of automation without consideration of human side consequences would be "abuse". To avoid disuse or misuse of automated vehicle systems, designing interfaces in a balanced way is required. Traditionally, trust in automation has been viewed as what the operator can have toward an automated system (e.g., supervisory control), whereas Chiou and Lee (2021) assert that trust is relational and so, it can be best understood through the interactions between entities (e.g., human-AI/robot teaming). Interactions will likely change users' trust calibration. Accordingly, the measures about trust can also be updated from reliance, acceptance, and decisions to use or not to use the automated technology in team performance.

9.2.6.2 Trust Measures

As with other HCI constructs, trust cannot be measured directly. Therefore, researchers measure participants' trust with subjective ratings (e.g., Jian, Bisantz, & Drury, 2000; Madsen & Gregor, 2000) and compare these ratings to trust-related behaviors, such as reliance on automation (Meyer, 2004). For example, the 12-item assessment scale uses trust and distrust as its subscale. Another scale (Körber, 2018) for trust in automation was developed and validated in the automated driving context. It has familiarity, intention of developers, propensity to trust, reliability/competence, understanding/predictability, trust in automation, understanding, and performance. Another automated driving-specific scale is the Situational Trust Scale for Automated Driving (STS-AD) (Holthausen et al., 2020), which measures situational trust (more dynamic compared to dispositional trust, which is rather static). As behavioral measures, speed and acceleration are often measured to represent reliance and compliance (Cotté, Meyer, & Coughlin, 2001; Yamada & Kuchar, 2006). Eye-tracking is another measure of trust in automation. Data from subjective and behavioral measures have been further validated with physiological measures (Bethel, Salomon, Murphy, & Burke, 2007; Montague, Xu, & Chiou, 2014). Another consideration is looking into cognitive trust and affective trust respectively. According to McAllister (1995), cognitive (cognition-based) trust is defined as trust based on the knowledge and evidence on someone's ability and achievements and affective (affect-based) trust is defined as trust based on the emotional bond with someone. Lee and Lee (2022) showed that in their driving simulation study, politeness strategies of the in-vehicle agent significantly influenced affective trust, but not cognitive trust in the autonomous driving context.

9.2.6.3 Trust in Automated Driving

Obviously, performance expectations and its reliability are the important adoption determinants for automated vehicles (Kaur & Rampersad, 2018). Research shows that drivers' demographics (e.g., culture, gender, age, and social norms) can influence their initial trust formation (Hoff and Bashir, 2015). Then, system performance and usability can also influence their trust level during and after using the automated vehicle technologies (Hoff and Bashir). In terms of interface design, transparency, and uncertainty have been the focus of trust research in automated vehicles. For example, Lorenz, Kerschbaum, and Schumann (2014) showed that their AR system, which suggests a corridor the driver can safely steer through in their takeover, resulted in higher consistency in driving trajectories. Another study (Wintersberger et al., 2017) showed that augmenting traffic objects on the windshield increased drivers' trust and technology acceptance. This study demonstrated that only showing relevant objects without explaining the specific driving actions can also enhance driver comprehension and thereby, trust. Large et al. (2019a) explored trust-related themes with a natural language-based conversational user interface, which used the Wizard-of-Oz method in the fully autonomous vehicle. Interestingly, they found that while participants engaged in a conversation with their natural language-based agents, they still considered that trust is still highly functional and task-based, rather than interpersonal or social. This might imply that people still have a mental model that the current interfaces respond to users' requests, rather than making actual conversation. Zang and Jeon (2022) showed the interaction effects between transparency and system reliability in the conditionally automated vehicle context. In their driving simulation study, with high reliability, participants rated proactive agents (i.e., the agents provide all information at once) significantly higher in intelligent and competent scales than the on-demand agents (i.e., participants can request further information after brief information provided), whereas they rated on-demand agents significantly higher in warmth and numerically higher in trust scales. Interestingly, maximum lateral acceleration in takeover showed significant interaction effects. With high reliability, the proactive agent led to significantly higher maximum lateral acceleration than the on-demand agent. With low reliability, the on-demand agent led to numerically higher maximum lateral acceleration than the proactive agent. This might imply that when the system is highly reliable, providing more information might increase the participants' confidence; whereas when the system is unreliable, providing less information might increase the participants' urgency perception. Much research predicts that when anthropomorphism increases, users' trust will increase (Eyssel et al., 2012), but the detailed effects should be further elaborated.

Researchers point out that more research can be conducted on dynamic aspects of trust, including trust development procedure, continuous measures, and how trust can influence interactions between a driver and a vehicle (or agent) as well as how interactions can influence their trust level (Chiou & Lee, 2021). As mentioned, there might be different mechanisms between cognitive trust and affective trust. Given that psychological research shows how emotions influence trust (Dunn & Schweitzer, 2005), this topic in the automated driving domain would be an important future research topic.

9.2.7 Motion Sickness

9.2.7.1 Non-Driving-Related Tasks and Automated Vehicles

Increased levels of vehicle automation (SAE Levels 3–5) change the way users tend to use their vehicles (SAE International 2021). One of the most noticeable changes is in the role of the driver; the *active driver* begins to take on the role of a *passenger*. This change in the driver's role is accompanied by a change in the tasks that the driver performs. With the driver free to take on the role of a passenger, they are able to engage in as many non-driving-related tasks (NDRTs) or passenger tasks (PTs) as possible within the vehicle cabin. A myriad of such NDRTs have already been identified which indicates that future users will want to engage in such NDRTs or PTs to make better use of their time during transportation (Lee et al., 2022; Shi & Frey, 2021). This is the promise of self-driving vehicles, where the driver is able to take on the role of a passenger. The challenge, however, in engaging in PTs effectively is motion sickness. Motion sickness is characterized by feelings of malaise, dizziness, nausea, and vomiting (Kuiper et al., 2020b; Reilhac et al., 2016). The human body is especially susceptible to motion sickness when performing tasks while in motion. If society were to truly enjoy the benefits of privately owned autonomous vehicles, it is paramount that manufacturers provide motion sickness mitigating solutions in their vehicles.

9.2.7.2 Research Framework

A survey of existing research led to the development of a framework that addresses the different aspects of motion sickness research – bird's-eye-view research framework (Dam & Jeon, 2021). Figure 9.3 shows the framework. The researcher must begin with an understanding of the Causation

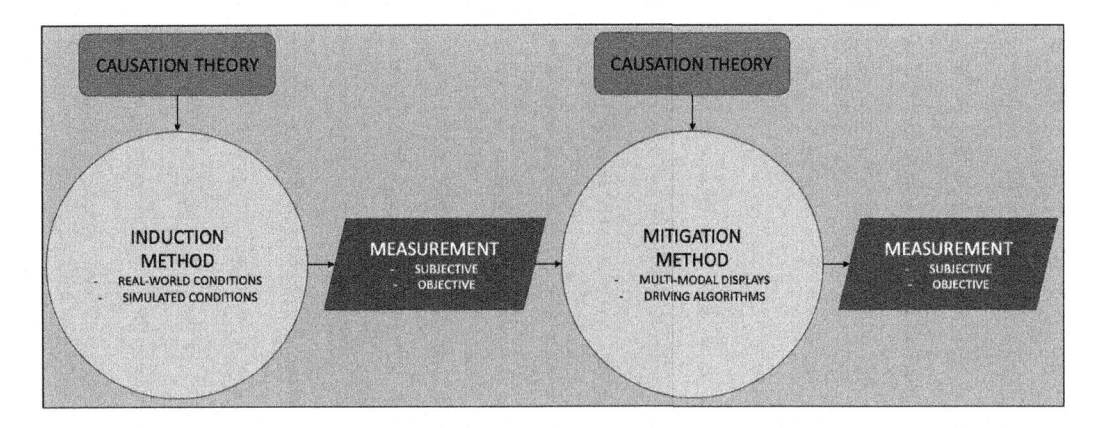

FIGURE 9.3 Bird's-eye-view research framework.

Theory behind motion sickness. Next, to further test the effectiveness of any potential solution, the researcher must first use a verified induction method, using real-world tracks or simulators to induce motion sickness in participants. Measurement of motion sickness symptoms is a challenging yet important aspect, since verification of the effectiveness of any solution truly depends on the measured constructs. Finally, the area of mitigating solutions needs to be explored with a human-centric approach, such that passengers can engage in NDRTs or PTs advantageously.

9.2.7.3 Causation Theory

To suggest mitigation strategies, it is important to understand what causes motion sickness. The most widely accepted theory is the sensory conflict theory, which suggests that when there is a mismatch between sensory information and bodily sensory state, motion sickness occurs (Kuiper et al., 2018). When the vestibular system (responsible for maintaining sense of balance and orientation) and the visual system (responsible for sense of sight) send conflicting information to the brain, it leads to symptoms of motion sickness. As the brain develops over time, it learns to accept a certain combination of postural orientation and visual information as normal; it is when conflicting information about orientation and vision is received, that the brain perceives it to be in a conflicting state of being, thus leading to motion sickness.

Another causation theory, that is not as widely accepted but is more applicable to the context of driving, is the Anticipation Theory (Diels & Bos, 2016; Kuiper et al., 2020a). The lack of knowledge about immediate upcoming motion can lead to motion sickness. This theory is able to explain why drivers are less likely to get motion sick when driving on a windy, hilly, curvy road, whereas passengers are more likely to experience motion sickness symptoms. Drivers remain in control of the vehicle at all times and have complete awareness about how the vehicle will move in the next moment, whereas passengers tend to simply experience the motion as it occurs. It has been demonstrated that even without sight, simply knowing the direction of upcoming motion through auditory cues, can help reduce feelings of motion sickness (Kuiper et al., 2020a). This theory is interesting because with the advent of automated driving, drivers will begin to take on the role of passengers, thus increasing their susceptibility to becoming motion sick. The Anticipation Theory remains to be tested and validated further, but presents a novel approach toward understanding motion sickness causation among passengers of automated vehicles.

9.2.7.4 Motion Sickness Induction

The first step in motion sickness research is the effective induction of motion sickness. Real-world driving and driving simulators (with and without the use of VR headsets) have both been shown to be verified induction methods. In real-world driving scenarios, participants take the role of a

FIGURE 9.4 'bdpq' NDRT

passenger in a fully autonomous vehicle, and are engaged in an NDRT or PT. Vehicles are instrumented accordingly to give the passenger the impression that the vehicle is self-driving as the Wizard-of-Oz strategy. The driving route is carefully planned to take place in real traffic (Yusof et al., 2020; Wang et al., 2020; Karjanto et al., 2017) or on a test track (Kuiper et al., 2018; Salter et al., 2019; Paddeu et al., 2020; Schartmüller & Riener, 2020; Bohrmann & Bengler, 2020) consisting of serpentine roads or slalom maneuvers. The driving characteristics are also manipulated to be more aggressive; lateral accelerations of up to 2.84 m/s^2 have been used to induce motion sickness when combined with a visual NDRT (Karjanto et al., 2017). A well-used visual NDRT has been the 'bdpq' test (Figure 9.4) where the participant is asked to pick out a pattern of two consecutive letters from a random matrix of "b", "d", "p", and "q" (Bohrmann & Bengler, 2020).

Due to greater costs of real-world methods, driving simulators are often a more popular choice. Driving simulators are also required to have at least 4 degrees of freedom (roll, pitch, yaw, and heave). The advantage of driving simulators is that a much more diverse set of conditions can be replicated with minimal cost. The driving characteristics and route can be similar to real-world methods, and must be designed specifically to increase sensory conflict; NDRTs and PTs should be designed to omit peripheral information to the extent possible.

9.2.7.5 Motion Sickness Measures

Measurement of motion sickness symptoms is the second challenge. This is critical to ensuring the effectiveness of the induction or mitigation methods. Most current measurement methods include subjective rating scales. Some of the most commonly used scales are: the Motion Sickness Assessment Questionnaire (MSAQ), the Fast Motion Sickness (FMS) Scale, and the Misery Index Scale (MISC). MSAQ is a nine-point multi-dimensional scale consisting of 16 questions that is administered pretest and posttest only (Gianaros & Stern; 2010). FMS, consisting of only one question, is a 20-point scale that can be administered every minute (Keshavarz & Hecht, 2011). The MISC scale is an 11-point rating scale that can be administered in a temporally varying nature, although ratings above a 6 indicate vomiting, which is a penultimate symptom of motion sickness (Reuten et al., 2020; Reuten et al., 2021; Winkel et al., 2022). These existing scales are not uniform in nature, and therefore, the validity of the measurements across studies can vary widely. Further, these scales have only been verified against other subjective scales such as the Simulator Sickness Questionnaire (SSQ), Pensacola Diagnostic Index, and Nausea Profile (Cramer et al., 1968; Kennedy et al., 1993; Muth et al., 1996).

Other measures of motion sickness include objective measures such as galvanic skin response, heart rate variability, and electrogastrography. The challenge with objective measures is their susceptibility to noise, inter-participant variability, and the likelihood of such measures not reaching statistical significance in studies. Nonetheless, electrogastrography (EGG) has still been found to be a reliable metric of motion sickness (Popović et al., 2019; Zhang et al., 2016). EGG is the measurement of myoelectrical activity in the stomach using non-intrusive electrodes. Factors such as the crest factor, dominant frequency, and amplitude increase have been shown to be sufficient in objectively capturing motion sickness symptoms (Gruden et al., 2021).

9.2.7.6 Mitigation Methods

Current mitigation methods have approached the problem from the perspective of existing causation theories. Since the Sensory Conflict Theory is widely accepted, most mitigation methods have attempted to bridge the information gap between the visual and vestibular sensory systems. It is recommended to locate visual displays at eye height of a seated passenger to allow for greater peripheral vision; displays at the height that require passengers to look down lead to greater motion sickness symptoms (Kuiper et al., 2018). Minimization of head movements has also been attempted to reduce the degree of inputs to the vestibular system (Kennedy et al., 1993; Lackner & Graybiel, 1984), by requiring passengers to fix their gaze on a distant object. This technique is also popular among dancers when spinning to minimize incongruence between the visual and vestibular systems. Further, displaying content on visual displays that are consistent with the vehicle's motion or trajectory has also been known to reduce sensory conflict (Bos et al., 2005).

Based on the Anticipation Theory, a few mitigation strategies have been developed that are dedicated to providing passengers with information on either the current or upcoming motion of the vehicle. In-vehicle LED strips, to provide visual cues three seconds before an upcoming motion (left or right 90 degree turns) to rear seat passengers, have been shown to reduce motion sickness symptoms and improve SA (Karjanto et al., 2017). Visual cues have been studied on personal devices, such as smartphones or e-readers; the motion of bubbles floating along the perimeter of the device's screen has been used to indicate acceleration characteristics of the vehicle in real time (Meschtscherjakov et al., 2019). VR headsets have also been used as a tool to immerse participants in fully autonomous vehicles; similar visual cues in the form of multiple small balls of light within the interior surfaces of the virtual vehicle, with as little as 500 ms prediction period, have shown to reduce motion sickness symptoms, too. (Winkel et al., 2021). Additionally, other modalities such as audio, haptic, and olfactory cues have been investigated based on the Anticipation Theory. Speech-based auditory cues, to inform participants of fore and aft motion, have been shown to reduce motion sickness symptoms in a proof-of-concept study by Kuiper et al. (2020a, b). Haptic feedback, provided at 183 Hz for a duration of three seconds with an ON and OFF time period of 0.6 seconds, displayed at the corresponding wrist for left and right turns, three seconds before the event has been shown to improve SA of participants when they are engaged in a visually intensive NDRT (Yusof et al., 2020). Finally, olfactory interventions such as lavender and ginger scents have been shown to increase engagement in NDRTs/PTs (Schartmüller & Reiner, 2020) while in an automated vehicle, but the data from these studies did not reach statistical significance.

9.3 DRIVING RESEARCH METHODS AND ENVIRONMENTS

9.3.1 Research Approaches

There are a great number of research approaches that can be used to evaluate automotive user interfaces. The most typical study approaches are driving simulator studies, test track studies, on-road vehicle studies, and naturalistic driving studies. The experimental control, validity, and costs vary across those approaches (Figure 9.5 and Table 9.2).

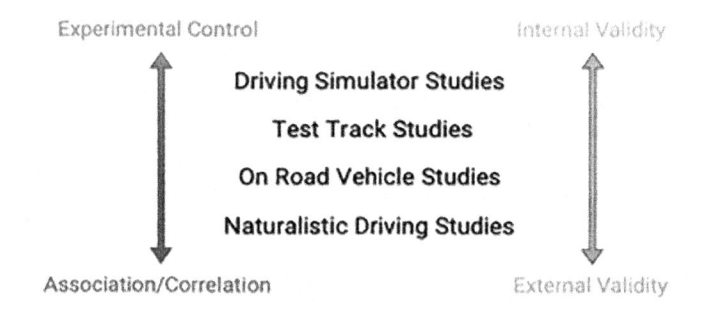

FIGURE 9.5 Experimental control and validity differences across research approaches.

TABLE 9.2
Comparison between Driving Simulator Studies and Naturalistic Driving Studies

	AutoUI Evaluation	Strengths	Weaknesses
Driving Simulator Studies	• Ideation • Prototyping	• Simulating dangerous scenarios • Ability to draw the causal-effect relationship	• Low external validity • Contrived situation • Simulation sickness • High initial costs
Naturalistic Driving Studies	• Beta testing • Aftermarket	• High external validity • Ability to capture real driving behaviors • Power of data after properly coded	• Time-consuming • Hard to draw the causal-effect relationship

9.3.2 Driving Simulator Studies

Driving simulators vary in terms of their fidelity. A display, a control panel (e.g., simply a keyboard), and a chair as the seat can formulate a basic simulator with the lowest fidelity. Each component can be advanced to enhance fidelity. The display can be upgraded to triple TV-size monitors, or a VR head-mounted display to enhance reality and immersion. The control panel can be upgraded to the steering wheel and pedal sets. Finally, the seat can be advanced to a fixed-based or motion-based driving seat, or even more realistically, to a cab that can accommodate different vehicle types. The National Advanced Driving Simulator (NADS-1 simulator, Figure 9.6) is one of the world's highest-fidelity simulators that provides the most realistic driving simulation experiences.

Although with limited external validity, the advantages of driving simulator studies cannot be underestimated. Dangerous scenarios (e.g., pedestrians crossing the road, near-crashes) or constrained situations (e.g., implementing a new intervention on the road) in real-life can be simulated in a driving simulator study with no more than minimal risks. For instance, Jamson et al. (2010) were able to use a driving simulator study to test out 20 different speed abatement techniques that can otherwise be challenging or infeasible using real-world testing. Researchers also have great control over the virtual environment components and keep the exposure consistent across participants, thus providing great internal validity to support the causal-effect relationship between certain variables and corresponding driving behaviors. Using the Wizard-of-Oz methods, researchers are able to test innovative prototypes for future technologies that do not exist yet, which largely contributes to the success of product development by introducing human factors and user-centered design at an early stage. Given the advancement of vehicle automation, much research has explored automotive user interfaces implemented in future vehicles with above Level 3 driving automation systems. Researchers have designed and evaluated different interfaces to facilitate the human-automation authority transition under conditional driving automation (e.g., Du et al., 2020; Hester et al., 2017; Hock et al., 2016; Roche & Brandenburg, 2020; Wang et al., 2022b). Innovative interfaces have also

FIGURE 9.6 NADS-1 simulator. (Photo credit: University of Iowa)

been proposed to address user experiences in highly automated vehicles (Large et al., 2019a; Lee et al., 2019; Wang et al., 2021), and eHMIs for AV and pedestrian communication (Bazilinskyy et al., 2022). Finally, data acquisition is also convenient for driving simulator studies since most of the simulators available to date provide data extraction fairly easily.

Nevertheless, there are debatable voices that challenge the validity of driving simulator studies. Considering that participants are aware of the simulation situation and know they cannot get into a crash and injure themselves they might carry out extreme behaviors that they would not do in real-life environments. Similarly, such contrived situations can also prevent drivers from exhibiting their natural behaviors, such as texting while driving. However, a systematic review that surveyed driving simulator studies indicates that at least 50% of studies reported equivalent with real drives (Wynne et al., 2019). Another challenge in using a driving simulator is the simulation sickness associated with the desynchronization of visual stimuli and physical movement.

Despite all these debates, the driving simulator study is still a powerful and cost-effective approach that helps researchers understand driver behaviors, especially their interaction with future automotive user interfaces in the brainstorming and prototyping stage.

9.3.3 Naturalistic Driving Studies

9.3.3.1 History of Naturalistic Driving Studies

Based on Heinrich's Triangle, for every injury accident, there are similar non-injured conflicts happened (Heinrich, 1941). Thus, if the engineers can reduce the number of non-injured conflicts, the number of injury accidents and fatalities can be reduced. Following Heinrich's Triangle, transportation researchers at General Motors developed a method to videotape traffic conflicts at an intersection (Parker & Zegeer, 1989). Thereafter, variations of the traffic conflict techniques have been developed for instrumented vehicles that capture driver behaviors in traffic flows, which is later referred to as Naturalistic Driving Study (NDS) (Neale et al., 2002).

9.3.3.2 Understanding Naturalistic Driving Studies, Strengths, and Weaknesses

Large-scale NDS can be used to better understand driver behaviors that result in crashes or to assess in-vehicle safety systems from the safety aspect. The latter is also known as a field operational test (FOT). In both NDS and FOT, driver behaviors are captured in the real-world setting: they drive on normal routes, are under normal daily pressures, and experience normal traffic conditions, which ensures a high external validity of such studies to a large extent. NDS does not include experimental control but instead recruits a large number of participants and collects data over an extended period of time to ensure adequate data covering a variety of traffic dynamics and environmental conditions. The Second Strategic Highway Research Program (SHRP2 NDS) is, to date, the largest NDS that covers 3,542 drivers at six data-collection sites across the United States from October 2010 to December 2013. It is a large-scale observational-type cohort study, whose data have been professionally coded and used to understand how driver characteristics (e.g., age, gender) and behaviors (e.g., cell phone use, distraction), vehicle parameters, and environmental factors impact crash risk (Dingus et al., 2016; Guo et al., 2017). SHRP2 data have also been used to develop driver-supporting systems for future automobiles (Mata-Carballeira et al., 2019; Papazikou et al., 2021).

Another significant strength of NDS is the high-quality, high-resolution driving performance data coupled with video recordings. However, such data power can be lost if experimenters frequently interact with participants as they are in the driving simulator studies. To ensure the data power, the vehicle instrumentation should be minimal to reduce the sense of intrusiveness. Additionally, participants should be required to do nothing but drive normally. Thus, it is very beneficial to carry out NDS/FOT to understand the impact of a particular automotive user interface on driving behaviors after its availability to the general market. For instance, Orlovska and her colleagues have conducted two naturalistic driving studies to understand the impact of driving context on the use of automated driver assistance systems (ADAS) (Orlovska, Novakazi, et al., 2020) and to evaluate the ADAS across Chinese, Swedish, and American markets (Orlovska, Wickman, et al., 2020).

Exactly due to its large-scale, naturalistic driving studies are usually expensive and time-consuming. The data collection can take up to years, and the data coding afterward is very consuming, which requires considerable training and reviewing. Recruitment, data acquisition equipment, and data storage all require large-scale funding. But once the data are collected and properly coded, they are very beneficial to the research community and can be used to answer a variety of research questions. Another limitation regarding the capability of assessing automotive user interfaces is that NDS is more applicable to the existing user interfaces considering that the participants are usually recruited to drive their personal vehicles.

9.4 MEASUREMENT

9.4.1 Behavioral Measures

To evaluate the DDT performance of active drivers and passive drivers, behavioral measures can be applied to assess drivers' performance on the vehicle motor control and the OEDR performance. Behavioral measures capture the physical and visual components of performing the DDT. Specifically, the physical components refer to drivers' motor control, aiming to assess the longitudinal and lateral vehicle control. The visual components indicate the gaze direction and visual scanning pattern that can help the researcher understand drivers' OEDR and might also infer their motor control actions. The following two sections introduce the specific measures related to these two components.

9.4.1.1 Measuring Drivers' Motor Control

Drivers' motor control of the subject vehicles can be measured from longitudinal and lateral perspectives. Within each aspect, the measures include capturing the subject vehicle's sole performance, or the metrics that are associated with other road users or traffic violations. The longitudinal aspect mainly captures drivers' speed maintenance and distance maintenance in relation to the leading vehicles (if any). The lateral aspect captures drivers' lane-keeping control and lane-changing behaviors (if any). Table 9.3 presents a list of primary behavioral measures that can gauge drivers' motor control. For a comprehensive list of driving performance measures, refer to McDonald et al. (2019)

TABLE 9.3
Primary Driver Motor Control Metrics

Category	Metric	Definition
Longitudinal Control	Speed	Minimum, maximum, average, or standard deviation of speed can be calculated and compared with the speed limits.
	Longitudinal acceleration	The acceleration rate along the straight line.
Lateral Control	Lane deviation	The distance between the center of the vehicle and the center of the traveling lane. Usually, the standard deviation of lane position is calculated to indicate the control stability.
	Lateral acceleration	The acceleration rate along the lateral axis.
	Steering wheel angle	The maximum or standard deviation of steering wheel angle in radians or degrees can be used to assess the control stability.
Relation to other road users	Minimum time headway to the lead vehicle	The shortest during between the subject vehicle and the lead vehicle in second. Shorter headway indicates unsafe car following behaviors.
	Time to collision (TTC)	The time took for a collision to occur at an instantaneous speed or distance.
Traffic violation	Over speed	The total time the participant exceeds the speed limit, or the percentage of speed exceeded according to the speed limit.

and SAE International (2015). These measures can be compared before and after the implementation of automotive user interfaces to assess the potential benefits or any unintended consequences of introducing advanced technologies in the vehicle. For instance, both emotion regulation prompts and SA prompts were found to encourage safer driving behaviors indicated by smaller lane deviation (Jeon et al., 2015). Similarly, an in-vehicle intelligent agent with a conversational speech style also yielded a stable takeover maneuver indicated by smaller standard deviation of lane position (Wang et al., 2022b).

In the driving simulator study and the naturalistic study, the subject vehicle's sole performance can be acquired straightforwardly either through the simulator data logging system or the data collection devices instrumented on the subject vehicle. SAE International has published operational definitions of driving performance measures (SAE International, 2015) that can be used to extract and preprocess data from driving simulator logs or from data acquisition equipment from instrumented vehicles.

9.4.1.2 Measuring Drivers' Attention

Understanding drivers' visual attention is also important in addition to their vehicle motor control. Eye-tracking, the methodology that assists researchers in understanding visual attention (Schall & Romano Bergstrom, 2014), has been widely used in HCI and usability research to study eye activities (Jacob & Karn, 2003) and has also been widely adopted in the driving research context. In general, eye activities include eye movement, blinking activity, and changes in pupil diameter.

9.4.1.2.1 Eye Movement Metrics

Eye movement is the most frequently used eye activity to understand driver visual attention. Eye movement indicates the location of the gaze, the duration of the fixation, and the movement from one fixation to another (i.e., saccades) or from one area to another new area (i.e., transition) (Schall & Romano Bergstrom, 2014). These components together formulate scanning patterns. Table 9.4 lists several frequently used eye-movement metrics. The eye-movement metrics can be further categorized into fixation-derived metrics, saccade-derived metrics, and scanpath-derived metrics (Joseph & Murugesh, 2020; Poole & Ball, 2006). A fixation is the extremely brief pause

TABLE 9.4
Frequently Used Eye-Movement Metrics and Their Definitions

Eye-Movement Metrics		Description
Fixation-derived	Fixation Count	The number of fixations within a time period or within a specific area.
	Fixation Duration	The duration of all fixations given a period.
	Fixations per AOI	Total fixation counts in a specific AOI
	Gaze (i.e., dwell, fixation cluster)	The sum of all fixation durations within a specific area.
	Fixation Spatial Density	The spatial distribution of fixations on a given area.
	Fixation Rate/Frequency	The frequency of fixating on an AOI over a fixed amount of time.
	Time to first fixation on-target	The time took to fixate on a target
	Visual Entropy	The measure of visual scanning randomness and uncertainty based on information theory.
Saccade-derived	Number of Saccades	The number of rapid movements from one fixation to another.
	Saccade Amplitude	Distance traveled from onset to offset.
Scanpath-derived	Scanpath duration	Duration of a complete sequence of saccade-fixate-saccade sequence.
	Scanpath length	Length of a complete sequence of saccade-fixate-saccade sequence
	Scanpath direction	The direction of fixations and interlinked saccades occur.

of eye movement on a specific area of the visual field (Carter & Luke, 2020; Schall & Romano Bergstrom, 2014). The length of fixations varies depending on the tasks but is generally between 180–300 ms (Rayner, 2009). Saccades are rapid movements of the eye from one fixation to another. The scanpath is the spatial and temporal complete sequence of fixations and interconnecting saccades (Joseph & Murugesh, 2020; Poole & Ball, 2006). Often, researchers define areas of interest (AOIs) to cover certain parts of the visual field (e.g., two-lane roads, road signs, and speedometers) and analyze only the eye movements that fall in those AOIs to help address their research questions.

Fixation count, fixation rate, and gaze were found to be positively correlated with driver SA measured by other widely adopted and validated SA measures such as the SAGAT (Endsley, 1988b) and SA rating technique (SART) (Zhang et al., 2020). Fixation duration and frequency can also be converted to eyes-off-road time or frequency to evaluate driver inattention. Subsequently, the time to the first fixation on the target can be considered as an automated takeover temporal measure, gaze reaction time, to evaluate the time required for drivers to redirect the gaze to the forward roadway after distraction (Eriksson et al., 2019). Finally, the scanpath-derived metrics are used to identify the search strategies, which can be further used to categorize experienced and novice drivers and develop training protocols.

9.4.1.2.2 Blinking Activity

Blinking is mostly involuntary. Blink rate and blink duration have been used frequently to indicate driver fatigue and sleepiness. Blink rate refers to the blink frequency per minute. The higher blink rate is associated with greater fatigue due to eye dryness. Blink duration refers to the closure time duration of a blink. Typically, the blink duration is 100–400 ms (Schiffman, 2001). A longer blink duration is usually associated with increased sleepiness (Anund et al., 2008; Ingre et al., 2006), which has been incorporated into some driver monitoring systems to monitor driver fatigue and perform interventions (Brandt et al., 2004; Danisman et al., 2010).

9.4.1.2.3 Changes in Pupil Size

Pupillometry is the measurement of changes in pupil diameter, which can be used to infer different types of brain activities. As an autonomic sympathetic nervous system response, pupil dilation can indicate attention, interest, or emotion (e.g., fear, anxiety) associated with mental workload and

Automotive User Interfaces

arousal (Ahern & Beatty, 1979; Iqbal et al., 2004; Romano Bergstrom et al., 2014). As a widely reported reliable reflection of mental workload, pupil diameter increases with the problem becomes more challenging (Hess & Polt, 1964; Krejtz et al., 2018), which can provide a "very effective index of the momentary load on a subject as they perform a mental task" (Kahneman & Beatty, 1966). However, since pupil dilation is associated with several internal cognitive or emotional activities, it has to be with contexts to provide a reasonable explanation of how pupil dilation changes relate to task variation.

Pupil diameter data have to be properly preprocessed before further inferential analysis. Mathôt et al. (2018) and Kret and Sjak-Shie (2019) have provided extensive explanations and guidance on why and how to preprocess the pupil size data. This guidance has been well practiced in previous driving-related research (He et al., 2022; Wang et al., 2021), and we encourage researchers to keep up the standards in pupillometry analyses.

9.4.1.2.4 Eye-Tracking Device Types

There are three main types of eye-tracking devices used in driving research. Wearable eye-tracking glasses, screen-mounted eye trackers, and cameras. The applicable research methods, advantages, and disadvantages of each type are listed in Table 9.5. No matter which type of eye-tracking device is used, eye-tracking data provide valuable information as long as the data are properly cleaned and coded.

TABLE 9.5
Applicable Research Methods, Advantages, and Disadvantages of Different Eye-Tracking Devices

	Research Methods	Advantages	Disadvantages
Wearable Glasses	• Driving simulator studies • Test track studies • On-road vehicle studies • Field operational tests	• Portable • Larger data acquisition range.	• Poor fitness from the anthropometry aspect and incompatibility with other personal wearable equipment (e.g., eyeglasses, masks), especially considering that some devices require recalibration once the glasses position changes. • Difficult in extracting spatial-related eye-tracking metrics. Extra eye-tracking makers are required. • Sense of intrusiveness.
Screen-Mounted	• Driving simulator studies	• Easy to calculate spatial-related eye movement metrics.	• Lost track of participants or requiring multiple mounted devices to reach a full capture. • Limited use contexts, at least one designated visual display is required.
Video recording	• Driving simulator studies • Test track studies • On-road vehicle studies • Field operational tests • Naturalistic driving studies	• Larger acquisition range. • Integrated with other gestures or postures to get a more comprehensive understanding of the scenarios.	• High cost and time-consuming in data cleaning and extraction. • Low precision for existing computer vision algorithms. • Sense of intrusiveness.

9.4.2 Neurophysiological Measures

The use of neural correlates of driving performance is a novel approach that has shown its validity over past research endeavors. The brain exhibits heightened levels of activity while driving, as the task requires the activation of neural pathways for attention, evaluation, planning, and monitoring among other driving tasks. As such, neurophysiological measures can be taken and show differences in brain oxygenation and neural activity as drivers perform different tasks under varying levels of stress and workload. These measures include functional magnetic resonance imaging (fMRI), electroencephalography (EEG), functional near-infrared spectroscopy (fNIRS), and magnetoencephalography (MEG) (Navarro et al., 2018).

fMRI can detect changes in blood flow and indicate areas of increased activity based on higher levels of blood flow (Heeger & Ress, 2002; Logothetis, 2008). fMRI activity can change based on the level of the road familiarity drivers have with (Mader et al., 2009). The comparison of readings between the familiar and unfamiliar routes shows significant activation for the unfamiliar route in the middle temporal and occipital cortex and in the cerebellum, indicating that drivers in unfamiliar settings have heightened levels of cautiousness. Neurological readings can help in identifying the deterioration of several brain circuits due to alcohol consumption (Calhoun et al., 2004; Carvalho et al., 2006; Meda et al., 2009). A study by Calhoun et al. (2004) showed that alcohol disrupted activity in the orbitofrontal and motor regions of the brain.

EEG technology enables researchers to analyze oscillatory brain activity by placing electrodes on the scalp surface and detecting changes in electrical activity. EEG can be used to detect changes in mental workload, with higher levels leading to changes in salient EEG components such as the alpha, beta, theta, and delta bands (Brookings et al., 1996; Di Flumeri et al., 2018; Hankins & Wilson, 1998). EEG can also be used to detect age-related differences in driving behavior (Getzmann et al., 2018). In their study, higher theta power, which reflects higher mental effort, was associated with better steering performance in older drivers.

fNIRS is another technique used to detect brain activity. fNIRS uses near-infrared-range light to measure the concentration changes of oxygenated hemoglobin (HbO) and deoxygenated hemoglobin (HbR) (Hoshi, 2016; Villringer et al., 1993). fNIRS data can be used to detect changes in workload in driving with passengers (Pradhan et al., 2015), prolonged driving (Xu et al., 2017), or at curves (Oka et al., 2015). fNIRS has also been shown to detect the effect of emotions in driving, with angry driving having significantly different readings than the neutral state in an exploratory study (FakhrHosseini et al., 2015b).

Brain-imaging methods offer different neurological data based on the method chosen. While fMRI offers the highest spatial resolution, EEG provides the highest temporal resolution (Laureys et al., 2002). fNIRS is in between the two in terms of spatial and temporal resolution. As such, researchers can utilize a range of methods simultaneously to detect and evaluate neurological responses to driving with the best range of data (Ahn et al., 2016).

9.4.3 Computational Modeling

Among different levels of approaches and methods in HCI research, most driving research has focused on behavioral level approaches (e.g., simulation experiments or naturalistic driving observations). As discussed, more and more neurophysiological approaches have been adopted in automotive user interface research and design. The last approach is modeling, which has often been used in HCI.

Modeling is used to represent complicated phenomena in an abstract format, to assess the relationship between design parameters and outcomes, and to predict future states (McClelland, 2009). Using models in HCI has diverse advantages. It can be more accessible. Students or junior researchers can readily update the parameters and check the products in real time. It can be more economical. For instance, human subjects research does not need to be conducted every time, but simulation by varying parameters is possible. Computational models also allow researchers to quantify how each parameter will influence human behaviors.

9.4.3.1 Qualitative Modeling of Driving Behaviors

Following Michon (1985), many hierarchical driving models have used the three-level hierarchy structure to describe a driver's behavior. For example, the Hierarchical Driver Model (Boer, 1998) is composed of an operational, tactical, and strategic decision-maker in a bottom-up manner. More dynamic theories have also been introduced to describe the relationship between a driver and tasks or contexts, going beyond the individual information processing level. To illustrate, motivational models have considered risk (Taylor, 1964) as a core role of the drivers' decision-making process to obtain driving safety. Those models involve risk compensation models (e.g., risk homeostasis theory (Adams, 1985; Fuller, 1984; Wilde, 1982)), risk threshold models (e.g., zero risk theory (Näätänen, & Summala, 1976)), and risk avoidance models (e.g., task-capability interface model (TCI)). In the TCI model, driving task difficulty is determined by the comparison between the demands of the driving task and the capability of the driver (Fuller, 2005). There has also been an effort to make an overarching affect-integrated driving behavior model (Jeon, 2015), but it is a qualitative modeling framework.

9.4.3.2 Computational Modeling in Manual Driving

There are different types of computational modeling approaches. Some researchers used cognitive architecture to model drivers' behaviors, which is a top-down approach. Salvucci (2006) developed a computational driver model based on the Adaptive Control of Thought-Rational (ACT-R) to predict driver lane keeping and lane changing behavior. He created a rapid prototyping and evaluation tool for secondary tasks using a similar architecture, called Distract-R (Salvucci, 2009). Zhang and Wu (2018) used the Queuing Network-Model Human Processor (QN-MHP) to model driver performance for speech warning systems in connected vehicles with different warning characteristics, such as warning timing, warning reliability, and warning style (i.e., notification and command warnings). Zhang, Wu and Wan (2016) also adopted the QN-MHP to predict driver performance for speech warnings by modeling warning loudness and semantics, given other empirical experiments. Jeong and Liu (2017) used the QN-MHP to predict drivers' eye glances and workload for four stimulus-response secondary tasks while driving. Other researchers used statistical and machine learning models, which is a bottom-up approach, to formalize driver maneuvers. For example, Wu, Boyle, and Marshall (2017) designed a logistic regression model to demonstrate that drivers' demographic information (e.g., age and location) can predict their choice between steering and braking. Hu et al. (2017) used a decision tree to predict driver maneuvers in a cut-in scenario.

9.4.3.3 Computational Modeling in Automated Driving

To capture the potential confusion about the different automation modes of the semi-automated vehicle, Janssen et al. (2019) introduced a Hidden Markov Model framework to formalize the beliefs that drivers may have about the mode. Gold, Happee, and Bengler (2018) modeled driver takeover performance in level 3 semi-automated vehicles using a regression modeling approach, but they did not model the specific impact of auditory cues on takeover performance. Recent studies built the models using the QN-MHP framework to predict drivers' response time to auditory takeover displays using drivers' perceived intuitiveness and urgency (Ko, Sanghavi et al., 2022) and acoustic characteristics of auditory displays (Ko, Kutchek et al., 2022; Sanghavi, Zhang, & Jeon, 2023).

9.5 INTERFACES FOR FUTURE MOBILITY – STATE OF THE ART

9.5.1 Controls – Voice Input, Text Input, and Gestures

At the first International Conference on Automotive User Interfaces (AutoUI), Kern and Schmidt (2009) provided a layout of design space for automotive user interfaces. Following Geiser (1985),

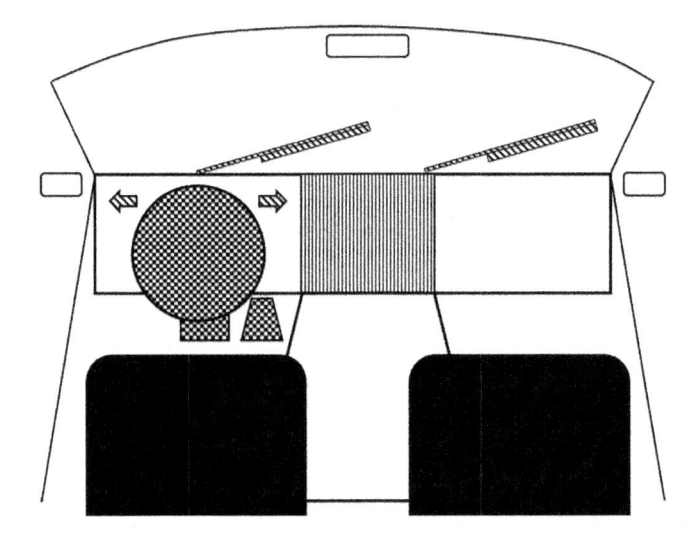

FIGURE 9.7 Primary (checkered), secondary (diagonal stripes), and tertiary (vertical stripes) control devices.

they classified three different categories of control devices: primary, secondary, and tertiary. The primary devices are used for maneuvering a vehicle, e.g., a steering wheel and pedals. The secondary devices are used to help the primary task, including turn signals or windscreen wipers. The tertiary devices are used for infotainment systems, such as BMW iDrive (https://www.bmw.com/en/index.html). These are depicted in Figure 9.7.

Kern and Schmidt included followings as input modalities: button, button with haptic feedback, discrete, continuous knobs, multifunctional controller, stalk control, slider, touchscreen, pedals, and thumbwheel, all of which were implemented in vehicles in 2009. Since then, a number of new control devices have been introduced in the community. One exceptional study (Large et al., 2017) showed the potential of a joystick as an alternative of primary driving control in automated vehicles. They showed that it is worthy of further exploration, but novice drivers rated it as higher in workload and collided with other vehicles. Other than that, research has focused more on secondary and tertiary tasks. Among those, speech recognition-based interfaces are the most common way of eyes-free interaction. Hua and Ng (2010) provided guidelines for speech recognition interfaces based on literature review and a case study. Winter et al. (2010) conducted language pattern analysis for more natural language speech applications in vehicles. Riener and Wintersberger (2011) introduced the proximity sensing device mounted near the gearshift, which monitored finger movements to control the mouse cursor on a screen in the dashboard. Hood et al. (2012) reported the implementation of a non-invasive electroencephalography (EEG)-based brain-computer interface (BCI) to control the functions of a car in a driving simulator. This exploratory attempt was innovative, but since the introduction of automated vehicles, research on BCI in the automotive area has been on the wane. Another novel approach was done by Murer et al. (2012) who explored the back of the steering wheel for text input when combined with a head-up display (HUD). Their user study showed that their approach was successful in terms of keeping drivers' hands on the wheel while typing. Gaze-based interaction was also designed. Riegler et al. (2020) investigated the effects of dwell times and feedback designs when using gaze-based interactions with windshield displays for SAE Level 3 vehicle environments.

There has been much research on hand gesture-based in-vehicle user interfaces. One of the first attempts by Ohn-Bar et al. (2012) categorized six classes of hand gestures using color and depth input. Riener et al. (2013) tried to make a standardization of spatial properties of gestures and gesture classes. Their evaluation study showed that most gestures are performed in a limited region in the car, bounded by the steering wheel, rear mirror, and gearshift regardless of the functions and individuals. May et al. (2017) created a recommended in-air gesture set based on a participatory elicitation study and online survey. There have been attempts to combine the gesture interfaces with other modalities. Pfleging et al. (2012) combined speech and gestures so that speech was used for function identification (e.g., mirror) and gestures for manipulation (e.g., left/right). Cui et al. (2021) explored gestural input on the steering wheel to improve the interaction efficiency of voice user interfaces. Studies have also integrated touch screens and gesture interactions (Burnett et al., 2013; Rümelin and Butz, 2013). There has also been research on in-between touch and gesture. Ahmad et al. (2016) introduced predictive in-car touchscreen with mid-air selection. It determines the item the user intends to select, early in the pointing gesture, and accordingly simplifies-expedites the target acquisition. Specific feedback has also been studied with respect to air gesture interactions, including ultrahaptic feedback (Shakeri et al., 2018; Harrington et al., 2018) and auditory feedback (Sterkenburg et al., 2019).

With the increase of automation, research has shifted its focus from manual control to monitoring of automated vehicle states. However, it is not clear if the control interfaces will completely disappear from the vehicle. At least in the near future, the control interfaces will remain for takeover in semi-automated vehicles. Given that automation will increase accessibility for people with disabilities or those who have traditionally not been allowed to drive (e.g., children), newer and more accessible control interfaces are expected to appear (Detjen et al., 2022; Jeon et al., 2016).

9.5.2 Displays – Uni/Multimodal Displays (HUD, Olfactory, and Auditory)

A variety of unimodal displays have been investigated for use within in-vehicle environments. Visual, auditory, vibrotactile, and olfactory displays can be used for the tasks of informing, alerting, and responding to drivers and users during driving. Visual displays include both HUDs and head-down displays (HDDs), and have a long history in the automotive industry (Liu & Wen, 2004). Auditory displays include earcons, speech messages, and auditory icons and have been shown to provide quick feedback to drivers (Xiang, Yan, Weng, & Li, 2016). Vibrotactile displays use haptic feedback from driver seats to inform drivers about driving changes and have been shown to have potential benefits through presenting driving information in a safe way, for example, eliciting faster reactions from the driver (Liu et al., 2012; Fitch, Kiefer, Hankey, and Kleiner, 2007). Static vibrotactile displays in the driver's seat cushion have been shown to be more effective than dynamic vibrotactile displays. Reaction time and hand-eye coordination movements were quicker for takeover requests and lane change operations, respectively; recognition rates were also higher for static vibration patterns than moving patterns, indicating their effectiveness for warning signals over direction cues (Petermeijer et al., 2017a, b). Another form of haptic feedback would be reactive forces applied to the gas pedal to suggest appropriate speed limits to the driver. In a simulator study making use of one-pedal driving, where lifting the foot off the gas pedal would cause the vehicle to decelerate at a maximum rate of $1.5\,m/s^2$, it was found that by increasing the force required to depress the gas pedal, drivers could be guided to adopt a lower speed; when the force required to depress the gas pedal was decreased, it would indicate to the driver that it was now safe to accelerate again, thus leading to safer driving (Saito and Raksincharoensak, 2019). Olfactory feedback can be induced through the addition of smells while driving and has been the subject of research (Okazaki, Haramaki, & Nishino, 2018).

Multimodal displays refer to a display that utilizes two or more display modalities to communicate with drivers. Multimodal displays have been shown to outperform unimodal displays in forward collision warnings using visual and vibrotactile displays (Haas & Van Erp, 2014) and combinations of visual, auditory, and vibrotactile displays (Politis, Brewster, & Pollick, 2013).

In automated vehicles, multimodal displays can support users in the takeover task. As the efficacy of multimodal displays has been shown in past studies, their application to takeover situations that usually employ auditory alerts (Stojmenova et al., 2020) has been explored. de Oliveira Faria et al. (2021) have explored the use of AR displays to alert drivers through AR cues. Another research group found that the addition of a vibrotactile modality helped in improving driving performance, reaction time, and workload in addition to being preferred by users (Hong & Yang, 2022).

While the use of multimodal displays can be of great interest, designers should consider that adding more is not always guaranteed to be better. A study that evaluated the use of multimodal displays using a combination of LED lights, an earcon, a speech message, and haptic feedback (Hong & Yang, 2022) found that the combination of the LED light and speech message led to higher stress and more unstable steering. It is also worth mentioning that multimodal displays can lead to greater annoyance (Politis et al., 2013), which can then negatively impact user compliance with the alert. It is thus important to design multimodal displays with a careful selection of modalities that can bring the best benefit, as efficient design can make it so multimodal displays have similar ratings of frustration when compared to unimodal displays while still providing better performance (Biondi, Strayer, Rossi, Gastaldi, & Mulatti, 2017).

9.5.3 Emerging Applications

9.5.3.1 In-Vehicle Intelligent Agents

Intelligent agents (IAs) that can autonomously and intelligently help users with assigned tasks (Padgham & Winikoff, 2005) have become one of mature smart home technologies as voice user interfaces, creating more convenient and accessible at-home experiences. Similarly, IAs can also help with both active and passive drivers whose visual resources are heavily occupied by the DDT. The roles of in-vehicle intelligent agents (IVIAs) largely depend on the driving automation and the resultant driver-automation function allocation (Wang et al., 2022a). Not only can IVIAs reduce drivers' workload when interacting with an in-vehicle infotainment system, but also they are capable of enhancing user trust and acceptance (Koo et al., 2015; Large et al., 2019b) and promoting overall user experience (Joo & Lee-Won, 2016; Wang et al., 2021), which is critical under future automated vehicle setting. However, until vehicles equipped with Level 4 or 5 driving automation systems are available to the general public, drivers are still required to complete the DDT and thus, can only tolerant minimal in-cabin distractions. Another challenge to have an intelligent agent in the vehicle is overreliance and complacency that is often problematic during human-automation interaction (Lee & See, 2004; Parasuraman & Manzey, 2010). In this way, the IVIAs should be carefully designed to minimize distraction and prevent unintended consequences.

Objective functions and design characteristics for IVIAs are two main considerations for IVIA development. Objective function is defined as the primary purpose of IVIA situated in a driving context. The objective functions of an IVIA can be classified into four categories: (1) directly supporting driving and NDRTs, (2) reinforcing general cognitive process regarding distraction, attention, and SA, (3) providing affective intervention and support, and (4) promoting overall experience and subjective evaluations (Lee & Jeon, 2022). The detailed objective function allocated to the IVIA needs to be adjusted based on the user context. For instance, IVIAs situated in vehicles with no driving automation system are expected to have advanced driving task-supporting function, while IVIAs fitted in the vehicles with Level 5 ADS are supposed to create better user experience. Affective support has been overlooked for a long time and has been brought up recently. Although driving under elevated emotional status is not prevalent, it increases the crash risk by 9.8 times, which is even higher than other well-known driving distractions (e.g., texting using handheld devices increases crash risks by 6.1) (Dingus et al., 2016). Affective IVIAs have also been proposed to mitigate the negative emotional effects and encourage safe driving behaviors (Dittrich & Mathew, 2021; Jeon, Walker, & Gable, 2015; Johnsson et al., 2005; Joo & Lee-Won, 2016).

The achievement and reliability of the objective functions mentioned above rely on the design considerations of IVIAs, which also inversely depend on their functions and the type of in-vehicle users they are serving. Visual representations and vocal characteristics are two main categories of design considerations that influence IVIAs' effects on driver performance and perception.

The visual representation, such as agent embodiment, determines users' first impression and contributes agent anthropomorphism. Agents with enhanced anthropomorphism (e.g., human-like appearance) induced positive user perception and social presence that facilitate user trust (Large et al., 2019b; Zhang et al., 2021; Niu et al., 2018; Zihsler et al., 2016). Such influence is critical especially in fully autonomous vehicles, where companionship, perceived safety, and user experience are valued most.

The vocal characteristics– features that pertain to the vocal outputs from IVIAs–can be further categorized into speech features and information contents. Speech features included but are not limited to speech style, voice attributes (e.g., age, gender), speech tone, and speed. The information contents of IVIAs can maintain different levels of information transparency, express urgency, deliver agent attitude, or perform affective intervention. These vocal characteristics are essential human characteristics and contribute greatly to the anthropomorphism of an agent, and thus, have greater impact on driver behaviors and perceptions. For instance, different speech styles of IVIAs have various impacts on drivers' attention and their maneuvers afterward. Messages in an assertive speech style were perceived as more urgent and yielded a faster reaction time by distracted passive drivers in semi-automated vehicles, compared to non-assertive speech style (Wong et al., 2019). Conversational speech style not only elicited higher user evaluation but can also encourage safer after-takeover maneuver compared to information style (Wang et al., 2022b).

9.5.3.2 VR and AR Applications to the In-Vehicle Context

Emerging technologies such as VR and AR have the possibility to redefine activities users can do within the vehicle. AR and VR applications have gained momentum in recent years, even though they were conceptualized back in 1994 by Milgram and Kishino (1994). They used a continuum or spectrum (Milgram's Reality-Virtuality Continuum) to describe such technologies. Real World was on the left end of the spectrum, and a completely Virtual World on the right; the entire region in between was termed Mixed Reality (MR). Within the MR region, the concept of AR was defined by Milgram et al. (1993) as the essence of computer graphic enhancement of video images of real scenes. More recently, Sikora (2018) defined AR as the combination of interactive virtual content, associated with real-world information, with real-world content. Due to the evolving nature of such technologies, updating the taxonomy can be beneficial to developers and practitioners. Skarbez et al. (2021) revisited Milgram's Reality-Virtuality Continuum to better define MR, using three dimensions–Extent of World Knowledge, or awareness of the immediate physical world around the user; Immersion, or the extent of illusion provided by the system; and Coherence, or the degree of plausibility of the system as experienced by the user.

Additional recent works have investigated alternative modalities to define AR; audio AR, or AAR is the use of the auditory modality instead of visual to provide additional information to the user, since system requirements for visually impaired users can be drastically different from those with normal vision (Blum et al., 2013). A study using Grounded Theory described three dimensions for AAR applications–extent of Immersion (as described by Skarbez et al.); user context or information designed to best assist the user in their primary task; and Customization, or ensuring the same sound, although audible to more than one user, is customized sufficiently to be meaningful only to the intended user (Dam et al., 2022b). AAR applications are already present in existing vehicles, but can be elevated through better implementation based on the latest taxonomies. For example, consider the humble turn signal. Through an intermittent clicking sound along with a visual component (green arrows on the instrument cluster), it provides the driver and all occupants of the vehicle with feedback of the system, meaning the turn lights outside the vehicle are active–this would be User Context and the only AAR dimension of the auditory signal since it does not provide sufficient Immersion or Customization. This application can be improved based on the two missing dimensions. Audio spatialization to indicate the left and right blinker (Jeon et al., 2009) would be a way to enhance Immersion, and even possibly eliminate the need for visual feedback; the option to Customize these sounds such that they may be perceived as a pleasant auditory stimulus by the drivers, but important system status feedback by the drivers would further make the humble turn signal a strong AAR application. Similarly, if users of an automated vehicle want to receive predictive auditory cues for upcoming motion as a way to mitigate motion sickness symptoms, the same principles can be applied to the design of the auditory displays for said cues.

9.5.3.3 Vulnerable Road Users and eHMI

In addition to focusing on the needs of vehicle users, the advent of highly automated vehicles presents the need to study users external to the vehicle, too. Such users include pedestrians, bicyclists, motorcyclists, or anyone on the road that is unprotected by a closed frame or shell; such road users are categorized as Vulnerable Road Users (VRUs) (Wegman and Aarts, 2006; van Haperen et al., 2019). Crosswalk safety is a challenge that is exacerbated by the presence of automated vehicles because pedestrians may not always have the option to communicate with the driver (Sucha et al., 2017; Liu et al., 2021). Hand gestures are a common way for pedestrians to explicitly communicate to drivers their intentions, but drivers' yielding rates can be poor (Zhuang & Wu, 2014). As a result, to design safer automated vehicles for VRUs, pedestrians' needs should be taken into account.

Unsignalized crosswalks present an even greater challenge since the pedestrian and the vehicle/driver must have a shared mental model to ensure safe crossings. More times than not, pedestrians are distracted; these distractions can be technological or social. Technological distractions can have a greater impact on pedestrians' SA, but it has been observed that pedestrians tend to compensate for it by adopting safer crossing behaviors (Dam et al., 2022a; Stavrinos et al., 2011). These include a greater likelihood to check left and right before crossing, or turning down or off their personal listening devices (PLDs). Nonetheless, this does not describe the entire situation since technologically distracted pedestrians are also more likely to resume their distractions once they have started crossing and while still on the crosswalk; compared to that, non-distracted pedestrians were observed to have a greater likelihood of continuing to check left and right even while traversing the crosswalk (Dam et al., 2022a). There is ongoing research to understand the extent of degradation in auditory SA due to the use of technological distractions such as PLDs at crosswalks (Dam et al., 2022c). The other form of distraction is social distractions, which include being engaged in a conversation while crossing, following or focusing on other VRUs instead of approaching traffic, and crossing as a group. Social distractions can be more prevalent than technological distractions, and thus, lead to more occurrences of risky behaviors by pedestrians (Thompson et al., 2013; Aghabayk et al., 2021; Dam et al., 2022a). These behaviors include not checking left and right before crossing, or blindly following the leading pedestrians. This could be due to pedestrians expecting other members of the group to remain vigilant, or receive cues of any danger from them instead of the actual element of danger (speeding vehicle). There is also the expectation that a group of pedestrians are more salient to drivers than a single pedestrian.

VRU safety is also influenced by the infrastructure of the location in question. Pedestrians' expectations for vehicles change if a crosswalk is located at an exit of a roundabout, or is on a straight road but their view is partially occluded by a parked vehicle, such as a bus. When a blind spot is present, both distracted and non-distracted pedestrians tend to be more cautious, but their willingness to cross before or after a vehicle can be ambiguous (Dam et al., 2022a). Such blind spots prevent drivers from being able to spot pedestrians from a distance. This creates the need to implement safety systems to ensure both the driver or vehicle, and the pedestrian have a shared understanding of the situation to ensure safe crossings. eHMIs can be effective in communicating the vehicle's intent to the pedestrian, especially in ambiguous situations to bridge the gap between the expectations of the pedestrian and the vehicle (Ackermans et al., 2020). Given our affinity to recognize human faces, it has also been shown that simply adding a set of "eyes" to a vehicle, leads to faster correct crossing decision-making on behalf of the pedestrian (Chang et al., 2017). It has also been demonstrated that eHMIs that incorporate contextual information in addition to yielding intent, such as location or distance of the pedestrian with respect to the approaching vehicle, or percentage completion of yielding, lead to more efficient interactions between VRUs and automated vehicles (Dey et al., 2020).

However, it is not always possible for the driver or vehicle to yield to pedestrians given the presence of other traffic or available stopping distance from the crosswalk; this creates the need for eHMIs to explicitly state its non-yielding intent especially when it begins to slow down as it approaches an awaiting pedestrian (Dey et al., 2021; Dey et al., 2022). Research also shows how continuous use of explicit cues such as flashing lights, text or displays with texts and symbols, may not only be distracting for pedestrians, but may also add to the cognitive workload of performing a street crossing (Moore et al., 2019). Moore et al.'s study showed that in instances where clear interaction between the vehicle and pedestrian is possible, implicit eHMIs such as slowing down of the vehicle or engine sound itself, is sufficient to indicate yielding intent of the automated vehicle. Such results indicate the eHMIs should be able to adapt their external displays to the context and provide explicit or implicit cues accordingly. Finally, it is also important to investigate the impact of erroneous eHMI messages, which could have severe consequences. Since pedestrians expect autonomous vehicles to be safer, the issue of over reliance on the accuracy of eHMIs should be considered carefully, along with the idea of standardizing eHMI messages (Holländer et al., 2019).

9.5.3.4 Accessibility and Various Types of Ride

The introduction of fully autonomous vehicles will bring a variety of ride styles as well as increase accessibility for many users. Jeon et al. (2016) discussed the challenges and opportunities of autonomous vehicles for diverse populations, including people with various difficulties/disabilities, older adults, and children. The potential includes allowing not only those who were less able to drive, but also those who were not allowed to drive, to ride themselves with fully autonomous vehicles. For these people, there are a number of research questions and practical implementation issues, such as emergency controllability, centralized meta-monitoring system, a new driver license, etc. One of the attempts already implemented and tested is a low-speed urban shuttle. Once well implemented, this type of system can provide improved mobility for short-distance trips and for first mile/last mile access to and from public transit services. For example, the Virginia Tech Transportation Institute operated a low-speed automated shuttle and surveyed shuttle riders and non-riders (Kim & Doerzaph, 2022). Results showed that shuttle riders reported more positive attitudes toward the shuttle operations than non-riders. Also, many participants strongly supported rules and restrictions governing shuttle operations on public roadways. In their user enactment study on shared automated vehicles with 30 older adults, Gluck et al. (2020) showed that older adults need assistive features for vehicle ingress and egress, storage for mobility aids, and emergency service contact system with concerns about privacy. Another study (Brinkley et al., 2020) with blind and visually impaired participants, showed that they are concerned about equipment failure and self-driving commercial vehicles such as heavy trucks. Most survey respondents (93.31%) expressed that they are interested in owning self-driving vehicle technology. Their focus group study identified many challenges, including parking guidance, orienting oneself to the desired destination upon arrival (e.g., building's front door), vehicle breakdown, etc. Another futuristic ride is urban air mobility. Given that this is a new area of research, little research has been conducted on this topic. Kim, Lim et al. (2022) hosted workshops at AutoUI Conference series and discussed the role of urban air mobility, people's expectations, and design considerations. The development of automated vehicles is also changing the vehicle into a mobile workplace. In his dissertation project, Schartmüller (2021) studied text entry, text comprehension, motion sickness, and speech-to-text in conditionally automated vehicles. There are still a number of challenges in implementing the mobile office in conditionally automated vehicles, but it will not be long after the introduction of full automation that the mobile office comes true.

9.6 CONCLUSION

This chapter introduces an overview of automotive user interfaces with a focus on HCI, including theoretical basis, methods, interfaces, and interactions. As with other HCI fields, the automotive user interface domain requires truly multidisciplinary approaches, including psychology, engineering, computer science, human factors, and design. Thus, how to integrate diverse approaches and methods in a balanced way is a key to its design success. With the introduction of automated vehicles, this area has been sharply changing, but the same core constructs are still important in research and design. This chapter is expected to be the entry point to further resources by providing the core concepts and methods as well as resources. The authors hope that this chapter can guide researchers, designers, engineers, and students to advance and expand this exciting research field.

REFERENCES

Ackermans, S., Dey, D., Ruijten, P., Cuijpers, R. H., & Pfleging, B. (2020). The effects of explicit intention communication, conspicuous sensors, and pedestrian attitude in interactions with automated vehicles. In *Proceedings of the 2020 CHI Conference on Human Factors in Computing Systems* (pp. 1–14). https://doi.org/10.1145/3313831.3376197

Adams, J. G. (1985). *Risk and Freedom: The Record of Road Safety Regulation*. London: Transport Publishing Projects.

Aghabayk, K., Esmailpour, J., Jafari, A., & Shiwakoti, N. (2021). Observational-based study to explore pedestrian crossing behaviors at signalized and unsignalized crosswalks. *Accident Analysis & Prevention*, 151, 105990. https://doi.org/10.1016/j.aap.2021.105990

Ahern, S., & Beatty, J. (1979). Pupillary responses during information processing vary with scholastic aptitude test scores. *Science*, 205(4412), 1289–1292.

Ahmad, B. I., Langdon, P. M., Godsill, S. J., Donkor, R., Wilde, R., & Skrypchuk, L. (2016, October). You do not have to touch to select: A study on predictive in-car touchscreen with mid-air selection. In *Proceedings of the 8th International Conference on Automotive User Interfaces and Interactive Vehicular Applications* (pp. 113–120). https://doi.org/10.1145/3003715.3005461

Ahn, S., Nguyen, T., Jang, H., Kim, J. G., & Jun, S. C. (2016). Exploring neuro-physiological correlates of drivers' mental fatigue caused by sleep deprivation using simultaneous EEG, ECG, and fNIRS data. *Frontiers in Human Neuroscience*, 10, 219.

Anund, A., Kecklund, G., Peters, B., Forsman, Å., Lowden, A., & Åkerstedt, T. (2008). Driver impairment at night and its relation to physiological sleepiness. *Scandinavian Journal of Work, Environment & Health*, 34(2), 142–150. https://doi.org/10.5271/SJWEH.1193

Bakhtiari, S., Zhang, T., Zafian, T., Samuel, S., Knodler, M., Fitzpatrick, C., & Fisher, D. L. (2019). Effect of visual and auditory alerts on older drivers' glances toward latent hazards while turning left at intersections. *Transportation Research Record*, 2673(9), 117–126.

Barkana, Y., Zadok, D., Morad, Y., & Avni, I. (2004). Visual field attention is reduced by concomitant hands-free conversation on a cellular telephone. *American Journal of Ophthalmology*, 138(3), 347–353.

Bazilinskyy, P., Kooijman, L., Dodou, D., Mallant, K., Roosens, V., Middelweerd, M., Overbeek, L., & de Winter, J. (2022). Get out of the way! Examining eHMIs in critical driver-pedestrian encounters in a coupled simulator. In *Proceedings of the 14th International Conference on Automotive User Interfaces and Interactive Vehicular Applications* (pp. 360–371). https://doi.org/10.1145/3543174.3546849

Bethel, C. L., Salomon, K., Murphy, R. R., & Burke, J. L. (2007). Survey of psychophysiology measurements applied to human-robot interaction. In *Proceedings - IEEE International Workshop on Robot and Human Interactive Communication* (pp. 732–737). https://doi.org/10.1109/ROMAN.2007.4415182

Biondi, F., Strayer, D. L., Rossi, R., Gastaldi, M., & Mulatti, C. (2017). Advanced driver assistance systems: Using multimodal redundant warnings to enhance road safety. *Applied Ergonomics*, 58, 238–244.

Blum, J. R., Bouchard, M., & Cooperstock, J. R. (2013). Spatialized audio environmental awareness for blind users with a smartphone. *Mobile Networks and Applications*, 18(3), 295–309. https://doi.org/10.1007/s11036-012-0425-8

Boer, E. R. (1998). A driver model of attention management and task scheduling: Satisficing decision making with dynamic mental models. In *Proceedings of the XVIIth European Annual Conference on Human Decision Making and Manual Control*.

Bohrmann, D., & Bengler, K. (2020). Reclined posture for enabling autonomous driving. In T. Ahram, W. Karwowski, S. Pickl, & R. Taiar (Eds.), *Human Systems Engineering and Design II* (Vol. 1026, pp. 169–175). Springer International Publishing. https://link.springer.com/10.1007/978-3-030-27928-8_26

Bos, J.E., MacKinnon, S.N., and Patterson, A. (2005). Motion sickness symptoms in a ship motion simulator: Effects of inside, *Outside, and No View*. 76(12), 9.

Brandt, T., Stemmer, R., & Rakotonirainy, A. (2004). Affordable visual driver monitoring system for fatigue and monotony. In *Conference Proceedings - IEEE International Conference on Systems, Man and Cybernetics* (Vol. 7, pp. 6451–6456). https://doi.org/10.1109/ICSMC.2004.1401415

Brinkley, J., Huff Jr, E. W., Posadas, B., Woodward, J., Daily, S. B., & Gilbert, J. E. (2020). Exploring the needs, preferences, and concerns of persons with visual impairments regarding autonomous vehicles. *ACM Transactions on Accessible Computing (TACCESS)*, 13(1), 1–34. https://doi.org/10.1145/3372280

Brookings, J. B., Wilson, G. F., & Swain, C. R. (1996). Psychophysiological responses to changes in workload during simulated air traffic control. *Biological Psychology*, 42(3), 361–377.

Burdett, B. R., Charlton, S. G., & Starkey, N. J. (2019). Mind wandering during everyday driving: An on-road study. *Accident Analysis & Prevention*, 122, 76–84.

Burnett, G., Crundall, E., Large, D., Lawson, G., & Skrypchuk, L. (2013, October). A study of unidirectional swipe gestures on in-vehicle touch screens. In *Proceedings of the 5th International Conference on Automotive User Interfaces and Interactive Vehicular Applications* (pp. 22–29). https://doi.org/10.1145/2516540.2516545

Calhoun, V. D., Pekar, J. J., & Pearlson, G. D. (2004). Alcohol intoxication effects on simulated driving: Exploring alcohol-dose effects on brain activation using functional MRI. *Neuropsychopharmacology*, 29(11), 2097–2107.

Cantin, V., Lavallière, M., Simoneau, M., & Teasdale, N. (2009). Mental workload when driving in a simulator: Effects of age and driving complexity. *Accident Analysis & Prevention*, 41(4), 763–771.

Carter, B. T., & Luke, S. G. (2020). Best practices in eye tracking research. *International Journal of Psychophysiology*, 155(October 2019), 49–62. https://doi.org/10.1016/j.ijpsycho.2020.05.010

Carvalho, K. N., Pearlson, G. D., Astur, R. S., & Calhoun, V. D. (2006). Simulated driving and brain imaging: Combining behavior, brain activity, and virtual reality. *CNS Spectrums*, 11(1), 52–62.

Chang, C.-M., Toda, K., Sakamoto, D., & Igarashi, T. (2017). Eyes on a car: An interface design for communication between an autonomous car and a pedestrian. In *Proceedings of the 9th International Conference on Automotive User Interfaces and Interactive Vehicular Applications* (pp. 65–73). https://doi.org/10.1145/3122986.3122989

Charles, R. L., & Nixon, J. (2019). Measuring mental workload using physiological measures: A systematic review. *Applied Ergonomics*, 74, 221–232.

Chiou, E. K., & Lee, J. D. (2021). Trusting automation: Designing for responsivity and resilience. *Human Factors*, 65(1), 137–165.

Cooper, P. J., Pinili, M., & Chen, W. (1995). An examination of the crash involvement rates of novice drivers aged 16 to 55. *Accident Analysis & Prevention*, 27(1), 89–104.

Cotté, N., Meyer, J., & Coughlin, J. F. (2001, October). Older and younger drivers' reliance on collision warning systems. In *Proceedings of the Human Factors and Ergonomics Society Annual Meeting* (Vol. 45, No. 4, pp. 277–280). Sage, CA: Los Angeles, CA: SAGE Publications.

Cramer, D. B., Graybiel, A., Miller, E. F., & Wood, C. D. (1968). *Diagnostic Criteria for Grading the Severity of Acute Motion Sickness (No. NAMI-1030)*. NASA.

Cui, Z., Gong, H., Wang, Y., Shen, C., Zou, W., & Luo, S. (2021, September). Enhancing interactions for in-car voice user interface with gestural input on the steering wheel. In *13th International Conference on Automotive User Interfaces and Interactive Vehicular Applications* (pp. 59–68). https://doi.org/10.1145/3409118.3475126

Dam, A., & Jeon, M. (2021). A review of motion sickness in automated vehicles. In *13th International Conference on Automotive User Interfaces and Interactive Vehicular Applications* (pp. 39–48). https://doi.org/10.1145/3409118.3475146

Dam, A., Duff, C., Jeon, M., & Patrick, R. N. C. (2022c). Effects of personal listening devices on pedestrians' acoustic situation awareness in a virtual reality environment. In *The 27th International Conference on Auditory Display (ICAD 2022)* (p. 4).

Dam, A., Oberoi, P., Pierson, J., Jeon, M., & Patrick, R. N. C. (2022a). Technological and social distractions at unsignalized and signalized campus crosswalks: A multi-stage naturalistic observation study. *SSRN Electronic Journal*. https://doi.org/10.2139/ssrn.4281911

Dam, A., Siddiqui, A., Leclerq, C., & Jeon, M. (2022b). Extracting a definition and taxonomy for audio augmented reality (AAR) using grounded theory. In *HFES 66th International Annual Meeting Technical Program* (p. 5).

Danisman, T., Bilasco, I. M., Djeraba, C., & Ihaddadene, N. (2010). Drowsy driver detection system using eye blink patterns. In *2010 International Conference on Machine and Web Intelligence, ICMWI 2010-Proceedings* (pp. 230–233). https://doi.org/10.1109/ICMWI.2010.5648121

de Oliveira Faria, N., Merenda, C., Greatbatch, R., Tanous, K., Suga, C., Akash, K., ... & Gabbard, J. (2021, September). The effect of augmented reality cues on glance behavior and driver-initiated takeover on SAE Level 2 automated-driving. In *Proceedings of the Human Factors and Ergonomics Society Annual Meeting* (Vol. 65, No. 1, pp. 1342–1346). Sage CA: Los Angeles, CA: SAGE Publications.

Deffenbacher, J. L., Deffenbacher, D. M., Lynch, R. S. and Richards, T. L. Anger, aggression, and risky behavior: A comparison of high and low anger drivers. *Behaviour Research and Therapy*, 41, 6, (2003), 701–718. https://doi.org/10.1016/S0005-7967(02)00046-3

Deffenbacher, J. L., Deffenbacher, D. M., Lynch, R. S., & Oetting, E. R. (2001). Further evidence of reliability and validity for the driving anger expression inventory. *Psychological Reports*, 89(3), 535–540. https://doi.org/10.2466/pr0.2001.89.3.

Dennehy, K. (1997). Cranfield situation awareness scale.

Detjen, H., Schneegass, S., Geisler, S., Kun, A., & Sundar, V. (2022, September). An emergent design framework for accessible and inclusive future mobility. In *Proceedings of the 14th International Conference on Automotive User Interfaces and Interactive Vehicular Applications* (pp. 1–12). https://doi.org/10.1145/3543174.3546087

Dey, D., Habibovic, A., Berger, M., Bansal, D., Cuijpers, R., & Martens, M. (2022). Investigating the need for explicit communication of non-yielding intent through a slow-pulsing light band (SPLB) eHMI in AV-pedestrian interaction. In *Proceedings of the 14th International Conference on Automotive User Interfaces and Interactive Vehicular Applications* (pp. 307–318).

Dey, D., Holländer, K., Berger, M., Eggen, B., Martens, M., Pfleging, B., & Terken, J. (2020). Distance-dependent eHMIs for the interaction between automated vehicles and pedestrians. In *12th International Conference on Automotive User Interfaces and Interactive Vehicular Applications* (pp. 192–204). https://doi.org/10.1145/3409120.3410642

Dey, D., Matviienko, A., Berger, M., Pfleging, B., Martens, M. & Terken, J. (2021). Communicating the intention of an automated vehicle to pedestrians: The contributions of eHMI and vehicle behavior. *it - Information Technology*, 63(2), 123–141. https://doi.org/10.1515/itit-2020-0025

Di Flumeri, G., Borghini, G., Aricò, P., Sciaraffa, N., Lanzi, P., Pozzi, S., Vignali, V., Lantieri, C., Bichicchi, A., & Simone, A. (2018). EEG-based mental workload neurometric to evaluate the impact of different traffic and road conditions in real driving settings. *Frontiers in Human Neuroscience*, 12, 509.

Diels, C., & Bos, J. E. (2016). Self-driving carsickness. *Applied Ergonomics*, 53, 374–382. https://doi.org/10.1016/j.apergo.2015.09.009

Dingus, T. A., Guo, F., Lee, S., Antin, J. F., Perez, M., Buchanan-King, M., & Hankey, J. (2016). Driver crash risk factors and prevalence evaluation using naturalistic driving data. *Proceedings of the National Academy of Sciences of the United States of America*, 113(10), 2636–2641. https://doi.org/10.1073/pnas.1513271113

Dittrich, M., & Mathew, N. (2021). Emotional feedback to mitigate aggressive driving: A real-world driving study. In *Lecture Notes in Computer Science (Including Subseries Lecture Notes in Artificial Intelligence and Lecture Notes in Bioinformatics)*, 12684 LNCS (pp. 88–101). https://doi.org/10.1007/978-3-030-79460-6_8

Dmitrenko, D., Maggioni, E., Brianza, G., Holthausen, B. E., Walker, B. N., & Obrist, M. (2020, April). Caroma therapy: Pleasant scents promote safer driving, better mood, and improved well-being in angry drivers. In *Proceedings of the 2020 Chi Conference on Human Factors in Computing Systems* (pp. 1–13).

Du, N., Zhou, F., Pulver, E. M., Tilbury, D. M., Robert, L. P., Pradhan, A. K., & Yang, X. J. (2020). Examining the effects of emotional valence and arousal on takeover performance in conditionally automated driving. *Transportation Research Part C: Emerging Technologies*, 112, 78–87.

Duivenvoorden, K., Hogema, J., Hagenzieker, M., & Wegman, F. (2015). The effects of cyclists present at rural intersections on speed behavior and workload of car drivers: A driving simulator study. *Traffic Injury Prevention*, 16(3), 254–259.

Dunn, J.R. and Schweitzer, M.E. (2005). Feeling and believing: The influence of emotion on trust. *Journal of Personality and Social Psychology*, 88(5), p.736.

Ekman, P. (1999). Basic emotions. In T. Dalgleish & M. J. Power (Eds.), *Handbook of Cognition and Emotion* (pp. 45–60). Sussex: John Wiley & Sons Ltd..

Endsley, M. R. (1988a). Design and evaluation for situation awareness enhancement. In *Proceedings of the Human Factors Society Annual Meeting*.

Endsley, M. R. (1988b). Situation awareness global assessment technique (SAGAT). In *Proceedings of the IEEE 1988 National Aerospace and Electronics Conference*.

Endsley, M. R. (1995). Towards a new paradigm for automation: Designing for situation awareness. *IFAC Proceedings*, 28(15), 365–370.

Endsley, M. R. (2018). Situation awareness in future autonomous vehicles: Beware of the unexpected. In *Congress of the International Ergonomics Association* (pp. 303–309). Cham: Springer International Publishing.

Eriksson, A., Petermeijer, S. M., Zimmermann, M., de Winter, J. C. F., Bengler, K. J., & Stanton, N. A. (2019). Rolling out the red (and Green) Carpet: Supporting driver decision making in automation-to-manual transitions. *IEEE Transactions on Human-Machine Systems*, 49(1), 20–31. https://doi.org/10.1109/THMS.2018.2883862

Eyssel, F., Kuchenbrandt, D., Bobinger, S., De Ruiter, L., & Hegel, F. (2012, March). 'If you sound like me, you must be more human' on the interplay of robot and user features on human-robot acceptance and anthropomorphism. In *Proceedings of the Seventh Annual ACM/IEEE International Conference on Human-Robot Interaction* (pp. 125–126). https://doi.org/10.1145/2157689.2157717

FakhrHosseini, M., Jeon, M., & Bose, R. (2015a). Estimation of drivers' emotional states based on neuroergonmic equipment: An exploratory study using fNIRS. In *Adjunct Proceedings of the 7th International Conference on Automotive User Interfaces and Interactive Vehicular Applications*.

Fakhrhosseini, S. M., & Jeon, M. (2017). Affect/emotion induction methods. In M Jeon (Ed.), *Emotions and Affect in Human Factors and Human-Computer Interaction* (pp. 235–253). Academic Press.

FakhrHosseini, S. M., & Jeon, M. (2019). How do angry drivers respond to self-selected music? A comprehensive perspective on assessing emotion, *Journal on Multimodal User Interfaces*, 13(2), 137–150.

FakhrHosseini, S. M., Jeon, M., & Bose, R. (2015b). Estimation of drivers' emotional states based on neuroergonmic equipment: An exploratory study using fNIRS. In *Proceedings of the 7th International Conference on Automotive User Interfaces and Vehicular Applications (AutomotiveUI'15)*, Nottingham, UK, September 1-3.

FakhrHosseini, S. M., Landry, S., Tan, Y-Y., Bhattarai, S., & Jeon, M. (2014). If you're angry, turn the music on: Music can mitigate anger effects on driving performance. In *Proceedings of the 6th International Conference on Automotive User Interfaces and Vehicular Applications (AutomotiveUI'14)*, Seattle, WA, USA, September 17–19.

Fan, X. A., Bi, L. Z., & Chen, Z. L. (2010, July). Using EEG to detect drivers' emotion with Bayesian Networks. In *2010 International Conference on Machine Learning and Cybernetics* (Vol. 3, pp. 1177–1181). IEEE.

Fayyad, J., Jaradat, M. A., Gruyer, D., & Najjaran, H. (2020). Deep learning sensor fusion for autonomous vehicle perception and localization: A review. *Sensors*, 20(15), 4220.

Fitch, G. M., Kiefer, R. J., Hankey, J. M., & Kleiner, B. M. (2007). Toward developing an approach for alerting drivers to the direction of a crash threat. *Human Factors*, 49(4), 710–720.

Fuller, R. (1984). A conceptualization of driving behaviour as threat avoidance. *Ergonomics*, 27(11), 1139–1155. https://doi.org/10.1080/00140138408963596

Fuller, R. (2005). Towards a general theory of driver behaviour. *Accident Analysis & Prevention*, 37(3), 461–472. https://doi.org/10.1016/j.aap.2004.11.003

Geiser, G. (1985). Man machine interaction in vehicles. *ATZ* 87, 74–77.

Getzmann, S., Arnau, S., Karthaus, M., Reiser, J. E., & Wascher, E. (2018). Age-related differences in pro-active driving behavior revealed by EEG measures. *Frontiers in Human Neuroscience*, 12, 321.

Gianaros, P. J., & Stern, R. M. (2010). A questionnaire for the assessment of the multiple dimensions of motion sickness. *Aviation, Space, and Environmental Medicine*, 72(2), 115.

Gluck, A., Boateng, K., Huff Jr, E. W., & Brinkley, J. (2020, September). Putting older adults in the driver seat: Using user enactment to explore the design of a shared autonomous vehicle. In *12th International Conference on Automotive User Interfaces and Interactive Vehicular Applications* (pp. 291–300). https://doi.org/10.1145/3409120.3410645

Gold, C., Happee, R., & Bengler, K. (2018). Modeling take-over performance in level 3 conditionally automated vehicles. *Accident Analysis & Prevention*, 116, 3–13. https://doi.org/10.1016/j.aap.2017.11.009

Greatbatch, R. L., Kim, H., & Doerzaph, Z. (2022). The effects of augmented reality head-up display graphics on driver situation awareness and takeover performance in driving automation systems. In *2022 IEEE Conference on Virtual Reality and 3D User Interfaces Abstracts and Workshops (VRW)*.

Gross, J. J. (1998). The emerging field of emotion regulation: An integrative review. *Review of General Psychology*, 2(3), 271–299.

Gruden, T., Popović, N. B., Stojmenova, K., Jakus, G., Miljković, N., Tomažič, S., & Sodnik, J. (2021). Electrogastrography in autonomous vehicles-an objective method for assessment of motion sickness in simulated driving environments. *Sensors*, 21(2), 550. https://doi.org/10.3390/s21020550

Gugerty, L. (2011). Situation awareness in driving. In J. Lee, M. Rizzo, D. Fisher, & J. Caird (Eds.), *Handbook for Driving Simulation in Engineering, Medicine and Psychology* (Vol. 1, pp. 265–272). Boca Raton, FL: CRC Press.

Guo, F., Klauer, S. G., Fang, Y., Hankey, J. M., Antin, J. F., Perez, M. A., Lee, S. E., & Dingus, T. A. (2017). The effects of age on crash risk associated with driver distraction. *International Journal of Epidemiology*, 46(1), 258–265. https://doi.org/10.1093/ije/dyw234

Haas, E. C., & Van Erp, J. B. (2014). Multimodal warnings to enhance risk communication and safety. *Safety Science*, 61, 29–35.

Hancock, P. A., Lesch, M., & Simmons, L. (2003). The distraction effects of phone use during a crucial driving maneuver. *Accident Analysis & Prevention*, 35(4), 501–514.

Hancock, P. A., Wulf, G., Thom, D., & Fassnacht, P. (1990). Driver workload during differing driving maneuvers. *Accident Analysis & Prevention*, 22(3), 281–290.

Hankins, T. C., & Wilson, G. F. (1998). A comparison of heart rate, eye activity, EEG and subjective measures of pilot mental workload during flight. *Aviation, Space, and Environmental Medicine*, 69(4), 360–367.

Harrington, K., Large, D. R., Burnett, G., & Georgiou, O. (2018, September). Exploring the use of mid-air ultrasonic feedback to enhance automotive user interfaces. In *Proceedings of the 10th International Conference on Automotive User Interfaces and Interactive Vehicular Applications* (pp. 11–20). https://doi.org/10.1145/3239060.3239089

Harris, H., & Nass, C. (2011, May). Emotion regulation for frustrating driving contexts. In *Proceedings of the SIGCHI Conference on Human Factors in Computing Systems* (pp. 749–752). https://doi.org/10.1145/1978942.1979050

Hart, S. G. (2006). NASA-task load index (NASA-TLX); 20 years later. In *Proceedings of the Human Factors and Ergonomics Society Annual Meeting*.

He, X., Stapel, J., Wang, M., & Happee, R. (2022). Modelling perceived risk and trust in driving automation reacting to merging and braking vehicles. *Transportation Research Part F: Traffic Psychology and Behaviour*, 86, 178–195. https://doi.org/10.1016/J.TRF.2022.02.016

Hedlund, J., Simpson, H., & Mayhew, D. (2005). *International Conference on Distracting Driving: Summary of Proceedings and Recommendations*.

Heeger, D. J., & Ress, D. (2002). What does fMRI tell us about neuronal activity? *Nature Reviews Neuroscience*, 3(2), 142–151.

Heinrich, H. W. (1941). *Industrial Accident Prevention. A Scientific Approach* (2nd ed.). New York: McGraw-Hill.

Hess, E. H., & Polt, J. M. (1964). Pupil size in relation to mental activity during simple problem-solving. *Science*, 143(3611), 1190–1192. https://doi.org/10.1126/science.143.3611.1190

Hester, M., Lee, K., & Dyre, B. P. (2017). "Driver take over": A preliminary exploration of driver trust and performance in autonomous vehicles. In *Proceedings of the Human Factors and Ergonomics Society Annual Meeting* (Vol. 61, No. 1, pp. 1969–1973). Sage CA: Los Angeles, CA: SAGE Publications. https://doi.org/10.1177/1541931213601971

Hock, P., Kraus, J., Walch, M., Lang, N., & Baumann, M. (2016). Elaborating feedback strategies for maintaining automation in highly automated driving. In *Proceedings of the 8th International Conference on Automotive User Interfaces and Interactive Vehicular Applications* (pp. 105–112). https://doi.org/10.1145/3003715.3005414

Hoff, K. A., & Bashir, M. (2015). Trust in automation: Integrating empirical evidence on factors that influence trust. *Human Factors*, 57(3), 407–434. https://doi.org/10.1177/001872081454

Holländer, K., Wintersberger, P., & Butz, A. (2019). Overtrust in external cues of automated vehicles: An experimental investigation. In *Proceedings of the 11th International Conference on Automotive User Interfaces and Interactive Vehicular Applications* (pp. 211–221). https://doi.org/10.1145/3342197.3344528

Holthausen, B. E., Wintersberger, P., Walker, B. N., & Riener, A. (2020, September). Situational trust scale for automated driving (STS-AD): Development and initial validation. In *12th International Conference on Automotive User Interfaces and Interactive Vehicular Applications* (pp. 40–47).

Hood, D., Joseph, D., Rakotonirainy, A., Sridharan, S., & Fookes, C. (2012, October). Use of brain computer interface to drive: preliminary results. In *Proceedings of the 4th International Conference on Automotive User Interfaces and Interactive Vehicular Applications* (pp. 103–106). https://doi.org/10.1145/2390256.2390272

Hoogendoorn, R. G., Hoogendoorn, S. P., Brookhuis, K. A., & Daamen, W. (2011). Adaptation longitudinal driving behavior, mental workload, and psycho-spacing models in fog. *Transportation Research Record*, 2249(1), 20–28.

Hong, S., & Yang, J. H. (2022). Effect of multimodal takeover request issued through A-pillar LED light, earcon, speech message, and haptic seat in conditionally automated driving. *Transportation Research Part F: Traffic Psychology and Behaviour*, 89, 488–500.

Horswill, M. S., & McKenna, F. P. (2004). Drivers' hazard perception ability: Situation awareness on the road. In S. Banbury & S. Tremblay (Eds.), *A Cognitive Approach to Situation Awareness: Theory and Application*, (pp. 155–175). Burlington, VT: Ashgate.

Hoshi, Y. (2016). Hemodynamic signals in fNIRS. *Progress in Brain Research*, 225, 153–179.

Hu, M., Liao, Y., Wang, W., Li, G., Cheng, B., & Chen, F. (2017). Decision tree-based maneuver prediction for driver rear-end risk-avoidance behaviors in cut-in scenarios. *Journal of Advanced Transportation*. https://doi.org/10.1155/2017/7170358

Hua, Z., & Ng, W. L. (2010, November). Speech recognition interface design for in-vehicle system. In *Proceedings of the 2nd International Conference on Automotive User Interfaces and Interactive Vehicular Applications* (pp. 29–33). https://doi.org/10.1145/1969773.1969780

Ingre, M., Åkerstedt, T., Peters, B., Anund, A., & Kecklund, G. (2006). Subjective sleepiness, simulated driving performance and blink duration: Examining individual differences. *Journal of Sleep Research*, 15(1), 47–53. https://doi.org/10.1111/J.1365-2869.2006.00504.X

Iqbal, S. T., Zheng, X. S., & Bailey, B. P. (2004). Task-evoked pupillary response to mental workload in human-computer interaction. In *Conference on Human Factors in Computing Systems - Proceedings* (pp. 1477–1480). https://doi.org/10.1145/985921.986094

Jacob, R. J. K., & Karn, K. S. (2003). Eye tracking in human-computer interaction and usability research: Ready to deliver the promises. In *The Mind's Eye: Cognitive and Applied Aspects of Eye Movement Research* (pp. 573–605). https://doi.org/10.1016/B978-044451020-4/50031-1

Jamson, S., Lai, F., & Jamson, H. (2010). Driving simulators for robust comparisons: A case study evaluating road safety engineering treatments. *Accident; Analysis and Prevention*, 42(3), 961–971. https://doi.org/10.1016/J.AAP.2009.04.014

Janssen, C. P., Boyle, L. N., Kun, A. L., Ju, W., & Chuang, L. L. (2019). A hidden markov framework to capture human-machine interaction in automated vehicles. *International Journal of Human-Computer Interaction*, 35(11), 947–955. https://doi.org/10.1080/10447318.2018.1561789

Jeon, J. Y., et al. (2009). Subjective evaluation of heavy-weight floor impact sounds in relation to spatial characteristics. *The Journal of the Acoustical Society of America*, 125(5), 2987–2994.

Jeon, M. (2015). Towards affect-integrated driving behaviour research. *Theoretical Issues in Ergonomics Science*, 16(6), 553–585. https://doi.org/10.1080/1463922X.2015.1067934

Jeon, M. (2016). Don't cry while you're driving: Sad driving is as bad as angry driving. *International Journal of Human-Computer Interaction*, 32(10), 777–790.

Jeon, M. (2017a). Emotions and affect in human factors and human-computer interaction: Taxonomy, theories, approaches, and methods. In M. Jeon (Ed.), *Emotions and Affect in Human Factors and Human-Computer Interaction*, San Diego: Academic Press.

Jeon, M. (2017b). Emotions in driving. In M. Jeon (Ed.), *Emotions and Affect in Human Factors and Human-Computer Interaction* (pp. 437–474). Academic Press.

Jeon, M., & Croschere, J. (2015). Sorry, I'm late; I'm not in the mood: Negative emotions lengthen driving time. In D. Harris (Ed.). *Engineering Psychology and Cognitive Ergonomics (EPCE) 2015, LNAI 9174* (pp. 1–8). Switzerland: Springer International Publishing.

Jeon, M., & Walker, B. N. (2011a). Emotion detection and regulation interface for drivers with traumatic brain injury. In *Workshop on Accessibility. ACM SIGCHI Conference on Human Factors in Computing Systems (CHI'11)*, Vancouver, BC, Canada, May 7-12, 2011.

Jeon, M., & Walker, B. N. (2011b). What to detect? Analyzing factor structures of affect in driving contexts for an emotion detection and regulation system. In *Proceedings of the Human Factors and Ergonomics Society Annual Meeting* (Vol. 55, No. 1, pp. 1889–1893). Sage CA: Los Angeles, CA: SAGE Publications. https://doi.org/10.1177/107118131155139

Jeon, M., Politis, I., Shladover, S. E., Sutter, C., Terken, J. M., & Poppinga, B. (2016, October). Towards life-long mobility: Accessible transportation with automation. In *Adjunct Proceedings of the 8th International Conference on Automotive User Interfaces and Interactive Vehicular Applications* (pp. 203–208). https://doi.org/10.1145/3004323.3004348

Jeon, M., Walker, B. N., & Gable, T. M. (2015). The effects of social interactions with in-vehicle agents on a driver's anger level, driving performance, situation awareness, and perceived workload. *Applied Ergonomics*, 50, 185–199. https://doi.org/10.1016/j.apergo.2015.03.015

Jeon, M., Walker, B. N., & Yim, J. B. (2014). Effects of specific emotions on subjective judgment, driving performance, and perceived workload. *Transportation Research Part F: Traffic Psychology and Behaviour*, 24, 197–209.

Jeong, H., & Liu, Y. (2017, June). Modeling of stimulus-response secondary tasks with different modalities while driving in a computational cognitive architecture. In *Driving Assessment Conference* (Vol. 9, No. 2017). University of Iowa. https://doi.org/10.17077/drivingassessment.1615

Jian, J.-Y., Bisantz, A. M., & Drury, C. G. (2000). Foundations for an empirically determined scale of trust in automated systems. *International Journal of Cognitive Ergonomics*, 4(1), 53–71. https://doi.org/10.1207/S15327566IJCE0401_04

Johnsson, I.-M., Nass, C., Harris, H., & Takayama, L. (2005). Matching in-car voice with driver state: Impact on attitude and driving performance. In *Driving Assessment 2005 : Proceedings of the 3rd International Driving Symposium on Human Factors in Driver Assessment, Training, and Vehicle Design* (pp. 173–180). https://doi.org/10.17077/drivingassessment.1158

Joo, Y. K., & Lee-Won, R. J. (2016). An agent-based intervention to assist drivers under stereotype threat: Effects of in-vehicle agents' attributional error feedback. *Cyberpsychology, Behavior, and Social Networking*, 19(10), 615–620. https://doi.org/10.1089/cyber.2016.0153

Jorna, P. (1993). Heart rate and workload variations in actual and simulated flight. *Ergonomics*, 36(9), 1043–1054.

Joseph, A. W., & Murugesh, R. (2020). Potential eye tracking metrics and indicators to measure cognitive load in human-computer interaction research. *Journal of Scientific Research*, 64(01), 168–175. https://doi.org/10.37398/JSR.2020.640137

Kahneman, D., & Beatty, J. (1966). Pupil diameter and load on memory. *Science*, 154(3756), 1583–1585. https://doi.org/10.1126/science.154.3756.1583

Kantowitz, B. H., & Simsek, O. (2001). Secondary-task measures of driver workload. In P.A. Hancock, & P.A. Desmond (Eds.), *Stress, Workload, and Fatigue* (pp. 395–408). Mahwah, NJ: L. Erlbaum.

Karjanto, J., Md. Yusof, N., Wang, C., Delbressine, F., Rauterberg, M., Terken, J., & Martini, A. (2017). Situation awareness and motion sickness in automated vehicle driving experience: A preliminary study of peripheral visual information. In *Proceedings of the 9th International Conference on Automotive User Interfaces and Interactive Vehicular Applications Adjunct* (pp. 57–61. https://doi.org/10.1145/3131726.3131745

Kaur, K., & Rampersad, G. (2018). Trust in driverless cars: Investigating key factors influencing the adoption of driverless cars. *Journal of Engineering and Technology Management*, 48, 87–96. https://doi.org/10.1016/j.jengtecman.2018.04.006

Kennedy, R. S., Lane, N. E., Berbaum, K. S., and Lilienthal, G. M. (1993). Simulator sickness questionnaire: An enhanced method for quantifying simulator sickness, *The International Journal of Aviation Psychology*, 3(3), 203–220, DOI: 10.1207/s15327108ijap0303_3

Kern, D., & Schmidt, A. (2009, September). Design space for driver-based automotive user interfaces. In *Proceedings of the 1st International Conference on Automotive User Interfaces and Interactive Vehicular Applications* (pp. 3–10). https://doi.org/10.1145/1620509.1620511

Keshavarz, B., & Hecht, H. (2011). Validating an efficient method to quantify motion sickness. *Human Factors: The Journal of the Human Factors and Ergonomics Society*, 53(4), 415–426. https://doi.org/10.1177/0018720811403736

Kim, H., & Doerzaph, Z. (2022, September). Road user attitudes toward automated shuttle operation: Pre and post-deployment surveys. In *Proceedings of the Human Factors and Ergonomics Society Annual Meeting* (Vol. 66, No. 1, pp. 315–319). Sage CA: Los Angeles, CA: SAGE Publications. https://doi.org/10.1177/10711813226610

Kim, Y. W., Lim, C., Ji, Y. G., Yoon, S. H., Colley, M., & Meinhardt, L. M. (2022, September). The 2nd workshop on user experience in urban air mobility: From ground to aerial transportation. In *Adjunct Proceedings of the 14th International Conference on Automotive User Interfaces and Interactive Vehicular Applications* (pp. 168–171). https://doi.org/10.1145/3544999.3550223

Klauer, S. G., Sayer, T. B., Baynes, P., & Ankem, G. (2016). Using real-time and post hoc feedback to improve driving safety for novice drivers. In *Proceedings of the Human Factors and Ergonomics Society Annual Meeting*.

Ko, S., Kutchek, K., Zhang, Y., & Jeon, M. (2022). Effects of non-speech auditory cues on control transition behaviors in semi-automated vehicles: Empirical study, modeling, and validation. *International Journal of Human-Computer Interaction*, 38(2), 185–20 https://doi.org/10.1080/10447318.2021.1937876

Ko, S., Sanghavi, H., Zhang, Y., & Jeon, M. (2022). Modeling the effects of perceived intuitiveness and urgency of≈various auditory warnings on driver takeover performance in automated vehicles. *Transportation Research Part F: Traffic Psychology and Behaviour*, 90, 70–83. https://doi.org/10.1016/j.trf.2022.08.008

Koo, J., Kwac, J., Ju, W., Steinert, M., Leifer, L., & Nass, C. (2015). Why did my car just do that? Explaining semi-autonomous driving actions to improve driver understanding, trust, and performance. *International Journal on Interactive Design and Manufacturing*, 9(4), 269–275. https://doi.org/10.1007/s12008-014-0227-2

Körber, M. (2018, August). Theoretical considerations and development of a questionnaire to measure trust in automation. In *Congress of the International Ergonomics Association* (pp. 13–30). Cham: Springer. https://doi.org/10.1016/j. apergo.2017.07.006

Krejtz, K., Duchowski, A. T., Niedzielska, A., Biele, C., & Krejtz, I. (2018). Eye tracking cognitive load using pupil diameter and microsaccades with fixed gaze. *PLoS One*, 13(9), e0203629. https://doi.org/10.1371/journal.pone.0203629

Kret, M. E., & Sjak-Shie, E. E. (2019). Preprocessing pupil size data: Guidelines and code. *Behavior Research Methods*, 51(3), 1336–1342. https://doi.org/10.3758/s13428-018-1075-y

Kuiper, O. X., Bos, J. E., & Diels, C. (2018). Looking forward: In-vehicle auxiliary display positioning affects carsickness. *Applied Ergonomics*, 68, 169–175. https://doi.org/10.1016/j.apergo.2017.11.002

Kuiper, O. X., Bos, J. E., Diels, C., Schmidt, E. A. (2020a). Knowing what's coming: Anticipatory audio cues can mitigate motion sickness. *Applied Ergonomics*, 68, 169–175. https://doi.org/10.1016/j.apergo.2020.103068

Kuiper, O. X., Bos, J. E., Schmidt, E. A., Deils, C., & Wolter, S. (2020b). Knowing what's coming: Unpredictable motion causes more motion sickness. *Human Factors: The Journal of the Human Factors and Ergonomics Society*, 62, 8, (December 2018), 1339–1348. https://doi.org/10.1177/0018720819876139

Kundinger, T., Wintersberger, P., & Riener, A. (2019). (Over) Trust in automated driving: The sleeping pill of tomorrow? In *Extended Abstracts of the 2019 CHI Conference on Human Factors in Computing Systems*.

Lackner, J.R., and Graybiel, A. (1984). Elicitation of motion sickness by head movements in the microgravity phase of parabolic flight maneuvers. *Aviation, Space, and Environmental Medicine*, 55(6), 513–520.

Large, D. R., Banks, V., Burnett, G., & Margaritis, N. (2017, September). Putting the joy in driving: Investigating the use of a joystick as an alternative to traditional controls within future autonomous vehicles. In *Proceedings of the 9th International Conference on Automotive User Interfaces and Interactive Vehicular Applications* (pp. 31–39). https://doi.org/10.1145/3122986.3122996

Large, D. R., Clark, L., Burnett, G., Harrington, K., Luton, J., Thomas, P., & Bennett, P. (2019a). "It's small talk, Jim, but not as we know it.": Engendering trust through human-agent conversation in an autonomous, self-driving car. In *Proceedings of the 1st International Conference on Conversational User Interfaces - CUI'19* (pp. 1–7). https://doi.org/10.1145/3342775.3342789

Large, D. R., Harrington, K., Burnett, G., Luton, J., Thomas, P., & Bennett, P. (2019b). To please in a pod: Employing an anthropomorphic agent-interlocutor to enhance trust and user experience in an autonomous, self-driving vehicle. In *Proceedings -11th International ACM Conference on Automotive User Interfaces and Interactive Vehicular Applications, AutomotiveUI 2019* (pp. 49–59). https://doi.org/10.1145/3342197.3344545

Laureys, S., Peigneux, P., & Goldman, S. (2002). Brain imaging. In H. D'haenen, J. A. den Boer, P. Willner (Eds.) *Biological Psychiatry* (pp. 155–166). New York: John Wiley & Sons Ltd.

Lee, J. D. (2008). Fifty years of driving safety research. *Human Factors*, 50(3), 521–528.

Lee, J. D., & See, K. A. (2004). Trust in automation: Designing for appropriate reliance. *Human Factors*, 46(1), 50–80. https://doi.org/10.1518/hfes.46.1.50_30392

Lee, J. D., Young, K. L., & Regan, M. A., 2008. Defining driver distraction. In M. A. Regan, J. D. Lee, & K. L. Young (Eds.), *Driver Distraction: Theory, Effects, and Mitigation* (pp. 31–40). Boca Raton, FL: CRC Press Taylor & Francis Group.

Lee, J. G., & Lee, K. M. (2022). Polite speech strategies and their impact on drivers' trust in autonomous vehicles. *Computers in Human Behavior*, 127, 107015.

Lee, S. C., & Jeon, M. (2022). A systematic review of functions and design features of in-vehicle agents. *International Journal of Human-Computer Studies*, 165, 102864. https://doi.org/10.1016/J.IJHCS.2022.102864

Lee, S. C., Nadri, C, Sanghavi, H., & Jeon, M. (2022). Eliciting user needs and design requirements for user experience in fully automated vehicles, *International Journal of Human-Computer Interaction*, 38(3), 227–239. https://doi.org/10.1080/10447318.2021.1937875

Lee, S. C., Yoon, S. H., & Ji, Y. G. (2021). Effects of non-driving-related task attributes on takeover quality in automated vehicles. *International Journal of Human-Computer Interaction*, 37(3), 211–219.

Lee, S., Ratan, R., & Park, T. (2019). The voice makes the car: Enhancing autonomous vehicle perceptions and adoption intention through voice agent gender and style. *Multimodal Technologies and Interaction*, 3(1). https://doi.org/10.3390/mti3010020

Liang, N., Yang, J., Yu, D., Prakah-Asante, K. O., Curry, R., Blommer, M., Swaminathan, R., & Pitts, B. J. (2021). Using eye-tracking to investigate the effects of pre-takeover visual engagement on situation awareness during automated driving. *Accident Analysis & Prevention*, 157, 106143.

Liao, Y., Li, G., Li, S. E., Cheng, B., & Green, P. (2018). Understanding driver response patterns to mental workload increase in typical driving scenarios. *IEEE Access*, 6, 35890–35900.

Lindemann, P., Lee, T.-Y., & Rigoll, G. (2018). Catch my drift: Elevating situation awareness for highly automated driving with an explanatory windshield display user interface. *Multimodal Technologies and Interaction*, 2(4), 71.

Liu, H., Hirayama, T., & Watanabe, M. (2021). Importance of instruction for pedestrian-automated driving vehicle interaction with an external human machine interface: Effects on pedestrians' situation awareness, trust, perceived risks and decision making. *arXiv preprint arXiv:2102.07958*. https://arxiv.org/abs/2102.07958

Liu, W. C., Jeng, M. C., Hwang, J. R., Doong, J. L., Lin, C. Y., & Lai, C. H. (2012). The response patterns of young bicyclists to a right-turning motorcycle: a simulator study. *Perceptual and Motor Skills*, 115(2), 385–402.

Liu, Y.-C., & Wen, M.-H. (2004). Comparison of head-up display (HUD) vs. head-down display (HDD): Driving performance of commercial vehicle operators in Taiwan. *International Journal of Human-Computer Studies*, 61(5), 679–697.

Logothetis, N. K. (2008). What we can do and what we cannot do with fMRI. *Nature*, 453(7197), 869–878.

Lorenz, L., Kerschbaum, P., & Schumann, J. (2014, September). Designing take over scenarios for automated driving: How does augmented reality support the driver to get back into the loop? In *Proceedings of the Human Factors and Ergonomics Society Annual Meeting* (Vol. 58, No. 1, pp. 1681–1685). Sage CA: Los Angeles, CA: Sage Publications.

Mader, M., Bresges, A., Topal, R., Busse, A., Forsting, M., & Gizewski, E. R. (2009). Simulated car driving in fMRI-cerebral activation patterns driving an unfamiliar and a familiar route. *Neuroscience Letters*, 464(3), 222–227.

Madsen, M., & Gregor, S. (2000). Measuring human-computer trust. In *Proceedings of the 11th Australasian Conference on Information Systems* (pp. 6–8).

Mata-Carballeira, Ó., Gutiérrez-Zaballa, J., del Campo, I., & Martínez, V. (2019). An FPGA-based neuro-fuzzy sensor for personalized driving assistance. *Sensors*, 19(18), 4011. https://doi.org/10.3390/S19184011

Mathôt, S., Fabius, J., van Heusden, E., & van der Stigchel, S. (2018). Safe and sensible preprocessing and baseline correction of pupil-size data. *Behavior Research Methods*, 50(1), 94–106. https://doi.org/10.3758/S13428-017-1007-2/FIGURES/6

Matthews, M. D., & Beal, S. A. (2002). *Assessing Situation Awareness in Field Training Exercises*. Research Report 1795, U.S. Army Research Institute for the Behavioural and Social Sciences.

May, K. R., Gable, T. M., & Walker, B. N. (2017, September). Designing an in-vehicle air gesture set using elicitation methods. In *Proceedings of the 9th International Conference on Automotive User Interfaces and Interactive Vehicular Applications* (pp. 74–83). https://doi.org/10.1145/3122986.3123015

Mayer, R.C., Davis, J.H., Schoorman, F.D. (1995). An integrative model of organizational trust. *Academy of Management Review*, 20(3), 709–734. Retrieved November 23, 2022 from https://www.jstor.org/stable/258792

McAllister, D.J., 1995. Affect-and cognition-based trust as foundations for interpersonal cooperation in organizations. *Academy of Management Journal*, 38(1), pp.24–59.

McClelland, J. L. (2009). The place of modeling in cognitive science. *Topics in Cognitive Science*, 1(1), 11–38. https://doi.org/10.1111/j.1756-8765.2008.01003.x

McDonald, A. D., Alambeigi, H., Engström, J., Markkula, G., Vogelpohl, T., Dunne, J., & Yuma, N. (2019). Toward computational simulations of behavior during automated driving takeovers: A review of the empirical and modeling literatures. *Human Factors*, 61(4), 642–688. https://doi.org/10.1177/0018720819829572

McKnight, A. J., & McKnight, A. S. (1993). The effect of cellular phone use upon driver attention. *Accident Analysis & Prevention*, 25(3), 259–265.

Meda, S. A., Calhoun, V. D., Astur, R. S., Turner, B. M., Ruopp, K., & Pearlson, G. D. (2009). Alcohol dose effects on brain circuits during simulated driving: An fMRI study. *Human Brain Mapping*, 30(4), 1257–1270.

Mehler, B., Reimer, B., Coughlin, J. F., & Dusek, J. A. (2009). Impact of incremental increases in cognitive workload on physiological arousal and performance in young adult drivers. *Transportation Research Record*, 2138(1), 6–12.

Meschtscherjakov, A., Strumegger, S., & Trösterer, S. (2019). Bubble margin: Motion sickness prevention while reading on smartphones in vehicles. In D. Lamas, F. Loizides, L. Nacke, H. Petrie, M. Winckler, & P. Zaphiris (Eds.), *Human-Computer Interaction - INTERACT 2019* (Vol. 11747, pp. 660–677). Springer International Publishing. https://link.springer.com/10.1007/978-3-030-29384-0_39

Meyer, J. (2004). Conceptual issues in the study of dynamic hazard warnings. *Human Factors*, 46(2), 196–204. https://www.ncbi.nlm.nih.gov/pubmed/15359670

Michon, J. A. (1985). A critical view of driver behavior models: What do we know, what should we do? In *Human Behavior and Traffic Safety* (pp. 485–524). Boston, MA: Springer. https://doi.org/10.1007/978-1-4613-2173-6_19

Milgram, Paul, et al. (1993). Applications of augmented reality for human-robot communication. In *Proceedings of 1993 IEEE/RSJ International Conference on Intelligent Robots and Systems (IROS'93)* (Vol. 3). IEEE.

Milgram, P. & Kishino, Fumio. (1994). A taxonomy of mixed reality visual displays. *IEICE Transactions on Information Systems*, 77(12), 1321–1329.

Miller, S. (2001). *Workload Measures*. Iowa City: National Advanced Driving Simulator.

Montague, E., Xu, J., & Chiou, E. K. (2014). Shared experiences of technology and trust: An experimental study of physiological compliance between active and passive users in technology-mediated collaborative encounters. *IEEE Transactions on Human-Machine Systems*, 44(5), 614–624. https://doi.org/10.1109/THMS.2014.2325859

Moore, D., Currano, R., Strack, G. E., & Sirkin, D. (2019). The case for implicit external human-machine interfaces for autonomous vehicles. In *Proceedings of the 11th International Conference on Automotive User Interfaces and Interactive Vehicular Applications* (pp. 295–307). https://doi.org/10.1145/3342197.3345320

Morel, M., Petit, C., Bruyas, M. P., Chapon, A., Dittmar, A., Delhomme, G., & Collet, C. (2006, January). Physiological and behavioral evaluation of mental load in shared attention tasks. In *2005 IEEE Engineering in Medicine and Biology 27th Annual Conference* (pp. 5526–5527). IEEE.

Mourant, R. R., & Rockwell, T. H. (1972). Strategies of visual search by novice and experienced drivers. *Human Factors*, 14(4), 325–335.

Muhundan, S., & Jeon, M. (2021, September). Effects of native and secondary language processing on emotional drivers' situation awareness, driving performance, and subjective perception. In *13th International Conference on Automotive User Interfaces and Interactive Vehicular Applications* (pp. 255–262). https://doi.org/10.1145/3409118.3475148

Murer, M., Wilfinger, D., Meschtscherjakov, A., Osswald, S., & Tscheligi, M. (2012, October). Exploring the back of the steering wheel: Text input with hands on the wheel and eyes on the road. In *Proceedings of the 4th International Conference on Automotive User Interfaces and Interactive Vehicular Applications* (pp. 117–120). https://doi.org/10.1145/2390256.2390275

Muth, E. R., Stern, R. M., Thayer, J. F., & Koch, K. L. (1996). Assessment of the multiple dimensions of nausea: The Nausea Profile (NP). *Journal of Psychosomatic Research*, 40(5), 511–520. https://doi.org/10.1016/0022-3999(95)00638-9

Näätänen, R., & Summala, H. (1976). *Road-User Behaviour and Traffic Accidents*. Amsterdam: North-Holland Publishing Company.

Nadri, C., Ko, S., Diggs, C., Winters, M., Sreehari, V., & Jeon, M. (2021a). Novel auditory displays in highly automated vehicles: Sonification improves driver situation awareness, perceived workload, and overall experience. In *Proceedings of the Human Factors and Ergonomics Society Annual Meeting*.

Nadri, C., Ko, S., Diggs, C., Winters, M., Vattakkandy, S., & Jeon, M. (2023). Sonification use cases in highly automated vehicles: Designing and evaluating use cases in level 4 automation. *International Journal of Human-Computer Interaction*, 1–11. https://doi.org/10.1080/10447318.2023.2180236

Nadri, C., Lee, S. C., Kekal, S., Li, Y., Li, X., Lautala, P., Nelson, D., & Jeon, M. (2021b). Investigating the effect of earcon and speech variables on hybrid auditory alerts at rail crossings. In *Paper Presented at the Proceedings of the 26th International Conference on Auditory Display (ICAD 2021)*.

Nagasawa, T., & Hagiwara, H. (2016). Workload induces changes in hemodynamics, respiratory rate and heart rate variability. In *2016 IEEE 16th International Conference on Bioinformatics and Bioengineering (BIBE)*.

Navarro, J., Reynaud, E., & Osiurak, F. (2018). Neuroergonomics of car driving: A critical meta-analysis of neuroimaging data on the human brain behind the wheel. *Neuroscience & Biobehavioral Reviews*, 95, 464–479.

Neale, V. L., Klauer, S. G., Knipling, R. R., Dingus, T. A., Holbrook, G. T., & Petersen, A. (2002). *The 100 Car Naturalistic Driving Study, Phase I - Experimental Design*. https://www.nhtsa.gov/DOT/NHTSA/NRD/Multimedia/PDFs/Crash%20Avoidance/2002/100CarPhase1Report.pdf

Niu, D., Terken, J., & Eggen, B. (2018). Anthropomorphizing information to enhance trust in autonomous vehicles. *Human Factors and Ergonomics in Manufacturing & Service Industries*, 28(6), 352–359. https://doi.org/10.1002/HFM.20745

Ohn-Bar, E., Tran, C., & Trivedi, M. (2012, October). Hand gesture-based visual user interface for infotainment. In *Proceedings of the 4th International Conference on Automotive User Interfaces and Interactive Vehicular Applications* (pp. 111–115). https://doi.org/10.1145/2390256.2390274

Oka, N., Yoshino, K., Yamamoto, K., Takahashi, H., Li, S., Sugimachi, T., Nakano, K., Suda, Y., & Kato, T. (2015). Greater activity in the frontal cortex on left curves: A vector-based fNIRS study of left and right curve driving. *PLoS One*, 10(5), e0127594.

Okazaki, S., Haramaki, T., & Nishino, H. (2019). A safe driving support method using olfactory stimuli. In *Complex, Intelligent, and Software Intensive Systems: Proceedings of the 12th International Conference on Complex, Intelligent, and Software Intensive Systems (CISIS-2018)* (pp. 958–967). Springer International Publishing.

Orlovska, J., Novakazi, F., Lars-Ola, B., Karlsson, M. A., Wickman, C., & Söderberg, R. (2020). Effects of the driving context on the usage of Automated Driver Assistance Systems (ADAS) -Naturalistic Driving Study for ADAS evaluation. *Transportation Research Interdisciplinary Perspectives*, 4, 100093. https://doi.org/10.1016/J.TRIP.2020.100093

Orlovska, J., Wickman, C., & Söderberg, R. (2020). Naturalistic driving study for Automated Driver Assistance Systems (ADAS) evaluation in the Chinese, Swedish and American markets. *Procedia CIRP*, 93, 1286–1291. https://doi.org/10.1016/J.PROCIR.2020.04.108

Paddeu, D., Parkhurst, G., & Shergold, I. (2020). Passenger comfort and trust on first-time use of a shared autonomous shuttle vehicle. *Transportation Research Part C: Emerging Technologies*, 115, 102604. https://doi.org/10.1016/j.trc.2020.02.026

Padgham, L., & Winikoff, M. (2005). Agents and multi-agent systems. In M. Wooldridge (Ed.), *Developing Intelligent Agent Systems* (pp. 1–6). John Wiley & Sons, Ltd. https://doi.org/10.1002/0470861223.ch1

Papazikou, E., Thomas, P., & Quddus, M. (2021). Developing personalised braking and steering thresholds for driver support systems from SHRP2 NDS data. *Accident Analysis & Prevention*, 160, 106310. https://doi.org/10.1016/J.AAP.2021.106310

Parasuraman, R., & Manzey, D. H. (2010). Complacency and bias in human use of automation: An attentional integration. *Human Factors*, 52(3), 381–410. https://doi.org/10.1177/0018720810376055

Parasuraman, R., & Riley, V. (1997). Humans and automation: Use, misuse, disuse, abuse. *Human factors*, 39(2), 230–253.

Parker, M., & Zegeer, C. (1989). *Traffic Conflict Techniques for Safety and Operations: Observers Manual* (No. FHWA-IP-88-027, NCP 3A9C0093). United States. Federal Highway Administration. https://rosap.ntl.bts.gov/view/dot/14308

Pauzié, A. (2008). A method to assess the driver mental workload: The driving activity load index (DALI). *IET Intelligent Transport Systems*, 2(4), 315–322.

Petermeijer, S. M., Cieler, S., & de Winter, J. C. F. (2017b). Comparing spatially static and dynamic vibrotactile take-over requests in the driver seat. *Accident Analysis & Prevention*, 99, 218–227. https://doi.org/10.1016/j.aap.2016.12.001

Petermeijer, S., Bazilinskyy, P., Bengler, K., & De Winter, J. (2017a). Take-over again: Investigating multimodal and directional TORs to get the driver back into the loop. *Applied Ergonomics*, 62, 204–215.

Pfleging, B., Schneegass, S., & Schmidt, A. (2012, October). Multimodal interaction in the car: Combining speech and gestures on the steering wheel. In *Proceedings of the 4th International Conference on Automotive User Interfaces and Interactive Vehicular Applications* (pp. 155–162). https://doi.org/10.1145/2390256.2390282

Plavšić, M., Klinker, G., & Bubb, H. (2010). Situation awareness assessment in critical driving situations at intersections by task and human error analysis. *Human Factors and Ergonomics in Manufacturing & Service Industries*, 20(3), 177–191.

Politis, I., Brewster, S., & Pollick, F. (2013, October). Evaluating multimodal driver displays of varying urgency. In *Proceedings of the 5th International Conference on Automotive User Interfaces and Interactive Vehicular Applications* (pp. 92–99).

Poole, A., & Ball, L. J. (2006). Eye tracking in HCI and usability research. In *Encyclopedia of Human Computer Interaction* (pp. 211–219). IGI Global. https://doi.org/10.4018/978-1-59140-562-7.ch034

Popović, N. B., Miljković, N., Stojmenova, K., Jakus, G., Prodanov, M., & Sodnik, J. (2019). Lessons learned: Gastric motility assessment during driving simulation. *Sensors*, 19(14), 3175. https://doi.org/10.3390/s19143175

Pradhan, A. K., Hu, B. G., Buckley, L., & Bingham, C. R. (2015). Pre-frontal cortex activity of male drivers in the presence of passengers during simulated driving: An exploratory functional near-infrared spectroscopy (fNIRS) study. In *Driving Assesment Conference*,

Pritzl, T.J., 2003. The effect of experimentally enhanced daydreaming on an electroencephalographic measure of sleepiness. Unpublished Dissertation. North Carolina State University, Raleigh, NC.

Rayner, K. (2009). Eye movements and attention in reading, scene perception, and visual search. *Quarterly Journal of Experimental Psychology*, 62(8), 1457–1506. https://doi.org/10.1080/17470210902816461

Reed, M. P., & Green, P. A. (1999). Comparison of driving performance on-road and in a low-cost simulator using a concurrent telephone dialling task. *Ergonomics*, 42(8), 1015–1037.

Reid, G. B., & Nygren, T. E. (1988). The subjective workload assessment technique: A scaling procedure for measuring mental workload. In *Advances in Psychology* (Vol. 52, pp. 185–218). Elsevier.

Reilhac, C. D., Bos, J. E., Katharina Hottelart, Patrice. (2016). Motion sickness in automated vehicles: The elephant in the room. *Road Vehicle Automation* 3, 121–129. https://doi.org/10.1007/978-3-319-40503-2_10

Reuten A., Bos J., Smeets J.B. (2020) The metrics for measuring motion sickness. In *Driving Simul Conf Europe* (Vol. 2020, pp. 1–4).

Reuten, A. J. C., Nooij, S. A. E., Bos, J. E., & Smeets, J. B. J. (2021). How feelings of unpleasantness develop during the progression of motion sickness symptoms. *Experimental Brain Research*, 239(12), 3615–3624. https://doi.org/10.1007/s00221-021-06226-1

Riegler, A., Aksoy, B., Riener, A., & Holzmann, C. (2020, September). Gaze-based interaction with windshield displays for automated driving: Impact of dwell time and feedback design on task performance and subjective workload. In *12th International Conference on Automotive User Interfaces and Interactive Vehicular Applications* (pp. 151–160). https://doi.org/10.1145/3409120.3410654

Riener, A., Jeon, M., & Alvarez, I. (Eds.). (2022). *User Experience Design in the Era of Automated Driving*. Springer.

Riener, A., & Wintersberger, P. (2011, November). Natural, intuitive finger based input as substitution for traditional vehicle control. In *Proceedings of the 3rd International Conference on Automotive User Interfaces and Interactive Vehicular Applications* (pp. 159–166). https://doi.org/10.1145/2381416.2381442

Riener, A., Ferscha, A., Bachmair, F., Hagmüller, P., Lemme, A., Muttenthaler, D., ... & Weger, F. (2013, October). Standardization of the in-car gesture interaction space. In *Proceedings of the 5th International Conference on Automotive User Interfaces and Interactive Vehicular Applications* (pp. 14–21). https://doi.org/10.1145/2516540.2516544

Roche, F., & Brandenburg, S. (2020). Should the urgency of visual-tactile takeover requests match the criticality of takeover situations. *IEEE Transactions on Intelligent Vehicles*, 5(2), 306–313. https://doi.org/10.1109/TIV.2019.2955906

Romano Bergstrom, J., Duda, S., Hawkins, D., & McGill, M. (2014). Physiological response measurements. In *Eye Tracking in User Experience Design* (pp. 81–108). Elsevier. https://doi.org/10.1016/B978-0-12-408138-3.00004-2

Roscoe, A. H. (1992). Assessing pilot workload. Why measure heart rate, HRV and respiration? *Biological Psychology*, 34(2–3), 259–287.

Rümelin, S., & Butz, A. (2013, October). How to make large touch screens usable while driving. In *Proceedings of the 5th International Conference on Automotive User Interfaces and Interactive Vehicular Applications* (pp. 48–55). https://doi.org/10.1145/2516540.2516557

Russell, J. A. (1980). A circumplex model of affect. *Journal of Personality and Social Psychology*, 39(6), 1161.

SAE International. (2015). *Operational Definitions of Driving Performance Measures and Statistics*. https://doi.org/10.4271/J2944_201506

SAE International. (2021). *Taxonomy and Definitions for Terms Related to Driving Automation Systems for On-Road Motor Vehicles*. Warrendale, PA: SAE International. https://doi.org/10.4271/J3016_202104

Saito, Y., & Raksincharoensak, P. (2019). Effect of risk-predictive haptic guidance in one-pedal driving mode. *Cognition, Technology & Work*, 21(4), 671–684. https://doi.org/10.1007/s10111-019-00558-3

Salmon, P. M., Lenne, M. G., Walker, G. H., Stanton, N. A., & Filtness, A. (2014). Exploring schema-driven differences in situation awareness between road users: An on-road study of driver, cyclist and motorcyclist situation awareness. *Ergonomics*, 57(2), 191–209.

Salmon, P. M., Young, K. L., & Cornelissen, M. (2013). Compatible cognition amongst road users: The compatibility of driver, motorcyclist, and cyclist situation awareness. *Safety Science*, 56, 6–17.

Salter, S., Diels, C., Herriotts, P., Kanarachos, S., & Thake, D. (2019). Motion sickness in automated vehicles with forward and rearward facing seating orientations. *Applied Ergonomics*, 78, 54–61. https://doi.org/10.1016/j.apergo.2019.02.001

Salvucci, D. D. (2006). Modeling driver behavior in a cognitive architecture. *Human Factors*, 48(2), 362–380. https://doi.org/10.1518/001872006777724417

Salvucci, D. D. (2009). Rapid prototyping and evaluation of in-vehicle interfaces. *ACM Transactions on Computer-Human Interaction (TOCHI)*, 16(2), 1–33.

Sanghavi, H., Zhang, Y., & Jeon, M. (2020). Effects of anger and display urgency on takeover performance in semi-automated vehicles. In *Proceedings of the 12th International Conference on Automotive User Interfaces and Interactive Vehicular Applications (AutomotiveUI'20)*, Washington, DC, September 21-22 (Virtual Conference).

Sanghavi, H., Zhang, Y., & Jeon, M. (2023). Exploring the influence of driver affective state and auditory display urgency on takeover performance in semi-automated vehicles: Experiment and modelling, *International Journal of Human-Computer Studies*, 171, 102979. https://doi.org/10.1016/j.ijhcs.2022.102979

Schall, A., & Romano Bergstrom, J. (2014). Introduction to eye tracking. In *Eye Tracking in User Experience Design* (pp. 3–26). Elsevier. https://doi.org/10.1016/B978-0-12-408138-3.00001-7

Schartmüller, C. (2021). Towards a mobile office: User interfaces for safety and productivity in conditionally automated vehicles. Doctoral Dissertation. Johannes Kepler University of Linz, Austria.

Schartmüller, C., & Riener, A. (2020). Sick of scents: Investigating non-invasive olfactory motion sickness mitigation in automated driving. In *12th International Conference on Automotive User Interfaces and Interactive Vehicular Applications* (pp. 30–39). https://doi.org/10.1145/3409120.3410650

Schiffman, H. R. (2001). *Sensation and Perception: An Integrated Approach* (5th ed.). John Wiley & Sons.

Scholtz, J. C., Antonishek, B., & Young, J. D. (2005). Implementation of a situation awareness assessment tool for evaluation of human-robot interfaces. *IEEE Transactions on Systems, Man, and Cybernetics-Part A: Systems and Humans*, 35(4), 450–459.

Scholtz, J., Antonishek, B., & Young, J. (2004). Evaluation of a human-robot interface: Development of a situational awareness methodology. In *37th Annual Hawaii International Conference on System Sciences, 2004. Proceedings of the*.

Shakeri, G., Williamson, J. H., & Brewster, S. (2018, September). May the force be with you: Ultrasound haptic feedback for mid-air gesture interaction in cars. In *Proceedings of the 10th International Conference on Automotive User Interfaces and Interactive Vehicular Applications* (pp. 1–10). https://doi.org/10.1145/3239060.3239081

Shi, E., & Frey, A. T. (2021). Non-driving-related tasks during level 3 automated driving phases - measuring what users will be likely to do. *Technology, Mind, and Behavior*, 2(2). https://doi.org/10.1037/tmb0000006

Sikora, M., et al. (2018). Soundscape of an archaeological site recreated with audio augmented reality. *ACM Transactions on Multimedia Computing, Communications, and Applications (TOMM)* 14(3), 1–22.

Skarbez, R., Smith, M., & Whitton, M. C. (2021). Revisiting Milgram and Kishino's reality-virtuality continuum. *Frontiers in Virtual Reality*, 2, 647997.

Smith, S. S., Horswill, M. S., Chambers, B., & Wetton, M. (2009). Hazard perception in novice and experienced drivers: The effects of sleepiness. *Accident Analysis & Prevention*, 41(4), 729–733.

Smolensky, M. (1993). Toward the physiological measurement of situation awareness: The case for eye movement measurements. In *Proceedings of the Human Factors and Ergonomics Society 37th Annual Meeting*.

Stavrinos, D., Byington, K. W., & Schwebel, D. C. (2011). Distracted walking: Cell phones increase injury risk for college pedestrians. *Journal of Safety Research*, 42(2), 101–107. https://doi.org/10.1016/j.jsr.2011.01.004

Sterkenburg, J., Landry, S., & Jeon, M. (2019). Design and evaluation of auditory-supported air gesture controls in vehicles, *Journal on Multimodal User Interfaces*, 13(2), 55–70. https://doi.org/10.1007/s12193-019-00298-8

Stojmenova, K., Sanghavi, H., Nadri, C., Ko, S., Tomažič, S., Sodnik, J., & Jeon, M. (2020). *Use of Spatial Sound Notifications for Takeover Requests in Semi-autonomous Vehicles-a Cross-Cultural Study*. ICIST2020.

Strayer, D. L., Drews, F. A., & Johnston, W. A. (2003). Cell phone-induced failures of visual attention during simulated driving. *Journal of Experimental Psychology: Applied*, 9(1), 23.

Sucha, M., Dostal, D.l, Risser, R. (2017). Pedestrian-driver communication and decision strategies at marked crossings. *Accident Analysis and Prevention*, 102, 41–50. https://doi.org/10.1016/j.aap.2017.02.018

Taylor, D. H. (1964). Drivers' galvanic skin response and the risk of accident. *Ergonomics*, 7(4), 439–451. https://doi.org/10.1080/00140136408930761

Thomas, A. M., & Jeon, M. (2020). Comparative analysis of facial affect detection algorithms. In *Proceedings of 2020 IEEE International Conference on Systems, Man, and Cybernetics (SMC2020)*, Toronto, Canada, October 11-14 (Virtual Conference).

Thompson, L. L., Rivara, F. P., Ayyagari, R. C., & Ebel, B. E. (2013). Impact of social and technological distraction on pedestrian crossing behaviour: An observational study. *Injury Prevention*, 19, 232–237. https://doi.org/10.1136/injuryprev-2012-040601

Tong, W., Hussain, A., Bo, W. X., & Maharjan, S. (2019). Artificial intelligence for vehicle-to-everything: A survey. *IEEE Access*, 7, 10823–10843.

US Department of Transportation, National Highway Traffic Safety Administration (2014a). *Fatality Analysis Reporting Encyclopedia*. Retrieved from fars.nhtsa.dot.gov/Main/index.aspx.

US Department of Transportation, National Highway Traffic Safety Administration (2014b). *Driver electronic device use in 2012*. Report No. DOT HS 811 884. Retrieved from nrd.nhtsa.dot.gov/Pubs/811884.pdf.

van Haperen, W., Riaz, M. S., Daniels, S., Saunier, N., Brijs, T., & Wets, G. (2019). Observing the observation of (vulnerable) road user behaviour and traffic safety: A scoping review. *Accident Analysis & Prevention*, 123, 211–221. https://doi.org/10.1016/j.aap.2018.11.021

van Leeuwen, P. M., Happee, R., & de Winter, J. C. (2015). Changes of driving performance and gaze behavior of novice drivers during a 30-min simulator-based training. *Procedia Manufacturing*, 3, 3325–3332.

Vasey, E., Ko, S., & Jeon, M. (2018, September). In-vehicle affect detection system: Identification of emotional arousal by monitoring the driver and driving style. In *Adjunct Proceedings of the 10th International Conference on Automotive User Interfaces and Interactive Vehicular Applications* (pp. 243–247). https://doi.org/10.1145/3239092.3267417

Villringer, A., Planck, J., Hock, C., Schleinkofer, L., & Dirnagl, U. (1993). Near infrared spectroscopy (NIRS): A new tool to study hemodynamic changes during activation of brain function in human adults. *Neuroscience Letters*, 154(1–2), 101–104.

Waag, W. L., & Houck, M. R. (1994). Tools for assessing situational awareness in an operational fighter environment. *Aviation, Space, and Environmental Medicine*, 65(5 Suppl), A13-9.

Walker, G. H., Stanton, N. A., & Young, M. S. (2008). Feedback and driver situation awareness (SA): A comparison of SA measures and contexts. *Transportation Research Part F: Traffic Psychology and Behaviour*, 11(4), 282–299.

Wang, C., Zhao, X., Fu, R., & Li, Z. (2020). Research on the comfort of vehicle passengers considering the vehicle motion state and passenger physiological characteristics: Improving the passenger comfort of autonomous vehicles. *International Journal of Environmental Research and Public Health*, 17(18), 6821. https://doi.org/10.3390/ijerph17186821

Wang, M., Hock, P., Chan Lee, S., Baumann, M., & Jeon, M. (2022a). Jarvis in the car: Report on characterizing and designing in-vehicle intelligent agents workshop. In *Proceedings of the Human Factors and Ergonomics Society Annual Meeting* (Vol. 66, No. 1, pp. 948–952). Sage CA: Los Angeles, CA: SAGE Publications. https://doi.org/10.1177/1071181322661445

Wang, M., Lee, S. C., Montavon, G., Qin, J., & Jeon, M. (2022b). Conversational voice agents are preferred and lead to better driving performance in conditionally automated vehicles. In *Proceedings of the 14th International Conference on Automotive User Interfaces and Interactive Vehicular Applications* (Vol. 1, No. 1, pp. 86–95). https://doi.org/10.1145/3543174.3546830

Wang, M., Lee, S. C., Sanghavi, H. K., Eskew, M., Zhou, B., & Jeon, M. (2021). In-vehicle intelligent agents in fully autonomous driving: The effects of speech style and embodiment together and separately. In *13th International Conference on Automotive User Interfaces and Interactive Vehicular Applications* (pp. 247–254). https://doi.org/10.1145/3409118.3475142

Wegman, F., & Aarts, L. (Eds.). (2006). *Advancing Sustainable Safety: National Road Safety Outlook for 2005-2020*. Leidschendam: SWOV Institute for Road Safety Research.

Wilde, G. J. (1982). The theory of risk homeostasis: Implications for safety and health. *Risk Analysis*, 2(4), 209–225. https://doi.org/10.1111/j.1539-6924.1982.tb01384.x

Winkel, K. N., Irmak, T., Kotian, V., Pool, D. M., & Happee, R. (2022). Relating individual motion sickness levels to subjective discomfort ratings. *Experimental Brain Research*. https://doi.org/10.1007/s00221-022-06334-6

Winkel, K. N., Pretto, P., Nooij, S. A. E., Cohen, I., & Bülthoff, H. H. (2021). Efficacy of augmented visual environments for reducing sickness in autonomous vehicles. *Applied Ergonomics*, 90, 103282. https://doi.org/10.1016/j.apergo.2020.103282

Winter, U., Grost, T. J., & Tsimhoni, O. (2010, November). Language pattern analysis for automotive natural language speech applications. In *Proceedings of the 2nd International Conference on Automotive User Interfaces and Interactive Vehicular Applications* (pp. 34–41). https://doi.org/10.1145/1969773.1969781

Wintersberger, P., von Sawitzky, T., Frison, A. K., & Riener, A. (2017, September). Traffic augmentation as a means to increase trust in automated driving systems. In *Proceedings of the 12th Biannual Conference on Italian Sigchi Chapter* (pp. 1–7). https://doi.org/10.1145/3125571.3125600

Wong, P. N. Y., Brumby, D. P., Babu, H. V. R., & Kobayashi, K. (2019). "Watch out!" Semi-autonomous vehicles using assertive voices to grab distracted drivers' attention. In *Conference on Human Factors in Computing Systems – Proceedings* (pp. 5–10). https://doi.org/10.1145/3290607.3312838

Wu, X., Boyle, L. N., & Marshall, D. (2017, September). Drivers' avoidance strategies when using a forward collision warning (FCW) system. In *Proceedings of the Human Factors and Ergonomics Society Annual Meeting* (Vol. 61, No. 1, pp. 1939–1943). Sage CA: Los Angeles, CA: SAGE Publications. https://doi.org/10.1177/1541931213601964

Wynne, R. A., Beanland, V., & Salmon, P. M. (2019). Systematic review of driving simulator validation studies. *Safety Science*, 117(March), 138–151. https://doi.org/10.1016/j.ssci.2019.04.004

Xiang, W., Yan, X., Weng, J., & Li, X. (2016). Effect of auditory in-vehicle warning information on drivers' brake response time to red-light running vehicles during collision avoidance. *Transportation Research Part F: Traffic Psychology and Behaviour*, 40, 56–67.

Xu, L., Wang, B., Xu, G., Wang, W., Liu, Z., & Li, Z. (2017). Functional connectivity analysis using fNIRS in healthy subjects during prolonged simulated driving. *Neuroscience Letters*, 640, 21–28.

Yamada, K., & Kuchar, J. K. (2006). Preliminary study of behavioral and safety effects of driver dependence on a warning system in a driving simulator. *IEEE Transactions on Systems, Man, and Cybernetics-Part A: Systems and Humans*, 36(3), 602–610.

Yoon, S. H., & Ji, Y. G. (2019). Non-driving-related tasks, workload, and takeover performance in highly automated driving contexts. *Transportation Research Part F: Traffic Psychology and Behaviour*, 60, 620–631.

Young, K. L., Salmon, P. M., & Cornelissen, M. (2013). Missing links? The effects of distraction on driver situation awareness. *Safety Science*, 56, 36–43.

Young, K., Regan, M., & Hammer, M. (2007). Driver distraction: A review of the literature. *Distracted Driving*, 2007, 379–405.

Yusof, M.N., Karjanto, J., Terken, J. M. B., Delbressine, F. L. M., & Rauterberg, G. W. M. (2020). Gaining situation awareness through a vibrotactile display to mitigate motion sickness in fully-automated driving cars. *International Journal of Automotive and Mechanical Engineering*, 17(1), 7771–7783. https://doi.org/10.15282/ijame.17.1.2020.23.0578

Zang, J., & Jeon, M. (2022). The effects of transparency and reliability of in-vehicle intelligent agents on driver perception, takeover performance, workload and situation awareness in conditionally automated vehicles. *Multimodal Technologies and Interaction*, 6(9), 82.

Zhang, L.L., Wang, J.Q., Qi, R.R., Pan, L.L., Li, M., & Cai, Y.L. (2016). Motion sickness: Current knowledge and recent advance. *CNS Neuroscience & Therapeutics*, 22(1), 15–24. https://doi.org/10.1111/cns.12468

Zhang, S., et al. (2021). Motivation, social emotion, and the acceptance of artificial intelligence virtual assistants—Trust-based mediating effects. *Frontiers in Psychology*, 12, 728495.

Zhang, T., Yang, J., Liang, N., Pitts, B. J., Prakah-Asante, K. O., Curry, R., Duerstock, B. S., Wachs, J. P., & Yu, D. (2020). Physiological measurements of situation awareness: A systematic review. *Human Factors: The Journal of the Human Factors and Ergonomics Society*, 3. https://doi.org/10.1177/0018720820969071

Zhang, Y., & Wu, C. (2018, September). Modeling the effects of warning lead time, warning reliability and warning style on human performance under connected vehicle settings. In *Proceedings of the Human Factors and Ergonomics Society Annual Meeting* (Vol. 62, No. 1, pp. 701–701). Sage CA: Los Angeles, CA: SAGE Publications. https://doi.org/10.1177/1541931218621158

Zhang, Y., Wu, C., & Wan, J. (2016). Mathematical modeling of the effects of speech warning characteristics on human performance and its application in transportation cyberphysical systems. *IEEE Transactions on Intelligent Transportation Systems*, 17(11), 3062–3074. https://doi.org/10.1109/TITS.2016.2539975

Zhuang, X., Wu, C. (2014). Pedestrian gestures increase driver yielding at uncontrolled mid-block road crossings. *Accident Analysis and Prevention* 70, 235–244.

Zihsler, J., Schwager, D., Hock, P., Szauer, P., Walch, M., Rukzio, E., & Dzuba, K. (2016). Carvatar: Increasing trust in highly-Automated driving through social cues. In *AutomotiveUI 2016-8th International Conference on Automotive User Interfaces and Interactive Vehicular Applications, Adjunct Proceedings* (pp. 9–14). https://doi.org/10.1145/3004323.3004354

10 Human–Robot Interaction

Connor Esterwood, Qiaoning Zhang, X. Jessie Yang, and Lionel P. Robert

10.1 INTRODUCTION

Organizations across our society are increasingly relying on robots to engage in interactions with humans. A robot can be defined as a sophisticated machine that is equipped with sensors, processing capabilities, and actuators that enable it to perceive, analyze, and interact with its surroundings in a physical manner (You et al., 2018; You and Robert Jr, 2018). Human–robot interaction (HRI) is an area of research that focuses on identifying and understanding the factors that promote or hinder human interaction with robots. The study of HRI is multidisciplinary and involves fields such as psychology, information science, computer science, engineering, and design and has the potential to transform various fields of human endeavor such as finance, manufacturing, health care, and education. At its core, HRI research seeks to design robots that are more responsive, engaging, and trustworthy to promote their acceptance by humans. This includes conducting research that identifies ways to build robots that are capable of interacting with humans in an intuitive, comfortable, and natural way to help promote collaborations between humans and robots.

The goal of this chapter is to provide a comprehensive overview of the most vital areas shaping HRI. To accomplish this, this chapter is organized in the following way. First, it presents and discusses the types of robots. The field of HRI has explored a diverse range of robots, revealing their impact on various outcomes (Robert, 2018; Robert Jr et al., 2020). Second, this chapter presents a scoping literature review that surveys trust in HRI. Trust is the foundation by which humans have sought, engaged, and benefited from one another. It is no surprise that trust has been shown to be vital for collaborative action between humans and robots. Third, this chapter identifies and discusses the literature on personality in HRI. Personality, both human and robot, has been shown to impact the interactions between humans and robots. Personality can be viewed as a representation of an individual human or robot's future behaviors, cognitions, and emotional reactions (Robert, 2018). To fully grasp the intricacies of personality in HRI, this chapter employs a multidisciplinary approach encompassing several views on both human and robot personality. Next, this chapter presents the literature on robot explanations. "Robot explanations" can be defined as the reasons that the robot provides to make its actions clear or easy to understand (Zhang et al., 2021). Robot explanations can decrease the uncertainty associated with the robot's actions by providing transparency. Finally, this chapter delves into the various metrics used to evaluate HRI. We also discuss the latest research on evaluation metrics of human interaction with robots. Traditionally, evaluation metrics of human interaction with robots have survey-based static measures. However, recent advances in sensors allow for real-time measures based on physiological changes, which can be obtained alongside or in place of traditional survey measures. In summary, this chapter provides an overview of important HRI areas shaping the field today.

10.2 ROBOTS USED IN HRI RESEARCH

10.2.1 Type of Robot

There are many definitions of the term "robot." The definition that best aligns with the use of robots in the HRI field is offered by You and Robert Jr (2018), who defined robots as technologies that can have either virtual or physical-embodied actions. As You and Robert Jr (2018) pointed out, embodiment and representation of embodied behaviors are what make robots different from other artificial intelligence (AI) technologies. Consistent with this definition, the field of HRI has utilized a wide range of robotic platforms, including custom one-off designs and standardized commercially available robots, with the latter being especially important for ensuring replication and reproducibility. Generally, at least 20 types of robots have been employed in HRI studies (Robert, 2018; Zimmerman et al., 2022), highlighting the diverse array of robotic platforms present in the literature. One way to categorize these robots is by their morphology, which can be classified as either humanoid or non-humanoid. Research has shown that differences between these morphologies can have a significant impact on HRI outcomes (Robert, 2018; Robert Jr et al., 2020). In the following sections, we dive deeper into these two types of robots and provide a brief overview of the most common robots used to represent each type.

10.2.1.1 Humanoid Robot

Humanoid robots, engineered to mimic human form and behavior, offer crucial insights for HRI research. These robots are defined by their human-like appearance, typically featuring a head, two arms, two legs, and a torso (Hirai et al., 1998; Ishida et al., 2001). Their configuration enables them to execute tasks that closely resemble human actions, although some humanoid robots only replicate specific body parts to concentrate on certain aspects of human-like interaction.

The expansive category of humanoid robots encompasses two primary subcategories: avatars and human-like robots. Avatars can manifest as virtual representations or physical platforms, conveying responses based on simulated facial expressions, gaze, or other cues (Zimmerman et al., 2022). These embodiments simulate human presence and interaction, enabling researchers to investigate human perceptions and responses to humanoid representations in various contexts. On the other hand, human-like robots are physical machines explicitly engineered to resemble humans in both form and behavior. Figure 10.1 shows that among the humanoid robots used in research, the Nao robot, Pepper robot, and Baxter robot have been identified as the most popular choices across multiple studies (Robert, 2018; Zimmerman et al., 2022).

10.2.1.2 Non-Humanoid Robot

Non-humanoid robots typically have simpler embodiments that are targeted toward specific tasks or domains (Cha et al., 2018; Coeckelbergh, 2011; Terada et al., 2007). As a result, these robots are utilized in a wide variety of settings and applications, such as health care, agriculture, space,

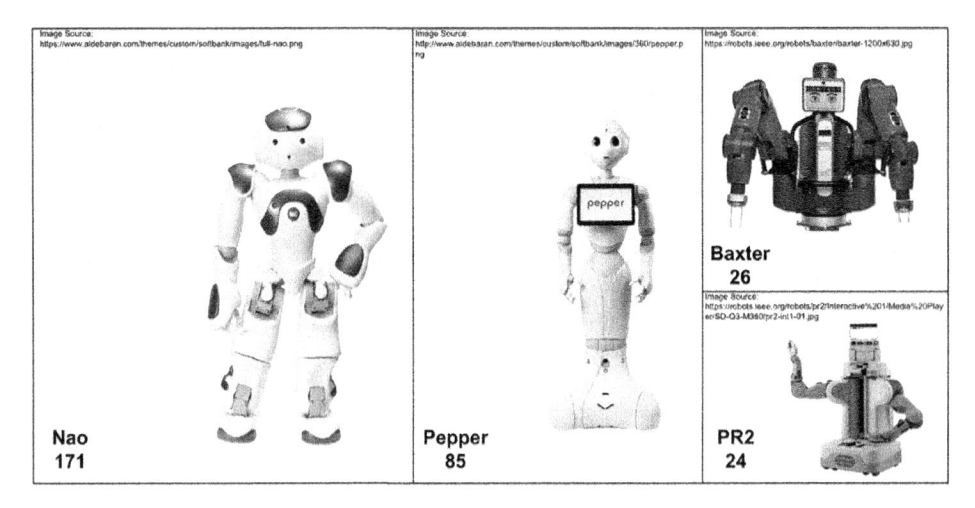

FIGURE 10.1 Figure illustrates the most popular humanoid robots identified by Zimmerman et al. (2022) and the number of times they were used across the human–robot interaction (HRI) literature between 2015 and 2021.

industry, automotive, and service (Cha et al., 2018). These robots encompass a variety of forms, shapes, and functionalities, with numerous robot types falling under this classification. As identified by Zimmerman et al. (2022), there are 15 distinct types of robots classified as non-humanoid robots. Among these are aquatic robots, which operate in aquatic environments (Georgiades et al., 2004; Long et al., 2006), and audio-only robots that interact with humans primarily through auditory means using voice commands and responses, without a physical or visual presence (Guinness et al., 2019; Raghunath et al., 2021). Other non-humanoid robot types include drones, which are capable of flying and performing tasks like aerial photography, surveillance, and delivery (Bhat et al., 2022; Suzuki, 2018); robotic hands, which are specialized robots that replicate human hand functionality (Piazza et al., 2019); non-industrial arms, which can be used in non-industrial settings (Millo et al., 2021); and image/video robots, which utilize visual information for communication or interaction (Raghunath et al., 2021). Industrial manipulators, also known as robotic arms, are employed in manufacturing and production settings for tasks (Zanchettin et al., 2013). Industrial mobile robots are autonomous machines designed to navigate complex industrial environments (Schneier et al., 2015). Mobile manipulators combine mobility and manipulation capabilities, enabling them to navigate and interact effectively within their environment (Bostelman et al., 2016). Mobile platforms, which move on wheels or tracks, navigate various environments for tasks such as transportation, exploration, or assistance (Sørensen et al., 2015). Mobility assistant robots are specifically designed to help individuals with mobility challenges, such as wheelchair users or those with physical impairments, by providing support and guidance (Geravand et al., 2016). Simulation/video game robots exist within virtual environments and serve as interactive characters or elements (Roitberg et al., 2021). Telepresence robots facilitate remote users' virtual presence in a different location, enabling interaction (Tsui et al., 2011). Toy robots, designed for entertainment and education, have various shapes, sizes, and functionalities (Michaud et al., 2000). Last, written vignettes are textual descriptions of robot behavior or interactions used in studies where the actual robot is not present or necessary for the experiment (Moyle et al., 2013).

In the HRI field, research focusing on non-humanoid robots frequently showcases prominent robot types such as mobile manipulators, virtual or gaming robots, and industrial manipulators. As illustrated in Figure 10.2, the UR5, Sawyer, and UR10 robots have been recognized as the leading choices across various HRI studies.

10.2.2 Physically Present V.S. Virtually Represented Robots

Within the HRI field, the traditional approach of presenting physical/real-world robots to subjects in laboratory and naturalistic settings is increasingly being challenged by the use of virtual representations of robots. These virtual representations include two-dimensional and three-dimensional videos, interactive game-based environments, and virtual reality, and have gained popularity as a result of recent advances in simulation and increased accessibility of game engines and computer graphics (Mara et al., 2021). Although the coronavirus disease 2019 (COVID-19) pandemic accelerated the adoption of virtual representations of robots, researchers have been exploring this alter-native method for various reasons, including cost, complexity, unpredictability, and difficulties in programming physical robots (Esterwood et al., 2023). Virtual representations provide greater methodological flexibility, allowing for more extensive opportunities for manipulating robot characteristics and behaviors, which can broaden the range of research questions that can be explored.

Studies have found minimal differences in humans' overall experience with robots between physical/real-world robots and the same robots presented in a virtual format. However, differences have been observed in humans' perceptions of a robot's utility, immediacy, perceptions and attitudes, and performance (Kamide et al., 2014; Liang and Nejat, 2022; Mara et al., 2021). The suitability of virtual representations of robots as proxies for physical/real-world robots in HRI is under investigation. Moderating factors, such as the type of robot and the context of the study, may play a role in determining the parity between the two formats (Liang and Nejat, 2022).

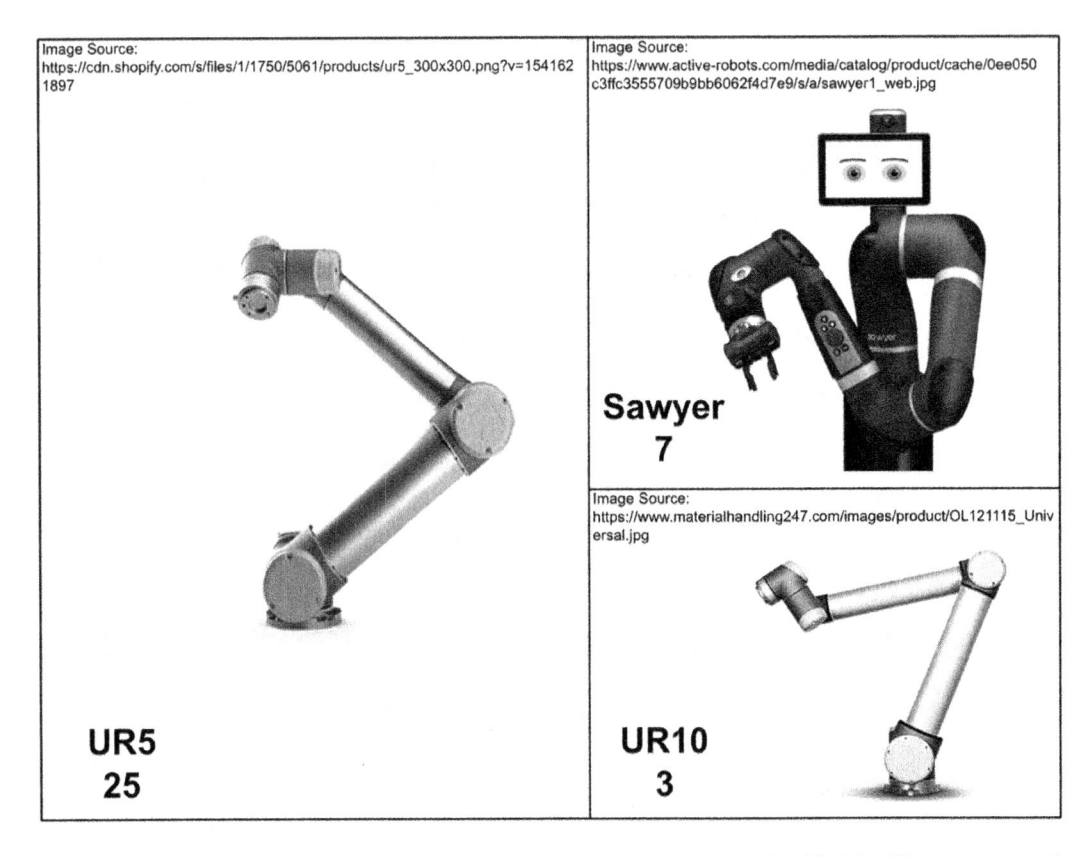

FIGURE 10.2 Figure illustrates the most popular non-humanoid robots identified by Zimmerman et al. (2022) and the number of times they were used across the human–robot interaction (HRI) literature between 2015 and 2021.

While some integrative work has been done in this space, such as Liang and Nejat (2022)'s investigation, the scope of this work has been limited to assistive robots in health care and well-being settings. Further research is needed to provide stronger support for or against the parity of physical/real-world robots and virtual representations of robots in the broader context of HRI research. Unfortunately, the current literature does not provide enough data for lower-level meta-analyses, making it challenging to draw firm conclusions. Therefore, it is crucial for researchers to continue exploring the potential benefits and limitations of both methods to better understand the use of virtual representations of robots in HRI research.

10.3 TRUST IN HRI

Trust is a vital component of any effective HRI. This is because without trust, humans fail to fully leverage robots. For example, if humans do not trust a robot, they are less likely to rely on that robot to perform the sorts of tasks that make robots useful. As a result, the benefits of robots are drastically reduced. This can lead to scenarios where work arrangements become not only unproductive but ultimately damaging to the overall productivity and well-being of workers. This is especially true when one considers recent shifts in the role of robots in working arrangements. Specifically, the roles that robots play in workplaces are shifting from tools and to teammates (You and Robert 2018), and, as a result, the various psychological and social aspects that lead to effective human–human teams are increasingly present in heterogeneous human–robot teams. Indeed, the importance of trust in this regard has not gone unnoticed, and a wealth of literature on this topic has begun to emerge from HRI literature. In response, this section introduces the concept of human–robot trust and provides a summary of the factors that impact when robots are seen as worthy of trust. In addition, we discuss recent computational work for estimating human–robot trust in real time. We close this section with an introduction to the emerging field of human–robot trust management and recovery. In doing so, we provide a starting point for those seeking to learn more on these topics.

10.3.1 Human–Robot Trust and Trustworthiness

Trust is a complex and multifaceted aspect of HRI. As a result, no single universally accepted definition for trust—either in general or in the context of HRI—has been established. Three common definitions of trust, however, have been gaining popularity across HRI literature. These definitions stem from different fields but overlap in several important places while diverging in others.

The oldest of these definitions is that of Mayer et al. (1995). They defined trust as "The **willingness** of a party to be **vulnerable** to the actions of another party based on the **expectation** that the other will perform a particular action important to the trustor, irrespective of the ability to monitor or control that other party" (Mayer et al., 1995, Pg.712), emphasis ours. This definition stems from Human–Human Interaction (HHI) literature and contains three distinct elements. First is the central concept of vulnerability or risk, second is the positioning of trust as pre-behavioral, and third is the role of expectations. The first of these (i.e., vulnerability) is a vital component of trust because vulnerability implies risk (Robert et al., 2009). Indeed, without risk or the potential for "something of importance to be lost," trust is ultimately unnecessary, unneeded, and relatively meaningless (Mayer et al., 1995).

The vulnerability element of trust is the most widely included in subsequent trust definitions, with the remaining two popular definitions of trust in HRI implying or explicitly including vulnerability. Specifically, Lee and See (2004) referenced vulnerability in their definition of trust, which is *"The attitude that an agent will help achieve an individual's goals in a situation characterized by uncertainty and **vulnerability**"*. Hancock et al. (2011), on the other hand, implies vulnerability in their trust definition by defining trust as: *"The reliance by an agent that actions **prejudicial to their well-being** will not be undertaken by influential others"* (emphasis ours in both statements). In both cases, the risk is ultimately incurred by a trustor in the form of the trustee's potential actions. This, therefore, makes the trustor vulnerable to the trustee. The second element of trust based on Mayer et al. (1995)'s definition is the distinction between trusting behaviors and trust itself. In their words, with our emphasis: *"The fundamental difference between trust and trusting behaviors is between a **willingness** to assume risk and actually **assuming** risk."* (Mayer et al., 1995, Pg.724). This distinction allows one to draw a line between trust and trust-related outcomes Robert et al. (2009). This is important because not all risk-taking behaviors are trust- dependent. Other definitions of trust in HRI have been less consistent on this point. For example, although Lee and See (2004)'s definition of trust places trust as an attitude that is conceptually closer to willingness and distinct from behavior, Hancock et al. (2011)'s definition of trust as "reliance" could be interesting as a form of risk- taking or behavior. This is not to say that Hancock et al. (2011)'s definition of trust is incorrect, however, but rather that it may be inconsistent with Mayer et al. (1995) and Lee and See (2004)'s definitions when examined at a deeper level.

Finally, the third element of trust based on Mayer et al. (1995)'s definition relates to expectations. These expectations are synonymous with the concept of trustworthiness. Trustworthiness can be defined as "a multifaceted construct that captures the competence and character of the trustee" (Colquitt et al., 2007, Pg.909). Trustworthiness is distinct from trust and largely precedes it (Mayer et al., 1995). It does so by influencing the expectations that one has of a trustee and therefore their willingness to trust. For example, a trustor is more disposed to trusting a trustee who seems trustworthy but not a trustee who appears untrustworthy. What makes someone or something trust- worthy, however, is more complex, but research in HHI has provided some useful frameworks.

Generally, the most popular framework for trustworthiness divides it into three components. These are ability, integrity, and benevolence. Ability is the skillfulness or competency that trustees are believed to have at their disposal (Mayer et al., 1995). In HRI, this is a human's belief that a robot can do what it has promised. Integrity is the degree to which the trustee is seen as honest and adherent to an acceptable set of principles (Kim et al., 2020, Pg.2). In HRI, this is a human's belief that a robot is honest and acts in a morally consistent manner. Finally, benevolence is "the extent to which a trustee is believed to want to do good to the trustor, aside from an egocentric profit motive" (Mayer et al., 1995, Pg.718). In HRI this is a human's belief that a robot is acting for the human's benefit and is free from conflicts of interest. In general, the lower these expectations are, the less disposed the trustor is to be vulnerable to and rely on (i.e., trust) the trustee (Colquitt et al., 2007). This has been found to be true as much for humans (Colquitt and Salam, 2012; Colquitt et al., 2007; Poon, 2013) as for robots (Esterwood and Robert, 2021). Many factors can influence trustworthiness and by extension trust, and a wealth of research in HRI exists on this topic.

Ultimately, a range of definitions and conceptualizations exist on trust. Each of these approaches differs and ongoing debate among these and other definitions persists. These debates are increasingly common and have begun to emerge not only in the field of HRI but across multiple domains. In this chapter, however, we focus on the three increasingly popular definitions of trust in HRI that we highlighted in italics. With these established, however, another question comes to mind. Specifically, what factors influence trust in robots? Fortunately, a great deal of research has been conducted and a series of meta-analyses hold special insight that may help us answer this question.

10.3.2 Factors Influencing Trust in HRI

Across the HRI literature, an ever-increasing number of studies have sought to determine which factors in HRI can influence human–robot trust. Generally, these studies can be grouped into three distinct categories (Hancock et al., 2011; Sanders et al., 2011). These are human-related factors, robot-related factors, and contextual-related factors (Hancock et al., 2011). A summary of these factors is presented in Figure 10.3 based on the conceptual model proposed by Sanders et al. (2011).

This model has been empirically examined via tow sequential meta-analysis (Hancock et al., 2011), and the analysis provided support for a handful of the factors proposed by Sanders et al. (2011). With regard to human-related factors, these factors appeared to significantly impact trust overall, but, on closer examination only one appeared significant (Hancock et al., 2021). In particular, only factors associated with a human's characteristics as opposed to abilities appeared to significantly impact trust. Furthermore, within this sub-factor, only satisfaction, expectancy, comfort, and personality appeared to be significantly influential (Hancock et al., 2021).

For robot-related factors, these factors can be subdivided into performance-based and attribute-based factors. Overall, both of these sub-factors appear to be significantly influential in combination and when examined individually (Hancock et al., 2021). Within these sub-factors, however, the impact is not equally distributed, nor is it always positive. In particular, only the performance-based factors of dependability and reliability significantly influence trust, with dependability actually having a negative impact (Hancock et al., 2021). Furthermore, for attribute-based factors, only robot personality significantly influences trust (Hancock et al., 2021).

Finally, for contextual factors, these factors can be subdivided into collaboration- based and tasking-based sub-factors. Overall, these factors do not appear to significantly influence trust in robots. When examined individually, however, collaboration- based sub-factors do (Hancock et al., 2021). Across Hancock et al. (2021)'s meta- analysis, however, relatively few studies in this category made firm conclusions on the true impact of such factors.

Taken together, Hancock et al. (2021)'s meta-analysis and review of the antecedents of trust in HRI point to a handful of significant factors that can influence trust. In particular, it appears that a human's comfort, expectancy, personality traits, and satisfaction are influential alongside a robot's personality, dependability, and reliability. Furthermore, contextual factors such as collaboration

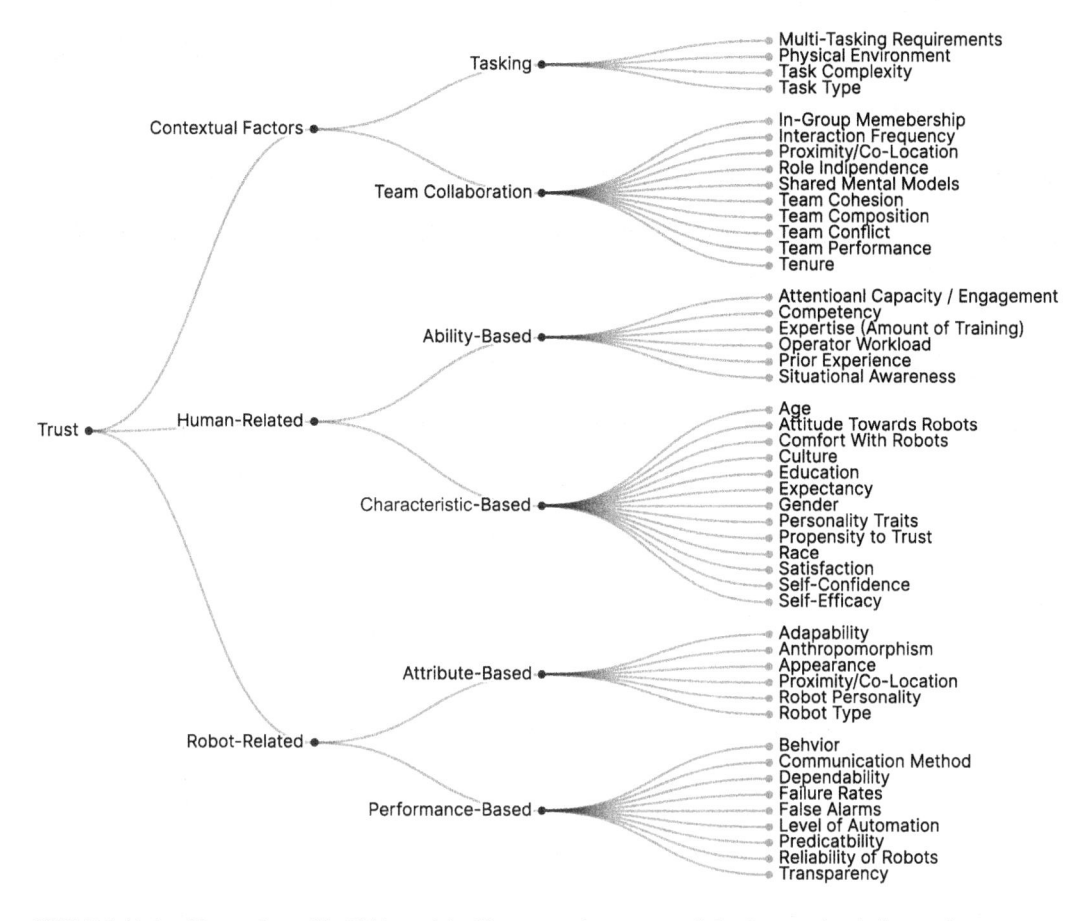

FIGURE 10.3 Hancock et al.'s 2021 model of human–robot trust and the factors that influence it.

may also be important. While these results highlight important trust-relevant factors for human–robot trust, they are by no means the only factors that may be useful. In addition, interactions among factors may exist, making some factors only relevant considering others. As a result, more research is needed, but the factors highlighted in Figure 10.4 may be of use to researchers and designers alike when considering this future work.

10.3.3 Trust Dynamics and Computational Trust Models in HRI

As described in Section 10.2.2, a growing body of research is identifying factors influencing one's trust in automation. The majority of this body of research adopts a snapshot view of trust and evaluates a person's trust at specific points in time, usually at the end of an experiment. This snapshot view, however, does not acknowledge that trust can change as a result of continual interactions with autonomy. As shown in Figure 10.5, at time t, agents A, B, and C have the same level of trust. However, their trust dynamics are different if examined over the time horizon.

Therefore, more recently another line of research has emerged that focuses on understanding the dynamics of trust formation and evolution when a person interacts with autonomy repeatedly (de Visser et al., 2020; Guo and Yang, 2021; Yang et al., 2021). Empirical studies have investigated how trust strengthens or decays as a result of moment-to-moment interactions with autonomy (Lee and Moray, 1992; Manzey et al., 2012; Moray et al., 2000; Yang et al., 2021). Based on these studies, Yang et al. (2023) summarized three major properties of trust dynamics: *continuity*, *negativity bias*, and *stabilization*. Continuity means that trust at the present moment is significantly associated with

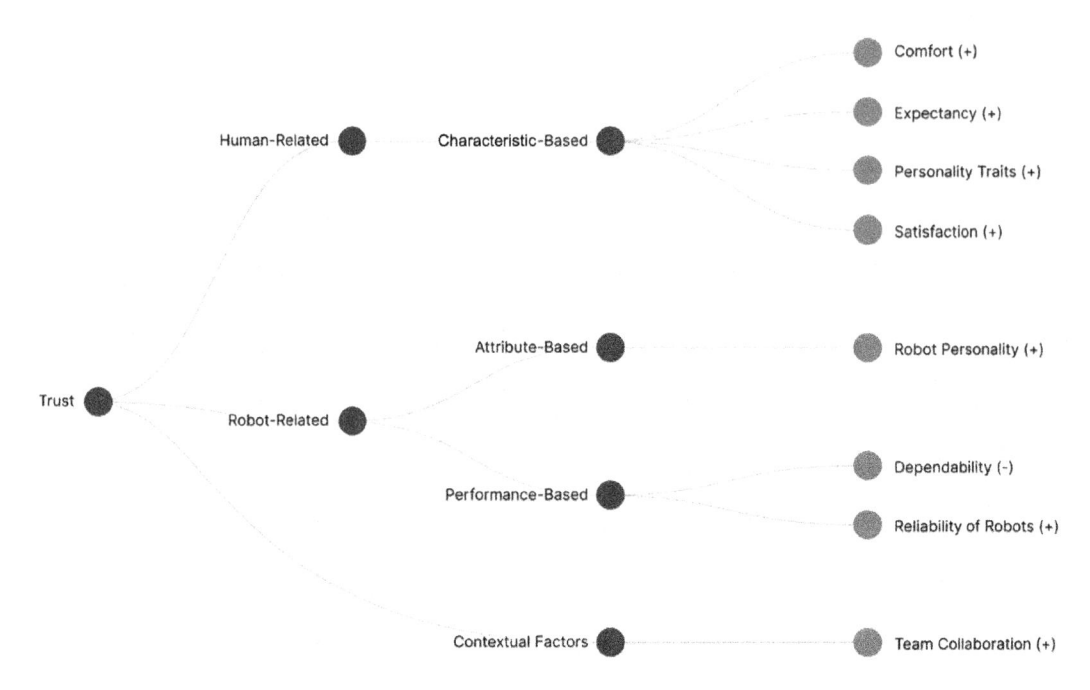

FIGURE 10.4 Hancock et al.'s 2021 model of human–robot trust and the factors that influence it where only significant effects are included. (+) indicates a positive impact on trust, (–) indicates a negative impact on trust.

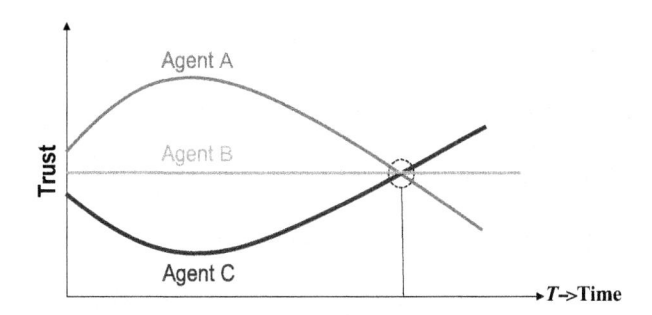

FIGURE 10.5 The static "snapshot" view versus the dynamic view of trust. At time t, Agents A, B, and C have the same level of trust. However, their trust dynamics are different.

trust at the previous moment and increases upon automation successes and decreases upon automation failures. Negativity bias means that negative experiences related to autonomy failures have a greater influence on trust than positive experiences related to autonomy successes. Stabilization means that a person's trust stabilizes over repeated interactions with the same autonomy.

Acknowledging that trust is a dynamic variable, several computational models of trust in automation have been developed. Lee and Moray (1992) proposed an auto-regressive moving average vector (ARMAV) time series model of trust that calculated trust at the present moment t as a function of trust at the previous moment $t-1$, task performance, and the occurrence of automation failures. Yang et al. (2017) examined how trust in automation evolved as an average human agent gained experience interacting with robotic agents. Their results showed that the average human agent's trust in automation stabilized over repeated interactions, and this process can be modeled using a first-order linear time-invariant dynamic system (Yang et al., 2017). Hu et al. (2016) proposed to predict trust as a dichotomy of trust vs. distrust by analyzing the human agent's electroencephalography (EEG) and galvanic skin response (GSR) data. Similarly, Lu and Sarter (2019) proposed the use of eye-tracking metrics, including fixation duration and scan path length, to infer a human's real-time trust. Their follow-up study (Lu, 2020) used three machine-learning techniques, logistic regression, *k*-nearest neighbors (kNN), and random forest to classify the human's real-time trust level.

Instead of using physiological signals, Xu and Dudek (2015) developed the online probabilistic trust inference model (OPTIMo) utilizing Bayesian networks to estimate human trust from automation's performance and human behaviors. Building from the three empirical properties of trust dynamics (i.e., continuity, negativity bias, and stabilization). Guo and Yang (2021) proposed to model trust as a beta random variable and predict trust value in a Bayesian framework. Guo and Yang (2021) compared their model prediction results against the two models developed by Lee and Moray (1992) and Xu and Dudek (2015) and showed that their model significantly outperformed the other two models. Moreover, given that the model complies with the empirically identified properties, it guarantees good explainability and generalizability. Using Gaussian processes, Soh et al. (2020) proposed a multi-task trust transfer model that can predict human trust in a robot's capabilities across multiple tasks. More recently, Guo et al. (2023a, b) proposed the trust inference and propagation (TIP) model, the first mathematical framework for computational trust modeling in multi-human multi-robot teams. The authors asserted that there are two types of experiences that any human agent has with any robot in a multi-human multi-robot team: direct and indirect experiences. The TIP model explicitly accounts for both types of experiences and successfully captures the underlying trust dynamics, significantly outperforming a baseline model (Guo et al., 2023a, b).

The ability of a robot to accurately estimate a human's trust level in real time has led to the research of trust-ware HRI, wherein a robot can adapt its behavior in accordance with a human's trust (Azevedo-Sa et al., 2021a). Similarly, Xu and Dudek (2016) proposed a framework for using an estimated trust for trust-aware conservative control (TACtiC), in which an autonomous agent can momentarily change its behavior to be more conservative if the human's trust falls too low. In Akash et al. (2019), a trust-workload partially observable Markov decision process (POMDP) model was trained and solved to generate optimal policies for a robot to control its transparency to improve the performance of the human–robot team. Further, Chen et al. (2020) presented a trust-POMDP model that can be solved to generate optimal policies for the robot to calibrate the human's trust and improve team performance. Guo et al. (2021) presented a reverse psychology model of human trust behavior and compared it with the disuse model.

The abovementioned work focuses on the human's trust in the autonomous/robotic agent. Recently, the first bi-directional trust model that encompasses both the human's trust in the robotics agent and the robotic agent's trust in the human was developed by Azevedo-Sa et al. (2021b). In the work, Azevedo-Sa et al. (2021b) introduced a novel capabilities-based bi-directional multi-task trust model that can be used for trust prediction from either a human or a robotic trustor agent. This model is useful for control authority allocation applications that involve both the human's trust in the robot and the robot's trust in the human.

10.3.4 Trust Management and Recovery in HRI

As mentioned, many factors influence trust in robots. Trust, however, is not fixed and changes over time (de Visser et al., 2020; Guo and Yang, 2021; Yang et al., 2023). Trust increases when robots meet expectations and decreases when they do not. Although increases in trust are relatively easy to manage, decreases can have lasting effects and are hard to recover from (Lewicki and Brinsfield, 2017). Generally, such decreases are the result of violations of trust. Trust violations can take multiple forms but generally result from a trustee failing to meet the expectations of a trustor (Esterwood and Robert, 2022b). This produces a reduction in the trustor's willingness to be vulnerable and therefore trust (Costa et al., 2018; Esterwood and Robert, 2022b; Esterwood and Robert, 2023a,b; Gillespie et al., 2021; Kim et al., 2004).

Trust violations in HRI can take on three distinct forms: ability-based, integrity- based, and benevolence-based (Grover et al., 2014). An ability-based trust violation occurs when a robot fails to meet a human's performance expectations or perform a task as assigned (Sebo et al., 2019). An integrity-based trust violation occurs when the robot breaches a human's expectation of honesty and ethical conduct (Sebo et al., 2019). Lastly, a benevolence-based trust violation arises when the robot is perceived as lacking care and fails to fulfill a human's expectation of its purpose (Esterwood and Robert, 2022b). Each of these types of violations can seriously undermine trust and result in a variety of negative outcomes.

Fortunately, trustees can use several strategies to mitigate trust decreases and restore trust. The most common of these are short-term verbal trust repair strategies, including apologies, promises, explanations, and denials (Esterwood and Robert, 2022b; Lewicki and Brinsfield, 2017). Apologies, expressions of regret or remorse for a perceived transgression, are the most widely used trust repair strategy across the literature (Esterwood and Robert, 2022b). They are believed to repair trust through encouraging forgiveness (Esterwood and Robert, 2023c). Promises, on the other hand, are statements of commitment to positive future performance (Schweitzer et al., 2006). They aim to restore trust by shifting the focus from past to future behaviors and are believed to work through encouraging forgetfulness (Esterwood and Robert, 2023).

Explanations, which are discussed in more detail in Section 10.4 of this chapter, are statements that provide clear reasoning behind a trust violation. They seek to establish a shared understanding between a trustor and trustee by conveying transparency (Esterwood and Robert, 2023; Ezenyilimba et al., 2022; Lewicki and Brinsfield, 2017; Lewicki et al., 2016; Rawlins, 2008). Explanations are hypothesized to repair trust through informing the trustor (Esterwood and Robert, 2023). Denials, on the other hand, are trust repair strategies that redirect blame or reject culpability for a trust violation (Baker et al., 2018). They aim to establish the complete innocence of the trustee by shifting blame away from them and onto another entity (Lewicki and Brinsfield, 2017). Denials are hypothesized to work through misinforming because they rely largely on inaccurate information provided by the trustee to the trustor (Esterwood and Robert, 2023).

Generally, the efficacy of these repair strategies in restoring human–robot trust is mixed, with some of these strategies appearing to be effective at certain times but not others (Esterwood and Robert, 2022b). This may largely relate to other factors acting as moderators. For example, timing (Kox et al., 2021; Robinette et al., 2015), violation type (Sebo et al., 2019; Zhang, 2021), anthropomorphism (Esterwood and Robert, 2021), attitude (Esterwood and Robert, 2022a) and severity (Correia et al., 2018) have each been shown to impact the efficacy of different trust repairs. The field of trust repair in HRI, however, is still young, and much remains unknown. In particular, additional moderators or previously unexamined main effects may more clearly explain these diverging results. Alternatively, human-related factors and individual differences might largely determine the efficacy of these repairs. Indeed, much work is ongoing, and the field of trust repair continues to expand.

10.4 PERSONALITY IN HRI

Human and robot personalities play a significant role in shaping how humans interact with and use robots (Esterwood et al., 2021b; Hancock et al., 2021; Robert Jr et al., 2020). These personalities have been found to impact not just trust, but also a wide range of other outcomes as well (Alarcon et al., 2021; Esterwood et al., 2021a; Robert Jr et al., 2020). Personality is therefore an important factor to consider in the context of HRI. In this section, we delve into the concept of personality in HRI and its impact on HRI. We start by defining personality and introducing some of the various ways in which it has been conceptualized both within the HRI literature and more generally across the personality psychology literature. We then summarize what has been found across the literature and how human personality, robot personality, and the match or mismatch between the two have impacted HRI.

10.4.1 Personality and Personality Traits

To begin this section, we must first define what personality is and briefly introduce the different ways that personality has been approached in the fields of personality psychology and HRI alike. Personality can be defined as an individual's "characteristic pattern of behavior in the broad sense (including thoughts, feelings, and motivation)" (Baumert et al., 2017, Pg.527). Generally, across the field of personality psychology, five distinct approaches to personality have garnered the most attention, and debate among these schools of thought is abundant (see McMartin (2016) for a full review). The most common approach to personality across HRI literature, however, is the trait-based approach to personality (Esterwood et al., 2021b).

The trait-based approach to personality posits that traits act as the foundational elements by which personalities are constructed (McMartin, 2016). Traits are "organized dispositions within the individual" (McMartin, 2016, Pg.30). These dispositions are seen as relatively stable and, as a result, they can be used to predict and explain various aspects of human behavior (Allport, 1937; McCrae and Costa Jr, 2008). The exact makeup and number of these traits in humans is a topic of lively debate within the personality psychology literature, and multiple sets of traits have garnered empirical support (McMartin, 2016). One commonality across each of these approaches, however, is that they see personality as multidimensional and as capable of being subdivided into specific operational variables. This has historically allowed researchers to precisely link specific traits to particular outcomes, which may partially explain the popularity of this approach (Esterwood et al., 2021b; Haslam, 2007; McMartin, 2016; Tasa et al., 2011).

The most common and widely supported approach to personality traits across many fields (Li et al., 2014; Robert, 2018; Robert Jr et al., 2020), including HRI (Esterwood and Robert, 2020; Lee and Nass, 2003; Pocius, 1991; Robert, 2018; Robert Jr et al., 2020; Völkel et al., 2019) is that of the Big Five. The Big Five personality traits are extraversion, neuroticism, openness, agreeableness, and conscientiousness (Goldberg, 1992; John et al., 2008; McCrae and Costa Jr, 2008). Extraversion is often conceived of as a spectrum with two poles, one being extraversion and the other introversion. Extraversion is the extent to which someone is outgoing, assertive, vocal, and sociable (Rhee et al., 2013), whereas introversion is the level to which an individual is timid, prefers quietness, and enjoys solitude (Driskell et al., 2006). Neuroticism is the degree to which someone is easily angered, not well-adjusted, insecure, or lacks self-confidence (Driskell et al., 2006). Openness to experience is typically described as the amount to which one is imaginative, curious, and open-minded (McCrae and Costa Jr, 1997). Agreeableness can be characterized by how cooperative and friendly someone is (Peeters et al., 2006). Finally, conscientiousness is the degree to which individuals are thorough, deliberate, and mindful of their actions (Tasa et al., 2011).

Each of these personality traits in sum comprises an individual's personality. This personality can have a direct impact on how humans think, behave, see others, and feel emotions (Hassabis et al., 2014; Peeters et al., 2006). This makes personality an informative factor to consider when examining differences between individuals and various patterns of behavior and cognition (Connelly et al., 2018). It is no surprise then that this topic has gained some attention in the field of HRI. The various ways that personality has been examined in HRI can be broadly categorized into three perspectives: (1) how humans' personalities impact HRI, (2) how humans' perceptions of robots' personalities impact HRI, and (3) how the degree of similarity or difference between humans' personalities and their perceptions of robots' personalities impacts HRI.

10.4.2 How Does a Human's Personality Impact HRI?

Humans' personalities have been shown to have direct and indirect impacts on a range of outcomes in HRI. In particular, two recent qualitative reviews of the literature on personality identified 20 sets of outcomes across the HRI literature that were examined in reference to personality (Robert, 2018; Robert Jr et al., 2020). These outcomes can be more concisely grouped into seven overarching categories. These are proximetrics (e.g., distance, touch), attitudes, anthropomorphism, performance, trust, emotional response, and acceptance. Across each of these categories, with the exception of anthropomorphism, results have been somewhat mixed (Robert, 2018; Robert Jr et al., 2020). In particular, studies examining how personality impacts proximetrics, attitudes, trust, emotional response, and acceptance each found significant results at some points and non-significant results at others.

Fortunately, subsequent work in this domain has accounted for one set of these mixed results. In particular, a meta-analysis based on a review of the literature examined the impact of personality on acceptance. This study (Esterwood et al., 2021b) showed that personality appeared to significantly impact acceptance. This impact, however, was not without moderators, and the authors identified other factors as influential. Additional analysis on other outcomes with mixed results, however, has not been conducted given the relatively small number of studies examining these relationships. Future work is therefore needed, but for the time being human personality does appear to directly impact humans' acceptance of robots and anthropomorphism.

10.4.3 How Does a Robot's Personality Impact HRI?

While the impact of humans' personalities in HRI has received the majority of attention in the HRI literature, there is growing research focused on a robot's perceived personality as well. A robot's personality can be defined in a similar manner to that of humans – i.e., a set of distinctive patterns of behavior – however, it is important to note that a robot's personality does not arise from the robot itself. Instead, robot's personalities emerge through the observations, interactions, and expectations of humans of a given robot (Esterwood et al., 2022; Tay et al., 2014). In this way, a robot's personality is predominantly shaped by human's own perceptions which, in turn, are influenced by multitudinous other factors. Although a comprehensive understanding of these factors is currently lacking in the literature, certain common aspects of a robot's design, such as voice cues, gestures, facial expressions, posture, and body movement, appear to play a role (Lee et al., 2006; Mileounis et al., 2015). It should be noted that these factors have not been thoroughly assessed regarding personality traits beyond extroversion. Nonetheless, studies have shown that faster speech, higher pitch, increased volume of speech, and more dynamic and rapid movements are often associated with perceptions of a robot as extroverted, while the opposite characteristics tend to make the robot appear introverted (Lee et al., 2006; Mileounis et al., 2015).

Regardless of these factors, the impact of robot's personalities – however they are manifested – appear significant. Indeed, recent reviews of the literature have uncovered 31 outcomes associated with robot personality in HRI. These can be grouped into three categories: attitudes (i.e., perceptions of robots), acceptance (i.e., intention to use robots), and interaction quality (i.e., enjoyment, fun, perceived control) (Robert, 2018; Robert Jr et al., 2020). Findings from this literature have generally shown mixed results, but most studies indicate that for each of these groups of outcomes robot personality is likely an important factor to consider. This is especially the case for acceptance because this outcome has benefited from meta-analysis. In particular, humans' willingness to accept robots was found to be significantly impacted by the type of personality humans see the robot as possessing (Esterwood et al., 2022). Taken together, this literature highlights that robot personality is an important consideration when designing robots and that this aspect of HRI can influence a range of outcomes.

10.4.4 What Impact Does Similarity Between Humans' and Robots' Personalities Have in HRI?

In the previous two sections, we introduced the effects that both human personality and robot personality have on HRI. Human and robot personalities, however, also interact. Namely, the similarities or differences between human personality and robot personality can have implications for a range of outcomes. To date, the outcomes examined in the literature have included humans' perceptions of the quality of a robot, their perceptions of the quality of their interactions with the robot, their perceptions of the robot's personality, and their willingness to accept robots (Robert, 2018; Robert Jr et al., 2020).

Generally, each of these outcomes has produced both significant and non-significant results, making firm conclusions difficult. Acceptance, however, has been examined via meta-analysis, which has allowed for closer examination of the impacts of similarity in personality on acceptance. In particular, results indicated that personality similarity between humans and robots appears to have a significant and positive relationship with acceptance (Esterwood et al., 2021a). These findings further highlight how both humans' and robots' personalities are important considerations when designing effective HRIs.

10.5 EXPLANATION IN HRI

The field of explainable artificial intelligence (XAI) has gained renewed interest as researchers seek to improve transparency, interpretability, and understandability of AI systems in response to ethical concerns and a lack of trust (Angwin et al., 2022; Miller, 2019; Stubbs et al., 2007). Research suggests that transparent decision-making processes and algorithms are crucial to building trust in AI systems (Hayes and Shah, 2017; Luo et al., 2022; Zhang et al., 2021).

Explanations, which refer to the reasoning or logic behind actions, can provide vital information that justifies the decisions of intelligence agents, ultimately leading to more trusting and efficient interactions (Hayes and Shah, 2017; Miller, 2019; Zhang et al., 2021). Explanations are necessary for various AI applications, including HRI. Sharing expectations about behavior and intentions between humans and robots is crucial for building trust and understanding the robot's behavior, which can improve communication and collaboration effectiveness (Setchi et al., 2020).

10.5.1 EXPLAINABLE AI IN HRI

Strategies that focus on conveying intention and resolving ambiguity in verbal interactions contribute to the effectiveness of HRI and collaboration. Explainable AI can help individuals without in-depth knowledge of AI understand, predict, and ultimately trust AI systems, as well as identify and address any potential issues or errors in the robot's decision-making processes (Miller, 2019). To achieve explainable AI in HRI, some techniques include natural language generation to explain the robot's actions (Bisk et al., 2016; McDonald, 2010; Tellex et al., 2020), visual explanations (Edmonds et al., 2019; Mishra et al., 2022), and causal reasoning to demonstrate the logic behind the robot's decisions (Alaieri and Vellino, 2016; Erdem et al., 2011; Mota et al., 2021).

Providing explanations about a robotic system's intention, state, capability, and upcoming actions can significantly aid users in developing an accurate mental model of the system (Kulesza et al., 2013). This model helps users continuously understand the explanation, anticipate future actions, and take necessary precautions when unforeseen circumstances arise (Naujoks et al., 2017a). Using a theory of mind to guide robot behaviors and information sharing, as demonstrated by Devin and Alami (2016), led to increased collaboration efficiency when knowledge was communicated appropriately. Additionally, robots capable of identifying and communicating relevant details about their behavior and reasoning make better teammates than those lacking these capabilities because users can then better understand the system and the reasoning behind its actions, as noted by Körber et al. (2018).

Explanations can help clarify the responsibilities of users and robot systems, particularly as robots become more autonomous and individuals tend to attribute blame to the robot instead of themselves or their coworkers (Kim and Hinds, 2006). Through cooperative perception, explanations can demonstrate that both parties are partners by explaining how the system operates and clarifying what users are expected to do. An understanding of whether it is the users or the AI system that determines the system's behavior enables more effective interaction with the system

(Naujoks et al., 2017a; Stanton and Young, 1998). This understanding is particularly important when the robot teammate performs an unexpected action because an explanation can help to increase the perceived responsibility of both parties.

Explanations play a significant role in shaping human attitudes toward and acceptance of robots. For instance, Ambsdorf et al. (2022) designed a HRI scenario to investigate the impact of robots using XAI to explain their actions and found that robots providing reasoning about their actions were perceived as more human-like and lively than those simply announcing their actions. Rational explanations can also improve trust and performance among users who are unfamiliar with a task (Schaffer et al., 2019). Example-based explanations have been shown to provide a better understanding of the system and have a positive impact on trust (Cai et al., 2019). This assertion was supported by Chiou et al. (2022), who demonstrated that increasing awareness of the purpose, process, and performance of robot teammates can help humans retain situational awareness.

10.5.2 Explanation Timing, Content, and Modality and HRI

The role of explanations in HRIs is pivotal, contributing significantly to the acceptance of robots. Robots are often engineered to mimic human intelligence and physicality. Their capacity to provide explanations assists humans in developing precise mental models. These explanations supply crucial information justifying the robot's behaviors, thereby enabling humans to comprehend and anticipate the robot's actions. An increasing number of studies are investigating the influence of robot explanations on behavioral and attitudinal outcomes (Du et al., 2019; Forster et al., 2017; Han et al., 2021; Lettl and Schulte, 2013; Lyons et al., 2023; Koo et al., 2016; Körber et al., 2018; Naujoks et al., 2017b; Ruijten et al., 2018; Shen et al., 2020, Stange and Kopp, 2021; Zhu and Williams, 2020). Notably, three substantial areas of robot explanations – timing, content, and modality – have been recognized and examined in the research field.

10.5.2.1 Explanation Timing

The timing of explanations, namely the point at which a robot offers explanation, plays a vital role in enhancing the efficacy of such clarifications. One can categorize explanation timing into three groups: pre-action (explanation offered before action), in situ (explanation concurrently with action), and post-action (explanation after action).

Various studies have delved into the impact of pre-action robot explanations. For example, Stange and Kopp (2021) scrutinized the repercussions of a social robot, Pepper, offering proactive self-explanations before versus after engaging in an undesirable behavior. Results suggested that while participants experienced less uncertainty regarding future events, they also felt less in control, and displayed diminished trust and lower intentions of interacting with a robot that proactively explained its actions. Similarly, Zhu and Williams (2020) found no compelling evidence linking proactive explanations to positive outcomes in human–robot team tasks. In fact, they noted potential drawbacks to such explanations, as participants perceived them as verbose and unnecessary. Contrarily, research in the realm of automated vehicles (AV), a subset of mobile robots, indicated that pre-action explanations tend to yield positive outcomes. Koo et al. (2016) discovered that explanations issued 1 second before an AV's action alleviated driver anxiety and heightened their sense of control, preference, and alertness. Consistent with this, Du et al. (2019) observed that explanations given 7 seconds before an AV's action garnered more trust and preference than those offered post-action or not at all, reducing anxiety and workload. Moreover, verbal messages describing an AV's impending action, relayed 7 seconds prior, generated higher levels of trust, anthropomorphism, and usability (Forster et al., 2017). Ruijten et al. (2018) found that an AV supplying pre-action explanations appeared more trustworthy, intelligent, human-like, and likable than those devoid of such clarifications.

The impact of in situ explanations during tasks in human–robot collaborations has also been explored. Han et al. (2021) employed a Rethink Robotics Baxter humanoid robot for a handover task in HRI, assessing participants' perceptions of the necessity and timing of robot explanations. Their findings demonstrated that participants universally acknowledged the need for robots to provide explanations and emphasized the importance of in situ timing.

Post-action explanations have likewise been subjected to investigation. Du et al. (2019) noted that explanations furnished 1 second after the AV's action resulted in the lowest levels of trust and preference, compared to pre-action explanations or none at all. Körber et al. (2018) conducted a mixed-design study which offered explanations 14 seconds post-action, revealing that while drivers believed they understood the system and the rationale for the AV's action, their trust in and acceptance of AVs did not significantly improve compared to when no explanation was provided. Shen et al. (2020), using AV driving videos, examined which driving scenarios necessitated explanations, and how the requirement for explanation differed according to the situation and driver type. The research identified a correlation between the need for an explanation, the driver type, and the driving scenario, with more aggressive drivers requiring fewer explanations. However, near-crash situations unequivocally demanded clear explanations (Shen et al., 2020).

10.5.2.2 Explanation Content

Explanation content pertains to the information about robotic actions provided to humans, and its influence on human reactions has been a key subject of past research. The content of explanations has been classified into three types: (1) 'what' – the actions executed by the robot (Stange and Kopp, 2021; Miller, 2019; Koo et al., 2015; Wiegand et al., 2019), (2) 'why' - the rationale behind the actions (Lyons et al., 2023; Han et al., 2021; Koo et al., 2016), and (3) 'what+why' – encompassing both actions and the reasons behind them (Koo et al., 2015; Du et al., 2019). Different types of explanation content have demonstrated varied effects on human attitudes and behavior.

'What-only' explanations convey solely the actions taken by the robot (i.e., what it will do or did do). Revealing a robot's intended actions can increase perceived understandability if the robot's activities appear ambiguous to users (Stange and Kopp, 2021). It might also positively influence user perceptions by suggesting that the robot is cognizant of potential misunderstandings and is proactively addressing them (Miller, 2019). However, this type of explanation also has its limitations. A study by Koo et al. (2015), using a fixed-base driving simulator with a realistic AV model, found that 'what-only' explanations led to lower acceptance and poorer driving performance compared to other explanation types ('what+why', 'why-only', and no explanation). 'What-only' explanations were deemed the least acceptable and most hazardous in terms of driving performance.

'Why-only' explanations deliver the logic behind robotic actions. Research by Han et al. (2021) discovered the significance of 'why' explanations (e.g., reasoning behind certain behaviors, failures, disobedience, etc.) in HRIs. Without an explanation for unclear behavior or task incompletion, participants inferred that there were issues with the robot needing resolution. Similarly, Lyons et al. (2023) used an autonomous search robot in an Urban Search and Rescue (USAR) scenario to investigate the role of explanations when the robot deviated from expected behavior. Explanations that emphasized the robot's awareness of the environment and why certain events occurred were found to effectively mitigate decreases in trust and trustworthiness. Moreover, 'why-only' explanations resulted in reduced anxiety and improved trust, preference, and driving performance, compared to other types of explanations in AV domain (Koo et al., 2015; Koo et al., 2016). These explanations help drivers anticipate events, maintain control, and enhance situational awareness. Wiegand et al. (2019) found that presenting 'why-only' explanations, including details about detected object movements and contextual information, significantly improved people's understanding of the situation and their situational awareness.

The 'what+why' explanation encompasses both the action performed by the robot and the logic behind it. Support for the efficacy of 'what+why' explanations has been found in various studies. For instance, Forster et al. (2017) discovered that this type of explanation bolsters trust, anthropomorphism, and usability in mobile robots (i.e., AVs). Additionally, Naujoks et al. (2017a) found that 'what+why' explanations help reduce visual workload, thereby making the automation more user-friendly as drivers do not need to constantly monitor the system's interface to decipher its intentions and actions. However, it is crucial to note that the influence of 'what+why' explanations on trust in AVs can vary based on factors like the driving event, vehicle action, driving environment, and the perspective of the explanation. In an experiment conducted by Ha et al. (2020), a driving simulator equipped with virtual reality technology was used to evaluate the effect of perceived risk and explanations on trust in AVs. The experiment featured four automated driving environments with varying weather conditions (clear day, snowy night) and speeds (fast, slow), and three explanation conditions: no explanation, 'what+why' explanation without a subject, and 'what+why' explanation from a third-person perspective. The findings demonstrated that both the perceived risk of the driving environment and the type of explanation played a vital role in influencing the impact of 'what+why' explanations on trust in AVs. Interestingly, in low-risk perceptions, third-person explanations were most effective in building trust. However, as the perceived risk escalated, the efficacy of third-person explanations diminished, and providing no explanation proved to be the most effective approach.

10.5.2.3 Explanation Modality

The communication approach adopted by robots to relay information to passengers and operators is referred to as their 'modality', which significantly affects both user experience and human perceptions. A modality can be understood as an independent channel through which automation and humans can exchange sensory data (Karray et al., 2008). Research on robot explanations has primarily employed two modalities: auditory and visual. Auditory explanations are typically delivered through a robot or robotic platform using a neutral tone (Lettl and Schulte, 2013; Du et al., 2019; Körber et al., 2018; Koo et al., 2015, 2016; Ruijten et al., 2018; Naujoks et al., 2017b). On the other hand, visual explanations often take the form of text-based natural language processing and annotations (Harbers et al., 2011; Wang et al., 2016), motion or light cues (Baraka et al., 2016; Anjomshoae et al., 2019), or graphical representations and images (Chen et al., 2018; Lim and Dey, 2011) integrated into the user interface to provide explanatory information. In the realm of HRI, expressive motions and lights have been identified as the most effective means of communicating the robot's internal state (Baraka et al., 2016; Anjomshoae et al., 2019).In the automated vehicle domain, while the influence of modality on explanations provided by level 4 and higher AVs is still under investigation, prior research has examined the effectiveness of alert modality in levels 1–3 driving automation, with a particular focus on vehicle display design. Studies generally indicate that auditory modality is a superior choice to visual modality, due to its less distracting nature and its enhanced ability to direct attention compared to visual warnings (Bernsen and Dybkjær, 2001; Cao et al., 2010; Wickens, 2008). This makes it a preferable choice for issuing warnings and promptly conveying potential danger levels (Wheatley and Hurwitz, 2001; Bernsen and Dybkjær, 2001; Cao et al., 2010; Wickens, 2008). However, auditory information has been found to trigger higher levels of perceived annoyance and surprise in drivers, leading to increased stress, delayed reactions, and incorrect responses (Nees et al., 2016; Dingus et al., 1997). In contrast, visual modalities such as icons displayed on a head-up display, boost perceptions of ease-of-use, transparency, and satisfaction (Du et al., 2021; Avetisyan et al., 2022). Visual warnings like texts and icons also support continuous awareness of the surrounding environment and require shorter recognition times for urgency compared to auditory warnings (Politis et al., 2015).

10.6 EVALUATION METRICS IN HRI

Research in the field of HRI has examined a wide breadth of outcomes. As a result, discussing all possible outcomes and the measures for each is outside of the scope of this section. Some outcomes in the HRI literature, however, are especially common. In particular, a recent examination of metrics and methods used in HRI (Zimmerman et al., 2022) as well as a range of systematic reviews and meta-analyses (Esterwood et al., 2021b; Hancock et al., 2011; Naneva et al., 2020;

Roesler et al., 2021) have uncovered a handful of common metrics and measures in HRI. These outcomes can be categorized into two different groups, namely, human-directed measurements and robot-directed measurements. For human-directed measurements, the most popular outcomes across this literature include trust (Hancock et al., 2011), acceptance (Naneva et al., 2020), workload (Prewett et al., 2009), human personality (Esterwood et al., 2022; Robert, 2018; Robert Jr et al., 2020), and attitude (Naneva et al., 2020). For robot-based measurements, the most common measures include anthropomorphism (Roesler et al., 2021), and robot personality (Esterwood et al., 2022; Robert, 2018; Robert Jr et al., 2020). While a complete account of each of the measures used for each of these outcomes is largely absent, some of these outcomes have received special attention. This allows us to examine their associated measures at a high level.

10.6.1 Common Human-Related Measurements

10.6.1.1 Trust

Of the outcomes common to HRI, trust measurements have perhaps received the most attention, with lively debate over the most suitable scales to use, when, and why being common throughout the literature (Chita-Tegmark et al., 2021; Kessler et al., 2017). Exacerbating this debate and the challenges with measuring trust in HRI in general is the frequency of custom instruments (Zimmerman et al., 2022). Although such scales are not inherently invalid and offer many benefits in the form of flexibility, they do limit the reproducibility of results because many of these measures are not fully reported (Zimmerman et al., 2022). Independent of these custom instruments, however, several common measures—referred to as "named surveys"—do exist (Zimmerman et al., 2022).

These named surveys include the Human–Robot Trust questionnaire (Schaefer, 2013), the Madsen and Gregor Measure of Human-Computer Trust (Madsen and Gregor, 2000), the Multi-Dimensional Measure of Trust (MDMT) (Ullman and Malle, 2018), and the Trust in Automation Scale (Jian et al., 2000). In addition, trust has been measured via the Muir Trust Scale (Muir and Moray, 1996), the Mayer Trust Scale (Mayer et al., 1995), and the Lee and See Trust Scale (Lee and See, 2004). Each of these scales purports to measure trust, but each varies in its respective approaches, with some emphasizing different aspects of trust than others. This divergence in measures likely stems from general disagreement in definitions of trust and the degree to which a robot can be considered like a human or instead akin to automation.

Recently, there has been increasing research attention on developing computational trust models, with several notable works (Xu and Dudek, 2015, Guo and Yang, 2021, Soh et al., 2020). Please refer to Section 10.2.3 for details.

10.6.1.2 Attitude

In addition to trust, a common outcome in the HRI literature is that of attitude. Generally, attitude can be considered as an overall construct or divided into various sub-components (Breckler, 1984; Naneva et al., 2020). Measures of attitude as an overall construct have largely relied on the Negative Attitudes toward Robots Scale (NARS) (Nomura et al., 2004), with additional measures such as the Robot Anxiety Scale (Nomura et al., 2008), Attitude toward Technology Scale (Chuttur, 2009), the Ezer Analogies Measure (Ezer, 2008), and the Attitudes toward Working with Robots scale (AWRO) (Robert, 2021) emerging as popular alternatives (Zimmerman et al., 2022).

When considering attitude at a sub-component level, the majority of studies in HRI focus on the affective sub-component of attitudes (Naneva et al., 2020). This has been measured most frequently via the two sub-scales (NARS-S1 and NARS-S2) of NARS (Naneva et al., 2020; Nomura et al., 2004; Zimmerman et al., 2022). Less common alternatives, however, include sub-scales from other measures such as the likeability component of the Godspeed Questionnaire Series (Bartneck et al., 2009) and – as with trust – myriad custom measures (Naneva et al., 2020).

Finally, the cognitive sub-dimension of attitude has largely been measured via one specific sub-scale of NARS (NARS-S2) (Naneva et al., 2020). Various other sub-scales included in measures of acceptance, however, have also been used (Naneva et al., 2020). Specifically, sub-components of the Almere model of robot acceptance (Heerink et al., 2010) and the Unified Theory of Acceptance and Use of Technology (UTAUT) (Venkatesh et al., 2003) have each been deployed to measure cognitive attitudes in an HRI context (Conti et al., 2017; Shin and Choo, 2011; Tay et al., 2014).

10.6.1.3 Human Personality

Personality has been increasingly examined in HRI (Robert, 2018; Robert Jr et al., 2020). As mentioned in Section 10.3, the most common measure of personality has historically been the Big Five index of personality traits. Unsurprisingly, this index has also been widely used in the context of HRI (Esterwood et al., 2021b; Robert, 2018; Robert Jr et al., 2020; Santamaria and Nathan-Roberts, 2017; Zimmerman et al., 2022). The Big Five index has been commonly used to measure extraversion, neuroticism, openness, agreeableness, and conscientiousness, but most commonly studies have used the index to measure extraversion exclusively (Esterwood and Robert, 2020; Santamaria and Nathan-Roberts, 2017).

While the Big Five personality index remains dominant, it is far from the only approach taken to measure personality. Indeed, measures such as the Myers-Briggs test, the NEO Personality Inventory (Costa and McCrae, 1992), and the Eysenck Personality Inventory (EPI) (Eysenck and Eysenck, 1975) have received attention (Esterwood and Robert, 2020; Santamaria and Nathan-Roberts, 2017). Each of these tests, however, relies on certain assumptions about what a personality is and how it can be used to predict behavior. In particular, these measures of personality exclusively take the trait-based approach to personality. This is but one of many approaches, and criticisms and critiques of the exclusive use of this approach have been presented across the personality psychology literature (McMartin, 2016). Such alternatives, however, have not fully manifested in the HRI domain, and personality remains measured mostly through the Big Five index and, by extension, via trait-based approaches to personality.

10.6.1.4 Workload

Finally, workload is another common outcome measured in the HRI literature (Prewett et al., 2010; Zimmerman et al., 2022). This outcome differs from the others because it is measured in a more standardized fashion. Specifically, workload has consistently been measured across studies with the NASA TLX questionnaire (Hart and Staveland, 1988). In addition, this measure has remained fairly unmodified across studies (Zimmerman et al., 2022). This is unusual because other measures have almost exclusively received some form of modification (Zimmerman et al., 2022). These modifications are mostly minimal alterations to wording to suit the study's design and subjects. As such, the NASA TLX is as close to a standard measure of an outcome as the HRI literature has seen. That is not to say, however, that the NASA TLX is the only measure of workload available to HRI researchers. Indeed, prior work has sought to create HRI-specific measures of workload stemming from the NASA TLX (Yagoda, 2010). This work, however, positions its measure as an accompaniment to the NASA TLX rather than a replacement.

10.6.2 COMMON ROBOT-RELATED MEASURES

10.6.2.1 Anthropomorphism

Another common outcome in the HRI literature is that of anthropomorphism. This outcome has been most commonly assessed through a measure named the Godspeed (Bartneck et al., 2009; Mara et al., 2022; Roesler et al., 2021; Zimmerman et al., 2022). Godspeed has long been the standard not only for measuring anthropomorphism itself but also, to a lesser extent, for assessing humans' attitudes toward robots (Roesler et al., 2021). The Godspeed, however, is not without limitations, and recent critiques have highlighted several major shortcomings of this measure (Carpinella et al., 2017; Ho and MacDorman, 2010; Kühne and Peter, 2022; Roesler et al., 2021). In particular, the Godspeed appears to suffer from confounded effects, poor factor loading, high correlation between dimensions, and issues with semantic differentiation in response formats (Ho and MacDorman, 2010).

In response to this criticism, recent work has re-conceptualized anthropomorphism in HRI to more clearly divide this concept from animacy and social presence (Kühne and Peter, 2022). Specifically, such re-conceptualizations have adopted a more multidimensional approach based on the theory of mind. In particular, new dimensions separate from those in the Godspeed have been proposed (Kühne and Peter, 2022). Such approaches, however, are in their infancy. As a result, validation and the establishment of formalized scales—much less their widespread adoption—have not fully emerged.

10.6.3 ROBOT PERSONALITY

The measurement of robot personality parallels that of human personality as measures of robot personality often rely on the trait-based approach to personality as exemplified via the Big Five personality index. Also similar to human personality measures, however, this research has also predominantly concentrated on extroversion, neglecting the exploration of other traits within the Big Five (Esterwood et al., 2022; Santamaria and Nathan-Roberts, 2017). Consequently, a comprehensive understanding of robot personality as a multifaceted construct remains limited, creating an exciting prospect for further investigation and advancement in this field. Exacerbating this issue is the lack of uniformity –and in some cases direct measurements—of robot personalities. Specifically, the HRI literature on this topic to date has diverged greatly in how they opt to measure robot personality (Esterwood et al., 2022). For example, of the seven studies examining extroversion identified by Esterwood et al., (2022) none used the same measurement instruments weakening the capability of future studies to build upon their results. Ultimately, more standardization in measures is clearly needed in regard to robot personality and future work should endeavor not only to use more consistent measures but also to report said measures in full.

10.7 CONCLUSION

In this chapter, we reflected on the HRI literature by examining trust, personality, explanation, and evaluation metrics. Each of these areas represents some of the most influential areas in the study of HRI. Yet, much work remains to deepen our understanding of HRI, with significant implications for scholarship and practice. As robots continue to advance, their use is expected to spread across various spheres of human life. Accordingly, the study of HRI will become increasingly important for our society. In closing, we hope this chapter propels research toward the next step in advancing our understanding of this area.

REFERENCES

Akash, K., Polson, K., Reid, T., and Jain, N. (2019). Improving human-machine collaboration through transparency-based feedback - part I: Human trust and workload model. *IFAC-PapersOnLine*, 51(34):315–321.

Alaieri, F., & Vellino, A. (2016). Ethical Decision Making in Robots: Autonomy, Trust and Responsibility: Autonomy Trust and Responsibility. In Social Robotics: 8th International Conference, ICSR 2016, Kansas City, MO, November 1–3, 2016 Proceedings 8 (pp. 159–168). Springer International Publishing.

Alarcon, G. M., Capiola, A., and Pfahler, M. D. (2021). The role of human personality on trust in human-robot interaction. In *Trust in Human-Robot Interaction* (pp. 159–178). Elsevier, Cambridge, MA.

Allport, G. W. (1937). *Personality: A Psychological Interpretation*. Henry Holt and Company, New York.

Ambsdorf, J., Munir, A., Wei, Y., Degkwitz, K., Harms, H. M., Stannek, S., Ahrens, K., Becker, D., Strahl, E., and Weber, T. (2022). Explain yourself! Effects of explanations in human-robot interaction. In *2022 31st IEEE International Conference on Robot and Human Interactive Communication (RO-MAN)* (pp. 393–400). IEEE.

Angwin, J., Larson, J., Mattu, S., and Kirchner, L. (2022). Machine bias. In Kirsten Martin (ed.) *Ethics of Data and Analytics* (pp. 254–264). Auerbach Publications, Boca Raton, FL.

Anjomshoae, S., Najjar, A., Calvaresi, D., and Främling, K. (2019). Explainable agents and robots: Results from a systematic literature review. In *18th International Conference on Autonomous Agents and Multiagent Systems (AAMAS 2019), Montreal, Canada, May 13-17, 2019* (pp. 1078–1088). International Foundation for Autonomous Agents and Multiagent Systems.

Avetisyan, L., Ayoub, J., and Zhou, F. (2022). Investigating explanations in conditional and highly automated driving: The effects of situation awareness and modality. *Transportation Research Part F: Traffic Psychology and Behaviour*, 89:456–466.

Azevedo-Sa, H., Jayaraman, S. K., Esterwood, C. T., Yang, X. J., Robert Jr, L. P., and Tilbury, D. M. (2021a). Real-time estimation of drivers' trust in automated driving systems. *International Journal of Social Robotics*, 13(8):1911–1927.

Azevedo-Sa, H., Yang, X. J., Robert, L. P., and Tilbury, D. M. (2021b). A unified bidirectional model for natural and artificial trust in human-robot collaboration. *IEEE Robotics and Automation Letters*, 6(3):5913–5920.

Baker, A. L., Phillips, E. K., Ullman, D., and Keebler, J. R. (2018). Toward an understanding of trust repair in human-robot interaction: Current research and future directions. *ACM Transactions on Interactive Intelligent Systems (TiiS)*, 8(4):1–30.

Baraka, K., Paiva, A., and Veloso, M. (2016). Expressive lights for revealing mobile service robot state. In *Robot 2015: Second Iberian Robotics Conference: Advances in Robotics, Volume 1* (pp. 107–119). Springer International Publishing.

Bartneck, C., Kulić, D., Croft, E., and Zoghbi, S. (2009). Measurement instruments for the anthropomorphism, animacy, likeability, perceived intelligence, and perceived safety of robots. *International Journal of Social Robotics*, 1:71–81.

Baumert, A., Schmitt, M., Perugini, M., Johnson, W., Blum, G., Borkenau, P., Costantini, G., Denissen, J. J., Fleeson, W., Grafton, Ben Jayawickereme, E., Kurzius, E., MacLeod, C., Miller, L., Read, S. J., Roberts, B., Robinson, M. D., Wood, D., Wrzus, C., and Mottus, R. (2017). Integrating personality structure, personality process, and personality development. *European Journal of Personality*, 31(5):503–528.

Bernsen, N. O. and Dybkjær, L. (2001). Exploring natural interaction in the car. In *CLASS Workshop on Natural Interactivity and Intelligent Interactive Information Representation* (Vol. 2, No. 1).

Bhat, S., Lyons, J. B., Shi, C., and Yang, X. J. (2022). Clustering trust dynamics in a human-robot sequential decision-making task. *IEEE Robotics and Automation Letters*, 7(4):8815–8822.

Bisk, Y., Yuret, D., and Marcu, D. (2016). Natural language communication with robots. In *Proceedings of the 2016 Conference of the North American Chapter of the Association for Computational linguistics: Human Language Technologies* (pp. 751–761).

Bostelman, R., Hong, T., and Marvel, J. (2016). Survey of research for performance measurement of mobile manipulators. *Journal of Research of the National Institute of Standards and Technology*, 121:342.

Breckler, S. J. (1984). Empirical validation of affect, behavior, and cognition as distinct components of attitude. *Journal of Personality and Social Psychology*, 47(6):1191.

Cai, C. J., Jongejan, J., and Holbrook, J. (2019). The effects of example-based explanations in a machine learning interface. In *Proceedings of the 24th International Conference on Intelligent User Interfaces* (pp. 258–262).

Cao, Y., Mahr, A., Castronovo, S., Theune, M., Stahl, C., and Müller, C. A. (2010). Local danger warnings for drivers: The effect of modality and level of assistance on driver reaction. In *Proceedings of the 15th International Conference on Intelligent User Interfaces* (pp. 239–248).

Carpinella, C. M., Wyman, A. B., Perez, M. A., and Stroessner, S. J. (2017). The robotic social attributes scale (rosas) development and validation. In *Proceedings of the 2017 ACM/IEEE International Conference on human-robot interaction* (pp. 254–262).

Cha, E., Kim, Y., Fong, T., and Mataric, M. J. (2018). A survey of nonverbal signaling methods for non-humanoid robots. *Foundations and Trends® in Robotics*, 6(4):211–323.

Chen, J. Y., Lakhmani, S. G., Stowers, K., Selkowitz, A. R., Wright, J. L., and Barnes, M. (2018). Situation awareness-based agent transparency and human-autonomy teaming effectiveness. *Theoretical Issues in Ergonomics Science*, 19(3), 259–282.

Chen, M., Nikolaidis, S., Soh, H., Hsu, D., and Srinivasa, S. (2020). Trust-aware decision making for human-robot collaboration: Model learning and planning. *ACM Transactions on Human-Robot Interaction (THRI)*, 9(2):1–23.

Chiou, E. K., Demir, M., Buchanan, V., Corral, C. C., Endsley, M. R., Lematta, G. J., Cooke, N. J., and McNeese, N. J. (2022). Towards human-robot teaming: Tradeoffs of explanation-based communication strategies in a virtual search and rescue task. *International Journal of Social Robotics*, 14:1117–1136.

Chita-Tegmark, M., Law, T., Rabb, N., and Scheutz, M. (2021). Can you trust your trust measure? In *Proceedings of the 2021 ACM/IEEE International Conference on Human-Robot Interaction* (pp. 92–100).

Chuttur, M. (2009). Overview of the technology acceptance model: origins, developments and future directions. *Sprouts: Working Papers on Information Systems.* 37(9):290.

Coeckelbergh, M. (2011). Humans, animals, and robots: A phenomenological approach to human-robot relations. *International Journal of Social Robotics*, 3:197–204.

Colquitt, J. A. and Salam, S. C. (2012). Foster trust through ability, benevolence, and integrity. In *Locke, Edwin (ed.) Handbook of Principles of Organizational Behavior: Indispensable Knowledge for Evidence-Based Management* (pp. 389–404). John Wiley & Sons, Inc., New York City.

Colquitt, J. A., Scott, B. A., and LePine, J. A. (2007). Trust, trustworthiness, and trust propensity: A meta-analytic test of their unique relationships with risk taking and job performance. *Journal of Applied Psychology*, 92(4):909.

Connelly, B. S., Ones, D. S., and Hülsheger, U. R. (2018). Personality in industrial, work and organizational psychology: Theory, measurement and application. In Ones, D. S., Anderson, N., Viswesvaran, C., and Sinangil, H. K. (Eds.) *The SAGE Handbook of Industrial, Work and Organizational Psychology* (Vol. 1, pp. 320–365). SAGE Inc., Thousand Oaks, CA.

Conti, D., Di Nuovo, S., Buono, S., and Di Nuovo, A. (2017). Robots in education and care of children with developmental disabilities: A study on acceptance by experienced and future professionals. *International Journal of Social Robotics*, 9:51–62.

Correia, F., Guerra, C., Mascarenhas, S., Melo, F. S., and Paiva, A. (2018). Exploring the impact of fault justification in human-robot trust. In *Proceedings of the 17th International Conference on Autonomous Agents and Multiagent Systems* (pp. 507–513).

Costa, A., Ferrin, D., and Fulmer, C. (2018). Trust at work. In Ones, D. S., Anderson, N., Viswesvaran, C., and Sinangil, H. K. (Eds.) *The Sage Handbook of Industrial, Work & Organizational Psychology* (pp. 435–467). SAGE Inc., Thousand Oaks, CA.

Costa, P. T. and McCrae, R. R. (1992). Normal personality assessment in clinical practice: The neo personality inventory. *Psychological Assessment*, 4(1):5.

de Visser, E. J., Peeters, M. M., Jung, M. F., Kohn, S., Shaw, T. H., Pak, R., and Neerincx, M. A. (2020). Towards a theory of longitudinal trust calibration in human-robot teams. *International Journal of Social Robotics*, 12(2):459–478.

Devin, S. and Alami, R. (2016). An implemented theory of mind to improve human-robot shared plans execution. In *2016 11th ACM/IEEE International Conference on Human-Robot Interaction (HRI)* (pp. 319–326). IEEE.

Dingus, T. A., McGehee, D. V., Manakkal, N., Jahns, S. K., Carney, C., and Hankey, J. M. (1997). Human factors field evaluation of automotive headway maintenance/collision warning devices. *Human Factors*, 39(2):216–229.

Do Quang, H., Manh, T. N., Manh, C. N., Tien, D. P., Van, M. T., Tien, K. N., and Duc, D. N. (2020). An approach to design navigation system for omnidirectional mobile robot based on ROS. *International Journal of Mechanical Engineering and Robotics Research*, 9(11):1502–1508.

Driskell, J. E., Goodwin, G. F., Salas, E., and O'Shea, P. G. (2006). What makes a good team player? Personality and team effectiveness. *Group Dynamics: Theory, Research, and Practice*, 10(4):249.

Du, N., Zhou, F., Tilbury, D., Robert, L. P., and Yang, X. J. (2021). Designing alert systems in takeover transitions: The effects of display information and modality. In *13th International Conference on Automotive User Interfaces and Interactive Vehicular Applications* (pp. 173–180), Leeds, UK: ACM.

Edmonds, M., Gao, F., Liu, H., Xie, X., Qi, S., Rothrock, B., Zhu, Y., Wu, Y. N., Lu, H., and Zhu, S.-C. (2019). A tale of two explanations: Enhancing human trust by explaining robot behavior. *Science Robotics*, 4(37):eaay4663.

Erdem, E., Haspalamutgil, K., Palaz, C., Patoglu, V., and Uras, T. (2011). Combining high-level causal reasoning with low-level geometric reasoning and motion planning for robotic manipulation. In *2011 IEEE International Conference on Robotics and Automation* (pp. 4575–4581). IEEE.

Esterwood, C. and Robert, L. P. (2020). Personality in healthcare human robot interaction (h-hri) a literature review and brief critique. In *Proceedings of the 8th International Conference on Human-Agent Interaction* (pp. 87–95).

Esterwood, C. and Robert, L. P. (2021). Do you still trust me? Human-robot trust repair strategies. In *2021 30th IEEE International Conference on Robot & Human Interactive Communication (RO-MAN)* (pp. 183–188). IEEE.

Esterwood, C. and Robert, L. P. (2022a). Having the right attitude: How attitude impacts trust repair in human-robot interaction. In *2022 17th ACM/IEEE International Conference on Human-Robot Interaction (HRI)* (pp. 332–341). IEEE.

Esterwood, C. and Robert, L. P. (2022b). A literature review of trust repair in hri. In *2022 31st IEEE International Conference on Robot and Human Interactive Communication (RO-MAN)* (pp. 1641–1646). IEEE.

Esterwood, C. and Robert, L. P. (2023a). Three strikes and you are out!: The impacts of multiple human-robot trust violations and repairs on robot trustworthiness. *Computers in Human Behavior*, 142:107658.

Esterwood, C., Essenmacher, K., Yang, H., Zeng, F., and Robert, L. P. (2021a). Birds of a feather flock together: But do humans and robots? A meta-analysis of human and robot personality matching. In *2021 30th IEEE International Conference on Robot & Human Interactive Communication (RO-MAN)* (pp. 343–348). IEEE.

Esterwood, C., Essenmacher, K., Yang, H., Zeng, F., and Robert, L. P. (2021b). A meta-analysis of human personality and robot acceptance in human-robot interaction. In *Proceedings of the 2021 CHI Conference on Human Factors in Computing Systems* (pp. 1–18).

Esterwood, C., Essenmacher, K., Yang, H., Zeng, F., and Robert, L. P. (2022). A personable robot: Meta-analysis of robot personality and human acceptance. *IEEE Robotics and Automation Letters*, 7(3):6918–6925.

Esterwood, C. and Robert, L. (2023b). The warehouse robot interaction sim: An open-source HRI research platform. In *2023 ACM/IEEE International Conference on Human-Robot Interaction*.

Esterwood, C., and Robert, L. P. (2023c). The theory of mind and human–robot trust repair. *Scientific Reports*, 13(1), 9877.

Eysenck, H. J. and Eysenck, S. B. G. (1975). *Manual of the Eysenck Personality Questionnaire (Junior & Adult)*. Hodder and Stoughton Educational, London.

Ezenyilimba, A., Wong, M., Hehr, A., Demir, M., Wolff, A., Chiou, E., and Cooke, N. (2022). Impact of transparency and explanations on trust and situation awareness in human-robot teams. *Journal of Cognitive Engineering and Decision Making*, 17(1):75–93.

Ezer, N. (2008). *Is a Robot an Appliance, Teammate, or Friend? Age-Related Differences in Expectations of and Attitudes towards Personal Home-Based Robots*. PhD thesis, Georgia Institute of Technology.

Forster, Y., Naujoks, F., and Neukum, A. (2017). Increasing anthropomorphism and trust in automated driving functions by adding speech output. In *2017 IEEE Intelligent Vehicles Symposium (IV)* (pp. 365–372).

Georgiades, C., German, A., Hogue, A., Liu, H., Prahacs, C., Ripsman, A., Sim, R., Torres, L.-A., Zhang, P., Buehler, M., et al. (2004). Aqua: An aquatic walking robot. In *2004 IEEE/RSJ International Conference on Intelligent Robots and Systems (IROS)(IEEE Cat. No. 04CH37566)* (Vol. 4, pp. 3525–3531). IEEE.

Geravand, M., Werner, C., Hauer, K., and Peer, A. (2016). An integrated decision-making approach for adaptive shared control of mobility assistance robots. *International Journal of Social Robotics*, 8:631–648.

Gillespie, N., Lockey, S., Hornsey, M., and Okimoto, T. (2021). Trust repair: A multilevel framework. In Gillespie, N., Fulmer. C., Lewicki, R., (Eds.) *Understanding Trust in Organizations* (pp. 143–176). Routledge, Oxfordshire.

Goldberg, L. R. (1992). The development of markers for the big-five factor structure. *Psychological Assessment*, 4(1):26.

Grover, S. L., Hasel, M. C., Manville, C., and Serrano-Archimi, C. (2014). Follower reactions to leader trust violations: A grounded theory of violation types, likelihood of recovery, and recovery process. *European Management Journal*, 32(5):689–702.

Guinness, D., Muehlbradt, A., Szafir, D., and Kane, S. K. (2019). Robographics: Dynamic tactile graphics powered by mobile robots. In *Proceedings of the 21st International ACM SIGACCESS Conference on Computers and Accessibility* (pp. 318–328).

Guo, Y. and Yang, X. J. (2021). Modeling and predicting trust dynamics in human-robot teaming: A Bayesian inference approach. *International Journal of Social Robotics*, 13:1899–1909.

Guo, Y., Shi, C., and Yang, X. J. (2021). Reverse psychology in trust-aware human-robot interaction. *IEEE Robotics and Automation Letters*, 6(3):4851–4858.

Guo, Y., Yang, X. J., and Shi, C. (2023a). TIP: A trust inference and propagation model in multi-human multi-robot teams. In *Companion of the 2023 ACM/IEEE International Conference on Human-Robot Interaction, HRI '23* (pp. 639–643), New York: Association for Computing Machinery.

Guo, Y., Yang, X. J., and Shi, C. (2023b). Enabling team of teams: A trust inference and propagation (TIP) model in multi-human multi-robot teams. In *Proceedings of Robotics: Science and Systems*. Daegu, South Korea.

Ha, T., Kim, S., Seo, D., and Lee, S. (2020). Effects of explanation types and perceived risk on trust in autonomous vehicles. *Transportation Research Part F: Traffic Psychology and Behaviour*, 73:271–280.

Han, Z., Phillips, E., and Yanco, H. A. (2021). The need for verbal robot explanations and how people would like a robot to explain itself. *ACM Transactions on Human-Robot Interaction (THRI)*, 10(4), 1–42.

Hancock, P. A., Billings, D. R., Schaefer, K. E., Chen, J. Y., De Visser, E. J., and Parasuraman, R. (2011). A meta-analysis of factors affecting trust in human-robot interaction. *Human Factors*, 53(5):517–527.

Hancock, P. A., Kessler, T. T., Kaplan, A. D., Brill, J. C., and Szalma, J. L. (2021). Evolving trust in robots: Specification through sequential and comparative meta-analyses. *Human Factors*, 63(7):1196–1229.

Harbers, M., van den Bosch, K., and Meyer, J. J. C. (2011, January). A theoretical framework for explaining agent behavior. In *SIMULTECH* (pp. 228–231).

Hart, S. G. and Staveland, L. E. (1988). Development of NASA-TLX (task load index): Results of empirical and theoretical research. In Hancock, P., and Meshkati, N,. (Eds.) *Advances in Psychology* (Vol. 52, pp. 139–183). Elsevier, Amsterdam.

Haslam, N. (2007). Trait psychology. In Haslam, N. (Ed.) *Introduction to Personality and Intelligence* (pp. 17–45). London: Sage Publications Ltd.

Hassabis, D., Spreng, R. N., Rusu, A. A., Robbins, C. A., Mar, R. A., and Schacter, D. L. (2014). Imagine all the people: How the brain creates and uses personality models to predict behavior. *Cerebral Cortex*, 24(8):1979–1987.

Hayes, B. and Shah, J. A. (2017). Improving robot controller transparency through autonomous policy explanation. In *Proceedings of the 2017 ACM/IEEE International Conference on Human-Robot Interaction, HRI '17* (pp. 303–312). New York: Association for Computing Machinery.

Heerink, M., Kröse, B., Evers, V., et al. (2010). Assessing Acceptance of Assistive Social Agent Technology by Older Adults: the Almere Model. *Int J of Soc Robotics*, 2, 361–375.

Hirai, K., Hirose, M., Haikawa, Y., and Takenaka, T. (1998). The development of Honda humanoid robot. In *Proceedings. 1998 IEEE International Conference on Robotics and Automation (Cat. No. 98CH36146)* (Vol. 2, pp. 1321–1326). IEEE.

Ho, C.-C. and MacDorman, K. F. (2010). Revisiting the uncanny valley theory: Developing and validating an alternative to the Godspeed indices. *Computers in Human Behavior*, 26(6):1508–1518.

Hu, W.-L., Akash, K., Jain, N., and Reid, T. (2016). Real-time sensing of trust in human-machine interactions. *IFAC-PapersOnLine*, 49(32):48–53.

Ishida, T., Kuroki, Y., Yamaguchi, J., Fujita, M., and Doi, T. T. (2001). Motion entertainment by a small humanoid robot based on open-r. In *Proceedings 2001 IEEE/RSJ International Conference on Intelligent Robots and Systems. Expanding the Societal Role of Robotics in the the Next Millennium (Cat. No. 01CH37180)* (Vol. 2, pp. 1079–1086). IEEE.

Jian, J.-Y., Bisantz, A. M., and Drury, C. G. (2000). Foundations for an empirically determined scale of trust in automated systems. *International Journal of Cognitive Ergonomics*, 4(1):53–71.

John, O. P., Naumann, L. P., and Soto, C. J. (2008). Paradigm shift to the integrative big five trait taxonomy. In O. P. John, R. W. Robins, & L. A. Pervin (Eds.), *Handbook of Personality: Theory and Research* (3rd ed., pp. 114–158). New York: Guilford Press.

Kamide, H., Mae, Y., Takubo, T., Ohara, K., and Arai, T. (2014). Direct comparison of psychological evaluation between virtual and real humanoids: Personal space and subjective impressions. *International Journal of Human-Computer Studies*, 72(5):451–459.

Karray, F., Alemzadeh, M., Abou Saleh, J., and Arab, M. N. (2008). Human-computer interaction: Overview on state of the art. *International Journal on smart Sensing and Intelligent Systems*, 1(1):137–159.

Kessler, T. T., Larios, C., Walker, T., Yerdon, V., and Hancock, P. (2017). A comparison of trust measures in human-robot interaction scenarios. In *Advances in Human Factors in Robots and Unmanned Systems: Proceedings of the AHFE 2016 International Conference on Human Factors in Robots and Unmanned Systems, July 27-31, 2016, Walt Disney World®, Florida, USA* (pp. 353–364). Springer.

Kim, P., Ferrin, D., Cooper, C., and Dirks, K. (2004). Removing the shadow of suspicion: The effects of apology versus denial for repairing competence-versus integrity-based trust violations. *Journal of Applied Psychology*, 89(1):104.

Kim, T., & Hinds, P. (2006). Who should I blame? Effects of autonomy and transparency on attributions in human-robot interaction. In *ROMAN 2006-The 15th IEEE international symposium on robot and human interactive communication* (pp. 80–85). IEEE.

Kim, W., Kim, N., Lyons, J. B., and Nam, C. S. (2020). Factors affecting trust in high-vulnerability human-robot interaction contexts: A structural equation modelling approach. *Applied Ergonomics*, 85:103056.

Koo, J., Kwac, J., Ju, W., Steinert, M., Leifer, L., and Nass, C. (2015). Why did my car just do that? Explaining semi-autonomous driving actions to improve driver understanding, trust, and performance. *International Journal on Interactive Design and Manufacturing (IJIDeM)*, 9(4):269–275.

Koo, J., Shin, D., Steinert, M., and Leifer, L. (2016). Understanding driver responses to voice alerts of autonomous car operations. *International Journal of Vehicle Design*, 70(4):377–392.

Körber, M., Prasch, L., and Bengler, K. (2018). Why do i have to drive now? Post hoc explanations of takeover requests. *Human Factors*, 60(3):305–323.

Kox, E. S., Kersholt, J. H., Hueting, T. F., and De Vries, P. W. (2021). Trust repair in human-agent teams: The effectiveness of explanations and expressing regret. *Autonomous Agents and Multi-Agent Systems*, 35(2):30.

Kühne, R. and Peter, J. (2022). Anthropomorphism in human-robot interactions: A multidimensional conceptualization. *Communication Theory*, 33(1), 42–52.

Kulesza, T., Stumpf, S., Burnett, M., Yang, S., Kwan, I., and Wong, W.-K. (2013). Too much, too little, or just right? Ways explanations impact end users' mental models. In *2013 IEEE Symposium on Visual Languages and Human Centric Computing* (pp. 3–10). IEEE.

Lee, J. D. and Moray, N. (1992). Trust, control strategies and allocation of function in human-machine systems. *Ergonomics*, 35(10):1243–1270.

Lee, J. D. and See, K. A. (2004). Trust in automation: Designing for appropriate reliance. *Human Factors*, 46(1):50–80.

Lee, K. M. and Nass, C. (2003). Designing social presence of social actors in human computer interaction. In *Proceedings of the SIGCHI Conference on Human Factors in Computing Systems* (pp. 289–296). ACM.

Lee, K. M., Peng, W., Jin, S. A., & Yan, C. (2006). Can robots manifest personality?: An empirical test of personality recognition, social responses, and social presence in human–robot interaction. *Journal of Communication*, 56(4):754–772.

Lettl, B. and Schulte, A. (2013). Self-explanation capability for cognitive agents on-board of UCAVs to improve cooperation in a manned-unmanned fighter team. In *AIAA Infotech@ Aerospace (I@ A) Conference* (p. 4898).

Lewicki, R. J. and Brinsfield, C. (2017). Trust repair. *Annual Review of Organizational Psychology and Organizational Behavior*, 4:287–313.

Lewicki, R. J., Polin, B., and Lount Jr, R. B. (2016). An exploration of the structure of effective apologies. *Negotiation and Conflict Management Research*, 9(2):177–196.

Li, N., Barrick, M. R., Zimmerman, R. D., and Chiaburu, D. S. (2014). Retaining the productive employee: The role of personality. *Academy of Management Annals*, 8(1):347–395.

Liang, N. and Nejat, G. (2022). A meta-analysis on remote HRI and in-person HRI: What is a socially assistive robot to do? *Sensors*, 22(19):7155.

Lim, B. Y. and Dey, A. K. (2011, August). Design of an intelligible mobile context-aware application. In *Proceedings of the 13th International Conference on Human Computer Interaction with Mobile Devices and Services* (pp. 157–166).

Long, J. H., Schumacher, J., Livingston, N., and Kemp, M. (2006). Four flippers or two? Tetrapodal swimming with an aquatic robot. *Bioinspiration & Biomimetics*, 1(1):20.

Lu, S., Zhang, M. Y., Ersal, T., and Yang, X. J. (2019). Workload management in teleoperation of unmanned ground vehicles: Effects of a delay compensation aid on human operators' workload and teleoperation performance. *International Journal of Human-Computer Interaction*, 35(19):1820–1830.

Lu, Y. (2020). *Detecting and Overcoming Trust Miscalibration in Real Time Using an Eye-tracking Based Technique*. PhD thesis, University of Michigan.

Luo, R., Du, N., and Yang, X. J. (2022). Evaluating effects of enhanced autonomy transparency on trust, dependence, and human-autonomy team performance over time. *International Journal of Human-Computer Interaction*, 38(18–20):1962–1971.

Lyons, J. B., aldin Hamdan, I., and Vo, T. Q. (2023). Explanations and trust: What happens to trust when a robot partner does something unexpected? *Computers in Human Behavior*, 138:107473.

Madsen, M. and Gregor, S. (2000). Measuring human-computer trust. In *11th Australasian Conference on Information Systems* (Vol. 53, pp. 6–8). Citeseer.

Manzey, D., Reichenbach, J., and Onnasch, L. (2012). Human performance consequences of automated decision aids: The impact of degree of automation and system experience. *Journal of Cognitive Engineering and Decision Making*, 6(1):57–87.

Mara, M., Appel, M., and Gnambs, T. (2022). Human-like robots and the uncanny valley: A meta-analysis of user responses based on the Godspeed scales. *Zeitschrift für Psychologie*, 230(1):33.

Mara, M., Stein, J.-P., Latoschik, M. E., Lugrin, B., Schreiner, C., Hostettler, R., and Appel, M. (2021). User responses to a humanoid robot observed in real life, virtual reality, 3d and 2d. *Frontiers in Psychology*, 12:633178.

Mayer, R. C., Davis, J. H., and Schoorman, F. D. (1995). An integrative model of organizational trust. *Academy of Management Review*, 20(3):709–734.

McCrae, R. R. and Costa Jr, P. T. (1997). Personality trait structure as a human universal. *American Psychologist*, 52(5):509.

McCrae, R. R. and Costa Jr, P. T. (2008). The five-factor theory of personality. In O. P. John, R. W. Robins, and L. A. Pervin (Eds.), *Handbook of Personality: Theory and Research* (3rd ed., pp. 159–181). The Guilford Press.

McDonald, D. D. (2010). Natural language generation. In Indurkhya, N., and Damerau, F., (Eds) *Handbook of Natural Language Processing* (Vol. 2, pp. 121–144). London.

McMartin, J. (2016). *Personality Psychology: A Student-Centered Approach.* Sage Publications, Thousand Oaks, CA.

Michaud, F., Clavet, A., Lachiver, G., and Lucas, M. (2000). Designing toy robots to help autistic children an open design project for electrical and computer engineering education. In *2000 Annual Conference* (pp. 5–205).

Mileounis, A., Cuijpers, R.H., Barakova, E.I. (2015). Creating Robots with Personality: The Effect of Personality on Social Intelligence. In: *Artificial Computation in Biology and Medicine. IWINAC 2015.* Lecture Notes in Computer Science (vol. 9107). Springer, Cham.

Miller, T. (2019). Explanation in artificial intelligence: Insights from the social sciences. *Artificial Intelligence*, 267:1–38.

Millo, F., Gesualdo, M., Fraboni, F., and Giusino, D. (2021). Human likeness in robots: Differences between industrial and non-industrial robots. In *Proceedings of the 32nd European Conference on Cognitive Ergonomics* (pp. 1–5).

Mishra, A., Soni, U., Huang, J., and Bryan, C. (2022). Why? Why not? When? Visual explanations of agent behaviour in reinforcement learning. In *2022 IEEE 15th Pacific Visualization Symposium (PacificVis)* (pp. 111–120). IEEE.

Moray, N., Inagaki, T., and Itoh, M. (2000). Adaptive automation, trust, and self-confidence in fault management of time-critical tasks. *Journal of Experimental Psychology Applied*, 6(1):44–58.

Mota, T., Sridharan, M., and Leonardis, A. (2021). Integrated commonsense reasoning and deep learning for transparent decision making in robotics. *SN Computer Science*, 2(4):242.

Moyle, W., Jones, C., Cooke, M., O'Dwyer, S., Sung, B., and Drummond, S. (2013). Social robots helping people with dementia: Assessing efficacy of social robots in the nursing home environment. In *2013 6th International Conference on Human System Interactions (HSI)* (pp. 608–613). IEEE.

Muir, B. M. and Moray, N. (1996). Trust in automation. Part II. Experimental studies of trust and human intervention in a process control simulation. *Ergonomics*, 39(3):429–460.

Naneva, S., Sarda Gou, M., Webb, T. L., and Prescott, T. J. (2020). A systematic review of attitudes, anxiety, acceptance, and trust towards social robots. *International Journal of Social Robotics*, 12(6):1179–1201.

Naujoks, F., Forster, Y., Wiedemann, K., and Neukum, A. (2017a). A human-machine interface for cooperative highly automated driving. In N. A. Stanton, S. Landry, G. Di Bucchianico, and A. Vallicelli, (Eds.), *Advances in Human Aspects of Transportation* (pp. 585–595), Cham: Springer International Publishing.

Naujoks, F., Forster, Y., Wiedemann, K., and Neukum, A. (2017b). Improving usefulness of automated driving by lowering primary task interference through HMI design. *Journal of Advanced Transportation*, 2017:6105087.

Nees, M. A., Helbein, B., and Porter, A. (2016). Speech auditory alerts promote memory for alerted events in a video-simulated self-driving car ride. *Human Factors*, 58(3):416–426.

Nomura, T., Kanda, T., Suzuki, T., and Kato, K. (2004). Psychology in human-robot communication: An attempt through investigation of negative attitudes and anxiety toward robots. In *RO-MAN 2004. 13th IEEE International Workshop on Robot and Human Interactive Communication (IEEE Catalog No. 04TH8759)* (pp. 35–40). IEEE.

Nomura, T., Kanda, T., Suzuki, T., and Kato, K. (2008). Prediction of human behavior in human-robot interaction using psychological scales for anxiety and negative attitudes toward robots. *IEEE Transactions on Robotics*, 24(2):442–451.

Peeters, M. A., Van Tuijl, H. F., Rutte, C. G., and Reymen, I. M. (2006). Personality and team performance: A meta-analysis. *European Journal of Personality*, 20(5):377–396.

Piazza, C., Grioli, G., Catalano, M., and Bicchi, A. (2019). A century of robotic hands. *Annual Review of Control, Robotics, and Autonomous Systems*, 2:1–32.

Pocius, K. E. (1991). Personality factors in human-computer interaction: A review of the literature. *Computers in Human Behavior*, 7(3):103–135.

Politis, I., Brewster, S., and Pollick, F. (2015). To beep or not to beep? Comparing abstract versus language-based multimodal driver displays. In *Proceedings of the 33rd Annual ACM Conference on Human Factors in Computing Systems*, CHI '15 (pp. 3971–3980), New York: Association for Computing Machinery.

Poon, J. M. (2013). Effects of benevolence, integrity, and ability on trust-in-supervisor. *Employee Relations*, 35(4):396–407.

Prewett, M. S., Johnson, R. C., Saboe, K. N., Elliott, L. R., and Coovert, M. D. (2010). Managing workload in human-robot interaction: A review of empirical studies. *Computers in Human Behavior*, 26(5):840–856.

Prewett, M. S., Saboe, K. N., Johnson, R. C., Coovert, M. D., and Elliott, L. R. (2009). Workload in human-robot interaction: A review of manipulations and outcomes. In *Proceedings of the Human Factors and Ergonomics Society Annual Meeting* (Vol. 53, pp. 1393–1397). Sage CA: Los Angeles, CA: SAGE Publications.

Raghunath, N., Myers, P., Sanchez, C. A., and Fitter, N. T. (2021). Women are funny: Influence of apparent gender and embodiment in robot comedy. In *Social Robotics: 13th International Conference, ICSR 2021, Singapore, Singapore, November 10-13, 2021, Proceedings 13* (pp. 3–13). Springer.

Rawlins, B. (2008). Measuring the relationship between organizational transparency and employee trust. *Public Relations Journal*, 2(2):1–21.

Rhee, J., Parent, D., and Basu, A. (2013). The influence of personality and ability on undergraduate teamwork and team performance. *SpringerPlus*, 2(1):16.

Robert Jr, L. P., Alahmad, R., Esterwood, C., Kim, S., You, S., Zhang, Q., et al. (2020). A review of personality in human-robot interactions. *Foundations and Trends® in Information Systems*, 4(2):107–212.

Robert, L. (2018). Personality in the human robot interaction literature: A review and brief critique. In *Proceedings of the 24th Americas Conference on Information Systems, Aug* (pp. 16–18).

Robert, L. P. (2021). A measurement of attitude toward working with robots (AWRO): A compare and contrast study of AWRO with negative attitude toward robots (NARS). In *Human-Computer Interaction. Interaction Techniques and Novel Applications: Thematic Area, HCI 2021, Held as Part of the 23rd HCI International Conference, HCII 2021, Virtual Event, July 24-29, 2021, Proceedings, Part II 23* (pp. 288–299). Springer.

Robert, L. P., Denis, A. R., and Hung, Y.-T. C. (2009). Individual swift trust and knowledge-based trust in face-to-face and virtual team members. *Journal of Management Information Systems*, 26(2):241–279.

Robinette, P., Howard, A. M., and Wagner, A. R. (2015). Timing is key for robot trust repair. In *International Conference on Social Robotics* (pp. 574–583). Springer.

Roesler, E., Manzey, D., and Onnasch, L. (2021). A meta-analysis on the effectiveness of anthropomorphism in human-robot interaction. *Science Robotics*, 6(58):eabj5425.

Roitberg, A., Schneider, D., Djamal, A., Seibold, C., Reiß, S., and Stiefelhagen, R. (2021). Let's play for action: Recognizing activities of daily living by learning from life simulation video games. In *2021 IEEE/RSJ International Conference on Intelligent Robots and Systems (IROS)* (pp. 8563–8569). IEEE.

Ruijten, P. A. M., Terken, J. M. B., and Chandramouli, S. N. (2018). Enhancing trust in autonomous vehicles through intelligent user interfaces that mimic human behavior. *Multimodal Technologies and Interaction*, 2(4):62.

Sanders, T., Oleson, K. E., Billings, D. R., Chen, J. Y., and Hancock, P. A. (2011). A model of human-robot trust: Theoretical model development. In *Proceedings of the Human Factors and Ergonomics Society Annual Meeting* (Vol. 55, pp. 1432–1436). Sage CA: Los Angeles, CA: SAGE Publications.

Santamaria, T. and Nathan-Roberts, D. (2017). Personality measurement and design in human-robot interaction: A systematic and critical review. In *Proceedings of the Human Factors and Ergonomics Society Annual Meeting* (Vol. 61, pp. 853–857). Sage CA: Los Angeles, CA: SAGE Publications.

Schaefer, K. (2013). *The Perception and Measurement of Human-Robot Trust*. PhD thesis, University of Central Florida.

Schaffer, J., O'Donovan, J., Michaelis, J., Raglin, A., and Höllerer, T. (2019). I can do better than your AI: Expertise and explanations. In *Proceedings of the 24th International Conference on Intelligent User Interfaces* (pp. 240–251).

Schneier, M., Schneier, M., and Bostelman, R. (2015). *Literature Review of Mobile Robots for Manufacturing*. US Department of Commerce, National Institute of Standards and Technology, Gaithersburg, MD.

Schweitzer, M. E., Hershey, J. C., and Bradlow, E. T. (2006). Promises and lies: Restoring violated trust. *Organizational Behavior and human Decision Processes*, 101(1):1–19.

Sebo, S. S., Krishnamurthi, P., and Scassellati, B. (2019). "I don't believe you": Investigating the effects of robot trust violation and repair. In *2019 14th ACM/IEEE International Conference on Human-Robot Interaction (HRI)* (pp. 57–65). IEEE.

Setchi, R., Dehkordi, M. B., and Khan, J. S. (2020). Explainable robotics in human-robot interactions. *Procedia Computer Science*, 176:3057–3066.

Shen, Y., Jiang, S., Chen, Y., Yang, E., Jin, X., Fan, Y., and Campbell, K. D. (2020). To explain or not to explain: A study on the necessity of explanations for autonomous vehicles. *arXiv preprint arXiv:2006.11684*.

Shin, D.-H. and Choo, H. (2011). Modeling the acceptance of socially interactive robotics: Social presence in human-robot interaction. *Interaction Studies*, 12(3):430–460.

Soh, H., Xie, Y., Chen, M., and Hsu, D. (2020). Multi-task trust transfer for human-robot interaction. *The International Journal of Robotics Research*, 39(2-3):233–249.

Sørensen, C., De Reuver, M., and Basole, R. C. (2015). Mobile platforms and ecosystems. *Journal of Information Technology, 30:195–197.*

Stange, S. and Kopp, S. (2021, November). Explaining before or after acting? How the timing of self-explanations affects user perception of robot behavior. In *Social Robotics: 13th International Conference, ICSR 2021, Singapore, Singapore, November 10-13, 2021, Proceedings* (pp. 142–153). Cham: Springer International Publishing.

Stanton, N. A. and Young, M. S. (1998). Vehicle automation and driving performance. *Ergonomics*, 41(7):1014–1028.

Stubbs, K., Wettergreen, D., and Hinds, P. (2007). Autonomy and common ground in human-robot interaction: A field study. *IEEE Intelligent Systems*, 22(2):42–50.

Suzuki, S. (2018). Recent researches on innovative drone technologies in robotics field. *Advanced Robotics*, 32(19):1008–1022.

Tasa, K., Sears, G. J., and Schat, A. C. (2011). Personality and teamwork behavior in context: The cross-level moderating role of collective efficacy. *Journal of Organizational Behavior*, 32(1):65–85.

Tay, B., Jung, Y., and Park, T. (2014). When stereotypes meet robots: The double-edge sword of robot gender and personality in human-robot interaction. *Computers in Human Behavior*, 38:75–84.

Tellex, S., Gopalan, N., Kress-Gazit, H., and Matuszek, C. (2020). Robots that use language. *Annual Review of Control, Robotics, and Autonomous Systems*, 3:25–55.

Terada, K., Shamoto, T., Ito, A., and Mei, H. (2007). Reactive movements of non-humanoid robots cause intention attribution in humans. In *2007 IEEE/RSJ International Conference on Intelligent Robots and Systems* (pp. 3715–3720). IEEE.

Tsui, K. M., Desai, M., Yanco, H. A., and Uhlik, C. (2011). Exploring use cases for telepresence robots. In *Proceedings of the 6th International Conference on Human-Robot Interaction* (pp. 11–18.

Ullman, D. and Malle, B. F. (2018). What does it mean to trust a robot? Steps toward a multidimensional measure of trust. In *Companion of the 2018 ACM/IEEE International Conference on Human-Robot Interaction* (pp. 263–264).

Venkatesh, V., Morris, M. G., Davis, G. B., and Davis, F. D. (2003). User acceptance of information technology: Toward a unified view. *MIS Quarterly*, 27:425–478.

Völkel, S. T., Schödel, R., Buschek, D., Stachl, C., Au, Q., Bischl, B., Bühner, M., and Hußmann, H. (2019). 2 opportunities and challenges of utilizing personality traits for personalization in HCI. In Augstein, M., Herder, E., and Worndl, W. (Eds.) *Personalized Human-Computer Interaction* (p. 31), De Gruyter, Berlin.

Wang, N., Pynadath, D. V., and Hill, S. G. (2016, May). The impact of pomdp-generated explanations on trust and performance in human-robot teams. In *Proceedings of the 2016 International Conference on Autonomous Agents & Multiagent Systems* (pp. 997–1005).

Wheatley, D. J. and Hurwitz, H. B. (2001). The use of a multi-modal interface to integrate in-vehicle information presentation. In *Driving Assessment Conference* (Vol. 1, No. 2001). University of Iowa.

Wickens, C. D. (2008). Multiple resources and mental workload. *Human Factors*, 50(3):449–455.

Wiegand, G., Schmidmaier, M., Weber, T., Liu, Y., and Hussmann, H. (2019). I drive - you trust: Explaining driving behavior of autonomous cars. In *Extended Abstracts of the 2019 CHI Conference on Human Factors in Computing Systems*, CHI EA '19 (pp. 1–6), New York: Association for Computing Machinery.

Xu, A. and Dudek, G. (2015). OPTIMo: Online probabilistic trust inference model for asymmetric human-robot collaborations. In *Proceedings of the Tenth Annual ACM/IEEE International Conference on Human-Robot Interaction* (pp. 221–228). ACM.

Xu, A. and Dudek, G. (2016). Maintaining efficient collaboration with trust-seeking robots. In *2016 IEEE/RSJ International Conference on Intelligent Robots and Systems (IROS)* (pp. 3312–3319).

Yagoda, R. E. (2010). Development of the human robot interaction workload measurement tool (hri-wm). In *Proceedings of the Human Factors and Ergonomics Society Annual Meeting*, volume 54 (pp. 304–308). Sage CA: Los Angeles, CA: Sage Publications.

Yang, X. J., Guo, Y., and Schemanske, C. (2023). From trust to trust dynamics: Combining empirical and computational approaches to model and predict trust dynamics in human-autonomy interaction. In V. G. Duffy, S. J. Landry, J. D. Lee, and N. A. Stanton, (Eds.), *Human-Automation Interaction: Transportation* (pp. 253–265). Springer, Cham, Berlin.

Yang, X. J., Schemanske, C., and Searle, C. (2021). Toward quantifying trust dynamics: How people adjust their trust after moment-to-moment interaction with automation. *Human Factors*, 65(5):862–878.

Yang, X. J., Unhelkar, V. V., Li, K., and Shah, J. A. (2017). Evaluating effects of user experience and system transparency on trust in automation. In *Proceedings of the 2017 ACM/IEEE International Conference on Human-Robot Interaction, HRI '17* (pp. 408–416), ACM, New York.

You, S. and Robert Jr, L. P. (2018). Emotional attachment, performance, and viability in teams collaborating with embodied physical action (EPA) robots. *Journal of the Association for Information Systems*, 19(5):377–407.

You, S., Kim, J.-H., Lee, S., Kamat, V., and Robert Jr, L. P. (2018). Enhancing perceived safety in human-robot collaborative construction using immersive virtual environments. *Automation in Construction*, 96:161–170.

Zanchettin, A. M., Bascetta, L., and Rocco, P. (2013). Achieving humanlike motion: Resolving redundancy for anthropomorphic industrial manipulators. *IEEE Robotics & Automation Magazine*, 20(4):131–138.

Zhang, Q., Yang, X. J., and Robert, L. P. (2021). What and when to explain? A survey of the impact of explanation on attitudes toward adopting automated vehicles. *IEEE Access*, 9:159533–159540.

Zhang, X. (2021). *"Sorry, it was My Fault": Repairing Trust in Human-Robot Interactions*. Thesis, University of Oklahoma.

Zhu, L. and Williams, T. (2020). Effects of proactive explanations by robots on human-robot trust. In *Social Robotics: 12th International Conference, ICSR 2020, Golden, CO, USA, November 14-18, 2020, Proceedings 12* (pp. 85–95). Springer International Publishing.

Zimmerman, M., Bagchi, S., Marvel, J., and Nguyen, V. (2022). An analysis of metrics and methods in research from human-robot interaction conferences, 2015-2021. In *2022 17th ACM/IEEE International Conference on Human-Robot Interaction (HRI)* (pp. 644–648). IEEE.

11 Human-Agent Teaming

Anthony L. Baker, Shan G. Lakhmani, and Jessie Y. C. Chen

11.1 INTRODUCTION

Teams are more than the sum of their parts, and the same goes for human-agent (H-A) teams. The development of faster and smarter computers, robotics, and artificial intelligence is inexorable, and incredible progress is made every year, so it is no surprise that the cutting edge of teamwork and productivity has long sought to integrate humans with intelligent machines to increase overall work capacity.

H-A teaming is a way to take advantage of the rapid information processing inherent to intelligent computer systems while also leveraging the flexibility and novel problem-solving capabilities of humans. Robots and intelligent agents have seen deployment in many contexts such as search and rescue, underwater repair/recovery operations, elderly home care, and the battlefield (Asakawa et al., 2002; Barnes & Evans, 2016; Kawamura & Iskarous, 1994; Ventura & Lima, 2012). In the optimal case, H-A teams can be faster, smarter, and more effective than humans alone, but in practice, the successful integration of humans and agents involves challenges that are in some cases unique from, and in other cases identical to, the challenges faced by human-only teaming. Before we delve into the details of H-A teaming, we will review some of its fundamental concepts, such as what makes a team and what defines an agent. Later sections of this chapter will expand on many of the topics introduced here.

Imagine a team. What do you think it looks like? It probably has multiple team members with different skills or roles. You might have imagined that the team has some job they work on together or a goal they are focused on achieving. The team members likely depend on each other to work on their individual tasks, so that the team can work toward their shared goal. Interdependence and shared goals are two hallmarks of how science has defined teams (Cohen & Bailey, 1997; Dyer, 1984; Hackman, 1987; Sundstrom, de Meuse, & Futrell, 1990) so to define a team, we might say that it's comprised of two or more people, all working toward a common goal, and relying on each other to complete some tasks that move them toward that goal. So, how do they move toward that goal? What *is* teamwork?

The answer is both simple and incredibly complicated. If you think of teamwork, you might imagine people talking about what they're going to do next, sharing tools that are needed for certain tasks, lifting something together to move it somewhere else, or coming up with a plan to stop an opposing team from scoring points. Teamwork is innate to the human experience; we are born with the capacity to work together toward common goals. However, there has been a tremendous amount of research done to try to understand how exactly it is that teams can translate knowledge and skills into performance outcomes. Consider that a team of experts is not necessarily an expert team; science has regularly shown that simply bringing together a group of experts is not the same (and not always as effective) as having a group that expertly works together (Burke, Salas, Wilson-Donnelly, & Priest, 2004; Eccles & Tenenbaum, 2004; Reyes & Salas, 2019; Salas, Cannon-Bowers, & Johnston, 1997). There is more to teamwork than simply knowing what to do. Effective teams can do a multitude of subtle things that allow them to work together most efficiently, sharing knowledge and resources when needed, and deftly adapting to new situations to eventually achieve their goals. A tremendous amount of research has been conducted to try to understand the nuances of what it is that makes teams successful or unsuccessful, and it appears that so long as there are teams being formed to solve new problems, there will be new questions about why and how they work together!

In this chapter, we use "agent" as a catch-all term that encompasses robots, autonomy, automation, or otherwise any computer-based entity that can serve as a team member in some capacity (O'Neill, McNeese, Barron, & Schelble, 2020). In other words, the agent is interdependent with the other team members (which may be agents or humans), and all team members share some common goal.

How are robots, autonomy, and automation different? The answer has been the subject of some debate. Robots are characterized by a physical form that allows them to interact with the environment (Baker, Phillips, Ullman, & Keebler, 2018), whereas automation and autonomy are the programs, algorithms, and computational intelligence systems by which data inputs are processed into decisions that may be acted upon. In general, automation is seen as less sophisticated, involving the rote execution of some functions and tasks that may have been previously held by a human (Parasuraman & Riley, 1997; Parasuraman, Sheridan, & Wickens, 2000). By contrast, autonomy is considered to involve a higher degree of machine intelligence and a greater capability for the system to make decisions independent from oversight.[1] Indeed, it is often argued that autonomy is a more advanced level of automation (Endsley, 2017; Hancock, 2017). As a result, within the last decade or so, there has been a noticeable evolution in how we think about agents in the context of a team, with some characterizing this change as "from tools to teammates" (Bradshaw et al., 2009; Phillips, Ososky, Grove, & Jentsch, 2011). The change has coincided with advancements in agents that allow them to make more complex decisions and plans and share those with humans. Regardless, not every application will require a highly intelligent agent, and there will always be a need for simpler autonomy that carries out repetitive but important tasks, such as package sorting or automotive assembly.

Many factors play into one's perception of an agent as a tool or a teammate, such as its appearance, its methods of communicating information, and whether we believe that the agent has some freedom of determining its own actions (Wynne & Lyons, 2018). Likewise, that perception can affect other things, such as how people interact with the agent, or whether they trust it to carry out certain functions. Consequently, *those* things can partly determine how well a human and an agent can work together to produce some outcomes. The intertwined nature of all these factors is not surprising and leads us to a key principle at the heart of this chapter.

The input-process-output (IPO) model is a way to explain how teams work (Figure 11.1). Inputs consist of things that the team "starts with", such as knowledge or certain skills. Processes are things that teams "do", such as resolving disputes, making decisions, or sharing information. Outputs are everything that comes out of the team, such as the number of widgets they produce, or how accurately they identified certain targets.

The IPO model represents how teamwork translates inputs into outputs through team processes. It was first studied in the 1970s and the 1980s when scientists began to realize that teamwork inputs do not simply lead straight to outputs, and that the processes of teamwork were a vital piece of the puzzle (Hackman, 1987; McGrath, 1984; Steiner, 1972). This model has been more recently refined by Ilgen et al. (2005) who argued that the IPO model is too simple to capture some of the nuances of team dynamics. They recognized that sometimes the factors that transmit the influence from inputs to outputs are not processes, but rather cognitive or affective states that emerge from the team's interactions, such as trust or a sense of team cohesion (and thus these are referred to as "emergent states"). They also argued that the IPO model implies that teamwork is inherently linear and ends at the output, whereas real team dynamics seem to be cyclical and may not always be in order from Inputs to Processes to Outputs. If you imagine working with a team to identify suspicious packages, each package you classify correctly or incorrectly might be considered an "output" that gives your team a small bit of information. That information builds your knowledge of what kinds of packages

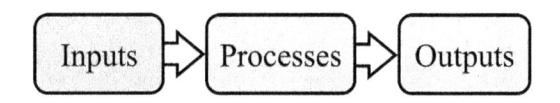

FIGURE 11.1 Input-Process-Output model of teamwork.

FIGURE 11.2 IMOI model of teamwork.

you can expect to see (your team's collective knowledge is considered an "input"). These points all illustrate that teamwork can perhaps be better characterized using a different model, and this one is called the Inputs-Mediators-Outputs-Inputs (IMOI) model (Ilgen et al., 2005) (Figure 11.2).

There are a few changes from the IPO model. First, the middle term is now called "Mediators", which consists of both processes and emergent states. Second, there is a link from outputs back to inputs. Third, there are bidirectional arrows between the main three steps, implying that the factors can sometimes affect each other in a nonlinear manner. This model is better able to represent the varied and often complex dynamics of teamwork and has seen wide support across the literature. The IMOI model has also served as a useful base for representing H-A team effectiveness. For example, it was recently applied to the context of human-robot teams, and key factors and relationships were more clearly distinguished (Esterwood & Robert, 2020; You & Robert, 2017). Figure 11.3 contains a version of that model, which has been adapted for the context of H-A teaming (recall that "agent" is broader than "robot").

In this model, the cyclical nature of the IMOI model is preserved, and inputs are now classified into several groups, such as team-level factors or those specific to just the human(s) or agent(s). Altogether, the model is better able to characterize some of the unique aspects of H-A teaming, such as the influence of distinct inputs that come from both the human and the agent.

We have so far reviewed some of the basic concepts of what makes a team as well as how teamwork can be modeled. The next sections of this chapter will detail some of the important factors and concepts in H-A teaming to both streamline our discussion and ground it in teamwork science. First we discuss Inputs, then Mediators, then Outputs. Taking good measurements is a high priority in this field, so we will also discuss Assessments to understand how and why some constructs are measured in H-A teaming. We note here that, according to the current and ever-evolving science of teams, some of the factors do not strictly fall under one category; for example, depending on the context, trust can be considered a relatively unchanging general disposition toward others

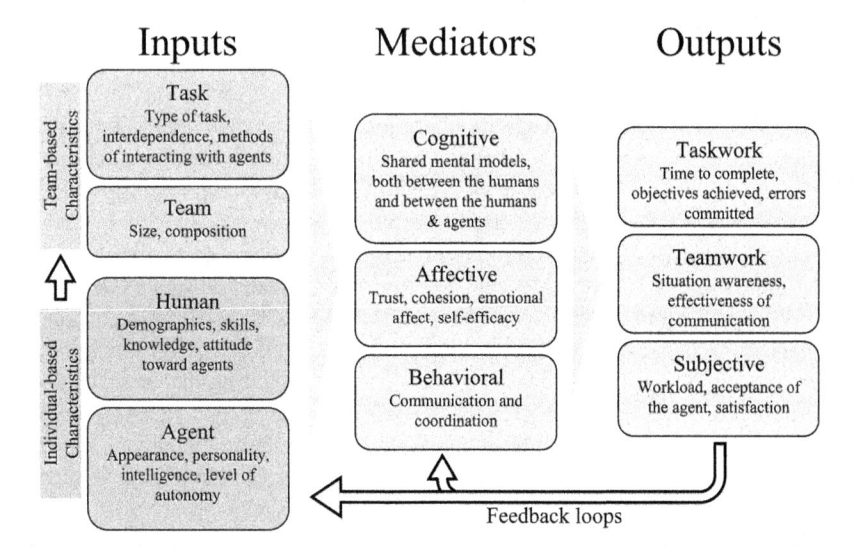

FIGURE 11.3 An IMOI model of human-agent teaming. (Adapted from You & Robert, 2017).

(i.e. an input), a belief that arises from an interaction with someone else (a mediator), or even an end state that a system is designed to elicit from a person (an output)! The topic of factors influencing multiple steps in the IMOI cycle is beyond the scope of our discussion, so to simplify the format of this chapter, we will present each factor under the step (input, mediator, or output) in which it might be commonly studied for H-A teaming.

11.2 INPUT FACTORS

When a team is formed, it never starts as a blank slate, even if the team members don't know anything about each other or the task they are supposed to complete. Every team member brings something to the table. These are known as inputs: properties of the individuals, the team, or their environment that exist when the team is formed. They are characteristics like a person's knowledge and skills, or a team's size, or its assignment of roles and tasks. In H-A teaming, the agent's inputs are also relevant to the team; these are features such as the agent's reliability, appearance, or how autonomous it is. In this section, we will review some of the commonly studied inputs that are associated with humans, agents, and the team as a whole.

11.2.1 HUMAN-BASED INPUTS

This first group of inputs relates to properties associated with the humans in a team. Here, we will describe expertise and training as well as individual and cultural differences.

11.2.1.1 Expertise and Training

Expertise is a body of well-organized, task-relevant knowledge (Sonnentag & Volmer, 2009). While expertise among individual team members is not a guarantee of team effectiveness, it can still have positive effects. At the most basic level, a team member who is an expert can accomplish a task under their purview as an individual (Burke, Salas, Wilson-Donnelly, & Priest, 2004; Sonnentag & Volmer, 2009). While this refers to an individual completing a task, it is the team that technically produces the outcome. Thus, a team of experts on the same topic can accomplish tasks associated with that topic more effectively. The distribution of expertise within a team can influence how well the team can handle certain tasks and situations. For example, knowledge homogeneity (where everyone has a high level of knowledge about the same topic) among team members can yield effective team performance (Cooke, Salas, Kiekel, & Bell, 2004). In contrast, more heterogenous teams (composed of individuals who all have different areas of expertise) can tackle a wider array of tasks. This functional diversity of expertise can lead to greater innovation and faster response to organizational change, though it can also lead to more conflict and reduced information sharing (Bunderson & Sutcliffe, 2002; Mathieu, Maynard, Rapp, & Gilson, 2008). Through cross-training—where each team member is trained on the tasks, duties, and responsibilities of their team members—team members can gain an understanding of what their team members know and what they need to know, even if they don't reach an expert-level body of knowledge (Blickensderfer, Cannon-Bowers, & Salas 1998). Furthermore, when a team member knows which of their teammates are experts in certain areas, they can use that information to coordinate more effectively (DeChurch & Mesmer-Magnus 2010; Fiore & Wiltshire, 2016).

Much like how cross-training with one's teammates can help a team member understand their teammate's knowledge base and responsibilities, training with an agent can help a teammate align their thought processes with the agent and reduce their uncertainty regarding the agent. Agents, like their human counterparts, can bring expertise to the team, with concomitant benefits and drawbacks. In one study looking at expertise in H-A teams, when the agent had expertise in an area in which the human teammate underperformed, teams had higher performance and the human teammate reported greater satisfaction with the agent, in comparison to H-A teams where agents held expertise in the same field as their human teammates (Abuhaimed & Sen, 2023). Like with human teammates, an accurate understanding of an agent team member's capabilities and expertise is useful for, if not critical to, the success of an H-A team (Salas, Shuffler, Thayer, Bedwell, & Lazzara, 2015; You & Robert, 2017).

11.2.1.2 Individual and Cultural Differences

Individual differences between people are another key input and play a considerable role in how other aspects of teamwork play out. Significant individual differences in human cognitive task performance and interaction with automation are well documented in the literature (Chen & Barnes, 2014; Manzey, Reichenbach, & Onnasch, 2012; Matthews, Lin, Panganiban, & Long, 2019; Ingram, Moreton, Gancz, & Pollick, 2021; Lyons & Guznov, 2019; Pop, Shrewsbury, & Durso, 2015; Rovira, Pak, & McLaughlin, 2017). Based on an extensive literature review, Chen and Barnes (2014) identified factors such as operator attentional control, spatial ability, and gaming experience that play important roles in H-A team performance. In addition to these factors, working memory capacity has also been found to impact human trust and task performance when working with intelligent agents, particularly those who are unreliable (Rovira et al., 2017). Similarly, individuals' differences in propensity to trust may impact their interactions with intelligent systems: those with a higher propensity to trust tend to trust unreliable agents more than those with lower propensity (Ingram et al., 2021). Research has also found that individual differences in cognitive abilities can sometimes have a greater impact on H-A teaming than the effects of any interface design manipulations (Rodes & Gugerty, 2012).

Individual difference factors related to the human have also been found to impact their interaction with machine agents (Lyons & Guznov, 2019; Matthews et al., 2019). For example, Lyons and Guznov (2019) investigated the relationship between the perfect automation schema (PAS; a measure of one's general attitudes toward robots) and human trust in intelligent agents in two different contexts: human–robot interaction and F-16 pilots' perception of the automatic ground collision avoidance system. Lyons and Guznov found that the PAS's high expectations facet had a positive relationship with trust. In other words, participants who reportedly held high expectations of the capabilities of agents were more likely to also have a higher degree of trust in the agent systems. Matthews and colleagues (2019) examined how individual differences in attitudes toward robots in general (measured by the PAS and the Negative Attitudes toward Robots Scale [NARS]) were related to their forming of different mental models of a robot's analytic performance in experimental scenarios. Matthews et al. found that participants' general attitudes toward robots (via the PAS) were associated with their trust perceptions of the robot in the experimental scenarios. On the other hand, the NARS was found to be predictive of trust, particularly the psychological aspects in the experimental contexts. The authors concluded that human biases (e.g. unreasonable expectations of robot capability or negative attitudes toward human-like robots) could potentially influence their trust and situation awareness during H-A teaming.

At the cultural level, Chien and colleagues (2019) examined H-A teaming in three distinct cultural backgrounds, based on the Cultural Syndromes Theory developed by the authors, which classifies cultures into Dignity, Face, or Honor based on the meaning and importance a culture assigns to "norms of exchange, reciprocity, punishment, honesty, and trustworthiness" (p. 2). Participants from the United States (Dignity), Taiwan (Face), and Turkey (Honor) experienced a simulation-based task in which they worked with an intelligent planning agent to manage a team of five unmanned aerial vehicles (UAVs) and, simultaneously, identify/attack hostile targets and reroute the UAVs when necessary. The intelligent agent's levels of automation (discussed in Section 2.2.1) and transparency (discussed in Section 3.3.2) were also manipulated. The experimental results showed that agent transparency had an impact on operators' interactions with the planning agent and their compliance with the agent's recommendations, but the effects of agent transparency were significantly influenced by participants' culture. For example, face culture participants had a higher tendency to accept recommendations from an opaque agent than did participants from the other two cultural backgrounds. These results suggest that when transitioning autonomy technologies from one culture to another, user interface modifications and training interventions may be required due to the effects of cultural differences on system reliance related to agent transparency.

11.2.2 AGENT-BASED INPUTS

This group of inputs relates to the characteristics of the agent that influence the team's mediators and outputs. Important agent characteristics that we will review are its level of autonomy and its reliability.

11.2.2.1 Level of Autonomy

Broadly, the agent's level of autonomy relates to how well it can complete tasks without human supervision or control. Partly owing to the wide variation in agent designs, functions, and intended applications, there are several different ways of understanding an agent's level of autonomy. Perhaps the earliest schema for describing levels of automation was proposed by Sheridan and Verplank (1978). They outlined ten levels of increasing agent capability, with the lowest level consisting of all decisions and actions being taken by the human with no computer assistance, and the highest level consisting of the computer handling all decisions and actions while ignoring the human. In the levels between those extremes, the agent goes from offering some decision/action alternatives, to executing some suggestions with approval, to executing things automatically, eventually doing so without informing the human unless requested otherwise.

A later model by Parasuraman and colleagues (2000) improved on this by representing core H-A activities as information processing, consisting of four steps: the acquisition of information, analysis of that information, a decision being made, and then an action being taken. In this way, each discrete step can have its own level of autonomy, making it easier to classify agents based on the kind of control they have over different steps in the process of gathering and acting on information. The autopilot system on a modern airliner is capable of a high level of autonomy for all information processing steps and is capable of controlling every part of a flight except for takeoff (autopilot systems on some airliners are capable of landing the plane, but this functionality is not commonly used). By comparison, many commercial and general planes also have Terrain Avoidance and Warning Systems (TAWS) that can calculate when a plane is potentially dangerously close to the ground, providing warnings to raise the altitude to avoid a possible crash. This system does not take any action on the plane, so it can be said that the TAWS has a high level of autonomy for information acquisition and analysis, a lower level for decision-making, and no autonomy for taking action.

A similar model of levels of autonomy was proposed by Endsley and Kaber (1999). This model also separates H-A activities into four functions: in this case, *monitoring* the environment, *generating* options and strategies, *selecting* from those options and strategies, and *implementing* the chosen selection. Then, the authors differentiate between ten possible levels of automation that each has increasing degrees of autonomy within those functions. At the lowest level, the humans are in full manual control, and from there, the system increases from serving as action support, to assuming shared control, providing decision support, eventually having supervisory control just before the highest level, at which the system is considered to have Full Automation.

Regardless of the model used to describe an agent's level of autonomy, the different responsibilities assigned to the agent can significantly affect the execution of teamwork. Having a "human in the loop" is a critical concept in H-A teaming: it is often crucial to have at least one human aware of the agent's activities, whether to maintain a shared understanding of the team, or to be able to take control from the agent in the event of a failure or other adverse event (Endsley, 1995, 2017). Improperly implementing high levels of automation can make it extremely difficult to respond to problems with the agent's activities, as the sudden transition of workload from agent to humans can place huge demands on a team structure that is not prepared to handle it (Onnasch, Wickens, Li, & Manzey, 2014). No matter the level of automation implemented for an agent, maintaining situation awareness[2] among the H-A team is key to ensuring that they are able to best take advantage of the agent's capabilities while being well-positioned to adapt to adverse events by transitioning workload between the agent and humans and executing appropriate backup behaviors (Endsley & Kaber, 1999; Salas, Sims, & Burke, 2005).

FIGURE 11.4 Would you trust your car to drive itself? (Photo courtesy Ian Maddox (2017), CC BY SA 4.0.)

The past decade has seen an explosion in the number of cars on the road that have at least some autonomous capabilities, such as adaptive cruise control that can speed up or slow down to match the car in front of you. Now that you know more about the (sometimes very blurry) differences between automation and autonomy, you might have a deeper understanding of the nuances of self-driving cars.

Earlier, you read about ways to characterize levels of autonomy. In cars, different levels of autonomy correspond to the number of functions that are allocated to the car versus to the driver. SAE International (an international organization that develops standards for engineering and the automotive industry) has established six levels of automation in cars, ranging from Level 0 to Level 5, with each level adding more capabilities to the car (SAE International, 2021). Levels 0–2 consist of "driver support features", meaning the driver is still considered to be in control and the autonomous capabilities must be constantly supervised to maintain the appropriate degree of safety. These levels include vehicle capabilities like automatic emergency braking, automatic lane centering, and adaptive cruise control.

For Levels 3–5, the available autonomous capabilities instead consider the driver as "not driving" when the capabilities are engaged. These automated driving features enable the car to maneuver in traffic jams or serve as local driverless taxis, and the car may possibly be sold without a steering wheel or pedals. At the highest level of autonomy, the car is expected to be able to drive everywhere in all conditions. Until we reach that highest level of autonomy, it is a wonder that there are infamous examples of drivers asleep at the wheel of the most advanced "self-driving" cars available today, which have only reached SAE Level 2 in the eyes of manufacturers and regulators.

With guidance from the National Highway Traffic Safety Administration, manufacturers are developing these autonomous driving capabilities and many new cars sold today include at least the lowest autonomy level features (automatic emergency braking, blind spot warnings, & lane departure warnings). Ultimately, these systems are promising in the fight to reduce traffic fatalities, which were estimated at almost 43,000 in 2022 alone (NHTSA, 2023).

11.2.2.2 Agent Performance and Reliability

Another agent-based input factor that can impact H-A teaming is the agent's performance, and in particular, the agent's competency, reliability, and integrity (Caldwell et al., 2022). A number of meta-analyses have identified agent performance as a key antecedent of human trust in the agent (Hancock et al. 2011; Schaefer, Chen, Szalma, & Hancock, 2016; Hancock, Kessler, Kaplan, Brill, & Szalma, 2021). In these analyses, several factors have been found to impact human trust in the agent: performance/capability (e.g., failure rate, false alarm rate), behavior, reliability, dependability, predictability, and so on. These analyses also show that performance-related factors are consistently found to have greater influence on human trust than factors related to agent attributes, human partners, and environments/contexts. We will focus on reliability and will discuss the effects of agent reliability on H-A team performance.

Empirical evidence shows that humans tend to "merge" their trust across system components with various levels of reliability, meaning their trust in higher-reliability components of the system can be degraded by components with lower reliabilities (Keller & Rice, 2010). In order to deal with this ineffective trust calibration, it has been suggested that greater system transparency (see Section 3.3.2) should be provided to the operator (e.g., raw information inspection, queries, etc.) so they can verify the system's output (Rovira et al., 2007). Additionally, "what-if" simulations (e.g., testing different options, adjusting weights of constraints) have been found to benefit H-A team performance (Cabour, Morales-Forero, Ledoux, & Bassetto, 2022; Clare, Cummings, How, Whitten, & Toupet, 2012).

Empirical evidence also shows that humans' loss of trust in unreliable machine agents can be influenced by the agents' degree of anthropomorphism (de Visser et al., 2016). Anthropomorphic agents (compared with the more machine-like agents) were found to support better trust resilience, especially when there was greater uncertainty associated with the tasking environment. Human-like trust repair behaviors (e.g., apologies, explanations, etc.) can be used to mitigate human operators' loss of trust in unreliable agents. In the context of automated driving, Kraus et al. (2020) also found that agent transparency (e.g., information about system unreliability) could mitigate trust degradations after system malfunctions. In contrast to the above findings, Wright, Chen, & Lakhmani (2020) found that reliability of a (robotic) agent, even with a higher level of transparency, had a profound effect on human teammates' perceptions of the agent – the unreliable agent was perceived as less animate, likable, intelligent, and safe, compared with the more reliable agent.

Continuing from our discussion of individual differences in Section 2.1.2, the effects of agent reliability on H-A team performance can be modulated by individual differences factors. For example, Ingram et al. (2021) found that participants with a higher propensity to trust (as measured by their responses to the Propensity to Trust Machines Questionnaire by Merritt et al. (2013)) were more likely to trust an image classifying agent than their lower-propensity counterparts, particularly when the agent was incorrect or under uncertain situations (e.g., poor image quality).

11.2.3 TEAMWORK-BASED INPUTS

While the previous sections have discussed individual-level inputs such as expertise and levels of autonomy, the input-mediator-output model also accounts for inputs that stem from team-level characteristics such as team structure and task interdependence (Mathieu, Maynard, Rapp, & Gilson, 2008; You & Robert, 2017). The following two sections will discuss task allocation and roles within a team.

11.2.3.1 Task Allocation

Important to teamwork is the allocation of tasks, roles, and responsibilities, either within a team or across multiple teams (Jobidon, Turcotte, Aubé, Labrecque, Kelsey, & Tremblay, 2017). When it comes to task allocation in a team structure, if the tasks that need to be done are highly interdependent and require specialized team members, then the task allocation structure is considered *functional*; however, if the tasks can be done more independently and team members have a broader set of responsibilities, then that task allocation is *divisional* (Hollenbeck, Ellis, Humphrey, Garza, & Ilgen, 2011). Given that agents are particularly well suited to specialized, highly structured, and repetitive tasks, they can fulfill a needed niche is certain teams (Sebo, Stoll, Scassellati, & Jung, 2020).

One of the progenitors of H-A teaming research is Fitts' (1951) list. Also known as HABA-MABA (Humans Are Better At; Machines Are Better At), Fitts' list attempts to characterize the general strengths and weaknesses of humans and machines (Bradshaw, Feltovich, & Johnson, 2012). By determining the tasks (also referred to as functions in this context) in which machines would excel and the tasks in which humans would excel, Fitts provided guidance for how early human-machine teams could allocate tasks, so that the team outcome could be better, reached faster, and done more frugally (Bradshaw, Feltovich, & Johnson, 2012; de Winter & Dodou, 2014). Both function allocation research and technology have progressed since Fitts' time, however. Advancements in technology have led to agents being able to take on a broader set of responsibilities and complete a wider variety of tasks, allowing agents to contribute to more tasks for their teams (O'Neill, McNeese, Barron, & Schelble, 2020).

As more advanced agents have been developed, agents have been able to respond to changes in dynamic, complex environments by changing their behavior or even adjusting their level of autonomy (discussed earlier in Section 2.2.1). An agent's ability to take on different tasks and adjust their level of automation in response to a changing environment is known as dynamic task allocation or adjustable autonomy (Johnson et al., 2014). This method of H-A interaction, where humans and agents dynamically adjust their task allocation depending on the needs of the team or the environment, allows the flexibility for each agent to contribute to tasks in which they perform best (Hearst, Allen, Horvitz, & Guinn, 1999). This strategy can facilitate collaborative problem solving, coordination, and shared cognitive decisions (Hearst, Allen, Horvitz, & Guinn, 1999; Bradshaw et al., 2003; Johnson et al., 2014). However, to enable teams to use such a dynamic task allocation strategy, it is critical to have thorough monitoring of many of the team's processes and interactions, as those data streams are needed to understand when or how to adapt to changing task conditions (DeCostanza et al., 2018). With sufficient data about human and agent team members, adaptive strategies can allow a H-A team to dynamically balance workload across team members, enabling the team to handle more complex situations (Mina, Kannan, Jo, & Min, 2020). For both human and H-A teams, the allocation of tasks and responsibilities can affect team mediators and outputs, leading to positive outcomes when all actors—human and agent—can contribute what they do best to the larger group effort.

11.2.3.2 Roles

In a team, a role can be defined as a cluster of related and goal-directed behaviors that someone takes on within a certain context (Mathieu, Tannenbaum, Kukenberger, Donsbach, & Alliger, 2015). Roles can be more *personal*, focusing on the social-psychological dynamics of a team, or more *positional*, focusing on the characteristics of the job and responsibilities of the position. With the advancement of technology, agents are growing better suited at taking on the responsibilities of human teammates and behaving in ways expected of those inhabiting those roles (Bradshaw et al., 2009). An agent, however, can't be considered a one-to-one replacement for a human teammate. First, to fill a role within a team, an agent must be recognized by its human teammates as a distinct entity acting in that role (O'Neill, McNeese, Barron, & Schelble, 2020). Second, the social-psychological dynamics of a team still exist. Once an agent is recognized as a team member with a role, it will be subject to the group dynamics within a team (Ososky, Schuster, Phillips, & Jentsch, 2013).

In a team dynamic, the agent must exist within the team's hierarchy as a subordinate, a peer, or a supervisor. An agent subordinate is the most common role distribution in H-A teams. In this team iteration, humans team members delegate specialized functions to a subordinate agent, usually with final authority over any decisions (Chen & Barnes, 2014); this approach can lead to human team members taking more responsibility for successful task completion but can in turn lead to amplified cognitive demands (Endsley, 2017; Hinds, Roberts, & Jones, 2004). While agent subordinates are more frequently in use currently, creating and working with an agent peer is a goal that many are pursuing. As a peer, an agent interacts on the same footing as their teammates and would thus—in addition to any specialized functionality—need to allocate tasking, coordinate responsibilities with their teammates, and share information (Hearst, Allen, Horvitz, & Guinn, 1999; Bradshaw et al. 2003; Salas, Grossman, Hughes, & Coultas, 2015). The final, and least used, H-A interaction pattern is that of the agent supervisor. In previous human-robot dyad research, humans have been found to be critical of robot supervisors, blame robot supervisors for failure, and cede task responsibility to robot supervisors (Hinds, Roberts, Jones, 2004; Lei & Rau, 2021). It seems that further development is needed before humans can be satisfied with agent supervisors. In the end, using prespecified roles, and the expectations that come with them, can be a way to integrate agents' inputs with human inputs in a way that leads to successful teamwork.

11.3 MEDIATORS

Mediators are all the things that teams think, know, and do during the course of teamwork. More accurately, mediators consist of two types of factors: processes and emergent states. Processes relate to the team's actual interactions, and are things that teams do, like make plans or resolve conflicts. Emergent states are the cognitive or affective states that emerge during teamwork, like a sense of trust or cohesion. For H-A teaming, processes and emergent states can be categorized into three groups: attitudes, behaviors, and cognition. This section will highlight some of the key mediators in H-A teaming that fall into each category.

11.3.1 ATTITUDES

The attitudes that a person holds toward others and toward the team can influence how well they can function within that team. Two of the most important attitudes that have received significant attention in teamwork science are trust and cohesion. What does it mean to trust someone else? As it turns out, the answer is quite complicated.

11.3.1.1 Trust

Across decades of trust research, this concept has usually been defined as a willingness to be vulnerable based on expectations about the behaviors or outcomes of another party, (Barber, 1983; Mayer, Davis, & Schoorman, 1995; Rempel, Holmes, & Zanna, 1985; Rotter, 1967). In other words, trust can be considered an emergent state of accepting some risk on the belief that another party will do something. In a group project, you trust your partners to complete their parts of the project when they say they will, at risk of missing the deadline or receiving a poor grade. Likewise, you trust your smartphone's map app to route you to your destination in a timely manner.

You don't trust your smartphone the same way you trust your group project teammates, however. Indeed, there are many different ways of characterizing trust. Two of the common trust types are called affective and cognitive trust. Affective trust is the "traditional" type that develops across many shared experiences and is based on emotion (McAllister, 1995; Rempel et al., 1985; Rousseau, Sitkin, Burt, & Camerer, 1998). In contrast, cognitive trust involves assessments that another party is competent and reliable and is based on more logic and reasoning (Erdem & Ozen, 2003; Lewis & Weigert, 1985; McAllister, 1995). Other ways of conceptualizing trust have involved how it quickly forms in new teams (swift trust; Meyerson, Weick, & Kramer, 1996) and how it can be multidimensional, such as how we are able to trust a close friend to keep secrets and support us but completely distrust their ability to competently drive a car (Lewicki, McAllister, & Bies, 1998).

Trust has received much attention because it plays a very important role in how team members interact (Krausman et al., 2022; Mathieu, Maynard, Rapp, & Gilson, 2008). Trust forms the foundation upon which teams can rely on each other. Interpersonal trust has been found to be important for maintaining effectiveness in individual relationships and on an organizational level (McAllister, 1995). Trust in one's team is positively associated with job satisfaction (Martins, Gilson, & Maynard,

2004) and team satisfaction (Costa, 2003). Most importantly, across many different fields of study, trust has repeatedly been associated with the performance of a team (Breuer, Hüffmeier, & Hertel, 2016; de Jong, Dirks, & Gillespie, 2015; Feitosa, Grossman, Kramer, & Salas, 2020).

Agents are obviously different than humans, and so the way we trust them (and define that trust) is different as well. Generally, the most important factor that decides whether you trust an agent is its reliability (Hancock et al., 2011; Lee & See, 2004; Schaefer, Chen, Szalma, & Hancock, 2016), so the more consistently an agent does what it is supposed to do, the more likely you are to trust it. To a lesser extent, other factors play into trust toward an agent, like its transparency and explainability (see Section 3.3.2), its appearance and personality, its behavior, the level of risk in the task, and one's general attitude toward robots and technology (Hancock et al., 2011; Kaplan, Kessler, Brill, & Hancock, 2021). External and organizational factors can also affect the trustworthiness of an intelligent agent; for example, sound software engineering principles, good organizational practices toward the use of agents, and external oversight and certification of systems can all foster agent trustworthiness (Shneiderman, 2020).

To ensure that an H-A team can perform well, people should have high trust in agents that are extremely reliable, and they should avoid trusting agents in situations when they are unreliable. This is a concept called trust calibration: human trust should be calibrated to the agent's range of capabilities (Lee & See, 2004). When a human user understands the limitations of an agent and interacts with it within the bounds of those limitations, then the human has appropriately calibrated trust (Lyons, 2013). However, a human's trust in an agent can be inappropriately calibrated, which can cause difficulties. If a human's trust in an agent's capabilities exceeds the agent's actual capabilities, then that is known as overtrust. Overtrust can lead to misuse of an agent—e.g. using an agent when it is inappropriate to do so—overreliance on the agent, and complacency (de Visser, Cohen, Freedy, and Parasuraman, 2014; Lee & See, 2004). If a human's trust in an agent is inappropriately calibrated in the other direction, where the human's trust in an agent falls below the agent's capabilities, then that is known as distrust. Distrust, or the expectation of agent incompetence, can lead to disuse—ignoring the agent and doing the task manually—thus removing the advantages of having an agent in the first place (Lee & See, 2004; Muir, 1994). Maintaining an appropriately calibrated trust in an agent is particularly important in H-A teams, because trust can influence a number of team processes, including cooperation, information sharing, the building of affective bonds, and communication (Mesmer-Magnus, DeChurch, Jimenez-Rodriguez, Wildman, & Shuffler 2011; Salas, Shuffler, Thayer, Bedwell, & Lazzara, 2015; Sebo, Stoll, Scassellati, & Jung, 2020). As a mediator, trust can impact team outcomes, influencing the effectiveness of team training on performance outcomes (Mathieu, Maynard, Rapp, & Gilson, 2008). Trust also mediates the relationship between psychosocial factors—psychological similarity and shared activities—and job satisfaction and team performance (Lu, 2015).

11.3.1.2 Cohesion

Like trust, cohesion is considered to be one of the most important properties of a group. Unsurprisingly, it is also one of the most widely studied characteristics of a team (Dion, 2000; Kozlowski & Ilgen, 2006). Over that period of study, cohesion has had numerous definitions. The earliest definitions emphasized the "field of forces" that keep members together in a group and inoculate the group against disruptive forces (Festinger, Schachter, & Back, 1950; Gross & Martin, 1952). Later definitions emphasized that cohesion reflects a group's commitment to an overall task (Goodman, Ravlin, & Schminke, 1987) or a group's tendency to stick together in pursuit of its objectives (Carron, 1982) and/or satisfying the group members' affective needs (Carron & Brawley, 2000). While older definitions reflected a more unidimensional approach to cohesion (Mudrack 1989), later definitions started to reflect a multidimensional concept of cohesion (Salas, Grossman, Hughes, & Coultas, 2015). While there are a number of different dimensions and subdimensions of cohesion (Lakhmani et al., 2022a), the two that tend to be examined most often include task cohesion—an attraction to the group based on a shared commitment toward completing a group's goal—and social cohesion—an attraction to the group based off of the social relationship within

FIGURE 11.5 Military robot being prepared to inspect a bomb.

Cohesion and trust are affective states that grow over time as teammates share experiences and learn more about each other. Humans can likewise gain new perspectives on their agent teammates after working together over a longer span of time. Interviews with explosive ordnance disposal (EOD) military personnel have explored how they feel toward the bomb disposal robots they work regularly with, finding that the EOD specialists often develop a surprising attachment toward their robot teammates.

Singer (2009) found that EOD squads regularly assigned names to their robots. Some squads gave their robots military promotions and even awarded them honorary medals for being "wounded" in combat. These behaviors are a way to give life and purpose to a robot, reflecting the robot's deep importance to its squad's safety and survival. Indeed, the EOD specialists often reported expressing feelings of gratitude toward their robots for dealing with an uncountable variety of dangerous situations. From interviews with 23 EOD specialists, Carpenter (2013) also found that they experienced feelings of loss and melancholy when the robots were damaged or destroyed while disarming explosives. In a similar vein, there have even been reports of specialists holding funerals for destroyed robots, as a final symbolic gesture. When a H-A team can build an important history of experiences together, the lines between human and agent can become blurred.

the group (Dion, 2000; Zaccaro, Gualtieri, & Minionis, 1995). In the current literature, multidimensional approaches to cohesion are more often used (Salas, Grossman, Hughes, & Coultas, 2015).

In H-A teams, agents can help team cohesion in one of three ways. First, an agent can help individual teammates with their actual tasks (Sycara & Sukthankar, 2006; Sukthankar, Shumaker, & Lewis, 2011). By helping individual teammates complete shared goals, agents can help the team's overall cohesion (Lu, 2015; Zaccaro, Gualtieri, & Minionis, 1995). The second way that an agent can help promote team cohesion is by helping human team members perform behaviors that lead to a more cohesive team (Sukthankar, Shumaker, & Lewis, 2011). An agent that makes it easier to share information, or that facilitates coordination, can yield a more cohesive team. The third way that an agent can help team cohesion is by acting as a team member and behaving in ways that promote team cohesion (Lakhmani et al., 2022a; Sycara & Sukthankar, 2006). While an agent can't feel cohesion (or much of anything really), it can promote those feelings in others through its actions and be the subject of others' feelings (Ososky, Schuster, Phillips, & Jentsch, 2013). However, introducing an agent as a teammate can complicate matters, as humans and agents can differently influence the team's cohesion.

One way that introducing an agent to a human team can inadvertently influence the team's cohesion is by affecting group dynamics. Recalling our previous discussion of roles (Section 2.3.2), the agent's position in the hierarchy can affect cohesion. One study found that people were more critical of a robot supervisor, giving it less credit and more blame than a robot that served as a peer or a subordinate (Hinds, Roberts, & Jones, 2004; Lei & Rau, 2021). Another study found that a participant, when working with both a human and a robot, attributed more blame for failures to the robot than the human (Lei and Rau, 2021).

Another way that an agent can influence a team's cohesion stems from how it interacts with its human teammates. Humans can interact with others using speech, non-verbal communication methods, and using technology. Agents *are* technology, and the ways in which they can interact with humans are constrained by their design and capabilities. If humans are able to use multiple media-rich methods to exchange information via multiple channels, while agents are limited to poorer communication methods and fewer channels, then of course the agent can have a negative influence on the team's cohesion (Lakhmani et al., 2022a; Wilson, 2014). Given the myriad of complex effects that introducing agents to a team can have on that team's cohesion, a great deal of work is being done to understand and measure potential effects (Abrams & Rosenthal-von der Pütten, 2020; Lakhmani et al., 2022a; Neubauer et al., 2021; Walliser et al., 2019; Wang et al., 2020).

11.3.2 BEHAVIORS

In a team, communication is a mediator that can be observed directly from peoples' behaviors. Arguably, it is the most crucial component of teams; one can barely imagine working together as a team without being able to communicate anything. Indeed, teams of divers often rely on dive slates or hand signals to exchange information without having to speak, and as a last resort, they coordinate dive plans ahead of time in order to execute their tasks in the event that one becomes unable to communicate.

You may have noticed from that example that communication involves both a behavioral component (the actions taken to share information, like speaking or using hand signals) and a cognitive component (the information that then changes how people think). We will first discuss communication as a behavior, setting the groundwork for the next section about team cognition.

Scientifically speaking, communication is the exchange of information between team members using verbal or non-verbal channels (Adams, 2007; Mesmer-Magnus & Dechurch, 2009). It allows teams to address conflicts, align toward specific goals, and synchronize information across different sources (Marks, Mathieu, & Zaccaro, 2001; Salas, Cooke, & Rosen, 2008; Salas et al., 2005). As information is shared, our knowledge and understanding evolve, shaping other team processes and outcomes (Kozlowski & Ilgen, 2006; Mathieu, Heffner, Goodwin, Salas, & Cannon-Bowers, 2000). Teams can communicate among themselves, with other teams, or with larger organizations above or outside of themselves, so there are myriad factors that can affect how well a team communicates and whether they experience good or poor outcomes as a result.

Communication can be broken down into simpler components. At its core, the act of communicating involves a sender (e.g., a person speaking), a message, a receiver (e.g., someone listening), and a channel through which the message is sent, such as a phone call, an e-mail, a whisper, or so on. This is every component needed to send a message from one party to another, and this process occurs back and forth as information is shared. We take turns in a predictable manner during conversations, in order to let conversation flow naturally between speakers (Barnlund, 2008; Jurafsky & Martin, 2007; Sacks, Schegloff, & Jefferson, 1974).

In H-A teams, the agent introduces unique capabilities to the team, but likewise also introduces challenges; many of these challenges stem from how the agent and the rest of the team interact. H-A team communication has to fit within the agent's capabilities, and the communication capabilities of agents are evolving extremely rapidly. That said, agents can communicate through many different forms of input in order to team effectively with humans. Most commonly, H-A communication is carried out via a computer interface or a remote controller, such as the controllers used for EOD

ChatGPT, a groundbreaking new technology, has been recently released by the research company OpenAI. The ChatGPT system (OpenAI, 2023) allows a user to have natural conversations with an AI-based conversational agent. ChatGPT can provide human-like responses to almost any prompt imaginable: it can answer simple and complex questions, write programming code, compose essays or song lyrics, and much more. ChatGPT is based on a Large Language Model (LLM), a type of AI that has been trained on a vast amount of data from internet and other media sources, such as discussion boards, blog posts, books, journals, and news articles.

The capabilities of ChatGPT are unprecedented and have been both transformative and disruptive in many spaces. Researchers have found that ChatGPT responses can achieve passing scores on collegiate and professional level examinations such as those required to obtain a medical license (Kung et al., 2023), and academic organizations have grappled with how to handle the threat of ChatGPT-based plagiarism. LLM-based capabilities will be significant for the advancement of H-A teaming, which has until now been hampered by the limited conversational capabilities of intelligent systems (as those who have spoken with smart voice assistants may attest).

Researchers have called for greater accountability from those who involve ChatGPT in their work and have identified the need for more transparency into the workings of conversational AI systems (van Dis et al., 2023). The current iteration of ChatGPT is still not completely reliable, sometimes committing "hallucinations" in which it confidently claims facts or information which are actually false or improperly reasoned (Azamfirei, Kudchadkar, & Fackler, 2023). This can be problematic if the user is trying to talk to ChatGPT about less-familiar subjects, as the user may not be aware that the information is subtly or significantly incorrect. In spite of these drawbacks, it is clear that the advantages offered by LLMs such as ChatGPT are tremendous, and an incredible amount of resources are being committed into developing these AI systems further, which will spur advances in fields such as robotics, medicine, computer science, academia, and many more.

robots (Army Technology, 2018). Some researchers have also investigated the use of gestures, hand signals, and tactile or haptic interface to send commands to robots (Barber et al., 2015; Barber, Lackey, Reinerman-Jones, & Hudson, 2013), which could offer more flexibility in how the robot's teammates can interact with it in the field, but these avenues are still in development and not as robust as using interfaces or controllers (Baker, Schaefer, & Hill, 2019).

Team communication is often in the service of developing a shared understanding of the team's environment, tasks, and activities. To coordinate effectively with agents, H-A teams need a shared understanding of their environment, their tasks, and their teamwork (Bisk, Yuret, & Marcu, 2016; Ososky et al., 2012). Humans and agents need to be able to communicate information back and forth in order to establish and maintain an ongoing understanding of what each team member is doing, and how they can most effectively work toward their goals (Chen & Barnes, 2014; Marathe, Schaefer, Evans, & Metcalfe, 2018). Thus, team communication is naturally linked with team cognition, which we explore in the following section.

11.3.3 TEAM COGNITION

Team cognition involves the interplay between a team's activities and the development of their knowledge and understanding of resources, tasks, the environment, and anything else they may do or be involved with. Through a team's interactions, they develop shared mental models and a sense of shared situation awareness. A shared mental model is a mental representation of knowledge that is shared by team members (Mathieu, Maynard, Rapp, & Gilson, 2008). Similarly, shared situation awareness is the degree to which team members possess the same *situation awareness*, which consists of a basic understanding of the situation relevant to the team's goals, how team members interpret that situation, and projections of future outcomes that are relevant to the team and its goals (Endsley, 2017). Intrateam similarities in knowledge structures, understanding of a situation at a point in time, and a collective awareness of knowledge distribution are all team-level mediators that can facilitate effective team performance (Cooke, Salas, Cannon-Bowers, & Stout 2000; Marks, Sabella, Burke, & Zaccaro 2002; Mathieu, Maynard, Rapp, & Gilson, 2008). The two sections below will discuss shared mental models and situation awareness, as well as how an agent's transparency can influence the rest of the team's cognition and shared mental models.

11.3.3.1 Shared Mental Models and Situation Awareness

Team cognition in H-A teams incorporates not only knowledge structures shared between humans but also those shared between humans and their agent counterparts. As with human teams, a shared understanding of required tasks, and the environments in which the tasks take place, is a prerequisite for the effective performance of group tasks (Marks, Sabella, Burke, & Zaccaro 2002; You & Robert, 2017). Human team members can build a shared mental model, but agents are not expected to be able to do so in the same way that humans can; instead, humans and agents can share mental models by having complementary information built from accurate assessments of the environment and the actors within it (Cannon-Bowers & Salas, 2001; You & Robert, 2017). Essentially, humans and agents can share mental models by gaining information from common sources and explicitly disclosing knowledge that has been created. Shared situation awareness is expected to work similarly. With a common set of information available, humans and agents can build a compatible model of the situation they inhabit, enabling coordination (Bradshaw et al., 2009). Coordination, in turn, is a necessary prerequisite for shared tasking in teams, especially with interdependent tasks (Bradshaw et al., 2009; You & Robert, 2017). In fact, coordination, communication, and other H-A interactions can allow individual team members to update the teams' knowledge structures (Cooke, Salas, Kiekel, & Bell, 2004; Sebo, Stoll, Scassellati, & Jung, 2020).

Some well-known H-A interaction findings can be relevant when discussing the maintenance of team cognition in H-A teams. Agent teammates can be designed to keep their human teammates "in the loop" (Fiore & Wiltshire, 2016). By doing so, they can prevent the deleterious effects of humans being divorced from their tasks, such as lower trust in the agent, lower situation awareness, and higher cognitive load (Grote, Weyer, & Stanton, 2014; Kilgore & Voshell, 2014; Stubbs, Hinds, & Wettergreen, 2007). An agent that acts to maintain its human teammates' shared mental models and shared situation awareness can prevent negative outcomes on the human's side, but it can have difficulty when maintaining its own information base; while agents can use sensors to gain information about their environments, sensing information about their human teammates and their cognitive states is more complicated, and agent designers must use innovative methods to incorporate agents into the creation and maintenance of shared mental models (Bradshaw et al., 2009).

11.3.3.2 Agent Transparency

In H-A teaming, characteristics of the agent can influence the team's cognition and shared mental models. Agent's transparency is an important characteristic, defined by Chen et al. (2014) as the "quality of an interface pertaining to its abilities to afford an operator's comprehension about an intelligent agent's intent, performance, future plans, and reasoning process" (p.2). This factor has received significant attention in recent years (see a review by Zhou, Li, Zhang, & Sun, 2022). One of the most cited agent transparency frameworks is the Situation awareness-based Agent Transparency (SAT) framework (Chen et al., 2014; see Figure 11.6) — based on the situation awareness (SA) model by Endsley (1995). The SAT framework consists of three levels of information requirements from one agent (e.g. a robot) to its partner (e.g. a human operator) to support the partner's *perception* of the agent's current actions and plans (Level 1), *comprehension* of its underlying logic (Level 2), and *projections* of future outcomes based on the agent's predicted end-states of current actions and plans (e.g. likelihood of success/failure), and any uncertainty associated with the projections (Level 3). The SAT framework was updated in 2018 (Chen et al. 2018) to incorporate bidirectional

> **Situation Awareness-based Agent Transparency**
>
> **Level 1: Goals & Actions**
>
> *Agent's current status/actions/plans*
>
> - Purpose: Desire (Goal selection)
> - Process: Intentions (Planning/Execution); Progress
> - Performance
> - Perception (Environment/Teammates)
>
> **Level 2: Reasoning**
>
> *Agent's reasoning process*
>
> - Reasoning process (Belief/Purpose)
> - Motivations
>   - Environmental & other constraints/affordances
>
> **Level 3: Projections**
>
> *Agent's projections/predictions; uncertainty*
>
> - Projection of future outcomes
> - Uncertainty and potential limitations; Likelihood of success/failure
> - History of Performance

FIGURE 11.6 Situation Awareness-based Agent Transparency (SAT) framework. (Adapted from Chen et al. (2014). Figure from Chen et al. (2018), CC BY.)

communications and additional teaming-related aspects between human and machine agents. The SAT framework has been adopted by numerous research groups with their own modifications and additions such as "what if" simulation as a Level 3 item (Cabour, Morales-Forero, Ledoux, & Bassetto, 2022; Sanneman & Shah, 2022).

The effects of agent transparency on operator performance have been investigated in multiple studies (Bhaskara, Skinner, & Loft, 2020; van de Merwe, Mallam, & Nazir, 2022). A meta-analysis by van de Merwe et al. (2022) shows the positive effects of agent transparency on human teammates' performance, especially in the contexts of H-A joint decision-making with humans playing the role of a supervisor. Mercado et al. (2016) and Stowers et al. (2020) found that when interacting with an intelligent planning agent in a multirobot management tasking environment, human operators performed better with a more transparent agent. The human operators were better able to calibrate their trust to the capabilities of the agent when the agent was more transparent, allowing them to make better decisions about whether to accept or reject the agent's recommendations.

While the benefits of agent transparency for operator performance are well documented, there have been inconsistent findings regarding the effects of transparency on operator workload. Helldin and colleagues (2014) found that there were costs associated with greater system transparency in terms of increased operator workload. Additionally, they found that decision times may have been impacted by the large amount of information to be processed associated with transparent interfaces. However, the same result of increased operator workload was not observed in other transparency studies (Mercado et al., 2016; Selkowitz, Lakhmani, & Chen, 2017; Stowers et al., 2020) except for a minor increase in response times reported by Stowers et al. (2020). A meta-analysis by van de Merwe et al. (2022) found no evidence of elevated workload associated with greater transparency (particularly in the context of H-A teaming where the human holds a supervisor role). It is worth noting that information about uncertainty could particularly impact the operator's workload, as suggested by Kunze and colleagues (2019).

Agent transparency also impacts humans' perceptions of the agent. For example, transparent agents tend to be perceived as more intelligent and human-like, compared with more opaque agents (Mercado et al., 2016; Roth, Schulte, Schmitt, & Brand, 2019). Empirical studies have shown that as agent transparency increases, human trust in the agent also tends to increase (Helldin et al., 2014;

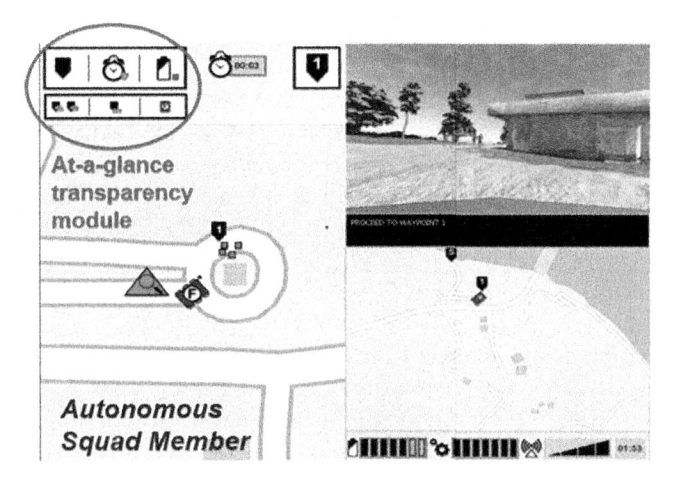

FIGURE 11.7 Human-machine interface for an Autonomous Squad Member, which includes an "at a glance" transparency module (top-left).

Mercado et al., 2016; Selkowitz et al., 2017; Stowers et al., 2020). However, there is also evidence that human teammates can exhibit complacency (overtrust) when interacting with highly transparent agents (Bhaskara et al., 2021; Loft et al., 2021).

Transparent interfaces for HAT have been implemented in various contexts such as human-robot interaction (HRI), multi-agent management systems, human-swarm interaction, autonomous driving, and explainable AI systems. In the context of HRI, a wide variety of systems have applied transparent principles toward designing their human-machine interfaces (HMI), from robots for military purposes (e.g., threat detection) to assistive robots for older adults. For example, the SAT framework (described in Section 3.3.2) was used to guide the design of the HMI for a military robot known as the Autonomous Squad Member (Figure 11.7). The HMI features an "at a glance" transparency module, where iconographic representations are used to indicate the ASM's current actions and plans (relating to Level 1 of the SAT model), top motivators (Level 2), and projected outcomes and uncertainties (Level 3) (Selkowitz et al., 2017; Wright et al., 2020).

Transparency principles have also been applied to the designs of H-A interfaces in multi-agent management contexts. For example, a transparent HMI based on the SAT framework was developed for a planning agent for multirobot management purposes (Mercado et al., 2016; Stowers et al., 2020). In a series of simulation-based experiments, it was found that participants performed better with a more transparent agent, and did not experience increased workload (Mercado et al., 2016; Stowers et al., 2020). Australian researchers developed a multi-agent management HMI based on the SAT framework and evaluated the interface's efficacy (Bhaskara et al., 2021; Loft et al., 2021). Interestingly, while Mercado et al. and Stowers et al. found that participants were able to calibrate their trust in the agent effectively, the studies by Bhaskara et al. and Loft et al. found that the humans exhibited complacency behaviors (i.e., too much trust) when interacting with a highly transparent agent. Finally, in the context of helicopter cockpit environments, there are efforts toward applying transparency principles based on the SAT framework to design the HMIs of multi-agent management systems (Roth et al., 2019; Hartnett, 2021). In a simulation-based experiment, Roth et al. (2019) found that participants performed better with a more transparent agent. This finding is consistent with Mercado et al. and Stowers et al. as well as a meta-analysis on agent transparency by van de Merwe et al. (2022).

Explainable AI (XAI) is an emerging area that seeks to make the data products of machine learning and artificial intelligence easier to understand. If the models, analyses, and data provided by AI systems can be meaningful and actionable, human operators can make better use of them during teamwork. Various techniques have been developed to deliver XAI capabilities, such as feature- or policy-based explanations, clarification of causal links, contrastive explanations, and more (Miller,

2019; Rawal et al., 2022). These capabilities can be embedded in augmented reality systems to enable projection of XAI contents directly into the HRI environment (Han et al., 2019), allowing the human to better understand the data coming from AI agents. XAI has also been used to enable natural language-based bidirectional communication in H-A teaming (Pynadath, Barnes, Wang, & Chen, 2018). In short, while the push for XAI is relatively new, it is building on transparency concepts developed within the SAT model and earlier transparency research (Chen et al., 2014; Chen et al., 2018; Cabour, Morales-Forero, Ledoux, & Bassetto, 2022; Holder and Wang, 2021; Chien, Yang, & Yu, 2022).

11.4 OUTPUTS

Teamwork inputs and processes interact to result in outputs: the products of a team's efforts, which can be as tangible as a wooden canoe that a team of carvers and carpenters build together or as intangible as the collective satisfaction they experience as a result of a job well done. There are many ways to conceptualize teamwork outputs; sometimes, a distinction is drawn between *effectiveness* (the extent to which a team achieves its intended purposes and goals) and *performance* (the successful interaction and coordination that allows the team to achieve its outcomes). In this manner, effectiveness could be described as "what did the team do", such as 'produce 25 widgets' or 'identify targets with an accuracy above 80%'. Performance could be described as "how did they do it", such as 'by engaging in certain communication behaviors to build a useful shared mental model' or 'by discussing disagreements early to avoid developing conflicts. Note that in this definition, performance does not solely consist of output-related aspects; it invokes aspects of the team's inputs and mediators to provide additional context for how the team functioned.

The team's purpose and organizational setting will determine which outputs are important. For a surgical team, key outputs might include patient recovery times or the occurrence rate of complications such as hospital-acquired infections. For military units that utilize unmanned aerial systems, key outputs might relate to the amount and quality of intelligence gathered for a geographic area, the efficiency of navigation routes used, or the number of targeted points of interest detected on video feeds. These can all be considered indicators of the teams' effectiveness.

Effectiveness and performance can be measured in many different ways. Surveys and questionnaires are extremely common due to their ease of implementation, but sometimes lack impartiality and detail. Behavioral observation involves trained observers who gather detailed information about team functions. Likewise, interviews can be developed and implemented to gain useful insights directly from team members. Objective measures can also be used to capture data directly from team members during the course of teamwork; such measures might include heart rate, respiration rate, galvanic skin response (an indicator of stress), or eye movements. The measurement of team effectiveness depends heavily on the team's context; while error rate is a useful factor to know, it relates differently to a surgical team than to a UAS reconnaissance team.

The following sections will discuss some of the assessment methods used to gather important information about H-A teamwork, shedding light on how key processes and emergent states result in various performance outcomes. We explore the measurement of trust, communication, and transparency; though these are not strictly "output" factors, they are critical to the outputs produced by H-A teams, and so they are often important for assessment.

11.4.1 TRUST

There are a variety of methods for conducting trust assessments in H-A teams. Subjective measures, such as surveys, reflect the conscious (and naturally, biased) views held by a person. There are numerous scales for different kinds of trust, such as the Negative Attitudes Towards Robots Scale (Nomura, Suzuki, Kanda, & Kato, 2006) and the Trust Perception Scale (Schaefer, 2016). Different scales can reflect different kinds of trust toward different subjects at different times, so selecting the most appropriate scale and implementing it at the appropriate time is key. Trust can also be assessed via behavior.

In team interactions where individuals speak, the analysis of speech content can reflect team trust (Lakhmani et al., 2022b). Reliance upon or compliance with an agent is another behavioral indicator of trust (Merritt, Heimbaugh, LaChapell, & Lee, 2013). Additionally, proxemics, or how close one stands to a human or robot, is another behavioral indicator of trust, with closer proximity indicating greater trust (MacArthur, Stowers, and Hancock, 2017). Finally, a person's physiological functions can correspond with trust in a system. Electrodermal activity, larger pupil sizes, and increased heart rate can indicate increased engagement, arousal, and trust when working with an agent (Krausman et al., 2022). Entrainment, or the synchronization of the physiological states between two or more people, can also indicate increased trust (Mønster, Håkonsson, Eskildsen, & Wallot, 2016).

Trust assessments are also shaped by the point in the team's lifespan in which the assessments are carried out. Teamwork is implied to be cyclical by the IMOI model, and trust is likewise cyclical. In the first stage of a team's existence, trust stems from any outside knowledge one brings with them and one's dispositional propensity to trust (Krausman et al., 2022). Following that initial stage of trust is the orientation stage. This is the stage where an individual becomes able to judge their team members' trustworthiness and competence. At this point, one has some experience with their teammates, so one can make some initial judgment of an agent's reliability and the overall level of team trust (Krausman et al., 2022; Lee & See, 2004). The trust calibration stage comes next. In this stage, an H-A team interacts in different contexts. Through multiple cycles of interaction in multiple contexts, one can ascertain how trustworthy the agent is, and under what circumstances. In the experience stage, the last stage of the trust cycle, one has established a history with their team, which can influence future trust in a particular agent, or in agents overall. Assessments at the beginning of the trust cycle may end up yielding more stable, dispositional reflections of trust, while assessments at the orientation cycle may be more volatile (Krausman et al., 2022). Thus, the relative timing of the trust assessments can influence their findings.

11.4.2 Communication

Good communication in H-A teams is crucial to their success, but because H-A teaming is a small subset of all teaming research, the study of H-A team communication is relatively limited compared to the vast amount of research into how human-only teams communicate. As a result, the field of H-A research has leveraged assessments of communication that were first designed with humans in mind. Baker et al. (2021) identified and described key communication assessments that can be applied to H-A teaming, some of which are explored below (see also Krausman et al., 2022).

The structure of team interactions can explain much about how well or poorly the team is communicating. Some assessments thus look at the patterns and back-and-forth exchanges in how team members interact. Social network analysis (SNA), for example, can be used to characterize the relationships between actors (i.e., humans or agents) in a network (Shaw, 1964; Wasserman & Faust, 1994). Using SNA, one can evaluate the connectedness of any given actor in the network, which can make it easier to identify points where communication may become overloaded, such as if an actor is too highly connected relative to their expected tasks. Team communication can also be assessed by comparing the extent to which team members push information to other teammates versus how much they request (i.e., pull) from teammates, providing unique insights into the team's cognition and shared SA (Johnson et al., 2020, 2021; McNeese et al. 2018).

Other assessments are aimed at inferring the emotional states of team members through features of their interactions. Facial expressions are a window into our emotions, and they can be captured using specialized equipment (e.g., computer vision, or facial electromyography to track muscle activity) or through live observation and manual coding. Then, different expressions can be calculated based on their association with emotional states such as anger, fear, happiness, or surprise (Ekman & Friesen, 1971), which can help explain the affective states that drive H-A interactions like trust (Neubauer et al., 2020). Emotional states like stress can also be inferred by comparing how vocal features such as pitch, volume, and the rate at which one speaks change from a person's baseline, and this has been done using neural networks to rapidly assess those features (Casale, Russo, Scebba, & Serrano, 2008; Stuhlsatz et al., 2011).

Finally, some communication assessments for H-A teaming are based on analyzing the content of the team's communication (i.e., what is being said). Recalling our earlier discussion on grounding and shared mental models, it has been noted that people engaged in a conversation will sometimes use similar word choices and sentence structures in their phrases, ostensibly as a way to make the communication more efficient (Heuer, Müller-Frommeyer, & Kauffeld, 2020; Semin, 2007). Language style matching (LSM) is an assessment method that leverages this to evaluate the degree of similarity between two speakers based on how similarly they use 'function words' such as articles, adverbs, or prepositions (Niederhoffer & Pennebaker, 2002). In a related approach, latent semantic similarity (LSS) is a method that looks at how speakers' utterances (words, sounds, or phrases) are used in relation to other utterances. The relationships between utterances can be analyzed using latent semantic analysis (Landauer, Foltz, & Laham, 1998) and evaluated for trends that indicate similarity between speakers. LSM and LSS can help characterize the extent to which speakers are engaging with each other and using discussion to lead to a deeper shared understanding of something.

11.4.3 Transparency and XAI

Because agent transparency is considerably linked to team cognition, it can be useful to evaluate aspects of the agent's transparency to gain a deeper understanding of their influences on human team members' plans, reasoning, and decisions. Several organizations have published their guidance on the design of systems with transparent HMIs; concurrently, some assessment methods have also been proposed to evaluate system transparency levels (IBM, 2018; Winfield et al., 2021). For example, multiple sets of metrics have been proposed by XAI researchers to assess the effectiveness of XAI systems (Rawal et al., 2022). The metrics largely focus on four aspects of the XAI systems: correctness and robustness; usefulness, understandability, and processing difficulty for the human; congruence between humans' and machines' mental models; and generalizability, adaptability, and versatility (Miller, 2019).

Another assessment method, the Explanation Satisfaction Scale, was developed to evaluate XAI users' experiences with a focus on assessing congruence between AI system outputs and the users' mental models of the problem space (Hoffman et al., 2018). The SAT framework (Chen et al., 2014, 2018), with its subcomponents in each SAT level, can also be used to evaluate a system's transparency level or whether any transparency principles are missing from an HMI. Similarly, Sanneman and Shah (2022) proposed the SA Framework for XAI (SAFE-AI) to assess the transparency of agent XAI aspects and how well they support a human's situation awareness. Finally, there are assessments that have been designed specifically to evaluate the XAI aspects of decision support and recommender systems (van der Waa et al., 2021).

11.5 CONCLUSION AND FRONTIERS FOR FUTURE WORK

H-A teams are in some ways quite similar to, and in other ways very different from, teams comprised of humans. The strengths and weaknesses of intelligent agents add a unique dimension to teams that incorporate them. As the future ushers in further advances to the processing power and capabilities of agents, and as the science of H-A teaming evolves, H-A teams will become more flexible, more intelligent, and more effective. In this chapter, we explored the fundamentals of teamwork via the model of inputs, mediators, and outputs. We discussed key factors in H-A teaming, such as levels of autonomy, trust, and agent transparency. We also presented assessment methods for a few important aspects of teamwork. While this chapter reviewed many topics within H-A teaming, our discussions remained at a high level, and there is more to be learned about the implementation and refinement of H-A teams. Table 11.1 summarizes the H-A teaming factors highlighted throughout this chapter.

TABLE 11.1
Summary of H-A Teaming Factors

Type	Subgroup	Factor	Key Info	References
Input	Human-based	Expertise and training	Well-organized, task-relevant knowledge. The distribution of team members' knowledge can affect how they perform.	Sonnentag and Volmer (2009); Blickensderfer, Cannon-Bowers, and Salas (1998); Salas, Shuffler, Thayer, Bedwell, and Lazzara (2015)
Input	Human-based	Individual and cultural differences	Individuals and cultures are innately different in what they know about agents and how they interact with them.	Chen & Barnes (2014); Lyons and Guznov (2019); Chien et al. (2019)
Input	Agent-based	Level of Autonomy	The extent to which tasks can be handled with or without human control. Good H-A teaming needs a balance between autonomy & supervision.	Sheridan and Verplank (1978); Parasuraman, Sheridan, and Wickens (2000); Endsley and Kaber (1999)
Input	Agent-based	Performance and reliability	Agent reliability induces human trust, but other factors like transparency or agent behaviors can affect that relationship.	Caldwell et al. (2022); Keller and Rice (2010); Rovira, McGarry, and Parasuraman (2007)
Input	Teamwork-based	Task allocation	Team members should have tasks that align with their capabilities. Advanced agents can dynamically adapt to changing tasks and situations.	Jobidon et al. (2017); Hollenbeck et al. (2011); Bradshaw, Feltovich, & Johnson (2012).
Input	Teamwork-based	Roles	Team roles dictate each member's hierarchy and interactions. H-A team roles usually depend on the agent's level of autonomy.	Mathieu et al. (2015); Bradshaw et al. (2009); O'Neill, McNeese, Barron, and Schelble (2020)
Mediator	Attitudes	Trust	Trust and reliability are interrelated, and human trust should be appropriately calibrated to an agent's actual capabilities.	Mayer, Davis, and Schoorman (1995); Hancock et al. (2011); Lee and See (2004)
Mediator	Attitudes	Cohesion	Agents can promote team cohesion by supporting humans' tasks and collective behaviors, but agents can also inadvertently affect group dynamics.	Dion (2000); Salas, Grossman, Hughes, and Coultas (2015); Lakhmani et al. (2022a)
Mediator	Behaviors	Communication	A team's communication can be a window into how they develop many other inputs, processes, and outputs.	Mesmer-Magnus and Dechurch (2009); Mavridis (2015); Marathe, Schaefer, Evans, and Metcalfe (2018)
Mediator	Team cognition	Shared mental models and situation awareness	Teams need a shared understanding of their tasks and environment, and agents must support that.	Mathieu, Maynard, Rapp, and Gilson (2008); Endsley (2017); Cannon-Bowers and Salas (2001)
Mediator	Team cognition	Agent transparency	Agents should be designed so that human teammates can understand what they are planning and doing.	Chen et al. (2014); Zhou, Li, Zhang, and Sun (2022); Endsley (1995)

(Continued)

TABLE 11.1 (*Continued*)
Summary of H-A Teaming Factors

Type	Subgroup	Factor	Key Info	References
Output		Trust	Trust assessments can help identify if humans are appropriately relying on agents and other human teammates throughout the team's life cycle.	Krausman et al. (2022); Schaefer (2016)
Output		Communication	Communication analysis is useful for studying other key factors such as shared mental models, coordination, trust, and so on.	Baker et al. (2021); Demir et al. (2015)
Output		Transparency and XAI	Assessing transparency and explainability is important to ensuring that agents can support a team's situation awareness and coordination.	Rawal et al. (2022); Miller (2019); IBM (2018); Sanneman and Shah (2022)

There are many frontiers of ongoing research into H-A teaming that will inform the future of the domain, and we will discuss four of them. One key area of study involves bidirectional communication. Because agents are still not able to communicate with humans as naturally as another human can, it is sometimes difficult for H-A teams to develop clear mental models about what the agent teammates are doing, thinking, or planning (Chen et al., 2018; Marathe, Schaefer, Evans, & Metcalfe, 2018). In the future, agents will need to be able to communicate their intentions more clearly, which can aid their transparency and allow them to better support human teammates (Lyons, Sycara, Lewis, & Capiola, 2021). However, it is not yet clear exactly how agents should communicate their intent in order to support the team's performance. If scientists can improve the ability of agents to communicate information naturally and reciprocally with human teammates, the teams will be able to interact and function far more efficiently.

Another important frontier for research and development in H-A teaming involves the management of trust. When an agent behaves in a way that runs counter to a teammate's expectations, it can damage the person's trust of the agent, leading to ineffective teaming. If the agent can identify when those trust violations occur, and can determine the appropriate method to work with their teammate to restore trust, it can allow the team to navigate complex situations or long-term partnerships more effectively (De Visser, Pak, & Shaw, 2018; Baker, Phillips, Ullman, & Keebler, 2018). However, it is still not clear whether certain ways of repairing trust (e.g., apologies, explanations, denials) are only appropriate for certain situations, or how to select the best method for repairing trust for a given situation, so it will take more work to understand how agents should attempt to repair their teammates' trust when it is breached.

A third area of active study in H-A teaming involves the development and validation of appropriate measures and metrics. Good measurement is the key to more completely understanding teamwork, but H-A teams are often studied in the context of a specialized purpose or a limited duration, which can complicate efforts to thoroughly understand their teamwork and generalize findings to other H-A teams. New measures of trust, resilience, mental models, and many other constructs are being developed and tested, but their validity needs to be evaluated (National Academies of Sciences, Engineering, and Medicine, 2021).

Those efforts dovetail with a fourth frontier of research in H-A teaming: the need to understand H-A teams over longer time horizons. Because much research is conducted using short-term, ad hoc teamwork between humans and agents, there is much to be learned about how H-A interaction dynamics evolve in the long term (National Academies of Sciences, Engineering, and Medicine,

2021). For example, it is relatively easy to study how an H-A team can adapt to a set of goals within the context of missions across one or a few days. If the team is expected to interact across weeks, months, and perhaps years (as might be expected in future contexts of long-duration spaceflight or multi-domain military operations, for example), a team's goals across that time will evolve considerably and the team will have to adapt appropriately, which will have unknown implications for the team's roles, task allocation, interdependencies, and more. Thus, upcoming research will seek to study H-A teams over longer durations to begin to understand how H-A teaming dynamics evolve beyond the short term.

The future of H-A teaming, and of research into it, is incredibly bright. The persistent pursuit of faster processors and more intelligent machines will feed new and exciting advances in how we can incorporate humans and agents into smarter, safer, more productive, and more effective teams. While there is yet much to understand at the cutting edge of H-A research, the way ahead sees humans and machines supporting each other long into the future.

NOTES

1. For more on the topic of defining autonomy, we refer the reader to Beer, Fisk, and Rogers (2014).
2. Situation awareness is discussed in more detail in Section 11.3.3, "Team Cognition".

REFERENCES

Abrams, A. M., & Rosenthal-von der Pütten, A. M. (2020). I-C-E framework: Concepts for group dynamics research in human-robot interaction. *International Journal of Social Robotics*, *12*(6), 1213–1229.

Abuhaimed, S., & Sen, S. (2023). Influence of expertise complementarity on ad hoc human-agent team effectiveness. In *International Conference on Principles and Practice of Multi-Agent Systems* (pp. 679–688). Cham: Springer.

Adams, S. K. (2007). *Disciplinarily Hetero- and Homogeneous Design Team Convergence: Communication Patterns and Perceptions of Teamwork*. (Thesis). Virginia Tech, Retrieved from https://vtechworks.lib.vt.edu/handle/10919/34802 Available from vtechworks.lib.vt.edu

Army Technology. (2018). *TALON Tracked Military Robot*. Retrieved from https://www.army-technology.com/projects/talon-tracked-military-robot/

Asakawa, K., Kojima, J., Kato, Y., Matsumoto, S., Kato, N., Asai, T., & Iso, T. (2002). Design concept and experimental results of the autonomous underwater vehicle AQUA EXPLORER 2 for the inspection of underwater cables. *Advanced Robotics*, *16*(1), 27–42. doi:10.1163/156855302317413727

Azamfirei, R., Kudchadkar, S. R., & Fackler, J. (2023). Large language models and the perils of their hallucinations. *Critical Care*, *27*(1), 120. doi:10.1186/s13054-023-04393-x

Baker, A. L., Fitzhugh, S. M., Huang, L., Forster, D. E., Scharine, A., Neubauer, C., ... & Cooke, N. J. (2021). Approaches for assessing communication in human-autonomy teams. *Human-Intelligent Systems Integration*, *3*(2), 99–128. doi:10.1007/s42454-021-00026-2

Baker, A. L., Phillips, E. K., Ullman, D., & Keebler, J. R. (2018). Toward an understanding of trust repair in human-robot interaction: Current research and future directions. *ACM Transactions on Interactive and Intelligent Systems*, *8*(4), 1–30. doi:10.1145/3181671

Baker, A. L., Schaefer, K. E., & Hill, S. G. (2019). *Teamwork and Communication Methods and Metrics for Human-Autonomy Teaming* (ARL-TR-8844). Aberdeen Proving Ground, MD, United States. Retrieved from https://apps.dtic.mil/sti/pdfs/AD1083400.pdf

Barber, B. (1983). *The Logic and Limits of Trust*. Rutgers University Press.

Barber, D. J., Abich, J., Phillips, E., Talone, A. B., Jentsch, F., & Hill, S. G. (2015). Field assessment of multimodal communication for dismounted human-robot teams. *Proceedings of the Human Factors and Ergonomics Society Annual Meeting 59*(1), 921–925. doi:10.1177/1541931215591280

Barber, D. J., Lackey, S., Reinerman-Jones, L., & Hudson, I. (2013). Visual and tactile interfaces for bi-directional human robot communication. In *Unmanned Systems Technology XV* (Vol. 8741, pp. 269–279).

Barnes, M. J., & Evans, A. W. (2016). Soldier-robot teams in future battlefields: An overview. In F. Jentsch & M. Barnes (Eds.), *Human-Robot Interactions in Future Military Operations* (pp. 29–50) London: CRC Press.

Barnlund, D. C. (2008). A transactional model of communication. In C. D. Mortensen (Ed.), *Communication Theory* (2nd ed., pp. 47–57). Piscataway, NJ: Routledge.

Beer, J. M., Fisk, A. D., & Rogers, W. A. (2014). Toward a framework for levels of robot autonomy in human-robot interaction. *Journal of Human Robot Interaction*, 3(2), 74–99. doi:10.5898/JHRI.3.2.Beer

Bhaskara, A., Skinner, M., & Loft, S. (2020). Agent transparency: A review of current theory and evidence. *IEEE Transactions on Human-Machine Systems*, 50, 215–224.

Bhaskara, A., Duong, L., Brooks, J., Li, R., McInerney, R., Skinner, M., ... & Loft, S. (2021). Effect of automation transparency in the management of multiple unmanned vehicles. *Applied Ergonomics*, 90, 103243.

Bisk, Y., Yuret, D., & Marcu, D. (2016). Natural Language Communication with Robots. In *Proceedings of the 2016 Conference of the North American Chapter of the Association for Computational Linguistics: Human Language Technologies* (pp. 751–761).

Blickensderfer, E., Cannon-Bowers, J. A., & Salas, E. (1998). Cross-training and team performance. In J. A. Cannon-Bowers & E. Salas (Eds.), *Making Decisions under Stress: Implications for Individual and Team Training* (pp. 299–311). American Psychological Association. doi:10.1037/10278-011

Bradshaw, J. M., Feltovich, P. J., & Johnson, M. (2012). Human-agent interaction. In G. A. Boy (Ed.), *The Handbook of Human-Machine Interaction: A Human-Centered Design Approach*. Farnham, UK: Ashgate Publishing.

Bradshaw, J. M., Feltovich, P. J., Jung, H., Kulkarni, S., Taysom, W., & Uszok, A. (2003). Dimensions of adjustable autonomy and mixed-initiative interaction. In *International Workshop on Computational Autonomy* (pp. 17–39). Berlin, Heidelberg: Springer.

Bradshaw, J. M., Feltovich, P., Johnson, M., Breedy, M., Bunch, L., Eskridge, T., ... & van Diggelen, J. (2009). *From Tools to Teammates: Joint Activity in Human-Agent-Robot Teams*. Paper presented at the International Conference on Human Centered Design, San Diego, CA.

Breuer, C., Hüffmeier, J., & Hertel, G. (2016). Does trust matter more in virtual teams? A meta-analysis of trust and team effectiveness considering virtuality and documentation as moderators. *Journal of Applied Psychology*, 101(8), 1151–1177. doi:10.1037/apl0000113

Bunderson, J. S., & Sutcliffe, K. M. (2002). Comparing alternative conceptualizations of functional diversity in management teams: Process and performance effects. *Academy of Management Journal*, 45(5), 875–893.

Burke, C. S., Salas, E., Wilson-Donnelly, K., & Priest, H. (2004). How to turn a team of experts into an expert medical team: Guidance from the aviation and military communities. *Quality and Safety in Health Care*, 13(suppl 1), i96–i104. doi:10.1136/qshc.2004.009829

Cabour, G., Morales-Forero, A., Ledoux, É., & Bassetto, S. (2022). An explanation space to align user studies with the technical development of Explainable AI. *AI & Society*. doi:10.1007/s00146-022-01536-6

Caldwell, S., Sweetser, P., O'Donnell, N., Knight, M. J., Aitchison, M., Gedeon, T., ... & Conroy, D. (2022). An agile new research framework for hybrid human-AI teaming: Trust, transparency, and transferability. *ACM Transactions on Interactive Intelligent Systems*, 12(3), 1–36.

Cannon-Bowers, J. A., & Salas, E. (2001). Reflections on shared cognition. *Journal of Organizational Behavior: The International Journal of Industrial, Occupational and Organizational Psychology and Behavior*, 22(2), 195–202.

Carpenter, J. (2013). *The Quiet Professional: An investigation of U.S. military Explosive Ordnance Disposal Personnel Interactions with Everyday Field Robots*. [Dissertation, University of Washington]. Retrieved from https://digital.lib.washington.edu/researchworks/bitstream/handle/1773/24197/Carpenter_washington_0250E_12154.pdf

Carron, A. V. (1982). Cohesiveness in sport groups: Interpretations and considerations. *Journal of Sport Psychology*, 4(2), 123.

Carron, A. V., & Brawley, L. R. (2000). Cohesion: Conceptual and measurement issues. *Small Group Research*, 31(1), 89–106.

Casale, S., Russo, A., Scebba, G., & Serrano, S. (2008). *Speech Emotion Classification Using Machine Learning Algorithms*. Paper presented at the 2008 IEEE International Conference on Semantic Computing.

Chen, J. Y., & Barnes, M. J. (2014). Human-agent teaming for multirobot control: A review of human factors issues. *IEEE Transactions on Human-Machine Systems*, 44(1), 13–29. doi:10.1109/THMS.2013.2293535

Chen, J. Y., Lakhmani, S. G., Stowers, K., Selkowitz, A. R., Wright, J. L., & Barnes, M. (2018). Situation awareness-based agent transparency and human-autonomy teaming effectiveness. *Theoretical Issues in Ergonomics Science*, 19(3), 259–282.

Chen, J. Y., Procci, K., Boyce, M., Wright, J. L., Garcia, A., & Barnes, M. (2014). *Situation Awareness-Based Agent Transparency* (ARL-TR-6905). Aberdeen Proving Ground, MD: U.S. Army Research Laboratory.

Chien, S. Y., Lewis, M., Sycara, K., Kumru, A., & Liu, J. S. (2019). Influence of culture, transparency, trust, and degree of automation on automation use. *IEEE Transactions on Human-Machine Systems*, 50, 205–214.

Chien, S., Yang, C., & Yu, F. (2022). Xflag: Explainable fake news detection model on social media. *International Journal of Human-Computer Interaction*, *38*, 1808–1827.

Clare, A. S., Cummings, M. L., How, J. P., Whitten, A. K., & Toupet, O. (2012). Operator objective function guidance for a real-time unmanned vehicle scheduling algorithm. *Journal of Aerospace Computing, Information, and Communication*, *9*, 161–173.

Cohen, S. G., & Bailey, D. E. (1997). What makes teams work: Group effectiveness research from the shop floor to the executive suite. *Journal of Management*, *23*(3), 239–290. doi:10.1016/S0149-2063(97)90034-9

Cooke, N. J., Salas, E., Cannon-Bowers, J. A., & Stout, R. J. (2000). Measuring team knowledge. *Human Factors*, *42*(1), 151–173.

Cooke, N. J., Salas, E., Kiekel, P. A., & Bell, B. (2004). Advances in measuring team cognition. In E. Salas & S. M. Fiore (Eds.), *Team Cognition: Understanding the Factors that Drive Process and Performance* (pp. 83–106). American Psychological Association. doi:10.1037/10690-005

Costa, A. C. (2003). Work team trust and effectiveness. *Personnel Review*, *32*(5), 605–622. doi:10.1108/00483480310488360

de Jong, B. A., Dirks, K. T., & Gillespie, N. (2015). Trust and team performance: A meta-analysis of main effects, contingencies, and qualifiers. In *Academy of Management Proceedings* (Vol. 2015, No. 1, p. 14561). Briarcliff Manor, NY: Academy of Management. doi:10.5465/AMBPP.2015.234

de Visser, E. J., Cohen, M., Freedy, A., & Parasuraman, R. (2014). A design methodology for trust cue calibration in cognitive agents. In *Virtual, Augmented and Mixed Reality. Designing and Developing Virtual and Augmented Environments: 6th International Conference, VAMR 2014, Held as Part of HCI International 2014, Heraklion, Crete, Greece, June 22-27, 2014, Proceedings, Part I 6* (pp. 251–262). Springer International Publishing.

de Visser, E. J., Monfort, S. S., McKendrick, R., Smith, M. A., McKnight, P. E., Krueger, F., & Parasuraman, R. (2016). Almost human: Anthropomorphism increases trust resilience in cognitive agents. *Journal of Experimental Psychology: Applied*, *22*, 331–349.

De Visser, E., Pak, R., & Shaw, T. (2018). From 'automation' to 'autonomy': The importance of trust repair in human-machine interaction. *Ergonomics* 61, 1409–1427. doi:10.1080/00140139.2018.1457725

de Winter, J. C., & Dodou, D. (2014). Why the Fitts list has persisted throughout the history of function allocation. *Cognition, Technology & Work*, *16*(1), 1–11.

DeChurch, L. A., & Mesmer-Magnus, J. R. (2010). The cognitive underpinnings of effective teamwork: A meta-analysis. *Journal of Applied Psychology*, *95*(1), 32.

DeCostanza, A., Marathe, A. R., Bohannon, A., Evans, A. W., Palazzolo, E. T., Metcalfe, J. S., & McDowell, K. (2018). *Enhancing Human-Agent Teaming with Individualized, Adaptive Technologies: A Discussion of Critical Scientific Questions* (ARL-TR-8359). DEVCOM Army Research Laboratory. Retrieved from https://apps.dtic.mil/sti/pdfs/AD1051552.pdf

Demir, M., McNeese, N. J., Cooke, N. J., Ball, J. T., Myers, C., & Frieman, M. (2015). Synthetic teammate communication and coordination with humans. *Proceedings of the Human Factors and Ergonomics Society Annual Meeting*, *59*(1), 951–955. doi:10.1177/1541931215591275

Dion, K. L. (2000). Group cohesion: From" field of forces" to multidimensional construct. *Group Dynamics: Theory, Research, and Practice*, *4*(1), 7.

Dyer, J. L. (1984). Team research and team training: A state-of-the-art review. In F. A. Muckler (Ed.), *Human Factors Review* (pp. 285–323). Santa Monica, CA: Human Factors and Ergonomics Society.

Eccles, D. W., & Tenenbaum, G. (2004). Why an expert team is more than a team of experts: A social-cognitive conceptualization of team coordination and communication in sport. *Journal of Sport and Exercise Psychology*, *26*(4), 542–560. doi:10.1123/jsep.26.4.542

Ekman, P., & Friesen, W. V. (1971). Constants across cultures in the face and emotion. *Journal of Personality and Social Psychology*, *17*(2), 124.

Endsley, M. R. (1995). Toward a theory of situation awareness in dynamic systems. *Human Factors*, *37*, 32–64.

Endsley, M. R. (2017). From here to autonomy: Lessons learned from human-automation research. *Human Factors*, *59*(1), 5–27. doi:10.1177/0018720816681350

Endsley, M. R., & Kaber, D. B. (1999). Level of automation effects on performance, situation awareness and workload in a dynamic control task. *Ergonomics*, *42*(3), 462–492. doi:10.1080/001401399185595

Erdem, F., & Ozen, J. (2003). Cognitive and affective dimensions of trust in developing team performance. *Team Performance Management: An International Journal*, *9*(5), 131–135. doi:10.1108/13527590310493846

Esterwood, C., & Robert, L. P. (2020). *Human Robot Team Design.* Paper presented at the Proceedings of the 8th International Conference on Human-Agent Interaction.

Feitosa, J., Grossman, R., Kramer, W. S., & Salas, E. (2020). Measuring team trust: A critical and meta-analytical review. *Journal of Organizational Behavior*, *41*(5), 479–501. doi:10.1002/job.2436

Festinger, L., Schachter, S., & Back, K. (1950). *Social pressures in informal groups; a study of human factors in housing* Stanford, CA: Stanford University Press.

Fiore, S. M., & Wiltshire, T. J. (2016). Technology as teammate: Examining the role of external cognition in support of team cognitive processes. *Frontiers in Psychology*, 7, 1531.

Fitts, P. M. (1951). *Human engineering for an effective air-navigation and traffic-control system*. Washington, DC: National Research Council.

Goodman, P. S., Ravlin, E., & Schminke, M. (1987). Understanding groups in organizations. *Research in Organizational Behavior*, 9, 121–173.

Gross, N., & Martin, W. E. (1952). On group cohesiveness. *American Journal of Sociology*, 57(6), 546–564.

Grote, G., Weyer, J., & Stanton, N. A. (2014). Beyond human-centred automation-concepts for human-machine interaction in multi-layered networks. *Ergonomics*, 57(3), 289–294.

Hackman, J. R. (1987). The design of work teams. In J. Lorsch (Ed.), *Handbook of Organizational Behavior* (pp. 315–342). Englewood Cliffs, NJ: Prentice Hall.

Han, Z., Allspaw, J., Norton, A., & Yanco, H. A. (2019). Towards a robot explanation system: A survey and our approach to state summarization, storage and querying, and human interface. In *Proceedings of the AI-HRI Symposium at AAAI-FSS*.

Hancock, P. A. (2017). Imposing limits on autonomous systems. *Ergonomics*, 60(2), 284–291. doi:10.1080/00140139.2016.1190035

Hancock, P. A., Billings, D. R., Schaefer, K. E., Chen, J. Y., de Visser, E. J., & Parasuraman, R. (2011). A meta-analysis of factors affecting trust in human-robot interaction. *Human Factors*, 53(5), 517–527. doi:10.1177/0018720811417254

Hancock, P. A., Kessler, T. T., Kaplan, A. D., Brill, J. C., & Szalma, J. L. (2021). Evolving trust in robots: Specification through sequential and comparative meta-analyses. *Human Factors*, 63(7), 1196–1229.

Hartnett, G. (2021). *Transparency at a Glance: A Concept for Transparency of Semiautonomous and Fully Autonomous (SAFA) and Artificial Intelligence (AI) Systems through Dynamic Infographics (DIG)* (Report number DAC-TR-2021-081). Aberdeen Proving Ground, MD: U.S. Army Data and Analysis Center.

Hearst, M. A., Allen, J., Horvitz, E., & Guinn, C. (1999). Mixed-initiative interaction. *IEEE Intelligent Systems*, 14(5), 14–23.

Helldin, T., Ohlander, U., Falkman, G., & Riveiro, M. (2014, June). Transparency of automated combat classification. In *International Conference on Engineering Psychology and Cognitive Ergonomics* (pp. 22–33). Cham: Springer.

Heuer, K., Müller-Frommeyer, L. C., & Kauffeld, S. (2020). Language matters: The double-edged role of linguistic style matching in work groups. *Small Group Research*, 51(2), 208–228.

Hinds, P. J., Roberts, T. L., & Jones, H. (2004). Whose job is it anyway? A study of human-robot interaction in a collaborative task. *Human-Computer Interaction*, 19(1–2), 151–181.

Hoffman, R., Mueller, S., Klein, G., & Litman, J. (2018). Metrics for explainable AI: Challenges and prospects. arXiv preprint, https://arxiv.org/abs/1812.04608.

Holder, E., & Wang, N. (2021). Explainable artificial intelligence (XAI) interactively working with humans as a junior cyber analyst. *Human-Intelligent Systems Integration*, 3, 139-153.

Hollenbeck, J. R., Ellis, A. P., Humphrey, S. E., Garza, A. S., & Ilgen, D. R. (2011). Asymmetry in structural adaptation: The differential impact of centralizing versus decentralizing team decision-making structures. *Organizational Behavior and Human Decision Processes*, 114(1), 64–74.

IBM. (2018). *IBM's Principles for Trust and Transparency*. Retrieved March 17, 2021, from https://www.ibm.com/blogs/policy/trust-principles.

Ilgen, D. R., Hollenbeck, J. R., Johnson, M., & Jundt, D. (2005). Teams in organizations: From input-process-output models to IMOI models. *Annual Review of Psychology*, 56, 517–543. Retrieved from https://search.proquest.com.ezproxy.libproxy.db.erau.edu/docview/205830055/abstract/8077237986914C26PQ/1

Ingram, M., Moreton, R., Gancz, B., & Pollick, F. (2021). Calibrating trust toward an autonomous image classifier. *Technology, Mind, and Behavior*, 2(1). doi:10.1037/tmb0000032

Jobidon, M. E., Turcotte, I., Aubé, C., Labrecque, A., Kelsey, S., & Tremblay, S. (2017). Role variability in self-organizing teams working in crisis management. *Small Group Research*, 48(1), 62–92.

Johnson, C. J., Demir, M., Zabala, G. M., He, H., Grimm, D. A., Radigan, C., … & Gorman, J. C. (2020). Training and verbal communications in human-autonomy teaming under degraded conditions. In *2020 IEEE Conference on Cognitive and Computational Aspects of Situation Management (CogSIMA)* (pp. 53–58). IEEE. doi:10.1109/CogSIMA49017.2020.9216061

Johnson, M., Bradshaw, J. M., Feltovich, P. J., Jonker, C. M., Van Riemsdijk, M. B., & Sierhuis, M. (2014). Coactive design: Designing support for interdependence in joint activity. *Journal of Human-Robot Interaction*, 3(1), 43–69.

Johnson, C. J., Demir, M., McNeese, N. J., Gorman, J. C., Wolff, A. T., & Cooke, N. J. (2021). The impact of training on human-autonomy team communications and trust calibration. *Human Factors*. doi:10.1177/00187208211047323

Jurafsky, D., & Martin, J. H. (2007). Dialogue and conversational agents. In *Speech and Language Processing: An Introduction to Natural Language Processing, Computational Linguistics, and Speech Recognition* (1st ed.), Upper Saddle River, NJ: Prentice Hall.

Kaplan, A. D., Kessler, T. T., Brill, J. C., & Hancock, P. A. (2021). Trust in artificial intelligence: Meta-analytic findings. *Human Factors*, 63(7), 1196–1229. doi:10.1177/00187208211013988

Kawamura, K., & Iskarous, M. (1994, 12–16 Sept. 1994). *Trends in Service Robots for the Disabled and the Elderly*. Paper presented at the Proceedings of IEEE/RSJ International Conference on Intelligent Robots and Systems (IROS'94).

Keller, D., & Rice, S. (2010). System-wide versus component-specific trust using multiple aids. The *Journal of General Psychology*, 137, 114–128.

Kilgore, R., & Voshell, M. (2014). Increasing the transparency of unmanned systems: Applications of ecological interface design. In *International Conference on Virtual, Augmented and Mixed Reality* (pp. 378–389). Cham: Springer.

Kozlowski, S. W., & Ilgen, D. R. (2006). Enhancing the effectiveness of work groups and teams. *Psychological Science in the Public Interest*, 7(3), 77–124. doi:10.1111/j.1529-1006.2006.00030.x

Kraus, J., Scholz, D., Stiegemeier, D., & Baumann, M. (2020). The more you know: Trust dynamics and calibration in highly automated driving and the effects of take-overs, system malfunction, and system transparency. *Human Factors*, 62(5), 718–736.

Krausman, A., Neubauer, C., Forster, D., Lakhmani, S., Baker, A. L., Fitzhugh, S. M., ... & Schaefer, K. E. (2022). Trust measurement in human-autonomy teams: Development of a conceptual toolkit. *ACM Transactions on Human-Robot Interaction*, 11(3), 1–58.

Kung, T. H., Cheatham, M., Medenilla, A., Sillos, C., De Leon, L., Elepaño, C., ... & Tseng, V. (2023). Performance of ChatGPT on USMLE: Potential for AI-assisted medical education using large language models. *PLOS Digital Health*, 2(2), e0000198. doi:10.1371/journal.pdig.0000198

Kunze, A., Summerskill, S. J., Marshall, R., & Filtness, A. J. (2019). Automation transparency: Implications of uncertainty communication for human-automation interaction and interfaces. *Ergonomics*, 62(3), 345–360.

Lakhmani, S. G., Neubauer, C., Krausman, A., Fitzhugh, S. M., Berg, S. K., Wright, J. L., ... & Schaefer, K. E. (2022a). Cohesion in human-autonomy teams: An approach for future research. *Theoretical Issues in Ergonomics Science*, 23(6), 687–724. doi:10.1080/1463922X.2022.2033876

Lakhmani, S. G., Pollard, K. A., Forster, D. E., Krausman, A. S., Julia, W. L., & McGhee, S. M. (2022b). *Guidelines for Collecting Laboratory Speech Data*. DEVCOM Army Research Laboratory. Retrieved from https://apps.dtic.mil/sti/pdfs/AD1161191.pdf

Landauer, T. K., Foltz, P. W., & Laham, D. (1998). An introduction to latent semantic analysis. *Discourse Processes*, 25(2–3), 259–284. doi:10.1080/01638539809545028

Lee, J. D., & See, K. A. (2004). Trust in automation: Designing for appropriate reliance. *Human Factors*, 46(1), 50–80. doi:10.1518/hfes.46.1.50_30392

Lei, X., & Rau, P. L. P. (2021). Effect of relative status on responsibility attributions in human-robot collaboration: Mediating role of sense of responsibility and moderating role of power distance orientation. *Computers in Human Behavior*, 122, 106820.

Lewicki, R. J., McAllister, D. J., & Bies, R. J. (1998). Trust And distrust: New relationships and realities. *Academy of Management Review*, 23(3), 438–458. doi:10.5465/AMR.1998.926620

Lewis, J. D., & Weigert, A. (1985). Trust as a social reality. *Social Forces*, 63(4), 967–985. doi:10.1093/sf/63.4.967

Loft, S., Bhaskara, A., Lock, B. A., Skinner, M., Brooks, J., Li, R., & Bell, J. (2021). The impact of transparency and decision risk on human-automation teaming outcomes. *Human Factors*. doi:10.1177/00187208211033445

Lu, L. (2015). Building trust and cohesion in virtual teams: The developmental approach. *Journal of Organizational Effectiveness: People and Performance*, 2(1), 55–72.

Lyons, J. & Guznov, S. (2019). Individual differences in human-machine trust: A multi-study look at the perfect automation schema. *Theoretical Issues in Ergonomics Science*, 20, 440–458.

Lyons, J. B. (2013). Being transparent about transparency: A model for human-robot interaction. In *2013 AAAI Spring Symposium Series*.

Lyons, J. B., Sycara, K., Lewis, M., & Capiola, A. (2021). Human-autonomy teaming: Definitions, debates, and directions. *Frontiers in Psychology*, 12. doi:10.3389/fpsyg.2021.589585

MacArthur, K. R., Stowers, K., & Hancock, P. A. (2017). Human-robot interaction: Proximity and speed-slowly back away from the robot!. In P. Savage-Knepshield, & J. Chen (Eds.), *Advances in Human Factors in Robots and Unmanned Systems* (pp. 365–374). Cham: Springer.

Maddox, I. (2017). Tesla Autopilot Engaged in Model X.jpg. CC-BY-SA 4.0. https://commons.wikimedia.org/wiki/File:Tesla_Autopilot_Engaged_in_Model_X.jpg

Manzey, D., Reichenbach, J., & Onnasch, L. (2012). Human performance consequences of automated decision aids: The impact of degree of automation and system experience. *Journal of Cognitive Engineering and Decision Making*, 6(1), 57–87.

Marathe, A. R., Schaefer, K. E., Evans, A. W., & Metcalfe, J. S. (2018). Bidirectional communication for effective human-agent teaming. In *Virtual, Augmented and Mixed Reality: Interaction, Navigation, Visualization, Embodiment, and Simulation: 10th International Conference, VAMR 2018, Held as Part of HCI International 2018, Las Vegas, NV, USA, July 15-20, 2018, Proceedings, Part I 10* (pp. 338–350). Springer International Publishing.

Marks, M. A., Mathieu, J. E., & Zaccaro, S. J. (2001). A temporally based framework and taxonomy of team processes. *The Academy of Management Review*, 26(3), 356–376. doi:10.2307/259182

Marks, M. A., Sabella, M. J., Burke, C. S., & Zaccaro, S. J. (2002). The impact of cross-training on team effectiveness. *Journal of Applied Psychology*, 87(1), 3.

Martins, L. L., Gilson, L. L., & Maynard, M. T. (2004). Virtual teams: What do we know and where do we go from here? *Journal of Management*, 30(6), 805–835. doi:10.1016/j.jm.2004.05.002

Mathieu, J. E., Heffner, T. S., Goodwin, G. F., Salas, E., & Cannon-Bowers, J. A. (2000). The influence of shared mental models on team process and performance. *Journal of Applied Psychology*, 85(2), 273–283.

Mathieu, J. E., Tannenbaum, S. I., Kukenberger, M. R., Donsbach, J. S., & Alliger, G. M. (2015). Team role experience and orientation: A measure and tests of construct validity. *Group & Organization Management*, 40(1), 6–34.

Mathieu, J., Maynard, M. T., Rapp, T., & Gilson, L. (2008). Team effectiveness 1997-2007: A review of recent advancements and a glimpse into the future. *Journal of Management*, 34(3), 410–476.

Matthews, G., Lin, J., Panganiban, A. R., & Long, M. D. (2019). Individual differences in trust in autonomous robots: Implications for transparency. *IEEE Transactions on Human-Machine Systems*, 50(3), 234–244.

Mavridis, N. (2015). A review of verbal and non-verbal human-robot interactive communication. *Robotics and Autonomous Systems*, 63, 22–35. doi:10.1016/j.robot.2014.09.031

Mayer, R. C., Davis, J. H., & Schoorman, F. D. (1995). An integrative model of organizational trust. *The Academy of Management Review*, 20(3), 709. doi:10.2307/258792

McAllister, D. J. (1995). Affect- and cognition-based trust as foundations for interpersonal cooperation in organizations. *The Academy of Management Journal*, 38(1), 24–59. doi:10.2307/256727

McGrath, J. E. (1984). *Groups: Interaction and Performance*: Englewood Cliffs, NJ: Prentice-Hall.

McNeese, N. J., Demir, M., Cooke, N. J., & Myers, C. (2018). Teaming with a synthetic teammate: Insights into human-autonomy teaming. *Human Factors*, 60, 262–273. doi:10.1177/0018720817743223PubMed

Mercado, J. E., Rupp, M. A., Chen, J. Y., Barnes, M. J., Barber, D., & Procci, K. (2016). Intelligent agent transparency in human-agent teaming for Multi-UxV management. *Human Factors*, 58(3), 401–415.

Merritt, S. M., Heimbaugh, H., LaChapell, J., & Lee, D. (2013). I trust it, but I don't know why: Effects of implicit attitudes toward automation on trust in an automated system. *Human Factors*, 55, 520–534.

Mesmer-Magnus, J. R., & Dechurch, L. A. (2009). Information sharing and team performance: A meta-analysis. *The Journal of Applied Psychology*, 94(2), 535–546. doi:10.1037/a0013773

Mesmer-Magnus, J. R., DeChurch, L. A., Jimenez-Rodriguez, M., Wildman, J., & Shuffler, M. (2011). A meta-analytic investigation of virtuality and information sharing in teams. *Organizational Behavior and Human Decision Processes*, 115(2), 214–225.

Meyerson, D., Weick, K. E., & Kramer, R. M. (1996). Swift trust and temporary groups. In R. M. Kramer, & T. R. Tyler (Eds.), *Trust in Organizations: Frontiers of Theory and Research* (pp. 166–195). Thousand Oaks, CA: Sage Publications, Inc.

Miller, T. (2019). Explanation in artificial intelligence: Insights from the social sciences. *Artificial Intelligence*, 267, 1–38.

Mina, T., Kannan, S. S., Jo, W., & Min, B. C. (2020). Adaptive workload allocation for multi-human multi-robot teams for independent and homogeneous tasks. *IEEE Access*, 8, 152697–152712. doi:10.1109/ACCESS.2020.3017659

Mønster, D., Håkonsson, D. D., Eskildsen, J. K., & Wallot, S. (2016). Physiological evidence of interpersonal dynamics in a cooperative production task. *Physiology & Behavior*, 156, 24–34.

Moses, K. L. (2006). *Military robot being prepared to inspect a bomb.* US Government Public Domain. https://commons.wikimedia.org/wiki/File:Military_robot_being_prepared_to_inspect_a_bomb.jpg

Mudrack, P. E. (1989). Group cohesiveness and productivity: A closer look. *Human Relations, 42*(9), 771–785.

Muir, B. M. (1994). Trust in automation: Part I. Theoretical issues in the study of trust and human intervention in automated systems. *Ergonomics, 37*(11), 1905–1922.

National Academies of Sciences, Engineering, and Medicine. 2021. *Human-AI Teaming: State of the Art and Research Needs.* Washington, DC: The National Academies Press. doi:10.17226/26355

Neubauer, C., Gremillion, G., Perelman, B. S., La Fleur, C., Metcalfe, J. S., & Schaefer, K. E. (2020). *Analysis of Facial Expressions Explain Affective State and Trust-Based Decisions during Interaction with Autonomy.* Paper presented at the International Conference on Intelligent Human Systems Integration.

Neubauer, C., Forster, D. E., Blackman, J., Lakhmani, S., Fitzhugh, S. M., Krausman, A., ... & Rovira, E. 2021. Developing a new human autonomy team cohesion scale. In *Proceedings of the 65th Annual Human Factors and Ergonomics Society* (Vol. 65, No. 1, pp. 801–806). Los Angeles, CA: SAGE Publications.

NHTSA. (2023). *Early Estimate of Motor Vehicle Traffic Fatalities in 2022.* Retrieved from https://crashstats.nhtsa.dot.gov/Api/Public/ViewPublication/813428

Niederhoffer, K. G., & Pennebaker, J. W. (2002). Linguistic style matching in social interaction. *Journal of Language and Social Psychology, 21*(4), 337–360.

Nomura, T., Suzuki, T., Kanda, T., & Kato, K. (2006). Measurement of negative attitudes toward robots. *Interaction Studies, 7*(3), 437–454.

O'Neill, T., McNeese, N., Barron, A., & Schelble, B. (2020). Human-autonomy teaming: A review and analysis of the empirical literature. *Human Factors, 64*(5), 904–938. doi:10.1177/0018720820960865

Onnasch, L., Wickens, C. D., Li, H., & Manzey, D. (2014). Human performance consequences of stages and levels of automation: An integrated meta-analysis. *Human Factors, 56*(3), 476–488. doi:10.1177/0018720813501549

OpenAI. (2023). *Introducing ChatGPT.* Retrieved from https://openai.com/blog/chatgpt

Ososky, S., Schuster, D., Jentsch, F., Fiore, S., Shumaker, R., Lebiere, C., ... & Stentz, A. (2012). The importance of shared mental models and shared situation awareness for transforming robots from tools to teammates. *Proc. SPIE, 8387,* 838710.

Ososky, S., Schuster, D., Phillips, E., & Jentsch, F. G. (2013, March). Building appropriate trust in human-robot teams. In *2013 AAAI Spring Symposium Series.*

Parasuraman, R., & Riley, V. (1997). Humans and automation: Use, misuse, disuse, abuse. *Human Factors, 39*(2), 230–253. doi:10.1518/001872097778543886

Parasuraman, R., Sheridan, T. B., & Wickens, C. D. (2000). A model for types and levels of human interaction with automation. *IEEE Transactions on Systems, Man, and Cybernetics - Part A: Systems and Humans, 30*(3), 286–297. doi:10.1109/3468.844354

Phillips, E., Ososky, S., Grove, J., & Jentsch, F. (2011). From tools to teammates: Toward the development of appropriate mental models for intelligent robots. In *Proceedings of the human factors and ergonomics society annual meeting* (Vol. 55, No. 1, pp. 1491–1495). Sage CA: Los Angeles, CA: SAGE Publications. doi:10.1177/1071181311551310

Pop, V., Shrewsbury, A., & Durso, F. (2015). Individual differences in the calibration of trust in automation. *Human Factors, 57,* 545–556. doi:10.1177/0018720814564422

Pynadath, D. V., Barnes, M. J., Wang, N., & Chen, J. Y. (2018). Transparency communication for machine learning in human-automation interaction. In J. Zhou, & F. Chen (Eds.), *Human and Machine Learning. Human-Computer Interaction Series* (pp. 75–90). Cham: Springer.

Rawal, A., McCoy, J., Rawat, D. B., Sadler, B. M., & St. Amant, R. (2022). Recent advances in trustworthy explainable artificial intelligence: Status, challenges, and perspectives. *IEEE Transactions on Artificial Intelligence, 3,* 852-866.

Rempel, J. K., Holmes, J. G., & Zanna, M. P. (1985). Trust in close relationships. *Journal of Personality and Social Psychology, 49*(1), 95–112. doi:10.1037/0022-3514.49.1.95

Reyes, D. L., & Salas, E. (2019). What makes a team of experts an expert team? In R. F. Subotnik, P. Olszewski-Kubilius, & F. C. Worrell (Eds.), *The Psychology of High Performance: Developing Human Potential into Domain-Specific Talent.* (pp. 141–159). Washington, DC, US: American Psychological Association.

Rodes, W., & Gugerty, L. (2012). Effects of electronic map displays and individual differences in ability on navigation performance. *Human Factors, 43,* 589–599.

Roth, G., Schulte, A., Schmitt, F., & Brand, Y. (2019). Transparency for a workload-adaptive cognitive agent in a manned-unmanned teaming application. *IEEE Transactions on Human-Machine Systems, 50*(3), 225–233.

Rotter, J. B. (1967). A new scale for the measurement of interpersonal trust. *Journal of Personality*, *35*(4), 651–665. doi:10.1111/j.1467-6494.1967.tb01454.x

Rousseau, D. M., Sitkin, S. B., Burt, R. S., & Camerer, C. (1998). Not so different after all: A cross-discipline view of trust. *Academy of Management. The Academy of Management Review; Briarcliff Manor*, *23*(3), 393–404. Retrieved from https://search.proquest.com.ezproxy.libproxy.db.erau.edu/docview/210973020/abstract/475A0B8534E84D90PQ/1

Rovira, E., McGarry, K., & Parasuraman, R. (2007). Effects of imperfect automation on decision making in a simulated command and control task. *Human Factors*, *49*, 76–87.

Rovira, E., Pak, R., & McLaughlin, A. (2017) Effects of individual differences in working memory on performance and trust with various degrees of automation. *Theoretical Issues in Ergonomics Science*, *18*, 573–591.

Sacks, H., Schegloff, E. A., & Jefferson, G. (1974). A simplest systematics for the organization of turn-taking for conversation. *Language*, *50*(4), 696–735. doi:10.2307/412243

SAE International. (2021). *SAE Levels of Driving Automation Refined for Clarity and International Audience*. Retrieved from https://www.sae.org/blog/sae-j3016-update

Salas, E., Cannon-Bowers, J. A., & Johnston, J. H. (1997). How can you turn a team of experts into an expert team?: Emerging training strategies. *Naturalistic Decision Making*, *1*, 359–370.

Salas, E., Cooke, N. J., & Rosen, M. A. (2008). On teams, teamwork, and team performance: Discoveries and developments. *Human Factors*, *50*(3), 540–547. doi:10.1518/001872008X288457

Salas, E., Grossman, R., Hughes, A. M., & Coultas, C. W. (2015). Measuring team cohesion: Observations from the science. *Human Factors*, *57*(3), 365–374.

Salas, E., Shuffler, M. L., Thayer, A. L., Bedwell, W. L., & Lazzara, E. H. (2015). Understanding and improving teamwork in organizations: A scientifically based practical guide. *Human Resource Management*, *54*(4), 599–622.

Salas, E., Sims, D. E., & Burke, C. S. (2005). Is there a "big five" in teamwork? *Small Group Research*, *36*(5), 555–599. doi:10.1177/1046496405277134

Sanneman, L., & Shah, J. (2022). The situation awareness framework for explainable AI (SAFE-AI) and human factors considerations for XAI systems. *International Journal of Human-Computer Interaction*, *38*, 1772–1788.

Schaefer, K. E. (2016). Measuring trust in human robot interactions: Development of the "Trust Perception Scale-HRI". In R. Mittu, D. Sofge, A. Wagner, & W. Lawless (Eds.), *Robust Intelligence and Trust in Autonomous Systems* (pp. 191–218). Boston, MA: Springer.

Schaefer, K. E., Chen, J. Y., Szalma, J. L., & Hancock, P. A. (2016). A meta-analysis of factors influencing the development of trust in automation: Implications for understanding autonomy in future systems. *Human Factors*, *58*(3), 377–400. doi:10.1177/0018720816634228

Sebo, S., Stoll, B., Scassellati, B., & Jung, M. F. (2020). Robots in groups and teams: A literature review. *Proceedings of the ACM on Human-Computer Interaction*, *4*(CSCW2), 1–36.

Selkowitz, A., Lakhmani, S., & Chen, J. (2017). Using agent transparency to support situation awareness of the Autonomous Squad Member. *Cognitive Systems Research*, *46*, 13–25.

Semin, G. R. (2007). Grounding communication: Synchrony. In A. W. Kruglanski & E. T. Higgins (Eds.), *Social Psychology: Handbook of Basic Principles* (2nd ed., pp. 630–649). New York: The Guilford Press.

Shaw, M. E. (1964). Communication networks. In L. Berkowitz (Ed.), *Advances in Experimental Social Psychology* (Vol. 1, pp. 111–147). Cambridge, MA: Academic press.

Sheridan, T. B., & Verplank, W. L. (1978). *Human and computer control of undersea teleoperators* (NR-196-152). Massachusetts Institute of Technology Cambridge, Man-Machine Systems Lab. Retrieved from https://apps.dtic.mil/sti/pdfs/ADA057655.pdf

Shneiderman, B. (2020). Bridging the gap between ethics and practice: Guidelines for reliable, safe, and trustworthy human-centered AI systems. *ACM Transactions on Interactive Intelligent Systems (TiiS)*, *10*(4), 26. doi:10.1145/3419764

Singer, P. W. (2009). *Wired for War: The Robotics Revolution and Conflict in the 21st Century*. London: Penguin.

Sonnentag, S., & Volmer, J. (2009). Individual-level predictors of task-related teamwork processes: The role of expertise and self-efficacy in team meetings. *Group & Organization Management*, *34*(1), 37–66.

Steiner, I. D. (1972). *Group Processes and Productivity*. New York: Academic Press.

Stowers, K., Kasdaglis, N., Rupp, M. A., Newton, O. B., Chen, J. Y., & Barnes, M. J. (2020). The IMPACT of agent transparency on human performance. *IEEE Transactions on Human-Machine Systems*, *50*(3), 245–253.

Stubbs, K., Hinds, P. J., & Wettergreen, D. (2007). Autonomy and common ground in human-robot interaction: A field study. *IEEE Intelligent Systems*, *22*(2), 42–50.

Stuhlsatz, A., Meyer, C., Eyben, F., Zielke, T., Meier, G., & Schuller, B. (2011). *Deep Neural Networks for Acoustic Emotion Recognition: Raising the Benchmarks.* Paper presented at the 2011 IEEE International Conference on Acoustics, Speech and Signal Processing (ICASSP).

Sukthankar, G., Shumaker, R., & Lewis, M. (2011). Intelligent agents as teammates. In E. Salas, S. M. Fiore, & M. P. Letsky (Eds.), *Theories of Team Cognition.* New York: Routledge.

Sundstrom, E., de Meuse, K. P., & Futrell, D. (1990). Work teams: Applications and effectiveness. *American Psychologist, 45*(2), 120–133. doi:10.1037/0003-066X.45.2.120

Sycara, K., & Sukthankar, G. (2006). *Literature Review of Teamwork Models.* Pittsburgh, PA: Carnegie Mellon University.

van de Merwe, K., Mallam, S., & Nazir, S. (2022). Agent transparency, situation awareness, mental workload, and operator performance: A systematic literature review. *Human Factors.* doi:10.1177/00187208221077804

van der Waa, J., Nieuwburg, E., Cremers, A., & Neerincx, M. (2021). Evaluating XAI: A comparison of rule-based and example-based explanations. *Artificial Intelligence, 291,* 103404.

van Dis, E. A. M., Bollen, J., Zuidema, W., va Rooij, R., & Bockting, C. L. (2023). ChatGPT: Five priorities for research. *Nature, 614*(7947), 224–226. Retrieved from https://www.nature.com/articles/d41586-023-00288-7

Ventura, R., & Lima, P. U. (2012, 5-7 Sept. 2012). *Search and Rescue Robots: The Civil Protection Teams of the Future.* Paper presented at the 2012 Third International Conference on Emerging Security Technologies.

Walliser, J. C., de Visser, E. J., Wiese, E., & Shaw, T. H. (2019). Team structure and team building improve human-machine teaming with autonomous agents. *Journal of Cognitive Engineering and Decision Making, 13*(4): 258–278. doi:10.1177/1555343419867563.

Wang, D., Churchill, E., Maes, P., Fan, X., Shneiderman, B., Shi, Y., & Wang, Q. (2020). *From Human-Human Collaboration to Human-AI Collaboration: Designing AI Systems That Can Work Together with People.* Paper presented at the Extended Abstracts of the 2020 CHI Conference on Human Factors in Computing Systems, Honolulu, HI, USA. doi:10.1145/3334480.3381069

Wasserman, S., & Faust, K. (1994). *Social network analysis: Methods and applications.* Cambridge: Cambridge University Press.

Wilson, J. R. 2014. Fundamentals of systems ergonomics/human factors. *Applied Ergonomics, 45*(1), 5–13.

Winfield, A. F., Booth, S., Dennis, L. A., Egawa, T., Hastie, H., Jacobs, N., ... & Watson, E. (2021). IEEE P7001: A proposed standard on transparency. *Frontiers in Robotics and AI, 8,* 665729.

Wright, J., Chen, J., & Lakhmani, S. (2020). Agent transparency and reliability in human-robot interaction: The influence on user confidence and perceived reliability. *IEEE Trans. Human-Machine Systems, 50,* 254–263.

Wynne, K. T., & Lyons, J. B. (2018). An integrative model of autonomous agent teammate-likeness. *Theoretical Issues in Ergonomics Science, 19*(3), 353–374. doi:10.1080/1463922X.2016.1260181

You, S., & Robert, L. P. (2017). Teaming up with robots: An IMOI (inputs-mediators-outputs-inputs) framework of human-robot teamwork. *International Journal of Robotic Engineering (IJRE), 2*(3). Retrieved from https://papers.ssrn.com/abstract=3308173

Zaccaro, S. J., Gualtieri, J., & Minionis, D. (1995). Task cohesion as a facilitator of team decision making under temporal urgency. *Military Psychology, 7*(2), 77–93.

Zhou, Z., Li, Z., Zhang, Y., & Sun, L. (2022). Transparent-AI blueprint: Developing a conceptual tool to support the design of transparent AI agents. *International Journal of Human-Computer Interaction, 38*(18–20), 1846–1873.

12 Ambient Assisted Living Solutions

Martina Ziefle, Julia Offermann, and Wiktoria Wilkowska

12.1 INTRODUCTION

The world's population is aging rapidly, with significant demographic shifts occurring in both developed and developing countries. This shift has resulted in the challenge of aging societies, with complex and multidimensional requirements for care and quality of life. As people age, they require more support and short or long-term care, including healthcare, chronic disease management, daily assistance, social services, and organizational assistance (Deusdad et al., 2016). Ensuring that these requirements are met is essential for maintaining the health and well-being of older adults and for creating sustainable and resilient societies that can support people of all ages (Oeppen & Vaupel, 2002). However, there are severe and pressing bottlenecks regarding the shortage of healthcare workers, including doctors, nurses, and other healthcare professionals (Pickard, 2015). The bottleneck in the care supply chain is still larger in rural compared to urban areas (Terschüren et al., 2012) and is aggravated by the fact that healthcare workers experience increasing burnout and stress due to the high demand for care, long working hours, and the emotional toll of caring for older adults with chronic conditions (Gustafsson et al., 2010; Bruria et al., 2022; Yıldız, 2023). The precarious situation in nursing is exacerbated by bottlenecks in the financing options for medical care within existing healthcare models (Little & Triest, 2001). Thus, there is considerable pressure to develop novel solutions to meet the bottleneck in the care of older adults, but, at the same time, to meet the core requirements of older adults in care (Bloom et al., 2010, Mynatt et al., 2004).

Ambient assisted living (AAL) technologies hold a large potential to meet these caring bottlenecks. AAL technologies assist older adults and their families in the care process at home and provide enhanced safety and security to detect severe health problems, fall detection, emergency response systems, and remote monitoring as valuable for maintaining users' safety and independence. Using lifelogging AAL technologies to monitor vital signs or provide reminders for medication, older adults are supported to manage their health and to improve their overall quality of life (Kleinberger et al., 2007, Mynatt et al., 2004). Furthermore, AAL technologies can also help to reduce social isolation and loneliness among older adults, e.g., by connecting users across homes and locations with family members and friends that live remotely (Jaschinski & Ben Allouch, 2015, Shah & Patel, 2018, Kasugai et al., 2010)

Still, on the other hand, not only older adults but also care personnel might see severe drawbacks that come with the use of AAL technologies (Offermann-van Heek & Ziefle, 2019a, Offermann et al., 2023, Messner et al., 2019). First and foremost, perceived disadvantages are related to privacy concerns and security issues of data protection (Krasnova et al., 2012; Schomakers et al., 2019a,b, 2020).

In addition, an essential barrier is the uncertainty of what could happen with the personal data by third parties (Calero Valdez & Ziefle, 2019). Especially lifelogging technologies and sensors that monitor vital parameters continuously but also the use of cameras in the intimacy of own homes are often reported as invasive and intrusive and, therefore, a threat to individuals' privacy and security (Himmel & Ziefle, 2016; Offermann et al., 2022). Another considerable concern relates to the perceived stigma associated with using AAL technologies, particularly if they are perceived as only being used by individuals who are frail or dependent (Ziefle & Schaar, 2014). This may lead to reluctance to use the technology, particularly if it is visible to others (Farrington 2016).

This chapter sheds light on the different aspects of AAL technology – its technical potential but also the requirements of human and social values required by users of AAL technology.

12.1.1 Aging Societies and their Multidimensional Requirements for Care and Quality of Life

Addressing the challenge of aging societies requires a comprehensive and multifaceted approach that considers the unique needs of older adults, as well as the broader social, economic, and cultural contexts in which they live (Morgan & Kunkel, 2001; Grenier, 2012; Murray & Laditka, 2010).

Aging refers to the natural process of getting older, which involves physical, psychological, and social changes that occur over time (Martinson & Berridge, 2014, Grenier, 2012). As people age, they may experience changes in their appearance, physical abilities, cognitive functioning, and social roles (Friedrich, 2001). Aging can also bring new opportunities for personal growth and development as well as new challenges and limitations (Wiles et al., 2012; Wilkowska et al., 2018).

Aging attitudes refer to the beliefs and attitudes that people hold about aging and older adults. These attitudes can be positive or negative and may affect how people perceive and interact with older individuals. Positive attitudes toward aging can include respect, appreciation, and recognition of the value and contributions of older adults (Troutman-Jordan & Staples, 2014). Negative attitudes toward aging can include stereotypes, ageism, and discrimination based on age (Reich et al., 2020, Reichstadt et al., 2010).

Ageism is a form of discrimination that involves prejudice or stereotyping against individuals or groups based on their age, particularly against older adults (Palmore, 2005). Ageism can take many forms, including age-based assumptions, stereotypes, and discriminatory practices in employment, healthcare, and other areas of life (Ivan & Cutler, 2021). Ageism can have negative effects on older adults, including reduced access to resources and opportunities, social isolation, and negative health outcomes. It is important to recognize and challenge ageism in all its forms to promote fairness and equity for people of all ages (Shallcross et al., 2013).

12.1.2 Key Social Requirements for the Design and Implementation of AAL Technologies

The following social requirements can be referred to which should be considered in the development and integration of AAL technologies in line with target groups' characteristics and claims for human-centered, technology-assisted care.

- *Aging and health:* As individuals age, their health and functional abilities may decline, making it difficult for them to perform daily living activities independently (Whitbourne, 2002; 2012). AAL solutions can help older adults manage their health, monitor chronic conditions, and prevent falls and injuries (Reddy et al., 2021).
- *Attitudes toward aging:* Attitudes toward aging can impact how older adults perceive and adopt AAL solutions. Negative attitudes toward aging can lead to reluctance in adopting new technologies and resistance to change (Biermann et al., 2018; Offermann-van Heek et al., 2021). Therefore, AAL solutions must be designed to be user-friendly, intuitive, and aligned with the needs and preferences of older adults (Wilkowska et al., 2018).
- *Independence and autonomy:* Maintaining independence and autonomy is crucial for older adults' well-being and quality of life. AAL solutions can help older adults live independently and age in place, promoting their sense of control and self-esteem (Ziefle & Schaar, 2014; Bedrov & Bulaj, 2018).

Social interaction and connection: Social isolation and loneliness can have adverse effects on older adults' mental health and well-being. AAL solutions can help older adults stay connected with friends and family, engage in social activities, and prevent social isolation (Demiris et al., 2004)

Age-friendly design: AAL solutions must be designed to meet the needs of older adults, incorporating age-friendly design principles such as clear and simple interfaces, large text, and appropriate color contrast (Pinto et al., 1997, Ziefle, 2010; Oetjen & Ziefle, 2009)

Ethics: AAL solutions must be designed with ethical principles in mind, including respect for autonomy, privacy, and dignity (Ziefle and Schaar, 2014; Offermann-van Heek et al., 2019b, c). For example, AAL solutions that collect personal data must ensure that individuals' privacy is protected, and that the data is used only for its intended purpose (Schomakers & Ziefle, 2023, Offermann-van Heek et al., 2019b, Florez-Revuelta et al., 2018).

Cultural diversity: AAL solutions must be designed to respect and accommodate cultural diversity (Offermann-van Heek et al., 2020, Wilkowska et al., 2021). This includes considerations such as language barriers, different cultural attitudes toward technology, and differences in living arrangements and family structures (Metallo et al., 2022; Wilkowska et al., 2021).

Understanding aging and attitudes toward aging is essential in designing effective and accepted AAL solutions (Köttl et al., 2022). Aging affects individuals differently, and affective as well as cognitive attitudes toward aging can influence how individuals perceive and adopt AAL technologies (Biermann et al., 2018; Offermann et al., 2022, Higgs & Gilleard, 2020).

12.2 THE TECHNOLOGICAL POTENTIAL: USE CASES, FUNCTIONS, AND PROTOTYPES

Innovations in active and assisted living technologies, environments, and services, such as *smart homes and Ambient Assisted Living (AAL)*, offer promising and technology-assisted solutions to meet these health and social needs while benefiting from economic opportunities. While there are also overlapping functionalities in both concepts, still there are also slightly different foci (for an overview, see Maskeliūnas et al., 2019, Kleinberger et al., 2007).

A *smart home* – as the more general concept in comparison to AAL – is a home equipped with devices and systems that can be controlled remotely or automatically to increase convenience, comfort, and security. It involves integrating various technologies and devices, such as sensors, cameras, voice assistants, smart thermostats, smart appliances, and automated lighting systems, to provide a more efficient and personalized living experience. Other benefits of a smart home include increased security through smart locks and security cameras, energy efficiency through automated systems that optimize energy consumption, and improved health and wellness through devices that monitor air quality, humidity levels, and sleep patterns (Santoso & Redmond, 2015; Vanus et al, 2017).

AAL is a specific case of a smart home that is designed to assist people in their daily activities, particularly the elderly or those with disabilities and increased care needs. AAL environments provide support, increase independence, and improve the quality of life of individuals who require assistance in their daily living activities (Messner et al., 2019). AAL technology in smart homes also integrates various technologies and devices, such as motion sensors, wearable devices, smart lighting systems, voice assistants, and emergency response systems, to enable individuals to live independently and safely. These devices can monitor the activities of individuals and help when needed, such as reminding individuals to take their medication, detecting falls, and alerting emergency services if necessary (Maskeliūnas et al., 2019, Nehmer et al., 2006).

Recent advances in wearable computing with a variety of products on the market (e.g., wearable cameras and smartwatches, wristbands, or glasses), the increasing functionality of mobile devices and apps for health and well-being, and the easier and less expensive installation of home automation systems are all contributing to the growing use of health care and assisted living services (Stavropoulos et al., 2020, Choi et al., 2020). By sensors that are integrated at different locations – be it at home or in care facilities – such as e.g., sensors in floors or carpets, furniture, walls, or ceilings – or in different textiles or as wearable devices (e.g., wristbands, watches, etc.), activities and behavioral patterns can be monitored (Vacher et al., 2011, Choukou et al., 2021). These lifelogging technologies can enable and motivate individuals to comprehensively record data about themselves, their environment, and the people with whom they interact (Wilkowska et al., 2022). The collection and processing of physiological signals, movement, location, and activities performed form the basis for providing a variety of state-of-the-art services to improve people's health, well-being, and independence (Offermann et al., 2023). These services include personalized healthcare, well-being monitoring (physical activity, dietary habits), support for people with memory impairment, social participation, mobility, support for formal and informal caregivers (medication management, management of care activities), and predictive systems (declining cognitive abilities, aggressive behavior, and fall prevention) (for an overview see Choukou et al., 2021).

In Table 12.1, a non-exhaustive, but a generic overview of technical solutions for AAL environments and care situations is given.

TABLE 12.1

Overview about Use Cases and Prototypes for AAL Environments and Care Situations

Function	Description	Examples
	Use Cases	
Monitoring health	Wearable devices, sensors, and smart home systems can be used to monitor an individual's health and detect any changes that may require medical attention	For example, sensors can monitor blood pressure, heart rate, and sleep patterns, while wearable devices can track physical activity and provide feedback on exercise and fitness routines. For an overview see Cicirelli et al. (2021). An example for a digital medication system is reported by Siek et al. (2011).
Safety and security	AAL systems can be used to improve the safety and security of individuals in different ways, daily assistance, but also critical life situations or hazards.	For example, lifelogging technologies can help by detecting falls, alerting emergency services in case of an accident or emergency, and monitoring the home for potential hazards. For an overview see Cicirelli et al. (2021). More details about privacy preservation technologies can be found in Ravi et al. (2023).
Social interaction	AAL solutions can also be used to facilitate social interaction and reduce loneliness and isolation.	For example, social robots can be used to provide companionship and engage individuals in activities, while video chat systems can be used to connect with friends and family members who live far away. For an overview see Latikka et al. (2021). As an example, a social interaction system with a telepresence robot was developed by Coradeschi et al. (2013) that has been implemented in several caring homes.
	Prototypes	
Mobility aids	Smart walkers for people with limited mobility and vision: Smart walkers equipped with sensors and cameras can monitor an individual's walking patterns, detect falls, and provide assistance with balance and stability.	Ultrasound-based, infrared depth cameras and RGB cameras that are equipped with computer vision processing, software systems and user interfaces equipped with haptic and audio signals. For an overview, see Mostafa et al. (2021).

(Continued)

TABLE 12.1 (*Continued*)
Overview about Use Cases and Prototypes for AAL Environments and Care Situations

Function	Description	Examples
Smart assistants	Smart home assistants: Companies such as Google, Amazon, and Apple but also specialized SMEs (small and medium enterprises) offer smart home systems that can be customized with technical solutions, such as voice assistants, smart thermostats, and security cameras.	Voice Assistant Prototypes for Personal Health, e.g., by Google Assistant and Amazon Echo voice assistant platforms. For an overview see Shalini et al. (2019).
Smart textiles	Smart textiles: Smart textiles, such as clothing and bedding, can be equipped with sensors and electronics to monitor an individual's health, detect falls, and provide feedback on sleep quality.	For example, Saponas and colleagues developed PocketTouch, a system that enables gesture recognition on flat surfaces in wearable textiles (Saponas et al., 2011). Another example that can be referred to in this context is a smart couch (Rus et al., 2017) equipped with capacitive sensing integrated into its textile surface to facilitate touch interaction. For an overview see Offerman-van Heek et al. (2018).
Wearable devices	Wearable devices: Wearable devices, such as smartwatches and fitness trackers, can be used to monitor an individual's health, track physical activity, and provide feedback on exercise routines.	Wearables are often used for Alzheimers disease and dementia to detect fall injuries and wandering around (Rodrigues et al, 2018; Algase et al., 2003). For an overview see Stavropoulos et al. (2020).

12.3 TECHNOLOGY ACCEPTANCE, PRIVACY, AND TRUST IN AAL TECHNOLOGY

With the ever-increasing use of technical devices and digital applications that are increasingly entering private environments and homes, the users' acceptance is an essential cornerstone for the success of AAL lifelogging technologies. To understand the use or rejection of novel digital applications and to predict human attitudes and behaviors regarding the usage of digital applications at home, two theoretical approaches need to be considered and combined as a base for the empirical procedure.

12.3.1 TECHNOLOGY ACCEPTANCE

Acceptance has been evolved as a central component for the success of technical innovations and is a central prerequisite for the successful development of AAL technologies. Technology (social) acceptance is defined by the active and positive reception and the willingness to use novel technologies (Dethloff, 2004). Social acceptance has an attitude- and a behavior-related dimension: The acceptance of a technology requires a positive attitude toward the technology and an intention to use the technology.

The research on technology acceptance has its origin in empirical social research. Ajzen and colleagues developed the "Theory of Reasoned Action" to study social behavior (Ajzen et al., 1980). The approach assumes that behavioral intentions are preceded by attitudes (e.g., previous experience) and social norms. The "Theory of Planned Behavior" as the successor model (Ajzen, 1991) added another component – the perceived behavioral control – and empirically validated relationships between predictors of behavioral intent. The most influential model is the technology acceptance model (TAM) (Davis, 1989; Davis et al., 1989). According to the TAM, the intention to use a technology depends on the attitude toward the technology, which is shaped by two factors: the perceived usefulness and the ease of using a device. TAM was further developed in 2000 (TAM 2, Venkatesh & Davis, 2000) by adding further predictors of usefulness (the subjective norm, the relevance to the activity, the quality of the result, and the demonstrability of the results). The next model adjustment resulted in the TAM 3 (Venkatesh & Bala, 2008) with the addition of person-related factors (control in handling technology, external controllability, fear in dealing with computers, and playful use of computers) and two usage factors (perceived fun and usability), which influence acceptance mediated by the simplicity of use. The next step in acceptance modeling was the "Unified Theory of Acceptance and Use of Technology" (UTAUT) which supplements elements from socio-cognitive theory and motivation theory (Venkatesh et al., 2003).

Social acceptance research puts the requirements of end users into the central focus and communicates users' needs to technical designers, employers, and responsible persons in education, policy, and governance. The early integration of acceptance helps to use socially accepted pathways for digitally assisted working settings and to reflect human factors' needs already from scratch. This way, innovative solutions can be better adapted to the needs of the stakeholders (companies, technical experts, and end users) from the very beginning of the technical development to deployment and roll-out.

This way, the following aspects should be considered from a social acceptance point of view.

- *Identifying barriers to adoption:* Understanding technology acceptance can help designers identify barriers to adoption, such as lack of familiarity, fear of technology, or difficulty using the technology. By identifying these barriers, designers can create AAL solutions that are user-friendly, intuitive, and easy to use, promoting adoption by older adults.
- *Tailoring solutions to user needs:* User-centered design involves the user in the design process, ensuring that AAL solutions are tailored to the needs and preferences of older adults (Gulliksen et al, 2003; Coleman et al, 2016). By involving older adults in the design process, designers can identify design features that are important to them, such as larger text or simplified interfaces. This can lead to more effective and efficient solutions that promote independence, autonomy, and social connection for older adults.
- *Improving usability and user satisfaction:* Incorporating user-centered design principles can lead to AAL solutions that are more usable and satisfying for older adults. By designing solutions that are intuitive, user-friendly, and align with the needs and preferences of older adults, designers can promote a positive user experience, increasing the likelihood of adoption and sustained use.
- *Enhancing overall effectiveness:* By designing AAL solutions that are appropriate for older adults, designers can enhance the overall effectiveness of the technology. This can lead to better health outcomes, increased independence and autonomy, and improved quality of life for older adults.

Even though technology acceptance is a core element of designing user-centered AAL technologies, none of the TAM-based models did consider perceptions of data and information privacy as a model component, which, however, is vital for the use of digital applications. This especially applies to the context of technology-assisted health care and as different illnesses or ailments might have different (perceived) repercussions, be it a stigma for having to deal with the ailment and the fear of losing face if someone would know about it (Lahlou, 2008; Ziefle & Schaar, 2014). Therefore, general acceptance models can only cautiously be applied to the healthcare context and extensions of the original predictors need to be reconsidered (Holden & Karsh, 2010; Vassli & Fatshchian, 2018). Empirical evidence shows that decision patterns differ across usage contexts of digital health technologies (Schomakers et al., 2019a) as well as for different types of mobile health apps (Schomakers et al., 2019b). Research indicates that trust, perceptions of privacy, and social influence have a stronger impact in a serious medical context than in lifestyle contexts (Burbach et al., 2019, Schomakers et al., 2020, 2022).

12.3.2 Perception of Privacy and Trust toward Medical Technology

Any lifelogging technology is collecting data about users, their actions, and behaviors (Koops et al., 2017), and this areawide data collection raises serious worries about the keeping of information privacy (European Commission 2015; Gantz & Reinsel, 2012; Smith et al., 2011). In line with Burgoon (1982), privacy is defined as the limitation of access to the self and distinguishes between physical, psychological, interactional, and informational privacy. Information privacy (Westin 1967) focuses on privacy as control over information about oneself (informational privacy), which has evolved to a key component of any digital behavior.

Concerns about information privacy reflect the discrepancy between the desired and actual state of privacy (Ziefle et al., 2016; Li, 2011). Privacy concerns, among other attitudes and context factors, influence behavioral intentions and actual behaviors, e.g., the provision of information and use of technologies of data represents the critical point concerning the acceptance of digital applications in general, and medical services in particular (e.g., Calero Valdez & Ziefle, 2019; Offermann-van Heek et al., 2020).

The willingness of users to tolerate broad data collection and the tolerance of (technical) surveillance (Karabey, 2012; Wilkowska et al., 2021) is of utmost importance in this context. Characteristically, trade-offs need to be negotiated on different levels and situations: For example, the trade-off between keeping personal privacy on the one hand, and the disclosure of data to healthcare providers on the other hand is of importance (Ziefle et al., 2016; Ziefle & Calero Valdez, 2018). Typically, users follow an affective or experience-based understanding of data or information sensitivity (Schomakers et al., 2019b). The mismatch between the perceived sensitivity of personal data and the technical (factual) data sensitivity can lead to misconceptions of users, and, as a consequence, careless user behaviors on the one hand (undertrust) and exaggerated concerns on the other hand (over-trust) (Ziefle et al., 2016; Ziefle & Calero Valdez, 2018).

When it comes to the question if users would be willing to share their data and allow the constant monitoring by means of smart devices, the temporary benefits of the novel services might be higher than the concerns of what could happen with the data. This is referred to as privacy calculus (Dinev & Hart, 2006; Schomakers et al., 2022). According to the privacy calculus theory, users weigh the perceived risks to their privacy and the potential benefits they gain (Gerber et al., 2018). If the perceived benefits outweigh the perceived risks, or, if the effort for privacy protection is perceived as higher than the expected risks, then users share their data and agree that their digital footprints are monitored. The privacy calculus has been validated for, e.g., social networks (Dienlin & Metzger, 2016), location-based services (Sun et al., 2015), and health data (Calero Valdez & Ziefle, 2019) but also for wearables (Li et al., 2016; Burbach et al., 2019) and mobile devices (Keith et al., 2013).

Recent research showed that most users are quite sensitive in the context of data exchange and privacy issues, especially when the data is used by third parties without public transparency (Burbach et al., 2019). Also, users' diversity has an impact on privacy decisions, such as minor technical knowledge and over-trust in having control of data (Schomakers et al., 2019a, Schomakers et al., 2020). Additionally, the individual disposition to privacy needs and the extent of trust in unknown technologies, but also the usage context, gender, culture, and individual differences influence privacy decision-making (Wilkowska et al., 2021; Kehr et al., 2015, Offermann-van Heek et al., 2021). Another critical issue for users concerns the question of how long data may be stored and which authority is responsible for the storage. The longer the data storage and the more data is stored on servers beyond the control of the users (e.g., central servers of companies or public health management), the lower the willingness to share data, independently of the type of data (Offermann-van Heek & Ziefle, 2019a; Offermann-van Heek et al., 2020). Concerns are also higher the more personal the information is and the higher the probability of being identifiable (Calero Valdez & Ziefle, 2019). However, there is also empirical evidence that people seem to be differently vulnerable to those concerns (Baek et al., 2014; Burbach et al., 2019; Schomakers et al., 2018; Offermann et al., 2023).

Overall, a general distrust in the public is prevailing toward authorities dealing with data in a careful, protective, and diligent manner (Offermann-van Heek et al., 2019c) and, in addition, archival concerns toward invasions of privacy (Schmidt et al., 2016; Maidhof et al., 2022).

12.4 USERS' PERSPECTIVES AND EMPIRICAL ASSESSMENT OF AAL LIFELOGGING

The potential of AAL technology and the multitude of AAL solutions which are already on the market as well as those that are in the developmental phase shows that innovative technology is a central cornerstone to tackling the care bottlenecks and the increasing numbers of older people in care and to allow them to stay at home as long as possible and to live independently assisted by technology.

It becomes evident that the advantages of using AAL technologies are strongly acknowledged by care receivers, family members as well as care personnel. But, at the same time, it is also clear that these technologies come along with considerable personal risks and individual costs, in terms of concerns about privacy and intimacy, dignity, and protection motivation (Nissenbaum, 2010). Also, the limited control over the technologies, especially in severe health situations, in which life-end decisions may play a role (Leung, 2007; Proulx & Jacelon, 2004), can lead to perceptions of a loss of self-determination, and these perceptions should be heard and understood by technical designers and people in charge of policy and governance in the medical sector.

These barrier perceptions still might be more sensible in the care sector, as frailness at older age and protection needs, but also the confrontation with physical decline and dying are highly prevalent in the care sector (Offermann-van Heek & Ziefle, 2019a, Lloyd, 2004).

Betz and colleagues (2010) reported that the factors of security, trust in technology but also the medical system as well as privacy issues and data protection predominate in the use of AAL technologies by older people, while a work-life balance is important for relatives of people in need of care (Offermann-van Heek et al., 2019c). Care professionals tend to prioritize aspects of usability but are also concerned about their privacy envisioning that the supervisors are judging their caring quality (Offermann-van Heek & Ziefle, 2019a). In addition, the willingness for using AAL technology at home is very individual and reasons for (non-)usage strongly relate to affective and cognitive attitudes toward aging and life quality. Attitudes toward aging may vary within the diverse group of older adults and across different cultures, and this can impact older adults' evaluations of AAL technologies (Biermann et al., 2018). For example, cultures that place a higher value on interdependence and family may view AAL technologies that promote independence as less desirable (Wilkowska et al., 2019). Beliefs about privacy may also vary interculturally, and this can influence older adults' perceptions of AAL technologies, particularly those related to food, housing, and social norms (Offermann-van Heek et al., 2021). The decision to accept AAL thus requires a sensible understanding of the balancing of the perceived advantages and disadvantages associated with the use of technology.

In the following, we report on recent empirical insights into the perceptions of the older adults' group. The insights do not reflect a systematic exploration of the wide area of AAL technologies. Rather, the findings show the diversity and complexity of the users' voice, their evaluation of Aging, and the question of which part AAL technologies play, the perceived benefits, but also the barriers of using AAL at home as well as the often-ignored critical question, which function or actions at home AAL technology is allowed to at all.

For these questions, empirical social science methodologies were applied: Prior to each of the quantitative studies, interviews and focus groups were used as qualitative research methods as a basis for each quantitative approach, which aimed at exploring the respective focal points from the perspective of the older adults' group, the target group of AAL technology. These arguments are not listed in detail, as the results were mainly used as a basis for the design of the quantitative instruments and can be retrieved from the original publications.

12.4.1 Attitudes toward Aging

The first study was focusing on the perceptions of aging and the evaluations of older adults toward positive and negative aspects that are related to the aging process (Wilkowska et al., 2021). The items used in the quantitative study, which was run online with 384 German users, were collected in two focus groups with older adults (N = 18, 50% female users 60–75 years of age).

In Figure 12.1, the descriptive results are given.

As taken from Figure 12.1, aging does imply considerable positive effects, showing that the picture of aging in the older group is related to wisdom, serenity, and socially valuable family situations in which older adults see themselves as integrated. Negative effects of aging – generally much lower agreed on in comparison to the positive aspects of aging, focus on a decrease of vitality, mobility and independence accompanied by loneliness. If the health status of participants on attitudes toward aging is integrated into the analysis, there was a significant impact: Healthy older adults had a more positive attitude toward aging and rated the positive effects of aging higher and negative effects lower in comparison to participants with a chronic age-related disease.

12.4.2 Perceived Benefits and Barriers of Camera-Based Assistive AAL Technologies

The next empirical study was concerned with the question of which specific benefits of AAL technology are acknowledged and which barriers were seen, a sample of 126 Turkish and German users was contacted. The study aimed at understanding the role of different cultural values in the acceptance of AAL technologies. As a specific type of AAL technology, participants had to evaluate the use of camera-based assisting technologies (Offermann-van Heek et al., 2021). The selection of both countries was arbitrary, in this case, the interdependence and individualistic nature of Germany and, in contrast, the much higher family-centric attitudes of Turkish participants. Figure 12.2 shows the evaluations of the benefits in both nationalities.

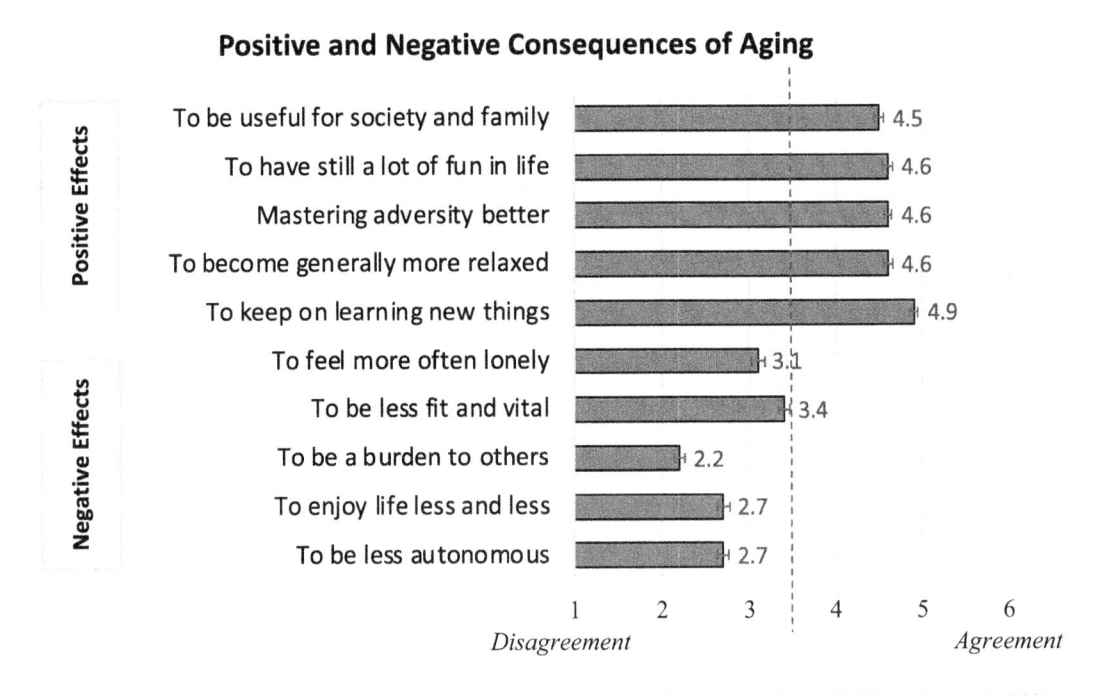

FIGURE 12.1 Evaluations toward negative and positive effects of aging (*N*=384). (Wilkowska et al., 2021.)

Ambient Assisted Living Solutions

	Turkish	German
Enabling of fast reactions in emergencies	5.0	4.4
Enabling fast access to health data	4.7	3.8
Ensuring a regular check of health from home	4.6	3.8
Relief of relatives	4.6	3.5
Trusting handling of health data	4.4	2.4
Relief in everyday life	4.1	3.0
Reducing dependency on others	4.1	3.0
Comparing life and health data with others	4.1	3.3
Increase in autonomy	3.9	3.1
Recording of memories	3.8	2.7

evaluation: rejection (1.0) — agreement (6.0)

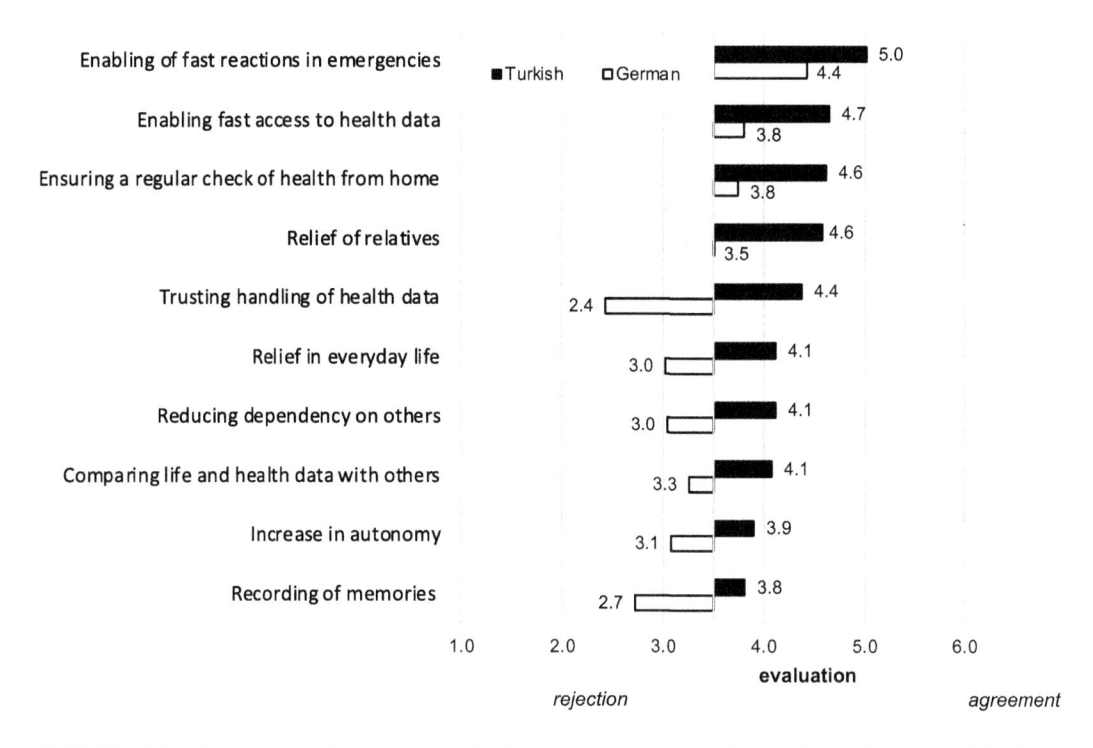

FIGURE 12.2 Evaluations of the benefits of using camera-based assisting technologies in Turkish (Black bars) and German (white bars) older users ($N = 126$). (Offermann-van Heek et al., 2021.)

What we see is a differentiated evaluation of AAL technologies, with both, concordant opinions between both nationalities, but also considerable differences.

First of all, the fast help in emergency situations, the fast access to health data as well enabling regular health checks as well as the relief of the care burden by relatives are seen as positive by both countries. However, the agreement is significantly lower in Germany as it is out of the perspective of the Turkish participants. However, there are also contradictory evaluations, in which German users do not see the benefits AAL technology could have as positive as the Turkish participants. Most remarkable is the difference between German and Turkish users in their trusting of the handling of health data, and their disliking of AAL technology in a private sphere (e.g., recording of memories).

How do German and Turkish participants evaluate the perceived barriers of using camera-based AAL technology at home? (Figure 12.3).

Turkish and German users do also evaluate the barriers to using camera-based assistive technologies differently. It becomes evident that German users have a much higher distrust in data handling and express higher concerns about the missing control over the data, and the uncertainty about what could happen with the data by third and unauthorized third parties in comparison to Turkish participants.

However, the picture is changing when the role of technology in humane care is under study. Here, the Turkish participants rate the concerns about loneliness and isolation, the missing human care (instead of technology), and the dependence on technologies much more negative in comparison to the German sample.

Evidently, cultural frames and different social and societal values the enormous impact of users' diversity on acceptance of AAL technology and should be considered in professional training but also in technology development (Offermann-van Heek et al, 2021).

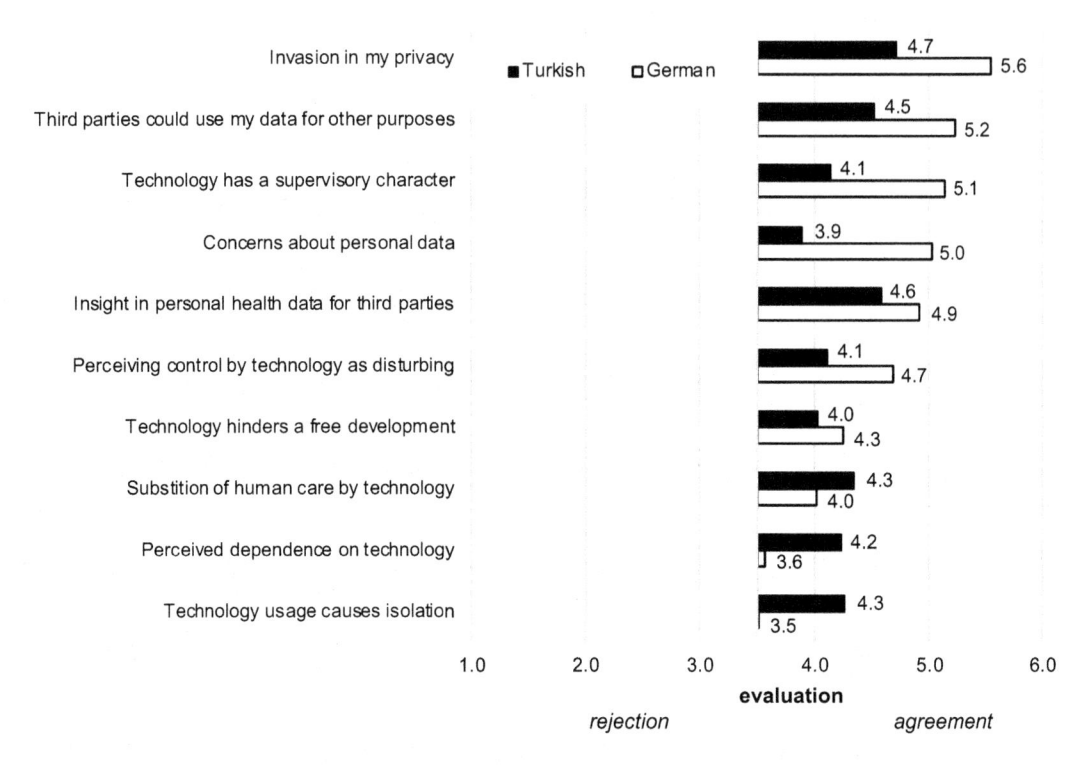

FIGURE 12.3 Evaluations of the barriers of using camera-based assisting technologies in Turkish (Black bars) and German (white bars) older users. (Offermann-van Heek et al., 2021.)

12.4.3 What is AAL Technology (Not) Allowed to Do?

Another online survey on ethical and social aspects in the context of the use of assistive lifelogging technologies focused specifically on what technological assistance may and may not be used in everyday life. The study was based on a preliminary expert interview study and two focus groups with potential AAL technology users (Offermann-van Heek et al., 2019b). The online survey was completed by $N=195$ German participants of different ages and with varying prior experience in health management and care.

The results (Figure 12.4) show that the respondents are significantly more positive about daily reminders, basic support in everyday life, recognition of speech, gestures or fingerprints, and video surveillance outside the private sphere in comparison to the tolerance toward video surveillance in the own living environment or the assumed replacing of human care and social contacts with AAL technology.

In addition, the respondents were also asked to rate which functions or activities assistive technology should not, or must not, take over (Figure 12.5).

Results show a clear picture: Lifelogging technology must not make users "dependent", must not make decisions on its own, and must not fail. Other potential barriers, as for example the complexity of technology usage, but also the potential privacy intrusion by the continuous monitoring as well as audio and video recordings were of lesser negative impact (Figure 12.5). Furthermore, lifelogging technologies should not prevent users from having the freedom to choose for help or not (self-determination), should not prevent social contacts, and should not take over too many tasks – as this again decreases the independence of users.

In addition, we explored participants' opinions on the general role and decision authority in the context of end-of-life decisions. It is essential to understand whether end users trust technology enough to allow it to take over decision-making authority in critical living conditions. We asked

Ambient Assisted Living Solutions 375

What is lifelogging technology allowed to do?

Function	Evaluation
Reminder (intake of medicine, appointments)	5.3
Support or help	5.0
Recognition of language and gestures	4.8
Video surveillance (outside the own house)	4.6
Recognition of fingerprints	4.7
Monitoring (measurement, recording of a process)	4.6
Storage of data when data is encrypted	4.4
Video surveillance (in the house)	3.7
Complement social contacts	3.7

Evaluation (min=1; max=6)
rejection — agreement

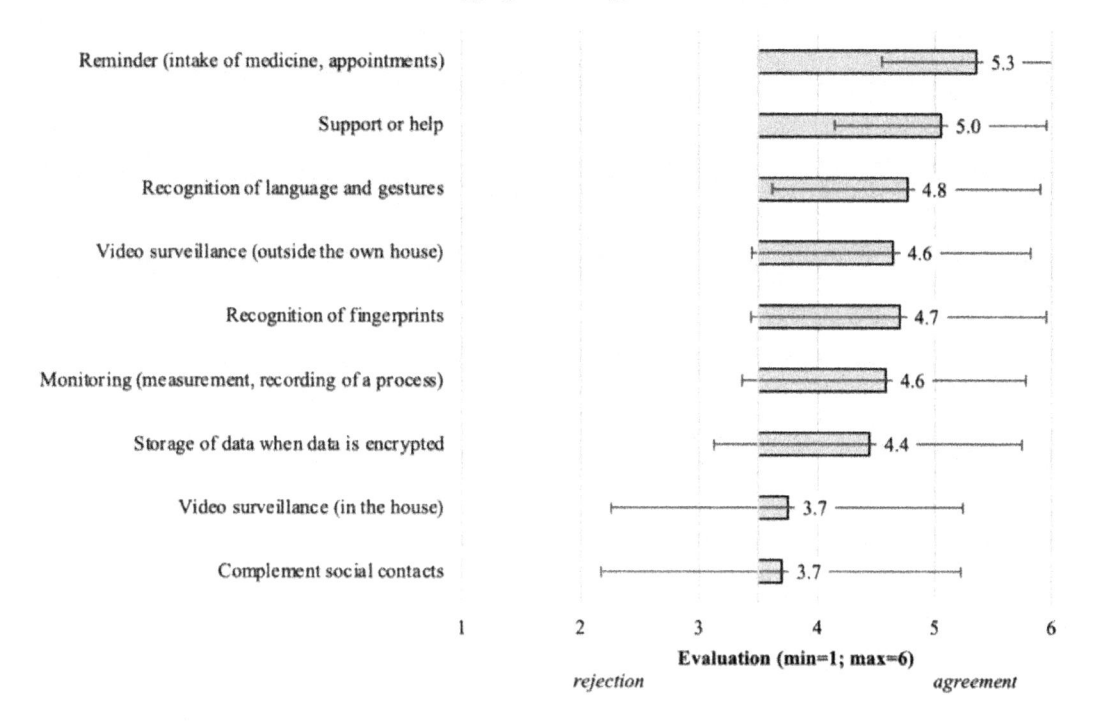

FIGURE 12.4 Evaluations of functions and actions AAL lifelogging technology is allowed to do out of the perspective of 195 German users. (Offermann-van Heek et al., 2019b.)

What is technology NOT allowed to do?

Function	Evaluation
Making dependent on technology	5.0
Making decisions independently	4.9
Technical failure	4.8
Restricting freedom of choice	4.6
Substitute social contacts	4.4
Taking over too many tasks	4.3
To be complex	4.1
Surveillance	4.0
Audio and video recordings	3.7

Evaluation (min=1; max=6)
rejection — agreement

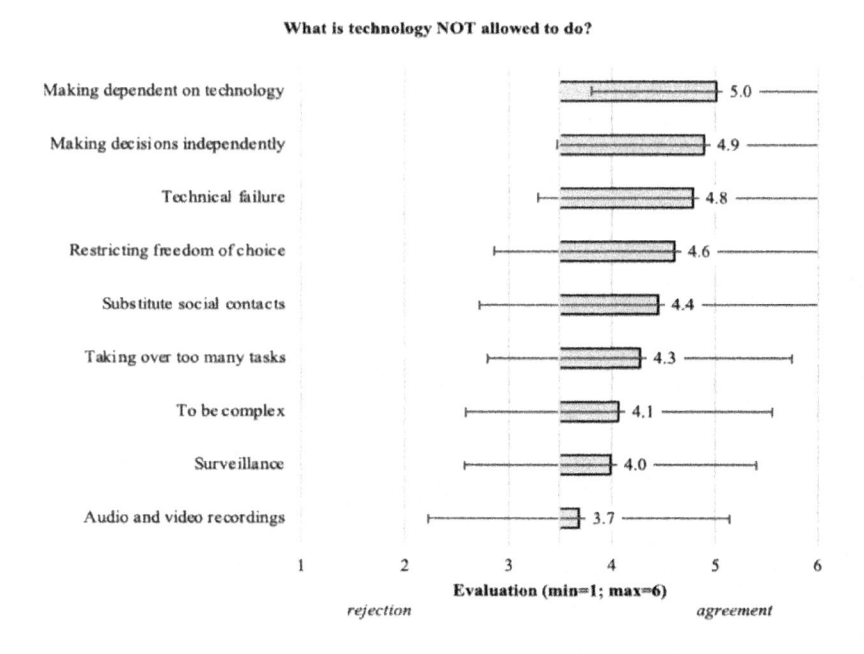

FIGURE 12.5 Evaluations of functions and actions AAL lifelogging technology is NOT allowed to do out of the perspective of 195 German users. (Offermann-van Heek et al., 2019b.)

"Technology is allowed to..."

- ... prolong life." — 4.1
- ... delay death." — 3.0
- ... decide between life and death." — 1.8

evaluation (min = 1; max = 6)
rejection — agreement

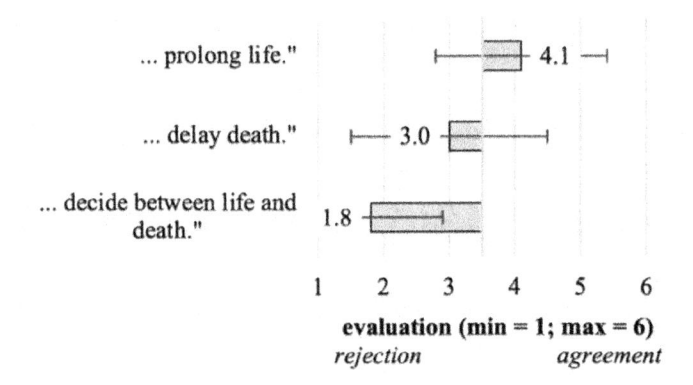

FIGURE 12.6 Evaluations of AAL lifelogging technology as end-of life-decision authority from the perspective of 195 German users. (Offermann-van Heek et al., 2019b.)

participants to assess whether technology is allowed to (1) prolong life, (2) delay end-of-life, and (3) decide between life and death. Participants gave their evaluations on a six-point Likert Scale (1=rejection, 6=confirmation). Results are shown in Figure 12.6.

Thus, technology is only allowed if it helps to prolong life. However, technology is not allowed to delay death and, most clearly, it must not have the decision authority between life and death. Of course, this question is morally extremely difficult to answer but it shows the deeply rooted concerns of the role of technology in a very vulnerable context.

12.5 GUIDELINES TO HUMAN-CENTERED DESIGN, USE AND MANAGEMENT OF AAL LIFELOGGING

In this final section, it is summarized that the understanding of technology acceptance and user-centered design in AAL technologies, their design, use, and management is an essential requirement for successful technical assistance. Many aspects of AAL technology acceptance, and thus the detailed investigation of and willingness to accept lifelogging AAL technologies, as well as the associated individual motives and barriers to use, have so far been largely ignored in the development and design of health-supporting technologies. Especially in the vulnerable application field of older adults' care, protection, independence, and dignity are key values in the context of AAL. Technical designers but also policy and governance bodies need to consider these values when developing AAL solutions. By promoting user-centered design principles and engaging with older adults during the design process, designers can create AAL technologies that are respectful, appropriate, and beneficial to older adults, and that align with their personal and social values and preferences for an appropriate quality of life at an older age.

Therefore, it is of great interest to identify acceptance-relevant factors to be able to prevent or reduce barriers to use at an early stage and to collect empirically the perspectives of those affected who already use the technology or are expected to use it in the future. This is even more important given the increasing importance of automated and AI-based AAL solutions that are applied for the care of older adults at home.

The following guidelines regarding design, data requirements in lifelogging AAL technologies as well as information and communication strategies are summarized (Table 12.2).

By integrating these principles into the design of AAL solutions, designers can develop products that cater to the specific needs and preferences of older adults, promoting independence, autonomy, and social connection. Simultaneously, these guidelines can serve as a reference for individuals not directly involved with the needs of older and frail patients.

TABLE 12.2

Guidelines for Human-Centered Design, Use, and Management of AAL Lifelogging Technologies (Offermann-van Heek et al., 2019b)

Dimensions	Guidelines
Design	**Participatory design** integrating users iteratively from initial phases
Decision making power	∂ users want to decide themselves: keep the user in control ∂ technology should not patronize and decide for users ∂ self-determined data sharing policy: sharing of data should be under control of users
Technology characteristics	∂ optimum: technology profiles with different integrated levels of privacy preservation: should be predefined but individually adaptable (ideal for diverse applications contexts and stakeholders) ∂ possibility of system deactivation: users should be in control of system's operation (e.g., emergency switch)
Interaction	∂ keep technologies easy to handle, as easy to learn as possible, as complex as necessary ∂ prevent feelings of heteronomy by well-defined and transparent data handling regulations ∂ design technologies that can be integrated in or combined with (well-known) existing systems (e.g., in care institutions, smart home systems)
Transparency	∂ enable a transparent overview of functions and performable actions ∂ ask beforehand if a specific action should be performed (depending on context, if possible) ∂ give feedback when a specific action is performed (e.g., transfer of data, activation of alarms or cameras)
Data Requirements	**Well-defined and transparent regulations of data handling** are needed
Type of data	∂ first, focus on safety-relevant functions: emergency detection and calls, monitoring of health parameters ∂ use highly aggregated data (e.g., binary room presence data instead of full sensor data) ∂ prevent a permanent recording of video- or audio-based data; if necessary, enable situational, authorized, and temporally limited recordings
Data processing	∂ prevent permanent storage, only temporally limited use (as shortly as useful possible; as long as necessary) ∂ enabling authorized and authenticated access to data which can be managed by the user
Information and Communication	**Providing open, transparent, and comprehensible information**
What?	∂ Users have to be informed about: **which** data are logged? **where** are data logged? **who** has access to data? **how long** is data stored? **why** is data stored?
How?	∂ providing **accessible information** (not too much, essential information first; detailed information on request; "speak the user's language") ∂ give the possibility to "turn off" the personal information gathering **at any time**

Communication Strategies	∂ **promote benefits** of technologies and **explain handling of potential barriers** (e.g., data security)
	∂ **short and comprehensible explanation** of technology usage and system functioning
	∂ **promote self-determined decisions** of technology handling (e.g., data access)
	∂ **clearly explain relationships between perceived barriers** and **benefits** (e.g., longer data storage enables deeper analyses of health development, changes in movements or behaviors)
	∂ **context- and stakeholder-tailored communication** as technologies and their benefits and barriers are perceived differently depending on diverse contexts and user groups
	  • **sports**: communicate data handling policy to prevent feeling of being controlled
	  • **medical home**: consultation and recommendations by doctors can facilitate technology adoption (as doctors are perceived as accepted and trustworthy)
	  • **medical home and professional care**: promote that technology usage aims for relief, support in everyday life, and autonomy enhancement; make clear that technology is *no substitute* for human attention and care personnel
	  • **age of users**: for younger users explaining handling of data- and privacy-related issues should be focused; for older users' potential benefits should be focused in communication
	  • **health status**: potential benefits of increased safety should be focused for people with chronic illnesses; clarifying the benefits of the technology for non-affected people by enabling them to empathize with living with chronic illnesses

It is crucial to recognize that technological development must involve end users, rather than being solely designed for them. This aligns with the concept of a paradigm shift in which the focus shifts from solely considering technical feasibility, legal and governance aspects, and financial consequences to prioritizing the central needs of users and defining what aspects are acceptable or not.

By placing users at the center of the development process, various expert perspectives can be acknowledged. This way, the different perspectives (technical designers, legal experts, end users, and family members) can be made visible, discussed, and balanced against each other. For instance, physicians might prioritize comprehensive data collection for optimal medical care, while patients or caregivers may be concerned about privacy and may not tolerate such extensive data collection for social reasons. In other cases, patients may willingly waive privacy protections for telemedicine applications, conflicting with existing legal frameworks.

Such conflicts and decision dilemmas are typical for the complex setting in which technology is applied in a highly sensible and vulnerable field of application (Offermann-van Heek et al., 2019b, Légaré et al., 2014). In line with ethical approaches (Stahl 2013; Stahl et al., 2013), the management of conflicts within and across the stakeholders involved in the technology-assisted care is a mandatory requirement of socially responsible technology development (Owen et al., 2012; Offermann-van Heek et al., 2019b, Wilkowska et al., 2018).

Future research and application endeavors should focus on developing integrative, interdisciplinary processes that aid in understanding how to cultivate tolerance toward trade-offs, negotiate conflicting conditions of use, and ultimately implement them effectively. This integrative approach requires the mental openness of all stakeholders involved, including experts, society, law, and politics. The aim is to foster and establish appropriate framework conditions that promote the design of technology that is accessible, intuitive, social, safe, diverse and involves the active participation of older adults in the design process.

For the successful implementation of the promising AAL concept and the social acceptance of AAL technologies that uphold dignity and self-determination in old age while enabling people to live longer at home, such an inclusive approach is indispensable (Owen et al., 2012; Rodríguez-Prat et al., 2016; Wilkowska et al., 2018, Leung, 2007).

ACKNOWLEDGMENTS

This work was partly funded by the German Federal Ministry of Education and Research project PAAL [6SV7955] and the project VisuAAL funded by the European Union's Horizon 2020 research and innovation programme under the Marie Skłodowska-Curie grant agreement No 861091.

REFERENCES

Ajzen, I. (1991). The theory of planned behavior. *Organizational Behavior and Human Decision*, 50(2), 179–211.

Ajzen, I., Fishbein, M., & Heilbroner, R.L. (1980). *Understanding Attitudes and Predicting Social Behavior* (Vol. 278). Englewood Cliffs, NJ: Prentice Hall.

Algase, D. L., Beattie, E. R., Leitsch, S. A., & Beel-Bates, C. A. (2003). Biomechanical activity devices to index wandering behaviour in dementia. *American Journal of Alzheimer's Disease & Other Dementias®*, 18, 85–92.

Baek, Y. M., Kim, E. M., & Bae, Y. (2014). My privacy is okay, but theirs is endangered: Why comparative optimism matters in online privacy concerns. *Computers in Human Behavior*, 31, 48–56.

Bedrov, A., & Bulaj, G. (2018). Improving self-esteem with motivational quotes: Opportunities for digital health technologies for people with chronic disorders. *Frontiers in Psychology*, 9, 2126.

Betz, D., Cieslik, S., Dinkelacker, P., Glende, S., Hartmann, C., Klein, P., ... & Zahneisen, A. (2010). Grundlegende Anforderungen an AAL-Technologien und-Systeme. In S. Meyer, & H. Mollenkopf (Eds.), *AAL in der alternden Gesellschaft. Anforderungen, Akzeptanz und Perspektiven*, Berlin (pp. 63–108).

Biermann, H., Offermann-van Heek, J., Himmel, S., & Ziefle, M. (2018) Ambient assisted living as support for aging in place: Quantitative users' acceptance study on ultrasonic whistles. *JMIR Aging*, *1*, e11825.

Bloom, D. E., Canning, D., & Fink, G. (2010). The greying of the global population and its macroeconomic consequences. *Twenty-First Century Society*, *5*(3), 233–242.

Bruria, A., Maya, S. T., Gadi, S., & Orna, T. (2022). Impact of emergency situations on resilience at work and burnout of Hospital's healthcare personnel. *International Journal of Disaster Risk Reduction*, *76*, 102994.

Burbach, L., Lidynia, C., Brauner, P., & Ziefle, M. (2019). Data protectors, benefit maximizers, or facts enthusiasts: Identifying user profiles for life-logging technologies. *Computers in Human Behavior*, *99*, 9–21.

Burgoon, J. K. (1982). Privacy and communication. In Burgoon M. (Ed.), *Communication Yearbook 6* (pp. 206–249). Beverly Hills, CA: SAGE.

Calero Valdez, A., & Ziefle, M. (2019). The users' perspective on the privacy-utility trade-offs in health recommender systems. *International Journal of Human-Computer Studies*, *121*, 108–121.

Choi, W, Wang, S, Lee, Y, Oh, H, & Zheng, Z. (2020). A systematic review of mobile health technologies to support self-management of concurrent diabetes and hypertension. *Journal of the American Medical Informatics Association*, *27*(6), 939–945.

Choukou, M. A., Shortly, T., Leclerc, N., Freier, D., Lessard, G., Demers, L., & Auger, C. (2021). Evaluating the acceptance of ambient assisted living technology (AALT) in rehabilitation: A scoping review. *International Journal of Medical Informatics*, *150*, 104461.

Cicirelli, G., Marani, R., Petitti, A., Milella, A., & D'Orazio, T. (2021). Ambient assisted living: A review of technologies, methodologies and future perspectives for healthy aging of population. *Sensors*, *21*(10), 3549. doi: 10.3390/s21103549

Coleman, R., Clarkson, J. O. H. N., & Cassim, J. (2016). *Design for Inclusivity: A Practical Guide to Accessible, Innovative and User-Centred Design*. Boca Raton, FL: CRC Press.

Coradeschi, S., Cesta, A., Cortellessa, G., Coraci, L., Gonzalez, J., Karlsson, L., ... & Ötslund, B. (2013,). Giraffplus: Combining social interaction and long term monitoring for promoting independent living. In *2013 6th International Conference on Human System Interactions (HSI)* (pp. 578–585). IEEE.

Davis, F. D. (1989). Perceived usefulness, perceived ease of use, and user acceptance of information technology. *MIS Quarterly*, *13*(3), 319. doi:10.2307/249008.

Davis, F., Bagozzi, P., & Warshaw, P. (1989). User acceptance of computer technology - A comparison of two theoretical models, *Management Science*, *35*(8), 982–1003.

Demiris, G., Rantz, M. J., Aud, M. A., Marek, K. D., Tyrer, H. W., Skubic, M., & Hussam, A. A. (2004). Older adults' attitudes towards and perceptions of 'smart home' technologies: A pilot study. *Medical Informatics and the Internet in Medicine*, *29*(2), 87–94.

Dethloff, C. (2004). *Akzeptanz und Nicht-Akzeptanz von technischen Produktinnovationen*. Lengerich, Germany: Pabst Science Publishers.

Deusdad, B. A., Pace, C., & Anttonen, A. (2016). Facing the challenges in the development of long-term care for older people in Europe in the context of an economic crisis. *Journal of Social Service Research*, *42*(2), 144–150.

Dienlin, T., & Metzger, M. J. (2016). An extended privacy calculus model for SNSs: Analyzing self-disclosure and self-withdrawal in a representative US sample. *Journal of Computer-Mediated Communication*, *21*(5), 368–383.

Dinev, T. & Hart, P. (2006). An extended privacy calculus model for e-commerce transactions. *Information Systems Research*, *17*(1), 61–80.

European Commission (2015). *Data Protection Eurobarometer* (Tech. Rep.). European Commission

Farrington, C. (2016). Wearable technologies and stigma in diabetes: The role of medical aesthetics. *The Lancet Diabetes & Endocrinology*, *4*(7), 566.

Florez-Revuelta, F., Mihailidis, A., Ziefle, M., Colonna, L., & Spinsante, S. (2018). Privacy-aware and acceptable lifelogging services for older and frail people: The PAAL project. In *2018 IEEE 8th International Conference on Consumer Electronics-Berlin (ICCE-Berlin)* (pp. 1–4). IEEE.

Friedrich, D. D. (2001). *Successful Aging: Integrating Contemporary Ideas, Research Findings, and Intervention Strategies*. Springfield, IL: Charles C Thomas Publisher.

Gantz, J., & Reinsel, D. (2012). The digital universe in 2020: Big data, bigger digital shadows, and biggest growth in the far east. *IDC iView: IDC Analyze the Future*, *2007*(2012), 1–16.

Gerber, N., Gerber, P., & Volkamer, M. (2018). Explaining the privacy paradox: A systematic review of literature investigating privacy attitude and behavior. *Computers & Security*, *77*, 226–261.

Grenier, A. (2012). *Transitions and the Lifecourse: Challenging the Constructions of 'Growing Old'*. Bristol: Policy Press.

Gulliksen, J., Göransson, B., Boivie, I., Blomkvist, S., Persson, J., & Cajander, Å. (2003). Key principles for user-centred systems design. *Behaviour and Information Technology*, *22*(6), 397–409.

Gustafsson, G., Eriksson, S., Strandberg, G., & Norberg, A. (2010). Burnout and perceptions of conscience among health care personnel: A pilot study. *Nursing Ethics*, *17*(1), 23–38.

Higgs, P., & Gilleard, C. (2020). The ideology of ageism versus the social imaginary of the fourth age: Two differing approaches to the negative contexts of old age. *Ageing & Society*, *40*(8), 1617–1630.

Himmel, S., & Ziefle, M. (2016). Smart home medical technologies: Users' requirements for conditional acceptance. *i-com*, *15*(1), 39–50.

Holden, R. J., & Karsh, B. (2010). The technology acceptance model: Its past and its future in health care. *Journal of Biomedical Informatics*, *43*(1):159–17.

Ivan, L., & Cutler, S. J. (2021). Ageism and technology: The role of internalized stereotypes. *University of Toronto Quarterly*, *90*(2), 127–139.

Jaschinski, C., & Ben Allouch, S. (2015). Why should I use this? Identifying incentives for using AAL technologies. In *Ambient Intelligence: 12th European Conference, AmI 2015, Athens, Greece, November 11-13, 2015, Proceedings 12* (pp. 155–170). Springer International Publishing.

Karabey, B. (2012). Big data and privacy issues. In *International Symposium on Information Management in a Changing World* (pp. 3–3). Berlin, Heidelberg: Springer.

Kasugai, K., Ziefle, M., Röcker, C., & Russell, P. (2010). Creating spatio-temporal contiguities between real and virtual rooms in an assistive living environment. In *Create10-The Interaction Design Conference* (pp. 1–6).

Kehr, F., Kowatsch, T., Wentzel, D., & Fleisch, E. (2015). Blissfully ignorant: The effects of general privacy concerns, general institutional trust, and affect in the privacy calculus. *Information Systems Journal*, *25*(6), 607–635.

Keith, M. J., Thompson, S. C., Hale, J., Lowry, P. B., & Greer, C. (2013). Information disclosure on mobile devices: Re-examining privacy calculus with actual user behavior. *International Journal of Human-Computer Studies*, *71*(12), 1163–1173.

Kleinberger, T., Becker, M., Ras, E., Holzinger, A., & Müller, P. (2007). Ambient intelligence in assisted living: Enable elderly people to handle future interfaces. In *International Conference on Universal Access in Human-Computer Interaction* (pp. 103–112). Berlin, Heidelberg: Springer.

Koops, B.-J., Newell, B. C., Timan, T., Skorvanek, I., Chorkrevski, T., & Galic, M. (2017). A typology of privacy. *University of Pennsylvania Journal of International Law*, *38*(2), 1–93.

Köttl, H., Allen, L. D., Mannheim, I., & Ayalon, L. (2022). Associations between everyday ICT usage and (self-) ageism: A systematic literature review. *Gerontologist*, *63*(7), 1172–1187.

Krasnova, H., Veltri, N. F., & Günther, O. (2012). Self-disclosure and privacy calculus on social networking sites: The role of culture. *Business & Information Systems Engineering*, *4*(3), 127–135.

Lahlou, S. (2008). Identity, social status, privacy and face-keeping in digital society. *Social Science Information*, *47*(3), 299–330.

Latikka, R., Rubio-Hernández, R., Lohan, E. S., Rantala, J., Nieto Fernández, F., Laitinen, A., & Oksanen, A. (2021). Older adults' loneliness, social isolation, and physical information and communication technology in the era of ambient assisted living: A systematic literature review. *Journal of Medical Internet Research*, *23*(12), e28022.

Légaré, F., Stacey, D., Brière, N., Robitaille, H., Lord, M. C., Desroches, S., & Drolet, R. (2014). An interprofessional approach to shared decision making: An exploratory case study with family caregivers of one IP home care team. *BMC Geriatr*, *14*, 83.

Leung, D. (2007). Granting death with dignity: Patient, family and professional perspectives. *International Journal of Palliative Nursing*, *13*(4), 170–174.

Li, H., Wu, J., Gao, Y., & Shi, Y. (2016). Examining individuals' adoption of healthcare wearable devices: An empirical study from privacy calculus perspective. *International Journal of Medical Informatics*, *88*, 8–17.

Li, Y. (2011). Empirical studies on online information privacy concerns: Literature review and an integrative framework. *Communications of the Association for Information Systems*, *28*(28), 453–496.

Little, J. S., & Triest, R. K. (2001). Seismic shifts: The economic impact of demographic change. An overview. In *Conference Series-Federal Reserve Bank of Boston* (Vol. 46, pp. 1–30). Federal Reserve Bank of Boston.

Lloyd, L. (2004). Mortality and morality: Ageing and the ethics of care. *Ageing & Society*, *24*(2), 235–256.

Maidhof, C., Hashemifard, K., Offermann, J., Ziefle, M., & Florez-Revuelta, F. (2022). Underneath your clothes: A social and technological perspective on nudity in the context of AAL technology. In *Proceedings of the 15th International Conference on PErvasive Technologies Related to Assistive Environments* (pp. 439–445).

Martinson, M., & Berridge, C. (2014) Successful aging and its discontents: A systematic review of the social gerontology literature. *The Gerontologist*, *55*(1):58–69

Maskeliūnas, R., Damaševičius, R., & Segal, S. (2019). A review of internet of things technologies for ambient assisted living environments. *Future Internet*, *11*(12), 259.

Messner, E. M., Probst, T., O'Rourke, T., Stoyanov, S., & Baumeister, H. (2019). mHealth applications: Potentials, limitations, current quality and future. In H. Baumeister, & C. Montag (Eds.), *Digital Phenotyping and Mobile Sensing* (pp. 235–248). Cham, Switzerland: Springer.

Metallo, C., Agrifoglio, R., Lepore, L., & Landriani, L. (2022). Explaining users' technology acceptance through national cultural values in the hospital context. *BMC Health Services Research*, *22*(1), 1–10.

Morgan, L., & Kunkel, S. (2001). *Aging: The Social Context*. Thousand Oaks, CA: Pine Forge Press.

Mostofa, N., Feltner, C., Fullin, K., Guilbe, J., Zehtabian, S., Bacanlı, S. S., … & Turgut, D. (2021). A smart walker for people with both visual and mobility impairment. *Sensors*, *21*(10), 3488.

Murray, L. M., & Laditka, S. B. (2010) Care transitions by older adults from nursing homes to hospitals: Implications for long-term care practice, geriatrics education, and research. *Journal of the American Medical Directors Association*, *11*(4), 231–238.

Mynatt, E. D., Melenhorst, A. S., Fisk, A. D., & Rogers, W. A. (2004). Aware technologies for aging in place: Understanding user needs and attitudes. *IEEE Pervasive Computing*, *3*(2), 36–41.

Nehmer, J., Becker, M., Karshmer, A., & Lamm, R. (2006). Living assistance systems: An ambient intelligence approach. In *Proceedings of the 28th International Conference on Software Engineering* (pp. 43–50).

Nissenbaum, H. (2010). *Privacy in Context: Technology, Policy, and the Integrity of Social Life*. Stanford, CA: Stanford University Press.

Oeppen, J, & Vaupel, J. W. (2002) Demography: Enhanced: Broken limits to life expectancy. *Science*, *296*, 1029–1031.

Oetjen, S., & Ziefle, M. (2009). A visual ergonomic evaluation of different screen types and screen technologies with respect to discrimination performance. *Applied Ergonomics*, *40*(1), 69–81.

Offermann, J., Wilkowska, W., & Ziefle, M. (2022). Interplay of perceptions of aging, care, and technology acceptance in older age. *International Journal of Human-Computer Interaction*, *39*(5), 1003–1015.

Offermann, J., Wilkowska, W., Maidhof, C., & Ziefle, M. (2023). Shapes of you? Investigating the acceptance of video-based AAL technologies applying different visualization odes. *Sensors*, *23*(3), 1143.

Offermann-van Heek, J., & Ziefle, M. (2019a). Nothing else matters! Trade-offs between perceived benefits and barriers of AAL technology usage. *Frontiers in Public Health*, *7*, 134.

Offermann-van Heek, J., Brauner, P., & Ziefle, M. (2018). Let's talk about TEX-Understanding consumer preferences for smart interactive textile products using a conjoint analysis approach. *Sensors*, *18*(9), 3152.

Offermann-van Heek, J., Schomakers, E. M., & Ziefle, M. (2019c). Bare necessities? How the need for care modulates the acceptance of AAL. *International Journal of Medical Informatics*, *127*, 147–156

Offermann-van Heek, J., Wilkowska, W., & Ziefle, M. (2020). Colors of aging: Cross-cultural perception of lifelogging technologies in older age. In *ICT4AWE* (pp. 38–49).

Offermann-van Heek, J., Wilkowska, W., & Ziefle, M. (2021). Cultural impact on perceptions of aging, care, and lifelogging technology: A comparison between Turkey and Germany. *International Journal of Human-Computer Interaction*, *37*(2), 156–168.

Offermann-van Heek, J., Wilkowska, W., Brauner, P., & Ziefle, M. (2019b). Guidelines for integrating social and ethical user requirements in lifelogging technology development. In *Proceedings of the 5th International Conference on Information and Communication Technologies for Ageing Well and e-Health - Volume 1: ICT4AWE* (pp. 67–79).

Owen, R., Macnaghten, P., & Stilgoe, J. (2012) Responsible research and innovation: From science in society to science for society, with society. *Science and Public Policy*, *39*, 751–60.

Palmore, E. (2005). Three decades of research on ageism. *Generations*, *29*(3), 87–90.

Pickard, L. (2015). A growing care gap? The supply of unpaid care for older people by their adult children in England to 2032. *Ageing & Society*, *35*(1), 96–123.

Pinto, M. R., De Medici, S., Zlotnicki, A., Bianchi, A., Van Sant, C., & Napou, C. (1997). Reduced visual acuity in elderly people: The role of ergonomics and gerontechnology. *Age and Ageing*, *26*(5), 339–344.

Proulx, K., & Jacelon, C. (2004). Dying with dignity: The good patient versus the good death. *American Journal of Hospice and Palliative Medicine®*, *21*(2), 116–120.

Ravi, S., Climent-Pérez, P., & Florez-Revuelta, F. (2023). A review on visual privacy preservation techniques for active and assisted living. *Multimedia Tools and Applications*, *83*(5), 14715–14755. https://doi.org/10.1007/s11042-023-15775-2

Reddy, A. R., Ghantasala, G. P., Patan, R., Manikandan, R., & Kallam, S. (2021). Smart assistance of elderly individuals in emergency situations at home. In *Internet of Medical Things: Remote Healthcare Systems and Applications* (pp. 95–115). Berlin: Springer Nature. https://link.springer.com/chapter/10.1007/978-3-030-63937-2_6

Reich, A. J., Claunch, K. D., Verdeja, M. A., Dungan, M. T., Anderson, S., Clayton, C. K., ... & Thacker, E. L. (2020). What does "successful aging" mean to you? - Systematic review and cross-cultural comparison of lay perspectives of older adults in 13 countries, 2010-2020. *Journal of Cross-Cultural Gerontology*, *35*, 455–478.

Reichstadt, J., Sengupta, G., Depp, C. A., Palinkas, L. A., & Jeste, D. V. (2010). Older adults' perspectives on successful aging: Qualitative interviews. *The American Journal of Geriatric Psychiatry*, *18*(7), 567–575.

Rodrigues, D., Luis-Ferreira, F., Sarraipa, J., & Goncalves, R. (2018). Behavioural monitoring of alzheimer patients with smartwatch based system. In *Proceedings of the 2018 International Conference on Intelligent Systems* (IS) (pp. 771–775). IEEE.

Rodríguez-Prat, A., Monforte-Royo, C., Porta-Sales, J., Escribano, X., & Balaguer, A. (2016). Patient perspectives of dignity, autonomy and control at the end of life: systematic review and meta-ethnography. *PloS one*, *11*(3), e0151435.

Rus, S., Braun, A., & Kuijper, A. (2017). E-textile couch: Towards smart garments integrated furniture. In A. Braun, R. Wichert, & A. Maña (Eds.), *Proceedings of the 13th European Conference on Ambient Intelligence* (pp. 214–224). Cham, Switzerland: Springer International Publishing.

Santoso, F., & Redmond, S. J. (2015). Indoor location-aware medical systems for smart homecare and telehealth monitoring: State-of-the-art. *Physiological Measurement*, *36*(10), R53.

Saponas, T. S., Harrison, C., & Benko, H. (2011). PocketTouch: Through-fabric capacitive touch input. In *Proceedings of the 24th Annual ACM Symposium on User Interface Software and Technology* (pp. 303–308). Santa Barbara, CA, New York: ACM.

Schmidt, T., Philipsen, R., Themann, P., & Ziefle, M. (2016). Public perception of V2X-technology-evaluation of general advantages, disadvantages and reasons for data sharing with connected vehicles. In *2016 IEEE Intelligent Vehicles Symposium (IV)* (pp. 1344–1349).

Schomakers, E. M., & Ziefle, M. (2023). Privacy vs. security: Trade-offs in the acceptance of smart technologies for aging-in-place. *International Journal of Human-Computer Interaction*, *39*(5), 1043–1058.

Schomakers, E. M., Lidynia, C., & Ziefle, M. (2019a). Listen to my heart? How privacy concerns shape users' acceptance of e-health technologies. In *2019 International Conference on Wireless and Mobile Computing, Networking and Communications (WiMob)* (pp. 306–311). IEEE.

Schomakers, E. M., Lidynia, C., & Ziefle, M. (2020). All of me? Users' preferences for privacy-preserving data markets and the importance of anonymity. *Electronic Markets*, *30*(3), 649–665.

Schomakers, E. M., Lidynia, C., Müllmann, D., & Ziefle, M. (2019b). Internet users' perceptions of information sensitivity. *International Journal of Information Management*, *46*, 142–150.

Schomakers, E. M., Lidynia, C., Vervier, L. S., Calero Valdez, A., & Ziefle, M. (2022). Applying an extended UTAUT2 model to explain user acceptance of lifestyle and therapy mobile health apps: Survey study. *JMIR mHealth and uHealth*, *10*(1), e27095.

Schomakers, E. M., Lidynia, C., Vervier, L., & Ziefle, M. (2018). Of guardians, cynics, and pragmatists a typology of privacy concerns and behavior. In *IoTBDS* (pp. 153–163). SCITEPRESS.

Shah, J. H., & Patel, A. D. (2018). Ambient assisted living system: The scope of research and development. In *2018 International Conference on Electrical, Electronics, Computers, Communication, Mechanical and Computing (EECCMC)*, Tamilnadu, India.

Shalini, S., Levins, T., Robinson, E. L., Lane, K., Park, G., & Skubic, M. (2019). Development and comparison of customized voice-assistant systems for independent living older adults. In *Human Aspects of IT for the Aged Population. Social Media, Games and Assistive Environments: 5th International Conference, ITAP 2019, HCII 2019, Proceedings, Part II 21* (pp. 464–479). Springer International Publishing.

Shallcross, A. J., Ford, B.Q., Floerke, V.A., & Mauss, I.B. (2013). Getting better with age: The relationship between age, acceptance, and negative affect. *Journal of Personality and Social Psychology*, *104*(4), 734.

Siek, K. A., Khan, D. U., Ross, S. E., Haverhals, L. M., Meyers, J., & Cali, S. R. (2011). Designing a personal health application for older adults to manage medications: A comprehensive case study. *Journal of Medical Systems*, *35*, 1099–1121.

Smith, H. J., Dinev, T., & Xu, H. (2011). Information privacy research: An interdisciplinary review. *MIS Quarterly*, *35*, 989–1015.

Stahl, B. C. (2013). Responsible research and innovation: The role of privacy in an emerging framework. *Science and Public Policy*, *40*(6), 708–716.

Stahl, B. C., Eden, G. & Jirotka, M. (2013). Responsible research and innovation in information and communication technology: Identifying and engaging with the ethical implications of ICTs. In *Responsible Innovation* (pp. 199–218). Hoboken, NJ: John Wiley & Sons, Ltd.

Stavropoulos, T. G., Papastergiou, A., Mpaltadoros, L., Nikolopoulos, S., & Kompatsiaris, I. (2020). IoT wearable sensors and devices in elderly care: A literature review. *Sensors, 20*(10), 2826.

Sun, Y., Wang, N., Shen, X. L., & Zhang, J. X. (2015). Location information disclosure in location-based social network services: Privacy calculus, benefit structure, and gender differences. *Computers in Human Behavior, 52*, 278–292.

Terschüren, C., Mensing, M., & Mekel, O. C. (2012). Is telemonitoring an option against shortage of physicians in rural regions? Attitude towards telemedical devices in the North Rhine-Westphalian health survey, Germany. *BMC Health Services Research, 12*(1), 1.

Troutman-Jordan, M., & Staples, J. (2014). Successful aging from the viewpoint of older adults. *Research and Theory for Nursing Practice, 28*(1), 87–104.

Vacher, M., Portet, F., Fleury, A., & Noury, N. (2011). Development of audio sensing technology for ambient assisted living: Applications and challenges. *International Journal of E-Health and Medical Communications (IJEHMC), 2*(1), 35–54.

Vanus, J., Belesova, J., Martinek, R., Nedoma, J., Fajkus, M., Bilik, P., & Zidek, J. (2017). Monitoring of the daily living activities in smart home care. *Human-Centric Computing and Information Sciences, 7*(1), 1–34.

Vassli, L.T., & Fatshchian, B.A. 2018. Acceptance of health-related ICT among elderly people living in the community: A systematic review of qualitative evidence. *International Journal of Human-Computer Interaction, 34*(2), 99–116.

Venkatesh, V., & Bala, H. (2008). Technology acceptance model 3 and a research agenda on interventions, *Decision Science, 39*(2), 273–315.

Venkatesh, V., & Davis, F. (2000). A theoretical extension of the technology acceptance model: Four longitudinal field studies, *Management Science, 46*(2), 186–204.

Venkatesh, V., Morris, M., Davis, F., & Davis, M. (2003). User acceptance of information technology - toward a unified view, *MIS Quarterly, 27*(3), 425–478.

Westin, A. F. 1967. Privacy and freedom. *American Sociological Review, 33*(1), 173.

Whitbourne, S. K. (2002). *The Aging Individual: Physical and Psychological Perspectives*. Springer Publishing Company. https://www.amazon.com/Aging-Individual-Physical-Psychological-Perspectives/dp/0826193617

Whitbourne, S. K. (2012). *The Aging Body: Physiological Changes and Psychological Consequences*. Springer Science & Business Media. https://link.springer.com/book/10.1007/978-1-4612-5126-2

Wiles, J. L., Leibing, A., Guberman, N., Reeve, J., & Allen, R. E. (2012). The meaning of "aging in place" to older people. *Gerontologist, 52*(3), 357–366.

Wilkowska, W., Brauner, P., & Ziefle, M. (2018). Rethinking technology development for older adults. A responsible research and innovation duty. In R. Pak, & A. Mc Laughlin (Eds.), *Aging, Technology, and Health* (pp. 1–30). Cambridge, MA: Academic Press.

Wilkowska, W., Offermann, J., Spinsante, S., Poli, A., & Ziefle, M. (2022). Analyzing technology acceptance and perception of privacy in ambient assisted living for using sensor-based technologies. *PLoS One, 17*(7), e0269642.

Wilkowska, W., Offermann-van Heek, J., Brauner, P., & Ziefle, M. (2019). Wind of change? Attitudes towards aging and use of medical technology. In *ICT4AWE* (pp. 80–91).

Wilkowska, W., Offermann-van Heek, J., Florez-Revuelta, F., & Ziefle, M. (2021). Video cameras for lifelogging at home: Preferred visualization modes, acceptance, and privacy perceptions among German and Turkish participants. *International Journal of Human-Computer Interaction, 37*(15), 1436–1454.

Yıldız, E. (2023). Psychopathological factors associated with burnout in intensive care nurses: A cross-sectional study. *Journal of the American Psychiatric Nurses Association, 29*(2), 122–135.

Ziefle, M. (2010). Information presentation in small screen devices: The trade-off between visual density and menu foresight. *Applied Ergonomics, 41*(6), 719–730.

Ziefle, M., & Calero Valdez, A. (2018). Decisions about medical data disclosure in the internet: An age perspective. In J. Zhou, & G. Salvendy (Eds.), *Human Aspects of IT for the Aged Population. Applications in Health, Assistance, and Entertainment. ITAP 2018*. Lecture Notes in Computer Science (Vol. 10927). Cham: Springer.

Ziefle, M., & Schaar, A. K. (2014). Technology acceptance by patients: Empowerment and stigma. In *Handbook of Smart Homes, Health Care and Well-Being* (pp. 1–10), Berlin.

Ziefle, M., Halbey, J., & Kowalewski, S. (2016) Users' willingness to share data on the internet: Perceived benefits and caveats. In *IoTBD 2016* (pp. 255–265). SCITEPRESS.

13 Urban Interfaces

Luke Hespanhol, Joel Fredericks, Martin Tomitsch, and Andrew Vande Moere

13.1 INTRODUCTION

One of the core tenets of urbanity is its inherent frictional state: cities bring together, within a well-defined geographical space, different aggregates of people and cultures, and with those, stories, values, wordviews, methods, and expectations about how life should be lived in relation to the surrounding peers and natural environment. The resulting built environment embeds those relationships and practices, expressed through spatial features and socially defined norms of behaviour constraining processes of engagement into what we often call civility. This continuously performative negotiation of competing – and often incompatible – priorities is translated into objects, services, and processes through which citizens interface both with one another and with the city itself (De Lange, Merx and Verhoeff, 2019).

Since the late 20th century, computing has emerged as an especially powerful condition shaping urban interfaces, from automation of various relationships and processes within the city, to autonomy (Cugurullo, 2020) over decision-making on urban matters, leading to a new class on non-human intelligent algorithmic citizens co-existing, co-managing, and mutually influencing human and non-human biological citizens, as well as the broader environment. As the set of urban relationships expands in complexity, so do the risks and opportunities associated with autonomous urbanity and the need for considerations about approaches to systems thinking and collective responsibility.

In this chapter, we provide an in-depth discussion into urban interfaces and their applications through the lens of human-computer interaction (HCI). In Section 13.2, we position urban interfaces in relation to other key related terms: urban informatics, media architecture, public displays, and smart cities. In Section 13.3, we introduce our typological model for urban interfaces, mapped to four city scales (personal, interpersonal, community, and metropolis) and four dimensions of concern (people, place, data, and systems). We then discuss those through case studies across the representative fields of governance (Section 13.4), digital placemaking (Section 13.5), urban data (Section 13.6), and digital infrastructure (Section 13.7). Following that, in Section 13.8, we articulate future directions for urban interfaces, including sustainability, hybridity, artificial intelligence (AI), and emerging methods, before concluding this chapter in Section 13.9.

13.2 DEFINITIONS

The notion of urban interfaces draws on a number of concepts that have been studied in the field of HCI, providing a design space for HCI solutions. This section introduces these concepts before offering a definition for urban interfaces.

13.2.1 Urban Informatics

The development of ubiquitous computing (Weiser, 1991) was the result of many waves of technological developments, including the advent of the Internet, pervasive displays, smartphones, wearable devices, Wi-Fi, various kinds of sensors, and social media, among others. Those technologies profoundly altered how we interact with each other, the range of activities we perform, and where those interactions and activities occur. In response to those dramatic changes in society, and in cities in particular, urban informatics thus emerged as a disciplinary domain *"situated at the intersection of notions, trends and considerations for place, technology, and people in urban environments"* (Foth, Choi & Satchell, 2011, p.2). It concerns itself with the investigation of the possibilities of technologies for the enhancement of urban life, empowerment of citizens, and the strengthening of civic debate (Foth, 2018), as well as the risks and challenges posed by the ill-thought design of those very same technologies and services, big data (Thakuriah, Tilahun & Zellner, 2017), and widespread surveillance (Bibri & Allam, 2022).

13.2.2 MEDIA ARCHITECTURE

The field of media architecture emerged around the increasing integration of digital media and technologies to the built environment in the second half of the 20th century, and the subsequent design of interactive environments since the early 2000s (McQuire, 2008). Media architecture investigates the intersection of technology, the city, and the social dynamics that unfold as a result. In that sense, media architecture and urban informatics do overlap significantly, yet the former tend to place greater emphasis on situated interventions in shared urban spaces, involving communities, and articulating shared narratives.

While originally focusing on the design of interactive urban media facades (Brynskov et al., 2009; Dalsgaard & Halskov, 2010; Wiethoff & Gehring, 2012), their usability (Fischer & Hornecker, 2012) and related architectural and contextual factors (Vande Moere & Wouters, 2012; Wouters et al., 2016a), the field of media architecture has over the years expanded to the design of interactive art installations (Hespanhol & Tomitsch, 2012), mediation of social interactions (Hespanhol et al., 2014; Hespanhol & Dalsgaard, 2015), civic participation (Hespanhol et al., 2015; Fredericks et al., 2018) and community engagement (Fredericks et al., 2015; Fredericks, Hespanhol & Tomitsch, 2016), tactical urbanism (Caldwell & Foth, 2014) and, more recently, social robotics (Hoggenmueller, Hespanhol & Tomitsch, 2020).

13.2.3 PUBLIC DISPLAYS

Public displays have been the subject of HCI studies as they provide a computational interface for people to interact with digital information in urban environments. Research in this area has focused on a range of fundamental aspects, such as how to support gestural interaction (Hespanhol et al., 2012; Ackad et al., 2015) and draw the attention of passers-by (Grace et al., 2013; Kukka et al., 2013). As an interface, public displays have many aspects in common with other display-based applications, while posing new challenges that are unique to an urban environment. For example, the way people interact with a public display is more opportunistic compared to personal devices (Michelis & Müller, 2011). Because of the public nature of this interaction, the design of public displays needs to carefully consider social aspects that are unique to urban environments, leading to phenomena such as display blindness (Müller et al., 2009), the honeypot effect (Wouters et al., 2016b), and performance (Dalsgaard & Hansen, 2008). Public displays have also been used as a component in urban informatics studies, for example, to support civic debate (Hespanhol et al., 2015). Specific implementations can further be considered a form of media architecture, though there is a subtle distinction between urban screens which are akin to public displays and media facades where there is a complete integration between the physical and digital elements (Tscherteu & Tomitsch, 2011).

13.2.4 SMART CITIES

The purview of a smart city is the intersection of technology and city processes. Effective governance of cities and delivery of services to citizens, enabling daily life to unfold smoothly and in accordance with societal norms, have been perennial concerns in city making. The definition of smartness, when applied to cities, has usually corresponded to an advanced state of integration, aided by digital technology, of its various elements so that they can work in tandem, with greater efficiency, and yielding benefits to citizens and government (Harrison & Donnelly, 2011; Batty, 2013).

With the fast growth in urbanisation, particularly since the turn of the 21st century, and concomitant expansion of affordable and mobile digital infrastructure, proposals about how to make cities smarter (Dirks & Keeling, 2009; Letaifa, 2015; Tomitsch, 2017) have gained significant traction, with smart city approaches being rolled out worldwide, with varying and often criticised outcomes: among other shortcomings, smart cities have been criticised for playing greater emphasis on efficiency and centralised, city-wide and top-down digital infrastructures, than qualitative sustainability frameworks (Ahvenniemi et al. 2017), or support for civic participation and citizen-led innovation (Hemment & Townsend, 2013; Shelton & Lodato, 2019). Yet, the notion of smart cities remains valuable when articulating the integration of pervasive technology in the design of intelligent urban systems capable of supporting more sustainable, equitable, and fair urban futures (Batty et al., 2012; Kumar & Dahiya, 2017; Silva, Khan & Han, 2018).

13.2.5 URBAN INTERFACES

Urban interfaces (De Lange, Merx & Verhoeff, 2019) represent the points of contact, enabled by digital technology, between the different stakeholders and elements in an urban environment. More importantly, they also describe how those points of contact invite interactions among those elements and co-shape often different perspectives, behaviours or controversies (Baibarac-Duignan & de Lange, 2021) around them. While they may include actual user interfaces for interactive urban systems, or 'city apps' (Tomitsch, 2016), the scope of urban interfaces we adopt here goes far beyond that, to encompass any aspect of urban living in which the exchange of information, knowledge or experience occurs, facilitated by digital technologies. In that sense, we can think of urban interfaces as a lens through which we can analyse and design urban life from the perspective of the *relationships* unfolding in the city (Jacobs, 2012). This relational character (Hespanhol, 2023b), in turn, allows urban interfaces to capture the way people, place, and technology co-constitute and mutually affect one another.

Urban interfaces can be enacted through urban informatics, may take the form of media architecture, enable interaction through public displays, and are intrinsic part of smart cities. Yet, they transcend those concepts by focusing on how technology can mediate the multitude of relationships between the various stakeholders of a city, including human, non-humans, the natural and built environments, and the implications of those relationships to norms of behaviour, value systems, culture, and civic life (Figure 13.1).

13.3 TYPOLOGICAL MODEL

As urban interfaces mediate relationships within the urban environment, two factors are worth of consideration when it comes to their design and application: (1) the parties involved in the relationship; and (2) the scale within the city those relationships are expressed at. Those factors suggest different lenses through which to approach urban interfaces and map out aspects relevant to their implementation.

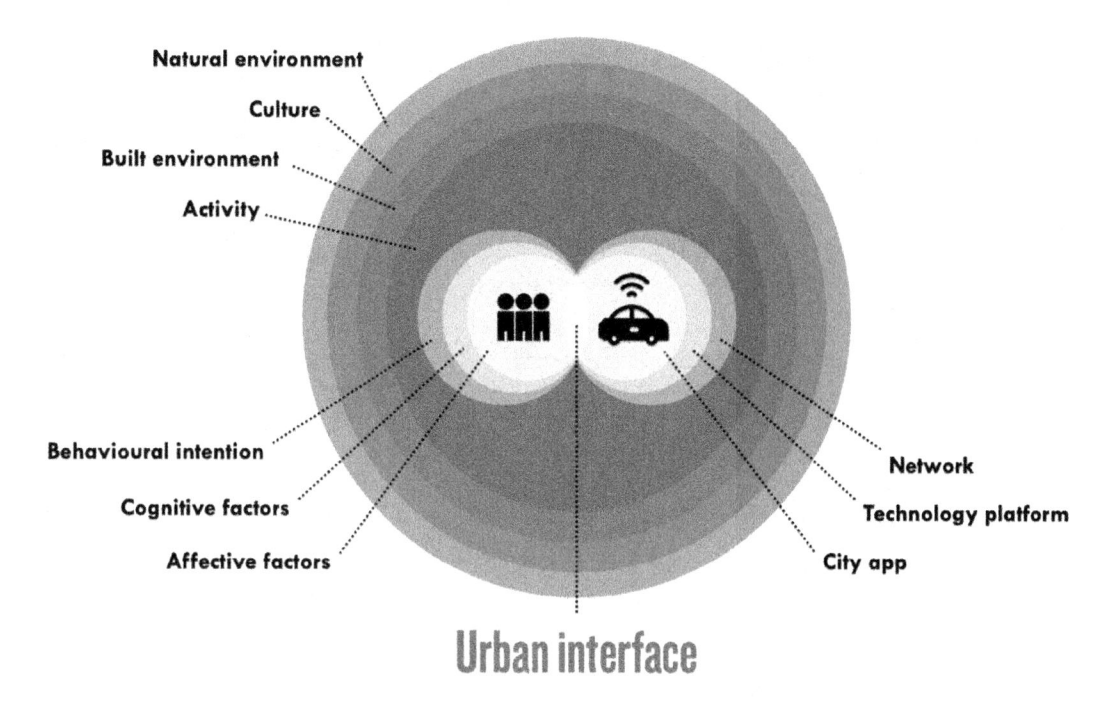

FIGURE 13.1 The scope of urban interfaces.

To assist with identifying the parties involved in the relationships addressed by an urban interface, we can take as a starting point the three core dimensions of urban informatics: people, place, and technology. However, given the acknowledged and growing importance of data gathering in smart cities (Thakuriah, Tilahun & Zellner, 2017), and the increasing relevance of AI systems for urban planning and management (Allam & Dhunny, 2019), we would argue that, rather than referring to 'technology' more broadly, it may be helpful instead to address (1) digital systems, and (2) the data fed to and produced by them, as separate concerns. In doing so, we end up with two dimensions related to the urban technological landscape – systems and data – and two dimensions which, we would argue, could be described as related to urban culture – people and place.

By *people*, in this context, we refer not only to the individual citizens inhabiting the city but also any other human factor at play, including culture, arts, institutions, legislation, ethical frameworks, and so on. Likewise, by *place*, we refer to the aggregate of the natural and built environments, their layouts and affordances, the activities performed in them by humans and non-humans alike, and shared notions of place identity, purpose, and value. We argue both dimensions as cultural insofar as they both encompass the range of relationships humans have traditionally developed with one another and their surrounding environments within societies. In contrast, the two dimensions of technology encompass new tools, practices, and methods afforded by ubiquitous urban computing, as well as the infrastructure required to support them. We therefore propose the dimension of data to capture not only the plethora of information gathered from citizens by government for use in the city administration, but also data exchanged across the various spheres of urban life, machine learning algorithms employed by urban services, and data models utilised for both operational and predictive government and policing. The last dimension, systems, captures the digital applications and infrastructure deployed in the city, whether or not directly interacting with citizens or used as part of smart city administration.

Importantly, each of those dimensions can also play out at different scales in the city, implying distinct relationships and levels of citizen or government accountability, as well as different levels of abstraction. Adapting the model proposed by Gardner & Hespanhol (2018) when discussing the

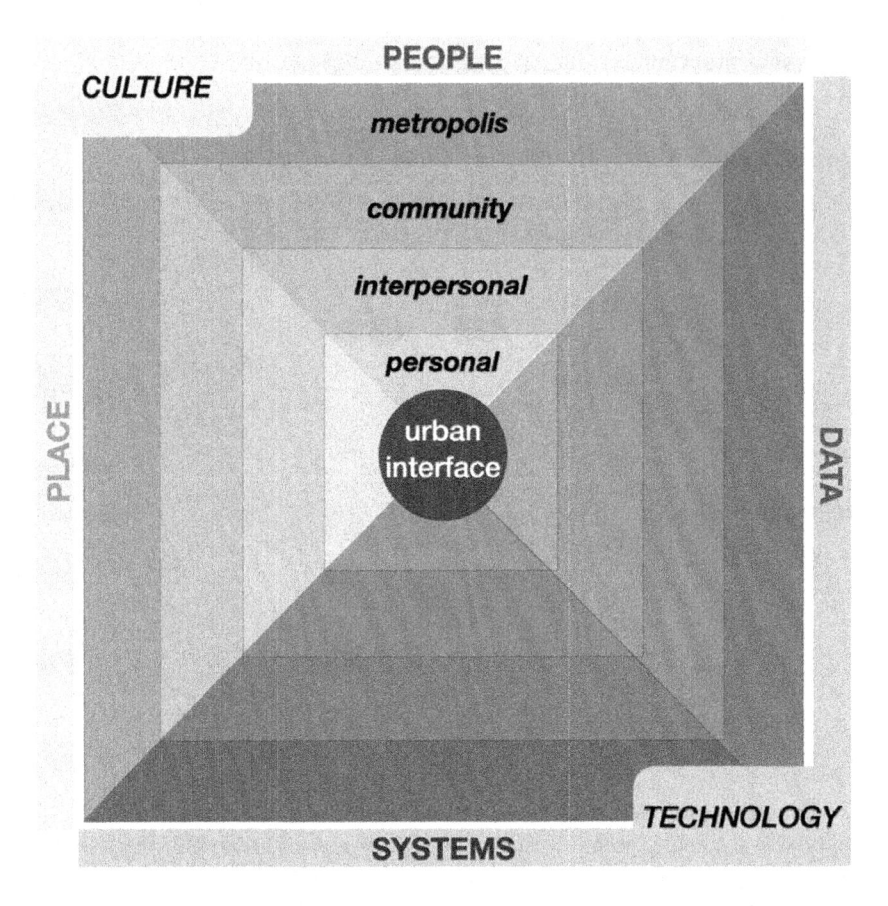

FIGURE 13.2 A typological model for urban interfaces.

different scales of a smart city, we propose four scales to aid the design and analysis of urban interfaces: *personal* (relationships at the level of a single individual interacting with the city), *interpersonal* (social interactions between multiple individuals within the city), *community* (relationships forged around shared factors of relevance, such as locations, activities, or causes), and *metropolis* (city-wide relationships).

Figure 13.2 illustrates our proposed typological model. Different combinations of the dimensions and scales described above may lead to different types of urban interfaces. Likewise, urban interfaces can be designed to work across dimensions and/or scales. In the following section, we discuss existing uses of urban interfaces that could be argued as exemplary of each dimension, namely governance (people), digital placemaking (place), data visualisation (data), and infrastructure (systems).

13.4 GOVERNANCE

Urban interfaces can act as a powerful tool for enhancing community participation in the governance of urban environments. By providing a platform for dialogue and collaboration, urban interfaces can play a key role in facilitating the development of effective policies, regulations and decision-making processes that reflect the needs, wants and aspirations of a variety of stakeholders. Urban environments are shaped by a complex interplay of cultural practices, social norms, and policy decisions, which have evolved over time in response to various factors, such as changing societal values, environmental impacts, and technological advancements.

To ensure the successful governance of urban environments, it is crucial to examine past outcomes, both successful and unsuccessful, and apply those insights to future policies, regulations, and decision-making processes. Collaboration among government agencies, private enterprises, community organisations, and local residents is essential for identifying and addressing potential risks or negative impacts associated with governance. By valuing the knowledge and expertise of various stakeholders and adopting a collaborative approach to governance, urban interfaces can help create more inclusive, sustainable, and equitable urban environments.

13.4.1 Community Engagement, Policies and Middle-Out

Community engagement is a critical aspect of effective governance within cities and regional centres. Typically, community engagement involves informing local communities about proposed projects and seeking their feedback through various channels, such as information sessions, online platforms, and official documents. However, while community engagement can be a valuable tool for obtaining feedback and ideas from local communities, it can also be reduced to tokenism or "engagement theatre" (Kamols et al., 2021) where top-down decision-makers ignore community input (Fredericks et al., 2020). As a result, grassroots activism movements have emerged worldwide, empowering local communities to voice their dissent and challenge top-down approaches to governance.

To address these issues, a collaborative approach involving a diverse representation of stakeholders is required for better governance outcomes. Drawing on their collective knowledge, skills and creativity would better help address the needs, wants and aspirations of people in urban environments. Implementing a middle-out engagement approach (Caldwell et al., 2021; Fredericks et al., 2023) would be a key driver for both the acquisition of skills and knowledge as well as creativity and innovation. The concept of middle-out engagement is aimed at fostering greater levels of collaboration and inclusion, connecting people across the multiple scales of the city (Figure 13.2). By tapping into the knowledge bank and expertise of top-down and bottom-up stakeholders, this approach seeks to reach a common ground where they can all work together as a representative coalition. Through active cooperation and dialogue, middle-out engagement encourages these stakeholders to jointly extend their efforts outward, resulting in more effective and representative outcomes. The concept of middle-out (Kinchla and Wolfe, 1979) has been used by researchers and industry professionals from various disciplines, such as biochemistry, engineering, social sciences, information technology, and interaction design.

Gemperle et al. (2023) showcased the effectiveness of the middle-out engagement approach in their urban community garden project. The project aimed to promote collaborative efforts in creating community governance for shared spaces and designing an interactive urban interface for local gardeners (Figure 13.3). This was achieved by engaging a diverse group of representatives from the top-down and bottom-up, who possessed valuable knowledge and expertise in community gardening. The representatives participated in interactive workshops to establish project requirements, identify user needs, and co-designed an urban interface for the community garden. Since the deployment of the urban interface, the representatives have been able to collaborate more effectively, resulting in the establishment of new procedures and practices that have enhanced the governance of the community garden and improved the overall user experience. The study results revealed that the middle-out approach facilitated effective communication, integration of objectives, and inclusive decision-making processes among the representatives.

The Chatty Bench Project (Gonsalves et al., 2021a) is another example of using a middle-out approach with an interactive urban interface. The project engaged property owners who were representatives from the top-down, while community members from a local neighbourhood centre and the village church represented people from the bottom-up. The project incorporated art, performance, and locative media to enable community members to create and participate in critical digital storytelling using a combination of physical and digital interactions. Overall, the project not only revealed critical stories but also activated public spaces that are often overlooked. It demonstrates how a collaborative approach can contribute to the governance of urban environments.

FIGURE 13.3 Representatives from the urban community garden project testing the interactive installation. (Photograph by John Marcel Gemperle.)

The project successfully brought together representatives from both the top-down and bottom-up, fostering an environment of cooperation and mutual understanding.

By adopting a collaborative approach, urban interfaces can better align with the needs and interests of local communities while addressing potential risks and negative impacts. Ultimately, collaborative approaches enable more equitable and sustainable urban environments that reflect the shared values and aspirations of their inhabitants.

13.4.2 Turning Actions into Governance

Urban environments incorporate a mix of diverse cultures, nationalities, and ethnicities where a variety of different types of people blend together. People all have unique requirements, perspectives, and needs for the areas they live, work, and socialise in. Meeting the diverse interests of all stakeholders and including their voices in the governance process is a complex undertaking. It requires strategies, such as middle-out engagement to increase participation and collaboration. By implementing this approach urban interfaces can be used as a mechanism to align the needs, wants, and desires of local communities, fostering more equitable and sustainable urban environments.

Collaboration among a variety of stakeholders can greatly benefit the governance of urban environments, where either government agencies or local communities can initiate conversations. In these cases, urban interfaces can play a vital role in facilitating the community engagement process in three key ways. Firstly, it can lower barriers between government and communities by providing a direct, approachable, safe, informal, yet credible and trustworthy channel for community dialogue. Secondly, it can set the context of the topics discussed, giving diverse stakeholders the opportunity to participate in the decision-making process. Lastly, it can improve the overall transparency of governance and accountability of all parties involved by providing clearer and up-to-date access to real-time discussion status. These opportunities for greater collaboration can be achieved by designing and rolling out initiatives and policies in continuous cooperation with top-down and bottom-up stakeholders. Urban interfaces can encourage participation from a variety of stakeholders and benefit the future governance of urban environments.

13.5 DIGITAL PLACEMAKING

13.5.1 Interfacing Citizens and the City

When urban interfaces are applied to the relationships between people and places, they tend to coalesce around processes of placemaking (Wyckoff, 2014), as practices aimed at forging purposes, qualities, affordances, and shared perceptions of public urban space have come to be known as. Original approaches to placemaking, as espoused by the Project for Public Spaces (2007), consisted in hands-on interventions into the built environment, collectively carried out by members of a community, to reimagine, reshape, and improve their neighbourhood. Those collective efforts would draw on civic participation to resolve competing viewpoints by finding common ground around shared values, iteratively testing possible visions for the urban spaces around them, and in the process also strengthening the bonds between community members themselves, bound by a shared sense of place. The resulting place would therefore be reflective of its community of users, feel safer and more inclusive, and help to further drive new shared stories, memories, and connections, in a continuous, citizen-driven process of self-improvement.

Urban interfaces, for their use of technology as the leading driver of such practices of urban transformation, qualify as *digital* placemaking. Because they often require dedicated (and potentially expensive) urban infrastructure, some digital placemaking approaches have often been criticised as departing from the traditional ethos of placemaking and not involving the place community in their design, following instead a top-down agenda of urban 'activations', and risking 'amplifying preexisting inequities, exclusions, or erasures in the ways that certain populations experience digital media' (Halegoua & Polson, 2021, p. 574). However, the very understanding about what digital placemaking consists of has become the subject of intense scrutiny in recent years, largely due to two factors. First, the different application of the term across otherwise isolated fields, such as urban studies and creative sectors, with lack of a shared theoretical framework (Basaraba, 2021), thus resulting in multiple definitions and approaches, as we discuss below. Second, the increasing pervasiveness of technology not only in cities, but across most areas of life. As some have argued, digital technology is so ingrained in the ways we interact in society at large, and in cities in particular, that previous attempts of labelling certain approaches as 'digital' become now rather pointless (Toland et al., 2020): all placemaking can be done with the aid of digital technologies, though opting out from using them surely remains an option (Hespanhol, 2022). Yet, employing technology certainly allows the potential for the design of certain placemaking activities which would be otherwise unlikely - through the use of media architecture and public displays, for example - while also facilitating their application across the various scales of a smart city (Figure 13.2). For that reason, placemaking carried out with the aid of urban interfaces tends to display particular features, challenges, and opportunities worthy of consideration.

13.5.2 The Making of Hybrid Places

One of the effects of ubiquitous computing (Weiser, 1991) has been the increasingly dual nature of placemaking as happening both through physical and digital means. The greater availability of digital infrastructure in urban spaces, as described in Section 13.7, also meant that it started to be increasingly used as part of many urban dynamics, including placemaking. That led, for example, to combinations of media facades and mobile devices or kiosks for situated civic participation (Schroeter & Foth, 2009; Behrens et al., 2014; Hespanhol et al., 2015), digitally enabled pop-up community engagement (Fredericks et al., 2018), city hacking (De Waal, De Lange & Bouw, 2017; Hespanhol & Tomitsch, 2019; Mulder & Kun, 2019), or location-based games to connect citizens to place (Pang et al., 2020). Likewise, digital services like social media have become instrumental for citizens to share stories, memories, and opinions, and in the process articulate common place and community identities (Breek et al., 2018; Waite, 2020; Wilken & Humphreys, 2021).

Beyond the use of technologies as tools for placemaking, another consequence of their ubiquity was a change in the nature of place itself (Hespanhol, 2022): when experiencing a place, citizens are often both present in a physical location and also partaking in interactions online. The nature of those interactions is, in many cases, intrinsically related to the place itself, yet decoupled from the built environment – they are *about* the place, *enabled* by it, and affecting it, however evolving entirely on digital channels or media. Virtual placemaking (Qabshoqa, 2018) has thus been proposed as a strategy to enable explorations on architecture and urban design using gamified mixed reality. Intensified in response to the COVID-19 lockdowns, and combined with other technologies such as QR codes, social media, video conference apps, projections, and displays, similar practices expanded to a range of interconnected platforms enabling continuous, yet fragmented conversations. Hespanhol (2022) defined *augmented placemaking* as this set of practices enabled by physical urban spaces equipped with digital infrastructure (superspaces) capable of enabling multiple digital place narratives (pluriplaces). Through those practices, a multiplicity of digitally enabled place narratives can potentially be made, and simultaneously coexist, around a single physical location, each of them accessed by different cohorts of citizens.

13.5.3 Approaches, Stakeholders, and Goals

In addition to the "standard" approach to placemaking proposed by the Project for Public Spaces (2007), Wyckoff (2014) identifies three others, each aligned with a particular agenda: strategic placemaking, tactical placemaking, and creative placemaking. Those three approaches also translate to digital counterparts. Strategic digital placemaking aims at solving a particular goal while improving a place, and it usually operates at the level of community and metropolis (Figure 13.2), driven by city makers responsible for shaping a particular urban programme. Examples include the use of urban screens for the development of cultural programmes to make a public precinct more vibrant and safer, such as the urban screen at Federation Square in Melbourne, Australia (Tomitsch et al., 2015), or the digital platforms at the Footbridge Gallery (Figure 13.4) at The University of Sydney campus (Hespanhol, 2023a). Likewise, recently emerging approaches to digital placemaking have aimed at strengthening the connections between citizens and urban nature (Fernandez de Osso Fuentes et al., 2023), thus creating urban interfaces capable of increasing health and well-being within the city. Those strategic urban interfaces are long-running, often involve many stakeholders, and may take years before starting to yield meaningful outcomes.

FIGURE 13.4 The Footbridge Gallery, at The University of Sydney, Australia. Artwork: "45 Rainbow". (Image courtesy of Xavier Ho.)

Tactical digital placemaking consists in grassroots tactical urbanism enacted through digital technologies and platforms. It may take the form of digital pop-ups (Fredericks et al., 2015; Fredericks et al., 2018), urban prototyping (Korsgaard & Brynskov, 2014; Hoggenmüller & Wiethoff, 2014), guerrilla urban projections (Hespanhol, 2022), or radical placemaking (Gonsalves et al., 2021b). It usually involves urban interfaces operating at the personal, interpersonal, and community scales of the city (Figure 13.2), and are designed as platforms for activism, citizen expression, or urban hacking.

Creative placemaking (Schupbach, 2015; Zitcer, 2020), in turn, can operate across all scales of the city. It involves the use of creative or artistic interventions rejuvenating and animating streetscapes (Wyckoff, 2014) and in the process addressing goals that may be either strategic or tactical. Likewise, it may also range from community-based digital storytelling (Wouters, Claes & Moere, 2018), to temporary interactive light installations (Hespanhol & Tomitsch, 2012; Hespanhol et al., 2014; Wouters et al., 2016a), to precinct-wide interventions, such as the 'Midnight Moment' arts program in Times Square, New York, USA (Tomitsch et al., 2015). A particular kind of creative placemaking, playful placemaking (Innocent, 2018; Chew, Hespanhol & Loke, 2021) adopts the hacking of street elements as the key approach to design place-based urban play and games as interfaces between people and the city.

13.6 GATHERING AND COMMUNICATING URBAN DATA

Interfaces can be more literally interpreted as boundary points of contact where data or information can be exchanged between two distinct systems. For an 'urban interface', these systems prototypically consist of official institutions and their citizens. Such data-driven urban interfaces appeared throughout the history of cities, as for example mediaeval street signage was invented to help guide citizens and visitors in the growing city environment, and notice boards were meant to update passers-by about upcoming events or official regulations. With the advancements in digital technology, including the development of interactive screens, low-energy electronics, and wireless networking, the urban interface has now gained the ability to convey more abstract, timely, and engaging information *to* citizens. Conversely, thanks to their interactive abilities, urban interfaces are now also increasingly used to gather data *from* citizens.

The exchange of data from and to citizens is closely related to the "smart city" concept, where data is believed to provide a more objective basis for making better decisions and automating many processes. Many smart cities therefore deploy urban interfaces to collect at least three different types of data (Lau et al., 2019). Physical data sources consist of sensors that capture information of a particular space or area, in terms of temperature, air quality, activities, and so on. Cyber data sources rather obtain information from the Internet domain such as via social media, web access logs or online opinion polls. Participatory data sources include the voluntary crowdsensing and crowdsourcing of data via personal or publicly accessible electronic devices. Whereas the first two types of urban interface often track people without their expressive consent by covertly hiding sensing devices within the cyber-physical urban infrastructure, participatory urban interfaces stand out as they require the conscious and deliberate engagement of citizens.

13.6.1 PARTICIPATORY URBAN INTERFACES

There have been various research initiatives investigating how the design and deployment of physical data sources can be more democratised via appropriate urban interface design methods, often under the umbrella of terms such as people-centric urban sensing (Campbell et al., 2006), bottom-up citizen sensing (Coulson et al., 2021), participatory sensing (Coulson et al., 2018a) or citizen science (Ali et al., 2021), among many others. As such, citizens have been invited to build and deploy humidity and temperature (Balestrini et al., 2017), air quality (Gabrys, Pritchard & Barrat, 2016; Jiang et al., 2016), noise (Coulson et al., 2018b) sensors, often by using highly specialised urban interface development toolkits and civic technologies (Daher et al., 2021). Other studies focused on how people could become a sensor themselves by creating new, or annotating existing, urban data with personal observations and viewpoints. Often, urban interfaces then take the shape of custom-developed smartphone applications or mobile websites with interactive mapping features (Boulos et al., 2011) that allow citizens to capture data in both time and space. These "mobile reports" (Crowley et al., 2012) describe environmental defects close to real-time, like potholes and damaged pavements (Pak, Chua & Vande Moere, 2017) or personal mobility habits (Lee & Sener, 2020), helping to create a so-called "wikification of GIS by the masses" (Boulos, 2005).

The voluntary participation via one's own personal smartphone as an urban interface not only requires a personal motivation that closely relates to social engagement (Reed et al., 2013) but also a dedicated commitment and effort to interact with the urban interface. In contrast, interactive public displays are promised to be more opportunistic, i.e., offering citizens to take part in urban data gathering while they are not deeply focused on some other task, such as when they wait around, linger around or simply pass by (Alt & Vehns, 2016). Moreover, public displays can be used as urban interfaces to specifically reach location-specific citizen groups (e.g., locals, tourists), while offering the opportunity to capture participatory urban data immediately and closely to the actual location of what is being surveyed. Accordingly, custom-developed public displays have been used as dedicated types of urban interfaces that can augment existing physical and cyber-physical data sources, such as to capture how citizens personally experience the noise levels and the number of people visiting a certain location (Houben et al., 2019), or whether and how they are impacted by local air quality issues (Claes, Coenen & Vande Moere, 2018). Other public display interventions derive completely new participatory data by surveying citizens with multiple-choice type of questionnaires. Some interventions combine an existing semi-permanent public display with a personal smartphone (Schroeter, Foth & Satchell, 2012), another touch-enabled screen (Hespanhol et al., 2015), a game-show-like button (Veenstra et al., 2015), dedicated plug-in push box (Hespanhol & Tomitsch, 2019) as a dedicated input modality. Other public displays use small or energy-saving screens to enable the surveying of citizens independently from existing urban infrastructure: custom-developed pin-like (Liu et al., 2019) or e-ink equipped displays with small push buttons (Coenen, Houben & Vande Moere, 2019) can be distributed over multiple locations simultaneously to survey different citizen subgroups or compare the distribution of answers over multiple locations (Coenen et al., 2021). Other types of urban interfaces tend to physically foreground their input modality by physically exaggerating the playful design of the input features, often in an attempt to increase their playfulness and lower their perceived accessibility. Such urban interfaces range from relatively simple carton board boxes (Koeman, Kalnikaité & Rogers, 2015), bespoke contraptions (Taylor et al., 2012) and public posters (Vlachokyriakos et al., 2014) with push buttons, to quite elaborate life-sized installations that are equipped with 'spinners', 'ball tubes', and large buttons (Golsteijn et al., 2015), or industrially looking dashboard with toggles, sliders, and rotary dials (Golsteijn et al., 2016), which all query citizens about their local experiences. Urban interfaces with more 'natural' input modalities specifically attempted to reach narrowly defined citizen subgroups through their input modality, as weight-sensitive floor mats solely invited pedestrians (Steinberger, Foth & Alt, 2014) or physically allowed bicyclists (Claes, Slegers & Vande Moere, 2016) to specifically respond to hyperlocal questions that were relevant to them.

13.6.2 Urban Data Analytics

The easily accessible, but also brief, direct, and opportunistic nature of all these urban interface technologies, however, limits the qualitative depth of what can be gathered to relatively summary or numerical types of participatory data, such as textual messages, Likert-scale opinions or smartphone images. Therefore, recent research has experimented on how direct questions could be replaced by more imaginative and open-ended use of storytelling, i.e. the method of combining facts and narratives to convey meaning. As such, several urban interface studies have investigated the use of storytelling to capture urban testimonials of neighbourhood residents. As such, public displays have been used with relatively limited success to collect input in the form of historical photos, audio-clips, and videos (Cheverst et al., 2017) or hyperlocal user-generated videos (Michielsen et al., 2020). More elaborate storytelling examples include an audio-recording old-style telephone that was passed around in a custom suitcase (Crivellaro et al., 2016), and a collection of custom-developed textual and auditory storytelling interfaces were mounted on the facades of houses (Wouters, Huyghe & Vande Moere, 2014).

13.6.3 URBAN DATA GATHERING LIMITATIONS

It has, however, been noted that participatory urban data interventions are often plagued with various financial, technological, and data reliability issues (Balestrini et al., 2015), and tend to result in 'just good enough data' (Gabrys, Pritchard & Barratt, 2016) of which the quality falls outside of the common scientific practices of legitimation and validation. Thus, while such data is often "meaningfully inefficient" (Gordon & Walter, 2019), it is yet sufficient to initiate discussions with other civic stakeholders, and contributes to new data practices that generate more open and democratic engagement with data in general (Gabrys, 2016), i.e. as citizens are actively involved into creating of data that addresses issues on a local level, the underlying technologies also become more relevant and trustworthy to their own lives (Coulson et al., 2021). However, there is a tendency for sampling bias as certain data dimensions are collected more readily than others, thereby contributing to issues about the quality of the data contributed by users (Löfgren & Webster, 2020). Moreover, it has been shown that most urban interfaces are still driven by utilitarian motives rather than facilitating placemaking, and are controlled by top-down gatekeepers who operate covertly and without accountability (Biedermann & Vande Moere, 2021).

13.6.4 URBAN DATA VISUALISATIONS

The smart city premise of collecting all these data sources is to combine or 'fuse' them together in order to create valuable insights into particular urban domains or applications. To ensure that these decisions are valid and feasible, these insights are often presented back to citizens in the form of data visualisations that entice citizens to collaboratively discuss and analyse the data they collect (Luther et al., 2009). Consequently, a closed feedback loop between citizens and city institutions is created: what was originally captured as participative data might lead to insightful decisions and regulations that in turn aim to maintain or influence the behaviour, opinion or attitude of citizens, and thus, expectedly, also leads to potentially noticeable changes in newly acquired urban data. Whereas most city institutions present their urban data through open data ecosystems (Zuiderwijk, Janssen & Davis, 2014) or dedicated websites that feature elaborate infographic dashboards (Alizadeh, Sarkar & Burgoyne, 2019), urban interfaces carry the intrinsic advantage to present the data back in a much more situated way (Vande Moere & Hill, 2012). While most urban data visualisations ignore the physical world, urban interfaces situated in the physical world can extend through the data's physical referent as well as the data's physical presence (Willett, Jansen & Dragicevic, 2016). Therefore, some studies investigated how urban interfaces can represent urban data within its own physical environment of origin, reaching the original citizen population but also situating the potential interpretation of this data within its own context. As such, potentially provocative local demographic data has been physically framed and situated as official street name boards (Claes & Vande Moere, 2013), whereas open government data has been represented as an elaborate physical installation that was revealed as a local community event (Perovich, Wylie & Bongiovanni, 2020). In that respect, research has shown the need to stage stepwise interactions to explore the visualisation, contextualise the data within the location, and to provide concrete ways allowing citizens to add their perspective to the visualisation (Schoffelen et al., 2015).

13.7 INFRASTRUCTURE

Urban interfaces utilise a range of technologies to support the implementation of applications for governance, placemaking and data visualisation. In that sense, these technologies represent the infrastructure for urban interfaces and the applications they support. Considering the underlying technologies as a networked infrastructure promotes the perspective that urban interfaces are not isolated but part of a wider system of components. "Zooming" in and out of that wider system reveals how the infrastructure supports urban interfaces at the personal, community and metropolis levels. For example, on a personal level, a pedestrian encountering an autonomous vehicle results in needs specific to that individual relationship, such as indicating whether it is safe for the pedestrian to cross the road in front of the vehicle. On a community level, the urban interface between autonomous vehicles and pedestrians needs to also consider other people and vehicles in the vicinity. On a metropolis level, all vehicles operating on a city's road need to interact with each other as well as with other urban elements such as traffic signals and other road users.

Technologies for infrastructure applications include media facades, public displays, wearables, robots, environmental sensors, and so on. The focus here is not on creating novel technologies, but to identify new ways for using technology to define how people interface with the infrastructure of cities. Akin to the notion of ubiquitous computing (Weiser, 1991), the technology used to implement the urban interfaces becomes part of the ambient environment – it is both the infrastructure on which urban interfaces operate and part of the city infrastructure itself. Using the example of autonomous vehicles, the vehicle as a technology provides an infrastructure for designing urban interfaces; at the same time, the vehicle as a robot becomes part of the city infrastructure along with roads, urban furniture, buildings, signage, etc.

13.7.1 DIGITAL INFRASTRUCTURE

The proliferation of digital technology has made it possible to weave digital information into the physical environment. Referred to as a city's digital layer (Kjeldskov & Paay, 2006), augmented space (Manovich, 2006) or info-structure (Tomitsch & Haeusler, 2015), this enables new forms of interaction between people and technology. In many cases, digital infrastructure applications draw also on urban data. By presenting people with data when and where it matters, they are able to make better informed decisions and hence better use of the city's infrastructure and resources. This is especially relevant given that urban populations are continuing to rise, which puts pressure on existing infrastructure, such as roads and the energy grid. For example, some cities have started to deploy digitally augmented road traffic signs, which provide real-time data about traffic flow (Figure 13.5). By providing motorists with this information, they are able to avoid congested roads, alleviating pressure on the transport network. The same effect can be achieved through wearable devices; for example, Google Maps provides information about road traffic and automatically suggests alternative routes to circumvent busy areas. Although Google Maps runs on a personal device, this also represents a form of digital infrastructure. Here, the infrastructure is augmented through the mobile device as an urban interface.

The notion of digital infrastructure extends to other urban elements, such as street furniture, including smart benches that allow people to interface with the infrastructure and charge their devices. Adding a digital layer can enhance urban experiences, which contributes to placemaking, and address urban issues. An example for such an intervention was the TetraBIN project (Figure 13.6), which wrapped a city bin in a low-resolution screen to gamify the act of putting rubbish in a bin (Tomitsch, 2014). The aim of the project was to demonstrate how an urban interface can playfully address the issue of littering in cities.

Although this is not commonly a concern, it is important that the design of infrastructure urban interfaces considers not only human but also non-human stakeholders. In a speculative design study

FIGURE 13.5 Digitally augmented traffic signs in Seoul, South Korea.

FIGURE 13.6 The TetraBIN project addresses the issue of littering by adding a digital layer to city bins.

that involved the design of a parklet, the researchers set out to account for possums as non-human stakeholders (Tomitsch et al., 2021). Through creating non-human personas and considering their perspectives when making design decisions, they discarded wireless phone charging stations as they were found to negatively interfere with urban wildlife. This shows opportunities for HCI beyond just creating the user interfaces but to also inform the design process in a way that leads to better outcomes. It also highlights the broader role of urban interfaces beyond HCI itself, toward also enabling and mediating broader social, cultural and ecological urban relationships.

13.7.2 Autonomous Infrastructure

Recent advancements in technology have led to robotic technologies becoming readily available and deployable even in city environments, which are less controlled and more complex compared to factories and warehouses. Corporations are investing in autonomous technology, from Amazon's delivery drones to Mercedes' driverless concept car. For the field of HCI, this presents new opportunities and challenges regarding the interaction between people and these robotic technologies. Urban interfaces are able to support this interaction and contribute to the adoption and acceptance of new applications like autonomous vehicles. Research findings suggest that people's trust in autonomous vehicles increases if they are able to understand what the vehicle sees and what it is going to do (Faas et al., 2021). In other words, autonomous vehicles need to communicate their awareness and their intent to pedestrians and other road users.

The field of HCI has turned its attention to this domain, focusing on the so-called external human-machine interfaces (eHMIs). These eHMIs embody urban interfaces as they enable people to interact with autonomous vehicles. Drawing on other HCI domains, like ambient displays (Wisneski et al., 1998), eHMIs make the internal operations of autonomous vehicles visible to pedestrians and other road users. HCI research has investigated, for example, the use of different modalities (Dou et al., 2021), shared-space scenarios (Burns et al., 2020), and how to design eHMIs for multiple pedestrian interactions (Wilbrink et al., 2021). The concept can be adopted for different kinds of autonomous vehicles, including small delivery robots to convey their predicted path to pedestrians (Yu, Hoggenmueller, & Tomitsch, 2023).

Autonomous city infrastructure extends beyond utilitarian applications such as driverless cars and delivery robots. Linking this domain with the notion of placemaking offers opportunities for new kinds of urban interfaces that activate spaces and alter the experience of passers-by. One such project is Woodie (Figure 13.7), a small autonomous robot capable of drawing with chalk, which was deployed as part of a light festival (Hoggenmueller, Hespanhol, & Tomitsch, 2020).

The field of HCI also provides methods for evaluating these kinds of complex urban interfaces. Deployment studies, like the Woodie project, can draw on established methods such as in-the-wild evaluations (Chamberlain et al., 2012). Often these kinds of interventions, like driverless cars, are expensive and complex to prototype and evaluating them in the wild could potentially put participants at risk. To circumvent those challenges, HCI studies have documented the benefits of virtual reality

FIGURE 13.7 The Woodie project, involving an urban robot capable of drawing with chalk. (Image courtesy of Marius Hoggenmueller.)

environments to simulate or replay 360-degree recordings of autonomous vehicles (Tran, Parker, & Tomitsch, 2021), allowing participants to experience their urban interfaces in a safe environment. Using such approaches makes it possible to prototype and evaluate speculative HCI applications for urban environments, which in turn has the ability to inform future technologies and deployment studies.

13.8 FUTURE DIRECTIONS

The nature, features, and capabilities of urban interfaces are continuously evolving in response to new developments in technology and society. The rapid urbanisation of communities across the world, coupled with omnipresent technology, has translated into incredibly complex networks of relationship between citizens, governments, institutions, big technology corporations, and the planet itself. By the second decade of the 21st century, three pressing concerns are of particular relevance when it comes to technology-mediated relationships between humans and the urban environment, thus urban interfaces: sustainability, hybridity, and AI. Those concerns cut across all four scales of the city, and play out across the four dimensions of governance, place, data, and technology included in our typological model above (Figure 13.2).

13.8.1 Sustainability

The climate crisis has prompted a rethink of the ways humans interact and the roles technology may play in supporting more sustainable approaches to urban living, including the design of technology itself. The release of the sustainable development goals (SDGs) by the United Nations Assembly in 2015 has highlighted the interconnectedness of many pressing issues faced by global populations, with one of the goals (SDG 11) focusing on promoting sustainable cities and communities. There have been calls by HCI scholars for a realignment of design efforts toward the SDGs (Hansson et al., 2021), with particular focus on sustainable communities (Fredericks et al., 2019). Some of the recent examples of urban interfaces discussed above illustrate that shift towards urban interfaces attentive to sustainable behaviour, such as the playful awareness-raising about littering of TetraBIN (Tomitsch, 2014), illustrated in Figure 13.6, or the urban community garden project (Figure 13.3) developed by Gemperle et al. (2023). Likewise, recent research efforts have sought to foster frameworks for urban resilience through an integrated approach to urban interfaces that combine human and other-than-human perspectives (Boon, De Waal, & Suurenbroek, 2023), or nurturing digital placemaking around urban fauna, flora, and natural systems (Fernandez de Osso Fuentes et al., 2023).

13.8.2 Hybridity

While every urban interface may be regarded as hybrid, in the sense of imposing a digital layer (Kjeldskov & Paay, 2006; Tomitsch & Haeusler, 2015) over the built environment of the city, the notion of hybridity has now permeated daily life to the point of becoming integral to urban life. Not only places are augmented (Manovich, 2006), but so is placemaking itself (Hespanhol, 2022), with human interactions being increasingly partaken simultaneously in built (physical) and programmed (virtual) environments as well as asynchronously in time. Accordingly, urban interfaces ought to also migrate from physical to virtual urban infrastructure, with direct implications for governance, placemaking, and management of data about the city and citizens. While many instances of hybrid urban interactions will continue to be driven by progress in mixed reality (Bibri & Allam, 2022), it is also worth noting that more prosaic technologies such as videoconference tools, popularised since the COVID-19 pandemic, have also spurred the normalisation of real-time hybrid interactions, translated into innovative urban interfaces in sectors such as cultural venues (Lai & Hespanhol, 2021), urban heritage (Morrison, 2022), and tourism (Coles, 2019).

13.8.3 Artificial Intelligence

The use of AI in cities is still incipient, with preliminary implementations consisting of urban interfaces targeting city-wide infrastructure and services, such as transport systems (Nikitas et al., 2020). Likewise, studies have investigated the potential of AI to perform an assistive role in urban planning, contributing to participatory decision-making for sustainable city development (Quan et al., 2019). A civic urban AI would dialogue with citizens across various stages of their urban lives, for example through a combination of predictive algorithms supporting preparedness to disruptions and thus urban resilience, and generative AI mediating governance and citizen approaches to sustainability. Given the infrastructure and technical skills required, most of those concepts would likely translate into top-down initiatives, operating largely at the level of the metropolis (Figure 13.2). However, the prospect of integrating AI into urban interfaces across the different scales of the city has become significantly more tangible since the release of open models and tools for public use, like conversational generative AI. Social urban robots like Woodie (Hoggenmueller, Hespanhol, & Tomitsch, 2020), illustrated in Figure 13.7, could be able to provide bespoke and site-specific engagement directly with citizens, and truly adaptable public spaces could be designed to support playful placemaking (Chew, Hespanhol & Loke, 2021).

13.8.4 Methods

Researchers have proposed new methods to assist with the design of urban interfaces in response to some of these new scenarios. For example, while developing personas is a widespread method to help determine the profiles for the stakeholders of a design solution, those stakeholders have traditionally been restricted to human beings. To address concerns about sustainability and the possible impact of technology on non-human stakeholders like animals and natural systems, scholars have proposed the development of non-human personas (Tomitsch et al., 2021), thus expanding the scope of relationships taken into account by the proposed solution.

The development of methods for the design of AI-driven urban interfaces remains an area to be further explored, but some hints may be taken from the methods addressing hybridity, which include approaches either for diagnosis or simulation. For the former, researchers have proposed methods like the Local-Social Index (Lai & Hespanhol, 2021), which provides an indication of the level of sociability and physical situatedness afforded by the hybrid design solution. Other methods explored hybridity itself as a means to support the simulation of challenging urban environments. For example, hyperreal prototyping (Hoggenmueller & Tomitsch, 2019) was proposed as an approach to simulate interactions between autonomous vehicles and pedestrians, too dangerous and costly to be tested in situ. Likewise, virtual urban prototyping (Fredericks et al., 2023) aims at the design and test of media architecture prior to its construction, or remote to where the test is being conducted.

Given the possibility of simulating possible futures, those methods also offer the possibility of contributing to sustainable agendas.

13.9 CONCLUSION

Urban interfaces capture a domain of increasing significance at the intersection of HCI, design, social sciences, and urban studies. While having digital technologies at their core, their scope goes far beyond the user interface itself, to encompass the myriad entangled relationships constituting urban life, their interdependence, and shared role co-constituting the situations, services, values, and environments, physical and digital, that come to be known as a city.

In this chapter, we discussed key concepts in urban HCI and how urban interfaces relate to them. We proposed a typological model to map the scales of a city in which urban interfaces can operate at (individual, interpersonal, community, and metropolis), and four core dimensions to support their design and implementation: governance, place, data, and infrastructure. We then discussed those four dimensions, illustrated through examples, and highlighted challenges and opportunities. Following this, we outlined future directions for the field, emerging concerns, concepts, applications, and methods.

As the field continues to evolve and urban interfaces become increasingly intelligent and hybrid, it is our hope that the HCI community can also take the necessary steps to ensure that they continue to spouse a duty of care for the diverse stakeholders within the city, humans and non-humans, and do so in ways that are sustainable, ethical, and considerate to the many systems and ecosystems sharing the urban realm.

REFERENCES

Ackad, C., Clayphan, A., Tomitsch, M., & Kay, J. (2015, September). An in-the-wild study of learning mid-air gestures to browse hierarchical information at a large interactive public display. In *Proceedings of the 2015 ACM International Joint Conference on Pervasive and Ubiquitous Computing* (pp. 1227–1238).

Ahvenniemi, H., Huovila, A., Pinto-Seppä, I., & Airaksinen, M. (2017). What are the differences between sustainable and smart cities? *Cities*, *60*, 234–245.

Ali, M. U., Mishra, B. K., Thakker, D., Mazumdar, S., & Simpson, S. (2021). Using citizen science to complement IoT data collection: a survey of motivational and engagement factors in technology-centric citizen science projects. *IoT*, *2*(2), 275–309.

Alizadeh, T., Sarkar, S., & Burgoyne, S. (2019). Capturing citizen voice online: Enabling smart participatory local government. *Cities*, *95*, 102400.

Allam, Z., & Dhunny, Z. A. (2019). On big data, artificial intelligence and smart cities. *Cities*, *89*, 80–91.

Alt, F., & Vehns, J. (2016, June). Opportunistic deployments: challenges and opportunities of conducting public display research at an airport. In *Proceedings of the 5th ACM International Symposium on Pervasive Displays* (pp. 106–117).

Baibarac-Duignan, C., & de Lange, M. (2021). Controversing the datafied smart city: Conceptualising a 'making-controversial' approach to civic engagement. *Big Data & Society*, *8*(2), 20539517211025557.

Balestrini, M., Diez, T., Marshall, P., Gluhak, A., & Rogers, Y. (2015). IoT community technologies: leaving users to their own devices or orchestration of engagement?. *EAI Endorsed Transactions on Internet of Things*, *1*(1). EAI.

Balestrini, M., Rogers, Y., Hassan, C., Creus, J., King, M., & Marshall, P. (2017, May). A city in common: a framework to orchestrate large-scale citizen engagement around urban issues. In *Proceedings of the 2017 CHI Conference on Human Factors in Computing Systems* (pp. 2282–2294). New York: ACM Press.

Basaraba, N. (2021). The emergence of creative and digital place-making: A scoping review across disciplines. *New Media and Society*, *25*(6), 1470–1497.

Batty, M., Axhausen, K. W., Giannotti, F., Pozdnoukhov, A., Bazzani, A., Wachowicz, M., ... & Portugali, Y. (2012). Smart cities of the future. *The European Physical Journal Special Topics*, *214*, 481–518.

Batty, M. (2013). Big data, smart cities and city planning. *Dialogues in Human Geography*, *3*(3), 274–279.

Behrens, M., Valkanova, N., gen. Schieck, A. F., & Brumby, D. P. (2014, June). Smart citizen sentiment dashboard: A case study into media architectural interfaces. In *Proceedings of the International Symposium on Pervasive Displays* (pp. 19–24)

Bibri, S. E., & Allam, Z. (2022). The metaverse as a virtual form of data-driven smart urbanism: On post-pandemic governance through the prism of the logic of surveillance capitalism. *Smart Cities*, *5*(2), 715–727.

Biedermann, P., & Vande Moere, A. (2021). A critical review of how public display interfaces facilitate placemaking. In *Proceedings of the Media Architecture Biennale 2020*, (pp. 170–181). New York: ACM Press.

Boon, B., de Waal, M., & Suurenbroek, F. (2023). *Public Spaces and Urban Resilience: State of Affairs in Dutch Cities and Exploring Human and Non-Human Perspectives*. Amsterdam: Hogeschool van Amsterdam.

Boulos, M. N. K. (2005). Web GIS in Practice III: creating a simple interactive map of England's strategic health authorities using google maps API, Google Earth KML, and MSN Virtual Earth Map Control. *International Journal of Health Geographics*, *4*(1), 1–8.

Boulos, M. N., Resch, B., Crowley, D. N., Breslin, J. G., Sohn, G., Burtner, R., ... & Chuang, K. Y. S. (2011). Crowdsourcing, citizen sensing and sensor web technologies for public and environmental health surveillance and crisis management: trends, OGC standards and application examples. *International Journal of Health Geographics*, *10*(1), 1–29.

Breek, P., Hermes, J., Eshuis, J., & Mommaas, H. (2018). The role of social media in collective processes of place making: A study of two neighborhood blogs in Amsterdam. *City & Community*, *17*(3), 906–924.

Brynskov, M., Dalsgaard, P., Ebsen, T., Fritsch, J., Halskov, K., & Nielsen, R. (2009, August). Staging urban interactions with media façades. In *INTERACT (1)* (pp. 154–167).

Burns, C. G., Oliveira, L., Hung, V., Thomas, P., & Birrell, S. (2020). Pedestrian attitudes to shared-space interactions with autonomous vehicles-A virtual reality study. In *Advances in Human Factors of Transportation: Proceedings of the AHFE 2019 International Conference on Human Factors in Transportation, July 24-28, 2019, 10* (pp. 307–316). Washington DC: Springer International Publishing.

Caldwell, G. A., & Foth, M. (2014, November). DIY media architecture: Open and participatory approaches to community engagement. In *Proceedings of the 2nd Media Architecture Biennale Conference: World Cities* (pp. 1–10).

Caldwell, G. A., Fredericks, J., Hespanhol, L., Chamorro-Koc, M., Barajas, M. J. S. V., & André, M. J. C. (2021). Putting the people back into the "smart": Developing a middle-out framework for engaging citizens. In A. Aurigi & N. Odendaal (Eds.), *Shaping Smart for Better Cities* (pp. 239–266). London: Academic Press.

Campbell, A. T., Eisenman, S. B., Lane, N. D., Miluzzo, E., & Peterson, R. A. (2006, August). People-centric urban sensing. In *Proceedings of the 2nd Annual International Workshop on Wireless Internet* (pp. 18-es). New York: ACM Press.

Chamberlain, A., Crabtree, A., Rodden, T., Jones, M., & Rogers, Y. (2012, June). Research in the wild: understanding 'in the wild' approaches to design and development. In *Proceedings of the Designing Interactive Systems Conference* (pp. 795–796).

Cheverst, K., Turner, H., Do, T., & Fitton, D. (2017). Supporting the consumption and co-authoring of locative media experiences for a rural village community: design and field trial evaluation of the SHARC2.0 Framework. *Multimedia Tools and Applications*, *76*, 5243–5274.

Chew, L., Hespanhol, L., & Loke, L. (2021). To play and to be played: Exploring the design of urban machines for playful placemaking. *Frontiers in Computer Science*, 3, 635949.

Claes, S., & Vande Moere, A. (2013, June). Street infographics: raising awareness of local issues through a situated urban visualization. In *Proceedings of the 2nd ACM International Symposium on Pervasive Displays* (pp. 133–138). New York: ACM Press.

Claes, S., Slegers, K., & Vande Moere, A. (2016, May). The bicycle barometer: design and evaluation of cyclist-specific interaction for a public display. In *Proceedings of the 2016 CHI Conference on Human Factors in Computing Systems* (pp. 5824-5835). New York: ACM Press.

Claes, S., Coenen, J., & Moere, A. V. (2018, September). Conveying a civic issue through data via spatially distributed public visualization and polling displays. In *Proceedings of the 10th Nordic Conference on Human-Computer Interaction* (pp. 597–608). New York: ACM Press.

Coenen, J., Houben, M., & Vande Moere, A. (2019, June). Citizen dialogue kit: public polling and data visualization displays for bottom-up citizen participation. In *Companion Publication of the 2019 on Designing Interactive Systems Conference 2019* (pp. 9–12). New York: ACM Press.

Coenen, J., Biedermann, P., Claes, S., & Moere, A. V. (2021, June). The stakeholder perspective on using public polling displays for civic engagement. In *C&T'21: Proceedings of the 10th International Conference on Communities & Technologies-Wicked Problems in the Age of Tech* (pp. 61–74). New York: ACM Press.

Coles, T. (2019). 10 Hidden in plain sight? AR apps and the sustainable management of urban heritage tourism. *Hidden Cities*, *5*(4), 209.

Coulson, S., Woods, M., Scott, M., & Hemment, D. (2018). Making sense: empowering participatory sensing with transformation design. *The Design Journal*, *21*(6), 813–833.

Coulson, S., Woods, M., Scott, M., Hemment, D., & Balestrini, M. (2018, June). Stop the noise! enhancing meaningfulness in participatory sensing with community level indicators. In *Proceedings of the 2018 Designing Interactive Systems conference* (pp. 1183–1192). New York: ACM Press.

Coulson, S., Woods, M., & Making Sense EU. (2021). Citizen sensing: An action-orientated framework for citizen science. *Frontiers in Communication*, 6, 629700.

Crivellaro, C., Taylor, A., Vlachokyriakos, V., Comber, R., Nissen, B., & Wright, P. (2016, May). Re-making places: hci, 'community building' and change. In *Proceedings of the 2016 CHI Conference on Human Factors in Computing Systems* (pp. 2958–2969). New York: ACM Press.

Crowley, D. N., Breslin, J. G., Corcoran, P., & Young, K. (2012, September). Gamification of citizen sensing through mobile social reporting. In *2012 IEEE International Games Innovation Conference* (pp. 1–5). IEEE.

Cugurullo, F. (2020). Urban artificial intelligence: From automation to autonomy in the smart city. *Frontiers in Sustainable Cities*, 2, 38.

Daher, E., Maktabifard, M., Kubicki, S., Decorme, R., Pak, B., & Desmaris, R. (2021). Tools for citizen engagement in urban planning. In: G.C. Lazaroiu, M. Roscia, & V.S. Dancu (eds) *Holistic Approach for Decision Making Towards Designing Smart Cities. Future City* (vol. 18, pp. 115–145). Cham: Springer.

Dalsgaard, P., & Halskov, K. (2010, April). Designing urban media façades: cases and challenges. In *Proceedings of the SIGCHI Conference on Human Factors in Computing Systems* (pp. 2277–2286).

Dalsgaard, P., & Hansen, L. K. (2008). Performing perception-staging aesthetics of interaction. *ACM Transactions on Computer-Human Interaction (TOCHI), 15*(3), 1–33.

De Lange, M., Merx, S., & Verhoeff, N. (2019). Urban interfaces: Between object, concept, and cultural practice. Introduction to urban interfaces: Media, art and performance in public space. *Leonardo Electronic Almanac, 22*(4).

De Waal, M., De Lange, M., & Bouw, M. (2017). The hackable city: Citymaking in a platform society. *Architectural Design, 87*(1), 50–57.

Dirks, S., & Keeling, M. (2009). *A Vision of Smarter Cities: How Cities can lead the Way into a Prosperous and Sustainable Future*. Somers, NY: IBM Global Business Services.

Dou, J., Chen, S., Tang, Z., Xu, C., & Xue, C. (2021). Evaluation of multimodal external human-machine interface for driverless vehicles in virtual reality. *Symmetry, 13*(4), 687.

Faas, S., Kraus, J., Schoenhals, A., & Baumann, M. (2021, May). Calibrating pedestrians' trust in automated vehicles: Does an intent display in an external HMI support trust calibration and safe crossing behavior? In *Proceedings of the 2021 CHI Conference on Human Factors in Computing Systems* (pp. 1–17).

Fernandez de Osso Fuentes, M. J., Keegan, B. J., Jones, M., & MacIntyre, T. (2023). Digital placemaking, health & wellbeing and nature-based solutions: A systematic review and practice model. *Urban Forestry and Urban Greening, 79*, 127796–127796.

Fischer, P. T., & Hornecker, E. (2012, May). Urban HCI: Spatial aspects in the design of shared encounters for media facades. In *Proceedings of the SIGCHI Conference on Human Factors in Computing Systems* (pp. 307–316).

Foth, M., Choi, J. H. J., & Satchell, C. (2011, March). Urban informatics. In *Proceedings of the ACM 2011 Conference on Computer Supported Cooperative Work* (pp. 1–8).

Foth, M. (2018). Participatory urban informatics: towards citizen-ability. *Smart and Sustainable Built Environment, 7*(1), 4–19.

Fredericks, J., Tomitsch, M., Hespanhol, L., & McArthur, I. (2015, December). Digital pop-up: Investigating bespoke community engagement in public spaces. In *Proceedings of the Annual Meeting of the Australian Special Interest Group for Computer Human Interaction* (pp. 634–642).

Fredericks, J., Hespanhol, L., & Tomitsch, M. (2016, June). Not just pretty lights: Using digital technologies to inform city making. In *Proceedings of the 3rd Conference on Media Architecture Biennale* (pp. 1–9).

Fredericks, J., Hespanhol, L., Parker, C., Zhou, D., & Tomitsch, M. (2018). Blending pop-up urbanism and participatory technologies: Challenges and opportunities for inclusive city making. *City, Culture and Society, 12*, 44–53.

Fredericks, J., Parker, C., Caldwell, G. A., Foth, M., Davis, H., & Tomitsch, M. (2019, December). Designing smart for sustainable communities: Reflecting on the role of HCI for addressing the sustainable development goals. In *Proceedings of the 31st Australian Conference on Human-Computer-Interaction* (pp. 12–15).

Fredericks, J., Tomitsch, M., & Haeusler, M. H. (2020). Redefining community engagement in smart cities: Design patterns for a smart engagement ecosystem. In C. Silva (Ed.), *Citizen-Responsive Urban E-Planning: Recent Developments and Critical Perspectives* (pp. 13–53). Hershey, PA: IGI Global.

Fredericks, J., Caldwell, G., Tomitsch, M., Haeusler, M. H., Colangelo, D., de Waal, M., ... & Tscherteu, G. (2023). *Media Architecture Compendium Volume 2: Concepts, Methods, Practice*. Stuttgart, Germany: AV Edition.

Gabrys, J. (2016). Citizen sensing: Recasting digital ontologies through proliferating practices. Theorizing the contemporary. *Cultural Anthropology*. Fieldsights, March 24. https://culanth.org/fieldsights/citizen-sensing-recasting-digital-ontologies-through-proliferating-practices

Gabrys, J., Pritchard, H., & Barratt, B. (2016). Just good enough data: figuring data citizenships through air pollution sensing and data stories. *Big Data & Society, 3*(2). https://doi.org/10.1177/2053951716679677

Gardner, N., & Hespanhol, L. (2018). SMLXL: Scaling the smart city, from metropolis to individual. *City, Culture and Society, 12*, 54–61.

Gemperle, J.M., Hoggenmueller, M., Fredericks, J. (2023). Exploring participatory design for urban community gardens. In *Proceedings of the 6th Conference on Media Architecture Biennale*.

Golsteijn, C., Gallacher, S., Koeman, L., Wall, L., Andberg, S., Rogers, Y., & Capra, L. (2015, January). VoxBox: A tangible machine that gathers opinions from the public at events. In *Proceedings of the Ninth International Conference on Tangible, Embedded, and Embodied Interaction* (pp. 201-208). New York: ACM Press.

Golsteijn, C., Gallacher, S., Capra, L., & Rogers, Y. (2016, June). Sens-Us: Designing innovative civic technology for the public good. In *Proceedings of the 2016 ACM Conference on Designing Interactive Systems* (pp. 39-49). New York: ACM Press.

Gonsalves, K., Foth, M., Caldwell, G., & Jenek, W. (2021a). Radical placemaking: An immersive, experiential and activist approach for marginalised communities. In H. Griffin (Ed.), *Connections: Exploring Heritage, Architecture, Cities, Art, Media* (Vol. 20.1, pp. 237–252). Canterbury, UK: AMPS (Architecture, Media, Politics, Society).

Gonsalves, K., Foth, M., & Caldwell, G. A. (2021b). Chatty bench project: Radical media architecture during COVID-19 pandemic. In *Media Architecture Biennale 20* (pp. 182–183).

Gordon, E., & Walter, S. (2019). Meaningful inefficiencies: Resisting the logic of technological efficiency in the design of civic systems. *The Playful Citizen, 1*(2019), 310–334.

Grace, K., Wasinger, R., Ackad, C., Collins, A., Dawson, O., Gluga, R., ... & Tomitsch, M. (2013, June). Conveying interactivity at an interactive public information display. In *Proceedings of the 2nd ACM International Symposium on Pervasive Displays* (pp. 19–24).

Halegoua, G., & Polson, E. (2021). Exploring 'digital placemaking'. *Convergence, 27*(3), 573–578.

Hansson, L. Å. E. J., Cerratto Pargman, T., & Pargman, D. S. (2021, May). A decade of sustainable HCI: Connecting SHCI to the sustainable development goals. In *Proceedings of the 2021 CHI Conference on Human Factors in Computing Systems* (pp. 1–19).

Harrison, C., & Donnelly, I. A. (2011, September). A theory of smart cities. In *Proceedings of the 55th Annual Meeting of the ISSS-2011, Hull, UK*.

Hemment, D., & Townsend, A. (2013). *Smart Citizens*. Manchester: FutureEverything.

Hespanhol, L., Tomitsch, M., Grace, K., Collins, A., & Kay, J. (2012, June). Investigating intuitiveness and effectiveness of gestures for free spatial interaction with large displays. In *Proceedings of the 2012 International Symposium on Pervasive Displays* (pp. 1–6).

Hespanhol, L., & Tomitsch, M. (2012, November). Designing for collective participation with media installations in public spaces. In *Proceedings of the 4th Media Architecture Biennale Conference: Participation* (pp. 33–42).

Hespanhol, L., Tomitsch, M., Bown, O., & Young, M. (2014, June). Using embodied audio-visual interaction to promote social encounters around large media façades. In *Proceedings of the 2014 Conference on Designing Interactive Systems* (pp. 945–954).

Hespanhol, L., Tomitsch, M., McArthur, I., Fredericks, J., Schroeter, R., & Foth, M. (2015, June). Vote as you go: Blending interfaces for community engagement into the urban space. In *Proceedings of the 7th International Conference on Communities and Technologies* (pp. 29–37).

Hespanhol, L., & Dalsgaard, P. (2015). Social interaction design patterns for urban media architecture. In *Human-Computer Interaction-INTERACT 2015: 15th IFIP TC 13 International Conference, Bamberg, Germany, September 14-18, 2015, Proceedings, Part III 15* (pp. 596–613). Springer International Publishing.

Hespanhol, L., & Tomitsch, M. (2019). Power to the people: Hacking the city with plug-in interfaces for community engagement. In M. de Lange & M. de Waal (Eds.), *The Hackable City: Digital Media and Collaborative City-Making in the Network Society* (pp. 25–50). Singapore: Springer.

Hespanhol, L. (2022). Augmented placemaking: Urban technologies, interaction design and public spaces in a post-pandemic world. *Interacting with Computers, 35*(5), 637–649.

Hespanhol, L. (2023a). Designing media architecture for digital placemaking on campus: Motivation, implementation and preliminary insights. In *Proceedings of the Media Architecture Biennale 2023, Online 14-15 June 2023, and Toronto, Canada, 21-23 June 2023*.

Hespanhol, L. (2023b). Human-computer intra-action: A relational approach to digital media and technologies. *Frontiers in Computer Science, 5*, 1083800.

Hoggenmüller, M., & Wiethoff, A. (2014, June). LightSet: Enabling urban prototyping of interactive media façades. In *Proceedings of the 2014 Conference on Designing Interactive Systems* (pp. 925–934).

Hoggenmueller, M., & Tomitsch, M. (2019, June). Enhancing pedestrian safety through in-situ projections: A hyperreal design approach. In *Proceedings of the 8th ACM International Symposium on Pervasive Displays* (pp. 1–2).

Hoggenmueller, M., Hespanhol, L., & Tomitsch, M. (2020, April). Stop and smell the chalk flowers: A robotic probe for investigating urban interaction with physicalised displays. In *Proceedings of the 2020 CHI Conference on Human Factors in Computing Systems* (pp. 1–14).

Houben, S., Bengler, B., Gavrilov, D., Gallacher, S., Nisi, V., Nunes, N. J., ... & Rogers, Y. (2019, June). Roam-IO: Engaging with people tracking data through an interactive physical data installation. In *Proceedings of the 2019 on Designing Interactive Systems Conference* (pp. 1157–1169). New York: ACM Press.

Innocent, T. (2018, November). Play about place: Placemaking in location-based game design. In *Proceedings of the 4th Media Architecture Biennale Conference* (pp. 137–143).

Jacobs, J. M. (2012). Urban geographies I: Still thinking cities relationally. *Progress in Human Geography*, 36(3), 412–422.

Jiang, Q., Kresin, F., Bregt, A. K., Kooistra, L., Pareschi, E., Van Putten, E., ... & Wesseling, J. (2016). Citizen sensing for improved urban environmental monitoring. *Journal of Sensors, 2016*. Hindawi.

Kamols, N., Foth, M., & Guaralda, M. (2021). Beyond engagement theatre: Challenging institutional constraints of participatory planning practice. *Australian Planner*, 57(1), 23–35.

Kinchla, R. A., & Wolfe, J. M. (1979). The order of visual processing: "Top-down," "bottom-up," or "middle-out". *Perception & Psychophysics*, 25, 225–231.

Kjeldskov, J., & Paay, J. (2006). Public pervasive computing: Making the invisible visible. *Computer*, 39(9), 60–65.

Koeman, L., Kalnikaité, V., & Rogers, Y. (2015, April). "Everyone Is Talking about It!" a distributed approach to urban voting technology and visualisations. In *Proceedings of the 33rd Annual ACM Conference on Human Factors in Computing Systems* (pp. 3127–3136). New York: ACM Press.

Korsgaard, H., & Brynskov, M. (2014, November). City bug report: Urban prototyping as participatory process and practice. In *Proceedings of the 2nd Media Architecture Biennale Conference: World Cities* (pp. 21–29).

Kukka, H., Oja, H., Kostakos, V., Gonçalves, J., & Ojala, T. (2013, April). What makes you click: Exploring visual signals to entice interaction on public displays. In *Proceedings of the SIGCHI Conference on Human Factors in Computing Systems* (pp. 1699–1708).

Kumar, T. V., & Dahiya, B. (2017). *Smart Economy in Smart Cities* (pp. 3–76). Berlin, Germany: Springer.

Lai, Z., & Hespanhol, L. (2021, November). An investigation on strategies for remote interactions with cultural spaces. In *Proceedings of the 33rd Australian Conference on Human-Computer Interaction* (pp. 170–175).

Lau, B. P. L., Marakkalage, S. H., Zhou, Y., Hassan, N. U., Yuen, C., Zhang, M., & Tan, U. X. (2019). A survey of data fusion in smart city applications. *Information Fusion*, 52, 357–374.

Lee, K., & Sener, I. N. (2020). Emerging data for pedestrian and bicycle monitoring: sources and applications. *Transportation Research Interdisciplinary Perspectives, 4*, 100095.

Letaifa, S. B. (2015). How to strategize smart cities: Revealing the SMART model. *Journal of Business Research*, 68(7), 1414–1419.

Liu, C., Bengler, B., Di Cuia, D., Seaborn, K., Nunes Vilaza, G., Gallacher, S., ... & Rogers, Y. (2018, June). Pinsight: a novel way of creating and sharing digital content through 'Things' in the wild. In *Proceedings of the 2018 Designing Interactive Systems Conference* (pp. 1169–1181). New York: ACM Press.

Löfgren, K., & Webster, C. W. R. (2020). The value of big data in government: The case of 'Smart Cities'. *Big Data & Society, 7*(1). Sage Journals. https://doi.org/10.1177/2053951720912775

Luther, K., Counts, S., Stecher, K. B., Hoff, A., & Johns, P. (2009, April). Pathfinder: An online collaboration environment for citizen scientists. In *Proceedings of the SIGCHI Conference on Human Factors in Computing Systems* (pp. 239–248). New York: ACM Press.

McQuire, S. (2008). *The Media City: Media, Architecture and Urban Space*. London: Sage.

Manovich, L. (2006). The poetics of augmented space. *Visual Communication*, 5(2), 219–240.

Michelis, D., & Müller, J. (2011). The audience funnel: Observations of gesture based interaction with multiple large displays in a city center. *International Journal of Human-Computer Interaction*, 27(6), 562–579.

Michielsen, D., Moere, A. V., Vannieuwenhuyze, J., Tsoumani, O., Van Der Graaf, S., Claes, S., & Libot, C. (2020, June). Hyperlocal user-generated video contributions on public displays. In *Proceedings of the 9TH ACM International Symposium on Pervasive Displays* (pp. 55–62). New York: ACM Press.

Morrison, J. (2022). Heritage, digital placemaking and user experience: An industry perspective. In Fabrizio Nevola, D. R. and Terpstra, N. (Eds.), *Hidden Cities: Urban Space, Geolocated Apps and Public History in Early Modern Europe*, Routledge Research in Digital Humanities, Chapter 2, pp. 39–58. London: Routledge.

Müller, J., Wilmsmann, D., Exeler, J., Buzeck, M., Schmidt, A., Jay, T., & Krüger, A. (2009). Display blindness: The effect of expectations on attention towards digital signage. In *Proceedings of the 7th International Conference on Pervasive Computing* (pp. 1–8). Berlin, Heidelberg: Springer.

Mulder, I., & Kun, P. (2019). Hacking, making, and prototyping for social change. In M. de Lange and M. de Waal (Eds.), *The Hackable City: Digital Media and Collaborative City-Making in the Network Society* (pp. 225–238). Singapore: Springer.

Nikitas, A., Michalakopoulou, K., Njoya, E. T., & Karampatzakis, D. (2020). Artificial intelligence, transport and the smart city: Definitions and dimensions of a new mobility era. *Sustainability, 12*(7), 2789.

Pak, B., Chua, A., & Vande Moere, A. (2017). FixMyStreet Brussels: socio-demographic inequality in crowdsourced civic participation. *Journal of Urban Technology, 24*(2), 65–87. https://doi.org/10.1080/10630732.2016.1270047

Pang, C., Neustaedter, C., Moffatt, K., Hennessy, K., & Pan, R. (2020). The role of a location-based city exploration game in digital placemaking. *Behaviour & Information Technology, 39*(6), 624–647.

Project for Public Spaces. (2007). *What is Placemaking?* Retrieved 20 April 2023 from https://www.pps.org/article/what-is-placemaking

Qabshoqa, M. (2018, September). *Virtual Place-Making - The Re-Discovery of Architectural Places through Augmented Play: A Playful Emergence between the Real and Unreal*. eCAADe. Lincoln: University of Lincoln..

Quan, S. J., Park, J., Economou, A., & Lee, S. (2019). Artificial intelligence-aided design: Smart design for sustainable city development. *Environment and Planning B: Urban Analytics and City Science, 46*(8), 1581–1599.

Reed, J., Raddick, M. J., Lardner, A., & Carney, K. (2013, January). An exploratory factor analysis of motivations for participating in Zooniverse, a collection of virtual citizen science projects. In *46th Hawaii International Conference on System Sciences* (pp. 610–619). IEEE.

Schoffelen, J., Claes, S., Huybrechts, L., Martens, S., Chua, A., & Moere, A. V. (2015). Visualising things. Perspectives on how to make things public through visualisation. *CoDesign, 11*(3–4), 179–192. https://doi.org/10.1080/15710882.2015.1081240

Schroeter, R., & Foth, M. (2009, November). Discussions in space. In *Proceedings of the 21st Annual Conference of the Australian Computer-Human Interaction Special Interest Group: Design: Open 24/7* (pp. 381–384).

Schroeter, R., Foth, M., & Satchell, C. (2012, June). People, content, location: Sweet spotting urban screens for situated engagement. In *Proceedings of the Designing Interactive Systems Conference* (pp. 146–155). New York: ACM Press.

Schupbach, J. (2015). Creative placemaking. *Economic Development Journal, 14*(4), 28–33.

Shelton, T., & Lodato, T. (2019). Actually existing smart citizens: Expertise and (non) participation in the making of the smart city. *City, 23*(1), 35–52.

Silva, B. N., Khan, M., & Han, K. (2018). Towards sustainable smart cities: A review of trends, architectures, components, and open challenges in smart cities. *Sustainable Cities and Society, 38*, 697–713.

Steinberger, F., Foth, M., & Alt, F. (2014, June). Vote with your feet: local community polling on urban screens. In *Proceedings of The International Symposium on Pervasive Displays* (pp. 44–49). New York: ACM Press.

Taylor, N., Marshall, J., Blum-Ross, A., Mills, J., Rogers, J., Egglestone, P., ... & Olivier, P. (2012, May). Empowering communities with situated voting devices. In *Proceedings of the SIGCHI Conference on Human Factors in Computing Systems* (pp. 1361–1370). New York: ACM Press.

Thakuriah, P., Tilahun, N. Y., & Zellner, M. (2017). Big data and urban informatics: Innovations and challenges to urban planning and knowledge discovery. In Thakuriah, P., Tilahun, N., & Zellner, M. (Eds), *Seeing Cities through Big Data: Research, Methods and Applications in Urban Informatics* (pp. 11–45). Springer, Cham, Switzerland.

Toland, A., Christ, M. C., & Worrall, J. (2020). *DigitalXPlace. In: Hes, D., Hernandez-Santin, C. (Eds.), Placemaking Fundamentals for the Built Environment* (pp. 253–274). Singapore: Springer.

Tomitsch, M. (2014). Towards the real-time city: An investigation of public displays for behaviour change and sustainable living. In *7th Making Cities Livable Conference*.

Tomitsch, M., McArthur, I., Haeusler, M. H., & Foth, M. (2015). The role of digital screens in urban life: New opportunities for placemaking. In *Citizen's Right to the Digital City: Urban Interfaces, Activism, and Placemaking* (pp. 37–54).

Tomitsch, M., & Haeusler, M. H. (2015). Infostructures: Towards a complementary approach for solving urban challenges through digital technologies. *Journal of Urban Technology, 22*(3), 37–53.

Tomitsch, M. (2016). City apps as urban interfaces. In A. Wiethoff & H. Hussmann (Eds.), *Media Architecture-Using Information and Media as Construction Material* (pp. 81–102). Berlin: De Gruyter Publishers.

Tomitsch, M. (2017). Making cities smarter. In *Making Cities Smarter*. JOVIS Verlag GmbH.

Tomitsch, M., Fredericks, J., Vo, D., Frawley, J., & Foth, M. (2021). Non-human personas: Including nature in the participatory design of smart cities. *Interaction Design and Architecture (s), 50*(50), 102–130.

Tran, T. T. M., Parker, C., & Tomitsch, M. (2021). A review of virtual reality studies on autonomous vehicle-pedestrian interaction. *IEEE Transactions on Human-Machine Systems*, *51*(6), 641–652.

Tscherteu, G., & Tomitsch, M. (2011, May). Designing urban media environments as cultural spaces. In *CHI 2011 Workshop on Large Urban Displays in Public Life*.

Vande Moere, A. V., & Hill, D. (2012). Designing for the situated and public visualization of urban data. *Journal of Urban Technology*, *19*(2), 25–46. https://doi.org/10.1080/10630732.2012.698065

Vande Moere, A. V., & Wouters, N. (2012, June). The role of context in media architecture. In *Proceedings of the 2012 International Symposium on Pervasive Displays* (pp. 1–6).

Veenstra, M., Wouters, N., Kanis, M., Brandenburg, S., te Raa, K., Wigger, B., & Moere, A. V. (2015, June). Should public displays be interactive? Evaluating the impact of interactivity on audience engagement. In *Proceedings of the 4th International Symposium on Pervasive Displays* (pp. 15–21). New York: ACM Press.

Vlachokyriakos, V., Comber, R., Ladha, K., Taylor, N., Dunphy, P., McCorry, P., & Olivier, P. (2014, June). PosterVote: expanding the action repertoire for local political activism. In *Proceedings of the 2014 Conference on Designing Interactive* Systems (pp. 795–804). New York: ACM Press.

Waite, C. (2020). Making place with mobile media: Young people's blurred place-making in regional Australia. *Mobile Media & Communication*, *8*(1), 124–141.

Weiser, M. (1991). The computer of the 21st century. *Scientific American*, *265*(3), 66–75.

Willett, W., Jansen, Y., & Dragicevic, P. (2016). Embedded data representations. *IEEE Transactions on Visualization and Computer Graphics*, *23*(1), 461–470.

Wisneski, C., Ishii, H., Dahley, A., Gorbet, M., Brave, S., Ullmer, B., & Yarin, P. (1998). Ambient displays: Turning architectural space into an interface between people and digital information. In *Cooperative Buildings: Integrating Information, Organization, and Architecture: First International Workshop, CoBuild'98 Darmstadt, Germany, February 25-26, 1998 Proceedings 1* (pp. 22–32). Berlin, Heidelberg: Springer.

Wiethoff, A., & Gehring, S. (2012, June). Designing interaction with media façades: A case study. In *Proceedings of the Designing Interactive Systems Conference* (pp. 308–317).

Wilbrink, M., Nuttelmann, M., & Oehl, M. (2021, September). Scaling up automated vehicles' eHMI communication designs to interactions with multiple pedestrians-putting eHMIs to the test. In *13th International Conference on Automotive User Interfaces and Interactive Vehicular Applications* (pp. 119–122).

Wilken, R., & Humphreys, L. (2021). Placemaking through mobile social media platform Snapchat. *Convergence*, *27*(3), 579–593.

Wouters, N., Huyghe, J., & Moere, A. V. (2014, October). StreetTalk: Participative design of situated public displays for urban neighborhood interaction. In *Proceedings of the 8th Nordic Conference on Human-Computer Interaction: Fun, Fast, Foundational* (pp. 747–756). New York: ACM Press.

Wouters, N., Downs, J., Harrop, M., Cox, T., Oliveira, E., Webber, S., … & Vande Moere, A. (2016a, June). Uncovering the honeypot effect: How audiences engage with public interactive systems. In *Proceedings of the 2016 ACM Conference on Designing Interactive Systems* (pp. 5–16).

Wouters, N., Keignaert, K., Huyghe, J., & Moere, A. V. (2016b, June). Revealing the architectural quality of media architecture. In *Proceedings of the 3rd Conference on Media Architecture Biennale* (pp. 1–4).

Wouters, N., Claes, S., & Moere, A. V. (2018, November). Hyperlocal media architecture: Displaying societal narratives in contested spaces. In *Proceedings of the 4th Media Architecture Biennale Conference* (pp. 76–83).

Wyckoff, M. A. (2014). Definition of placemaking: Four different types. *Planning & Zoning News*, *32*(3), 1.

Yu, X., Hoggenmueller, M., & Tomitsch, M. (2023, March). Your way or my way: Improving human-robot co-navigation through robot intent and pedestrian prediction visualisations. In *Proceedings of the 2023 ACM/IEEE International Conference on Human-Robot Interaction* (pp. 211–221).

Zitcer, A. (2020). Making up creative placemaking. *Journal of Planning Education and Research*, *40*(3), 278–288.

Zuiderwijk, A., Janssen, M., & Davis, C. (2014). Innovation with open data: Essential elements of open data ecosystems. *Information Polity*, *19*(1–2), 17–33.

14 Sustainability and Citizen Science

Yao Sun and Ann Majchrzak

14.1 INTRODUCTION

14.1.1 Sustainable Development

Sustainable development is defined as "development that meets the needs of the present without compromising the ability of future generations to meet their own needs" (World Commission on Environment and Development, 1987, p.43). In 2015, world leaders at the historic United Nations Sustainable Development Summit adopted 17 Sustainable Development Goals of the 2030 Agenda for Sustainable Development and 169 targets pertaining to global challenges including poverty, inequality, climate, environmental degradation, prosperity, and peace and justice (United Nations, 2015). It was the first time that environmental concerns were addressed from economic, social, and political perspectives rather than solely from a scientific viewpoint. Sustainable development is often framed as having three supporting pillars (i.e., economy, environment, and society) and is defined differently in different disciplines (see Elliot, 2001 for a review). It has been studied in a verities of areas such as mathematics, economy, climate, natural resources, energy, transportation, ecological and agricultural systems, local hazards and disasters, urban growth, etc. (Munasinghe, 2009). The theoretical frameworks regarding the meaning and ideologies of sustainable development are continuously evolving as strategies in practice continue to be reshaped by global development progress (Potter et al., 2008).

Citizen science offers great potential for monitoring and implementing sustainable development goals. It outperforms traditional survey which is often expensive and fails to include the marginalized and hard-to-reach population (Klopp & Petretta, 2017; Friesen et al., 2019) and can collect data on a wide range of demographic basis and from geographic locations that are otherwise inaccessible as well as engage most vulnerable groups and increase their representation in datasets (Amano et al., 2016; Danielsen et al., 2014). Citizen science can also help address gaps, for example, in the coverage of official environmental datasets by generating longitudinal data and overcoming the shortage of data for many environmental indicators (Pocock et al., 2015; United Nations, 2019). In other social scientific realms, citizen science can be leveraged to verify data sources and tracking international indicators for decreasing political manipulation (Klopp & Petretta, 2017; Lu et al., 2015) or maintaining biodiversity (Hecker et al., 2019; Hayhow et al., 2019). To facilitate achieving sustainable development goals, citizen science advances the public's understanding of societal problems and the development of scientific and technological solutions, as well as fosters policy-making and local action design (Kullenberg & Kasperowski, 2016; Lidskog, 2008; Rome & Lucero, 2019; Sauermann et al., 2020; Turbé et al., 2020). It leads to behavioral changes in the implementation of innovations to address sustainability challenges through promoting community engagement, educating stakeholders, as well as offering opportunities for the public to gain information and develop skills (Branchini et al., 2015; Cervantes & Hong, 2018; Stedman et al., 2009; West et al., 2020). Citizen science can promote the localization of sustainable development goals to cities and regions, complementing national datasets and adding richness and contextual information around sustainability indicators (Jameson et al., 2017; Pocock et al., 2018; van den Homberg & Susha, 2018). It can be helpful in setting local action agendas or identifying gaps between national goals and locally relevant targets (Sauermann et al., 2020; West & Pateman, 2017).

14.1.2 CITIZEN SCIENCE

Citizen science is a collaborative research technique that involves the public in gathering scientific information for solving scientific problems (Bhattacharjee, 2005; Cohn, 2008). It can be traced back to 1900 when the Christmas Bird Count was run by the National Audubon Society in the US, and then in 1932 when the British Trust for Ornithology designed nature conservation strategies with the help of amateur naturalists (Silvertown, 2009). Citizen scientists, accordingly, refer to these non-professional "volunteers who participate as field assistants in scientific studies" (Cohn, 2008, p.193).

Citizen science often acts as informal science outreach that can promote public understanding of science, as participants of citizen science projects deeply engage in scientific work rather than merely acting as research subjects or respondents (Brossard, Lewenstein & Bonney, 2005). Over the past decades, the concept and practice of citizen science has evolved from documenting natural occurrences such as breeding birds or new starts (Droege, 2007; Bonney et al., 2009) to the integration of data vetting, data collection protocol testing, and inclusive measures for public education (Bonney, 2007; Cohn, 2008). Citizen science has become an established approach to advancing scientific knowledge in many other areas, such as wildlife protection (Cannon et al., 2005; Hochachka & Dhondt, 2000; Prysby & Oberhauser, 2004), climate change (Groulx et al., 2017; Pecl et al., 2019), as well as disease prevention, biomedical research, and health equity (King et al., 2019; Rosas & King, 2021; Wiggins & Wilbanks, 2019). In the time of COVID-19 pandemic, for example, citizen science has played a critical role in tracking the spread of the disease, predicting outbreak locations, guiding population measures, and helping in the allocation of healthcare resources (Birkin, Vasileiou & Stagg, 2021).

Empowered by the integration of the Internet into everyday life, citizen science projects have become increasingly visible and accessible for people who are passionate and ready to contribute (Bonney et al., 2014). Implementing citizen science requires "creating the academic, project management, and informatics infrastructures to carrying out all the (task) elements and obtaining buy-in from citizens so that there is sufficient participation in the process" (Cooper et al., 2007, p. 16). Although citizen science projects widely adopt the structure of tasks of peer production and open communities, the involvement of the public can vary for different projects or tasks. Cooper et al. (2007) distinguished citizen science research from scientific consulting research, co-management research, and participatory action research, indicating that citizen science research model often accompanies with broad geographic scope as well as high research and education priority. Wilderman et al. (2007) identified the citizen science research model as a comprehensive integration of community consulting, community-defined research, contributions from community workers, and community-based participatory research. In the conceptualization developed by Bonney et al. (2009), citizen science projects are configured by contributory, collaborative, or co-creating dynamics occurring between scientists and members of the public. Contributory projects are "designed by scientists and for which members of the public primarily contribute data"; collaborative projects are "designed by scientists and for which members of the public contribute data but also may help to refine project design, analyze data, or disseminate findings"; and co-created projects are "designed by scientists and members of the public working together and for which at least some of the public participants are actively involved in most or all steps of the scientific process" (p. 11).

Citizen science projects also vary in terms of the nature of the tasks to be completed (see Wiggins & Crowston, 2011, for a review). In action-oriented projects, the tasks usually require citizens' long-term engagement in both planning and undertaking the tasks. In conversational projects, citizens are involved for the purpose of outreach or practicality, and citizens' engagement is needed mostly for a better understanding of the public. Investigative projects, however, often aim at providing real-world data for scientific studies, meaning that citizens are part of scientific studies. Educational projects often involve informal or formal educational activities that seek to improve scientific literacy among citizens. As information and communication technology leverages itself to citizen science, an increasing number of citizen science projects can be completed through collecting responses beyond geographical boundaries with open-source data management and analytics tools (Bonney et al., 2014).

14.1.3 Leveraging the Wisdom of Crowds

Leveraging the wisdom of crowds, crowdsourcing has been recognized as an effective tool for tackling broadly defined societal challenges in relation to sustainability. Crowdsourcing refers to an action that "harnesses the creative solutions of distributed individuals" that are generally large and undefined (Brabham, 2008, p. 76). According to Howe (2006), in crowdsourcing, a critical prerequisite is the open call broadcasted to a large network of potential labors. Crowdsourcing has been applied to various disciplines (see Estellés-Arolas & González-Ladrón-de-Guevara, 2012, for a review) and has evolved to a full-fledged conceptualization of "a type of participative online activity in which an individual, an institution, a non-profit organization, or company proposes to a group of individuals of varying knowledge, heterogeneity, and number, via a flexible open call, the voluntary undertaking of a task" (Estellés-Arolas & González-Ladrón-de-Guevara, 2012, p. 197). When members of the crowd participate in the undertaking of the task, they contribute their diverse experience, knowledge, or skills for mutual benefit, simultaneously satisfying their needs, such as economic or monetary needs, social recognition, self-esteem, or the development of individual skills and careers. Crowd-based innovation takes the forms of competitive and collaborative crowdsourcing. In competitive crowdsourcing, members of the crowd participate in ideation contests or tournaments to contribute ideas and compete for rewards (Afuah & Tucci, 2012; Blohm et al., 2013). In collaborative crowdsourcing, however, participating individuals generate solutions collectively via knowledge sharing and co-creation (Geiger & Schader, 2014). Crowdsourcing activities also vary in terms of functions and goals. Tasks such as voting or rating do not involve much intellectual effort and knowledge sharing (Surowiecki, 2005), whereas tasks such as solving scientific problems usually require participants to create more comprehensive and complex knowledge embodiment (Geiger & Schader, 2014).

Unleashing the crowd features the dynamic management of knowledge. Dynamic knowledge management includes managing the actors, infrastructure, as well as the processes in relation to knowledge creation, sharing, and application (Lee & Choi, 2003; Yeh et al., 2006). Research has shown that innovative solutions result from successfully managing heterogeneous sources of knowledge artifacts made explicit by a diverse range of knowledge contributors (Malhotra & Majchrzak, 2019; Santoro et al., 2018; Sun et al., 2020). When crowd members collectively generate innovative solutions to ill-structured societal and scientific challenges, they need to collaborate with others and rely on others' knowledge, as they "lack sufficient expertise to innovate alone when the knowledge frontier is complex and expanding" (Dahlander & Frederiksen, 2012, p. 988). In this regard, collaborative crowdsourcing is especially promising for addressing broadly defined scientific or societal challenges as it connects members of the crowd who would not otherwise come together and thus provides opportunities for these crowd members to share their different interpretations of the problems and collectively understand the challenge in its entirety (Head, 2019; McGahan et al., 2021; Roberts, 2000). As crowd members make their perspectives explicit, they develop a comprehensive understanding of the tacit knowledge relevant to the challenge, and thus are capable of collectively revealing previously unacknowledged facets of the challenge (Porter et al., 2020). Knowledge sharing and co-creation facilitate the construction of shared values regarding sustainability issues within the crowd as well as between the public and institutions, which further stimulates inclusivity and long-term involvement of the crowd (Di Fatta et al., 2018; Pfitzer et al., 2013).

In tackling scientific or societal challenges, citizens co-create scientific research through practices such as crowd science. Crowd science builds upon crowdsourcing and refers to "scientific research done in an open and collaborative fashion" (Franzoni & Sauermann, 2014, p. 1). Crowd science generates several advantages over conventional science by eliminating resource limits. Harnessing the wisdom of the crowd helps expand areas of scientific inquiry, arrive at results more efficiently, as well as support the surfacing of the most capable problem solvers and best solutions through a participatory self-selection process (Lakhani et al. 2007; Tucci et al., 2018). Members of the crowd get involved in all stages of the scientific research process, from the collection of raw data, data analysis, to critically evaluating potential social consequences, disseminating scientific findings, and clarifying misunderstandings (Ganna et al., 2021), such that they become mutually responsive in considering the societal implications of scientific research (Owen et al., 2020). Crowd-powered science acts as a research accelerator in many areas of sustainability development, such as health promotion (e.g., English et al., 2018), geo-tracking (e.g., See et al., 2016), software development (e.g., Bazilian et al., 2012), drug development (e.g., Thompson & Bentzien, 2020), biosciences (e.g., Curtis, 2014), astronomy (e.g., Simpson et al., 2014), etc.

Crowdsourcing is not only a way to seek for citizens' contribution but also a way to appreciate their concerns (Leighninger, 2011). It is increasingly being appropriated by NGOs, civil society organizations, and private sectors when dealing with sustainability development and research, because sustainability is a pressing imperative that calls for large-scale contribution from strong networks of academics, entrepreneurs, non-profit sectors, and other businesses, as well as highlights citizen behaviors when measuring total impact at the societal level (Certoma et al., 2015). Using crowdsourcing, citizens can be involved in planning new policies or formulating new legislation. For example, citizens' civic engagement has motivated constitutional changes (Aitamurto et al., 2017), solutions to climate change (Dawson et al., 2020; Kammermann & Dermont, 2018; Kythreotis et al. 2019), and other policy reforms in different nations (Hudson, 2018; Maboudi & Nadi, 2016).

14.2 CROWD-BASED OPEN APPROACHES TO TACKLING GRAND CHALLENGES

14.2.1 ROBUST ACTION

Tackling grand challenges such as eradicating poverty or combating climate change requires novel solutions that integrate diverse intellectual resources (Ferraro et al., 2015; George et al., 2016; United Nations, 2017). To this end, Ferraro et al. (2015) introduced an approach of robust action that can facilitate scaling locally generated solutions to address broadly defined challenges and identified three strategies for successfully harnessing the power of robust action. The first strategy is *participatory architecture*, which builds upon the theorization of hybrid forms (Callon et al., 2009) and refers to creating a structure that encourages contributing actors' prolonged engagement beyond the initial engagement. Such architecture also allows the engagement of actors with heterogenous knowledge background as well as supports the employment of diverse evaluative criteria and lateral accountability (Stark, 2009). With participatory architecture, interested actors are better capable of meaningfully engaging with counterparts in an 'even playing field' (Bartley, 2007; Callon, et al., 2009; Furnari, 2014; Mair & Hehenberger, 2014). To promote action, another important strategy is *multivocal inscription* (Ferraro et al., 2015; Furnari, 2014). Inscription, according to Latour (1999), refers to guidelines, norms, or routines developed in societal activities. Multivocal inscriptions can function as generic boundary objects to sustain diverse interpretations and various viewpoints contributed by actors (Star & Griesemer, 1989). In the context of grand challenges, multivocal inscriptions offer flexibility (Pinch & Bijker, 1987) and can provide "common ground for discussion among a range of development and environmental actors who are frequently at odds" (Sneddon et al., 2006, p. 254), thus allowing actors with multidisciplinary background to coordinate based on common grounds without requiring explicit agreement (Bechky, 2003; Bowker & Star, 1999). The third strategy of robust action deals with iterative and continued learning, which was referred to as *distributed experimentation* (Ferraro et al., 2015). Successful large-scale interventions, as Ferraro and colleagues argued, often build upon successive bottom-up actions leading to 'small wins' that make the "next solvable problem more visible" (Weick, 1984, p. 43). The small successes for different actors can have different meanings, such as personal satisfaction, career opportunities, monetary incentives, etc. Distributed experimentation outperforms centralized governance approaches as it allows distinctive problem-solving pathways to emerge and encourages actors to pursue additional experimentation (Dietz et al., 2003; Simon, 1996). Synthesizing the three robust action strategies enables individual organizations to fuel innovative solutions by attracting the engagement of heterogenous stakeholders.

Robust action has been applied to tackling many sustainability challenges. For example, to promote the adoption of wind power as clean energy, Denmark followed a path to construct participatory architecture, implement multivocal inscription, and conduct distributed experimentation (Garud & Karnøe, 2003; Karnøe, 1990). To put robust action to work, starting in the 1970s, a wide variety of Danish actors, including politicians, insurance companies, community consultants, engineers and engineering associations, and other stakeholders gathered for a series of informal meetings, which resulted in "numerous iterations of designs, construction, tests, and problem-solving prior to the approval of a new wind turbine" (Garud & Karnøe, 2003, p. 290) as well as "a specialized industrial network of users and suppliers" (Karnøe, 1990, p. 120). In the phase of multivocal inscription, Denmark's wind power sector managed to officially consolidate a design process involving a wide variety of communities to contribute diverse ways to tackle the challenge of refining wind turbine. Benefiting from the participatory architecture of the Danish wind sector and the multivocal inscription, Danish firms "deployed prototypes designed with simple engineering heuristics to engender a process of trial-and-error learning instead of pursuing a design-intensive R&D approach" (Garud & Karnøe, 2003, p. 282) and "carried out hundreds of experiments" (Karnøe, 1990, p. 113). These small-scale experimentations then stimulated more iterative discussions and experiments, ultimately cultivating a serious of scaled successes (Karnøe & Garud, 2012).

When applied to tackling more recent sustainability challenges, robust action is further empowered by Internet-based online crowdsourcing platforms. For example, in their Save Our Oceans study (Porter et al., 2020), the researchers employed an online crowdsourcing platform to facilitate large-scale robust action and found that robust action principles of participatory architecture, multivocal inscription, and distributed experimentation can be applied to each phase of crowdsourcing process. Specifically, the research team started with involving founding partners of Save Our Oceans in collaborative problem formulation, followed by launching a crowdsourcing platform in which members of the public, invited experts, and employees from partner organizations all joined to post ideas as well as enrich others' ideas. Idea evaluation and a match making between partner organizations and top idea holders happened after the ideation phase, before the partner organizations finally worked on developing and implementing the top ideas initiated by the crowd and enriched by the experts. This research extends the robust action framework by emphasizing the importance of generating continuous engagement, keeping novel ideas alive, as well as connecting actors and ideas across different stages of crowdsourcing.

14.2.2 OPEN INNOVATION IN SCIENCE

Open innovation refers to "a distributed innovation process involving purposive knowledge flows across organizational boundaries for monetary or non-monetary reasons" (Chesbrough & Bogers, 2014, p. 12). It describes a process in which actors transcend knowledge boundaries to access new knowledge to innovate, while at the same time sharing existing knowledge with other actors to facilitate innovation (Chesbrough & Bogers, 2014). Building on the distributed nature of innovation sources, various scholarly perspectives have emerged in studying open-source collaborative innovation in other social scientific fields (Baldwin & von Hippel, 2011; Raasch et al., 2009; von Hippel, 2005). Essentially, openness advances communal knowledge by empowering knowledge exchange, unleashes diverse professionals and allows them to collaborate across temporal or geographical boundaries, as well as leverages the human capital available to tackle complex societal problems (Chesbrough, 2020).

In science, open innovation is defined as "a process of purposively enabling, initiating, and managing knowledge flows and inter/transdisciplinary collaboration across organizational den disciplinary boundaries in scientific research" (Beck et al., 2022, p. 140). It functions in all scientific research phases and is configured by various collaboration forms. For example, in interdisciplinary research projects, scientists from different disciplines borrow, contrast, integrate, or transcend their knowledge, techniques, perspectives, and theories to answer important research questions lying at the intersection of multiple disciplines (Miller, 1982; Nowotny et al., 2006). Key to successful interdisciplinary collaborations is the breadth, depth, and innovativeness of knowledge integration, which depend upon knowledge diversity and coherence (Rafols & Meyer, 2010). Open innovation also fuels the evolution of research instruments and collaboration culture in scientific research, permitting the emergence of new collaborative connections, an advanced science education, and more efficient use of expensive instruments (Bos et al., 2007; D'Ippolito & Rüling, 2019; Finholt, 2002; Heck et al., 2018; Heradio et al., 2016; Kraut et al., 1988; Teasley & Wolinsky, 2001; Tuertscher et al., 2014; Waldrop, 2013). Openness in science also takes the form of open data and materials sharing, as well as open access publication approaches, benefiting real-time storing and sharing of scientific research findings (Beck et al., 2022).

Open innovation in science often involves co-creators other than academics in the scientific research process. The general public can co-create scientific research and facilitate the dissemination of scientific findings "in an open and collaborative fashion" (Franzoni & Sauermann, 2014, p. 1). Volunteers who may not be academics can help gather, code, or analyze data, as well as conduct experiments and collect measurements (English et al., 2018; Scheliga et al., 2018; Silvertown, 2009). Such deep interactions can motivate scientists to engage in innovation activities, close the gap between academia and the public, promote the democratization of science, as well as catalyze creativity and potentially give rise to more valuable social and scientific outcomes (Anderson, 1994; Bergen, 2009; Bonney et al., 2009; Llopis & D'Este, 2016; Tsai, 2012). Empowered by information technologies, recent open innovation in science has applied new approaches such as computational or gamified tools to sustain the public's motivation as well as bridging human and machine intelligence (Heck et al., 2018; Tinati et al., 2017). In addition, industry actors are increasingly co-creating scientific research through university-industry connections (Perkmann et al., 2013), which foster mutually meaningful knowledge exchange as well as reinforce trust among collaborators, paving the way for long-term scientific collaborations (Cohen et al., 2002; D'Este & Perkmann, 2011; Feller & Feldman, 2010; Perkmann & Walsh, 2008). Policymakers at various levels of government can, likewise, co-create in the scientific research process through open and collaborative policy-making actions (Attard et al., 2015). Such co-creation also takes the form of tackling mission-oriented grand societal challenges that involve scientific research in addressing complex ill-structured societal problems (Borrás & Edler, 2014; Kuhlmann & Rip, 2018; Mazzucato, 2018).

While open innovation in science yields significant scientific and social impact, it is also contingent upon the individual, organizational, disciplinary, or societal circumstances. At the individual level, whether open innovation in scientific practices can be successful depends on scientists' knowledge, background and characteristics. Interdisciplinary work experience is found to be related to higher intellectual productivity (Abreu & Grinevich, 2013; Dietz & Bozeman, 2005), and a higher level of education or academic training often pertains to the adoption of open science strategies (Ding, 2011). Personal traits, as well as individual abilities to seek, recognize, apply, and assimilate new knowledge are also important drivers of open and collaborative science practices (Batey & Furnham, 2006; Cohen & Levinthal, 1990; Heinström, 2003; Linek et al., 2017). Scholarly reputation also plays an important role in open science practices, which is usually referred to as Matthew Effect or 'star scientist' effect in attracting resources and building collaboration ties (Merton, 1968; Zucker et al., 2002). At the team or group level, team sizes, as well as collaborators' heterogeneity of mental models, expertise, and skills can increase the diversity of the shared knowledge and result in knowledge reuse and synthesis (Curral et al., 2001; Fleming & Sorenson, 2004; Gruber et al., 2013; Mote et al., 2016; van Noorden, 2015; Wuchty et al., 2007). Within a team, collaborators influence each other through practicing normative collaboration styles and adjusting to the prevailing work culture (Bercovitz & Feldman, 2008; Kenney & Goe, 2004; Tartari et al., 2014). At the organizational level, open science practices are often constrained by the infrastructures of research organizations and organizational policies and protocols (Baldini et al., 2007; Geuna & Muscio, 2009). At the society level, open science practice can be influenced by the evolution of technology, policy changes, as well as distinctive values and beliefs toward openness and collaboration (Berman, 2011; Etzkowitz & Leydesdorff, 2000; Felin & Zenger, 2014; Kukutai & Taylor, 2016; Lakhani et al., 2013; Shane, 2004).

14.2.3 Open Strategy Formulation

Open strategy formulation is defined as the process by which organizations include a wide variety of internal and external stakeholders to formulate innovation strategies that derive from a melding of diverse perspectives (Hautz et al., 2017; Stieger et al., 2012; Whittington et al., 2011). Open strategy formulation can increase organizations' inclusivity and transparency by promoting stakeholders' participation and co-development of strategies as well as fueling the exchange of stakeholder-contributed knowledge (Whittington et al., 2011). Adopting this approach can help organizations "widen the search for strategy ideas and improve commitment and understanding during implementation" (Whittington et al., 2011, p. 535) and facilitate collective sensemaking and understanding of the problem (Gioia & Chittipeddi, 1991; Gioia et al., 1994; Mintzberg, 1994). Such a collaborative and participatory approach has been applied to tackling a wide range of complex societal challenges and has been found effective in building trust among participating stakeholders, enabling scientific, technical, local and indigenous knowledge to be incorporated, helping citizens manage their inability and unwillingness to embrace socio-ecological transformation, as well as stimulating social learning and triggering collective systemic thinking (Blühdorn & Deflorian, 2019; Johnson et al., 2012; Malhotra et al., 2017; Ryan et al., 2022).

Collaborative crowdsourcing outperforms competitive crowd-based contests for open strategy formulation. In collaborative crowdsourcing, participating stakeholders engage in dialectical conversations to share diverse perspectives, discover perceptive similarities and differences, and thus co-create new knowledge without worrying about winning or losing (Haefliger et al., 2011; Marabelli & Newell, 2012; Stieger et al., 2012). As participants make their tacit knowledge explicit, the evolution of collective knowledge and the surfacing of different perspectives become transparent and publicly readable (Leonardi, 2014), and hence foster the emergence of commonly acceptable solutions. To better understand stakeholders' motivation to participate in formulating strategies and contributing solutions for tackling grand challenges, Clary et al. (1998) summarized six functions of voluntary work that may act as driving forces for individuals' participation in voluntary activities. The first function is values, meaning that individuals view voluntary work as a way of expressing their values toward altruistic and humanitarian concerns for others. The second and third functions indicate that volunteering offers opportunities for individuals to acquire new knowledge and skills, as well as potential career-related benefits. The fourth and fifth functions deal with individuals' impression management in social life, as volunteering is often viewed positively by others and can help reduce negative emotions or feelings. The last function suggests that taking part in voluntary work helps individuals to achieve self-realization and ego enhancement. In a similar vein, Batson et al. (2002) studied community-based voluntary work and categorized four types of motivation. Egoism, altruism, collectivism refer to the welfare of the volunteer, others, and the group or community, whereas principlism highlights the need to maintain moral principles such as justice. Lakhani and Wolf (2007) concluded that participants in open-source communities are driven by intrinsic and extrinsic factors. Intrinsic factors include enjoyment, satisfaction, and other positive feelings, whereas extrinsic motivation pertains to realistic outcomes such as monetary rewards or career opportunities. Other studies have also demonstrated intrinsic and extrinsic factors that motivate individuals to participate in crowdsourcing or open-source practices (Budhathoki & Haythornthwaite, 2013; Ryan & Deci, 2000).

Applying open innovation strategy formulation includes three stages (Hautz, 2017). In the first stage, the organization grants autonomy to participating stakeholders to contribute all kinds of ideas and suggestions, without rejecting any ideas out of hand. In the second stage, the organization reviews the contributed ideas, builds shared understanding, as well as selects the ideas that fit into the organization and have the potential to enhance the strategy. In the last stage, organizations need to work on integrating the selected ideas into their knowledge base to implement the ideas (Pappas & Wooldridge, 2007). In the past decade, rapid advances in information technology have made it easier to tap into the ideas generated by various stakeholders and afford more opportunities for traditionally low-power stakeholders to engage. The rise of social media, for example, acts as an open communication platform that facilitates collective innovation strategy formulation (Baptista et al., 2017). In online idea generation, anonymity increases contributors' willingness and volume of participation, especially among low-power stakeholders, by removing social barriers and reducing the pressure for conformity (Dennis & Garfield, 2003; Dennis et al., 2001; Nunamaker et al., 1991). Parallelism enables peripheral stakeholders to obtain equal access to the conversation by mitigating the 'production blocking' driven by high-power or centralized stakeholders in traditional offline communication, such that low-power stakeholders do not have to wait turn to contribute (Dennis, 1996; Gallupe et al., 1994; Valacich et al., 1994). Online open strategy development also fosters equal participation of stakeholders by reducing powerful stakeholders' control of the conversation content occurring in offline environment and supporting the common ground for collective idea generation and knowledge exchange (Dennis & Garfield, 2003; Tyran et al., 1992).

14.3 MANAGING COMPLEXITIES IN CITIZEN SCIENCE FOR SUSTAINABILITY

14.3.1 Complexities Associated with Knowledge Gaps

Complexities in crowd-based citizen science may arise associated with the knowledge-sharing process, especially knowledge gaps. Knowledge gaps occur when crowd members coming from heterogenous social contexts find it difficult to understand each other (Loebbecke et al., 2016; Trkman & Desouza, 2012). In tackling grand sustainability challenges, members of the crowd often need to solve broadly defined strategic problems and thus to share knowledge that has ambiguous boundaries and is difficult to be accurately perceived (Tsoukas, 2009). Moreover, in crowd-based innovation challenges, as conversations often occur asynchronously, participants' various degrees of involvement impede the development of shared mental models, commonly accepted norms and meanings of words, as well as a shared understanding of the challenge (Faraj et al., 2011). Cronin and Weingart (2007) defined that a knowledge representational gap "arises as a function of the cognition of individuals working together to solve a problem" (p. 762) at the collective level when individuals work together to solve a problem. Knowledge representational gaps result from crowd participants' different beliefs and values that are normally shaped by acculturation. Based on such beliefs and values, collaborators tend to generate different perceptions of the problem, different preferences of relevant information, as well as different strategies for resolving 'how' and 'why' issues related to the problem (Nonaka, 1994). Knowledge representational gaps can cause complexities for people collaboratively solving problems as they "increase the misunderstanding and misuse of information, decrease coordination both explicitly and implicitly, and lead to conflict that can be detrimental" (Cronin & Weingart, 2007, p. 768). Additionally, knowledge sharing in crowd-based challenges is often influenced by complexities pertaining to the transfer of tacit knowledge. Tacit knowledge is a composition of skills, expertise, and abilities that are difficult to be articulated and "is rooted in an individual's experience and values" (Holste & Fields, 2010, p. 129). Barriers may arise as transferring tacit knowledge requires collaborators' awareness of tacit knowledge, willingness and capacity to share and learn, as well as mental and physical actions for expressing and applying tacit knowledge (Argote, 1999; Fahey & Prusak, 1998; Nidumolu et al., 2001; Nonaka & Takeuchi, 1995; Stenmark, 2000, 2002).

Research has suggested several mitigation strategies to address the problem of knowledge gaps and knowledge sharing. Developing explicit routines, for example, can help manage complexities in knowledge sharing (Loebbecke et al., 2016; Trkman & Desouza, 2012). Involving mediators or facilitators for the intervention may also play a mitigating role (Marabelli & Newell, 2012). To bridge knowledge representational gaps, perspective making and perspective taking need to be encouraged. Perspective making is "a social practice in a form of life" that is "lifelike and plausible" as well as "fits the culturally bound demands of a form of life" within communities of knowing (Boland & Tenkasi, 1995, p. 357). Narratives help frame collective experience and retain collective memory (Bruner, 1990). Perspective making facilitates a collective understanding of "how actors socially construct their accounts of action and how actors constitute the character of their actions" (Mulkay et al, 1983, p. 195). Accordingly, perspective taking is configured by the appreciation and synthesis of each other's knowledge in collaboration (Dougherty, 1992; Henderson, 1994; Nonaka, 1994; Purser & Pasmore, 1992; Purser et al., 1992), and refers to a process of "exchange, evaluation, and integration of knowledge" that is "comprised of the interactions of individuals and not their isolated behavior" (Duncan & Weiss, 1979, p. 86). Perspective taking facilitates the deliberate adoption of others' viewpoints (Caruso et al., 2006), and can facilitate collaborators' understanding of "the thoughts, motives, and/or feelings" of others and "why they think and/or feel the way they do" (Parker et al., 2008, p.151). As such, taking others' perspectives fosters social interaction as it reduces conflict perceptions and stereotyping (Galinsky & Moskowitz, 2000; Sessa, 1996), as well as improves emotional regulation (Parker et al., 2008). Perspective taking also promotes the sharing and integration of diverse knowledge through motivating collaborators to comprehend, analyze, and elaborate on others' viewpoints, so that collaborators can undertake a cognitive reframing of the problem (Hargadon & Bechky, 2006; Titus, 2000). Collaborators adopting others' perfectives demonstrate more cooperative behaviors and less favoritism (Galinsky & Moskowitz, 2000; Parker & Axtell, 2001). Such actions can help reduce uncertainties, build trust, and develop mutual understanding in interpersonal relationships, thereby enabling the successful transfer of tacit knowledge that is often embedded in social and cultural contexts (Roberts, 2000). Frequent interactions stimulated by perspective taking can empower learning in collaboration (Nonaka & Takeuchi, 1995) and further trigger comprehensive decoding, explication of tacit knowledge, as well as exchange of tacit knowledge among collaborators (Ryan & O'Connor, 2013; Senker, 1995). As crowd members close their knowledge gaps and advance their understanding of the problem or challenge, they can revisit their own knowledge and recombine others' knowledge to collectively generate innovative solutions.

14.3.2 POTENTIAL RISK OF CONTENTIOUS CONFLICT

Collaborations in online communities suffer from potential contentious conflict among the members, which occurs frequently in large-scale online collaborations (Buriol et al., 2006). Individuals engaging in online contentious conflict tend to personally attack others using negative and destructive languages (Gebauer et al., 2013; Kittur et al., 2007; Papacharissi, 2004), and such back-forth activities hamper collaborative innovation to a large extent. Contentious posts may lead individuals to focus on defending their positions instead of contributing creative ideas and may potentially escalate hostile personal attacks (Grant & Berry, 2011; Parker & Axtell, 2001). Such negative interactions reduce participants' psychological safety (Edmondson, 1999) and prevent them from making innovative contributions. Psychological safety is a collective-level concept that describes a work climate manifested by "shared belief held by members of a team that the team is safe to interpersonal risk taking" (Edmondson, 1999, p. 350). When team members or collaborating actors feel psychologically safe, they undertake greater knowledge sharing and generate more learning behaviors (Bstieler & Hemmert, 2010; Mu & Gnyawali, 2003; Ortega et al., 2010; Siemsen et al., 2009). Feeling psychologically safe in collaboration also reduces membership turnover and facilitates collective goal achievement (Baer & Frese, 2003; Chandrasekaran & Mishra, 2012). Contentious conflict hampers trust and collective thinking and thus dampens collaborators' psychological safety (Brueller & Carmeli, 2011; Schulte et al., 2012). When contentious conflict arises, collaborators engage in less honest communication (Hirak et al., 2012; Nemanich & Vera, 2009; Walumbwa & Schaubroeck, 2009), and may become unwilling to spend time reading others' shared posts thoroughly or making novel contributions based on critical thinking, ultimately hindering the collective production of innovative solutions (Lee, 2005).

Contentious conflict differs from legitimate disagreements. Contentious conflict stems from interpersonal hostility and can develop into hostile relationships. Such kind of hostility occurs in interpersonal interactions usually when individuals have conflicts over their personal preferences regarding nonwork issues, such as social presentations, religious preferences, etc. (Jehn, 1995, 1997). Legitimate disagreements, however, revolve around task structure, task content, or task processes such as how to delegate resources (Jehn et al., 1997). Contentious conflict makes collaborators less likely to retain their engagement as it leads to lower morale in task completion process (Amason, 1996). When knowledge contributors devote their time to attacking others' positions and defending their own positions, they pay less attention to the problem or the exchange of useful knowledge and are less motivated to understand others' meanings or refine their own perspectives (Malhotra et al., 2017).

Algorithmic or human interventions need to be introduced to mitigate the potential risk of contentious conflict. For example, Wikipedia applied a three-revert rule to detecting and managing any pair of contributors intentionally deleting each other's edits to a Wikipedia article (Buriol et al., 2006). Once three reverts are detected, the contributor pair is automatically blocked from making further edits until a corresponding editor completes investigation and resolves the conflict. Reddit utilizes both human and machine moderators to facilitate discussions and create a platform that all users enjoy (Stephen, 2018). Applying moderators facilitates constructive discussions in the platform and helps host valuable user-generated content (Jones et al., 2019). In crowd-based open challenges, likewise, one common practice is to involve human moderators to facilitate online discussions and foster knowledge sharing (Nicolini, 2011). A guided open discussion can help avoid not only the risk of contentious conflict but also off-topic discussions (Stieger et al., 2012). Malhotra et al. (2017) applied a two-phase guided approach to crowd collaboration for an environmental innovation challenge. In the first phase, crowd collaborators were guided to share knowledge elements to collectively form a brief knowledge collage, without proposing full-fledged solutions. Participants were guided to share, for example, facts such as statistics, examples such as personal stories, tradeoffs describing inconsistent circumstances, and idea seeds that might trigger further discussions. Doing so encouraged participants to build a common knowledge ground and focus on understanding others' contributed knowledge fragments instead of defending their own positions. In the second phase, crowd collaborators were guided to propose solutions based on intentionally integrating others' shared knowledge elements and undertake iterations through bridging various knowledge elements without applying a win-lose negotiation strategy (Majchrzak et al., 2012). Precisely framing the challenge context also helps engaging the right participants with the desired expertise to join and contribute (Aitamurto et al., 2011; Jeppesen & Lakhani, 2010; King & Lakhani, 2013; Lopez-Vega et al., 2016; Mergel & Desouza, 2013). Challenge questions that are broadened to include different scenarios and encourage multiple types of contributions can effectively minimize contentious conflict among crowd participants (Malhotra et al., 2017). Moreover, a carefully designed timeframe plays a critical role in mitigating the risk of contentious conflict and managing online crowd-based interaction. Problems framed in a future-focused format can stimulate greater learning and innovative ideas as contributors are geared toward envisioning the future rather than focusing on current constraints or emotional complaints (Blomqvist & Levy, 2006; Gray, 1999; Trommsdorff, 1983).

14.3.3 Potential Risk of Self-Promotion

The degree to which crowd participants engage in self-promotion poses a risk to crowd-based open challenges. Research has found that when participants perceive the open collaboration platform as a broadcast platform for promoting themselves, knowledge sharing tends to be ineffective (Denyer et al., 2011). Participants are also likely to develop coalitions when they prioritize their individual concerns about receiving more votes and winning against others on such platforms (Dellarocas, 2010; Kittur et al., 2009). Self-promotion impedes participants' cognitive elaboration in their collaboration. Cognitive elaboration consists of illustrating one's own opinion and comparing it with others' opinions through conversations (O'Donnell & O'Kelly, 1994). Cognitive elaboration suggests that cooperative learning benefits from peer interaction in which collaborators undertake active information processing (Dansereau, 1988; O'Donnell & Dansereau, 1992). Elaboration not involving self-promotion can prompt learning activities such as deliberate imaging of the application, connecting to prior knowledge, restructuring existing cognitive structures, or adding new fragments to existing structures (O'Donnell & O'Kelly, 1994; Wittrock, 1986). Elaborative interactions that are not self-focused play a key role in knowledge collaboration as participants can reorganize their understanding by including new knowledge learned from others (Webb, 1989, 1991). Such conversational elaboration can spur the emergence of inter-subjective meaning that facilitates knowledge collaboration for innovation (Tsoukas, 2009). Inter-subjective meaning results from other-focused interactions and its completion is "dependent on the flow of the subsequent dialogue" (Sawyer, 2003, p.43). In other words, knowledge sharing in crowd-based open challenges is essentially an other-focused cognitive process that requires collaborators to be attentive to others' shared knowledge, provide responses, and collectively refine the communal knowledge. Self-focused promotion hampers this cognitive process as it diverges from the interactive cognitive elaboration and leads to narrow-focused collective outcomes. In open challenges, when engaging in such self-focused activities, crowd members devote time to promoting their own ideas rather than carefully reading others' shared knowledge, or they may even overwhelm the crowd with non-knowledge posts seeking for support to their own ideas. In Wikipedia, for example, self-promotion actions occur occasionally and are considered short-sighted, unfair to other contributors, and undermining the goal of Wikipedia in promoting knowledge collaboration (Robichaud, 2016). Contributors focusing on self-promotion tend not to make substantial or meaningful suggestions and fail to pay attention to others' knowledge or learn about others' perspectives. Consequently, critical knowledge gaps cannot be properly addressed while potential knowledge similarities fail to surface, impairing the collective production of comprehensive solutions.

To overcome the barriers caused by participants' self-promotion, a facilitator can be involved to monitor the discussion and help the crowd avoid posting self-promoted content (Stieger et al., 2012). Ensuring anonymity in crowd-based open discussion can also reduce self-promotion as participants are freed from the pressure to advocate or argue against anyone and can behave altruistically without seeking status in the online community (Lampel & Bhalla, 2007). Altruistic behaviors can further induce more positive interactions among crowd participants (Baytiyeh & Pfaffman, 2010). Anonymous posting also allows the involvement of facilitators to intervene in the discussion and make sure that the open discussion is geared toward a desired outcome (Kannangara & Uguccioni, 2013). Malhotra et al. (2017) employed a two-phase guideline to mitigate the risk of self-promotion. In the first stage, crowd participants were instructed to anonymously contribute knowledge that is relevant to the challenge question and not to seek for votes or any kind of support to their own posts. In the second stage, using explicit priming cues (Fujita & Trope, 2014), participants were particularly guided to integrate others' posts with specifying others' posts that they attempted to integrate. To motivate participants to constructively integrate others' posts, the platform also regularly identified participants who posted most comments to others' posts as the most collaborative participants. Such a strategy offers intrinsic incentives so that participants can feel recognized and therefore be motivated to maintain a focus on others' shared knowledge rather than focusing on promoting their own perspectives. Invitations sent to participants were crafted to direct participants' attention to developing collaborative solutions based on integrating others' shared knowledge and to explicitly promote constructive knowledge sharing by encouraging open, honest, comprehensive, and thought-provoking conversations. In Wikipedia, contributors are encouraged to disrupt homogenous narrowly focused viewpoints and diversify their perspectives (Bergen, 2016; Boboltz, 2015; Wagner et al., 2016) to maintain the alignment with public interest.

14.3.4 Temporal Complexity and Temporal Coordination

Tackling grand challenges often involves multiparty collaborations in which all involved parties need to coordinate their efforts. Different parties have different temporal structures in terms of how and when to perform their activities. Orlikowski and Yates (2002) indicated that "through their everyday action, actors produce and reproduce a variety of temporal structures which in turn shape the temporal rhythm and form of their ongoing practices" (p. 684). When various parties adjust actions to accomplish their common goals, they often need to deal with temporal complexity that emerges from their temporal structure incongruences (Fjeldstad et al., 2012; Garud et al., 2013; Gulati et al., 2012). Scientific collaborations involving a mix of public and private sectors may experience tensions such as goal inconsistency, as private organizations seek to fulfill short-term efficiency goals and public sectors focus more on long-term societal impacts (Caldwell et al., 2017; Slawinski & Bansal, 2015). In dealing with sustainability challenges, organizations from various industries often need to collaboratively develop solutions over sustained periods of time, resulting in misalignment in terms of their time horizons and innovation elements (Ansari & Garud, 2009). Such temporal complexity impedes collaboration as involving parties often find it difficult to reach an agreement on the timeline of collaboration progress. Complexities engendered by differences in collaborative rhythms also occur at the team or group level, when the research team faces challenges posed by involving collaborators' different career trajectories, shifting roles or identities, as well as life transitions that may restructure the collaboration pace (Jackson et al., 2011). Temporal complexities can emerge from the phenomenal rhythms of scientific studies as well. For example, climate change research often requires long-term collaborative efforts devoted to the projects and thus carries deep and challenging implications for scientific collaborations (Jackson et al., 2011). Reddy et al. (2006) found that different workgroups within the organization tend to generate distinctive understandings of the temporal structures in coordinating their activities. In online settings, long-term communication patterns regarding temporal issues eventually shape the nature of collaborations (Jones, 2003; Whittaker et al., 2002).

To overcome the barriers caused by temporal complexities, collaborating actors need to perform temporal coordination, defined as "effortful attempts to create or maintain alignment for engaging in joint action in the face of temporal complexities" (Hilbolling et al., 2022, p. 138). Temporal coordination can facilitate the alignment of rhythms and paces to support congruent collaborative activities (Bluedorn, 2002; Zerubavel, 1979). Collaborators can synchronize activities or use breakdown structures to build interdependent temporal structures and make joint progress (Eisenhardt & Tabrizi, 1995; Lindkvist et al., 1998; Lundin & Söderholm, 1995). To synchronize temporal structures, collaborators often undertake entrainment, which refers to "the adjustment of the pace and cycle of an activity to match or synchronize with that of another activity" (Ancona & Chong, 1996, p. 253). Such action facilitates collaborative work, especially innovation, as it effectively addresses the interdependencies between different temporal structures in collaborations (Bonneau, 2007; Dibrell et al., 2015; Hopp & Greene, 2017; Khavul et al., 2010; Pérez-Nordtvedt et al., 2008). Additionally, collaborating actors may also align the timing of activities through the mutual adaptation of incongruent temporal structures (Das, 2006; Davis, 2014; Standifer & Bluedorn, 2006). Involving parties may negotiate collaboration timelines and progress, create temporal boundary objects, or undertake shared project organizations to build congruency for collaborators to engage in joint action (Lindkvist et al., 1998; Lundin & Söderholm, 1995; Manning, 2008). In solving broadly defined sustainability problems, collaborating parties may follow three mechanisms to deal with temporal complexities (Hilbolling et al., 2022). The first mechanism is to leverage serendipitous alignment to create opportunities for collaborating parties to find connections and thus take joint action. Serendipitous alignment allows collaborators to build a common ground for interactions and set shared expectations. The second mechanism is temporary excluding parties, which refers to "the strategic decoupling of organizations with incongruent temporal structures that otherwise might hinder taking joint action" (Hilbolling et al., 2022, p. 149). By temporarily excluding a subset of the parties, the rest of the collaborating parties can avoid misaligning with the incongruent temporal structures and thus increase the work efficiency and make satisfactory progress without synchronizing various temporal structures. Based on this temporary exclusion mechanism, accordingly, the third mechanism refers to the fact that multiple collaborating parties identify the foundation for future reconnection or alignment and thus jointly envision the future. Such "aligning on the future" (Hilbolling et al., 2022, p. 150) creates a space for heterogenous parties to look into a shared future and fosters meaningful interactions about the future without tapping into temporal complexities. Taken together, the three mechanisms advance temporal coordination approaches to address temporal complexities in tackling grand sustainability challenges.

14.3.5 Ethical Issues, Data Bias, and Quality

Key to citizen science is the role of citizens and the data they contribute. One of the major issues pertains to research integrity in study design, data collection and sharing, misconduct prevention, promoting objectivity, and publishing research findings (Shamoo & Resnik, 2015). Various ethical and legal standards also embody the principles for protecting human subjects, governing the scope of research, scientific rigor, risk minimization, confidentiality, safety monitoring, protection of vulnerable subjects, as well as independent ethics review (Emanuel et al., 2000; Shamoo & Resnik, 2015). The White House memorandum offers three core principles that highlight full voluntariness of participation, meaningful and beneficial to participants, and participation being acknowledged (Holdren, 2015). Maintaining transparency and accountability is critical to preserving research integrity in participatory studies. Shrik et al. (2012) indicated that ethical participatory projects require "careful, intentional, and transparent employment of participation strategies to achieve targeted outcomes" (p. 3). Institutions and scientists should set accurate expectations for participation, make the expectations of different stakeholders explicit as well as balance these diverse expectations based on a dynamic evaluation of evolving interests (Shirk et al., 2012). Additionally, it is researchers' responsibility to disclose research outcomes and offer timely and frequent feedback to citizen collaborators. For example, the Mapping for Change project initiated by the University College London has benefited from working with the 'community champion' who is perceived as trustworthy by the local community in constructing long-term trust relationships with citizen collaborators and maintaining citizens' trust and engagement (Balestrini et al. 2017; Eleta et al., 2019). To address other concerns such as exploitation of human labor, some citizen science projects (e.g., Galaxy Zoo) have developed their unique guidelines emphasizing volunteers being treated as collaborators, volunteers' time being valued, and volunteers not being asked to perform tasks for which artificial intelligence is better fit (Prestopnik & Crowston, 2012).

The other major ethical issue lies in intellectual property and data governance. Based on the level of citizen involvement, citizen science projects can be classified into citizen-initiated projects, community-based participatory research, and citizen-assisted projects (Resnik, 2019). In the current networked environment, citizen science projects often require the collection of personal data or metadata contained in pictures, identifying information or other digital pieces of information, raising concerns about privacy issues. To address these concerns, Cavoukian (2010) suggested Privacy by Design principles that highlight embedded-in-design privacy, transparency and visibility, as well as user-centered design for protecting users' privacy. Crucial to data governance is data protection, as well as the openness of access and ownership. Essentially, data protection focuses on securing data against unauthorized use, whereas openness of access and ownership is about who defines, owns, and uses data (Bietz et al., 2019). The governance of data is of particular importance for biomedical research which often needs a large amount of personal data to make scholarly advances. Private organizations, for example, collect user-generated data by providing new products for patients, whereas many governmental research programs suffer from the lack of transparency or value reciprocity (Carter et al., 2015). Concerns were thus raised about the whole value chain being controlled by professionals and very rare citizen control allowed (Eleta et al., 2019). To tackle such issues, collective governance models have emerged that are increasingly employed to enable citizen involvement in the governance of science and allow citizens to participate in scientific studies not only as collaborators but also as the real instigators of the projects (Buyx et al., 2017; Evans, 2016). Smith et al. (2019) argued that the recognition of patient collaborators' contribution needs to be considered on a long-term basis from personal, scientific, and financial perspectives.

A relevant concern is data bias. Due to the availability of Internet of Things devices, data collected online can be contributed by a non-representative group of the population that fails to include older, geographically distant, and minority members, resulting in demographically or geographically skewed data (Acer et al., 2019; Havinga et al., 2020). As such, citizen science researchers will need to be attentive to the socio-demographic backgrounds of study respondents. Data quality can be a concern as well. Studies have reported discarding data due to the nature of volunteered information or substantial variances in data quality (Aoki et al., 2009; Andersson & Sternberg, 2016; Barzyk et al., 2018; Elwood et al., 2012; Ferster et al., 2013; Wylie et al., 2014; Weir et al., 2019; Vesnic-Alujevic et al., 2018). To address these pitfalls, research has suggested that citizen science studies need to consider the implications of data-empowered global citizens, and that government agencies need to publish protocols or guidelines on data quality (Barzyk et al., 2018; Black & White, 2016). Other scholars suggest approaches such as collecting longitudinal data or integrating participants into the protocol to overcome data quality challenges (Dema et al., 2019). Improving the training for volunteers and focusing on certain areas may also help increase data quality (Ferster et al., 2013; Heiss & Matthes, 2017).

14.4 CONCLUSION

Unleashing the wisdom of crowds, citizen science has been found to be an effective approach to tackling grand sustainability challenges. This chapter reviews theoretical frameworks and empirical studies on crowd-powered citizen science and sustainability, demonstrating the advantages and complexities in managing citizen science for sustainability. Featuring crowdsourced knowledge collaboration, theories of robust action, open innovation, and open strategy formulation highlight the importance of affording space and opportunities that allow members of the crowd to share, exchange, reuse, and integrate knowledge (Beck et al., 2022; Malhotra et al., 2017; Porter et al., 2020). In citizen science, crowd participants can engage in back-and-forth conversations to make their diverse tacit knowledge explicit (Leonardi, 2014; Malhotra & Majchrzak, 2019; Santoro et al., 2018; Sun et al., 2021), critically view others' knowledge and revisit their own knowledge (Dahlander & Frederiksen, 2012; Head, 2019; Majchrzak et al., 2012; McGahan et al., 2021), undertake distributed experimentation to make parallel attempts for problem solving (Dietz et al., 2003; Ferraro et al., 2015), and synthesize others' knowledge to collectively generate optimal solutions to the challenge (Dougherty, 1992; Galinsky & Moskowitz, 2000; Hargadon & Bechky, 2006; Henderson, 1994; Nonaka,1994; Parker & Axtell, 2001). The issues and complexities suggest that theories of crowd-based knowledge sharing in citizen science need to be further developed to capture the theoretical nuances regarding how to guide crowd participants to deal with timelines and temporal boundaries (Fjeldstad et al., 2012; Garud et al., 2013; Gulati et al., 2012), how to motivate participants to engage in other-focused constructive and non-promotional interactions (Dellarocas, 2010; Denyer et al., 2011; Kittur et al., 2009), as well as how to avoid ethical barriers and obtain high-quality data in citizen science research (Bietz et al., 2019; Buyx et al., 2017; Cavoukian, 2010; Emanuel et al., 2000; Evans, 2016; Shamoo & Resnik, 2015; Shrik et al., 2012).

Different from other sustainability approaches such as common pool resource management (Dietz et al., 2003; Garud & Gehman, 2012; Nisbet, 2009; Ostrom, 1990) and the strategic sustainable development framework (Ny et al., 2006; Robèrt et al., 2002), crowd-based open approaches are beyond reshaping governance structures or adherence to traditional domain frames and grant high degrees of freedom to crowd participants such that innovative solutions can emerge from their active knowledge sharing, revisiting, synthesizing, and co-creating. Research on citizen science for sustainability also sheds light on scientific practices and policy-making by suggesting that policymakers and science practitioners are expected to devote more resources to improve the openness, transparency, interoperability, and adaptability of citizen science (Certoma et al., 2015). Doing so can encourage open collaboration and knowledge co-creation, foster the flow of information, improve public decision-making, as well as prevent infrastructural centralization. Further, organizations that launch citizen science projects may need to develop strategic agility and absorptive capacity to better implement crowd-generated strategic solutions (Cohen & Levinthal, 1990; Doz & Kosonen, 2010).

The theoretical frameworks and empirical studies set the stage for future research on using crowd-powered open approaches to tackle sustainability grand challenges. Table 14.1 presents a summary of key social aspects involved in advancing sustainability using citizen science, roles

TABLE 14.1
Advancing Sustainability through Citizen Science and Crowdsourcing

Key Social Aspects and Components	Corresponding Roles of Crowdsourcing Activities	Various Forms of Governance Involved[a]	Features of Citizen Science	Exemplary Studies
Participating entities include citizens and communities, organizations, policymakers and regulators, as well as other constituencies	Involving public participation in promoting scientific advancements and decision-making on sustainable strategies	• Adaptive governance that implements rules and norms as well as maintains common knowledge resources • Strategic and sustainable governing framework that supports participatory backcasting and predetermined systemic goals at organizational and supra-organizational levels • Robust action governing approach that is configured by weak formal authority, changing constituencies, and emergent goals, allowing sustained novelty generation in tackling grand challenges	• Crowdsourcing platforms that are freely accessible for all citizens • Architecture and procedures that facilitate citizens' shared perceptions of the challenges • Interactivity that promotes knowledge sharing and collaboration • Open access data and knowledge that allows citizens' voices to be heard and encourages innovative solutions to emerge • Interconnectedness between local and global challenges that motivates shared practices and continuous engagement of citizens	• Bailey and Grossardt (2010) • Brabham (2009, 2012) • Campbell (1996) • Daffara (2011) • Dietz et al. (2003) • Eames and Egmose (2011) • Estellés-Arolas & González-Ladrón-de-Guevara (2012) • Ferraro et al. (2015) • Fricker (1998) • Hage et al. (2010) • Haughton and Hunter (2003) • Holden & Linnerud (2007) • Kallio et al. (2007) • Keirstead and Leach (2008) • Larsen et al. (2011) • Leifer (1991) • Lemke, Casper, and Moore (2011) • Marres and Rogers (2005) • Ny et al. (2006) • Ostrom (1990, 2009) • Padgett and Ansell (1993) • Robert et al. (2002, 2005, 2013) • Willett et al. (2012) • Ratcliffe and Krawczyk (2011) • Street (1997) • Stren et al. (1992) • Tonn (2007)
Multidimensional goals that include balanced and optimal measures of ecological, environmental, and societal sustainability	Enhancing proactive interconnections between the ecological system and social actors at system and individual levels			
Efficient implementation of strategies and principles that balance the fulfillment of human needs, environment protection, and ecosystem evolution	Connecting citizens, scientists, and decision-makers using technology-facilitated approaches, as well as promoting direct interactions and collaborative knowledge sharing			
Systemic equality that promotes justice at social and environmental levels	Optimizing power distribution in the process of data production, collection, aggregation, analysis, and interpretation			
Transparency of decision-making and knowledge management	Establishing effective institutionalized models for managing technology, infrastructure, data, and information			
Adaptation to contextual complexity that pertains to progressive and dynamically evolving social contexts	Crowd is involved in solving wicked problems that are configured by modularity and segmentation so that individual citizens can match their knowledge, skills, and experience with the tasks			
Interorganizational collaboration based on common understandings shared by institutions, governments, non-government organizations, and businesses	Offering a dialogic space and a common ground for various organizations to interact and negotiate on a non-discriminatory base			
International cooperation between actors from multiple nations to achieve common global sustainable development goals	Providing a public space beyond physical boundaries so that international and transnational actors can collaborate without mediation			

[a] The content presented in the columns of 'Various forms of governance involved', 'Features of citizens', and 'Exemplary studies' apply to all the content displayed in the first two columns, 'Key social aspects and components', and 'Corresponding roles of crowdsourcing activities'.

played by crowdsourcing activities, various forms of governance, and characteristics of citizen science. Future research needs to be attentive to issues such as how to reduce risks pertaining to organizations' competitive intelligence and make sure organizations' strategic plans are not disclosed to competitors when using such crow-based open approaches (Kannangara & Uguccioni, 2013; Marjanovic et al., 2012). Meanwhile, the precise mechanism by which participants' knowledge reuse facilitates knowledge integration and bridges knowledge gaps deserves closer examination. Future studies should investigate the boundary conditions under which knowledge reuse takes place based on developing shared knowledge or newly added knowledge or a combination of both (Bechky, 2003; Carlile, 2004; Rodan & Galunic, 2004; Star & Griesemer, 1989). The process by which crowd-generated innovation influences the development of organizational capabilities is not yet well-studied. To facilitate incorporating crowd-generated solutions and maintaining competitiveness in the competition, an organization needs to strengthen its complementary capabilities such that organizational inertia can be overcome to embrace evolutionary changes (Kelly & Amburgey, 1991). As information technology advances rapidly, newly developed participatory architectures can be employed to provide documentation of crowds' dynamic reflections such that previously inactive or newly joined participants can conveniently obtain a refreshed or nuanced understanding of the knowledge shared (Dietz et al., 2003; Etzion et al., 2017; Garud & Gehman, 2012). Finally, future research could study individual differences of crowd participants, such as professional and socioeconomic background, as well as how such differences yield cognitive and behavioral variations when participants collaborate on generating innovative solutions to grand sustainability challenges. Unpacking how participants behave differently will largely advance our understanding of how to connect crowd-based large-scale innovation with small-scale organizational or team-level scientific collaborations.

REFERENCES

Abreu, M. & Grinevich, V. (2013). The nature of academic entrepreneurship in the UK: Widening the focus on entrepreneurial activities. *Research Policy*, 42(2), 40822.

Acer, U. G., Broeck, M. V. D., Forlivesi, C., Heller, F. & Kawsar, F. (2019). Scaling crowdsourcing with mobile workforce: A case study with Belgian postal service. *Proceedings of the ACM on Interactive, Mobile, Wearable and Ubiquitous Technologies*, 3(2), 1–32.

Afuah, A. & Tucci, C. L. (2012). Crowdsourcing as a solution to distant search. *Academy of Management Review*, 37(3), 355–375.

Aitamurto, T., Landemore, H. & Saldivar Galli, J. (2017). Unmasking the crowd: Participants' motivation factors, expectations, and profile in a crowdsourced law reform. *Information, Communication & Society*, 20(8), 1239–1260.

Aitamurto, T., Leiponen, A. & Tee, R. (2011). The promise of idea crowdsourcing-benefits, contexts, limitations. *Nokia Ideasproject White Paper*, 1, 1–30.

Amano, T., Lamming, J. D. L. & Sutherland, W. J. (2016). Spatial gaps in global biodiversity information and the role of citizen science. *Bioscience*, 66, 393–400.

Amason, A. C. (1996). Distinguishing the effects of functional and dysfunctional conflict on strategic decision making: Resolving a paradox for top management teams. *Academy of Management Journal*, 39(1), 123–148.

Ancona, D. G. & Chong, C.-L. (1996). Entrainment: Pace, cycle, and rhythm in organizational behavior. *Research in Organizational Behavior*, 18, 251–284.

Anderson, J. V. (1994). Creativity and play: A systematic approach to managing innovation. *Business Horizons*, 37(2), 80–85.

Andersson, M. & Sternberg, H. (2016). Informating transport transparency. In *Proceedings of 49th Hawaii International Conference on System Sciences (HICSS)* (pp. 1841–1850).

Ansari, S. & Garud, R. (2009). Inter-generational transitions in socio-technical systems: The case of mobile communications. *Research Policy*, 38(2), 382–392.

Aoki, P. M., Honicky, R. J., Mainwaring, A., Myers, C., Paulos, E. & Subramanian, S. (2009). A vehicle for research: Using street sweepers to explore the landscape of environmental community action. In *Proceedings of the SIGCHI Conference on Human Factors in Computing Systems CHI* (pp. 375–384).

Argote, L. (1999). *Organizational Learning: Creating, Retaining, and Transferring Knowledge*. Boston, MA: Kluwer Academic.

Attard, J., Orlandi, F., Scerri, S. & Auer, S. (2015). A systematic review of open government data initiatives. *Government Information Quarterly*, 32(4), 399–418.

Baer, M. & Frese, M. (2003). Innovation is not enough: Climates for initiative and psychological safety, process innovations, and firm performance. *Journal of Organizational Behavior*, 24, 45–68.

Bailey, K. & Grossardt, T. (2010). Toward structured public involvement: Justice, geography and collaborative geospatial/geovisual decision support systems. *Annals of the Association of American Geographers*, 100, 57–86.

Baldini, N., Grimaldi, R. & Sobrero, M. (2007). To patent or not to patent? A survey of Italian inventors on motivations, incentives, and obstacles to university patenting. *Scientometrics*, 70(2), 333–354.

Baldwin, C. & Von Hippel, E. (2011). Modeling a paradigm shift: From producer innovation to user and open collaborative innovation. *Organization Science*, 22(6), 1399–1417.

Balestrini, M., Rogers, Y., Hassan, C., Creus, J., King, M. & Marshall, P. (2017). A city in common: A framework to orchestrate large-scale citizen engagement around urban issues. In *Proceedings of the 2017 CHI Conference on Human Factors in Computing Systems* (pp. 2282–2294).

Baptista, J., Wilson, A. D., Galliers, R. D. & Bynghall, S. (2017). Social media and the emergence of reflexiveness as a new capability for open strategy. *Long Range Planning*, 50(3), 322–336.

Bartley, T. (2007). Institutional emergence in an era of globalization: The rise of transnational private regulation of labor and environmental conditions. *American Journal of Sociology*, 113(2), 297–351.

Barzyk, T. M., Huang, H., Williams, R., Kaufman, A. & Essoka, J. (2018). Advice and frequently asked questions (FAQs) for citizen-science environmental health assessments. *International Journal of Environmental Research and Public Health*, 15(5), 960–980.

Batey, M. & Furnham, A. (2006). Creativity, intelligence, and personality: A critical review of the scattered literature. *Genetic, Social, and General Psychology Monographs*, 132(4), 355–429.

Batson, C. D., Ahmad, N. & Tsang, J. A. (2002). Four motives for community involvement. *Journal of Social Issues*, 58(3), 429–445.

Baytiyeh, H. & Pfaffman, J. (2010). Open-source software: A community of altruists. *Computers in Human Behavior*, 26(6), 1345–1354.

Bazilian, M., Rice, A., Rotich, J., Howells, M., DeCarolis, J., Macmillan, S., ... & Liebreich, M. (2012). Open-source software and crowdsourcing for energy analysis. *Energy Policy*, 49, 149–153.

Bechky, B. A. (2003). Object lessons: Workplace artifacts as representations of occupational jurisdiction. *American Journal of Sociology*, 109(3), 720–752.

Beck, S., Bergenholtz, C., Bogers, M., Brasseur, T. M., Conradsen, M. L., Di Marco, D., ... & Xu, S. M. (2022). The open innovation in science research field: A collaborative conceptualisation approach. *Industry and Innovation*, 29(2), 136–185.

Bercovitz, J. & Feldman, M. (2008). Academic entrepreneurs: Organizational change at the individual level. *Organization Science*, 19(1), 69–89.

Bergen, D. (2009). Play as the learning medium for future scientists, mathematicians, and engineers. *American Journal of Play*, 1(4), 413–428.

Bergen, S. (2016). Linking in: How historians are fighting Wikipedia's biases. *Perspectives on History*. Retrieved from https://www.historians.org/publications-and-directories/perspectives-on-history/september-2016/linking-in-how-historians-are-fighting-wikipedias-biases

Berman, E. P. (2011). *Creating the Market University: How Academic Science Became an Economic Engine*. New York: Princeton University Press.

Bhattacharjee, Y. (2005). Citizen scientists supplement work of Cornell researchers. *Science*, 308, 1402–1403.

Bietz, M, Patrick, K. & Bloss, C. (2019). Data donation as a model for citizen science health research. *Citizen Science: Theory and Practice*, 4(1):6, 1–11.

Birkin, L. J., Vasileiou, E. & Stagg, H. R. (2021). Citizen science in the time of COVID-19. *Thorax*, 76(7), 636–637.

Black, I. & White, G. (2016). Citizen science, air quality, and the internet of things. In Reis, C. I., & Maximiano, M. D. S. (Eds.), *Internet of Things and Advanced Application in Healthcare* (pp. 138–169). IGI Global.

Blohm, I., Leimeister, J. M. & Krcmar, H. (2013). Crowdsourcing: How to benefit from (too) many great ideas. *MIS Quarterly Executive*, 12(4), 199–211.

Blomqvist, K. & Levy, J. (2006). Collaboration capability–a focal concept in knowledge creation and collaborative innovation in networks. *International Journal of Management Concepts and Philosophy*, 2(1), 31–48.

Bluedorn, A. C. (2002). *The Human Organization of Time: Temporal Realities and Experience*. Stanford, CA: Stanford University Press.

Blühdorn, I. & Deflorian, M. (2019). The collaborative management of sustained unsustainability: On the performance of participatory forms of environmental governance. *Sustainability*, 11(4), 1189–1206.

Boboltz, S. (2015). Editors are trying to fix Wikipedia's gender and racial bias problem. *The Huffington Post*. Retrieved from https://www.huffpost.com/entry/wikipedia-gender-racial-bias_n_7054550.

Boland Jr, R. J., & Tenkasi, R. V. (1995). Perspective making and perspective taking in communities of knowing. *Organization Science*, 6(4), 350–372.

Bonneau, L. (2007). Inter-organisational time: The example of Quebec biotechnology. *International Journal of Innovation Management*, 11(1), 139–164.

Bonney, R. (2007). Citizen science at the Cornell lab of ornithology. In Yager, R. E. & Falk, J. H. (Eds), *Exemplary Science in Informal Education Settings: Standards-based Success Stories* (pp. 213–229). NSTA Press.

Bonney, R., Cooper, C. B., Dickinson, J., Kelling, S., Phillips, T., Rosenberg, K. V. & Shirk, J. (2009). Citizen science: A developing tool for expanding science knowledge and scientific literacy. *BioScience*, 59(11), 977–984.

Bonney, R., Shirk, J. L., Phillips, T. B., Wiggins, A., Ballard, H. L., Miller-Rushing, A. J. & Parrish, J. K. (2014). Next steps for citizen science. *Science*, 343(6178), 1436–1437.

Borrás, S. & Edler, J. (2014). *The Governance of Socio-technical Systems: Explaining Change*. Cheltenham, UK: Edward Elgar Publishing.

Bos, N., Zimmerman, A., Olson, J., Yew, J., Yerkie, J., Dahl, E. & Olson, G., (2007). From shared databases to communities of practice: A taxonomy of collaboratories. *Journal of Computer-Mediated Communication*, 12(2), 652–672.

Bowker, G. C. & Star, S. L. (1999). *Sorting Things Out: Classification and its Consequences*. Cambridge: MIT Press.

Brabham, D. C. (2008). Crowdsourcing as a model for problem solving: An introduction and cases. *Convergence*, 14(1), 75–90.

Brabham, D. C. (2009). Crowdsourcing the public participation process for planning projects. *Plan Theory*, 8, 242–262.

Brabham, D. C. (2012). Motivations for participation in a crowdsourcing application to improve public engagement. *Transit Planning*, 40, 307–328.

Branchini, S., Meschini, M., Covi, C., Piccinetti, C., Zaccanti, F. & Goffredo, S. (2015). Participating in a citizen science monitoring program: Implications for environmental education. *PLoS One*, 10, e0131812.

Brossard, B. D., Lewenstein, B. & Bonney, R. (2005). Scientific knowledge and attitude change: The impact of a citizen science project. *International Journal of Science Education*, 27(9), 1099–1121.

Brueller, D. & Carmeli, A. (2011). Linking capacities of high-quality relationships to team learning and performance in service organizations. *Human Resource Management*, 50, 455–477.

Bruner, J. S. (1990). *Acts of Meaning*. Cambridge, MA: Harvard University Press.

Bstieler, L. & Hemmert, M. (2010). Increasing learning and time efficiency in interorganizational new product development teams. *Journal of Product Innovation Management*, 27, 485–499.

Budhathoki, N. R. & Haythornthwaite, C. (2013). Motivation for open collaboration: Crowd and community models and the case of OpenStreetMap. *American Behavioral Scientist*, 57(5), 548–575.

Buriol, L.S., Castillo, C., Donato, D., Leonardi, S. & Millozzi, S. (2006). Temporal analysis of the wikigraph. In *Proceedings of Web Intelligence WI IEEE/WIC/ACM International Conference* (pp. 45–51). IEEE.

Buyx, A, Del Savio, L, Prainsack, B & Voelzke, H. (2017). Every participant is a PI: Citizen science and participatory governance in population studies. *International Journal of Epidemiology*, 46(2), 377–384.

Caldwell, N.D., Roehrich, J.K. & George, G. (2017). Social value creation and relational coordination in public- private collaborations. *Journal of Management Studies*, 54(6), 906–928.

Callon, M., Lascoumes, P. & Barthe, Y. (2009). *Acting in an Uncertain World: An Essay on Technical Democracy*. Cambridge: MIT Press.

Campbell, S. (1996). Green cities, growing cities, just cities? Urban planning and the contradictions of sustainable development. *Journal of the American Planning Association*, 62, 296–312.

Cannon, A. R., Chamberlain, D. E., Toms, M. P., Hatchwell, B. J. & Gaston, K. J. (2005). Trends in the use of private gardens by wild birds in Great Britain 1995-2002. *Journal of Applied Ecology*, 42(4), 659–671.

Carlile, P. R. (2004). Transferring, translating, and transforming: An integrative framework for managing knowledge across boundaries. *Organization Science*, 15(5), 555–568.

Carter, P, Laurie, G. T. & Dixon-Woods, M. (2015). The social licence for research: Why *care.data* ran into trouble. *Journal of Medical Ethics*, 41(5), 404–409.

Caruso, E. M., Epley, N. & Bazerman, M. H. (2006). The good, the bad, and the ugly of perspective taking in groups. *Research on Managing Groups and Teams*, 8, 201–224.

Cavoukian, A. (2010). Privacy by design: The definitive workshop. *Identity in the Information Society*, 3(2), 247–251.

Certoma, C., Corsini, F. & Rizzi, F. (2015). Crowdsourcing urban sustainability. Data, people and technologies in participatory governance. *Futures*, 74, 93–106.

Cervantes, M. & Hong, S.J. (2018). STI policies for delivering on the sustainable development goals. In *OECD Science, Technology and Innovation Outlook 2018: Adapting to Technological and Societal Disruption*. Paris, France: OECD Publishing.

Chandrasekaran, A. & Mishra, A. (2012). Task design, team context, and psychological safety: An empirical analysis of R&D projects in high technology organizations. *Production and Operations Management*, 21, 977–996.

Chesbrough, H. & Bogers, M. (2014). Explicating open innovation: Clarifying an emerging paradigm for understanding innovation. In *New Frontiers in Open Innovation* (pp. 3–28). Oxford: Oxford University Press.

Chesbrough, H. (2020). To recover faster from Covid-19, open up: Managerial implications from an open innovation perspective. *Industrial Marketing Management*, 88, 410–413.

Clary, E. G., Snyder, M., Ridge, R. D., Copeland, J., Stukas, A. A., Haugen, J. & Miene, P. (1998). Understanding and assessing the motivations of volunteers: A functional approach. *Journal of Personality and Social Psychology*, 74(6), 1516–1530.

Cohen, W. M. & Levinthal, D. A. (1990). Absorptive capacity: A new perspective on learning and innovation. *Administrative Science Quarterly*, 35(1), 128–152.

Cohen, W. M., Nelson, R. R. & Walsh, J. P. (2002). Links and impacts: The influence of public research on industrial R&D. *Management Science*, 48(1), 1–23.

Cohn, J. P. (2008). Citizen science: Can volunteers do real research? *BioScience*, 58(3), 192–197.

Cooper, C. B., Dickinson, J., Phillips, T. & Bonney, R. (2007). Citizen science as a tool for conservation in residential ecosystems. *Ecology and Society*, 12(2), 11–22.

Cronin, M. A. & Weingart, L. R. (2007). Representational gaps, information processing, and conflict in functionally diverse teams. *Academy of Management Review*, 32 (3), 761–773.

Curral, L. A., Forrester, R. H., Dawson, J. F. & West, M. A. (2001). It's what you do and the way that you do it: Team task, team size, and innovation-related group processes. *European Journal of Work and Organizational Psychology*, 10(2), 187–204.

Curtis, V. (2014). Online citizen science games: Opportunities for the biological sciences. *Applied & Translational Genomics*, 3(4), 90–94.

Daffara, P. (2011). Rethinking tomorrow's cities: Emerging issues on city foresight. *Futures*, 43, 680–689.

Dahlander, L. & Frederiksen, L. (2012). The core and cosmopolitans: A relational view of innovation in user communities. *Organization Science*, 23(4), 988–1007.

Danielsen, F., Topp-Jørgensen, E., Levermann, N., Løvstrøm, P., Schiøtz, M., Enghoff, M. & Jakobsen, P. (2014). Counting what counts: Using local knowledge to improve Arctic resource management. *Polar Geography*, 37(1), 69–91.

Dansereau, D. F. (1988). Cooperative learning strategies. In Weinstein, C. E., Goetz, E. T. & Alexander, P. A. (Eds.), *Learning and Study Strategies: Issues in Assessment, Instruction, and Evaluation* (pp. 103–120). San Diego: Academic Press.

Das, T.K. (2006). Strategic alliance temporalities and partner opportunism. *British Journal of Management*, 17(1), 1–21.

Davis, J.P. (2014). The emergence and coordination of synchrony in organizational ecosystems. In Adner, R, Oxley, J. E. & Silverman, B. S. (Eds.), *Collaboration and Competition in Business Ecosystems Advances in Strategic Management* (Vol. 30, pp. 197–237). Bingley: Emerald Group Publishing Limited.

Dawson, T., Hambly, J., Kelley, A., Lees, W. & Miller, S. (2020). Coastal heritage, global climate change, public engagement, and citizen science. *Proceedings of the National Academy of Sciences*, 117(15), 8280–8286.

Dellarocas, C. (2010). Online reputation systems: How to design one that does what you need. *MIT Sloan Management Review*, 51(3), 33–37.

Dema, T., Brereton, M. & Roe, P. (2019). Designing participatory sensing with remote communities to conserve endangered species. In *Proceedings of the 2019 CHI Conference on Human Factors in Computing Systems* (pp. 1–16).

Dennis, A. R. & Garfield, M. J. (2003). The adoption and use of GSS in project teams: Toward more participative processes and outcomes. *MIS Quarterly*, 27(2), 289–323.

Dennis, A. R. (1996). Information exchange and use in group decision making: You can lead a group to information, but you can't make it think. *MIS Quarterly*, 20(4), 433–457.

Dennis, A. R., Wixom, B. H. & Vandenberg, R. J. (2001). Understanding fit and appropriation effects in group decision support systems via meta-analysis. *MIS Quarterly*, 25(2), 169–173.

Denyer, D., Parry, E. & Flowers, P. (2011). "Social", "Open" and "Participative"? Exploring personal experiences and organisational effects of enterprise 2.0 use. *Long Range Planning*, 44(5), 375–396.

D'Este, P. & Perkmann, M. (2011). Why do academics engage with industry? The entrepreneurial university and individual motivations. *Journal of Technology Transfer*, 36(3), 316–339.

Di Fatta, D., Caputo, F. & Dominici, G. (2018). A relational view of start-up firms inside an incubator: The case of the ARCA consortium. *European Journal of Innovation Management*, 21(4), 601–619.

Dibrell, C., Fairclough, S. & Davis, P. S. (2015). The impact of external and internal entrainment on firm innovativeness: A test of moderation. *Journal of Business Research*, 68(1), 19–26.

Dietz, J. S. & Bozeman, B. (2005). Academic careers, patents, and productivity: Industry experience as scientific and technical human capital. *Research Policy*, 34(3), 349–367.

Dietz, T., Ostrom, E. & Stern, P. C. (2003). The struggle to govern the commons. *Science*, 302, 1907–1912.

Ding, W. W. (2011). The impact of founders' professional-education background on the adoption of open science by for-profit biotechnology firms. *Management Science*, 57(2), 257–273.

D'Ippolito, B. & C.-C. Rüling. (2019). Research collaboration in large scale research infrastructures: Collaboration types and policy implications. *Research Policy*, 48(5), 1282–1296.

Dougherty, D. (1992). Interpretive barriers to successful product innovation in large firms. *Organization Science*, 3(2), 179–202.

Doz, Y. L. & Kosonen, M. (2010). Embedding strategic agility: A leadership agenda for accelerating business model renewal. *Long Range Planning*, 43(2), 370–382.

Droege, S. (2007). Just because you paid them doesn't mean their data are better. In *Citizen Science Toolkit Conference* (pp. 13–26). Cornell Laboratory of Ornithology.

Duncan, R. & A. Weiss (1979). Organizational learning: Implications for organizational design. In Cummings, L. L. & Staw, B. M. (Eds.), *Research in Organizational Behavior* (Vol. 1). Greenwich, CT: JAI Press.

Eames, M. & Egmose, J. (2011). Community foresight for urban sustainability: Insights from the Citizens Science for Sustainability (SuScit) project. *Technological Forecasting and Social Change*, 78, 769–784.

Edmondson, A. (1999). Psychological safety and learning behavior in work teams. *Administrative Science Quarterly*, 44 (2), 350–383.

Eisenhardt, K.M. & Tabrizi, B. (1995). Accelerating adaptive processes: Product innovation in the global computer industry. *Administrative Science Quarterly*, 40(1), 84–110.

Eleta, I., Clavell, G. G., Righi, V. & Balestrini, M. (2019). The promise of participation and decision-making power in citizen science. *Citizen Science: Theory and Practice*, 4(1–8), 1–9.

Elliot, J. A. (2001). *An Introduction to Sustainable Development* (4th ed.). New York: Routledge.

Elwood, S., Goodchild, M. F. & Sui, D. Z. (2012). Researching volunteered geographic information: Spatial data, geographic research, and new social practice. *Annals of the Association of American Geographers*, 102(3), 571–590.

Emanuel, E.J., Wendler, D. & Grady, C. (2000). What makes clinical research ethical? *JAMA*, 283(20), 2701–2711.

English, P. B., Richardson, M. J. & Garzón-Galvis, C. (2018). From crowdsourcing to extreme citizen science: Participatory research for environmental health. *Annual Review of Public Health*, 39(1), 335–350.

Estellés-Arolas, E. & González-Ladrón-de-Guevara, F. (2012). Towards an integrated crowdsourcing definition. *Journal of Information Science*, 38(2), 189–200.

Etzion, D., Gehman, J., Ferraro, F. & Avidan, M. (2017). Unleashing sustainability transformations through robust action. *Journal of Cleaner Production*, 140, 167–178.

Etzkowitz, H. & Leydesdorff, L. (2000). The dynamics of innovation: From national systems and "mode 2" to a triple helix of university-industry-government relations. *Research Policy*, 29(2), 109–123.

Evans, B. J. (2016). Barbarians at the gate: Consumer-driven health data commons and the transformation of citizen science. *American Journal of Law & Medicine*, 42(4), 651–685.

Fahey, L. & Prusak, L. (1998). The 11 deadliest sins of knowledge management. *California Management Review*, 40(3), 265–276.

Faraj, S., Jarvenpaa, S. L. & Majchrzak, A. (2011). Knowledge collaboration in online communities. *Organization Science*, 22 (5), 1224–1239.

Felin, T. & Zenger, T. R. (2014). Closed or open innovation? Problem solving and the governance choice. *Research Policy*, 43(5), 914–925.

Feller, I. & Feldman, M. (2010). The commercialization of academic patents: Black boxes, pipelines, and rubik's cubes. *Journal of Technology Transfer*, 35(6), 597–616.

Ferraro, F., Etzion, D. & Gehman, J. (2015). Tackling grand challenges pragmatically: Robust action revisited. *Organization Studies*, 36, 363–90.

Ferster, C., Coops, N., Harshaw, H., Kozak, R. & Meitner, M. (2013). An exploratory assessment of a smartphone application for public participation in forest fuels measurement in the wildland-urban interface. *Forests*, 4, 1199–1219.

Finholt, T. A. (2002). Collaboratories. *Annual Review of Information Science and Technology*, 36(1), 73–107.

Fjeldstad, Ø. D., Snow, C. C., Miles, R. E. & Lettl, C. (2012). The architecture of collaboration. *Strategic Management Journal*, 33(6), 734–750.

Fleming, L. & O. Sorenson. (2004). Science as a map in technological search. *Strategic Management Journal*, 25(8–9), 909–928.

Franzoni, C. & H. Sauermann. (2014). Crowd science: The organization of scientific research in open collaborative projects. *Research Policy*, 43(1), 1–20.

Fricker, A. (1998). Measuring up to sustainability. *Futures*, 30, 367–375.

Friesen, J., Taubenböck, H., Wurm, M. & Pelz, P. F. (2019). Size distributions of slums across the globe using different data and classification methods. *European Journal of Remote Sensing*, 52(2), 99–111.

Fujita, K. & Trope, Y. (2014). Structured versus unstructured regulation: On procedural mindsets and the mechanisms of priming effects. In *Understanding Priming Effects in Social Psychology* (pp. 70–89). Daniel C. Molden (Ed.), The Guilford Press, New York.

Furnari, S. (2014). Interstitial spaces: Micro-interaction settings and the genesis of new practices between institutional fields. *Academy of Management Review*, 39(4), 439–462.

Galinsky, A. D. & Moskowitz, G. B. (2000). Perspective-taking: Decreasing stereotype expression, stereotype accessibility, and in-group favoritism. *Journal of Personality and Social Psychology*, 78, 708–724.

Gallupe, R. B., Cooper, W. H., Grise, M. L. & Bastianutti, L. (1994). Blocking electronic brainstorms. *Journal of Applied Psychology*, 79(1), 77–86.

Ganna, A., Verweij, K. J., Nivard, M. G., Maier, R., Wedow, R., Busch, A. S., ... & Zietsch, B. P. (2021). Response to Comment on "Large-scale GWAS reveals insights into the genetic architecture of same-sex sexual behavior". *Science*, 371(6536), eaba5693.

Garud, R. & Gehman, J. (2012). Metatheoretical perspectives on sustainability journeys: Evolutionary, relational and durational. *Research Policy*, 41, 980–995.

Garud, R. & Karnøe, P. (2003). Bricolage versus breakthrough: Distributed and embedded Agency in technology entrepreneurship. *Research Policy*, 32, 277–300.

Garud, R., Tuertscher, P. & van de Ven, A. H. (2013). Perspectives on innovation processes. *Academy of Management Annals*, 7(1), 775–819.

Gebauer, J., Füller, J. & Pezzei, R. (2013). The dark and the bright side of co-creation: Triggers of member behavior in online innovation communities. *Journal of Business Research*, 66 (9), 1516–1527.

Geiger, D. & Schader, M. (2014). Personalized task recommendation in crowdsourcing information systems-Current state of the art. *Decision Support Systems*, 65, 3–16.

George, G., Howard-Grenville, J. & Joshi, A. (2016). Understanding and tackling societal grand challenges through management research. *Academy of Management Journal*, 59, 1880–1895.

Geuna, A. & Muscio, A. (2009). The governance of university knowledge transfer: A critical review of the literature. *Minerva*, 47(1), 93–114.

Gioia, D. A. & Chittipeddi, K. (1991). Sensemaking and sensegiving in strategic change initiation. *Strategic Management Journal*, 12(6), 433–448.

Gioia, D. A., Thomas, J. B., Clark, S. M. & Chittipeddi, K. (1994). Symbolism and strategic change in academia: The dynamics of sensemaking and influence. *Organization Science*, 5(3), 363–383.

Grant, A. M. & Berry, J. W. (2011). The necessity of others is the mother of invention: Intrinsic and prosocial motivations, perspective taking, and creativity. *Academy of Management Journal*, 54(1), 73–96.

Gray, J. (1999). A bias toward short-term thinking in threat-related negative emotional states. *Personality and Social Psychology Bulletin*, 25(1), 65–75.

Groulx, M., Brisbois, M. C., Lemieux, C. J., Winegardner, A. & Fishback, L. (2017). A role for nature-based citizen science in promoting individual and collective climate change action? A systematic review of learning outcomes. *Science Communication*, 39(1), 45–76.

Gruber, M., MacMillan, I. C. & Thompson, J. D. (2013). Escaping the prior knowledge corridor: What shapes the number and variety of market opportunities identified before market entry of technology start-ups? *Organization Science*, 24(1), 280–300.

Gulati, R, Wohlgezogen, F. & Zhelyazkov, P. (2012). The two facets of collaboration: Cooperation and coordination in strategic alliances. *Academy of Management Annals*, 6(1), 531–583.

Haefliger, S., Monteiro, E., Foray, D. & Von Krogh, G. (2011). Social software and strategy. *Long Range Planning*, 44 (5), 297–316.

Hage, M., Leroy, P. & Petersen, A. C. (2010). Stakeholder participation in environmental knowledge production. *Futures*, 42, 254–264.

Hargadon, A. B. & Bechky, B. A. (2006). When collections of creatives become creative collectives: A field study of problem solving at work. *Organization Science*, 17, 484–500.

Haughton, G. & Hunter, C. (2003). *Sustainable Cities*. Taylor & Francis. New York.

Hautz, J. (2017). Opening up the strategy process—A network perspective. *Management Decision*, 55(9), 1956–1983.

Hautz, J., Seidl, D. & Whittington, R. (2017). Open strategy: Dimensions, dilemmas, dynamics. *Long Range Planning*, 50(3), 298–309.

Havinga, I., Bogaart, P. W., Hein, L. & Tuia, D. (2020). Defining and spatially modelling cultural ecosystem services using crowdsourced data. *Ecosystem Services*, 43, 101091.

Hayhow, D., Eaton, M., Stanbury, A., Burns, F., Kirby, W., Bailey, N., ... & Symes, N. (2019). State of Nature Report 2019. Retrieved from https://nbn.org.uk/wp-content/uploads/2019/09/State-of-Nature-2019-UK-full-report.pdf.

Head, B.W. (2019). Forty years of wicked problems literature: Forging closer links to policy studies. *Policy and Society*, 38, 180–197.

Heck, R., Vuculescu, O., Sørensen, J. J., Zoller, J., Andreasen, M. G., Bason, M. G., Ejlertsen, P., Elíasson, O., Haikka, P., Laustsen, J. S. & Nielsen, L. L. (2018). Remote optimization of an ultracold atoms experiment by experts and citizen scientists. *Proceedings of the National Academy of Sciences*, 115 (48), E11231–E11237.

Hecker, S., Wicke, N., Haklay, M. & Bonn, A. (2019). How does policy conceptualise citizen science? A qualitative content analysis of international policy documents. *Citizen Science: Theory and Practice*, 4(1), 32, 1–16.

Heinström, J. (2003). Five personality dimensions and their influence on information behaviour. *Information Research*, 9(1), 1–24.

Heiss, R. & Matthes, J. (2017). Citizen science in the social sciences: A call for more evidence. *GAIA - Ecological Perspectives for Science and Society*, 26, 22–26.

Henderson, R. (1994). Managing innovation in the information age. *Harvard Business Review*, 72(1), 100–105.

Heradio, R., L. De La Torre, D. Galan, F. J. Cabrerizo, E. Herrera-Viedma & S. Dormido. (2016). Virtual and remote labs in education: A bibliometric analysis. *Computers & Education*, 98(6), 14–38.

Hilbolling, S., Deken, F., Berends, H. & Tuertscher, P. (2022). Process-based temporal coordination in multiparty collaboration for societal challenges. *Strategic Organization*, 20(1), 135–163.

Hirak, R., Pang, A. C., Carmeli, A. & Schaubroeck, J. M. (2012). Linking leader inclusiveness to work unit performance: The importance of psychological safety and learning from failures. *The Leadership Quarterly*, 23, 107–117.

Hochachka, W. M. & Dhondt, A. A. (2000). Density-dependent decline of host abundance resulting from a new infectious disease. *Proceedings of the National Academy of Sciences*, 97(10), 5303–5306.

Holden, E. & Linnerud, K. (2007). The sustainable development area: Satisfying basic needs and safeguarding ecological sustainability. *Sustainable Development*, 15, 174–187.

Holdren, J. (2015). *Memorandum to the Heads of Executive Departments and Agencies: Addressing Societal and Scientific Challenges through Citizen Science and Crowdsourcing*. White House Office of Science and Technology Policy.

Holste, J. S. & Fields, D. (2010). Trust and tacit knowledge sharing and use. *Journal of Knowledge Management*, 14(1), 128–140.

Hopp, C. & Greene, F.J. (2017). In pursuit of time: Business plan sequencing, duration and intraentrainment effects on new venture viability. *Journal of Management Studies*, 55(2), 320–351.

Howe, J. (2006). The rise of crowdsourcing. *Wired Magazine*, 14(6), 1–4.

Hudson, B. (2018). Citizen accountability in the 'New NHS' in England. *Critical Social Policy*, 38(2), 418–427.

Jackson, S. J., Ribes, D., Buyuktur, A. & Bowker, G. C. (2011). Collaborative rhythm: Temporal dissonance and alignment in collaborative scientific work. In *Proceedings of the ACM 2011 Conference on Computer Supported Cooperative Work*, pp. 245–254.

Jameson, S., Lämmerhirt, D. & Prasetyo, E. (2017). *Acting Locally, Monitoring Globally? How to Link Citizen-Generated Data to SDG Monitoring*. DataShift, Open Knowledge International: Cambridge, UK.

Jehn, K. A. (1995). A multimethod examination of the benefits and detriments of intragroup conflict. *Administrative Science Quarterly*, 256–282.

Jehn, K. A. (1997). A qualitative analysis of conflict types and dimensions in organizational groups. *Administrative Science Quarterly*, 530–557.

Jehn, K. A., Chadwick, C. & Thatcher, S. M. (1997). To agree or not to agree: The effects of value congruence, individual demographic dissimilarity, and conflict on workgroup outcomes. *International Journal of Conflict Management*, 8, 287–306.

Jeppesen, L.B. & Lakhani, K.R. (2010). Marginality and problem-solving effectiveness in broadcast search. *Organization Science*, 21(5), 1016–1033.

Johnson, K. A., Dana, G., Jordan, N. R., Draeger, K. J., Kapuscinski, A., Olabisi, L. K. S. & Reich, P. B. (2012). Using participatory scenarios to stimulate social learning for collaborative sustainable development. *Ecology and Society*, 17(2).

Jones, Q. (2003). Applying cyber-archeology. In *Proceedings of Computer-Supported Cooperative Work*, Dordrecht: Kluwer.

Jones, R., Colusso, L., Reinecke, K. & Hsieh, G. (2019). r/science: Challenges and opportunities in online science communication. In *Proceedings of the 2019 CHI Conference on Human Factors in Computing Systems* (pp. 1–14).

Kallio, T. J., Nordberg, P. & Ahonen, A. (2007). Rationalizing sustainable development - A critical treatise. *Sustainable Development*, 15, 41–51.

Kammermann, L. & Dermont, C. (2018). How beliefs of the political elite and citizens on climate change influence support for Swiss energy transition policy. *Energy Research & Social Science*, 43, 48–60.

Kannangara, S. N. & Uguccioni, P. (2013). Risk management in crowdsourcing-based business ecosystems. *Technology Innovation Management Review*, 3(12), 32–38.

Karnøe, P. & Garud, R. (2012). Path creation: Co-creation of heterogeneous resources in the emergence of the Danish wind turbine cluster. *European Planning Studies*, 20(5), 733–752.

Karnøe, P. (1990). Technological innovation and industrial organization in the Danish wind industry. *Entrepreneurship & Regional Development*, 2(2), 105–124.

Keirstcad, J. & Leach, M. (2008). Bridging the gaps between theory and practice: A service niche approach to urban sustainability indicators. *Sustainable Development*, 16, 329–340.

Kelly, D. & Amburgey, T. L. (1991). Organizational inertia and momentum: A dynamic model of strategic change. *Academy of Management Journal*, 34(3), 591–612.

Kenney, M. & Goe, W. R. (2004). The role of social embeddedness in professorial entrepreneurship: A comparison of electrical engineering and computer science at UC Berkeley and Stanford. *Research Policy*, 33(5), 691–707.

Khavul, S., Pérez-Nordtvedt, L. & Wood, E. (2010). Organizational entrainment and international new ventures from emerging markets. *Journal of Business Venturing*, 25(1), 104–119.

King, A. & Lakhani, K. R. (2013). *Using Open Innovation to Identify the Best Ideas*. MIT Press. Cambridge, MA.

King, A. C., Winter, S. J., Chrisinger, B. W., Hua, J. & Banchoff, A. W. (2019). Maximizing the promise of citizen science to advance health and prevent disease. *Preventive Medicine*, 119, 44.

Kittur, A., Lee, B. & Kraut, R.E. (2009). Coordination in collective intelligence: The role of team structure and task interdependence. In *Proceedings of the SIGCHI Conference on Human Factors in Computing Systems* (pp. 1495–1504). ACM.

Kittur, A., Suh, B., Pendleton, B. A. & Chi, E. H. (2007). He says, she says: Conflict and coordination in Wikipedia. In *Proceedings of the SIGCHI Conference on Human Factors in Computing Systems* (pp. 453–462). ACM.

Klopp, J. M. & Petretta, D. L. (2017). The urban sustainable development goal: Indicators, complexity and the politics of measuring cities. *Cities*, 63, 92–97.

Kraut, R., Egido, C. & Galegher, J. (1988). Patterns of contact and communication in scientific research collaboration. In *Proceedings of the 1988 ACM conference on Computer-Supported Cooperative Work, Portland, Oregon*.

Kuhlmann, S. & Rip, A. (2018). Next-generation innovation policy and grand challenges. *Science and Public Policy*, 45(4), 448–454.

Kukutai, T. & J. Taylor. (2016). *Indigenous Data Sovereignty: Toward an Agenda* (Vol. 38). Australia: Anu Press.

Kullenberg, C. & Kasperowski, D. (2016). What is citizen science? - A scientometric meta-analysis. *PLoS One*, 11, e0147152.

Kythreotis, A. P., Mantyka-Pringle, C., Mercer, T. G., Whitmarsh, L. E., Corner, A., Paavola, J., … & Castree, N. (2019). Citizen social science for more integrative and effective climate action: A science-policy perspective. *Frontiers in Environmental Science*, 7, 10.

Lakhani, K. & Wolf, R.G. (2007). Why hackers do what they do: Understanding motivation and effort in free/open source software projects. In *Perspectives on Free and Open Source Software* (pp. 3–21). MIT Press: Cambridge, MA.

Lakhani, K. R., Jeppesen, L. B., Lohse, P. A. & Panetta, J. A. (2007). *The Value of Openess in Scientific Problem Solving* (pp. 7–50). Boston, MA: Division of Research, Harvard Business School.

Lakhani, K. R., Lifshitz-Assaf, H. & Tushman, M. L. (2013). Open innovation and organizational boundaries: task decomposition, knowledge distribution and the locus of innovation. In Grandori, A. (Ed.), *Handbook of Economic Organization* (pp. 355–382). Cheltenham, UK: Edward Elgar Publishing.

Lampel, J. & Bhalla, A. (2007). The role of status seeking in online communities: Giving the gift of experience. *Journal of Computer-Mediated Communication*, 12(2), 434–455.

Larsen, K., Gunnarsson-Ostling, U. & Westholm, E. (2011). Environmental scenarios and local-global level of community engagement: Environmental justice, jams, institutions and innovation. *Futures*, 43, 413–423.

Latour, B. (1999). *Pandora's Hope*. Cambridge: Harvard University Press.

Lee, H. & Choi, B. (2003). Knowledge management enablers, processes, and organizational performance: An integrative view and empirical examination. *Journal of Management Information Systems*, 20(1), 179–228.

Lee, H. (2005). Behavioral strategies for dealing with flaming in an online forum. *The Sociological Quarterly*, 46(2), 385–403.

Leifer, E. M. (1991). *Actors as Observers: A Theory of Skill in Social Relationships*. New York: Garland.

Leighninger, M. (2011). Citizenship and governance in a wild, wired world: How should citizens and public managers use online tools to improve democracy? *National Civic Review*, 100(2), 20–30.

Lemke, T., Casper, M. J. & Moore, L. J. (2011). *Biopolitics: An Advanced Introduction*. NYU Press. New York.

Leonardi, P. M. (2014). Social media, knowledge sharing, and innovation: Toward a theory of communication visibility. *Information Systems Research*, 25(4), 796–816.

Lidskog, R. (2008). Scientised citizens and democratised science. Re-assessing the expert-lay divide. *Journal of Risk Research*, 11(1–2), 69–86.

Lindkvist, L., Soderlund, J. & Tell, F. (1998). Managing product development projects: On the significance of fountains and deadlines. *Organization Studies*, 19(6), 931–951.

Linek, S. B., Fecher, B., Friesike, S. & Hebing, M. (2017). Data sharing as social dilemma: Influence of the researcher's personality. *PLoS One*, 12(8), e0183216.

Llopis, O. & D'Este, P. (2016). Beneficiary contact and innovation: The relation between contact with patients and medical innovation under different institutional logics. *Research Policy*, 45(8), 1512–1523.

Loebbecke, C., van Fenema, P. C. & Powell, P. (2016). Managing inter-organizational knowledge sharing. *Journal of Strategic Information Systems*, 25, 4–14.

Lopez-Vega, H., Tell, F. & Vanhaverbeke, W. (2016). Where and how to search? Search paths in open innovation. *Research Policy*, 45 (1), 125–136.

Lu, Y. L., Nakicenovic, N., Visbeck, M. & Stevance, A. S. (2015). Five priorities for the UN sustainable development goals. *Nature*, 520, 432–433.

Lundin, R. A. & Söderholm, A. (1995). A theory of the temporary organization. *Scandinavian Journal of Management*, 11(4), 437–455.

Maboudi, T. & Nadi, G. P. (2016). Crowdsourcing the Egyptian constitution: Social media, elites, and the populace. *Political Research Quarterly*, 69(4), 716–731.

Mair, J. & Hehenberger, L. (2014). Front-stage and backstage convening: The transition from opposition to mutualistic coexistence in organizational philanthropy. *Academy of Management Journal*, 57(4), 1174–1200.

Majchrzak, A., More, P. H. & Faraj, S. (2012). Transcending knowledge differences in cross-functional teams. *Organization Science*, 23(4), 951–970.

Malhotra, A. & Majchrzak, A. (2019). Greater associative knowledge variety in crowdsourcing platforms leads to generation of novel solutions by crowds. *Journal of Knowledge Management*, 23(8), 1628–1651.

Malhotra, A., Majchrzak, A. & Niemiec, R. M. (2017). Using public crowds for open strategy formulation: Mitigating the risks of knowledge gaps. *Long Range Planning*, 50(3), 397–410.

Manning, S. (2008). Embedding projects in multiple contexts-A structuration perspective. *International Journal of Project Management*, 26(1), 30–37.

Marabelli, M. & Newell, S. (2012). Knowledge risks in organizational networks: The practice perspective. *Journal of Strategic Information Systems*, 21(1), 18–30.

Marjanovic, S., Fry, C. & Chataway, J. (2012). Crowdsourcing based business models: In search of evidence for innovation 2.0. *Science and Public Policy*, 39(3), 318–332.

Marres, N. & Rogers, R. (2005). Recipe for Tracing the Fate of Issues and their Publics on the Web. https://research.gold.ac.uk/id/eprint/6548/1/Marres_05_Rogers_recipe_copy.pdf

Mazzucato, M. (2018). Mission-oriented innovation policies: Challenges and opportunities. *Industrial and Corporate Change*, 27(5), 803–815.

McGahan, A. M., Bogers, M, Chesbrough, H. & Holgersson, M. (2021). Tackling societal challenges with open innovation. *California Management Review*, 63(2), 49–61.

Mergel, I. & Desouza, K.C. (2013). Implementing open innovation in the public sector: The case of Challenge. gov. *Public Administration Review*, 73(6), 882–890.

Merton, R. K. (1968). The Matthew effect in science. *Science*, 159(3810), 56–63.

Miller, R. C. (1982). Varieties of interdisciplinary approaches in the social sciences: A 1981 overview. *Issues in Interdisciplinary Studies*, 1(1), 1–37.

Mintzberg, H. (1994). *The Rise and Fall of Strategic Planning: Reconceiving Roles for Planning, Plans, Planners*. New York: Free Press.

Mote, J., Jordan, G., Hage, J., Hadden, W. & Clark, A. (2016). Too big to innovate? Exploring organizational size and innovation processes in scientific research. *Science and Public Policy*, 43(3), 332–337.

Mu, S. H. & Gnyawali, D. R. (2003). Developing synergistic knowledge in student groups. *Journal of Higher Education*, 74, 689–711.

Mulkay, M., Potter, J. & Yearley, S. (1983). Why an analysis of scientific discourse is needed. In Knorr-Cetina, K. D. & Mulkay, M. (Eds.), *Science Observed*. London, UK: Sage Publications.

Munasinghe, M. (2009). *Sustainable Development in Practice: Sustainomics Methodology and Applications*. London: Cambridge University Press.

Nemanich, L. A. & Vera, D. (2009). Transformational leadership and ambidexterity in the context of an acquisition. *The Leadership Quarterly*, 20, 19–33.

Nicolini, D. (2011). Practice as the site of knowing: Insights from the field of telemedicine. *Organization Science*, 22 (3), 602–620.

Nidumolu, S., Subramani, M. & Aldrich, A. (2001). Situated learning and the situated knowledge web: Exploring the ground beneath knowledge management. *Journal of Management Information Systems*, 18(1), 115–150.

Nisbet, M. C. (2009). Communicating climate change: Why frames matter for public engagement. *Environment: Science and Policy for Sustainable Development*, 51(2), 12–23.

Nonaka, I. & Takeuchi, H. (1995). *The Knowledge-Creating Company: How Japanese Companies Create the Dynamics of Innovation?* Oxford, UK: Oxford University Press.

Nonaka, I. (1994). A dynamic theory of organizational knowledge creation. *Organization Science*, 5(1), 14–37.

Nowotny, H., Scott, P. & Gibbons, M. (2006). Re-thinking science: Mode 2 in societal context. In Carayannis, E. G. & Campbell, D. F. J. (Eds.), *Knowledge Creation, Diffusion, and Use in Innovation Networks and Knowledge Clusters. A Comparative Systems Approach across the United States, Europe and Asia* (pp. 39–51). London: Praeger Publishers.

Nunamaker, J. F., Dennis, A. R., Valacich, J. S., Vogel, D. R. & George, J. F. (1991). Electronic meeting systems to support group work. *Communications of the ACM*, 34(7), 40–61.

Ny, H., MacDonald, J. P., Broman, G., Yamamoto, R. & Robért, K. H. (2006). Sustainability constraints as system boundaries: An approach to making life-cycle management strategic. *Journal of Industrial Ecology*, 10(1–2), 61–77.

O'Donnell, A. M. & Dansereau, D. F. (1992). Scripted cooperation in student dyads: A method for analyzing and enhancing academic learning and performance. In Hertz-Lazarowitz, R. & Miller, N. (Eds.). *Interaction in Cooperative Groups: The Theoretical Anatomy of Group Learning*, pp. 121–140. New York: Cambridge University Press.

O'Donnell, A.M. & O'Kelly, J. (1994). Learning from peers: Beyond the rhetoric of positive results. *Educational Psychology Review*, 6(4), 321–349.

Orlikowski, W.J. & Yates, J. (2002). It's about time: Temporal structuring in organizations. *Organization Science*, 13(6), 684–700.

Ortega, A., Sanchez-Manzanares, M., Gil, F. & Rico, R. (2010). Team learning and effectiveness in virtual project teams: The role of beliefs about interpersonal context. *Spanish Journal of Psychology*, 13, 267–276.

Ostrom, E. (1990). *Governing the Commons: The Evolution of Institutions for Collective Action*. New York: Cambridge University Press.

Ostrom, E. (2009). A general framework for analyzing sustainability of social-ecological systems. *Science*, 325(5939), 419–422.

Owen, R., Macnaghten, P. & Stilgoe, J. (2020). Responsible research and innovation: From science in society to science for society, with society. In Marchant, G. E., & Wallach, W. (Eds.) *Emerging Technologies: Ethics, Law and Governance* (pp. 117–126). Routledge: New York.

Padgett, J. F. & Ansell, C. K. (1993). Robust action and the rise of the Medici, 1400-1434. *American Journal of Sociology*, 98, 1259–1319.

Papacharissi, Z. (2004). Democracy online: Civility, politeness, and the democratic potential of online political discussion groups. *New Media & Society*, 6(2), 259–283.

Pappas, J. M. & Wooldridge, B. (2007). Middle managers' divergent strategic activity: An investigation of multiple measures of network centrality. *Journal of Management Studies*, 44(3), 323–341.

Parker, S. K. & Axtell, C. M. (2001). Seeing another viewpoint: Antecedents and outcomes of employee perspective taking. *Academy of Management Journal*, 44, 1085–1100.

Parker, S. K., Atkins, P. W. B. & Axtell, C. M. (2008). Building better work places through individual perspective taking: A fresh look at a fundamental human process. In Hodgkinson, G. & Ford, K. (Eds.), *International Review of Industrial and Organizational Psychology* (pp. 149–196). Chichester, England: Wiley.

Pecl, G. T., Stuart-Smith, J., Walsh, P., Bray, D. J., Kusetic, M., Burgess, M., ... & Moltschaniwskyj, N. (2019). Redmap Australia: challenges and successes with a large-scale citizen science-based approach to ecological monitoring and community engagement on climate change. *Frontiers in Marine Science*, 6, 349.

Pérez-Nordtvedt, L., Payne, G. T., Short, J. C. & Kedia, B. L. (2008). An entrainment-based model of temporal organizational fit, misfit, and performance. *Organization Science*, 19(5), 785–801.

Perkmann, M. & Walsh., K. (2008). Engaging the scholar: Three types of academic consulting and their impact on universities and industry. *Research Policy*, 37(10), 1884–1891.

Perkmann, M., Tartari, V., McKelvey, M., Autio, E., Broström, A., D'Este, P., Fini, R., Geuna, A., Grimaldi, R. & Hughes, A. (2013). Academic engagement and commercialisation: A review of the literature on university-industry relations. *Research Policy*, 42(2), 423–442.

Pfitzer, M., Bockstette, V. & Stamp, M. (2013). Innovating for shared value. *Harvard Business Review*, 91(9), 100–107.

Pinch, T.J. & Bijker, W.E. (1987). The social construction of facts and artifacts: Or how the sociology of science and the sociology of technology might benefit each other. In Bijker, W. E., Hughes, T. P. & Pinch, T. J. (Eds.), *Social Construction of Technological Systems* (pp. 17–50). Cambridge: MIT Press.

Pocock, M. J., Roy, H. E., Preston, C. D. & Roy, D. B. (2015). The Biological Records Centre: A pioneer of citizen science. *Biological Journal of the Linnean Society*, 115(3), 475–493.

Pocock, M.J., Chandler, M., Bonney, R., Thornhill, I., Albin, A., August, T., August, T., ... & Danielsen, F. (2018). A vision for global biodiversity monitoring with citizen science. In David, A., Bohan, A. J. D., Woodward, G. & Jackson, M. (Eds.) *Advances in Ecological Research* (Vol. 59, pp. 169–223). Cambridge, MA: Academic Press.

Porter, A. J., Tuertscher, P. & Huysman, M. (2020). Saving our oceans: Scaling the impact of robust action through crowdsourcing. *Journal of Management Studies*, 57(2), 246–286.

Potter, R. B., Binns, J. A., Elliott, J. A. & Smith, D. (2008). *Geographies of Development* (3rd ed.). Harlow: Pearson Education Limited.

Prestopnik, N. R. & Crowston, K. (2012). Citizen science system assemblages: Understanding the technologies that support crowdsourced science. In *Proceedings of the* 2012 iConference, pp. 168–176.

Prysby, M. & Oberhauser, K. (2004). Temporal and geographical variation in monarch densities: Citizen scientists document monarch population patterns. In Oberhauser, K. S. & Solensky, M. J. (Eds.), *The Monarch Butterfly: Biology and Conservation* (pp. 9–20). Cornell University Press, Ithaca, NY.

Purser, R. E. & Pasmore, W. A. (1992). Organizing for Learning. In Woodman, R. & Pasmore, W. A. (Eds.), *Research in Organizational Change and Development* (Vol. 6). Greenwich, CT: JAI Press.

Purser, R. E., Pasmore, W. A. & Tenkasi, R. V. (1992). The influence of deliberations on learning in new product development teams. *Journal of Engineering and Technology Management*, 9(1), 1–28.

Raasch, C., Herstatt, C. & Balka, K. (2009). On the open design of tangible goods. *R&D Management*, 39(4), 382–393.

Rafols, I. & Meyer, M. (2010). Diversity and network coherence as indicators of interdisciplinarity: Case studies in bionanoscience. *Scientometrics*, 82(2), 263–287.

Ratcliffe, J. & Krawczyk, E. (2011). Imagineering city futures: The use of prospective through scenarios in urban planning. *Futures*, 43, 642–653.

Reddy, M. C., Dourish, P. & Pratt, W. (2006). Temporality in medical work: Time also matters. *Proceedings of Computer Supported Cooperative Work (CSCW)*, 15(1), 29–53.

Resnik, D.B. (2019). Citizen scientists as human subjects: Ethical issues. *Citizen Science: Theory and Practice*, 4(1–11), 1–7,

Shirk, J. L., Ballard, H. L., Wilderman, C. C., Phillips, T., Wiggins, A., Jordan, R., ... & Bonney, R. (2012). Public participation in scientific research: a framework for deliberate design. *Ecology and Society*, 17(2): 29. http://dx.doi.org/10.5751/ES-04705-170229

Robèrt, K. H., Schmidt-Bleek, B., De Larderel, J. A., Basile, G., Jansen, J. L., Kuehr, R., ... & Wackernagel, M. (2002). Strategic sustainable development-selection, design and synergies of applied tools. *Journal of Cleaner Production*, 10(3), 197–214.

Robèrt, K.-H., Basile, G., Broman, G., Byggeth, S., Cook, D., ... & Missimer, M. (2005). *Strategic Leadership towards Sustainability* (2nd ed.). Karlskrona, Sweden: Blekinge Institute of Technology.

Robèrt, K.-H., Broman, G. I. & Basile, G. (2013). Analyzing the concept of planetary boundaries from a strategic sustainability perspective: How does humanity avoid tipping the planet? *Ecology and Society*, 18 (2), 80–88.

Robichaud, D. (2016). Wikipedia edit-a-thons: Thinking beyond the warm fuzzies. *Partnership: The Canadian Journal of Library and Information Practice and Research*, 11(2), 1–10.

Rodan, S. & Galunic, C. (2004). More than network structure: How knowledge heterogeneity influences managerial performance and innovativeness. *Strategic Management Journal*, 25(6), 541–562.

Rome, C. & Lucero, C. (2019). Wild carrot (*Daucus carota*) management in the Dungeness Valley, Washington, United States: The power of citizen scientists to leverage policy change. *Citizen Science: Theory and Practice*, 4(1), 36.

Rosas, L. G. & King, A. C. (2021). The role of citizen science in promoting health equity. *Annual Review of Public Health*, 43, 215–234.

Ryan, E. J., Owen, S. D., Lawrence, J., Glavovic, B., Robichaux, L., Dickson, M., ... & Blackett, P. (2022). Formulating a 100-year strategy for managing coastal hazard risk in a changing climate: Lessons learned from Hawke's Bay, New Zealand. *Environmental Science & Policy*, 127, 1–11.

Ryan, R. M. & Deci, E. L. (2000). Intrinsic and extrinsic motivations: Classic definitions and new directions. *Contemporary Educational Psychology*, 25(1), 54–67.

Ryan, S. & O'Connor, R. V. (2013). Acquiring and sharing tacit knowledge in software development teams: An empirical study. *Information and Software Technology*, 55(9), 1614–1624.

Santoro, G., Vrontis, D., Thrassou, A. & Dezi, L. (2018). The Internet of Things: Building a knowledge management system for open innovation and knowledge management capacity. *Technological Forecasting and Social Change*, 136, 347–354.

Sauermann, H., Vohland, K., Antoniou, V., Balázs, B., Göbel, C., Karatzas, K., Mooney, P., Perello, J., Ponti, M, Samson, R. & Winter, S. (2020). Citizen science and sustainability transitions. *Research Policy*, 49(5), 103978.

Sawyer, R. K. (2003). *Improvised Dialogues*. Westport, CT: Ablex Publishing.

Scheliga, K., Friesike, S., Puschmann, C. & Fecher, B. (2018). Setting up crowd science projects. *Public Understanding of Science*, 27(5), 515–534.

Schulte, M., Cohen, N. A. & Klein, K. J. (2012). The coevolution of network ties and perceptions of team psychological safety. *Organization Science*, 23, 564–581.

See, L., Mooney, P., Foody, G., Bastin, L., Comber, A., Estima, J., ... & Rutzinger, M. (2016). Crowdsourcing, citizen science or volunteered geographic information? The current state of crowdsourced geographic information. *ISPRS International Journal of Geo-Information*, 5(5), 55.

Senker, J. (1995). Tacit knowledge and models of innovation. *Industrial and Corporate Change*, 4(2), 425–447.

Sessa, V. I. (1996). Using perspective taking to manage conflict and affect in teams. *Journal of Applied Behavioral Science*, 32, 101–115.

Shamoo, A. E. & Resnik, D. B. (2015). *Responsible Conduct of Research* (3rd ed). New York: Oxford University Press.

Shane, S. A. (2004). *Academic Entrepreneurship: University Spinoffs and Wealth Creation*. Cheltenham, UK: Edward Elgar Publishing.

Shirk, J. L., Ballard, H. L., Wilderman, C. C., Phillips, T, Wiggins, A., Jordan, R., McCallie, E., Minarchek, M., Lewenstein, B. V., Krasny, M. E. & Bonney, R. (2012). Public participation in scientific research: A framework for deliberate design. *Ecology and Society*, 17(2), 29.

Siemsen, E., Roth, A. V., Balasubramanian, S. & Anand, G. (2009). The influence of psychological safety and confidence in knowledge on employee knowledge sharing. *Manufacturing and Service Operations Management*, 11, 429–447.

Silvertown, J. (2009). A new dawn for citizen science. *Trends in Ecology & Evolution*, 24(9), 467–471.

Simon, H. A. (1996). *The Sciences of the Artificial* (3rd ed.). MIT Press. Cambridge, MA.

Simpson, R., Page, K. R. & De Roure, D. (2014). Zooniverse: Observing the world's largest citizen science platform. In *Proceedings of the 23rd International Conference on World Wide Web* (pp. 1049–1054).

Slawinski, N. & Bansal, P. (2015). Short on time: Intertemporal tensions in business sustainability. *Organization Science*, 26(2), 531–549.

Smith, E., Bélisle-Pipon, J. C. & Resnik, D. (2019). Patients as research partners; how to value their perceptions, contribution and labor? *Citizen Science: Theory and Practice*, 4(1–8), 1–9.

Sneddon, C., Howarth, R. B. & Norgaard, R. B. (2006). Sustainable development in a post-Brundtland world. *Ecological Economics*, 57(2), 253–268.

Standifer, R. L. & Bluedorn, A. C. (2006). Alliance management teams and entrainment: Sharing temporal mental models. *Human Relations*, 59(7), 903–927.

Star, S. L. & Griesemer, J. R. (1989). Institutional ecology, translations and boundary objects: Amateurs and professionals in Berkeley's Museum of Vertebrate Zoology, 1907-39. *Social Studies of Science*, 19, 387–420.

Stark, D. (2009). *The Sense of Dissonance*. Princeton: Princeton University Press.

Stedman, R., Lee, B., Brasier, K., Weigle, J. L. & Higdon, F. (2009). Cleaning up water? Or building rural community? Community watershed organizations in Pennsylvania. *Rural Sociology*, 74(2), 178–200.

Stenmark, D. (2000). Leveraging tacit organisational knowledge. *Journal of Management Information Systems*, 17(3), 9–24.

Stenmark, D. (2002). Sharing tacit knowledge: A case study at Volvo. In Barnes, S. (Ed.), *Knowledge Management Systems: Theory and Practice*. London, UK: Thomson Learning.

Stephen, B. (2018). Reddit updates its quarantine policy with an appeals process. Retrieved from https://www.theverge.com/2018/9/28/17914240/reddit-update-quarantine-policy-appeals-process.

Stieger, D., Matzler, K., Chatterjee, S. & Ladstaetter-Fussenegger, F. (2012). Democratizing strategy: How crowdsourcing can be used for strategy dialogues. *California Management Review*, 54(4), 44–68.

Street, P. (1997). Scenario workshops: A participatory approach to sustainable urban living? *Futures*, 29, 139–158.

Stren, R. E., White, R. R. & Whitney, J. B. (1992). *Sustainable Cities: Urbanization and the Environment in International Perspective*. Westview Press. Boulder, CO.

Sun, Y., Majchrzak, A. & Malhotra, A. (2021). Serial integration, real innovation: Roles of diverse knowledge and communicative participation in crowdsourcing. In *Proceedings of the 54th Hawaii International Conference on System Sciences* (pp. 4921–4929).

Sun, Y., Tuertscher, P., Majchrzak, A. & Malhotra, A. (2020), Pro-socially motivated interaction for knowledge integration in crowd-based open innovation. *Journal of Knowledge Management*, 24(9), 2127–2147.

Surowiecki, J. (2005). *The Wisdom of Crowds*. New York: Anchor.

Tartari, V., Perkmann, M. & Salter, A. (2014). In good company: The influence of peers on industry engagement by academic scientists. *Research Policy*, 43(7), 1189–1203.

Teasley, S. & Wolinsky, S. (2001). Scientific collaborations at a distance. *Science*, 292(5525), 2254–2255.

Thompson, D. C. & Bentzien, J. (2020). Crowdsourcing and open innovation in drug discovery: Recent contributions and future directions. *Drug Discovery Today*, 25(12), 2284–2293.

Tinati, R., Luczak-Roesch, M., Simperl, E. & Hall, W. (2017). An investigation of player motivations in eyewire, a gamified citizen science project. *Computers in Human Behavior*, 73, 527–540.

Titus, P. A. (2000). Marketing and the creative problem-solving process. *Journal of Marketing Education*, 22, 225–235.

Tonn, B. E. (2007). Futures sustainability. *Futures*, 39, 1097–1116.

Trkman, P. & Desouza, K.C. (2012). Knowledge risks in organizational networks: An exploratory framework. *Journal of Strategic Information Systems*, 21(1), 1–17.

Trommsdorff, G. (1983). Future orientation and socialization. *International Journal of Psychology*, 18(1–4), 381–406.

Tsai, K. C. (2012). Play, imagination, and creativity: A brief literature review. *Journal of Education and Learning*, 1(2), 15–20.

Tsoukas, H. (2009). A dialogical approach to the creation of new knowledge in organizations. *Organization Science*, 20(6), 941–957.

Tucci, C. L., Afuah, A. & Viscusi, G. (Eds.). (2018). *Creating and Capturing Value through Crowdsourcing*. Oxford University Press. Oxford, United Kingdom.

Tuertscher, P., Garud, R. & Kumaraswamy, A. (2014). Justification and interlaced knowledge at ATLAS, CERN. *Organization Science*, 25(6), 1579–1608.

Turbé, A., Barba, J., Pelacho, M., Mudgal, S., Robinson, L.D., Serrano-Sanz, F., Sanz, F., Tsinaraki, C., Rubio, J.-M. & Schade, S. (2020). Correction: Understanding the citizen science landscape for European environmental policy: An assessment and recommendations. *Citizen Science: Theory and Practice*, 5(1), 5.

Tyran, C. K., Dennis, A. R., Vogel, D. R. & Nunamaker, J. F. (1992). The application of electronic meeting technology to support strategic management. *MIS Quarterly*, 16(3), 313–334.

United Nations (2015). A/RES/70/1 UN general assembly transforming our world: The 2030 agenda for sustainable development. *Seventieth session of the General Assembly*.

United Nations (2017). *Goal 17: Revitalize the Global Partnership for Sustainable Development*. https://www.un.org/sustainabledevelopment/globalpartnerships/.

United Nations (2019). *Measuring Progress: Towards Achieving the Environmental Dimension of the SDGs*. https://wedocs.unep.org/handle/20.500.11822/27627.

Valacich, J. S., Dennis, A. R. & Connolly, T. (1994). Idea generation in computer-based groups: A new ending to an old story. *Organizational Behavior and Human Decision Processes*, 57(3), 448–467.

Van den Homberg, M. & Susha, I. (2018). Characterizing data ecosystems to support official statistics with open mapping data for reporting on sustainable development goals. *ISPRS International Journal of Geo-Information*, 7(12), 456.

Van Noorden, R. (2015). Interdisciplinary research by the numbers. *Nature*, 525(7569), 306–307.

Vesnic-Alujevic, L., Breitegger, M. & Guimarães Pereira, Â. (2018). 'Do-it-yourself' healthcare? Quality of health and healthcare through wearable sensors. *Science and Engineering Ethics*, 24(3), 887–904.

Von Hippel, E. (2005). *Democratizing Innovation*. MIT Press. Cambridge, MA

Wagner, C., Graells-Garrido, E., Garcia, D. & Menczer, F. (2016). Women through the glass ceiling: Gender asymmetries in Wikipedia. *EPJ Data Science*, 5, 1–24.

Waldrop, M. M. (2013). Education online: The virtual lab. *Nature News*, 499(7458), 268–270.

Walumbwa, F. O. & Schaubroeck, J. (2009). Leader personality traits and employee voice behavior: Mediating roles of ethical leadership and work group psychological safety. *Journal of Applied Psychology*, 94, 1275–1286.

Webb, N. M. (1989). Peer interaction and learning in small groups. *International Journal of Educational Research*, 13(1), 21–39.

Webb, N. M. (1991). Task-related verbal interaction and mathematics learning in small groups. *Journal for Research in Mathematics Education*, 22(5), 366–389.

Weick, K. E. (1984). Small wins: Redefining the scale of social problems. *American Psychologist*, 39(1), 40–49.

Weir, D., McQuillan, D. & Francis, R. A. (2019). Civilian science: The potential of participatory environmental monitoring in areas affected by armed conflicts. *Environmental Monitoring and Assessment*, 191(10), 1–17.

West, S. & Pateman, R. (2017). *How Could Citizen Science Support the Sustainable Development Goals?* Stockholm, Sweden: Stockholm Environment Institute. Available online: https://mediamanager.sei.org/documents/Publications/SEI-2017-PB-citizen-science-sdgs.pdf.

West, S. E., Büker, P., Ashmore, M., Njoroge, G., Welden, N., Muhoza, C., Osano, P., Makau, J., Njoroge, P. & Apondo, W. (2020). Particulate matter pollution in an informal settlement in Nairobi: Using citizen science to make the invisible visible. *Applied Geography*, 114, 102133.

Whittaker, S., Jones, Q., Nardi, B., Terveen, L., Creech, M., Isaacs, E. & Hainsworth, J. (2002). ContactMap: Using personal social networks to organize communication in a social desktop. In *Proceedings of ACM Conference on Computer-Supported Cooperative Work (CSCW)*. New York: ACM.

Whittington R, Basak-Yakis, B & Cailluet L. (2011). Opening strategy: Evolution of a precarious profession. *British Journal of Management*, 22(3), 531–544.

Wiggins, A. & Crowston, K. (2011). From conservation to crowdsourcing: A typology of citizen science. In *Proceedings of 44th Hawaii International Conference on System Sciences* (pp. 1–10). IEEE.

Wiggins, A. & Wilbanks, J. (2019). The rise of citizen science in health and biomedical research. *The American Journal of Bioethics*, 19(8), 3–14.

Wilderman, C. C., McEver, C., Bonney, R., Dickinson, J., Kelling, S. & Rosenberg, K. (2007, June). Models of community science: Design lessons from the field. In McEver, C., Bonney, R., Dickinson, J., Kelling, S., Rosenberg, K. & Shirk, J. L. (Eds.), *Citizen Science Toolkit Conference* (Vol. 1, No. 2, pp. 1–3). Ithaca, NY: Cornell Laboratory of Ornithology.

Willett, W., Heer, J. & Agrawala, M. (2012, May). Strategies for crowdsourcing social data analysis. In *Proceedings of the SIGCHI Conference on Human Factors in Computing Systems* (pp. 227–236).

Wittrock, M. C. (1986). Student thought processes. In Wittrock, M. C. (Ed.), *Handbook of Research on Teaching* (3rd ed., pp. 297–314). New York: Macmillan.

World Commission on Environment and Development (1987). World commission on environment and development. *Our Common Future*, 17(1), 1–91.

Wuchty, S., Jones, B. F. & Uzzi, B. (2007). The increasing dominance of teams in production of knowledge. *Science*, 316 (5827), 1036–1039.

Wylie, S. A., Jalbert, K., Dosemagen, S. & Ratto, M. (2014). Institutions for civic technoscience: How critical making is transforming environmental research. *The Information Society*, 30(2), 116–126.

Yeh, Y., Lai, S. & Ho, C. (2006). Knowledge management enablers: A case study. *Industrial Management & Data Systems*, 106(6), 793–810.

Zerubavel, E. (1979). *Patterns of Time in Hospital Life. A Sociological Perspective*. Chicago, IL: University of Chicago Press.

Zucker, L. G., Darby, M. R. & Armstrong, J. S. (2002). Commercializing knowledge: University science, knowledge capture, and firm performance in biotechnology. *Management Science*, 48(1), 138–153.

Index

Note: **Bold** page numbers refer to tables, *italic* page numbers refer to figures and page numbers followed by "n" refer to end notes.

ability 270, 310, 313, 320
abstraction, low-level of **175**, 177–180
acceptance 305, 316–319, 321, 322
accessibility 5, 12, 24, 263, 285, 289
    artificial intelligence 126
    by user interface adaptation 49–51
accountability, responsible AI 115–116
ACTA rules 60, *61*, 66
acting devices 173–174
action 408–412, 415, 416, 418, 419, 421
action-oriented projects 409
activity recognition 68
actors 410–413, 416, 418, 419, **422**
Adam Smith's rational choice theory 84
adaptability 44
adaptable widget library, adaptation features of widgets in **56–58**
adaptation-based approach, adoption of 70–71
adaptation rule 50, 54, 59, 64–66, 71
adaptive component hierarchies 60, *61*
Adaptive Control of Thought-Rational (ACT-R) 283
adaptive multimodal human robot interaction 59–63
adaptive systems 3
adaptive UI prototyping *55*, 55–59, **56–58**, *59*
adaptivity 44
ADAS *see* advanced driver assistance systems (ADAS); automated driver assistance systems (ADAS)
ADAS & ME applications 66–67
Adobe Creative Cloud 164n1
ADS *see* automated driving system (ADS)
advanced driver assistance systems (ADAS), personalized interaction with 64–66, *65*
affect 269
affective trust 271, *272*
after-task questionnaire (ASQ) 25
age-friendly design 366
ageism 365
agents 333–355
agent transparency, H-A teaming *347,* 347–350
aging 364–366, 371, 372, *372,* 376
    attitudes 365
aging societies 364, 365
agreeableness 315, 322
AI *see* artificial intelligence (AI)
Alan Cooper's theory 143
Alan Turing 108
Alexa Skills 209
algorithmic bias 121, 127
algorithmic interventions 417
Algorithm Watch 257
Amazon GO 246
Amazon's Alexa 253–254
Amazon's decommissioned HR recruiting tool 253
ambient assisted living (AAL) 364–378
ambient assisted living (AAL) lifelogging

attitudes toward aging 372, *372*
    human-centered design, use and management of 376–378, **377**
    users' perspectives and empirical assessment of 371–376
ambient assisted living (AAL) technologies 364–365, 374–375
    environments and care situations 367, **367–368**
    perceived benefits and barriers of camera-based assistive 372–376, *375, 376*
    perception of privacy and trust toward medical technology 370–371
    social requirements for design and implementation of 365–366
    technological potential 365–366, **367**
    technology acceptance 368–369
ambient intelligence (AmI) 1, 3
Ambient Intelligence Programme of ICS-FORTH 1
AmI Design Process 1, 4, 10, *11,* 12, 15, 23, 27, 37
AmI RTD Programme of ICS-FORTH 4
AmI-Solertis system 27–28, *28*
AmITest 29
Amper Music 243
anonymity 79
antecedents of trust 310
anthropomorphic Cues 222, 225
anthropomorphism 255
anticipation theory 273, 274
apologies 314
a posteriori adaptation 47–49
Application Programming Interface (API) 29
applications
    artificial intelligence 110
    Internet of Things 172–174, *173*
    to in-vehicle context VR 287–288
applied 410, 412–414, 417
apps for health and well-being 366
AR *see* augmented reality (AR)
architecture 9, 28
areas of interest (AOIs) 280
ARgus Designer tool 31
ARgus Workstation 31
articulated work 205, 206
artificial environments 127
artificial intelligence (AI) 1, 68, 201, 384, 400; *see also* responsible AI
    accelerating capabilities 108–109, *109*
    accessibility 126
    along the customer journey 246, *246*
    applications 110
    assistants 147, 153, 157–160
    benefits of 142
    birth of 108
    capabilities to human well-being 111
    for customer services **245,** 245–246

439

artificial intelligence (AI) (cont.)
    DEEP-MAX scorecard rating for **132**
    as design collaborator 152–153, *153*
    develops as a field 108
    for enhancing UX 154–163
    ethical 118
    explainability 111, 127
    human–computer interaction develops as a field 109–110
    integration of AI and HCI design processes 124, *124*
    new design material 150–152
    as new design material 151
    problems with AI emerge 110–111
    recommender system 118, 120, 121, 125
    technical foundations of 242–243
    for UI design 149–150
    urban interfaces 400
    usability 126–127
    winter 109
Artificial Intelligence Act (AIA) 257
artificial intelligence-aided design (AIAD) technology 148
artificial intelligence (AI) algorithms 127
    interpretability of 119
artificial intelligence (AI)-based systems 241, 242
    challenges of 253–257
    for customer services 257–253
    overarching consequences resulting from risks 256–257
    overarching regulatory frameworks and prevention measures 256
    potential negative effects of 253–254
artificial intelligence (AI)-based systems for customer services, predominant risks of
    algorithmically enforced filter bubbles 254–255
    anthropomorphism 255
    discrimination 255
    insufficient service quality 255
    lack of transparency 254
    personalization 256
    privacy 256
    proactivity 256
artificial intelligence (AI)-enabled IAs 159
artificial intelligence (AI)-enabled RS 162
artificial intelligence (AI) in HCI/UX research and evaluation 141, **142,** 143
    benefits of **144**
    challenges and future work 153–154
    digital personas 143–145
    qualitative analysis 146–147
    UX evaluation 147–148
artificial intelligence (AI) technology 37, 126, 141, 163
    characteristics and human factors issues of **156**
    in UX evaluation 148
ASQ *see* after-task questionnaire (ASQ)
assessment 335, 342, 347, 350–352, **354**
assistive technologies 46–47, 50, 70
attacker
    tools and techniques 85
    types 81
attacker's motivation 81
    curiosity 83
    denunciations and revelation 83
    fake news shaping people's opinions 82
    financial gain 81–82
    profiting by deceit 83
    revenge 83
    sabotage, cyberwar, and intelligent gathering 82
    solving crimes 83
    terrorist organizations 83
attention, factors shaping vulnerable behaviors 87
attitudes 337, 342, 343, 350, **353**
attitudes towards aging 365
Attitudes toward Working with Robots scale (AWRO) 321
Attitude toward Technology Scale 321
audio augmented reality (AAR) 287
auditability **132**
auditory displays *267,* 283, 285, 288
augmented reality (AR) 9, 67, 263
    applications to in-vehicle context 287–288
authentication, cyberattacks 94
automated driver assistance systems (ADAS) 278
automated driving
    situation awareness in 268–269
    trust in 271–272
    workload in 267
automated driving system (ADS) 121, 264
automated vehicles (AVs) 318
automatic reauthoring 45
automatic speech recognition (ASR) module 210
automation 35, 171
    bias 127
    and efficiency 241
    of human tasks 241
    industrial **3**
automation system, driving 263
autonomous infrastructure 396, 398
autonomous squad member 349, *349*
    human-machine interfaces for 349, *349*
autonomous system 116, 118, 129
    deployment of 118
autonomous vehicles 289
autonomy 159, 334, 337–340, 352, 365
auto-regressive moving average vector (ARMAV) 312
    time series model of trust 312
awareness
    education and, cyber protection 80–81
    training and awareness program effectiveness 98–99

B2B customer services 258
B2C customer services 258
'bdpq NDRT' test 274, *274*
behavior 338–343, 345–346, 350–351, **353**
    protective **101–102**
    vulnerable **101–102**
behavioral-based cyber technologies 80
behavioral trust 309
behavior pattern-schema 89
benevolence 310, 314
BERT 210, 213
bias 110, 111, 116, 117, 119
    algorithmic 121, 127
    automation 127
    potential 126
Big Five personality traits 315
Bing 243
Bitcoin 79
blackmail 81, 92
blinking activity, drivers' attention 280

Index

Book of Ideas 33–34
brainstorming 17, 18, 34
    ideas for smart kitchen environment 18
bugs 171, 184, 185
business-to-business (B2B) 241
business-to-customer (B2C) services 241

CAMELEON reference framework 50
camera-based assistive AAL technologies 373
Cana Brave Resort, customer service chatbot of 216–217, *217*, 219, *222*
card sorting 207
care and quality of life, multidimensional requirements for 365
caring bottlenecks 364
CASA paradigm *see* Computers Are Social Actors (CASA) paradigm
causation theory 273
CCA *see* Cybersecurity Countermeasures Awareness (CCA)
challenges 408, 410–419, 421, **422,** 423
    Internet of Things 178–186, 190
Chatbots
    applications for customer services and associated benefits 251–252
    purpose and technical functioning 250–251
ChatGPT 163–164, 201, 229, 243, 346
Chatty Bench Project 389
Chaves and Gerosa's social cues 214
CHI *see* co-evolutionary hybrid intelligence (CHI)
cities 384–387, 389, 392, 393, 396, 400
citizen 408–423
citizen science 409–410, 420
    for sustainability 415–421
    sustainability through **422**
city apps 386
city scales 384
civic participation 385, 386, 391
CJA *see* customer journey analytics (CJA)
cloud services 173–174
CNN *see* convolutional neural network (CNN)
co-evolutionary hybrid intelligence (CHI) 158
cognition 342, 345, 346–350, 351, 352
cognitive abilities of conversational agents 213–214
cognitive elaboration 417
cognitive reappraisal 270
cognitive (cognition-based) trust 271, 272
cohesion 334, 342, 343–345, **353**
collaboration 4, 5, 9, 10, 29, 412–414, 416–419, 421, **422,** 423
    tools 27
collaborative 409, 410, 412–416, 418, 419, **422**
collaborative design assistance 152
collaborators 413, 415–420
collective 414–418, 420
combinatorial optimization approach 50, 51, 67
communicating with computers 202–203
    application domains 203–204
    conversational agents and HCI 204–205
    definitions and classification 203
communication 320, 343–345, 347–354, **353**
community 384, 385, 388–393, 395, 399–401
    engagement 385, 389–390
COMPAS *see* Correctional Offender Management Profiling for Alternative Sanctions (COMPAS)

complexities 415–421
computational modeling
    drivers' attention 282–283
    in manual driving 283
computational models of trust in automation 312
Computers Are Social Actors (CASA) paradigm 214
Computer System Usability Questionnaire (CSUQ) 223
computer vision-based AI systems 126
conflict 415–417
connected devices 175
connection 366
conscientiousness 315, 322
consequences, AI-based system 256–257
constraints 67, 69
content-based filtering (CBF) algorithms 161
contentious conflict, potential risk of 416–417
context-aware assistance 160
context model 44
contextual and stakeholder-discovering, approaches and techniques 205–206
contextual factors 67, 310
contextual-related factors 310
continuity 311, 313
contribute 409–412, 414, 415, 417, 418, 420
conversational agent evaluation *222,* 222–223
    usability and user experience **223,** 223–224
    user perceptions and acceptance 224–225
conversational agents **182,** 183, 195, 203, 205
    cognitive abilities of 213–214
    design of 205
    and HCI 204–205
    language design for 204
    for social good 225–228, *227*
    task-oriented 210
conversational agents, building 208–209
    cognitive abilities of conversational agents 213–214
    end-to-end architecture *212,* 212–213
    modular architecture *210,* 210–212, **211**
conversational elaboration 418
conversational intelligence 214, **215,** 218, 221
conversational interfaces (CIs) 29
conversational tone 204
conversational user interface 206–207
conversation design 205–208
convolutional neural networks (CNNs) 147, 148, 161, 213
cookies, human behavior and privacy 92–93
coordination 341, 344, 347, 350, **354**
core AI-based systems for customer services 248, **248**
    Chatbots 250–252
    recommender systems 249–250
    virtual assistants 252–253
coronavirus disease 2019 (COVID-19) pandemic 307
Correctional Offender Management Profiling for Alternative Sanctions (COMPAS) 110, 121
CPS *see* cyber-physical systems (CPS)
creative placemaking 393
crosswalk safety 288
crowd 410–412, 415–418, 421, **422,** 423
crowd-based innovation challenges 415
crowd-based open approaches, to tackling grand challenges 411–415
    open innovation in science 412–414
    open strategy formulation 414–415
    robust action 411–412

crowd-powered science 411
crowd science 410
crowdsourcing 410–412, 414, **422,** 423
    sustainability through **422**
cryptocurrency 79
cultural differences 336, 337, **353**
cultural diversity 366
cultural frames 373
curiosity, attacker's motivation 83
custom-developed public displays 394
customer interface 246, *246*
customer journey 241, 242
    digital transformation of 243–245
customer journey analytics (CJA) 143
customer service chatbot, of Cana Brave Resort 216–217, *217,* 219, *222*
customer services 241, 242
    artificial intelligence for **245,** 245–246
    core AI-based systems for 257–253
    implementation and management of artificial intelligence for *247,* 247–248
customizability opportunities 171
cyberattacker 77
    motivation of 81, *82*
cyberattacks 93
    factors contributing to the increase in 79
cyberattacks, behavior used in 94
    authentication 94
    healthcare 96–97
    Internet of Things 96
    manufacturing 96
    phishing 94–95
    power of free 95
    scarcity 95–96
    scare tactics 96
cybercrime 77
    attackers 78
    ease of attacking 78
    number 78
    punishment 78
    scale 77–78
    significance 78
cyber hygiene 97
    changing people's behavior 99
    complexity and limitation 100
    cyber training programs 98
    employee training 98
    motivation 99–100
    training and awareness program effectiveness 98–99
    training delivery methods 98
cyber-physical systems (CPS) 1, 2
cyber protection
    behavioral-based cyber technologies 80
    defensive technologies 79–80
    education and awareness 80–81
    human behavior 80
    laws and regulations 80
cybersecurity 79
    effective decision-making in 88
    revenge in 90
Cybersecurity Countermeasures Awareness (CCA) 98
cybersecurity hygiene 97–100
    awareness 97
    training 98

cybersecurity privacy 77–102
Cybersecurity Ventures 78
cyberspace, self-disclosure in 89
cyber technologies, behavioral-based 80
cyber training programs 98
cyberwar 82

DAMSL (dialog act markup in several layers) 210
DARLENE AR system
    for Law Enforcement Agents 68
    UIs 68
DARLENE context-aware personalized intelligent UIs 68–70, *69, 70*
data 408–411, 413, 420–421, **422**
    bias 420–421
    gathering 387, 393, 394–395
    protection 364, 371
    quality 420
    storage 370, **377**
    visualisation 388, 395
    visualization tools 27
data-driven personas 143
DDT *see* dynamic driving task (DDT)
debugging tools **184,** 185, *186, 187*
decision authority 374, 376
decision-making
    factors shaping vulnerable behaviors 88
    logic, defined 51
    mechanisms 72
Decision-Making Specification Language (DMSL) 51–52, *52*
deep learning (DL) 108, 121, 242, 243
    classifier 145
DEEP-MAX scorecard rating for AI systems **132**
defensive technology, cyber protection 79–80
demographic change 364
denials 314
denunciations and revelation 83
deoxygenated hemoglobin (HbR) 282
design 384–386, 389, 391, 393, 394, 398–401
    approach 8
    challenge 16
    for diversity 49, 71
    guidelines 52, 53, 61, 151
    ideation 152
    iteration 24, 26
    materials 151
    methodology 5
    principles 1, 6, 10
    team 13–16, 24–26, 34, 35
    tools 27, 37
design and development 20, 37
    process 15
    team 12
design collaborator, AI as 152–153
designer-AI collaborative design activities *153*
Design for All 45, 46–48, **48,** 71
    standardization on 48
design framework, for Intelligent Environments 10-27
designing intelligent artifacts *vs.* designing intelligent services 5, **7–8,** *8*
design process 141, 152, 153
    Intelligent Environments and 1–4
Design Thinking 1, 10

Index

developer perspective, Internet of Things 188–194
development methodologies
    custom software engineering 190, **192**
    for IoT application 191
devices
    acting 173–174
    connected 175
    sensing 173
dialog management (DM) 208, 209
dialogue acts 210
dialogue design methods and techniques 206–208
dialogue flow 207
dialogue policy 211
digital channels 244
digital health technologies 369
digital infrastructure 384, 386, 387, 392, 396, 397
digital keys and certificates 90
digital personas 143
    benefits of 144
    challenges for 145
digital placemaking 384, 388, 391–393, 399
    urban interfaces 391–393
digital pop-ups 393
digital storytelling 389, 393
digital technologies 386, 391, 393, 400
digital transformation 201, 244
    of customer journey as basis 243–245
dignity 366, 371, 376, 378
disabilities 46, 48, 51, 62, 66, 70
disclosure of data 370
discrimination, AI-based systems for customer services 254
Distract-R 283
distributed experimentation 411–412
diverse 410–412, 414, 416, 420, 421
diversity of users' characteristics 126
diversity score (D) **132**
DM *see* dialog management (DM)
DMSL *see* decision-making specification language (DMSL)
drivers
    behavioral measurement 278
    distraction 263
    emotions 269–270
    fundamental understanding of 263–275
    inattention 264, 280
    in manual and automated driving modes 263–264
    motion sickness 272–275
    motor control metrics 278–279
    situation awareness 267
    trust 270–272
    workload 266
    workload in automated driving 267
    workload in manual driving 266
    workload measures 266
drivers' attention, measurement 279
    blinking activity 280
    changes in pupil size 280–281
    computational modeling 282–283
    eye movement metrics 279–280, **280**
    eye-tracking device types 281, **281**
    neurophysiological measurement 282
driving 263
    research approaches 275, **276**, *276*

    simulator studies 275–276
driving automation system 263
    human-automation function allocation on levels of **265**
driving behaviors, qualitative modeling of 283
driving modes, drivers in manual and automated 263–264
driving performance 266, 267, 270, 278, 279, 282, 286
driving simulator studies **275,** 276–278
dual-task performance 266
dual UIs 49
dynamic driving task (DDT) 263, 264
dynamic knowledge management 410

EAGER Designs Repository 52–54, *53,* **54**
education 114, 133
    and awareness, cyber protection 80–81
educational projects 409
effective decision-making, in cybersecurity 88
efficiency 122, 125, 127
Egoki System 50
eHMIs *see* external human-machine interfaces (eHMIs)
electroencephalography (EEG) 282, 284
    brain-computer interface 284
electrogastrography (EGG) 275
ELIZA 250
embodiment 203
emergency response systems 364, 366
emotional intelligence **216,** 221
emotionally aware assistance 160
emotional sharing 94
emotions 263, 264, 269–270, 272, 279, 280, 282
emotions, drivers
    effects on driving 270
    induction 269
    measurement 270
    mitigation methods 270
    situation awareness in 269
Emotion Value Assessment (EVA) 205–206
end-of-life decisions 374
end-to-end architecture, conversational agents *212,* 212–213
end-to-end (E2E) conversational AI systems 206
end-user debugging tools *187,* 188
End-User Development (EUD), Internet of Things 175
end-user perspective, Internet of Things (IoT) **175,** 175–188
end users 15, 26, 171–175
engage 409, 410, 414–418, 420, 421
engagement 408, 409, 411, 412, 417, 420, **422**
    customer **3**
    user **7**
enhanced situational awareness 68–70, *69, 70*
entity extraction 210
environments
    artificial 127
    physical 127
EPI *see* Eysenck Personality Inventory (EPI)
epidemics 111, 114, 133
equality 111, 114, 125, 133
equity 119, **132**
Ethereum 79
ethical AI 118
ethical considerations **6, 8,** 9, 121, 122, 124
ethics 366
ethics score (E) **132**

EUDebug *186*
European Union Agency For Network and Information Security Cybersecurity (ENISA) 99
EVA *see* Emotion Value Assessment (EVA)
evaluation
    expert-based 24
    methods 222–224
    user-based 25–26
expert-based evaluation 24
expertise 336, 340, **353**
explainable artificial intelligence (XAI) 111, 119, 127, 317, 349–350, 352
explanation
    content 319–320
    modality 320
    timing 318–319
explosive ordnance disposal (EOD) 344
extended reality (XR) environments, personalization and adaptation in 68–70
external human-machine interfaces (eHMIs) 263, 398
    Vulnerable Road Users and 288–289
extraversion 315, 322
eye-tracking 268, 271, 279
    device types 281, **281**
Eysenck Personality Inventory (EPI) 322

facilitate scaling 411
facilitate innovation 412
factors shaping vulnerable behaviors 85, *86*
    attention 86
    behavior pattern-schema 89
    decision-making 88
    human error 86–87
    human visual perception 87
    memory 88
    perception and cognition 87–89
    revenge 90
    self-disclosure 89
fairness, responsible AI 116–117
fake news shaping people's opinions 82
fall detection 364
fall prevention 367
fast access to health data 373
fast motion sickness (FMS) scale 274
field operational test (FOT) 277
filtering 11, 27, 37
financial gain, attacker's motivation 81–82
fine-grained adaptation of UIs 71
FIRMA 59, 60, *61, 62*
fixation-derived metrics 279
fluid layout, UI version with 45
formal education 98
FOT *see* field operational test (FOT)
frequently asked questions (FAQ) 62
functional limitations 45, 46, 49, 62, 70
function allocation 341
functional magnetic resonance imaging (fMRI) 282
functional near-infrared spectroscopy (fNIRS) 282
future mobility, interfaces for 283–285

galvanic skin response 275
gateways 173
General Data Protection Regulation (GDPR) 256
generative AI 249

generative models 209
Generative Pre-trained Transformer (GPT) 121
generative UI design 150
gesture-based in-vehicle user interfaces 284
gesture interactions 285
gesture recognition 60, 63
gestures 283–285
Global Positioning System (GPS) 93
good old-fashioned artificial intelligence (GOFAI) 242
governance 384, 386, 388, 388–390, 396, 399–400
    AI design and 128
    of AI systems 111
    frameworks 112
    levels of 123
GPT *see* Generative Pre-trained Transformer (GPT)
grand 411–415, 418, 419, 421, **422,** 423
graphical user interfaces (GUIs) 29, 109, 148, 207
    personalization 64
    personalization and adaptation *67*
grassroots activism movements 389
GRETA agent 213
groups 333, 335, 336, 338, 341–345, 347, 348
GUIs *see* graphical user interfaces (GUIs)

hand-eye coordination movements 285
haptic feedback 275
H-A teaming *see* human-agent teaming (H-A teaming)
HAT, transparent interfaces for 349
HCAI *see* human-centered AI (HCAI)
HCD *see* human-centered design (HCD)
head-down displays (HDDs) 285
head-up display (HUD) 66, 268, 284
healthcare, cyberattacks 96–97
heart rate variability 275
hedonic aspects 223
help 408, 409, 413–418, 421
Her (movie) 201
heuristic approach 67
Hidden Markov Model 283
Hierarchical Driver Model 283
hi-fi prototypes 21
HIR
    communication 350
    transparency and XAI 352
HomeShare project 125
HRI *see* Human–robot interaction (HRI)
HTTP *see* Hypertext Transfer Protocol (HTTP)
HTTPS *see* Hypertext Transfer Protocol Secure (HTTPS)
human-agent teaming (H-A teaming) 333–355
    agent-based inputs 338–340
    agent performance and reliability 340
    agent transparency *348,* 347–350
    attitudes 342
    behaviors 345–346
    cohesion 343–345
    expertise and training 336
    factors **353–354**
    human-based inputs 336–338
    IMOI model of 335, *335*
    individual and cultural differences 337
    input factors 336–342
    level of autonomy 338–340
    mediators 342–350
    outputs 350–352

Index

roles 341–342
shared mental models and situation awareness 347
task allocation 340–341
team cognition 346–347
teamwork-based inputs 340
trust 342–343, 350–351
human-agent (H-A) teams 333
human–AI interaction 112, **113**, 114, 118, 123
    definitions of 128
    economy and business 129–131
    history of 129
    levels of 128–129, *129*
    nature of risk in **130**
    social interactions 132
    work and the future of work 131
human-automation function allocation, on levels of driving automation systems **265**
human behavior and privacy
    cookies 92–93
    self revelation on social media 93–94
human behavior, cyber protection 80
human-centered AAL design 376–378
human-centered AI (HCAI) 111
    challenges in wide-scale adoption of 112
    community 122
    design and evaluation 122–123, 127–128
    future research directions for **112–113**
    HCI principles adapted to 123–125
    HCI professionals 114, *114*
    principles 111–112
    research – current limitations 125
    stakeholders 114
human-centered design (HCD) 46, 124
human-centered RS 162
human-centred, technology-assisted care 378
human–computer Interaction (HCI) 5, 384, 385, 398–400
    conversational agents and 204–205
    practitioner guidelines 121–122, 128, 132
    principles adapted to HCAI 123–125
    professionals 114, *114*
human computer interaction (HCI)/UX designers 151
human computer interaction (HCI)/UX professionals 143, 146, 148, 150, 155, 156, 162
    AI/ML-based algorithms allow 144
    classical activity for 159
    conduct UX evaluations 147
human computer interaction (HCI)/UX work, critical issues for **156**
Human–Computer Trust 321
human error, factors shaping vulnerable behaviors 85–86
human factors 5
    issues of AI technology **156**
Human–Human Interaction (HHI) 309
human intelligence 108, 109, 128, 132
human interventions 417
human-in-the-loop 347
human likeness 204, 216, 218, 221
human–machine interaction (HMI) elements 64
human-machine interfaces (HMI) 349
    for Autonomous Squad Member 349, *349*
humanoid robot 306, *306*
human personality 322
human-related factors 310
human-related measurements, trust 321

human–robot collaborations 318–319
human–robot interaction (HRI) 305–323, 337, 349
    adaptive multimodal 59–64
    anthropomorphism 322–323
    attitude 321–322
    content, and modality and 318
    evaluation metrics in 320–323
    explainable AI in 317–318
    explanation content 319–320
    explanation in 317–320
    explanation modality 320
    explanation timing 318–319
    factors influencing trust in 310–310, *311*
    human personality 322
    human-related measurements 321–322
    human's personality impact 315–316
    personality in 314–317
    research, robots 305–308
    robot personality 323
    robot-related measurement 322–323
    trust 321
    trust dynamics and computational trust models in 311–313, *312*
    trust in 308–309
    trust management and recovery in 313–314
    trust violations in 314
    workload 322
human–robot trust 309–310, *311, 312,* 314, 321
    questionnaire 321
    and trustworthiness 309–310
human's personality impact HRI 315–316
    vs. robot's personality impact HRI 316–317
human supervision 338
human vulnerable behaviors **101–102**
human well-being, shifting focus back from AI capabilities to 111
hybridity 384, 399–400
    urban interfaces 399–400
hybrid places 384–385
Hypertext Transfer Protocol (HTTP) 91
Hypertext Transfer Protocol Secure (HTTPS) 91

IA *see* intelligent assistant (IA)
IAs *see* intelligent agents (IAs)
ICS-FORTH's Intelligent Home 37
IDE *see* integrated development environment (IDE)
ideas 410, 412, 414–418
ideation
    method 18
    stage 16, 19, 21, 35
IDEs *see* integrated environments (IDEs)
IEs *see* Intelligent Environments (IEs)
IF-THEN rules 176, *177,* 178, *180,* 181, 182, *184*
    strategy **184, 182,** 183–184, **184**
IFTTT 177, *177,* 178, 181
IMOI model *see* Inputs-Mediators-Outputs-Inputs (IMOI) model
implementation
    of design process 4
    of intelligent living room 35
incidental affect 269
inclusion, conversational agents 225
independence 365
individual costs 371

individual differences 337, 340
individuals 410, 412–417, **422,** 423
industrial automation **3**
industrial design 22, 35
Industrial Internet Reference Architecture (IIRA) 190
industrial manipulators 307
inference
    causal 116
    incorrect 126
    robust 108
information 408, 409, 413, 415, 417, 420, 421, **422,** 423
Information and Communication Technologies (ICT) 46
information overload **175,** 178, 181
information sensitivity 370
Information Technology (IT) professional 87
initiative models 208
innovation 17, 35, 408, 410, 412–419, 421, 423
    balance 11
    opportunity for 16
    and research 10
    technological 13
input-mediator-output-input model 335, 340
input-process-output (IPO) model 334, *334*
inputs 334–342, 345, 350, 352, **353**
Inputs-Mediators-Outputs-Inputs (IMOI) model of H-A teaming 335, *335*
instant messaging tools 202
integral affect 269
integrated development environment (IDE) 28
integrated environments (IDEs) 55
integration 19
    of AI 3
    challenges 193
    of computational and physical components 2
    of diverse services 28
    heterogeneous subsystems 191
    issues 175, **175**
    of physical and digital components 5
    of smart technologies 13
    technology **6**
"Integration & Test" stage 35–36
integrity 310, 314
intelligence 108
intelligent agents (IAs) 286
intelligent artifacts *vs.* intelligent services 5, **7–8,** 9
intelligent assistant (IA) 157, 158
    human users' needs of 159
    UX of 159
Intelligent Environments (IEs)
    AmI Solertis: streamlining IE behavior definition 27–28, *28*
    application domains of **3,** 3–4, 10–12
    ARgus Designer 31
    creating services for 27
    design aspects of **6**
    design aspects specific to 5
    designing intelligent artifacts *vs.* designing intelligent services 5, **7–8,** 8
    and design process 1–4
    frameworks for dictating the behavior of 29
    and objectives 1, **2**
    and relevant concepts 1–3
    supporting tools 27–34
    UInify 31–32
    UXAmI observer 32–33

Wizard of AmI 30–31
Intelligent Environments (IEs), design for 4–5
    challenges, problems, and barriers 4–10
    intelligent spaces 9
    larger IEs 9
    motivation and approach 9–10
    similar design approaches 8
    smart artifacts 9
Intelligent Environments (IEs), design framework for 10–27
    affinity diagrams 15
    cognitive walkthroughs and heuristics 24–25
    expert-based evaluations 24
    field study-observations 14
    frame the design challenge 16
    interviews 14
    market research 14
    personas 14–15
    questionnaires 14
    scientific research 13–14
    stage 1: understand 12–15
    stage 2: define 16
    stage 3: ideate 17–19
    stage 4: filter and plan 19–20
    stage 5: design 20–22
    stage 6: evaluate design 23–26
    stage 7: (implement - integrate) & test 26–27
    stage 8: evaluate integrated solution 27
    SWOT 16–17
    understand and empathize with users 10–27
    understand technological and market trends 13
    understand the context of use 13
    user-based evaluation 25–26
    validating IEs behavior 29–30
'Intelligent Home' 4
intelligent living room 34–36, *36*
Intelligent Personal Assistants (IPAs) 181
intelligent user interfaces (IUI) 154–157
interaction 413, 416–419, 421, **422**
    design 209, 389
    toolkits 52–55
interaction technique 159, **179,** 181, 183, 187, 188
interdependence 333, 340
interior designer 15, 22, 34, 35
Internal Revenue Service (IRS) 96
International Conference on Automotive User Interfaces (AutoUI) 263, 283
international standardization organization 70
Internet of Everything (IoE) 172
Internet of Tangible Things 174
Internet of Things (IoT) 1
    challenges 178–187, 190
    challenges and opportunities 174–175
    components poses **192**
    context 175–178, *177,* 188–189
    cyberattacks 96
    definitions and application examples 172–174, *173*
    developer perspective 188–194
    devices 420
    devices and systems 171
    enabled environments 171
    End-User Development 175
    end-user perspective **175,** 175–186
    gap between perspectives 194
    guidance for future research 187–188, 193–194

Index

methodologies 193
personalization 184, 187, 195
system development 193
systems' characteristics and development process **192**
understanding of **192**
Internet of Things-Architecture (IoT-A) 190
interoperability 178, 189, 191–193
interpersonal 384, 388, 393, 400
interpretability of AI algorithms 119
intersectionality 228
interview 141
    transcripts 143, 146
intimacy 364, 371
in-vehicle context, VR and AR applications to 287–288
in-vehicle intelligent agents (IVIAs) 279, 286–287
involving parties 419
IoE *see* Internet of Everything (IoE)
IoT *see* Internet of Things (IoT)
IPAs *see* Intelligent Personal Assistants (IPAs)
IRS *see* Internal Revenue Service (IRS)
iterative design 23, 35
IUI *see* intelligent user interfaces (IUI)
IVIAs *see* in-vehicle intelligent agents (IVIAs)

Jigsaw metaphor 179
JMorph adaptable widget library 54–55, **56–58**
job loss 131

KJ method 15
knowledge 409–416, 418, 421, **422**, 423
knowledge gaps, complexities associated with 415–416
knowledge representational gaps 415

labeling theory 85
labor 128
language design, for conversational agents 204
language style matching (LSM) 352
lapses (skill-based errors) 86–87
Large Language Model (LLM) 346
large-scale language models 203
latent semantic similarity (LSS) 352
lateral control **279**
Law Enforcement Agents (LEAs), DARLENE AR system for 68
laws and regulations, cyber protection 80
learning 411, 412, 414, 416, 417
    deep 109, 121
    machine learning 131
LECTORstudio 29
levels of automation 337–339
levels of autonomy 338–340, 352
Levels of Detail (LoD) 68
leveraging the wisdom of crowds 410–411
lifelogging technologies 364, 367, **367**, 368, 370, 374, *375, 376,* **377**
    and sensors 364
light installations 393
linear temporal logic (LTL) 185
Litecoin 79
LLM *see* Large Language Model (LLM)
Local-Social Index 400
location sharing, self revelation on social media 93
location-tracking technologies 93
lo-fi prototypes 21

loneliness 364, 366, **367,** 372, 373
longitudinal control **265, 279**
long short-term memory (LSTM) network 147, 209, 212, 213
long-term 409, 410, 413, 419, 420
LSM *see* language style matching (LSM)

machine learning (ML) 131, 141, 203, 242
    classifier 210
    decisions 205
    as a service 143
machine-to-machine (M2M) communication standards, patterns of 190
management
    of AAL-lifelogging 376, **377**
    of care activities 367
manual driving
    computational modeling in 283
    situation awareness in 268
    workload in 266
Mapping for Change project 420
MASA paradigm *see* Media are Social Actors (MASA) paradigm
Mayer Trust Scale 321
MDMT *see* Multi-Dimensional Measure of Trust (MDMT)
measurement
    of anthropomorphism in HRI 322–323
    of attitudes in HRI 321–322
    of personality in HRI 322
    of trust in HRI 321 of workload in HRI 322
media architecture 384, 385, 386, 391, 400
Media are Social Actors (MASA) paradigm 214
mediators 335, 336, 338, 342–345, 350, 352, **353**
medical outcomes 111, 114, 133
medication management 367
members 409–412, 415, 416, 420, 421
memory, factors shaping vulnerable behaviors 88
mental models 337, 346–347, 350–352, **353**
Meta's chatbot BlenderBot 3 253
methods 384, 387, 393, 398, 400
metrics 352, 354
metropolis 384, 388, 392, 396, 400
Microsoft Speech Synthesis 60
middle-out 389–390
milgram's reality-virtuality continuum 287
mind perception theory 205
misery index scale (MISC) 274
mixed reality (MR) 67, 287
    algorithm 248
    reference framework 9
ML *see* mixed reality (MR)
motion sensors 366
motion sickness 263, 272–275, 288, 289
Motion Sickness Assessment Questionnaire (MSAQ) 274
motion sickness, drivers 272–275
    causation theory 273
    induction 273–274, *274*
    measurement 274–275
    mitigation methods 275
    non-driving-related tasks and automated vehicles 272
    research framework 272–273, *273*
motivation
    of cyber attackers 81, *82*
    cyber hygiene 99–100

motivation, attacker 81
    curiosity 83
    denunciations and revelation 83
    fake news shaping people's opinions 82
    financial gain 81–82
    profiting by deceit 83
    revenge 83
    sabotage, cyberwar, and intelligent gathering 82
    solving crimes 83
    terrorist organizations 83
motor control metrics, drivers 278–279
MR *see* Mixed Reality (MR)
MRT *see* multiple-resource theory (MRT)
MSAQ *see* Motion Sickness Assessment Questionnaire (MSAQ)
Muir Trust Scale 321
Multi-Dimensional Measure of Trust (MDMT) 321
multimodal displays 285–286
multimodality 65
multiple-resource theory (MRT) 87
multivocal inscriptions 411
Myers-Briggs test 322
My IoT Puzzle tool 179–180, *179*
MyUI (Mainstreaming Accessibility through Synergistic User Modeling and Adaptability) 50

NADS-1 simulator *see* National Advanced Driving Simulator (NADS-1 simulator)
NARS *see* Negative Attitudes toward Robots Scale (NARS)
NASA Task Load Index (NASA-TLX) 223, 266
NASA Task Load Index (TLX) workload scale 266
National Advanced Driving Simulator (NADS-1 simulator) 276, *277*
National espionage organizations and security agencies 82
national security agencies 80
natural interaction 71
naturalistic driving studies 275, **276**, 277–278
    history of 277
    strengths, and weaknesses 277–278
natural language generation (NLG) 208, 209, 212
natural language inference (NLI) 146
natural language processing (NLP) 29, 128, 143, 158, 188, 208–209, 212, 243
NDRTs *see* non-driving-related tasks (NDRTs)
Negative Attitudes toward Robots Scale (NARS) 321, 337
negativity bias 311–313
net promoter score (NPS) 25
neuroergonomic sensors 270
neuroscience 132
neuroticism 315, 322
Nielsen's heuristics 223
NLP *see* natural language processing (NLP)
non-driving related tasks (NDRT) 272–275, 286
non-humanoid robots 306–307, *308*
NPS *see* net promoter score (NPS)

object and event detection and response (OEDR) 263, 278
observed situational awareness 69
OEDR *see* object and event detection and response (OEDR)
olfactory displays 285
"one-size-fits-all" approach 70
online 410, 412, 415–420

online communities, collaborations in 416
online probabilistic trust inference model (OPTIMo) 313
online survey 93, **101**
on-road vehicle studies 275
ontology modeling 68
open 409–415, 417, 418, 421, **422,** 423
openness 315, 322
operational design domain (ODD) 263–264
opportunities, for human–computer interaction 174
Organisation for Economic Co-operation and Development (OECD) 111
organizations 410, 414, 419, 423
outputs 334, 335, 338, 340, 341, 350–352, **354**
over-trust 370
oxygenated hemoglobin (HbO) 282

PA *see* Persona Analytics (PA)
ParlAmI 29
partially observable Markov decision process (POMDP) model 313
participants 409, 410, 414–418, 420, 421, 423
participatory 409, 411, 412, 414, 420, **422,** 423
parties 418, 419
passenger tasks (PTs) 272
pattern-based frameworks 50
Pensacola Diagnostic Index 274
people commit cybercrime 83–84
    labeling theory 85
    rational choice theory 84
    social control theory 84
    social disorganization theory 84
    social learning theory 84
    strain theory 84
perceived risks 370
perceived situational awareness 69
perceptions
    of aging 372
    of privacy 369
persona 143, 145
Persona Analytics (PA) 144
personal 384, 385, 388, 392–396
personality 305
    in HRI 314–317
    and personality traits 315
    similarity 318
    trait-based approach to 315
    traits 310, 315, 316, 322
personalization 44, 49, 50, 171–172
    AI-based systems for customer services 257
    GUI 64, 67
    intentions 183, *184*
    IoT 184, 187, 195
    of IoT ecosystems 180
    logic 65, 66
    rule 66
personalized adaptive intelligent UIs 71
personal listening devices (PLDs) 288
personally identifiable information (PII) 90
personal privacy 370
personas 14–15
personification 214, 216, **216,** 221
perspective making 415–416
pervasive displays 384
phishing, cyberattacks 94–95

Index

physical environments 127
physically present *vs.* virtually represented robots 307–308
placemaking 384, 388, 391–393, 395, 396–400
planning
    project 11, 17
    strategic 17, 18
platform 412, 415, 417, 418, **422**
PLDs *see* personal listening devices (PLDs)
policy
    human-centered policy guidelines for AI 112, **112–113**
    National Artificial Intelligence Initiative Act of 2020 129
post-study system usability questionnaire (PSSUQ) 25
potential bias 126
poverty, reduce 111, 114, 133
power of free, cyberattacks 95
pragmatic aspects 223
prevention measurement, AI-based systems 257
privacy 91–92, 111, 112, 115, 120, 205
    AI-based systems for customer services 256
    calculus 370
    concerns 364, 370
    human behavior and 92–94
    policies 229
    protective operations 218
    users 225
proactive approaches 70–71
problem 408–411, 413–417, 419, 421, **422**
process 409–418, **422**, 423
process analysis
    development **192**
    IoT development 190
processes 334–336, 341–343, 345, 350, **353**
product-level adaptation 47
professional training 98
profiting by deceit, attacker's motivation 83
program effectiveness, training and awareness 98–99
programmers 172, 188, 194
programming framework 171, 191, **192**
Project for Public Spaces 392
projects 409, 413, 419–421
promises 314
promoting 408, 414, 417, 418, 420, **422**
protection motivation 370
protective behavior **101–102**
prototypes, lo-fi and hi-fi 21
prototyping 207, 209
    for IEs 31
    rapid 18, 20, 22
    tools 27
proximetrics 316
PSSUQ *see* post-study system usability questionnaire (PSSUQ)
psychological safety 416
PTs *see* passenger tasks (PTs)
public 408–410, 412, 413, 418, 419, 421, **422**
public displays 384, 386, 387, 392, 394, 395, 397
    custom-developed 395
public transparency 370
pupil diameter 281
pupillometry 280, 281
Puzzle 179, 186

QN-MHP *see* Queuing Network-Model Human Processor (QN-MHP)
qualitative analysis, AI in HCI/UX research and evaluation 146–147
qualitative modeling, of driving behaviors 283
quality of life 364–366, 369, 376
    multidimensional requirements for 364
Queuing Network-Model Human Processor (QN-MHP) 283

RAMCIP Robot UI *63,* 62–63, **64**
rational choice theory 84
reasoned action theory 99
reasoning 121, 127
    logical 128
    mechanism 44, 46
recommendation
    of IF-THEN rules *182*
    TAPrec 182
    techniques 181
recommender system (RS) 118, 120, 121, 125, 160–163
    applications for customer services and associated benefits 250
    purpose and technical functioning 250–251
RecRules 182
recurrent neural networks (RNN) 161
regulations 118
    laws and, cyber protection 80
regulatory frameworks, AI-based systems 257
reinforcement learning 51, 108, 109
reliability 336, 338, 340, 342, 351, **353**
Representational State Transfer (REST) 28
representation models **179,** 180–181, 183, 187
research 409–412, 415, 417, 419–421, 423
research gaps 194–195
resource-oriented web services (REST web services) 192
responsible AI
    accountability 115–116
    automated decision-making systems 120–121
    definition of 115
    ethical AI 118
    examples of 120–121
    fairness 116–117
    group fairness 117
    individual fairness 116
    large language models 121
    recommender system 120
    trustworthiness 118–120
responsive design 45
REST *see* Representational State Transfer (REST)
revelation
    denunciations and 83
    self revelation on social media 93
revenge
    attacker's motivation 83
    in cybersecurity 90
ride, accessibility and types of 289
risks, of AI-based systems for customer services 254–256
RNN *see* recurrent neural networks (RNN)
road rage 269
robotic arms 307
robot interaction, adaptive multimodal human 59–64
robot-related factors 310

robots 333, 334, 337, 342, 344, 345, 347, 348, 350
    defined 305
    in HRI research 305–308
    humanoid 306, *306*
    non-humanoid 306–307, *308*
    types 305–306
robot's personality impact HRI 316
    vs. human's personality impact HRI 316–317
robust 411, 412, 421, **422**
robust action 411–412
robustness 115, 129
role distribution 341
RS *see* recommender systems (RS)
rule-based approach 50
Rule Bot 181
rules discovery, promoting IF-THEN **189,** 181, 183
run-time adaptations 62
run-time interaction monitoring 71

SA *see* situational awareness (SA)
sabotage 82
    and terrorist attacks 78
saccade-derived metrics 279
safety 121–123, 127
SA Framework for XAI (SAFE-AI) 352
SAGAN model 149
SAGAT *see* situation awareness global assessment technique (SAGAT)
SA rating technique (SART) 280
SART *see* SA rating technique (SART)
Satisfiability Modulo Theories (SMT) solvers 185
Save Our Oceans study 412
SBE *see* smart building environments (SBE)
scalability 5, 10, 29
scanpath-derived metrics 279, 280
scarcity advertising, effectiveness of 95–96
scarcity marketing tactics 95–96
scare tactics 96
schemas/schemata 89
science 408–423
scientific 408–414, 419–421, **422,** 423
scientists 409, 413, 420, **422**
SCT *see* social cognitive theory (SCT)
SDGs *see* sustainable development goals (SDGs)
Second Strategic Highway Research Program (SHRP2 NDS) 278
Security Education, Training, and Awareness (SETA) Program 98
self-adaptation 44
    UI design 52, 59
    UIs capable of 44
self-determination 371, 374
    in old age 378
self-disclosure 89
    in cyberspace 89
self-promotion, potential risk of 417–418
self revelation on social media 93
    emotional sharing 94
    location sharing 93
Semantic Colored Petri Net (SCPN) approach 185
semantic encoding 210
sensing devices 173
sensors
    lifelogging technologies and 364
    neuroergonomic 270

sensory conflict theory 273, 275
Service-Oriented Architectures (SOA) 192
service provision 245–246
service quality, AI-based systems for customer services 255
sexism 228
shared mental models 346–347, 350, **354**
    human-agent teaming 346–347
sharing 410, 412, 413, 415–418, 420, 421, **422**
Simon's optimization-based design 149
simulation tools 27
Simulator Sickness Questionnaire (SSQ) 274
single authoring 45
SioT *see* Social Internet of Things (SioT)
situational awareness (SA) 68, 155
    enhanced 68–70, *69, 70*
    observed 69
    perceived 69
situational trust 271
Situational Trust Scale for Automated Driving (STS-AD) 271
situation awareness 263, 267–268, 337, 338, 346–347, *348,* 352, **353**
Situation Awareness-based Agent Transparency (SAT) framework 347, *348*
situation awareness (SA), drivers 267
    automated driving 268–269
    emotions 269
    manual driving 268
    measurement 268
    model 267–268
situation awareness global assessment technique (SAGAT) 268, 280
situation awareness, H-A teaming 346–347
slips (automatic behavior) 86–87
slot-filling 210
smart building environments (SBE) 9
smart cities 384, 386, 393
smart environments 174, 175
smart home 171, 172, 176, 181, 187, 366, **367, 368**
smartwatches 366, **368**
SNA *see* social network analysis (SNA)
social 408, 410–416, 421, **422**
social acceptance 368, 369, 378
social and societal values 373
social cognitive theory (SCT) 100
social control theory 84
social cues 214, **215–216,** 216
social dimension 214
    social cues 214, **215–216,** 216
    user evaluation after interaction 219–221, *220, 221*
    user expectations prior to conversation 216–217, *217*
    user experience during interaction 217–219, *218, 219*
social disorganization theory 84
social good, challenges and opportunities for 228–229
social influence 369
social interactions 366, **367,** 358, 388
    human–AI interaction 132
Social Internet of Things (SioT) 172
social isolation 364–366
social learning theory 84
socially accepted pathways 369
socially responsible technology development 378
social media-based personas 145
social media, self revelation on 93–94
social network analysis (SNA) 351

Index

social norms 368, 371
social recommender systems 249
social sciences 389, 400
societal 408, 410–414, 419, **422**
software architectures 190
software engineering 185, 189, 191
    development methodologies custom 190, **192**
solving crimes, attacker's motivation 83
specification, OpenAPI 28
speech recognition-based interfaces 284
SSQ *see* Simulator Sickness Questionnaire (SSQ)
stabilization 311–313
Stages of Change Model 99
stakeholders 369, **377,** 378, 408, 412, 414, 415, 420
    human-centered AI 114
standardization on Design for All 48
standards 128
static vibrotactile displays 285
storyboards 208
strain theory 84
strategic digital placemaking 392
STS-AD *see* Situational Trust Scale for Automated Driving (STS-AD)
Subjective Workload Assessment Technique (SWAT) 266
successful technical assistance 376
supervisory control 338
SUPPLE 51
support vector machine (SVM) 210
SUS *see* system usability questionnaire (SUS)
sustainability 384, 386, 399, 400, 408–423
    through citizen science and crowdsourcing **422**
    urban interfaces 399–400
sustainable development 408
sustainable development goals (SDGs) 399
SWAT *see* Subjective Workload Assessment Technique (SWAT)
symbolic AI 242
system usability questionnaire (SUS) 25
System Usability Scale (SUS) 223

tacit 410, 414–416, 421
tackle 412, 420, 421
tactical digital placemaking 393
tactical urbanism 385, 393
takeover 267–272, 279, 280, 283, 285
TAM *see* technology acceptance model (TAM)
TAPrec 182, *182*
task allocation 340–341, **353,** 354
task-capability interface model (TCI) 283
task-oriented conversational agents 209
task-oriented oriented systems 209
task-oriented systems 209
TAWS *see* Terrain Avoidance and Warning Systems (TAWS)
TCI *see* task-capability interface model (TCI)
team
    dynamics 334, 341
    effectiveness 335, 336, 350
    performance 336, 337, 340, 343, 347
    structure 338, 340
team cognition, H-A teaming 345
teamwork 333–335, *335,* 337, 338, 340, 342, 346, 349, 350, 351, **353,** 354
technology acceptance 368–371, 376
technology acceptance model (TAM) 225, 368–369
"technology-centered design" approach 122

temporal 412, 418–419, 421
temporal complexity 418–419
temporal coordination 418–419
Terrain Avoidance and Warning Systems (TAWS) 338
terrorism 81
terrorist attacks, sabotage and 78
testing 23–24
test track studies 275
TetraBIN project *397*
text input 283–285
text-to-speech synthesis (TTS) 212
theory of Reasoned Action (TRA) 99
think aloud 225
Think Aloud sessions 225
time to collision **279**
tools
    debugging **184,** 185, *186, 187*
    end-user debugging tools 186, *187*
    My IoT Puzzle tool 179–180, *180*
TRA *see* theory of Reasoned Action (TRA)
traditional UI techniques 154
traffic violation **279**
training 336, 337, 343, **353**
trait-based approach to personality 315
transformer-based model 209
transparency 254, 271, 272, 287, 337, 340, 343, 346, 347–349, *349,* 350, 352, **353, 354**
transparent algorithm 119
transportation 111, 114, 130, 133
trigger-action programming 176
trust 90, 213, 214, 227, 263, 270–272, 286, 287, 305, 335–337, *339,* 342–343, 347–351, **353,** 354, **354,** 368–371, 374
    calibration 340, 343, 351
    change 311
    digital keys and certificates 90
    'HTT PS' and 'HTTP' 91
    in human–robot interaction 308–309
    user 217
    user behavior and 91
trust-aware conservative control (TACtic) 313
trust, drivers 270–272
    in automated driving 271–272
    measurement 271
trust dynamics, and computational trust models in HRI 311–313, *312*
Trust in Automation Scale 321
trust inference and propagation (TIP) model 313
trust management and recovery in HRI 313–314
Trust Perception Scale 350
trust repair 314
trust violation 313, 314
trust-workload partially observable Markov decision process (trust-POMDP) 313
trustworthiness 115, 309–310, 319
    and human–robot trust 309–310
    responsible AI 118–119
turing test 108, 202
turn-taking 207, 217
typological model 384, 386–388, 399, 400

UAVs *see* unmanned aerial vehicles (UAVs)
ubiquitous computing 384, 391, 396
ubiquitous environments 1, 2
UCD *see* user-centered design (UCD)

UEQ 25
UInify 31–32
UMUX *see* usability metric for user experience (UMUX)
understanding 408–411, 414–419, 423
    of IoT environments 194–195
    of IoT system **192**
undertrust 370
Unified Theory of Acceptance and Use of Technology (UTAUT) 225, 368
unified UI methodology 49
uniform resource locator (URL) 91
unimodal displays 285
United Nations Conference on Trade and Development (UNCTD) 79
United Nations Educational Scientific and Cultural Organization (UNESCO) 111
United States Federal Trade Commission 81–82
United States National Artificial Intelligence (AI) Initiative Act of 2020 129
Universal Access 45, 49, 71
Universal Design 46–48
Universal Usability 45
unmanned aerial vehicles (UAVs) 337
urban
    data 384, 393–395
    hacking 393
    HCI 400–401
    informatics 384–385
    interaction design 389
    interfaces 384–401
    play 393
    prototyping 393, 400
    studies 391, 400
    technologies 387
urban data
    analytics 394
    gathering and communicating 393–395
    gathering limitations 395
    visualisations 395
urban informatics 384–385
urban interfaces 386, *387*
    artificial intelligence 400
    autonomous infrastructure 398–399
    community engagement 389–390
    digital infrastructure *397*, 396–397
    digital placemaking 391–393
    governance 388–389
    hybridity 399
    infrastructure 396
    interfacing citizens and the city 391
    making of hybrid places 391–392
    middle-out engagement approach 389
    participatory 393
    policies 390
    standard approach 392
    sustainability 399
    turning actions into governance 390
    typological model 386–388, *388*
urbanity 384
Urban Search and Rescue (USAR) scenario 319
URL *see* uniform resource locator (URL)
usability 24–26, 369
    AI 125–126
    maximum 32
    post-study system usability questionnaire 25
    system usability questionnaire 25
    usability metric for user experience 25
    and user experience **223**, 223–224
usability metric for user experience (UMUX) 25
usability testing 143, 47, 148
US Department of Defense Ethical Principles for Artificial Intelligence 116
user-based evaluation 25–26
user behavior and trust 91
user-centered design (UCD) 8, 13, 155, 224, 369, 376–378
user engagement 217, 220
user evaluation, after interaction 219–221, *220, 221*
user expectations, prior to conversation 216–217, *217*
user experience (UX) 2, 45, 46, 68, 126
    AI for enhancing 154–163
    design 13
    of intelligent assistant 159
    during interaction 217–219, *218, 219*
    usability and **223**, 223–224
user experience (UX) designers 15
user experience (UX) evaluation 141, 146–148
    AI technology in 148
    automatic 148
user feedback **2, 6, 7**
user interaction 1, **7**
user interface (UI) 141, **144,** 171, 178, 182, 183
    customization 49
    DARLENE context-aware personalized intelligent 68–70
    designers 22
    graphical user interfaces 148
    intelligent 154–157
    intelligent user interfaces 154, 155, 157
    prototype 149
    prototype, adaptive *55,* 55–59, **56–58,** *59*
    wireframe 149
user interface (UI) adaptation 44
    accessibility by 49–51
user interface (UI) design
    AI for 149–150
    self-adaptation 52, 58
user interface (UI) tools 51–59
    adaptive UI prototyping *55,* 55–59, **56–58,** *59*
    decision-making specification language 51–52, *52*
    EAGER Designs Repository 52–54, *53, 54*
    interaction toolkits 52–55
    JMorph 54–55, **56–58**
user modeling 71
user needs 3, 5, 14, 16, 23, 24
user perceptions 206, 224, 225
    and acceptance 223–224
user privacy 159
user requirement 14–17
user research 143, 146, 163
user satisfaction 207, 214, 369
users' communication journey 214–216
users' diversity 370, 373
users need 175, 176, 195
users' perspectives
    and empirical assessment of AAL lifelogging 371–376
UsiXML 50
UTAUT *see* Unified Theory of Acceptance and Use of Technology (UTAUT)
UXAmI observer 32–33, *33*

Index

value-sensitive design (VSD) methodology 205
vibrotactile displays 285
virtual assistants 201
    applications for customer services and associated benefits 253
    purpose and technical functioning 252
virtually represented robots 307–308
    vs. physically present 307–308
virtual rapid prototyping 37
virtual reality (VR) 9, 263
    applications to in-vehicle context 287–288
vision, factors shaping vulnerable behaviors 87
ViSiT 180
visualisation, urban data 395
visualization priority 68
voice assistants 202, 207, 366, **367**
voice-based interactions 254
voice input 283–285
Voice-User Interface (VUI) practitioners 206
VR *see* virtual reality (VR)
vulnerable behavior **101–102**
Vulnerable Road Users (VRUs) 288–289
    and eHMI 288–289

wearable cameras 366
wearable computing 366
wearable devices 366, 367, **367, 368**
Web Accessibility Initiative (WAI) 48
web apps 174
web-based dashboards 174
web-based tool 179
Web Content Accessibility Guidelines (WCAG) 48
well-being monitoring 367
White House memorandum 420
widget library, JMorph adaptable 54–55, **56–58**
willingness to use 368
Wizard of AmI tool 21, 30–31, 34
Wizard of Oz (WoZ) method 224, 272, 274, 276
Woodie project 398, *398*
work 409, 412–416, 420
workload 263, 266–267, 280–284, 286, 289
World Wide Web Consortium (W3C) 48
WoZ experiments 207
WYSIWYG ("What You See Is What You Get") design approach 55, 59

XAI *see* explainable artificial intelligence (XAI)
XR environments *see* extended reality (XR) environments

Zapier 177, *177*
Zimbardo's Stanford Prison experiment 96